Methods in Enzymology

Volume XVIII
VITAMINS AND COENZYMES
Part A

METHODS IN ENZYMOLOGY

EDITORS-IN-CHIEF

Sidney P. Colowick   Nathan O. Kaplan

Methods in Enzymology

Volume XVIII

Vitamins and Coenzymes

Part A

EDITED BY

Donald B. McCormick and Lemuel D. Wright

GRADUATE SCHOOL OF NUTRITION AND
SECTION OF BIOCHEMISTRY AND MOLECULAR BIOLOGY
CORNELL UNIVERSITY
ITHACA, NEW YORK

1970

ACADEMIC PRESS   New York and London

COPYRIGHT © 1970, BY ACADEMIC PRESS, INC.
ALL RIGHTS RESERVED
NO PART OF THIS BOOK MAY BE REPRODUCED IN ANY FORM,
BY PHOTOSTAT, MICROFILM, RETRIEVAL SYSTEM, OR ANY
OTHER MEANS, WITHOUT WRITTEN PERMISSION FROM
THE PUBLISHERS.

ACADEMIC PRESS, INC.
111 Fifth Avenue, New York, New York 10003

United Kingdom Edition published by
ACADEMIC PRESS, INC. (LONDON) LTD.
Berkeley Square House, London W1X 6BA

LIBRARY OF CONGRESS CATALOG CARD NUMBER: 54-9110

PRINTED IN THE UNITED STATES OF AMERICA

Table of Contents

Section V. Biotin and Derivatives

Section VI. Pyridoxine, Pyridoxamine, and Pyridoxal: Analogs and Derivatives

Preface

The expansion of information with respect to vitamins and their analogs and, particularly, the advent of newer chemical and biological techniques for studying the coenzyme forms and the enzymes involved in their biosynthesis and breakdown are reflected in the scope of Volume XVIII. It soon became apparent in the preparation of this work that some subdivision of the material was essential. Consequently, Volume XVIII of *Methods in Enzymology* will appear in three parts. A somewhat arbitrary division of the subject matter had to be made, and we make no apologies for the logic or lack of logic in the way this division was made. Part A covers the vitamin and coenzyme forms of ascorbate, thiamine, lipoate, pantothenate, biotin, and pyridoxine. Part B will cover nicotinate, flavins, and pteridines. Part C will cover the B_{12} group, ubiquinone, tocopherol, and vitamins A, K, and D.

For each vitamin–coenzyme group, detailed descriptions of current laboratory methods are given. Included are chemical and physical, enzymatic, and microbiological analyses; isolation and purification procedures for the coenzymes and derivatives and for the enzymes involved in their metabolism; chemical synthesis and reactions of natural forms, analogs, and radioactively labeled compounds; general metabolism including biosynthesis and degradation. In addition, information on properties and biochemical functions of the vitamins, coenzymes, and relevant enzymes are included.

We wish to thank the numerous contributors for their cooperation and patience. Though a few omissions of value to experimentalists in the area may have been made inadvertently, we believe that the subject has been adequately covered for the purposes intended. Even occasional overlaps, such as modifications of an assay procedure, were deliberately included to offer some flexibility in choice and to represent fairly the different researchers involved. We also wish to express our gratitude to Mrs. Patricia MacIntyre for her excellent secretarial assistance and to the numerous persons at Academic Press for their efficient and kind guidance.

DONALD B. MCCORMICK
LEMUEL D. WRIGHT

July, 1970

Contributors to Volume XVIII, Part A

Article numbers are shown in parentheses following the names of contributors.
Affiliations listed are current.

YASUSHI ABIKO (53, 57, 58, 59), *Research Laboratories, Daiichi Seiyaku Company Ltd., Tokyo, Japan*

ARTHUR F. ABT (1, 4), *1101 S. Paul Street, Baltimore, Maryland*

H. AHRENS (79, 81), *Department of Experimental Therapeutics, Roswell Park Memorial Institute, Buffalo, New York*

R. L. AIRTH* (14, 41, 42), *Cell Research Institute and Department of Botany, The University of Texas, Austin, Texas*

A. ALEXANDROV (13), *Department of Analytical Chemistry, High Institute of Natural Science and Mathematics, Plovdiv, Bulgaria*

CHARALAMPOS ARSENIS (9, 10, 11), *Department of Biological Chemistry, University of Illinois, Chicago, Illinois*

M. M. BAIG (2), *Florida State University, Tallahassee, Florida*

ROBERT N. BRADY (71), *Department of Biochemistry, Vanderbilt University, Nashville, Tennessee*

MYRON BRIN (23, 87), *Hoffmann-La Roche Inc., Nutley, New Jersey*

GENE M. BROWN (26, 34), *Department of Biology, Massachusetts Institute of Technology, Cambridge, Massachusetts*

RICHARD W. BURG (98) *Merck Sharp and Dohme, Research Laboratories, Rahway, New Jersey*

P. H. W. BUTTERWORTH (60, 61), *Department of Biochemistry, University College of London, London, England*

DONALD E. CADWALLADER (12), *School of Pharmacy, University of Georgia, Athens, Georgia*

A. F. CARLUCCI (62), *Scripps Institution of Oceanography, University of California at San Diego, La Jolla, California*

I. B. CHATTERJEE (3), *University College of Science, Department of Biochemistry, Calcutta, India*

C. J. CHESTERTON (60, 61), *Department of Virology, Royal Postgraduate Medical School, London, England*

JAMES E. CHRISTNER (65), *Ames Products Division Laboratory, Ames Company, Elkhart, Indiana*

MINOR J. COON (65), *Department of Biological Chemistry, The University of Michigan, Ann Arbor, Michigan*

J. R. COOPER (16, 39, 40), *Department of Pharmacology, Yale University School of Medicine, New Haven, Connecticut*

D. B. COURSIN (85), *St. Joseph Hospital, Lancaster, Pennsylvania*

V. DAS GUPTA (12), *College of Pharmacy, University of Houston, Houston, Texas*

S. DAVID (36), *Faculté des Sciences, Orsay, France*

D. C. DeJONGH (80), *Department of Chemistry, Wayne State University, Detroit, Michigan*

WALTER B. DEMPSEY (95), *Veterans Administration Hospital, Dallas, Texas*

BRUNO DEUS (38, 45), *Department of Internal Medicine, University of Freiburg, Freiburg, West Germany*

B. ESTRAMAREIX (36), *Faculté des Sciences, Orsay, France*

V. L. FLORENTIEV (89), *Institute of Molecular Biology, Academy of Sciences of the USSR, Moscow, USSR*

G. ELIZABETH FOERSTER (14), *Cell Research Institute and Department of Botany, The University of Texas, Austin, Texas*

WERNER FÖRY (72), *J. R. Geigy, AG., Basel, Switzerland*

H. FRICK (92), *Technical Department, F. Hoffmann-La Roche & Co., Basel, Switzerland*

SATORU FUJII (84), *Second Department of Internal Medicine, Osaka City University, Medical School, Osaka, Japan*

* Deceased.

SABURO FUKUI (33, 90, 91), *Laboratory of Industrial Biochemistry, Department of Industrial Chemistry, Kyoto University, Kyoto, Japan*

N. M. GREEN (73, 74), *National Institute for Medical Research, London, England*

C. J. GUBLER (20, 21, 22, 37), *Department of Chemistry, Brigham Young University, Provo, Utah*

SIGMUNDUR GUDBJARNASON (17), *Wayne State University, School of Medicine, Detroit, Michigan*

J. R. GUEST (46), *Department of Microbiology, University of Sheffield, Sheffield, England*

KIYOSHI HARADA (29), *Pharmaceutical Research Laboratory, Tanabe Seiyaku Company Ltd., Osaka, Japan*

BETTY E. HASKELL (86), *Department of Food Science, University of Illinois, Urbana, Illinois*

TARO HAYAKAWA (51), *Department of Pathological Biochemistry, Atomic Disease Institute, Nagasaki University School of Medicine, Nagasaki-shi, Japan*

A. A. HERBERT (46), *Department of Microbiology, University of Sheffield, Sheffield, England*

HELMUT HOLZER (45), *Department of Biochemistry, School of Medicine, University of Freiburg, Freiburg, Germany*

HIROSHI HIRANO (25), *Chemical Research Laboratories, Research and Development Division, Takeda Chemical Industry Ltd., Osaka, Japan*

H. HIRSHFELD (36), *The Weizmann Institute, Rehovoth, Israel*

MIYOSHI IKAWA (88), *Department of Biochemistry, University of New Hampshire, Durham, New Hampshire*

Y. ITOKAWA (16, 39), *Department of Hygiene, Kyoto University, Kyoto, Japan*

V. I. IVANOV (89), *Institute of Molecular Biology, Academy of Sciences of the USSR, Moscow, USSR*

SHOJIRO IWAHARA (70), *Department of Agricultural Biochemistry, Kagawa University, Kagawa, Japan*

JEAN-CLAUDE JATON (93), *Cardiac Unit, Massachusetts General Hospital, Boston, Massachusetts*

L. R. JOHNSON (22), *Endocrine Laboratory and Department of Medicine, Latter Day Saints Hospital, Salt Lake City, Utah*

ROBERT J. JOHNSON (77), *Departments of Biochemistry and Biophysics, Iowa State University, Ames, Iowa*

ELLIOT JUNI (43), *Department of Microbiology, The University of Michigan, Medical School, Ann Arbor, Michigan*

YASUO KAGAWA (5, 6), *Division of Biological Sciences, Section of Biochemistry and Molecular Biology, Cornell University, Ithaca, New York*

M. YA. KARPEISKY (89), *Institute of Molecular Biology, Academy of Sciences of the USSR, Moscow, USSR*

KEIICHI KOHNO (29), *Pharmaceutical Research Laboratory, Tanabe Seiyaku Company Ltd., Osaka, Japan*

HIROYUKI KOIKE (18), *Central Research Laboratories, Sankyo Company Ltd., Tokyo, Japan*

MASAHIKO KOIKE (50, 51), *Department of Pathological Biochemistry, Atomic Disease Institute, Nagasaki University School of Medicine, Nagasaki-shi, Japan*

S. G. KORENMAN (76), *University of Iowa School of Medicine, Iowa City, Iowa*

W. KORYTNYK (79, 80, 81, 83, 88), *Department of Experimental Therapeutics, Roswell Park Memorial Institute, Buffalo, New York*

FRANKLIN R. LEACH (48, 49), *Department of Biochemistry, Oklahoma State University, Stillwater, Oklahoma*

IRWIN G. LEDER (27, 28, 35), *National Institute of Arthritis and Metabolic Diseases, National Institutes of Health, Bethesda, Maryland*

HENG-CHUN LI (67, 69), *Department of Biochemistry, Massachusetts Institute of Technology, Cambridge, Massachusetts*

F. LOEWUS (2), *Department of Biology, State University of New York at Buffalo, Buffalo, New York*

DONALD B. McCORMICK (47, 63, 64, 67, 69, 70, 71, 72, 94), *Graduate School of Nutrition and Section of Biochemistry and*

Molecular Biology, Cornell University, Ithaca, New York

YOSHITAKE MANO (7), Department of Biochemistry, Faculty of Medicine, University of Tokyo, Tokyo, Japan

FUMINORI MASUGI (91), Kyoto University, Laboratory of Industrial Chemistry, Kyoto, Japan

TAIZO MATSUKAWA (25), Wako Pure Chemical Industry Ltd., Osaka, Japan

HENRY G. MAUTNER (56), Department of Biochemistry and Pharmacology, Tufts University School of Medicine, Boston, Massachusetts

DAVID E. METZLER (77), Departments of Biochemistry and Biophysics, Iowa State University, Ames, Iowa

A. V. MOREY (43), Department of Microbiology, Case Western Reserve University, Cleveland, Ohio

YOSHIHARU NAKAI (91), Laboratory of Industrial Biochemistry, Department of Industrial Chemistry, Kyoto University, Kyoto, Japan

R. A. NEAL (24), Department of Biochemistry, School of Medicine, Vanderbilt University, Nashville, Tennessee

P. C. NEWELL (30), Microbiology Unit, Department of Biochemistry, University of Oxford, Oxford, England

KOICHI OGATA (66, 68, 97), Department of Agricultural Chemistry, Kyoto University, Kyoto, Japan

NOBUKO OHISHI (33, 90), University of Nagoya, Institute of Biochemistry, Faculty of Medicine, Nagoya, Japan

KIYOSHI OKUDA (84), Clinical Chemistry Laboratory, Osaka City University Medical School, Osaka City, Japan

B. W. O'MALLEY (76), Department of Biology, Vanderbilt University, Nashville, Tennessee

RAÚL N. ONDARZA (54), Departamento de Bioquimica, Facultad de Medicina, Universidad Nacional Autónoma de México, México, D. F.

C. W. M. ORR (8), McCollum-Pratt Institute, Johns Hopkins University, Baltimore, Maryland

JOHN W. PORTER (60, 61), Department of Physiological Chemistry, University of Wisconsin, Madison, Wisconsin and Lipid Metabolism Laboratory, Veterans Administration Hospital, Madison, Wisconsin

ALAN B. PRITCHARD (47), General Foods Corporation, Technical Center, White Plains, New York

A. R. PROSSER (52, 82), Division of Nutrition, Bureau of Foods, Pesticides and Product Safety, Food and Drug Administration, Washington, D.C.

EDWARD F. ROGERS (44), Merck Sharp and Dohme, Research Laboratories, Merck and Company, Rahway, New Jersey

JEROME A. ROTH (63), Section of Biochemistry and Molecular Biology, Cornell University, Ithaca, New York

HELMUT RUIS (64, 71), Division of Biochemistry, Organisch-Chemisches Institüt der Universität, Vienna, Austria

S. F. SCHAEREN (92), Research Department, F. Hoffmann-La Roche & Co., Basel, Switzerland

A. J. SHEPPARD (52, 82), Division of Nutrition, Bureau of Foods, Pesticides and Product Safety, Food and Drug Administration, Washington, D.C.

NORIO SHIMAZONO (5, 6, 7), Department of Biochemistry, Tokyo Medical College, Tokyo, Japan

MASAO SHIMIZU (55), Research Laboratories, Daiichi Seiyaku Company Ltd., Tokyo, Japan

SHOICHI SHIMIZU (90), Laboratory of Industrial Biochemistry, Department of Industrial Chemistry, Kyoto University, Kyoto, Japan

P. SINAŸ (36), Faculté des Sciences, Orléans-la Source, France

ESMOND E. SNELL (86, 94), Department of Biochemistry, University of California, Berkeley, California

P. R. SUNDARESAN (85), Lipids Laboratory, Research Institute, St. Joseph Hospital, Lancaster, Pennsylvania

KANTARO SUZUKI (50), Department of Biochemistry, Nihon University School of Dentistry, Tokyo, Japan

S. TAKANASHI (78), *Drug Metabolism Laboratory, Research Laboratories, Chugai Pharmaceutical Co., Ltd., Tokyo, Japan*

Z. TAMURA (78), *Faculty of Pharmaceutical Sciences University of Tokyo, Tokyo, Japan*

YOSHIKI TANI (97), *Department of Agricultural, Chemistry, Kyoto University, Kyoto, Japan*

JUDITH P. TEPPER (67), *Section of Biochemistry and Molecular Biology, Cornell University, Ithaca, New York*

HOWARD TIECKELMANN (31), *State University of New York at Buffalo, Buffalo, New York*

RAYMOND V. TOMLINSON (31), *Syntex, Palo Alto, California*

PAUL F. TORRENCE (31), *National Institutes of Health, Bethesda, Maryland*

OSCAR TOUSTER (9, 10, 11), *Department of Molecular Biology, Vanderbilt University, Nashville, Tennessee*

GORO TSUKAMOTO (29), *Pharmaceutical Research Laboratory, Tanabe Seiyaku Company Ltd., Osaka, Japan*

R. G. TUCKER (30), *Public Health Laboratories, Radcliffe Infirmary, Oxford, England*

JOHANNES ULLRICH (19, 45), *Department of Biochemistry, School of Medicine, University of Freiburg, Freiburg, Germany*

HANNA UNGAR-WARON (93), *Department of Chemical Immunology, The Weizmann Institute of Science, Rehovoth, Israel*

ISAMU UTSUMI (29), *Pharmaceutical Research Laboratory, Tanabe Seiyaku Company Ltd., Osaka, Japan*

P. VASSILEVA-ALEXANDROVA (13), *Department of Analytical Chemistry, Institute of Food Industry, Plovdiv, Bulgaria*

SUSANNE L. VON SCHUCHING (1, 4), *Nuclear Medicine Research Department, Veterans Administration Center, Martinsburg, West Virginia*

HIROSHI WADA (96), *Department of Biochemistry, Osdka University, School of Medicine, Kitaku, Osaka*

MASAHISA WADA (84), *Second Department of Internal Medicine, Osaka City University Medical School, Osaka City, Japan*

P. P. WARING (15, 32), *U.S. Army Medical Research and Nutrition Laboratory, Denver, Colorado*

RU-DONG WEI (75), *Department of Biochemistry and the Kohlberg Laboratory, National Defense Medical Center, Medical Research Laboratory, Veterans General Hospital, Taipei, Taiwan*

ROSEANN WHITE (95), *Florida Technological University, Orlando, Florida*

JAMES L. WITTLIFF (41, 42), *Department of Biochemistry, University of Rochester, School of Medicine and Dentistry, Rochester, New York*

J. H. WITTORF (20, 21, 22), *Chemical Abstracts Service, Columbus, Ohio*

LEMUEL D. WRIGHT (47, 64, 67, 69, 70, 71), *Graduate School of Nutrition and Section of Biochemistry and Molecular Biology, Cornell University, Ithaca, New York*

J. WURSCH (92), *Research Department, F. Hoffmann-La Roche & Co., Basel, Switzerland*

SHOJIRO YURUGI (25), *Chemical Research Laboratories, Research and Development Division, Takeda Chemical Industry Ltd., Osaka, Japan*

TAKASHI YUSA (18), *Central Research Laboratories, Sankyo Company Ltd., Tokyo, Japan*

Z. Z. ZIPORIN (15, 32), *U.S. Army Medical Research and Nutrition Laboratory, Denver, Colorado*

METHODS IN ENZYMOLOGY

EDITED BY

Sidney P. Colowick and Nathan O. Kaplan

VANDERBILT UNIVERSITY
SCHOOL OF MEDICINE
NASHVILLE, TENNESSEE

DEPARTMENT OF CHEMISTRY
UNIVERSITY OF CALIFORNIA
AT SAN DIEGO
LA JOLLA, CALIFORNIA

I. Preparation and Assay of Enzymes
II. Preparation and Assay of Enzymes
III. Preparation and Assay of Substrates
IV. Special Techniques for the Enzymologist
V. Preparation and Assay of Enzymes
VI. Preparation and Assay of Enzymes (*Continued*)
Preparation and Assay of Substrates
Special Techniques
VII. Cumulative Subject Index

METHODS IN ENZYMOLOGY

EDITORS-IN-CHIEF

Sidney P. Colowick   Nathan O. Kaplan

VOLUME VIII. Complex Carbohydrates
Edited by ELIZABETH F. NEUFELD AND VICTOR GINSBURG

VOLUME IX. Carbohydrate Metabolism
Edited by WILLIS A. WOOD

VOLUME X. Oxidation and Phosphorylation
Edited by RONALD W. ESTABROOK AND MAYNARD E. PULLMAN

VOLUME XI. Enzyme Structure
Edited by C. H. W. HIRS

VOLUME XII. Nucleic Acids (Parts A and B)
Edited by LAWRENCE GROSSMAN AND KIVIE MOLDAVE

VOLUME XIII. Citric Acid Cycle
Edited by J. M. LOWENSTEIN

VOLUME XIV. Lipids
Edited by J. M. LOWENSTEIN

VOLUME XV. Steroids and Terpenoids
Edited by RAYMOND B. CLAYTON

VOLUME XVI. Fast Reactions
Edited by KENNETH KUSTIN

VOLUME XVII. Metabolism of Amino Acids and Amines (Parts A and B)
Edited by HERBERT TABOR AND CELIA WHITE TABOR

VOLUME XVIII. Vitamins and Coenzymes (Parts A, B, and C)
Edited by DONALD B. MCCORMICK AND LEMUEL D. WRIGHT

VOLUME XIX. Proteolytic Enzymes
Edited by GERTRUDE E. PERLMANN AND LASZLO LORAND

VOLUME XX. Nucleic Acids (Part C)
Edited by KIVIE MOLDAVE AND LAWRENCE GROSSMAN

Methods in Enzymology

Volume XVIII
VITAMINS AND COENZYMES
Part A

Section I

L-Ascorbic Acid

[1] Syntheses of Specifically Labeled Intermediates Used in the Preparation of Carbon- and Tritium-Labeled L-Ascorbic Acid

By Susanne L. von Schuching and Arthur F. Abt

I. Introduction

A study of the structure of L-ascorbic acid reveals that the spatial requirements of the hydroxyl groups on carbons 4 and 5 together with the chemical instability of the compound have precluded the development of a total synthesis. Both principal routes to L-ascorbic acid—(1) from L-sorbose, first described by Reichstein and Grüssner[1] (Fig. 1), and (2) from L-xylose and cyanide described by Reichstein et al.[2] and Ault et al.[3] (Fig. 2)—make use of available carbohydrates as building blocks. Both syntheses involve isomerization and lactonization of a 2-keto-L-hexonic acid or 3-keto-L-hexonic acid as the final step in the preparation of L-ascorbic acid.

The first investigators preparing ^{14}C-labeled L-ascorbic acid have adapted both of the original syntheses on a semimicro scale, incorporating improvements to raise the isotopic yield of L-ascorbic acid-[14]C. The radioactive carbon was introduced either by preparing labeled glucose as the starting material[4,5] or by introducing ^{14}C as labeled cyanide.[6,7]

A list of intermediates of the two syntheses is given in order to clarify the various approaches for the preparation of labeled L-ascorbic acid as well as the position of the labeled carbon or tritium in the molecule.

II. Outline of Syntheses

Synthesis 1.[1] L-Ascorbic Acid via L-Sorbose Pathway (Fig. 1)

Sequence for the preparation of L-ascorbic acid from 2-keto-L-gulonic acid:

D-Glucose
D-Glucitol (sorbitol)

[1] T. Reichstein and A. Grüssner, *Helv. Chim. Acta* **17**, 311 (1934).
[2] T. Reichstein and A. Grüssner, and R. Oppenauer, *Helv. Chim. Acta* **17**, 510 (1934).
[3] R. D. Ault, D. K. Baird, H. C. Carrington, W. N. Haworth, R. W. Herbert, E. L. Hirst, E. G. Percival, F. Smith, and M. Stacy, *J. Chem. Soc. (London)* p. 1419 (1933).
[4] H. L. Frush and H. S. Isbell, *J. Res. Natl. Bur. Std.* **59**, 289 (1957).
[5] P. G. Dayton, *J. Org. Chem.* **21**, 1535 (1956).
[6] J. J. Burns and C. G. King, *Science* **111**, 257 (1950).
[7] L. L. Salomon, J. J. Burns, and C. G. King, *J. Am. Chem. Soc.* **74**, 5161 (1952).

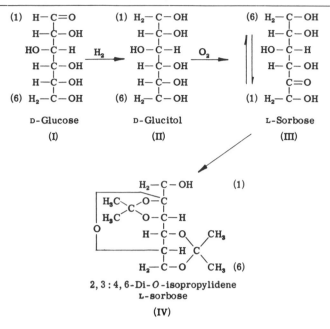

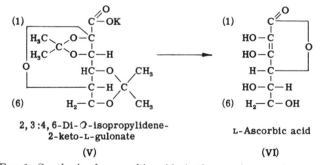

FIG. 1. Synthesis of L-ascorbic acid via the L-sorbose pathway.

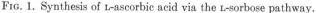

L-Sorbose
2,3:4,6-Di-*O*-isopropylidene-L-sorbose
Potassium 2,3:4,6-di-*O*-isopropylidene-L-sorbose
2-Keto-L-gulonate
L-Ascorbic acid

An apparent inversion of the carbon chain occurs during the oxidation of D-glucitol to L-sorbose because of the numbering of the carbon atoms.

Synthesis 2.[2,3] L-Ascorbic Acid via L-Xylosone and Cyanide (Fig. 2)

Sequence for the preparation of L-ascorbic acid from 3-keto-L-gulonic acid:

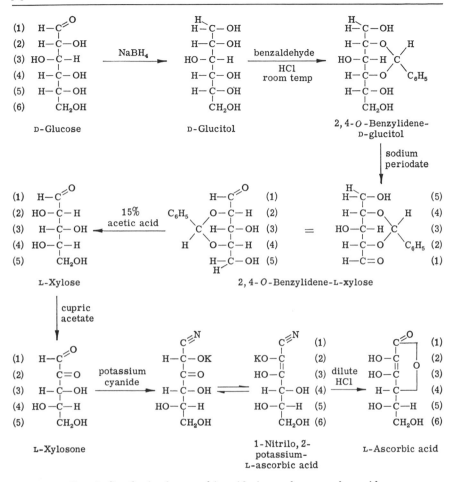

FIG. 2. Synthesis of L-ascorbic acid via L-xylosone and cyanide.

D-Glucose
D-Glucitol (sorbitol)
2,4-O-Benzylidene-D-glucitol
2,4-O-Benzylidene-L-xylose
L-Xylose
L-Xylosone + KCN
Imino-L-ascorbic acid
3-Keto-L-gulonic acid
L-Ascorbic acid

Apparent inversion of the carbon chain with loss of carbon 6 of glucose occurs during the glycol cleavage of 2,4-O-benzylidene-D-glucitol to 2,4-O-benzylidene-L-xylose because of the system of numbering carbon atoms.

III. Location of [14]C Label

L-Ascorbic Acid-1- through 6-[14]C Using D-Glucose-1- through 6-^{14}C as Starting Material (Synthesis 1; Fig. 1)

Ascorbic acid synthesized from D-glucose labeled in any of its carbon positions with ^{14}C yields ascorbic acid labeled at the opposite site of the chain due to the apparent inversion of the molecule during the oxidation of D-glucitol to L-sorbose, i.e., D-glucose-1-^{14}C is converted to L-ascorbic acid-6-^{14}C and D-glucose-2-^{14}C to L-ascorbic acid-5-[14]C.

L-Ascorbic Acid-2- through 6-[14]C Using D-Glucose-1- through 5-^{14}C as Starting Material (Synthesis 2; Fig. 2)

D-Glucose loses carbon 6 during the glycol cleavage of 2,4-mono-*O*-benzylidene-D-glucitol to 2,4-mono-*O*-benzylidene-L-xylose with an apparent inversion of the carbon chain; i.e., carbon 1 of glucose becomes carbon 6 of ascorbic acid; carbon 5 of glucose becomes carbon 2 of ascorbic acid.

L-Ascorbic Acid-2- through 6-[14]C Using L-Xylose-1- through 5-^{14}C as Starting Material (Synthesis 2; Fig. 2)

L-Xylose labeled in any of its 5 carbon positions yields L-ascorbic acid-^{14}C labeled in the 2 through 6 positions, due to the lengthening of the chain at carbon 1 of xylose; i.e., L-xylose-5-^{14}C yields L-ascorbic acid-6-^{14}C, and L-xylose-1-^{14}C yields L-ascorbic acid-2-[14]C.

L-Ascorbic Acid-1-^{14}C from L-Xylosone and Cyanide-[14]C (Synthesis 2; Fig. 2)

L-Ascorbic acid-1-^{14}C is obtained by coupling cyanide-^{14}C to L-xylosone, followed by hydrolysis of the resulting nitrilo-L-ascorbic acid-1-[14]C.

The use of cyanide-^{14}C to introduce labeled carbon into the C-1 position of L-ascorbic acid is the most widely used method for preparing labeled vitamin C.

IV. Preparation of Intermediates

Cyanide-[14]C

This intermediate is used for introducing labeled carbon by lengthening the carbon chain.

A number of syntheses for the reduction of barium carbonate-^{14}C to cyanide-^{14}C in the presence of ammonia have been discussed in Volume IV of this series.[8,9] An additional synthesis based on the same principle has

---

[8] S. Abraham and W. Z. Hassid, Vol. IV, p. 489.
[9] H. S. Anker, Vol. IV, p. 779.

been described.[10] All the syntheses require special equipment and experience.[10a] Na[14]CN is used in the preparation of glucose-[14]C.

D-Glucose-1-^{14}C obtained by the methods outlined below has been used in the preparation of L-ascorbic acid-1-^{14}C via the sorbose pathway on a semimicro scale.[4,5] The improvements introduced by these investigators make the synthesis of the chemically labile ascorbic acid by this pathway feasible. D-Glucose labeled with ^{14}C in any of its carbon positions may be employed without any change in the procedure.[10b]

It is realized that glucose labeled with ^{14}C in other than the end positions of the chain requires more intricate syntheses, which in turn reduce the radiochemical yield of the desired end product, labeled vitamin C. Nevertheless this disadvantage might be small in comparison to the gain of new information obtained by employing a differently tagged L-ascorbic acid in a metabolic study.

D-Glucose-1-^{14}C from D-Arabinose and Cyanide-^{14}C for the Preparation of L-Ascorbic Acid-6-[14]C[4,5]

The original cyanohydrin synthesis of Kiliani[11] consists of the addition of cyanide to an aldose, hydrolysis of the resulting epimeric cyanohydrins to aldonic acids, lactonization, and reduction to aldoses. The mixture of epimers has to be separated; this is done at the aldonic acid stage.

Isbell et al.[12,13] have modified the synthesis to favor the formation of D-glucose over its epimer D-mannose from D-arabinose and Na[14]CN. A radiochemical yield of 45% of D-glucose-^{14}C has been achieved on the basis of the Na[14]CN used.

D-Glucose-1-^{14}C from D-Arabinose and Nitromethane-[14]C

The procedure developed by Sowden[14,15] has been carried out with radioactive carbon. The yield of nitromethane from methanol-^{14}C is low (16%), reducing the final radiochemical yield of D-glucose-1-[14]C.

[10] S. von Schuching and T. Enns, *J. Am. Chem. Soc.* **78**, 4255 (1956).

[10a] It might be more economical to purchase cyanide-^{14}C from a commercial laboratory for occasional use.

[10b] H. L. Frush, L. T. Sniegoski, N. B. Holt, and H. S. Isbell, *J. Res. Natl. Bur. Std.* **69A**, 535 (1965).

[11] H. Kiliani, *Chem. Ber.* **19**, 767, 1128 (1886).

[12] H. S. Isbell, J. V. Karabinos, H. L. Frush, N. B. Holt, A. Schwebel, and T. T. Galkowski, *J. Res. Natl. Bur. Std.* **48**, 163 (1952).

[13] S. Abraham and W. Z. Hassid, Vol. IV, p. 512.

[14] J. C. Sowden, *Science* **109**, 229 (1949).

[15] S. Abraham and W. Z. Hassid, Vol. IV, p. 522.

D-Glucose-2-^{14}C from D-Arabinose-1-^{14}C for the Preparation of L-Ascorbic Acid-5-[14]C[4,5]

D-Arabinose-1-^{14}C has been obtained by Frush and Isbell[16] by lengthening the chain of D-erythrose with cyanide-^{14}C in 60% radiochemical yield. D-Arabinose-^{14}C may then be converted to D-glucose-2-^{14}C with inactive cyanide.[12,13]

The procedure for the preparation of D-arabinose according to these authors is carried out as follows:

Potassium D-*Arabonate-1-*[14]*C.* To a tube containing a solution of 3.1 millimoles of sodium cyanide-^{14}C, 5 millimoles of sodium hydroxide and 10 ml of water, frozen on the sides by cooling in a dry-ice bath, is added a small lump of carbon dioxide and a solution of 5 millimoles of sodium bicarbonate and 4 millimoles of D-erythrose in 10 ml of water. The mixture is thawed by shaking the loosely stoppered tube in an ice bath, then stored at 0° for 1 day and at room temperature for 5 days. The nitriles are hydrolyzed by heating the mixture on a steam bath for 7 hours in a stream of air, with addition of water as required. The residue is dissolved in water and passed through a cation-exchange column with thorough washing. The combined effluent and washings are concentrated under reduced pressure to about 25 ml and neutralized with potassium hydroxide. The mixture is diluted with 2.0 g of carrier, concentrated to 4 ml, diluted with methanol, and stored in a refrigerator to crystallize. The product is collected, recrystallized from water–ethanol, washed with cold, aqueous ethanol (1:4), and dried in a vacuum desiccator. The yield is 2.23 g, 42.7%, on a radiochemical basis.

D-*Arabono-γ-lactone-1-*[14]*C.* An aqueous solution containing 0.446 g (2.19 millimoles) of potassium D-arabonate-1-^{14}C is passed through a cation-exchange column of Amberlite IR-120 with thorough washing. The effluent and washings are evaporated to a syrup under reduced pressure, and the D-arabonic acid-1-^{14}C is transferred with a few drops of water to a reduction tube.[13] The solution is diluted with 5 ml of glacial acetic acid and heated for 5 hours in a boiling water bath, under an air stream. At 1 hour, the mixture is treated with 1 ml of water and 5 ml of glacial acetic acid; three 5-ml portions of acetic acid are added at 1-hour intervals. The pale yellow residue is dissolved in a few drops of methanol, seeded, and stored over calcium chloride for 1 day. The yield is 94%.

D-*Arabinose-1-*[14]*C.* To an ice-cooled, vigorously stirred mixture of 6.4 g of crystalline sodium acid oxalate, 20 ml of water and the D-arabono-γ-lactone-1-^{14}C is added 9.2 g of 5% sodium amalgam pellets.[13] After 3 hours of stirring, the mercury is removed, the solution is diluted with 5 volumes

[16] H. L. Frush and H. S. Isbell, *J. Res. Natl. Bur. Std.* **51**, 307 (1953).

of methanol, and the precipitated salts are filtered off and washed with methanol. The filtrate is cooled, made just basic to phenolphthalein with aqueous sodium hydroxide, concentrated under reduced pressure to 10 ml, and diluted with 5 volumes of methanol. The precipitated salts are filtered off and washed with methanol. The alcohol is evaporated under diminished pressure, and the aqueous solution is passed through a deionizing column with thorough washing. The combined solution is concentrated to 10 ml under reduced pressure (below 40°) and filtered through a bed of decolorizing charcoal into a 50-ml flask, where the filtrate and washings are lyophilized. The residue is dissolved in several drops of methanol and treated with isopropyl alcohol to incipient turbidity, then seeded and stored for several days to crystallize. The mother liquor is removed, and the product is washed with a few drops of methanol:propanol (2:1) and recrystallized from methanol solution by treatment with isopropyl alcohol. The yield is 0.185 g, 56.7%, on a radiochemical basis. The yield may be raised to 60% by treating the mother liquor with 0.1 g of carrier.

L-Xylose-1-^{14}C for the Preparation of L-Ascorbic Acid-2-^{14}C[6,7]

The preparation of L-xylose-1-^{14}C may be achieved by following the outline for the preparation of D-xylose-1-^{14}C from D-threose and Na^{14}CN given by Isbell et al.[17] using L-threose.

Imino-L-ascorbic Acid-1-^{14}C from L-Xylosone and Na^{14}CN for the Preparation of L-Ascorbic Acid-1-^{14}C[6,7,18]

The application of the xylosone-cyanide route[2,3] to the preparation of L-ascorbic acid-1-^{14}C was first described by Burns and King.[6]

The most important modification introduced by the first investigators[6,7] was the purification step in which ascorbic acid was separated from mineral salts after the hydrolysis of imino-L-ascorbic acid using an ion-exchange column.

In carrying out this synthesis, it was determined that the greatest loss of radioactive isotope was due to a low yield of L-ascorbic acid during the ion-exchange purification step. The difficulty was caused by large amounts of mineral salts and other impurities in the hydrolysis solution which interfered with the absorption of the small amounts of ascorbic acid on the anion-exchange column. It may be mentioned that ascorbic acid solutions containing minor amounts of contamination with mineral salts are absorbed and recovered quantitatively from an anion-exchange column. It was

[17] H. S. Isbell, H. L. Frush, and N. B. Holt, J. Res. Natl. Bur. Std. **53**, 325 (1954).

[18] S. L. von Schuching and G. H. Frye, Biochem. J. **98**, 652 (1966).

possible to improve the yield of radioactive ascorbic acid at this point by the isolation of ascorbic acid as the 5,6-mono-O-cyclohexylidene derivative.[18] After hydrolysis of this derivative, crystalline L-ascorbic acid was obtained.

Preparation of 5,6-Mono-O-cyclohexylidene-L-ascorbic Acid

Two methods are described for the preparation of the cyclohexylidene derivative of ascorbic acid and its hydrolysis to free ascorbic acid.

Method I

L-Ascorbic acid (1.0 g, 5.7 millimoles) is covered with 10 ml of redistilled cyclohexanone containing 0.14 ml of concentrated H_2SO_4. The solution becomes clear after shaking for 10 minutes and is allowed to stand for an additional 12 hours. The solution is then diluted with 50 ml of methanol and neutralized with solid $NaHCO_3$. After removal of the precipitated Na_2SO_4 by filtration, the solution is concentrated to a small volume *in vacuo* at 65°, and n-heptane (500 ml) is added. The resulting precipitate is twice crystallized from acetone (10 ml) and n-heptane (500 ml). The yield is 1.1 g (77%).

Method II

L-Ascorbic acid (1.0 g, 5.7 millimoles) is covered with 10 ml of redistilled cyclohexanone and shaken with 1 g of anhydrous copper sulfate for 4 hours. After filtration, the cyclohexanone is reduced to a small volume *in vacuo* at 65°, and 500 ml of n-heptane is added. Recrystallization of the precipitate is carried out as above. The yield is 1.28 g (90%).

Hydrolysis of 5,6-Mono-O-cyclohexylidene Ascorbic Acid

This compound may be hydrolyzed to give free ascorbic acid with 10% formic acid for 24 hours at room temperature. It may also be treated with a 10-fold excess of Dowex 50 (H^+ form) for 1 hour in water. L-Ascorbic acid is crystallized after evaporation of the solvent.

Synthesis of L-Ascorbic Acid-1-^{14}C with 5,6-Mono-O-cyclohexylidene-L-ascorbic Acid as an Intermediate[18]

L-Xylosone (1.2 g, 8 millimoles) is condensed with $Na^{14}CN$ (0.294 g, 6 millimoles, 10 mCi, of ^{14}C) according to the procedure of Salomon *et al.*,[7] but no carrier ascorbic acid is added before hydrolysis of the imino-L-ascorbic acid. The solution after hydrolysis is evaporated *in vacuo* at 40°. The syrup is thoroughly dried at room temperature with the aid of an oil pump. Next, 10 ml of redistilled cyclohexanone and 0.14 ml of concentrated

H_2SO_4 are added to the residue, and the mixture is shaken for 10 minutes and kept at room temperature for another 12 hours. Ethanol (50 ml) is added, and also enough $NaHCO_3$ to neutralize the solution. The resulting salts are filtered off, the solution is concentrated *in vacuo* at 65° to 10 ml, and *n*-heptane (500 ml) is added. After 24 hours at $-20°$, 5,6-mono-*O*-cyclohexylidene-L-ascorbic acid-1-^{14}C settles partly at the bottom of the flask as a sticky mass, and another part floats in the heptane layer. The *n*-heptane layer is filtered, and the precipitates are obtained. Further purification is carried out by reprecipitation from acetone and *n*-heptane and from acetone and methylcyclohexanone. Separation from inorganic ions, which interfere with the crystallization of ascorbic acid, is thus achieved. The air-dried material is hydrolyzed to free ascorbic acid by stirring the material in 50 ml of water and 5 g of Dowex 50 (H$^+$ form) resin, for 1 hour. The resin is filtered off, and 700 mg of L-ascorbic acid-1-^{14}C, specific activity 7μCi/mg (43% radiochemical yield), is obtained.

Paper chromatography is carried out using butan-1-ol:acetic acid: water (4:1:5 v/v/v) on Whatman 3MM paper.[19,20] Radioactive purity is established by scanning a paper chromatogram with a windowless scanner. A single peak coincides with the position of unlabeled ascorbic acid on the paper strip.

V. Location of Tritium Label in L-Ascorbic Acid-T

A tritiated ascorbic acid is suitable for metabolic studies if it contains the isotope in a stable position. The hydrogen atoms attached to carbons 4, 5, and 6 (one each on carbons 4 and 5, and two on carbon 6), are stable and may be replaced by tritium. By preparing a suitably labeled carbohydrate as an intermediate, the two basic syntheses outlined in Figs. 1 and 2 are applicable for the purpose.

In determining the position of the tritium label in the ascorbic acid molecule, the same considerations as those outlined for the position of the carbon label also apply. First, the apparent inversion of the chain has to be taken into account in determining the position of the isotope; and second, attention has to be paid to the possible loss of the isotope during the sequence of reactions.

The number of approaches to a tritiated L-ascorbic acid is smaller when compared to the number of approaches to L-ascorbic acid-^{14}C, since only three positions are available, compared to six positions available for the introduction of ^{14}C.

[19] S. Partridge and R. Westall, *J. Biol. Chem.* **42**, 238 (1948).
[20] L. W. Mapson and S. Partridge, *Nature* **164**, 479 (1949).

POSITION OF TRITIUM LABEL IN L-ASCORBIC ACID SYNTHESIZED
FROM TRITIATED INTERMEDIATES

L-Ascorbic Acid-6-T from D-Glucitol-1-T (Synthesis 1; Fig. 1)

D-Glucitol-1-T yields L-ascorbic acid-6-T.

L-Ascorbic Acid-4 -through 6-T from L-Xylose-3 through 5-T (Synthesis 2; Fig. 2)

L-Xylose labeled with tritium in positions 3, 4, or 5, yields L-ascorbic acid-T labeled in positions 4, 5, or 6.

The most important route for introducing tritium into a specific location of the carbohydrate chain is the reduction of lactone, keto, or aldehydic group of a carbohydrate to either an aldose or alditol. The use of sodium amalgam in acid solution for reduction has been introduced by Fischer.[21] Sodium amalgam in tritiated water has been employed by Isbell and his group.[22, 23]

Procedures originally developed for the application of borohydrides in the sugar series[24-29] served as models for the isotopic applications achieving higher specific activity. $LiBT_4$ and $NaBT_4$ are the most widely used agents.[30-32]

Preparation of Tritiated Intermediates

Lithium Borotritide

These compounds have been prepared by direct exchange of $LiBH_4$ in the dry state with tritium gas at a temperature of 200° for 97 hours.[33]

Detailed instructions for preparing tritiated lithium borohydride in a safe manner have been given.[34]

[21] E. Fischer, *Chem. Ber.* **24**, 2133, 3684, 1890.

[22] S. Abraham and W. Z. Hassid, Vol. IV, p. 514.

[23] H. S. Isbell, H. L. Frush, N. B. Holt, and J. D. Moyer, *J. Res. Natl. Bur. Std.* **64A**, 177 (1960).

[24] M. L. Wolfrom and H. B. Wood, *J. Am. Chem. Soc.* **73**, 2933 (1951).

[25] M. Abdel-Akher, J. K. Hamilton, and F. Smith, *J. Am. Chem. Soc.* **73**, 4691 (1951).

[26] M. L. Wolfrom and K. Anno, *J. Am. Chem. Soc.* **74**, 5583 (1952).

[27] Bibliography: Borohydrides in Cellulose and Sugar Chemistry. Metal Hydrides, Inc., Beverly, Massachusetts, 1959.

[28] H. L. Frush and H. S. Isbell, *J. Am. Chem. Soc.* **78**, 2844 (1956).

[29] J. K. Hamilton and F. Smith, *J. Am. Chem. Soc.* **76**, 3543 (1954).

[30] G. Moss, *Arch. Biochem. Biophys* **90**, 111 (1960).

[31] H. S. Isbell, H. L. Frush, and J. D. Moyer, *J. Res. Natl. Bur. Std.* **64A**, 359 (1960).

[31a] H. L. Frush, H. S. Isbell, and A. J. Fatiadi, *J. Res. Natl. Bur. Std.* **64A**, 433 (1960).

[32] M. Urguiza and N. N. Lichtin, *Tappi* **44**, 221 (1961).

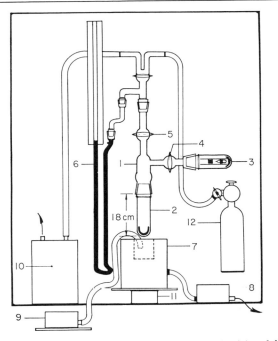

FIG. 3. Exchange apparatus used in the preparation of tritiated borohydrides. *1*, Three-way manifold; *2*, exchange vessel lined with platinum crucible; *3*, vessel containing glass wool, tritium gas vial, and iron bar for breaking the vial; *4* and *5*, 6-mm stopcocks; *6*, manometer; *7*, Hoskins electric furnace; *8*, Variac; *9*, galvanometer; *10*, vacuum pump; *11*, lab jack; *12*, hydrogen cylinder.

Sodium Borotritide

This compound is also prepared by direct exchange of $NaBH_4$ with tritium gas in the dry state, but higher temperatures are required; 350° and 24 hours.[34]

Figure 3 presents the diagram of the complete vacuum line used in conjunction with the exchange vessel in the authors' laboratory. The tritium gas is contained in a glass ampoule which is broken at the start of the exchange reaction. The assembly consists of a three-way manifold (*1*) which carries two 6-mm stopcocks (*4* and *5*) and three different standard tapers. A standard taper 29/42 connects to the reaction vessel, which consists of a Pyrex tube, 18 by 3.0 cm. A platinum crucible is placed inside the tube to protect the glass during the heating period (*2*). A standard taper 24/40 connects to a second round-bottom Pyrex tube, 14 by 2.5 cm (*3*).

[33] W. G. Brown, L. Kaplan, and K. E. Wilzbach, *J. Am. Chem. Soc.* **74**, 1343 (1952).
[34] H. S. Isbell and J. D. Moyer, *J. Res. Natl. Bur. Std.* **63A**, 177 (1959).

This tube contains the vial, an iron bar for breaking the vial, and glass wool to protect the glass bottom. The remaining components of the assembly are a small hydrogen tank (*12*), a manometer (*6*), a vacuum pump (*10*), a Hoskins furnace containing a thermocouple (*7*) connected to a galvanometer (*9*), and a Variac for exact temperature control (*8*). The furnace is placed on a lab jack (*11*) so that it may easily be raised and lowered.

The exchange reaction is carried out in the following manner: The apparatus is placed in a well-ventilated hood and is dried thoroughly by evacuating, flaming all-glass parts with a microburner, and cooling the assembly under vacuum. The apparatus is next filled with dry air; 1 g of NaBH$_4$ is placed in the platinum crucible in tube (*2*), and the tritium vial with its tip toward the standard taper and an iron bar are placed carefully in tube (*3*). The connections are made air tight with Apiezon grease N.[34a]

The apparatus is evacuated and filled with hydrogen gas from the tank. This is repeated three times before hydrogen is pumped out to a pressure of 0.75 atmosphere. The furnace then is raised, and the top of the furnace is covered with a layer of asbestos board to prevent melting of the stopcock grease from the heat.

The borohydride is conditioned by heating for 1 hour to 350°.

After cooling, the apparatus is filled with hydrogen to 0.33 atmosphere and the tritium-containing vial is broken with an induction coil and heated at 350° for 24 hours. After the apparatus has cooled to room temperature, it is filled and evacuated several times with tank hydrogen to remove unreacted tritium with the vent of the vacuum pump leading directly into the hood.

D-Glucitol-1-T

Directions have been given for the reduction of D-gluconolactone or D-gluconic-γ-lactone to D-glucose with LiBT$_4$ in water.[31] Further reduction leads to D-glucitol-1-T as described below.

D-Xylose-5-T

The D-isomer has been prepared by reduction of 5-aldo-1,2-O-isopropylidene-D-xylopentofuranose with LiBT$_4$ which was hydrolyzed to D-xylose-5-T[31] (Fig. 4). The substitution of the L-isomer would yield L-xylose-5-T as required for the synthesis of L-ascorbic acid-6-T by the osone cyanide route.

5-Aldo-1,2-O-isopropylidene-D-xylopentofuranose has been obtained from D-glucose. Acetonation of D-glucose leads to 1,2:5,6-O-di-O-iso-

[34a] Associated Electrical Industries. James G. Biddle Company, 1316 Arch Street, Philadelphia, Pennsylvania.

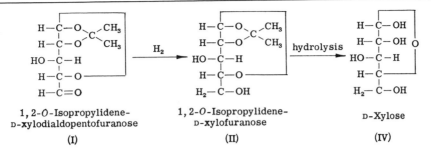

FIG. 4. Synthesis of D-xylose from 1,2-O-isopropylidene-D-xylodialdopentofuranose.

propylidene-D-glucose[35] which is then hydrolyzed preferentially to 1,2-mono-O-isopropylidene-D-glucose.[36] Cleavage with HIO_5 yields the substituted dialdehyde of D-xylose.[36a] The substitution of L-glucose as the starting material will yield the L-isomer.

L-Xylose-5-T

Two different methods are employed on a semimicro scale to introduce tritium into the 5 position of L-xylose using $NaBT_4$.

The first method requires three steps from the introduction of tritium to the completion of the synthesis replacing one hydrogen atom with tritium and results in a low radiochemical yield.

The second method gives higher radiochemical yields, since only one step is required following introduction of the isotope, and two hydrogen atoms are replaced by tritium.

Method I

D-Glucose is reduced with tritiated sodium borohydride to D-glucitol-1-T. Addition of benzaldehyde forms 2,4-O-benzylidene-D-glucitol-1-T. Glycol cleavage with periodate yields 2,4-O-benzylidene-L-xylose-5-T.

After hydrolysis to free L-xylose-5-T, the remaining steps to the formation of L-ascorbic acid-6-T are identical to those described for the synthesis of L-ascorbic acid-1-^{14}C, except that an excess of cyanide is used.[2,7,18]

Procedure

Reduction of D-Glucose to D-Glucitol-1-T. D-Glucose (1.8 g; 10 millimoles) is dissolved in 20 ml of water; 3 portions of 19 mg each of $NaBT_4$ (1.5 millimoles, 100 mCi total) are added with rapid stirring at 10-minute inter-

[35] C. L. Mehltretter, B. H. Alexander, R. L. Mellies, and C. E. Rost, *J. Am. Chem. Soc.* **73**, 2424 (1951).

[36] R. I. Gramera, A. Park, and R. L. Whistler, *J. Org. Chem.* **28**, 3230 (1963).

[36a] R. Schaffer and H. S. Isbell, *J. Res. Natl. Bur. Std.* **56**, 191 (1956).

vals. Next, 380 mg (10 millimoles) of nonradioactive $NaBH_4$ is added to provide an excess of reducing material, and the mixture is stirred for an additional 30 minutes.

A small portion (5 drops) of the reaction mixture diluted with 2 ml of water and acidified with dilute acetic acid is boiled for 1 minute to expel hydrogen and tested for reducing material with Fehling's solution. The reduction is quantitative at this point.

The reaction mixture is diluted to 100 ml with water, and the excess $NaBH_4$ is destroyed by addition of 20.6 g of dry Dowex 50 (H^+) in small increments. The mixture is stirred in contact with the resin for 45 minutes, until pH 4–5 is reached.

The resin is removed by filtration and thoroughly washed with water. The combined filtrates and washings are concentrated *in vacuo* at 50° to a syrup. Boron is removed as volatile methyl borate by repeatedly reconcentrating the solution to a syrup at 60° after addition of methanol. This process is continued until a negative test for boron is obtained with curcumin test paper.

The syrup is taken up in methanol and reconcentrated in a tared 100-ml boiling flask. The syrup is thoroughly dried under vacuum from an oil pump yielding 2.4 g of semisolid, amorphous mass. It is used without further purification.

Condensation of D-*Sorbitol-1-T with Benzaldehyde to 2,4-O-Benzylidene-*D-*glucitol-1-T.* This procedure follows the general outline of Ness.[37] The syrup is dissolved in 2.5 ml of water and the reaction vessel is flushed with dry purified nitrogen for 10 minutes. Freshly distilled benzaldehyde (1.02 ml, 10 millimoles) is added followed by 0.254 ml of concentrated HCl. Solid material begins to form after 30 minutes and the mixture completely solidifies after 2.5 hours. It is held at 4° for 16 hours, after which the mass is broken up with a spatula and washed from the flask into a sintered-glass filter with three 5-ml portions of ice water. The cake is washed with 5 ml of cold absolute ethanol and 5 ml of cold petroleum ether. After 20 hours in a desiccator over anhydrous $CaCl_2$, 1.75 g (65%) of white crystals (m.p. 160–164°) is obtained.

*Glycol Cleavage of 2,4-O-Benzylidene-*D-*glucitol-1-T.* The 2,4-O-benzylidene-D-glucitol-1-T (1.75 g, 7 millimoles), is suspended in 4 ml of water, and the suspension is cooled in an ice bath. A hot solution of 1.59 g (7.45 millimoles) of $NaIO_4$ is slowly added. After 15 minutes the character of the suspension changes because of the oxidation of the organic material and the precipitation of inorganic salt. At this time the inorganic salt is further precipitated by the addition of 150 ml of absolute alcohol, stirring

[37] R. K. Ness, *in* "Methods in Carbohydrate Chemistry" (R. L. Whistler and M. L. Wolfrom, eds.), Vol. 1, p. 90. Academic Press, New York, 1962.

and cooling being continued for 10 more minutes. The salt is removed by suction filtration, and the cake is washed with 40 ml of absolute alcohol. The combined filtrates are concentrated *in vacuo* to a thick syrup which, when dissolved in 150 ml of absolute ethanol, yields a small additional amount of salt. After filtration the clear solution is reconcentrated to a thin syrup and is hydrolyzed without further purification.

Hydrolysis of 2,4-O-Benzylidene-L-xylose-5-T to L-Xylose-5-T. The concentrated syrup is dissolved in 100 ml of 5% acetic acid and boiled under reflux for 3 hours. Benzaldehyde is removed by shaking the solution with ether. The solution is then concentrated under reduced pressure at 70°C to a syrup. The syrup is dissolved with warming in about 20 ml of absolute alcohol; after it has cooled to room temperature, crystallization generally begins easily when the inside of the flask is scratched with a stirring rod. After crystals had been obtained several times in this laboratory, the crystallization of L-xylose took place spontaneously. A yield of 0.51 g (53%) of white crystalline material, m.p. 145–147°, is obtained. Assay of radioactivity shows a radiochemical yield of 11%. This low incorporation of tritium despite complete reduction of the aldehyde group of glucose as demonstrated by a negative Fehling test may be caused by exchange of tritium by labile hydrogen atoms.

Method II (Fig. 5)

1,2:3,5-Di-O-cyclohexylidene-L-xylofuranose (I) is preferentially hydrolyzed to 1,2-mono-O-xylofuranose (II). Catalytic oxidation of (II) gives 1,2-mono-O-cyclohexylidene-L-xyluronic acid (III).[38]

The methyl ester of 1,2-mono-O-cyclohexylidene-L-xyluronic acid (IV) is prepared, and the ester group is reduced to 1,2-mono-O-cyclohexylidene-L-xylofuranose-5-T. L-Xylose-5-T is obtained after acid hydrolysis (Fig. 5).[39]

Procedure

Preparation of Catalyst.[35] To a solution of 1 g (0.017 mole) of chloroplatinic acid ($H_2PtCl_6 \cdot 6H_2O$, 400 mg of Pt) in 60 ml of water in a beaker is added 9 g of activated charcoal. The mixture is stirred mechanically and neutralized by the addition of sodium bicarbonate. It is then heated to 80° in a wax bath, and 5.5 ml of formaldehyde is added dropwise. Sodium hydrogen carbonate is added in portions to neutralize the formic acid formed. Care is taken to maintain the mixture slightly alkaline. Heating and stirring are continued for another 2 hours; during this time the level of the liquid is held constant by the periodic addition of water. After 2

---

[38] K. Heyns and J. Lenz, *Chem. Ber.* **94,** 346 (1961).
[39] S. L. von Schuching and G. H. Frye, *J. Org. Chem.* **30,** 1288 (1965).

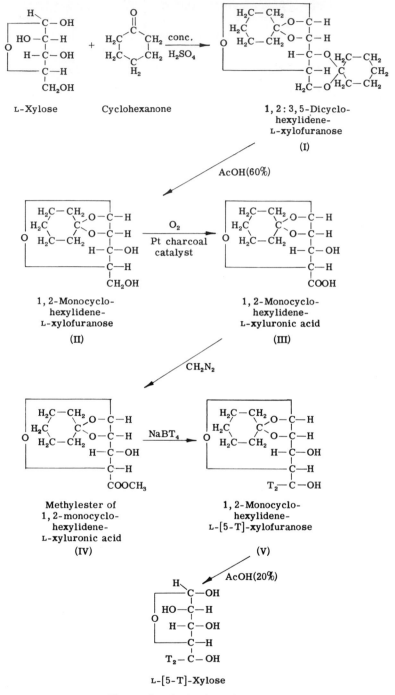

FIG. 5. Synthesis of L-xylose-5-T.

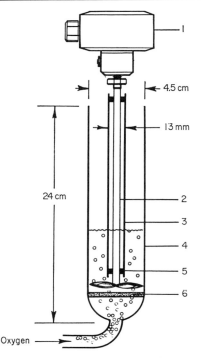

FIG. 6. Oxidation apparatus. *1*, High speed stirrer, 7000 rpm; *2*, stirring rod with butterfly stirrer; *3*, 13-mm diameter tube; *4*, reaction vessel; *5*, Teflon bushing; *6*, sintered-glass plate.

hours the mixture is cooled and filtered. The catalyst is washed well with water, then dried in air or under vacuum. The dried catalyst contains 4.5% platinum by weight.

Apparatus for Catalytic Oxidation (Fig. 6). A standard chromatographic column, diam. 45 mm and containing a fused coarse sinter at its lower end above the constriction, is cut off 240 ml above the disk (*6*) at a distance of 24 cm. The top of the tube is left open, and a brushless mechanical stirrer, maximum speed 7000 rpm, is centered above it (*1*). The stirring rod (7 mm in diameter), whose end is drawn into a butterfly shape only slightly shorter than the diameter of the column (*2*), extends to a few millimeters above the disk. In order to keep it steady at this high speed, the rod is enclosed by an outer glass tube of 13-mm o.d. (*3*). Two soft bushings are cut from a piece of Teflon tubing and placed around the rod inside this guard tube (*5*). The oxygen enters from the bottom of the sintered glass tube.

Heating is effected by wrapping an electric heating tape regulated by a rheostat around the lower part of the tube to the height of the liquid.

*1,2:3,5-Di-O-cyclohexylidene-*L-*xylofuranose (I).* L-Xylose (25 g, 0.167

mole) is suspended in cyclohexanone (264 ml, 2.5 mole) in a 1-liter long-neck flask, and concentrated sulfuric acid (3.61 ml, specific gravity 1.84) is added dropwise. After the addition of sulfuric acid, the reaction mixture is shaken at room temperature for 24 hours.

The sulfuric acid is neutralized by addition of sodium bicarbonate (11.1 g) with cooling; the sodium sulfate which is formed is filtered off. The solution is concentrated *in vacuo* at 65° to a thick syrup. The syrup is diluted with methanol and cooled at −20°. Crystallization begins after a few minutes, and 47 g (95% of theory) of crude I is recovered after 3 days at −20°. The material is recrystallized from alcohol, m.p. 103–104°; yield 41 g (79% of theory).

The compound is soluble in methanol, ethanol, ether, petroleum ether (b.p. 60–110°), dioxane, ethyl acetate, and acetone, but it is insoluble in water. It does not reduce hot Fehling's solution: $[\alpha]_D^{20} - 2.88°$ (c. 9.1, acetone).

1,2-Mono-O-cyclohexylidene-L-xylofuranose (*II*). A suspension of 1,2:3,5-di-*O*-cyclohexylidene-L-xylofuranose (I) (5 g, 0.016 mole) in 150 ml of 60% acetic acid (v/v) is vigorously shaken in a 300-ml Kjeldahl flask for 24 hours at room temperature.

The resulting clear solution is concentrated *in vacuo* at 50° to a syrup. A hundred milliliters of water is added and removed *in vacuo*. The syrup is thoroughly dried on a high-vacuum pump. It is then dissolved in 10 ml of acetone, and the acetone solution is added dropwise to 700 ml of rapidly stirred petroleum ether. This effects the separation of any unreacted dicyclo compound, since the dicyclo compound (I) is soluble in petroleum ether and the monocyclo compound (II) is insoluble in this solvent. After 24 hours at 4°, white crystals of (II) (4.3 g, 85% of theory) are obtained. This product melts at 83–84°. It is soluble in water, methanol, ethanol, ether, dioxane, ethyl acetate, and acetone. It is insoluble in petroleum ether and does not reduce hot Fehling's solution: $[\alpha]_D^{20} + 0.41°$ (c. 9.1, acetone).

1,2-Mono-O-cyclohexylidene-L-xyluronic Acid (*III*). To 2.3 g (0.01 mole) of 1,2-mono-*O*-cyclohexylidene-L-xylofuranose (II), 300 mg of sodium bicarbonate, and 1.7 g of 4.5% platinum–charcoal catalyst (680 mg of Pt) is added 100 ml of water in the oxidation apparatus described above. The stirrer is started at maximum speed, and oxygen is blown in the suspension at 50 ml/minute. The reaction mixture is maintained at 50°. After 30 minutes of reaction time, another 1.7 g of the catalyst together with 300 mg of sodium bicarbonate are added. The total reaction time is 2 hours. A few minutes after the start of the reaction the solution begins to foam considerably.

The oxidation is stopped after 2 hours. The catalyst is removed by suction filtration and washed well with water. The filtrate is concentrated *in vacuo* at 50° to about 50 ml and adjusted to pH 2 with dilute hydrochloric acid to destroy any excess bicarbonate. The solution is then brought to pH 5 with the addition of dilute ammonium hydroxide. Ten milliliters of 10% calcium chloride is added, and the precipitation of the calcium salt of (III) begins after 30 seconds. The mixture is allowed to stand overnight. The precipitate is filtered and washed with water and acetone, yielding 1.73 g (70% of theory).

The calcium salt of (III) (1 g) is dissolved in 50 ml of 1 N hydrochloric acid, and 1,2-mono-O-cyclohexylidene-L-xyluronic acid (III) is extracted with three 50-ml portions of ethyl acetate. The extract is concentrated *in vacuo* at 50° to dryness, dissolved in 5 ml of acetone, and the solution is then added dropwise to 800 ml of rapidly stirred petroleum ether. A quantitative yield of (III) is obtained (m.p. 159–161°). The compound is soluble in water and acetone, but insoluble in ether and petroleum ether: $[\alpha]_D^{20} + 3.67°$ (c. 9.1, acetone).

Methyl Ester of 1,2-Mono-O-cyclohexylidene-L-xyluronic Acid (IV). 1,2-Mono-O-cyclohexylidene-L-xyluronic acid (1 g) is thoroughly dried, and an excess of ethereal solution of diazomethane is poured over it. Vigorous evolution of nitrogen occurs, and the compound dissolves immediately. The solution is allowed to decolorize, and after spontaneous evaporation of the ether, the ester crystallizes. The crude crystals are dissolved in 10 ml of acetone and dropped slowly into 500 ml of petroleum ether. The solution becomes cloudy, and after 3 days at −20°, white fluffy crystals of (IV) are obtained weighing 0.93 g (88% of theory): m.p. 93–95°, $[\alpha]_D^{25} + 1.5°$ (c. 9.1, acetone).

L-Xylose by Reduction of Methyl 1,2-Mono-O-cyclohexylidene-L-xyluronate (IV) with NaBT₄

The run with tritiated sodium borohydride is carried out under a well-ventilated hood. A 516-mg quantity (0.002 mole) of the ester (IV) is dissolved in 40 ml of water in a 25-ml, two-neck, round-bottom flask under the hood. The outlet of one opening of the flask is connected through a piece of rubber tubing to the back of the hood. To the magnetically stirred solution at room temperature is added 34 mg of NaBT₄ in two portions (0.001 mole) containing 90 mCi of tritium. After 30 minutes, 38 mg of inactive NaBH₄ is added twice at 15-minute intervals. After 1 hour, the volume is brought to 100 ml with water and 8 g of cation-exchange resin is slowly added from an Erlenmeyer flask which is connected by a piece of rubber tubing directly to the second outlet of the reaction flask. The Erlenmeyer flask is operated remotely by lifting it with a pair of long-handled tongs (50 cm) in order to avoid any breathing of the unreacted tritium. Stirring is continued for 1 hour after the evolution of gas stops. The solution is then filtered, and boron is removed as methyl borate, as described above. Removal of the cyclohexanone group is carried out by hydrolysis of the residue with 150 ml of 25% acetic acid on the steam bath. The solution is concentrated and the residue is dissolved in 50 ml of water. Cyclohexanone is removed by extraction with ether. The solution is then alternately concentrated at 50° and reconstituted with distilled water to remove all labile tritium, which requires five cycles. The residue is taken up in 10 ml of water, and 1 g of inactive L-xylose is added as a carrier. The solution is again brought to dryness, taken up in alcohol, and ether is added to beginning turbidity. After 3 days at 4°, 670 mg of crystallized L-xylose-5-T is obtained (m.p. 143°). Assay of radioactivity shows 47.2 mCi (51.5%) total radioactivity in 670 mg of L-xylose-5-T. The supernatant solution contains an additional 28.2% of radioactivity.

L-Ascorbic Acid-6-T from L-Xylose-5-T

In the oxidation of L-xylose-5-T to L-xylosone-5-T[7,18] condensation with KCN and hydrolysis is carried out as described above using an excess of cyanide. The radiochemical yield is 42% based on the L-xylose-5-T used.

[2] Biosynthesis and Degradation of Isotopically Labeled Ascorbic Acid (Plants)

By F. Loewus and M. M. Baig

Studies of L-ascorbic acid biosynthesis,[1] in which specifically labeled sugars are used, have revealed that higher plants, unlike animals, form L-ascorbic acid from D-glucose or D-galactose by a path of conversion that conserves the 6-carbon chain and oxidizes C-1 of hexose to form the carboxyl group of L-ascorbic acid. When specifically labeled D-glucurono-γ-lactone or D-galacturonate or their corresponding C-1 reduced products, L-gulono-γ-lactone or L-galactono-γ-lactone, is provided, conversion to L-ascorbic acid proceeds with retention of the preformed carboxyl function as C-1 of L-ascorbic acid. Inasmuch as the pathway from D-glucose to D-glucuronic acid and D-galacturonic acid in plants involves a path of conversion in which the 6-carbon chain of D-glucose is conserved but is oxidized at C-6, the two L-ascorbic acid pathways of plants appear to be mutually exclusive pathways. Since the ultimate 6-carbon precursor in plants is hexose, the pathway in which C-1 of hexose is oxidized while C-6 remains reduced (a primary alcohol function) appears to represent the major route of L-ascorbic acid biosynthesis in plants.

[1] F. Loewus, *Phytochemistry* **2**, 109 (1963).

Biosynthesis of Labeled L-Ascorbic Acid in Intact Plant Tissues

D-*Glucose*. If label is located specifically in a single carbon, for example, in D-glucose-1-^{14}C, the L-ascorbic acid of plants fed this sugar will have a ^{14}C distribution pattern similar to patterns contained by hexose moieties of sucrose, starch, and cellulose isolated from the same tissue. That is, about 70–80% of the radioactivity is retained in C-1 and about 20–30% is found in C-6, a pattern that reflects movement of label through the triose phosphate pool.[2] This distribution takes place shortly after the labeled sugar is given to the plant, and there is little change with time in the pattern even if experiments are prolonged.[1] In the conversion of D-glucose-6-^{3}H to L-ascorbic acid in plant tissue, tritium is conserved at C-6, evidence that this carbon does not undergo oxidation. Net conversion of D-glucose to L-ascorbic acid is determined by the conditions under which label is given. As much as 2% conversion has been achieved in the case of parsley leaves pulsed with D-glucose-1-^{14}C followed by a 40-hour period in which the stems were kept in a solution of 1% *myo*-inositol,[3] however, redistribution of ^{14}C from C-1 into C-6 is much greater than that observed with water-immersed leaves.

D-*Glucurono-γ-lactone*. Plant tissues convert as much as 5% of this compound to L-ascorbic acid when dilute solutions are given to detached plant parts through cut surfaces. If a specific carbon of the lactone is labeled, conversion proceeds without randomization of [14]C. Thus D-glucurono-γ-lactone-1-^{14}C is converted to L-ascorbic acid-6-^{14}C and D-glucurono-γ-lactone-6-^{14}C is converted to L-ascorbic acid-1-[14]C. D-Galacturonic acid and its methyl ester are also converted in part to L-ascorbic acid without randomization of specific label. As in the case of the D-glucurono-γ-lactone, the carboxyl at C-6 of the uronic acid becomes C-1 of L-ascorbic acid.

L-*Gulono-γ-lactone*. Plant tissues convert as much as 10% of this compound to L-ascorbic acid when dilute solutions are given to detached plant parts. Conversion proceeds without randomization of specific ^{14}C label. Thus L-gulono-γ-lactone-1-^{14}C is converted to L-ascorbic acid-1-^{14}C while L-gulono-γ-lactone-6-^{14}C is converted to L-ascorbic acid-6-[14]C.

L-*Galactono-γ-lactone*. A wide variety of plants readily convert this compound to L-ascorbic acid.[4] Parsley leaves maintained in 0.5% L-galactono-γ-lactone for 190 hours showed an increase in L-ascorbic acid over 7-fold greater than in water-fed controls. In experiments with detached ripening strawberries or 7-day-old bean apices, as much as 80% of the administered L-galactono-γ-lactone-2-^{14}C is recovered as L-ascorbic acid-2-[14]C.

[2] J. Edelman, V. Ginsburg, and W. Z. Hassid, *J. Biol. Chem.* **213,** 843 (1955).

[3] F. Loewus, *Federation Proc.* **24,** 855 (1965).

[4] G. A. D. Jackson, R. B. Wood, and M. V. Prosser, *Nature* **191,** 282 (1961).

Procedure for Labeling Plant Tissues and for Isolating L-Ascorbic Acid

Detached plant parts readily take up small amounts of water through severed stems or petioles. Uptake is most efficient when precaution is taken to cut the stem while it is submerged in water and then to transfer the detached part to the labeling solution. When stem surfaces are first rinsed with a saturated solution of $Ca(ClO)_2$ and subsequent operations are kept sterile, interference from microbial infections is substantially diminished. Young leaves of parsley (*Petroselinum*) or apices of 7-day-old bean plants (*Phaseolus*) take up 0.2 to 2 ml per gram fresh weight through the cut surface. A current of air across the leaf accelerates uptake. Fleshy parts such as the detached ripening strawberry fruit (*Fragaria*) take up water or dilute sugar solutions much more slowly (0.1–0.2 ml per gram fresh weight in 24 hours). As solute concentration increases, uptake is reduced. Normally, a concentration of 0.5% sugar or neutral substance has little effect on uptake.

Labeled tissues are thoroughly macerated at 0° in 0.1% oxalic acid (10 ml per gram fresh weight) in a motor-driven blender. Insoluble residue is removed by centrifugation (10,000 g). Water-clear supernatant solutions can be analyzed for ascorbic acid by titrating aliquots, in 3% metaphosphoric acid (freshly prepared), with $NaHCO_3$-buffered 0.025% 2,6-dichlorophenolindophenol. The first faint persistent pink color is taken as the end point. Colored solutions or plant extracts that contain dye-reactive compounds other than L-ascorbic acid[5] should be assayed by gas chromatography after conversion of dried aliquots to their trimethylsilyl derivatives.[6–8] Loss of labeled L-ascorbic acid due to oxidation is minimized if carrier L-ascorbic acid is added to the supernatant solution immediately after aliquots have been removed for assay. Excess oxalic acid is removed as the insoluble calcium salt by adding an equivalent amount of calcium formate. The solution is passed through a short column of Dowex 50 H^+ resin and then loaded on a 1 × 15 cm column of Dowex 1 formate (8% cross-linked, 200–400 mesh) resin. Traces of neutral label are removed with water washes and acidic constituents are eluted with a dilute formic acid gradient that is formed by adding, at the rate of column flow, 500 ml of 0.06 N formic acid to a mixing chamber of constant volume that initially contains 200 ml of water. Ascorbic acid appears in the eluate volume between 290 and 370 ml.[9] Crystalline L-ascorbic acid is prepared by

[5] E. Epstein, M. W. Nabors, and B. B. Stowe, *Nature* **216,** 547 (1967).
[6] See C. C. Sweeley, W. W. Wells, and R. Bentley, Vol. XIII [7].
[7] M. Vecchi and K. Kaiser, *J. Chromatog.* **26,** 22 (1967).
[8] K. Pfeilsticker, *Z. Anal. Chem.* **237,** 97 (1968).
[9] F. Loewus and S. Kelly, *Arch. Biochem. Biophys.* **102,** 96 (1963).

evaporating combined fractions under reduced pressure at 40° and dissolving the residue in warm glacial acetic acid. On standing, the L-ascorbic acid crystallizes. Recrystallization can be done in glacial acetic acid or by dissolving L-ascorbic acid in absolute ethanol (20 mg/ml) followed by addition of ethyl ether. To facilitate recovery of labeled L-ascorbic acid, it is often advisable to add carrier to the combined fractions prior to recovery as a crystalline product.

Derivatives of L-Ascorbic Acid Useful in Tracer Studies

Di-o-aminoanil of L-*Dehydroascorbic Acid*.[10] To L-ascorbic acid (40 mg) in water (3 ml) in a small separatory funnel, add *p*-quinone (40 mg) in ethyl ether (5 ml). Shake the mixture for 4 minutes to oxidize the ascorbic acid. The aqueous layer is washed once with fresh ether (2 ml) to remove traces of *p*-quinone and then transferred to a 12-ml conical centrifuge tube containing *o*-diaminobenzene (55 mg). After aeration to remove traces of ether, the dark brown solution is heated gently until a color change to a light shade of brown is noted. At this point the volume is reduced, either with a stream of N_2 or under reduced pressure, to about 1 ml; at this point the product crystallizes in the acid form. To convert the anil to the lactone form, the dried crystalline residue is dissolved in methanol (1 ml) and then left to crystallize. After one recrystallization from methanol, the yield is 35–40 mg of yellow crystals, melting point, 180–181°.

*5,6-O-Isopropylidene-*L-*ascorbic Acid*.[11] L-Ascorbic acid (62 mg) is added to a flask containing dry acetone (5 ml) and anhydrous cupric sulfate (200 mg). The mixture is stoppered, shaken for 24 hours at 25°, and centrifuged. The supernatant solution is evaporated and the residue is extracted with warm *n*-heptane. The product (50–55 mg) crystallizes from *n*-heptane. Further recrystallization from absolute ethanol yields pure crystals melting at 119–120° with decomposition. Attempts to prepare 5,6-O-isopropylidene-D-araboascorbic acid by the same procedure results in a product having the same melting point and elemental analysis as D-araboascorbic acid. When mixed with authentic D-araboascorbic acid, this product gave no melting point depression.[12]

*2,3-Diphenacyl-*L-*ascorbic Acid*.[13] L-Ascorbic acid (70 mg) in water (1 ml) is carefully adjusted to pH 6 with $NaHCO_3$ (34 mg). To this is added ethanol (2 ml) and α-bromoacetophenone (159 mg). The reaction is refluxed for 2.5 hours, then diluted with water (1 ml), and left to crystallize. Moist

[10] H. Hasselquist, *Arkiv Kemi* **4**, 369 (1952).
[11] L. von Vargha, *Nature* **130**, 847 (1933).
[12] F. Loewus and S. Kelly, *Nature* **191**, 1059 (1961).
[13] C. S. Vestling and M. C. Rebstock, *J. Biol. Chem.* **161**, 285 (1945).

crystals are dissolved in ethanol (1 ml) and recrystallized by the addition of water (2 ml). The yield varies from 20 to 50 mg, melting point, 137–138°.

Degradation of L-Ascorbic Acid

A systematic procedure for the degradation of L-ascorbic acid that yields C-1 + C-2, C-3, C-4 + C-5, and C-6 has been described.[14, 15] More often, knowledge of the amount of label present in a specific carbon is all that is needed. Procedures outlined below permit one to estimate the amount of ^{14}C in C-1, C-2 (difference between C-1 + C-2 and C-1), C-5, and C-6. Tritium attached to C-5 or C-6 can also be determined by these methods. C-3 and C-4 (difference between C-4 + C-5 and C-5) can be determined if the threonic acid fragment, released in the course of C-1 + C-2 recovery, is isolated and degraded.

Carbon 1. Decarboxylation of L-ascorbic acid or the di-*o*-aminoanil of L-dehydroascorbic acid in 8 N H_2SO_4 yields C-1 as CO_2 in a yield of 90% or better.[14] The reaction can be run at 100° for 3 hours with a slow stream of N_2 passing through the reaction mixture to sweep out CO_2 or at 80° for 8 hours using the convenient reaction flask system of Katz, Abraham, and Chaikoff.[16] If the latter system is chosen, introduction of base into the center well of the evacuated flask should be postponed until completion of the decarboxylation.

Carbon 1 + Carbon 2. This fragment is obtained as crystalline oxalic acid by cleavage of L-ascorbic acid with NaOI.[14] To L-ascorbic acid (25 mg) in 0.1 N acetic acid (0.5 ml) is added 0.1 N I_2 (7.5 ml) followed by 1 N NaOH (2.5 ml). After standing 5 minutes, the solution is adjusted to pH 5 with glacial acetic acid. Oxalate is precipitated by the addition of 1 M calcium acetate (0.5 ml). The precipitate is washed with dilute ammonia, redissolved in 5 N HCl, and evaporated to dryness. Pure crystalline oxalic acid is recovered from the dried residue by sublimation at 40° under reduced pressure.

Carbon 5. When the di-*o*-aminoanil of L-dehydroascorbic acid is treated with periodate, C-5 is recovered as formic acid.[17] Anil (100 mg) dissolved in water (20 ml) is cooled to 0° and adjusted to pH 3 with N HCl. To this is added 0.1 M $NaIO_4$ (10 ml, freshly prepared). After 10 minutes, the reddish-brown precipitate is filtered off and the solution is run through a short column of Dowex 50 H^+ (8% cross-linked, 200–400 mesh) resin. Upon addition of $BaCO_3$ (0.5 g), formic acid precipitates along with other

[14] H. H. Horowitz, A. P. Doerschuk, and C. G. King, *J. Biol. Chem.* **199,** 193 (1952).

[15] H. H. Horowitz and C. G. King, *J. Biol. Chem.* **200,** 125 (1953).

[16] See S. Abraham and W. Z. Hassid, Vol. IV [22], especially pp. 531–532.

[17] F. Loewus, R. Jang, and C. G. Seegmiller, *J. Biol. Chem.* **222,** 649 (1956).

acids as its barium salt. Addition of a small amount of ethanol completes the precipitation. The supernatant solution, containing C-6 as formaldehyde, is removed. Recovery of formate as its crystalline salt follows a procedure recently described by Gabriel.[18] Barium formate in the residue is dissolved in water (3 ml) to which is added 1 M $(NH_4)_2SO_4$ (0.5 ml). After removal of the $BaSO_4$ precipitate, the solution is evaporated to dryness at 25° and the residue is extracted repeatedly with methanol. The methanolic extract is brought to dryness, and ammonium formate is recovered from the residue by sublimation at 40° under reduced pressure.

Carbon 6. Periodate oxidation of L-ascorbic acid or the di-*o*-aminoanil of L-dehydroascorbic acid releases C-6 as formaldehyde which can be recovered as its dimedon derivative.[19] L-Ascorbic acid (36 mg) is dissolved in water (2 ml). To this is added 0.5 M sodium phosphate (4.0 ml, pH 7.5) and freshly prepared 0.5 M $NaIO_4$ (2 ml, 107 mg/ml). After 1 hour, dimedon (5,5-dimethyl-1,3-cyclohexanedione) in ethanol (2 ml, 80 mg/ml) is added and the mixture is left overnight. Solids are recovered, washed with water (2.5 ml), and dried. Dimedon-formaldehyde is recovered from the insoluble residue by extraction with hot absolute ethanol (3.5 ml). Material insoluble in hot ethanol is discarded. Crystalline dimedon-formaldehyde is recovered upon the addition of water (1.5 ml) to the ethanolic solution.

Biosynthesis of Labeled L-Ascorbic Acid from L-Galactono-γ-lactone

Preparation of L-*Galactono-γ-lactone-2-*[14]*C*. This compound is readily prepared from *myo*-inositol-2-^{14}C by way of biosynthetic D-galacturonic acid.[20] Detached ripening strawberry fruits or other plant tissue in which pectin biosynthesis occurs will convert *myo*-inositol-2-^{14}C to D-galacturonic acid-5-^{14}C units of pectin in a radiochemical yield of 50–70%. The labeled tissue is macerated in 70% ethanol, centrifuged to recover ethanol-insoluble residue, and washed repeatedly with fresh portions of 70% ethanol to remove all traces of soluble radioactivity. Insoluble residue is suspended in 0.2% pectinase (EC 3.2.1.15) plus 0.1% EDTA (20 ml/0.1 g dry weight of residue). After 24 hours of incubation under toluene at 28°, D-galacturonic acid is recovered from the soluble portion by separation on Dowex 1 formate resin as described above for L-ascorbic acid. D-Galacturonic acid is located in the eluate volume between 240 and 350 ml. Combined fractions are reduced to dryness, taken up in a small volume of water, and reduced with an excess of $NaBH_4$.[21] L-Galactonic acid is recovered from the reaction

[18] O. Gabriel, *Carbohydrate Res.* **6**, 319 (1968).
[19] B. Bloom, *Anal. Biochem.* **3**, 85 (1962).
[20] F. Loewus, *Advan. Tracer Methodol.* **2**, 163 (1965).
[21] M. L. Wolfrom and K. Anno, *J. Am. Chem. Soc.* **41**, 1141 (1952).

mixture by passage through columns of Dowex 50 H+ and Dowex 1 formate resins. L-Galactonic acid is eluted from the latter column with the same gradient used to recover L-ascorbic acid and appears in the eluate volume between 220 and 260 ml. Combined fractions are evaporated to dryness and heated for 2 hours at 60° to convert the acid to its γ-lactone.

Conversion to L-Ascorbic Acid-2-¹⁴C. The apical portion of a 7-day-old bean plant (approximately 1 g) is detached above the cotyledons while the plant is held under water and then transferred to a vial containing 0.5% L-galactono-γ-lactone-2-¹⁴C (0.25 ml). As this solution is taken up through the cut stem, water is added in small increments to avoid exposing the cut surface to air. Radioactivity is taken up within 1 hour. The plant is held in a vial of water for another 23 hours during which time it takes up another 4 ml. At the end of this period, L-ascorbic acid is recovered as outlined above. The yield (1.6 mg), compared to a water-fed control (0.3 mg), represents almost stoichiometric conversion to L-ascorbic acid. Over 99% of the ¹⁴C is located in C-2.

Enzymatic Conversion of L-Galactono-γ-lactone to L-Ascorbic Acid. An enzyme oxidizing L-galactono-γ-lactone to L-ascorbic acid has been extracted from cauliflower particulate preparation and partially purified.[22]

[22] L. W. Mapson and E. Breslow, *Biochem. J.* **68**, 395 (1958).

[3] Biosynthesis of L-Ascorbate in Animals

By I. B. CHATTERJEE

L-Ascorbic acid is synthesized by most species of animals except the guinea pig, primates, the fruit bat, the pipistrelle, and some birds[1] of the Passeriformes order.

Evidence presented from this laboratory[2] and elsewhere[3] indicates that formation of L-ascorbic acid from D-glucurono-γ-lactone or D-glucuronate takes place according to the following route:

[1] Sixteen species of birds of the Passeriformes order are incapable of synthesizing L-ascorbic acid. C. R. Chaudhuri and I. B. Chatterjee, *Science* **164**, 435 (1969).
[2] I. B. Chatterjee, G. C. Chatterjee, N. C. Ghosh, J. J. Ghosh, and B. C. Guha, *Biochem. J.* **74**, 193 (1960).
[3] Y. Mano, K. Yamada, K. Suzuki, and N. Shimazono, *Biochim. Biophys. Acta* **34**, 563 (1959).

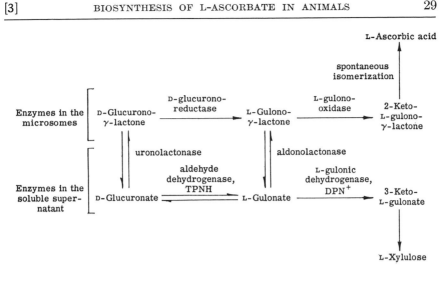

Biosynthesis of L-Ascorbate in the Microsomes

Since D-glucuronoreductase (L-gulono-γ-lactone:NADP oxidoreductase; EC 1.1.1.20) has not been separated from L-gulonooxidase (L-gulono-γ-lactone:oxygen oxidoreductase; EC 1.1.3.8), the result will represent a combined assay of both the enzymes.

Source of Enzyme

Sources include liver of mammals and those birds of the Passeriformes order that are capable of synthesizing the vitamin; kidney of amphibians, reptiles, and birds having a position lower than the Passeriformes in the phylogenetic tree. A potent source is liver from male young goat or calf. Enzymatic activity of liver from fasted or female animals[4] or liver infiltrated with fat is low.

Microsomes prepared from the tissue homogenate in 0.25 M sucrose[5] solution are washed once by resuspension in sucrose solution and recentrifugation at 105,000 g for 1 hour. The washed microsomes are finally dispersed in 0.25 M sucrose solution by gentle homogenization, so that 1 ml of the dispersion is equivalent to 1 g of wet tissue. Since the microsomal enzyme can convert only the lactone form of the substrate, contamination of microsomes by soluble supernatant solution containing lactonase results in an inhibited synthesis.

---

[4] I. B. Chatterjee and R. W. McKee, *Arch. Biochem. Biophys.* **109**, 62 (1965).
[5] P. Siekevitz, Vol. V [5].

Method

Reagents

Sodium phosphate buffer, 0.2 M, pH 7.4

D-Glucurono-γ-lactone semicarbazone, 0.01 M, freshly prepared

Sodium pyrophosphate, 0.05 M, pH adjusted to 7.4, freshly prepared

Potassium cyanide, 0.02 M, pH adjusted to 7.4, freshly prepared

Preparation of D-*Glucurono-γ-lactone Semicarbazone.* $NH_2NHCONH_2 \cdot HCl$ (2 g) and $CH_3COONa \cdot 3H_2O$ (2.45 g) are taken in a 100-ml conical flask and dissolved in 3 ml water by heating in a boiling water bath for about a minute. The content of the flask is cooled to about 70°, and to this is added a hot solution (approximately 70°) of D-glucurono-γ-lactone (3.14 g in 5 ml of water). The mixture is shaken by swirling, and the whole mass sets solid within 30 seconds. To this is added 20 ml of water; the mixture is stirred thoroughly with a glass rod for about 5 minutes, and the white silky lustrous precipitate is filtered off with suction. The precipitate is washed 5 times with 6-ml portions of water (until a silver nitrate test for chloride is negative), and then successively with ethanol and ether and finally dried over $CaCl_2$ under reduced pressure. The yield is 3.4 g of D-glucurono-γ-lactone semicarbazone; recrystallized from water, m.p. 198–199° with decomposition.

Assay Procedure. To a 20-ml conical flask are added 0.25 ml each of sodium phosphate buffer, sodium pyrophosphate, and potassium cyanide, 1 ml of D-glucurono-γ-lactone semicarbazone, 0.25 ml of tissue microsomes (approximately equivalent to 5 mg of protein), and 0.5 ml of water to a final volume of 2.5 ml. The reaction is carried out with shaking in air at 37° for 1.5 hours. The reaction is stopped by addition of 0.5 ml of 30% metaphosphoric acid solution. The precipitated protein is removed by centrifugation or filtration, and ascorbic acid is determined titrimetrically[2] with 2,6-dichlorophenolindophenol reagent. The ascorbic acid can also be estimated by the method of Roe and Kuether,[6] but in this case cyanide should be omitted from the incubation medium.[7]

Definition of Unit and Specific Activity. A unit of enzyme is defined as the amount of enzyme synthesizing 1 micromole of L-ascorbic acid per hour. Specific activity is expressed as units per milligram of protein.

[6] J. H. Roe and C. A. Kuether, *J. Biol. Chem.* **147,** 399 (1943).

[7] Estimation of ascorbic acid by the method of Roe and Kuether gives low values in the presence of cyanide and reduced glutathione. I. B. Chatterjee, N. C. Ghosh, J. J. Ghosh, and B. C. Guha, *Science* **126,** 608 (1957).

Properties

Stability. The enzymatic activity is stable for about 3 months when the intact liver or the microsomes are stored at −18°. Lyophilized microsomes kept in sealed tubes under vacuum and stored at −18° are stable for at least a year.

Specificity. The microsomal enzyme catalyzes the conversion of the following compounds at the relative rates shown: D-glucurono-γ-lactone semicarbazone, 100; D-glucurono-γ-lactone cyanohydrin, 60; D-glucurono-γ-lactone oxime, 50; D-glucurono-γ-lactone, 2; D-mannurono-γ-lactone, 1; L-idurono-γ-lactone, 1; L-gulono-γ-lactone, 100.[8]

pH Optimum. The optimal pH is 7.4, with rates 90% of maximal at pH 7.0 and 7.8.

Michaelis Constant. Using goat liver microsomes, the K_m for D-glucurono-γ-lactone semicarbazone in air is $2 \times 10^{-3}\ M$.

Inhibitors. The enzyme is inhibited by aldehydes.[9] It is not inhibited by cyanide, azide, ethylenediaminetetraacetate, or antimycin A.

ASSAY OF L-GULONO-γ-LACTONE OXIDASE

The microsomal enzyme catalyzes the conversion of L-gulono-γ-lactone according to the following reaction:[10]

$$\text{L-Gulono-γ-lactone} + O_2 \xrightarrow{\text{L-gulonooxidase}} \text{2-keto-L-gulono-γ-lactone} + H_2O_2$$

$$\text{2-Keto-L-gulono-γ-lactone} \xrightarrow{\text{spontaneous isomerization}} \text{L-ascorbic acid}$$

Since the enzyme has not been purified, the oxidase is preferentially assayed by measuring the formation of ascorbic acid rather than the oxygen uptake. The dehydrogenase component of the enzyme can be assayed using phenazine methosulfate as the electron carrier.

Assay Method

The assay method is similar to that described above using D-glucurono-γ-lactone semicarbazone except that D-glucurono-γ-lactone semicarbazone is replaced by L-gulono-γ-lactone as the substrate, and the pH of the incubation medium is kept at 7.0 instead of 7.4.

Solubilization of Microsomal Enzyme. The microsomal enzyme can be

---

[8] See also J. Kanfer, J. J. Burns, and G. Ashwell, *Biochim. Biophys. Acta* **31,** 556 (1959).

[9] I. B. Chatterjee and S. Dutta Gupta, *Abstr. Proc. 7th Intern. Congr. Biochem.*, p. 827 (1967).

[10] I. B. Chatterjee, J. J. Ghosh, N. C. Ghosh, and B. C. Guha, *Naturwissenschaften* **46,** 580 (1959).

solubilized by sodium deoxycholate, snake venom, or digitonin or by ultrasonic treatment. The method using sodium deoxycholate is as follows:

Goat liver microsomal suspension in 0.25 M sucrose (10 ml, equivalent to 250 mg protein) is added during 2 minutes to a solution of sodium deoxycholate (200 mg) in 5 ml of sodium phosphate buffer (0.2 M, pH 7.4) at 0° with constant swirling. A further 20 ml of sodium phosphate buffer (0.2 M, pH 7.4) is added and the mixture is centrifuged at 105,000 g for 40 minutes. The residue is discarded and the supernatant solution is subjected to ammonium sulfate treatment.

Ammonium Sulfate Treatment. Twenty milliliters of a saturated ammonium sulfate solution is added slowly with swirling to 20 ml of the deoxycholate-treated supernatant solution. The precipitate is gathered by centrifugation, dissolved in 10 ml of sodium phosphate buffer (0.1 M, pH 7.2), and dialyzed against 500 ml of sodium phosphate buffer (0.01 M, pH 7.2) with mechanical stirring for 1 hour. The precipitate which forms during dialysis is removed by centrifugation and the brownish yellow supernatant fluid containing the active enzyme is preferably assayed without delay. The temperature throughout the procedure is kept between 0° and 3°.

Properties

Inhibitors. The enzyme is inhibited by sodium borate, Antabuse (Disulfiram), and reversibly by p-chloromercuribenzoate. An apparent inhibition is observed when a high concentration (0.01 M or above) of cyanide or other aldehyde agents are used. This inhibition is probably due to trapping of the 2-keto intermediate, since oxygen uptake is not reduced.

Interfering Reactions. EFFECT OF SOLUBLE SUPERNATANT SOLUTION.[2] Irrespective of the species studied, the soluble supernatant solution obtained after sedimentation of microsomes at 105,000 g inhibits the microsomal conversion of L-gulono-γ-lactone into L-ascorbic acid. This inhibition is apparently due to the presence of aldonolactonase in the soluble supernatant solution which reversibly hydrolyzes the lactone into the corresponding free acid.

FORMATION OF LIPID PEROXIDE.[11] The synthesis of L-ascorbic acid in the microsomes is accompanied by a parallel formation of microsomal lipid peroxide. Lipid peroxide irreversibly inhibits the dehydrogenase component of L-gulonooxidase. Unless sodium pyrophosphate is used in the assay medium to prevent lipid peroxide formation, inhibited synthesis results. Besides sodium pyrophosphate, α,α'-dipyridyl, 8-hydroxyquinoline, or α-tocopherol may be used to prevent lipid peroxidation. Lipid peroxide is not formed when the incubation is carried out in the absence of air.

[11] I. B. Chatterjee and R. W. McKee, *Arch. Biochem. Biophys.* **110**, 254 (1965).

Assay for Dehydrogenase

Method

The method is based on the use of phenazine methosulfate in place of oxygen as the electron carrier and carrying out the incubation in the absence of air. An aerobic assay for the dehydrogenase has been previously described.[12]

Reagents

Sodium phosphate buffer, 0.2 M, pH 7.2
L-Gulono-γ-lactone, 0.1 M, freshly prepared
Phenazine methosulfate, 20 mg/ml in 0.02 M sodium phosphate buffer, pH 7.2, prepared just before use
Potassium cyanide, 0.02 M, freshly prepared and pH adjusted to 7.2

Procedure. To the main compartment of a Thunberg tube kept in crushed ice are added 0.25 ml each of sodium phosphate buffer, potassium cyanide, and microsomal dispersion and 1.55 ml of water. To the side arm are added 0.1 ml each of L-gulonolactone and phenazine methosulfate. A control Thunberg tube is kept without substrate. After addition, the Thunberg tube is connected to a high vacuum pump for exactly 1 minute. The contents of the tube are then mixed by tilting, and the incubation is carried out for 15 minutes at 37°. The reaction is stopped by addition of 0.5 ml of 30% metaphosphoric acid; the precipitated protein is removed, then the ascorbic acid is estimated titrimetrically[4] with 2,6-dichlorophenol-indophenol reagent.

Biosynthesis of L-Ascorbate in the Tissue Extract

Biosynthesis of L-ascorbic acid in the tissue extract containing both soluble supernatant solution and microsomes represents a combined assay of aldehyde dehydrogenase, aldonolactonase, and L-gulonooxidase.

Synthesis from Sodium d-Glucuronate

Method

Reagents

TPN, 5×10^{-3} M, dissolved in water and kept frozen
Sodium phosphate buffer, 0.2 M, pH 7.2
MnSO$_4$ solution, 0.1 M
Sodium glucuronate, 0.1 M, freshly prepared
Glucose 6-phosphate, 0.1 M, freshly prepared

---

[12] See C. Bublitz and A. L. Lehninger, Vol. VI [42].

Tissue Extract. A potent source is goat liver. The tissue is homogenized in 3 volumes of 0.04 M sodium phosphate buffer, pH 7.2. The homogenate is centrifuged at 10,000 g for 20 minutes, the sediment is discarded, and the supernatant fluid is used as the tissue extract.

Assay Method. To a 20-ml conical flask are added 0.75 ml of sodium phosphate buffer, 0.15 ml each of sodium glucuronate and glucose 6-phosphate, 0.1 ml of TPN, 0.1 ml of $MnSO_4$, 1 ml of tissue extract, and 0.25 ml water to a final volume of 2.5 ml. The reaction is carried out with shaking in air at 37° for 2.5 hours. The reaction is stopped by adding 0.5 ml of 30% metaphosphoric acid, the precipitated protein is removed and the ascorbic acid is estimated titrimetrically[2] with 2,6-dichlorophenolindophenol reagent or by Roe and Kuether[6] method.

SYNTHESIS FROM SODIUM L-GULONATE

Reagents

Sodium phosphate buffer, 0.2 M, pH 6.8
Sodium L-gulonate 0.04 M. Prepared by neutralizing 17.8 mg of L-gulono-γ-lactone with 1 ml of 0.1 N NaOH at 37° for 24 hours and diluting to 2.5 ml.
$MnCl_2$ solution, 0.02 M

Procedure. To a 20-ml conical flask are added 0.25 ml each of sodium phosphate buffer, sodium L-gulonate, and $MnCl_2$ solution, 1 ml of tissue extract, and 0.75 ml water to make a final volume of 2.5 ml. It is useful to measure ascorbic acid destruction by keeping a control incubation flask containing 1–2 micromoles of L-ascorbic acid in place of sodium L-gulonate. After incubation for 2 hours at 37° in air, the reaction is stopped by adding 0.5 ml of 30% metaphosphoric acid solution, and ascorbic acid is estimated in the protein-free filtrate by the method of Roe and Kuether.[6]

[4] Catabolism of Ascorbic Acid Labeled with Radioactive Carbon to CO_2. Excretory Pathway of $^{14}CO_2$ by Respiratory Exhalation

By ARTHUR F. ABT and SUSANNE L. VON SCHUCHING

I. Introduction

Radioactive tracer studies have shown that one of the catabolic end products of ascorbic acid is carbon dioxide. As noted by Burns, the entire carbon chain contributes to this catabolic product, the major portion of CO_2 being derived from the C-1 position.[1] It has been further demonstrated that the amount of $^{14}CO_2$ exhaled after administration of L-ascorbic acid-1-^{14}C is dependent on the previous dietary intakes of vitamin C; the lower the ascorbic acid intake, the lower the $^{14}CO_2$ excretion.[2] The difference in metabolic pathways in different species of animals and the human led to an investigation of the catabolism of L-ascorbic acid-1-^{14}C to $^{14}CO_2$ in guinea pigs, monkeys, and human subjects. It was found that the difficulty of separating expired $^{14}CO_2$ from other gases increases with the size of the experimental subject. This is especially true in the human, which expires about 12 liters of air per minute containing about 4% CO_2. A number of investigators have reported procedures for the fixation and assay of $^{14}CO_2$ from expired air, using an ionization chamber[3-5] or liquid scintillation system.[6] The list of references is not complete, but it represents a few of the basic methods.

The following outline describes easily applicable methods for the collection and determination of radioactivity of $^{14}CO_2$ in expired air from small and large animals and from human subjects after administration of labeled ascorbic acid. The separation of $^{14}CO_2$ from the remaining gases and the transfer of CO_2 to either a 1-liter ionization chamber or the preparation of the sample for counting in a liquid scintillation system is given, as well as a one-step determination of the specific radioactivity of expired $^{14}CO_2$ from human subjects.

CO_2 is a weak acid and may be trapped in an excess of alkaline absorber provided that the flow is slow enough to allow complete trapping. The stable CO_2 may be determined in one aliquot of the alkaline solution, and the radioactive CO_2 in another aliquot. In experiments where only the knowledge of the specific radioactivity of CO_2 at various intervals after administration of the labeled tracer is required, the absorption procedure may be simplified. A measured amount of the alkaline absorber is neutralized by an excess of exhaled CO_2 and the radioactivity of this sample is determined.

Determination of Radioactivity

Ionization Chamber. A 1-liter ionization chamber is evacuated and filled to atmospheric pressure with $^{14}CO_2$. The radioactivity is determined by measuring the rate of ion production with an electrometer.[7,8]

[1] J. J. Burns, H. B. Burch, and C. G. King. *J. Biol. Chem.* **191,** 501 (1951).
[2] S. von Schuching, T. Enns, and A. F. Abt, *Am. J. Physiol.* **199,** 423 (1960).
[3] F. J. Dominques, K. J. Gildner, R. R. Baldwin, and J. R. Lowy, *Intern. J. Appl. Radiation Isotopes* **7,** 77 (1959).
[4] B. M. Tolbert, M. Kirk, and F. Upman, *Rev. Sci. Instr.* **30,** 116 (1959).

Liquid Scintillation System. CO_2 absorbed in hyamine hydroxide is mixed with a solution of the appropriate fluor prior to the determination of radioactivity in a liquid scintillation spectrometer. The radioactivity is determined by measuring the photons produced by the passage of charged particles through the organic fluor.[9]

New developments in instrumentation and efficient fluor and solvent combinations resulting in higher counting efficiency have made the liquid scintillation system the method of choice.[10]

II. Collection and Fixation of Carbon Dioxide in Expired Air

Collection of CO_2 in Expired Air from Small Animals[2]

The following conditions must be fulfilled in the construction of a metabolic train for animals. The flow of air must be steady and sufficient for normal breathing of the animal and must be slow enough to allow complete absorption of CO_2 in the alkaline absorber. There must be a provision for the collection of excreta. Examples may be found in the literature for constructing metabolic chambers.[11,12] We describe here the techniques that we have used and found to be advantageous.

Figure 1 shows a metabolic train suitable for animals up to 2 pounds. A desiccator serves as the metabolic chamber (F).

The incoming air passes through the metabolic train by suction from a rotary air blast and suction pump (K). The air is rendered carbonate free and dry by passage through a 2-liter Mariotte bottle (A) filled with soda lime and a 500-ml Mariotte bottle (B) filled with ascarite. The flowmeter (C) is constructed from an absorption tower, height 15 inches, and controls the rate of proper flow of incoming air. Details of construction of the flowmeter (E) are shown in Fig. 2.

Following the flowmeter an empty flask (D) is put in line to prevent splattering. A mercury manometer (E) is attached with a "T" tube, just

[5] G. V. LeRoy, G. T. Okita, E. C. Tocus, and D. Charles, *Intern. J. Appl. Radiation Isotopes* **7**, 273 (1960).

[6] D. S. Frederickson and K. Ono, *J. Lab. Clin. Med.* **51**, 147 (1958).

[7] D. Steinberg and S. Udenfriend, Vol. IV, p. 426.

[8] F. P. Chinard, W. R. Taylor, M. F. Noland, and T. Enns, *Am. J. Physiol.* **196**, 535 (1959).

[9] D. Steinberg and S. Udenfriend, Vol. IV, p. 433.

[10] E. Rapkin, *Packard Tech. Bull. No. 7*, Jan., 1962.

[11] S. S. Jackel, E. H. Mosbach, J. J. Burns, and C. G. King, *J. Biol. Chem.* **186**, 569 (1950).

[12] G. A. Edwards, C. H. Edwards, and E. L. Gadsden, *Intern. J. Appl. Radiation Isotopes* **4**, 264 (1959).

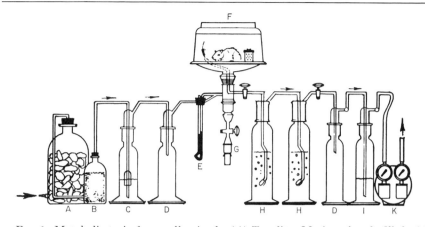

FIG. 1. Metabolic train for small animals. (*A*) Two-liter Mariotte bottle filled with soda lime; (*B*) 500-ml Mariotte bottle filled with ascarite; (*C*) flowmeter; for details, see Fig. 2; (*D*) empty absorption tower, 15 inches overall height; (*E*) mercury manometer; (*F*) metabolic chamber, for details, see Fig. 3; (*G*) standard taper 29/42; (*H*) two absorption towers, overall height 15 inches, filled with 250 ml of 2.5 N NaOH; (*I*) absorption tower, 15 inches overall height, filled with 100 ml of sulfuric acid for checking the flow of air through the system; (*K*) rotary air blast and suction pump.

before the metabolic chamber, and held at a negative of 8 mm below atmospheric pressure, which corresponds to 80 ml of air per minute.

The metabolic chamber consists of an inverted desiccator and is provided with a trap for the collection of urine. Details of construction are seen in Fig. 3. Next in line are two absorption towers (*H*), height 15 inches, which are filled with 250 ml of 2.5 N sodium hydroxide. The inlet tubes are equipped with coarse fritted disks which break up the incoming air into small bubbles, thus facilitating the absorption of CO_2. Another empty absorption tower (*D*) to avoid splattering is placed in front of an absorption tower (*I*) filled with 100 ml of concentrated sulfuric acid serving as a bubble counter. The rotary air blast and suction pump (*K*) complete the metabolic train.

Metabolic Experiment Using a Guinea Pig

An intramuscular injection of L-ascorbic acid-1-¹⁴C containing 4.2 μCi of ¹⁴C is given to a guinea pig weighing 300 g. Immediately after injection the animal is placed in the metabolic chamber and the passage of air through the train is started. It is possible to leave the animal in the chamber for a period of 23 hours, removing it for 1 hour for feeding and watering and replacing it again in the chamber. This procedure is continued for 8 days. The amount of alkali in the absorption tower is 500 ml of 2.5 N

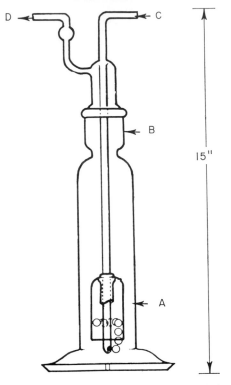

Fig. 2. Flowmeter used for regulating the air inlet through the metabolic train. The flowmeter consists of an absorption tower, 15 inches in overall height. It is filled with concentrated sulfuric acid and is placed in line directly before an empty flask. The inner dome (A) will lift to the desired level which can be marked on the outside of the tower. (A) Glass dome with inner seal; (B) standard taper 29/42; (C) air inlet; (D) air outlet.

NaOH equivalent to 0.625 mole of CO_2. This is sufficient excess of alkali to neutralize a guinea pig's CO_2 output for 24 hours. The guinea pig expires 0.01 mole of CO_2/hour.

If determinations of radioactive CO_2 are required at shorter intervals a parallel row of absorption towers is set up. A three-way stopcock allows switching the flow of air to the second set of absorption towers. This permits the alternate determination of radioactivity at intervals without interrupting the flow of air through the metabolic train.

Collection of CO_2 in Expired Air from Large Animals

Metabolic experiments with monkeys require special precautions because of their unruly behavior and the danger of severe infections that

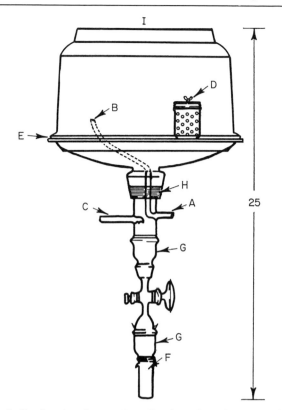

FIG. 3. Metabolic chamber for use in collection of respiratory carbon dioxide and urine. (A) Air inlet tube connected to (B) a piece of Tygon tubing; (C) air outlet tube; (D) perforated plastic can with metal cover filled with anhydrous calcium chloride (the can is inverted to picture a wing nut, which is soldered on the cover and allows the can to be wired to the wire mesh); (E) zinc-coated wire mesh placed on the lower portion of the desiccator; (F) urine collection flask; (G) standard taper 29/42; (H) one-hole rubber stopper; (I) an inverted 250-mm Scheibler-type desiccator with a rubber stopper. The overall height is 25 inches.

may result from scratches or bites obtained during handling. The metabolic train is the same as that described for smaller animals except for the size of the metabolic chamber; and the alkaline absorber is exchanged at 3-hour intervals, since a 6-pound monkey exhales 0.1 mole of CO$_2$ per hour.

Figure 4 shows the metabolic chamber used for monkeys. A 10-liter bell jar with tubulations and a ground flange are used. The jar is inverted, and the capacity is increased by placing the lower case of a desiccator, 12 inches in diameter, on the ground flange. The air inlet is on the bottom, and the outlet through the tubulations at the side.

The difficulty in handling a lively rhesus monkey may be circumvented

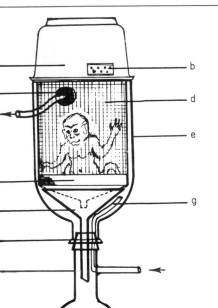

FIG. 4. (a) Lower case of a desiccator 12 inches in diameter; (b) perforated plastic can with metal cover filled with anhydrous calcium chloride to absorb excessive moisture; (c) one-hole rubber stopper with air outlet tube; (d) inner cage constructed from wire mesh; the top is enclosed by a wire mesh to which a wire handle is soldered (The monkey may be handled by lifting the whole wire cage at any time during the metabolic run.); (e) 10-liter bell jar with ground flange and tubulation; height, 19 inches; diameter, 10 inches; (f) metal funnel, used for collection of excreta, taken from a standard metabolic cage; (g) polyethylene plastic tubing serving as the air inlet; (h) glass funnel to collect excreta; (i) two-hole rubber stopper; (k) 500-ml long-neck flask with side arm. A piece of plastic tubing is pushed through the side arm to serve as the air inlet.

by the use of an inner cage constructed from wire mesh (d in Fig. 4). The lower end fits into a metal funnel taken from a metabolic cage. The top is enclosed by a wire mesh to which a wire handle is soldered. The wire cage fits into the bell jar assembly. The monkey is transferred to the inner cage at the beginning of the metabolic experiment and remains in the cage throughout the duration of the run.

Metabolic Experiment Using a Monkey

A rhesus monkey, 5.78 pounds, is given 85 μCi of L-ascorbic acid-1-^{14}C dissolved in 3 ml of water by stomach tube. The animal is placed im-

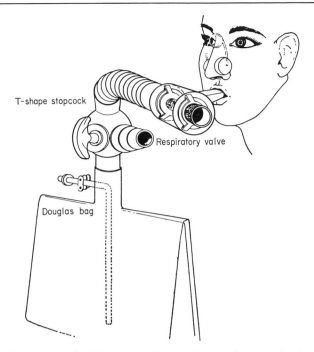

T-shape stopcock

Respiratory valve

Douglas bag

FIG. 5. Collection of exhaled air through a respiratory flutter valve into a modified Douglas bag.

mediately in the wire cage and then in the metabolic chamber (Fig. 4). The metabolic train is maintained as described above and allows intermittent determinations of CO_2 during a 15-day period.

Collection of CO_2 in Expired Air from Human Subjects

The large amounts of air involved in rhythmic inhalation and exhalation creates a technical difficulty in preparing carbon dioxide into a form readily available for the measurement of its radioactivity. The average normal human at rest exhales 12 liters of air per minute, which contains approximately 300 ml of CO_2. At an average rate of 18 breaths per minute, this means single exhalation of 670 ml of expired air containing 16 ml of CO_2 every 3 seconds.

We have found that the best results in human studies were obtained by frequent sampling of the expired CO_2 rather than continuous collection, after ingestion of 100 μCi of L-ascorbic acid-1-^{14}C.

Method 1

This method is based on the collection and fixation of expired $^{14}CO_2$ for assay in an ionization chamber or liquid scintillation system.

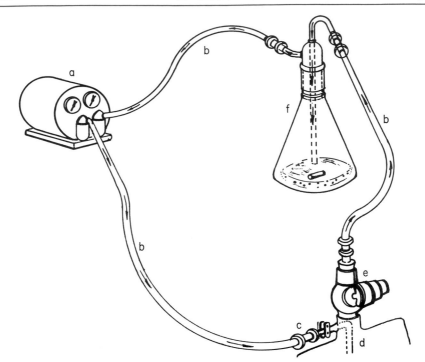

Fig. 6. Flow diagram for the fixation of carbon dioxide in an alkaline absorber. Seventy-five milliliters of 2.5 N NaOH or 100 ml of hyamine hydroxide may be used. (a) Rotary air blast and suction pump, (b) Tygon tubing, (c) Polyethylene tubing coupling, (d) Douglas bag, (e) T-shape stopcock, (f) 250-ml Erlenmeyer flask.

For Use with an Ionization Chamber. Expired air is collected in a modified Douglas bag for 3-minute periods at set intervals following the oral administration of radioactive ascorbic acid. The subject exhales through an McKerrow–Otis low-resistance valve.[13] The Douglas bag is equipped with a piece of Tygon tubing pushed through the side arm and extending nearly the whole length of the bag to ensure good mixing (Fig. 5).[14]

A 250-ml Erlenmeyer flask containing 75 ml of 2.5 N NaOH is used for the fixation of CO_2 from the Douglas bag. The flask, whose inlet was placed 1 cm above the surface of the liquid, contained a magnetic stirrer. Figure 6 demonstrates the fixation of CO_2 with the aid of a rotary air blast. The air from the Douglas bag circulates through the absorption vessel until all CO_2 is absorbed. Thirty minutes suffice for complete absorption.

[13] C. B. McKerrow and A. B. Otis, *J. Appl. Physiol.* **9,** 497 (1956).
[14] S. L. von Schuching and A. F. Abt, *in* "Methods in Tracer Methodology" (S. Rothchild, ed.), Vol. 2, p. 293. Plenum Press, New York, 1965.

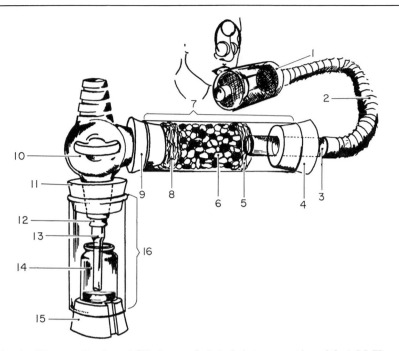

FIG. 7. Direct collection of CO_2 from exhaled air into a counting vial. *1*, McKerrow-Otis breathing valve; *2*, flexible hose; *3*, glass tubing, OD 36 mm; *4*, rubber stopper, No. 12; *5*, glass wool; *6*, anhydrous calcium chloride, No. 4 mesh, with indicator; *7*, glass tubing, length 20 cm and OD 6.5 cm; *8*, glass wool; *9*, rubber stopper, No. 12; *10*, Douglas bag stopcock; *11*, rubber filter adapter; *12*, rubber tubing; *13*, glass tubing, OD 12 mm; *14*, glass counting vial; *15*, rubber stopper, No. 10, indented to secure the vial and with two sections removed for venting; *16*, glass tubing, length 15 cm and OD 5 cm.

For Use with a Liquid Scintillation System. When a liquid scintillation system is used, the absorption vessel may be filled with 100 ml of hyamine hydroxide.[15] A drying tower must be placed before the absorption vessel, since water in the scintillation fluid causes quenching.

Method 2

In this method, CO_2 is collected for the determination of specific activity.[16]

The human subject exhales through a drying tube into a counting vial charged with 1 ml of 1 M hyamine hydroxide and a drop of phenolphthalein as indicator until the alkaline absorbent is completely neutralized by CO_2.

---

[15] J. M. Passman, N. S. Radin, and J. A. Cooper, *Anal. Chem.* **28**, 484 (1956).
[16] A. F. Abt and S. L. von Schuching, *Bull. Johns Hopkins Hosp.* **119**, 316 (1966).

Hyamine hydroxide is a monobasic alkali, and the determination of radioactivity in the sample is therefore at the same time a measure of the specific radioactivity of CO_2, provided that the normality of hyamine hydroxide is known.[17] Figure 7 shows the details of the technique.

Metabolic Experiment—Human Subjects

After oral ingestion of 100 μCi of L-ascorbic acid-1-^{14}C, the subject exhales at set intervals in Douglas bags of 30-liter capacity. The first sample is obtained 30 minutes after ingestion and thereafter at hourly intervals through the seventh hour. The collection times are then spaced at longer intervals and obtained until the fifteenth day. From this point, any procedure for the fixation and determination of [14]CO_2, as described above, may be used.

Determination of Radioactivity in the Ionization Chamber

The determination of radioactive CO_2 in the ionization chamber may be carried out by either of two methods: (1) CO_2 is liberated directly from 25 ml of the absorbent alkali solution; or (2) $BaCO_3$ is prepared, and CO_2 is liberated from the carbonate. There is no preferred method, the choice depends on the radioactivity in the expired CO_2, which might be very low at the end of a metabolic run, thus making mandatory the use of the maximum amount of [14]CO_2. A 1-liter ionization chamber accommodates a maximum of 0.0446 mole of CO_2.

Liberation of Carbon Dioxide from 10% NaOH–NaHCO₃ Solution into the Ionization Chamber. Figure 8[18] shows the vacuum line and ionization chamber. Twenty-five milliliters of the absorbent fluid is pipetted into a 100-ml round-bottom flask which contains a magnetic stirring bar. The contents of the flask are frozen by swirling the flask in an acetone–dry-ice

---

[17] *Titration of CO₂ absorbed in excess of NaOH.* Ten milliliters of 2.5 N NaOH absorber is neutralized with 1 N HCl to the phenolphthalein end point and then titrated with 1 N HCl to the bromocresyl green end point. The difference, in milliliters of HCl used between the end point obtained with phenolphthalein and the bromocresyl green point, represents the amount of CO₂ in the solution.

    CO₂ absorbed in excess of hyamine hydroxide. Ten milliliters of 1 M hyamine hydroxide is titrated with 1.0 N HCl to the phenolphthalein end point.

    Precipitation of BaCO₃ from NaHCO₃-NaOH solution. The contents of the absorption towers containing 2.5 N NaOH are transferred to a glass-stoppered Erlenmeyer flask For each 100 ml of the solution, 13.2 g of ammonium chloride is added. The ammonium chloride is used in order to reduce the hydroxyl concentration sufficiently to avoid precipitation of barium hydroxide. Barium chloride, 52 g, is added, and the Erlenmeyer flask is stoppered. The contents of the flask are mixed by swirling and warmed in a water bath at 80°. The resulting BaCO₃ is easily filtered on a sintered-glass funnel and dried by washing with acetone.

[18] S. L. von Schuching and C. W. Karickhoff, *Anal. Biochem.* **5,** 93 (1963).

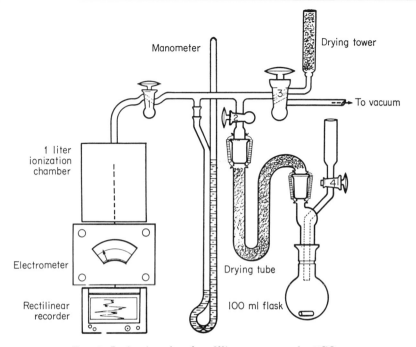

Fig. 8. Ionization chamber filling apparatus for $^{14}CO_2$.

bath and then attached to the vaccum line. The dry-ice bath is kept around the flask to keep it frozen until the system is evacuated with the help of a mechanical pump. Stopcock *3* is then closed, but the pump is kept running.

Ten milliliters of lactic acid are added through stopcock *4*. A seal of lactic acid is left above stopcock *4*. It has beeń found that two precautions assure a vacuum-tight system: (1) keeping the pump running and (2) keeping a seal of lactic acid. The dry-ice bath is removed, the mixture is allowed to thaw, and the liberated CO$_2$ fills the ionization chamber. The reaction is allowed to continue with stirring until there is no further rise in pressure shown by the manometer; this indicates complete liberation of CO$_2$. When no more CO$_2$ is developed, a slow stream of air is allowed to enter through stopcock *3* until the ionization chamber is at atmospheric pressure. Next, stopcock *1*, between the ionization chamber and the flask, is closed, the ion collector is disconnected, and the radioactivity is recorded in divisions per minute.

Liberation of CO$_2$ from BaCO$_3$ into the Ionization Chamber. Up to 8.3 g of BaCO$_3$ equivalent to 0.0446 mole of CO$_2$, may be used. Solid BaCO$_3$ is added to a 100-ml flask (Fig. 8), and the system is evacuated. Twenty-five milliliters of lactic acid is added through the stopcock, and the same procedure is followed as that described before.

Determination of Radioactivity in the Liquid Scintillation System. The high sensitivity of the liquid scintillation system makes this instrument the method of choice. One milliliter of the absorber, hyamine hydroxide, is mixed with 14 ml of liquid scintillation fluid, containing 4 g of 2,5-diphenyloxazole (PPO) and 0.5 g of 2,2-*p*-phenylene-bis(5-phenyloxazole) (POPOP) per liter of toluene. A high voltage setting of 3–900 is used. The radioactivity in the vial is then measured in a Tri-Carb liquid scintillation spectrometer (Packard Instrument Co.). The counting efficiency has been found to be 58%.

[5] Catabolism of L-Ascorbate in Animal Tissues

By YASUO KAGAWA and NORIO SHIMAZONO

I. Hydration of Dehydro-L-ascorbic Acid

Dehydro-L-ascorbic acid + H_2O → 2,3-diketo-L-gulonic acid

Assay Method[1]

Principle. The diketo-L-gulonic acid produced by the hydration of dehydro-L-ascorbic acid is determined with 2,4-dinitrophenylhydrazine by the modified method of Roe *et al.*,[2] after the remaining dehydro-L-ascorbic acid has been reduced by H_2S to L-ascorbic acid, which is more stable than the substrate. As the decarboxylation of diketo-L-gulonic acid by a contaminating enzyme is strongly inhibited by polycarboxylic acids, maleate is used as the buffer. Other assay methods to determine lactone hydrolysis, such as polarimetry, determination by hydroxamate formation, and ultraviolet spectroscopy, are inaccurate. The hydrolysis of dehydro-L-ascorbic acid is caused by an aldonolactonase; the purification procedure and properties of the enzyme are described in article [6] of this volume.

Reagents

Dehydro-L-ascorbic acid, 0.1 *M* (use within 5 minutes). The solution of dehydro-L-ascorbic acid is prepared by the oxidation of 0.2 *M* sodium-L-ascorbate solution with bromine, and a small amount of

[1] Y. Kagawa and H. Takiguchi, *J. Biochem.* (*Tokyo*) **51**, 197 (1961).
[2] J. H. Roe, M. B. Mills, M. J. Oesterling, and C. M. Damron, *J. Biol. Chem.* **174**, 201 (1948).

excess bromine, judged by the yellow color, is removed by the addition of thiourea.

$MgSO_4$, 0.1 M

Glutathione, 0.03 M

Tris(hydroxymethyl)aminomethane–maleate buffer, 0.2 M, pH 6.8

Metaphosphate, 3.5% (w/v), containing 0.35% (w/v) of $SnCl_2$

H_2S generator

2,4-Dinitrophenylhydrazine, 2%, in 9 N H_2SO_4

Thiourea, 1% (w/v), in metaphosphate, 5% (w/v)

H_2SO_4, 85% (w/v), Reagent grade

Procedure. The following components are mixed in a 10-ml stoppered centrifuge tube: 0.1 ml of $MgSO_4$, 0.01 ml of glutathione, the enzyme, and buffer to the final volume of 3.0 ml. The reaction is started by the addition of dehydro-L-ascorbic acid solution at 37°. Nonenzymatic control is always required. After 5 minutes the reaction is stopped by 0.5 ml of HPO_3 solution containing $SnCl_2$, and the mixture is rapidly reduced with bubbling H_2S at 0° for 15 minutes. The black precipitate is removed by centrifugation, and aliquot samples containing 20–500 millimicromoles of diketo-L-gulonate are assayed. To 1 ml of the sample solution, 1 ml of thiourea solution and 0.5 ml of dinitrophenylhydrazine solution are added, and then the mixture is incubated at 50° for 1 hour. The resulting hydrazone suspension is dissolved with gradual addition of 2.5 ml of 85% H_2SO_4 at 0°, and the optical density at 530 mμ is determined with a spectrophotometer.

Definition of Unit and Specific Activity. One unit is defined as 1 micromole of diketo-L-gulonate produced enzymatically (the amount of nonenzymatic hydrolysis is subtracted) in 1 minute in the condition described above. The specific activity is indicated by units per milligram of protein. The amount of protein is determined by the method of Warburg and Christian.[3]

Purification Procedure and Enzyme Properties

These are described in article [6] on catabolism of uronate and aldonate.

II. Decarboxylation of 2,3-Diketo-L-gulonic Acid

2,3-Diketo-L-gulonic acid + H_2O → L-xylonic acid (or L-lyxonic acid) + CO_2

Assay Method[4]

Principle. Carbon dioxide produced from diketo-L-gulonic acid is measured manometrically under nitrogen in a Warburg flask.[4] The substrate

---

[3] O. Warburg and W. Christian, see Vol. III, p. 451.

[4] Y. Kagawa, *J. Biochem. (Tokyo)* **51**, 134 (1962).

remaining after the enzymatic reaction may be determined by the method of Roe *et al.*,[5] since the major reaction products are L-xylonic and L-lyxonic acids.[4,6] However, as the nonenzymatic formation of oxalic and threonic acid from the substrate also occurs, the direct measurement of decarboxylation is preferable.

Reagents

Potassium 2,3-diketo-L-gulonic acid, 0.05 M. To prepare this compound, L-ascorbic acid (8.9 g) is added to KIO_3 solution (3.7 g in 50 ml of water), and the resulting colorless solution is gradually neutralized with 50 ml of 1 N KOH within 30 minutes (final pH 5). The solution is added to 800 ml of cold ethanol ($-20°$), and the resulting white precipitate is filtered off, dried under reduced pressure at 5°, and stored at $-20°$. Dissolve this hygroscopic white powder (183 mg) in 20 ml of redistilled water.
Phosphate buffer, 0.1 M, pH 6.8
Trichloroacetic acid, 5 M

Procedure. Potassium 2,3-diketo-L-gulonic acid (10.0 micromoles) is incubated in a Warburg flask with the enzyme and 25 micromoles of phosphate buffer (pH 6.8) in the final volume of 2.5 ml under nitrogen atmosphere at 37° for 30 minutes. The CO_2 evolution is determined after the addition of trichloroacetic acid (1 millimole). Nonenzymatic decarboxylation is corresponded to 6–9% of the substrate during the incubation in the above condition.

Definition of Unit and Specific Activity. A unit of the enzyme activity is defined as micromoles of diketo-L-gulonic acid decarboxylated in 1 minute under the assay condition described above. The specific activity is indicated by a unit per milligram of protein determined by Warburg and Christian's method.[3]

Purification Procedure

Step 1. Preparation of Supernatant Fraction. Rat liver is homogenized with 3 volumes of 0.15 M KCl, and the supernatant fraction is obtained after centrifugation at 105,000 g for 1 hour.

Step 2. Ammonium Sulfate Precipitation. Solid $(NH_4)_2SO_4$ is added to the supernatant fraction, and the precipitate between 60 and 80% saturation is dissolved in a small amount of distilled water. The solution is dialyzed against distilled water at 0° for 12 hours. The enzyme activity of

[5] J. H. Roe, M. B. Mills, M. J. Oesterling, and C. M. Damron, *J. Biol. Chem.* **174,** 201 (1948).
[6] J. Kanfer, G. Ashwell, and J. J. Burns, *J. Biol. Chem.* **235,** 2518 (1960).

the fraction obtained between 0 and 60% saturation is less than 27% of the total activity.

Step 3. Cellulose Column Chromatography. The dialyzed fraction is diluted with phosphate buffer (0.005 M, pH 7.5) to a final protein concentration of 5 mg/ml. A column of DEAE-cellulose (1 g per 50 mg of protein) is buffered with the same phosphate buffer. During the absorption of the diluted fraction on the column, the eluate contains 1 mg of protein per milliliter without the enzyme activity. After the elution with the same buffer and with 0.02 M NaCl in the buffer, the enzyme is eluted with 0.08 M NaCl in the buffer. The yields and steps in the method are summarized in the table.

PURIFICATION OF ENZYME FROM RAT LIVER

Fraction	Specific activity (milliunits/mg)	Activity recovery (%)	Total protein (mg)
Rat liver supernatant	3.6	100	6080
Ammonium sulfate fraction, dialyzed	19.6	55	606
DEAE-cellulose column fraction	204.0	12	12.6

Properties

Time Course of the Reaction. Diketo-L-gulonic acid lost and CO_2 produced are almost proportional to the incubation time.

Effect of pH. Optimal pH is 7.0 in phosphate buffer. Owing to the non-enzymatic decomposition of the substrate, an accurate measurement of activity is not possible in alkaline solution.

Stability of Enzyme. Ninety percent of the activity is lost on heating at 65° for 5 minutes. The activity of the dialyzed $(NH_4)_2SO_4$ fraction can be kept for months in a deep freezer ($-20°$), and that of the fractions from DEAE-cellulose, for a week at 0°.

Activators and Inhibitors. The enzyme activity is inhibited by poly-carboxylic acids, such as maleate, succinate, citrate, or phthalate (85–97% inhibition at 0.04 M). The activity of the purified enzyme is not influenced by the addition of Mg^{2+}, Mn^{2+}, or Co^{2+}. The activity is inhibited by metal chelating agents, especially by 8-hydroxyquinoline (90% inhibition at 10^{-4} M). The inhibition is restored after dialysis or by the addition of Mg^{2+} or Mn^{2+}. The enzymatic as well as nonenzymatic decomposition of diketo-L-gulonate is inhibited by borate (90%, at 10^{-3} M). Partial inhibition is observed with p-chloromercuribenzoate, monoiodoacetate, and diisopropyl-fluorophosphate in relatively high concentrations.

Effect of Substrate Concentration. The Michaelis constant determined with varying amounts of the substrate (0.033 mM to 40 mM) is estimated to be 1.5 × 10⁻².

Substrate Specificity. 2,3-Diketo-D- and L-gulonic and gluconic acids and 2,3-diketo-D-glucoheptonic acid are decarboxylated. 2,3-Diketo-D- and L-pentonic acid are also decarboxylated. α-Keto acids, such as pyruvate, α-ketoglutarate, 2-ketogluconate, 2-ketogulonate, and mesoxalate are not decarboxylated. It is important to notice that Winkelman *et al.* purified β-ketogulonate decarboxylase, which may be this enzyme.[7]

Distribution of Decarboxylase. The enzyme is found to be distributed in liver and kidney of the mammals tested (dog, hog, ox, guinea pig, rabbit, rat, monkey, and man).[4] The enzyme activity in other organs is hardly detected with manometry, although ¹⁴CO₂ evolution from dehydroascorbic acid-1-¹⁴C is slightly increased by the supernatant fractions.

[7] J. Winkelman and G. Ashwell, *Biochim. Biophys. Acta* **52,** 170 (1961).

[6] Catabolism of Uronate and Aldonate: Lactonases

By YASUO KAGAWA and NORIO SHIMAZONO

The preparation of aldonolactonase was described in Volume VI (p. 337); more recently, the separation of aldonolactonase and 6-phosphoglucono-lactonase has been performed.[1] Aldonolactonase shows hydrolyzing activity against L-dehydroascorbic acid.[2,3] The procedure described here is suitable for a large-scale preparation of this enzyme.

I. Preparation of Aldonolactonase[2]

[D(or L)-Gulono-γ-lactone Hydrolase, EC 3.1.1.18]
(Dehydroascorbate Hydrolyzing, D-Glucono-δ-lactone Hydrolyzing)

Principle, reagents, procedure, and definition of unit are described in article [5] of this volume.

Purification Procedure

Bovine liver acetone powder is preferably adopted as starting material, since the high activity of the water extract of the powder is without

[1] M. Kawada, Y. Kagawa, H. Takiguchi, and N. Shimazono, *Biochim. Biophys. Acta* **57,** 404 (1962).

[2] Y. Kagawa and H. Takiguchi, *J. Biochem.* (*Tokyo*) **51,** 197 (1961).

[3] M. Kawada, H. Takiguchi, Y. Kagawa, K. Suzuki, and N. Shimazono, *J. Biochem.* (*Tokyo*) **51,** 405 (1962).

diketoaldonate decarboxylase activity. The acetone powder is obtained from fresh or frozen bovine liver ($-20°$) treated with the solvent in a Waring blendor for 3 minutes and centrifuged. The resulting precipitate is again blended, centrifuged twice, desiccated by grinding in a mortar, and then stored in a desiccator at room temperature. By such treatment, 5.8 kg of bovine liver yields 1.1 kg of desiccated acetone powder.

The purification method is essentially the same as Yamada's for lactonase I,[4] which is identified as aldonolactonase.[5]

Step 1. Extraction. Ten grams of the acetone powder is extracted with 70 ml of distilled water (near $0°$) with stirring for 30 minutes.

Step 2. Heat Treatment. The extract is heated at $63°$ for 3 minutes with stirring, the temperature being raised within 5 minutes.

Step 3. Fractionation with Ammonium Sulfate. Solid ammonium sulfate is added to the extract; the precipitate obtained from 0.52 to 0.76 saturation is dissolved in a small amount of water, and the solution is dialyzed against distilled water.

TABLE I

PURIFICATION OF BOVINE LIVER ALDONOLACTONASE

Fraction	Specific activity (μmoles DKG[a] per mg protein)	Recovery (%)	Total protein (mg)
Acetone powder (water extract)	0.82	100	785
Heat treatment	1.54	88	367
Ammonium sulfate fraction, dialysis	3.94	40	67
DEAE-cellulose column treatment[b]	8.00	8	6.4

[a] 2,3-Diketo-L-gulonic acid.
[b] Value of one tube fraction.

Step 4. Cellulose Column Chromatography. Diethylaminoethylcellulose buffered with 0.01 M tris(hydroxymethyl)aminomethane(Tris)-HCl buffer of pH 7.75 is packed in a 2×10 cm column. The dialyzed fraction is absorbed on the column and eluted with 0.02 M NaCl in 0.01 M Tris-HCl buffer (pH 7.35). The specific activity, yield, and total protein at each step are summarized in Table I. The purification is about 10-fold from the water extract of the acetone powder, and about 22-fold from the bovine liver supernatant.

[4] K. Yamada, *J. Biochem. (Tokyo)* **46**, 361 (1959).
[5] J. Winkelman and A. L. Lehninger, *J. Biol. Chem.* **233**, 293 (1958).

Properties

Time Course of Delactonization. The amount of dehydroascorbic acid delactonized is proportional to the incubaton period in the presence of the enzyme, as long as the amount of the substrate remaining is large enough. On the other hand, nonenzymatic delactonization follows a first-order reaction.

Effect of pH. The enzyme has an optimum at pH 7.0. The enzyme activity at pH values higher than 7.6 is not measured owing to the rapid nonenzymatic hydrolysis. The k values of the nonenzymatic hydrolysis in the presence of Mg^{2+} ($6.7 \times 10^{-3}\ M$) are as follows:

pH	6.8	7.0	7.2	7.4	7.6
k (sec^{-1} $\times$ 10^{-3})	0.77	1.15	1.76	2.41	3.18

Enzyme Stability. The activity is completely lost on heating in a boiling water bath for 5 minutes. The purified enzyme solution is stable at 5° for weeks in the eluate obtained from diethylaminoethylcellulose column chromatography.

Metal Requirement. The purified enzyme is not active without Mn^{2+}, Mg^{2+}, or Co^{2+}.

Inhibitors. The enzyme activity is completely lost by the addition of p-chloromercuribenzoate ($10^{-6}\ M$), and is recovered (71%) with the addition of glutathione ($10^{-5}\ M$) in the presence of Mg^{2+}. Monoiodoacetate is also inhibitory (25% inhibition at $10^{-2}\ M$).

Substrate Specificity. The enzyme is effective on dehydro-L-ascorbic acid solution obtained from dehydro-L-ascorbic acid crystals and dehydro-L-ascorbic acid methanol complex. The enzyme hydrolyzes dehydro-D-araboascorbic acid at almost the same rate. The enzyme is also active on other lactones such as D-glucono-δ-lactone, D- and L-gulono-γ-lactone and D-glucurono-γ-lactone. The lactone ring of L-ascorbic acid is not hydrolyzed by this enzyme.

Contamination by Other Hydrolyzing Enzymes in the Preparation. Attempts to separate aldonolactonase and the dehydroascorbic acid delactonizing activity have not been successful. The preparation does not contain 6-phosphogluconolactonase activity.

II. Preparation of 6-Phosphogluconolactonase[6]

6-Phosphoglucono-δ-lactone Hydrolase

6-Phosphoglucono-δ-lactone + H_2O → 6-phosphogluconic acid

Principle. The remaining lactone is converted into hydroxamate, and the resulting brown-violet color in the presence of $FeCl_3$ is determined spectrophotometrically. Use of 6-phosphogluconate dehydrogenase and NADP, followed by spectrophotometry at 340 mμ, is easier but not accurate.

Reagents

6-Phosphoglucono-δ-lactone,[7] 0.1 M (use within 5 minutes). Free 6-phosphogluconic acid solution, which is obtained by passing the solution of barium 6-phosphogluconate through Dowex 50 (H+ form), is converted into δ-lactone by lyophilization and concentration in Methyl Cellosolve under reduced pressure as described by Isbell et al.[7] An aliquot weight of amorphous gum is dissolved into redistilled water in a stoppered test tube with vigorous shaking.

$MgSO_4$, 0.01 M

Phosphate buffer, 0.1 M, pH 7.0

Hydroxylamine reagent (1.5 M, pH 6.4), which is freshly prepared by mixing 1.0 ml of 3 M hydroxylamine hydrochloride, 0.1 ml of saturated NaOH and 0.9 ml of water

$FeCl_3$ solution, 20% in 4% trichloroacetic acid and 12% HCl

Procedure. The following components are mixed in a 10-ml stoppered centrifuge tube: 0.1 ml of 6-phosphoglucono-δ-lactone, 0.1 ml of $MgSO_4$, 0.5 ml of phosphate buffer, enzyme, and water to a final volume of 1.0 ml. The reaction is started by the addition of 6-phosphoglucono-δ-lactone and is continued at 37° for 5 minutes. A blank test is carried out without the enzyme. To the reaction mixture, 2 ml of freshly prepared hydroxylamine reagent is added and mixed. After the 20-minute incubation at 37°, 1.5 ml of $FeCl_3$ solution is added at 0°. If a precipitate is formed, it is removed by centrifugation, and after 20 minutes the colored supernatant solution is measured with a colorimeter at 540 mμ. A standard is necessary with each experiment.

Definition of Unit and Specific Activity. A unit of enzyme activity is defined as the activity that hydrolyzes enzymatically 1 micromole of lactone in 1 minute (subtract nonenzymatic hydrolysis). Owing to the nonenzymatic hydrolysis of the substrate (1.2 micromoles per 5 minutes at pH 7), the activity measurements both in the first-order range of the substrate concentration and in the highly reduced enzyme concentration are inaccurate. Usually the enzyme solution is diluted to a concentration which hydrolyzes about 2 micromoles of the substrate in 5 minutes. The specific activity of the enzyme is indicated by units per milligram of protein.

[6] M. Kawada, Y. Kagawa, H. Takiguchi, and N. Shimazono, *Biochim. Biophys. Acta* **57**, 404 (1962).

[7] H. S. Isbell, J. V. Karabinos, H. L. Frush, N. B. Holt, A. Schwebel, and T. T. Galkowski, *J. Res. Natl. Bur. Std. U.S.* **48**, 163 (1952).

Purification Procedure

Step 1. Preparation of the Supernatant Fraction. Rat liver is homogenized with 4 volumes of 0.25 M sucrose solution, and the supernatant fraction is obtained after centrifugation at 105,000 g for 1 hour.

Step 2. Ammonium Sulfate Precipitation. The supernatant fraction is heated at 50° for 10 minutes and cooled; solid ammonium sulfate is added. The precipitate between 30 and 90% saturation is collected. After the pellet is dissolved with the smallest amount of 0.01 M Tris-HCl, pH 7.75, the solution is dialyzed against the same buffer at 0° for 18 hours.

Step 3. Calcium Phosphate Gel Treatment. The dialyzed solution is stirred with calcium phosphate gel[8] (2 mg per milligram of protein in the solution) and allowed to stand for 30 minutes at 0°.

Step 4. Cellulose Column Chromatography. The gel is removed by centrifugation, and then the supernatant solution is absorbed on a DEAE-cellulose column, which is equilibrated with 0.01 M Tris-HCl, pH 7.75. The fraction eluted at 0.05 M NaCl in the same buffer contains aldonolactonase activity, and the following fraction eluted at 0.5 M NaCl contains 6-phosphogluconolactonase activity (Table II).

TABLE II
PURIFICATION OF RAT LIVER 6-PHOSPHOGLUCONOLACTONASE

Fraction	Specific activity (unit/mg protein)	Recovery (%)	Total protein (mg)
Rat liver supernatant	7.3	100	960
Ammonium sulfate fractionation and calcium phosphate gel treatment	17.6	18	70.3
DEAE-cellulose column treatment	59.5	3.4	4.0

Purification from Yeast Extract. This enzyme is also prepared from dried bakers' yeast, by autolysis in 3 volumes of 0.1 M phosphate buffer (pH 7.0) at 37° for 3 hours, and heating at 50° for 5 minutes, followed by ammonium sulfate fractionation (0.40–0.90 saturation), dialysis, negative absorption on calcium phosphate gel (2.55 mg/mg protein), and, finally, column chromatography on DEAE-cellulose. The same column described above is used, and the fraction eluted at 0.05 M NaCl shows 6-phosphogluconolactonase activity (131-fold purification).

---

[8] Prepared by the method described in Vol. II, p. 214.

Properties

Specificity. The enzyme is highly specific for 6-phospho-D-glucono-δ-lactone. Other lactones, including 6-phospho-D-glucono-γ-lactone, D-glucono-δ-lactone, D-galactonolactone, D-gulonolactone, L-gulonolactone, ascorbic acid, and dehydroascorbic acid are not hydrolyzed.

Activators and Inhibitors. Unlike aldonolactonase, divalent cations, such as Mg^{2+} or Mn^{2+}, are not required in either the animal or yeast enzyme. The activity is not inhibited by a chelating agent ($10^{-2} M$ EDTA), but is inhibited by *p*-chloromercuribenzoate and monoiodoacetate.

Effect of pH. The enzyme has an optimum at pH 7.2. Since the non-enzymatic hydrolysis of the lactone is 3 micromoles per 5 minutes at pH 8, the measurement of the activity beyond this range is impossible.

Distribution. The enzyme occurs in all the tissues tested, including liver, kidney, brain, spleen, and muscle of the rat. There is also a distribution among species. In the cell fractions, 97.2% of the total activity is localized in the supernatant fraction.

Reverse Reaction. Hydroxamate formation from 6-phospho-D-gluconic acid is slightly detected in the presence of hydroxylamine.

[7] Catabolism of Uronate and Aldonate: TPN-L-Hexonate Dehydrogenase (Liver)

By YOSHITAKE MANO and NORIO SHIMAZONO

$$\text{D-Glucuronolactone} \atop \text{or D-glucuronate} + \text{TPNH} + \text{H}^+ \rightleftharpoons {\text{L-gulonolactone} \atop \text{or L-gulonate}} + \text{TPN}^+$$

Although this enzyme has been classified on the basis of its action on two different substrates, glucuronate reductase and glucuronolactone reductase, both activities are thought to be due the same enzyme.[1,2] The enzyme is also called TPN-L-gulonate dehydrogenase.[1,3] It has been purified from rat liver,[1,2] pig kidney,[4] and yeast (*Schwanniomyces occidentalis*).[5] An assay method and purification procedure from pig kidney have been described in Volume VI,[6] but the following procedure will lead to a much greater purity with high yield.

[1] Y. Mano, K. Yamada, K. Suzuki, and N. Shimazono, *Biochim. Biophys. Acta* **34**, 563 (1959).
[2] Y. Mano, K. Suzuki, K. Yamada, and N. Shimazono, *J. Biochem. (Tokyo)* **49**, 618 (1961).
[3] S. Ishikawa and K. Noguchi, *J. Biochem. (Tokyo)* **44**, 465 (1957).
[4] J. L. York, A. P. Grollman, and C. Bublitz, *Biochim. Biophys. Acta* **47**, 298 (1961).
[5] A. Sivak and O. Hoffmann-Ostenhof, *Biochim. Biophys. Acta* **53**, 426 (1961).
[6] See Vol. VI [42].

Assay Method

Principle. With most substrates, the equilibrium is favored at neutral pH toward reduction of an aldehyde group of the hexose derivative to form L-hexonate. Thus, the enzyme is conveniently assayed spectrophotometrically by measuring the rate of change of absorbancy at 340 mμ of TPNH during the reduction of D-glucuronolactone or D-glucuronate.

Reagents

Potassium phosphate buffer, 0.05 M, pH 7.0

TPNH, 0.01 M (chemically[7] or enzymatically[8] reduced; the latter is much better in studies on the stoichiometry of reaction)

D-Glucurono-γ-lactone or sodium D-glucuronate (Chugai Pharmaceutical Co., Tokyo), 0.1 M

Procedure. To a cuvette with a 1-cm light path are added 0.1 ml of D-glucuronolactone or D-glucuronate, 0.05 ml of TPNH, 1.0 ml of the buffer, and water to give a final volume of 3.0 ml. Prior to the start of the reaction, all components except substrate are kept for 5 minutes in a water bath at 37°; the reaction is started by addition of the substrate solution, also at 37°. The determination is completed within 10 minutes. Rapid mixing of the components at the start can be effected by the use of a small glass rod. The decrease in absorbancy at 340 mμ can be followed conveniently by the use of a self-recording spectrophotometer. Under these conditions, no significant hydrolysis of lactones or lactonization of free acids takes place.

Application of Assay Method to Crude Tissue Extracts. When assaying crude undialyzed preparations, it is convenient to use a blank which contains the complete system except substrate and to record the difference in the rate of change of absorbancy between the cuvettes containing the substrate and the blank. Since crude preparations contain lactonase I (aldonolactonase), D-glucuronate must be used as substrate in these systems. The use of tris(hydroxymethyl)aminomethane (Tris) as a buffer should be avoided because possible contamination by an enzyme which catalyzes the reduction of this compound would interfere with the result of the assay.

Definition of Unit and Specific Activity. The unit of activity of the enzyme has been defined as a thousand times the change of optical density at 340 mμ in 10 minutes under the above conditions, and the specific activity of the enzyme is expressed as units per milligram of protein as determined spectrophotometrically.[9]

[7] See Vol. III [126].
[8] See Vol. III [127].
[9] See Vol. III [73].

Purification Procedure

Rats weighing about 150–200 g are killed by decapitation under ether anesthesia. Livers from 10 rats are minced and homogenized with 4 volumes of 0.25 M sucrose, using a Potter-Elvehjem glass homogenizer. The homogenate is immediately subjected to centrifugation at 105,000 g for 60 minutes at 0°. The resulting precipitate is discarded, and a clear supernatant solution (about 300 ml) is pooled. All the subsequent steps are carried out at 0° to 5°.

Ammonium Sulfate Fractionation. To each 100 ml of the supernatant solution, 23 g of ground solid ammonium sulfate is added slowly with stirring, giving a final saturation of 0.38. After standing for about 2 hours, the mixture is centrifuged; the precipitate is discarded. To the supernatant solution, more ammonium sulfate is added (12.2 g to each 100 ml of the original extract) to a saturation of 0.55. The solution is again centrifuged after standing for 2 hours, and the supernatant solution is discarded. The precipitate is dissolved in 300 ml of chilled water, 72.9 g of ammonium sulfate is added to the solution (to about 0.40 saturation), and after centrifugation another portion of ammonium sulfate (20.2 g) is added to the supernatant to reach 0.50 saturation. The mixture is allowed to stand for 2 hours. The precipitate resulting after centrifugation of this material is dissolved in 300 ml of 0.40 saturated ammonium sulfate solution, and another portion of ammonium sulfate (18.9 g) is added to the supernatant solution to reach a saturation of 0.50. The half-saturated solution is centrifuged, and the precipitates are dissolved in a minimum amount of water (about 3 ml). The enzyme solution is dialyzed against about 2 liters of distilled water until sulfate ions in the dialyzate cannot be detected by barium acetate. The resulting insoluble matter is spun down and discarded. The clear solution is diluted with water and Tris-HCl buffer (pH 7.5) until the protein concentration is below 0.2% and the final concentration of the buffer is 0.02 M; the solution is then subjected to diethylaminoethyl-(DEAE)-cellulose column chromatography.

Fractionation on DEAE-Cellulose.[10] DEAE-cellulose used for the chromatography is treated with 1 N NaOH and 1 N HCl and buffered with 0.02 M Tris-HCl buffer of pH 7.50. The diluted enzyme solution is passed through a DEAE-cellulose column (2 × 15 cm) with pumping by means of suitably controlled air pressure, at a flow rate of 10–15 ml per 10 minutes. Each fraction (5 ml) is collected by a fraction collector. The column is then washed with about 200 ml of the same buffer, and the enzyme is eluted usually as the second peak as monitored by the absorbancy at 280 mμ (tube number, about 30). However, the position of this second peak is

[10] See Vol. V [1].

somewhat variable, probably influenced by the exchange capacity. Each active fraction is combined, and, after the pH is adjusted to 7.0, the solution is stored in a frozen state at around $-20°$.

A summary of this purification is given in the table.

PURIFICATION OF TPN-L-HEXONATE DEHYDROGENASE

Enzyme preparation	Total volume (ml)	Total units	Total protein (mg)	Specific activity (units/mg protein)	Yield (%)
Extract	300	29,700	5589	5.3	100
Ammonium sulfate fraction	45.2	19,090	266.7	71.6	64
DEAE-cellulose fraction	40.0	13,480	5.3	2543	45

Properties

Stability. The enzyme can be stored in a frozen state, where it is fairly stable without detectable loss of activity for 3 weeks at $-20°$.

Activators and Inhibitors. The enzyme is activated by 10^{-2} M sulfate to about 150–200%. Hg^{2+} and Cu^{2+} completely inhibit the enzyme at 10^{-4} M, but p-chloromercuribenzoate hardly inhibits the activity even at concentrations of 10^{-4} M or more. Barbiturate (5,5'-diethylbarbiturate) at 3.3×10^{-3} M inhibits the reaction almost completely.

Specificity.[2,11] The enzyme catalyzes reversibly both oxidation and reduction of various sugars and derivatives between aldehyde and alcohol groups of the compounds. Direct reversible conversion between D-glucurono-γ-lactone and L-gulono-γ-lactone has been established. Other similar lactones can serve as substrate. Under the set conditions, the reduction reaction is broad with compounds having a free aldehyde group except for uronate, while in the oxidation process the enzyme acts exclusively upon L-hexonate and its lactone. The substrate specificity of the enzyme is very similar to that of aldose reductase or aldehyde reductase.[12,13] The specificity of the latter may vary with tissues from which the enzyme originates.

Kinetic Properties. The K_m is 6.9×10^{-4} M for D-glucuronolactone, and 3.3×10^{-4} M for D-glucuronate at 1.66×10^{-4} M TPNH. The equilibrium constant is 2.6×10^{-2} for the D-glucuronate system under the conditions given.

[11] Y. Mano, K. Suzuki, K. Yamada, and N. Shimazono, *Biochem. Biophys. Res. Commun.* **3**, 136 (1960).

[12] H. G. Hers, "Le Métabolisme du Fructose," p. 133. Editions Arscia, Bruxelles, 1957.

[13] H. G. Hers, *Biochim. Biophys. Acta* **37**, 120 (1960).

[8] The Inhibition of Catalase (Hydrogen-peroxide:Hydrogen Peroxide Oxireductase, EC 1.11.1.6) by Ascorbate

By C. W. M. Orr

$$\text{Catalase:} \quad 2\,H_2O_2 \xrightarrow{} 2\,H_2O + O_2$$
$$\text{ascorbate}$$

The ascorbate-mediated inhibition of catalase was first described by Foulkes and Lemberg.[1] These authors concluded that trace contamination by Cu^{2+} was partly responsible for the inhibition. Chance[2] also described the inhibition of catalase by ascorbate. He found that at concentrations of $10^{-3}\,M$ ascorbate, a spectrophotometrically identifiable, catalytically inactive species, designated complex II, was formed. Complex II is the result of the interaction of H_2O_2 with the heme moieties of the catalase molecule.[2] At the concentration of ascorbate used by Chance, it can be assumed that sufficient H_2O_2 is generated during the auto-oxidation of ascorbate to account for the appearance of complex II. More recently, it has been shown[3–5] that ascorbate inhibits catalase at concentrations as low as $2 \times 10^{-6}\,M$. Under these conditions no complex II was detected. A second mechanism, involving the inactivation of catalase (due to disintegration of the protein) by free radicals generated during the auto-oxidation of ascorbate, has been proposed.[5] The presence of divalent cations (particularly Cu^{2+} and Fe^{2+}) and Fe^{3+} during the auto-oxidation greatly increases this inhibition. Consequently, in demonstrating the effect of micromolar concentrations of ascorbate on catalase, particular care must be taken to ensure the absence of contaminating cations.

Reagents

H_2O, deionized and then glass-distilled is used at all times. This includes glassware, which should be rinsed with glass-distilled H_2O at least six times.

Ascorbic acid, reagent grade recrystallized from hot absolute alcohol. Prepare immediately before use as a $0.1\,M$ solution adjusted to pH 7.0 with N NaOH.

Tris, reagent grade recrystallized from hot absolute alcohol

[1] E. C. Foulkes and R. Lemberg, *Australian J. Exptl. Biol. Med. Sci.* **26,** 307 (1948).
[2] B. Chance, *Biochem. J.* **46,** 387 (1950).
[3] C. W. M. Orr, *Biochem. Biophys. Res. Commun.* **23,** 854 (1966).
[4] C. W. M. Orr, *Biochemistry* **6,** 2995 (1967).
[5] C. W. M. Orr, *Biochemistry* **6,** 3000 (1967).

HCl, the middle fraction from redistilled reagent grade HCl is used to neutralize the Tris buffers

Tris-chloride buffer, 0.05 M, pH 7.0

H_2O_2, reagent grade, 30%. To make 0.088 M H_2O_2, add 1 ml of 30% H_2O_2 to 99 ml of 0.01 M Tris-chloride, pH 7.0.

Preparation of Catalase

Crystalline catalase is obtained from a reliable supply house.[6] In order to remove low molecular weight contaminants, an aliquot containing 10 mg of protein is made up to 2 ml with 0.05 M Tris-chloride, pH 7.0, and applied to a 50 × 2 cm column of Sephadex G-100 previously equilibrated at 4° in 0.05 M Tris-chloride, pH 7.0. Catalase is eluted from the column with the same buffer, and all fractions in the protein peak with an absorbance ratio of 405 mμ:280 mμ greater than 0.75 are pooled. Before use, an aliquot is diluted appropriately (see below) in 0.01 M Tris-chloride, pH 7.0.

Catalase Assay

Catalase is assayed by a slight modification of the method of Chance.[7] Catalactic activity is determined by measuring the decrease in absorbance at 240 mμ when catalase is incubated with H_2O_2. The measurements can be made manually[8] or, more simply, by a recording spectrophotometer (Gilford or SP 800). In the latter case a chart speed of 1 inch/10 seconds is optimal, and a continuous recording is made for 20 seconds. The rate is determined by measuring the slope of the linear part of the curve (normally the first 10 seconds). The amount of catalase used in the assay should be adjusted so that there is no measureable absorption from the added enzyme solution itself and that a reduction of about 0.2 at 240 mμ occurs in 10 seconds; normally, 1–2 μg of catalase should be adequate. This rate would correspond to an activity of 4 units of catalase with a specific activity of 2000–4000.[9]

Assay Procedure. To a 3-ml cuvette is added 3 ml of 0.088 M H_2O_2 in 0.01 M Tris-chloride, pH 7.0. The cuvette is tilted, and 0.02 ml of a catalase-containing solution is pipetted onto the side wall so that it remains as a hanging drop as long as the cuvette is held at that angle. The cuvette is covered with parafilm, mixed by inversion, and inserted into the spectro-

---

[6] All chemicals used in 3, 4, and 5 were from British supply houses. Catalase was obtained from Koch-Light Ltd., Bucks, England, Lot No. 26388.

[7] B. Chance, *Methods Biochem. Anal.* **1**, 412.

[8] B. Chance, Vol. II, p. 766.

[9] A unit of catalase activity is defined as that amount of protein required to catalyze the reduction of 1 micromole of H_2O_2 in 1 minute.

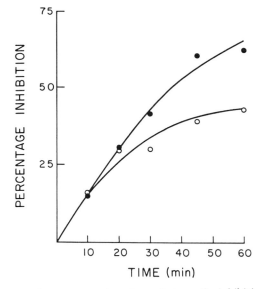

FIG. 1. The effect of the concentration of ascorbate on the inhibition of catalase when preincubated at $37°$: ●——●, $2 \times 10^{-4} M$ ascorbate; ○——○, $2 \times 10^{-6} M$ ascorbate.

photometer. Recording, against a blank of $0.01 M$ Tris-chloride, pH 7.0, is started immediately.

Detection of Inhibition of Catalase by Ascorbate. When catalase is pre-incubated with ascorbate the following procedure is used. Tris-chloride, pH 7.0 (0.5 micromole), 100 units of catalase, and 0.2 micromole of ascorbate are added to a tube in an ice bath, in that order. The tube, together with a control containing only Tris-chloride and catalase, is incubated at $37°$; 0.02 ml aliquots are removed at 5, 10, 15, 20, and 30 minutes and immediately assayed. The difference in the rate of H_2O_2 decomposition is obtained from the slopes of the control and ascorbate-containing tubes at each time point, and the percentage of inhibition is calculated (Fig. 1).

Factors Affecting the Inhibition. The inhibition of catalase by ascorbate is not an instantaneous process, but is time dependent. It has been shown that neither pH nor the concentration of H_2O_2 (within the spectrophotometrically useful range, i.e., 0.5–1.5 OD units at 240 mμ) affect the rate of inhibition. The rate of inhibition of catalase by ascorbate is strongly temperature dependent; at $0°$ there is little inhibition, but it increases as the temperature is raised. Also, as indicated in Fig. 1, the rate and extent of the inhibition of catalase are dependent on the initial concentration of ascorbate in the incubation mixture. It has been mentioned that divalent cations (such as Cu^{2+} and Fe^{2+}) or Fe^{3+}, which catalyze the auto-

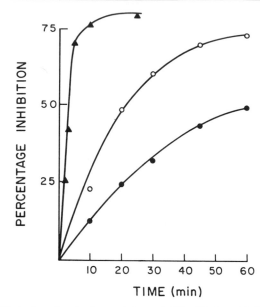

Fig. 2. The effect of preincubating catalase at 37° with $2 \times 10^{-4} M$ ascorbate alone ($\bullet$——$\bullet$), $2 \times 10^{-4} M$ ascorbate and $2 \times 10^{-4} M$ NaEDTA pH 7.0 ($\bigcirc$——$\bigcirc$), $2 \times 10^{-4} M$ ascorbate and $2 \times 10^{-6} M$ Cu^{2+} ($\blacktriangle$——$\blacktriangle$).

oxidation of ascorbate, profoundly enhance the rate of inhibition. These effects can be demonstrated by the addition of $2 \times 10^{-6} M$ of either cation to the reaction mixture (the effect of Cu^{2+} is shown in Fig. 2). In the absence of ascorbate, Cu^{2+} at this concentration does not inhibit catalase. By contrast, iron strongly inhibits the enzyme without ascorbate. Further, EDTA cannot be used to chelate these cations, since it itself potentiates the ascorbate-mediated inhibition of catalase (Fig. 2).

[9] Separation and Solubilization of the Mitochondrial NAD- and NADP-Linked Xylitol Dehydrogenases

By Charalampos Arsenis and Oscar Touster

$$\text{Xylitol} + \text{NADP}^+ \rightleftharpoons \text{L-xylulose} + \text{NADPH} + \text{H}^+$$
$$\text{Xylitol} + \text{NAD}^+ \rightleftharpoons \text{D-xylulose} + \text{NADH} + \text{H}^+$$

The two mitochondrial xylitol dehydrogenases [xylitol: NAD oxido-reductase (D-xylulose-forming), EC 1.1.1.9; and xylitol: NADP oxidoreductase (L-xylulose-forming), EC 1.1.1.10] of guinea pig liver have been located in different compartments of the organelle.[1] This finding provides a way to separate the two enzymes. Based upon the generally accepted sites of several marker enzymes,[2,3] it appears that NAD-xylitol (D-xylulose) dehydrogenase is weakly bound to the outer membrane or the outside of the inner membrane of the mitochondrion, or is present between the two membranes, whereas the NADP-xylitol dehydrogenase appears to be located in the matrix confined by the inner membrane or is bound rather strongly to the inner membrane.

Mitochondria can be readily resolved, by centrifugation of the disrupted particles in a discontinuous sucrose gradient, into three fractions representing (a) a soluble fraction consisting of enzymes usually found in the space between the two membranes or easily detached from the membranes, (b) a fraction consisting of the outer membranes, and (c) a pellet consisting primarily of the core of the mitochondria. The procedure by Parsons and Williams[2] as modified by Sottocasa et al.[3] was found to yield the best results and will be described below.

Assay Method[4]

Principle. The formation of NADH or NADPH is followed by the change in the absorbance at 340 mμ with the use of a Gilford recording spectrophotometer (Vol. V [34], p. 317).

Reagents

    Xylitol, 0.067 M
    NAD or NADP, 0.0067 M
    Tris-HCl buffer, 0.04 M, pH 9.0
    L-Cysteine, 0.01 M
    MgCl$_2$, 0.05 M
    Sodium malonate, 0.10 M, pH 9.0

Procedure. NAD- or NADP-dependent xylitol dehydrogenase is assayed in mixtures containing 0.1-ml portions of buffer, cysteine, MgCl$_2$, coenzyme, sodium malonate, and enzyme in a total volume of 0.9 ml. The reaction is initiated by addition of 0.1 ml of xylitol solution to the experimental cuvette.

Definition of Unit and Specific Activity. One unit of enzyme is defined as that amount which catalyzes the formation of 1 millimicromole of reduced

---

[1] C. Arsenis and O. Touster, *J. Biol. Chem.* **243**, 4396 (1968).
[2] D. F. Parsons and C. R. Williams in Vol. X, p. 443.
[3] G. L. Sottocasa, B. Kuylenstierna, L. Ernster, and A. Bergstrand, Vol. X, p. 448.
[4] O. Touster and G. Montesi in Vol. V, p. 317.

DISTRIBUTION OF XYLITOL DEHYDROGENASES IN MITOCHONDRIAL SUBFRACTIONS

Fraction	Volume (ml)	Total protein (mg)	NAD-xylitol (D-xylulose) dehydrogenase			NADP-xylitol (L-xylulose) dehydrogenase		
			Total activity (units)	Specific activity		Total activity (units)	Specific activity	
				Units/mg	Units/g		Units/mg	Units/g
Soluble	8.3	6.88	79.8	11.6	19.2	6.88	1.0	1.66
Interface zone	1.7	0.63	2.2	3.5	0.5	0.34	0.55	0.08
Pellet	10.0	41.0	0	0	0	31.98	0.78	7.7
Supernatant solution after sonic treatment and centrifugation	10.0	35.00	0	0	0	84.96	2.36	20.4
Resuspended pellet after sonic treatment and centrifugation	10.0	7.10				8.02	1.13	1.92

pyridine nucleotide per minute. Specific activity is expressed as units per milligram of protein.

Preparation of Mitochondrial Fractions. Male guinea pigs are fasted for 18–20 hours before they are killed; their livers are quickly removed and homogenized in 3 volumes of cold 0.25 M sucrose. All subsequent operations are carried out at 0–4°. The mitochondrial fraction is prepared essentially as described by de Duve *et al.*[5] The mitochondrial preparation (approximately 50 mg of protein) was suspended in 7.5 ml of 10 mM Tris-phosphate buffer, pH 7.5, containing 0.02% bovine serum albumin. After 5 minutes at 0°, shrinking of the particles is accomplished by adding 2.5 ml of 1.8 M sucrose, which was 2 mM in $MgSO_4$ and 2 mM in ATP. After centrifugation in a discontinuous sucrose gradient as described (Vol. X [72c], p. 448), the clear, yellow supernatant solution and the zone at the interface of the two sucrose layers are separated from the pellet. The pellet is resuspended in 0.45 M sucrose containing 0.015% bovine serum albumin, in a volume (10 ml) equal to that of the sucrose suspension subjected to the gradient centrifugation. Another resuspended mitochondrial pellet derived from the above fractionation process is treated for 10 seconds at 0° and 3 A with an S-110 Branson Sonifier (step horn with 0.5-inch tip) and centrifuged in a Spinco No. 40 rotor for 60 minutes at 40,000 rpm. The pellet is resuspended in 0.45 M sucrose containing 0.015% bovine serum albumin, in a volume (10 ml) equal to the volume of the original sample used in the sucrose gradient. The distribution of the two mitochondrial xylitol dehydrogenases is shown in the table.

[5] C. de Duve, B. C. Pressman, R. Gianetto, R. Wattiaux, and F. Appelmans, *Biochem. J.* **60,** 604 (1955).

[10] Cytoplasmic NADP-Linked Xylitol Dehydrogenase [Xylitol:NADP Oxidoreductase (L-Xylulose-Forming), EC 1.1.1.10]

By CHARALAMPOS ARSENIS and OSCAR TOUSTER

$$\text{Xylitol} + \text{NADP}^+ \rightleftharpoons \text{L-xylulose} + \text{NADPH} + \text{H}^+$$

When NADP-linked xylitol dehydrogenase, which has long been known to occur in mitochondria,[1] is assayed in solutions of acetone powder preparations made from whole liver, much higher activities are found than when

[1] O. Touster, V. H. Reynolds, and R. M. Hutcheson, *J. Biol. Chem.* **221,** 697 (1956).

mitochondria or mitochondrial subfractions are assayed.[2,3] Further studies have recently revealed the existence of high NADP-linked xylitol dehydrogenase activity in the cytosol of guinea pig liver homogenate.[4]

Assay Method

Principle. The formation of NADPH was followed by the change in the absorbance at 340 mμ with the use of a Gilford recording spectrophotometer.[5]

Reagents

Glycine-NaOH, 0.1 M, pH 9.0
Xylitol, 0.53 M
NADP, 0.007 M
MgCl$_2$, 0.05 M

Procedure. Mixtures for assaying NADP-xylitol dehydrogenase contain 10 micromoles of glycine-NaOH buffer, pH 9.0, 0.7 micromoles of NADP, 5 micromoles of MgCl$_2$ and enzyme in a total volume of 0.9 ml. The reaction is initiated by the addition of 0.1 ml of xylitol solution to the experimental cuvette.

Definition of Unit and Specific Activity. One unit of enzyme is defined as that amount which catalyzes the formation of 1 millimicromole of NADPH per minute. Specific activity is defined as units per milligram of protein determined by the method of Lowry *et al.*[6]

Purification Procedure

Step 1. Preparation of the High Speed Supernatant Solution. Guinea pig liver (22 g) is homogenized in 2 volumes of cold 1.1% KCl solution and centrifuged at 800 g for 10 minutes. The supernatant solution is separated, and the pellet is resuspended in 1.1% KCl solution and rehomogenized. After centrifugation, the supernatant solutions are combined and centrifuged in a Spinco No. 30 rotor for 60 minutes at 29,000 rpm. The supernatant solution is brought to 88 ml by the addition of 1.1% KCl solution (fraction I).

Step 2. Alumina C$_\gamma$ Fractionation. The high speed supernatant solution (fraction I, 1460 mg of protein) is mixed with Alumina C$_\gamma$ (292 mg dry weight). After standing for 10 minutes at 0°, the suspension is centrifuged at 12,000 g for 10 minutes. The supernatant solution (fraction II) contains

[2] S. Hollmann and G. Lauman, *Z. Physiol. Chem.* **348**, 1073 (1967).
[3] C. Arsenis, T. Maniatis, and O. Touster, *J. Biol. Chem.* **243**, 4395 (1968).
[4] C. Arsenis and O. Touster, *J. Biol. Chem.* **244**, 3895 (1969).
[5] O. Touster and G. Montesi in Vol. V, p. 317.
[6] O. H. Lowry, N. J. Rosebrough, A. L. Farr, and R. J. Randall, *J. Biol. Chem.* **193**, 265 (1951).

Fractionation of Cytoplasmic NAD- and NADP-Linked Xylitol Dehydrogenases from Guinea Pig Liver

Fraction	Treatment	Volume (ml)	Total protein (mg)	Total activity (units)		Specific activity (units/ml)		Yield (%)	
				NADP dehydrogenase	NAD dehydrogenase	NADP dehydrogenase	NAD dehydrogenase	NADP dehydrogenase	NAD dehydrogenase
I	High speed supernatant solution	88	1460	49,700	57,200	34	39	100	100
II	Alumina C$_\gamma$ supernatant solution	84	756	3,400	36,200	11	47.9	16.9	63.3
III	Alumina C$_\gamma$ gel eluate	44	150	31,700	700	210	4	63.7	1.2

most of the NAD-linked xylitol dehydrogenase activity and about 17% of the NADP-linked xylitol dehydrogenase activity. The gel is washed by suspending it in half the original volume of water and centrifuging as above. Most of the NADP-linked xylitol dehydrogenase is then eluted by resuspending the gel in half the original volume of 0.1 M sodium phosphate, pH 7.5, and centrifuging as before (fraction III). A summary of the enzyme purification procedure is shown in the table (p. 67).

[11] A Lysosomal Acid Phosphatase of Rat Liver with High Activity toward α-D-Glucuronic Acid 1-Phosphate

By CHARALAMPOS ARSENIS and OSCAR TOUSTER

$$\alpha\text{-D-Glucuronic acid 1-phosphate} + H_2O \rightarrow \text{D-glucuronic acid} + P_i$$

Since UDP-glucose is readily oxidized to UDP-glucuronic acid, the latter nucleotide appears to be a likely physiological source of free D-glucuronic acid, a precursor of L-ascorbic acid. One pathway for formation of glucuronic acid from its nucleotide derivative involves the sequential hydrolysis of UDP-glucuronic acid to α-D-glucuronic acid 1-phosphate and then to free glucuronic acid. However, even a moderately specific glucuronic acid 1-phosphatase has not as yet been found in liver, although a lysosomal acid phosphatase has been purified[1-3] that has relatively high activity toward this phosphate. Although, in view of present assumptions about the degradative function of lysosomes, it is questionable whether this phosphatase would be directly involved in a biosynthetic pathway such as the synthesis of ascorbic acid, the preparation and the assay conditions of this lysosomal phosphatase are given below.

Assay Methods

Principle. Inorganic phosphate produced by the action of phosphatase is determined by a modification[3] of the Lowry–Lopez method,[4] and *p*-nitrophenol is determined by its absorbance at 410 mμ at pH 10–11.

[1] C. Arsenis, S. Hollmann, and O. Touster, *Abstr. Am. Chem. Soc. Meeting*, C-286 September, 1966, New York.
[2] C. Arsenis and O. Touster, *J. Biol. Chem.* **242**, 3400 (1967).
[3] C. Arsenis and O. Touster, *J. Biol. Chem.* **243**, 5702 (1968).
[4] L. F. Leloir and C. E. Cardini Vol. III, p. 845.

Reagents

Sodium malonate, 0.2 M, pH 4.0

α-D-Glucuronic acid 1-phosphate, tripotassium salt adjusted to pH 4.0 with HCl, 0.1 M

p-Nitrophenyl phosphate, 0.1 M

Procedure. The standard assay for phosphatase contains, in a total volume of 1.0 ml, 10 micromoles of substrate, 50 micromoles of sodium malonate, pH 4.0, and an amount of enzyme which hydrolyzes less than 20% of the substrate. After 30 minutes of incubation at 37°, the reaction is stopped by the addition of 1.0 ml of 0.25 M sodium acetate buffer, pH 3.1, and then heating in a boiling water bath for 1 minute. Inorganic phosphate produced is determined by addition of 0.2 ml each of 1% solutions of ammonium molybdate and ascorbic acid in water. The absorbancy at 700 mμ is proportional to the inorganic phosphate concentration. When p-nitrophenyl phosphate is the substrate, the p-nitrophenol produced is determined by stopping the reaction with 10 ml of 0.04 M NaOH and determining the absorbance at 410 mμ.

Definition of Unit and Specific Activity. One unit of enzyme corresponds to the release of 1 micromole of product per hour. Specific activity is defined as units per milligram of protein determined by the method of Lowry *et al.*[5]

Experimental Procedure

Step 1. Preparation of Lysosomal Extract. The lysosomal fraction, prepared from 15 g of rat liver as described by Ragab *et al.*,[6] is suspended in cold water containing 0.1% (v/v) Triton X-100 and dialyzed overnight against 0.001 M Tris-HCl buffer, pH 7.4. After removal of insoluble material by centrifugation, the supernatant solution is used for column chromatography.

Step 2. Column Chromatography. The supernatant solution obtained above is poured over a 1.9 × 20 cm DEAE-cellulose column equilibrated with 0.01 M Tris-HCl buffer, pH 7.4. Phosphatase activity in eluates from the column is determined with 10 mM p-nitrophenyl phosphate as the substrate in 0.05 M sodium malonate buffer, pH 4.5. Application of 100 ml of 0.01 M Tris-HCl buffer, pH 7.4, to the column elutes the first acid phosphatase peak. Elution with 100 ml of 0.01 M 5'-dAMP in the same Tris buffer then yields the lysosomal acid nucleotidase.[3] Finally, the phosphatase with high activity toward α-D-glucuronic acid 1-phosphate is

[5] O. H. Lowry, N. J. Rosebrough, A. L. Farr, and R. L. Randall, *J. Biol. Chem.* **193**, 265 (1951).

[6] H. Ragab, C. Beck, C. Dillard, and A. L. Tappel, *Biochim. Biophys. Acta* **148**, 501 (1967).

eluted with 100 ml of 0.01 M Tris-HCl buffer, pH 7.4, which is 0.2 M in NaCl. The most active fractions (fraction 26 through 29) are pooled and concentrated by ultrafiltration. Salient features of the purification are indicated by the data in the table. The extent of purification of the enzyme is much greater than is apparent from this table, since the preparation of the lysosomal extract involves separation of enzyme from over 97% of the liver protein.

FRACTIONATION OF PHOSPHATASE ACTIVITY FROM RAT LIVER LYSOSOMES

Fraction	Volume (ml)	Total protein (mg)	Total activity (units)	Specific activity (units/mg)
Lysosomal extract	6.5	19.1	105	55
DEAE-cellulose				
Fractions 6 through 9	40	0.58	124	214
Fractions 16 through 19	40	0.17	82	476
Fractions 26 through 29	40	3.58	294	82

Properties

This phosphatase appears to be a nonspecific sugar phosphate phosphohydrolase,[2] hydrolyzing a wide variety of sugar phosphates including nucleotides. Its pH optimum is about 4.0.

Section II

Thiamine: Phosphates and Analogs

[12] Acid Dye Method for the Analysis of Thiamine

By V. Das Gupta and Donald E. Cadwallader

Thiamine is an essential nutritional or therapeutic component of a great many oral and injectable pharmaceutical preparations. Many biological and chemical methods for the quantitative determination of thiamine have been proposed. Chemical assay methods are usually preferred to biological methods since analyses can be performed rapidly and economically, and they are usually more reliable for routine determinations.

Although a number of chemical assay methods of thiamine are known, only the coupling reaction with diazotized 6-aminothymol[1] and thiochrome reaction[2] have been developed as precise methods of assay. In recent years, the thiochrome method has been used almost exclusively; however, close attention to details makes this method unsuitable for rapid control work.

An accurate, precise, and simple colorimetric method of analysis has been developed.[3] This method is based on the salt formation of thiamine with an acid dye, bromothymol blue. The resulting thiamine dye salt is distinguished by its solubility in organic solvents and thus can be separated, and the dye component be measured spectrophotometrically.

Assay Procedure

Reagents and Materials

Bromothymol blue (Fisher Scientific Co. or W. H. Curtin Co.) used without further purification
Chloroform N. F., USP grade, as the organic solvent
Thiamine hydrochloride or mononitrate, USP grade

Preparation of Solutions. A chloroform solution of dye ($4 \times 10^{-4} M$) is prepared by dissolving 62.44 mg of bromothymol blue in enough chloroform to make 250 ml of solution.

A pH 6.6 buffer solution is prepared by taking 50 ml of 0.2 M potassium dihydrogen phosphate solution, adding 16.4 ml of 0.2 M sodium hydroxide solution and then enough distilled water to make 200 ml.

Aqueous solutions of thiamine are prepared by dissolving 40 mg of either thiamine hydrochloride or thiamine mononitrate in enough pH 6.6

[1] K. J. Hayden, *Analyst* **82**, 61 (1957).
[2] "United States Pharmacopeia." 17th rev., p. 888. Mack Publ., Easton, Pennsylvania, 1965.
[3] V. Das Gupta and D. E. Cadwallader, *J. Pharm. Sci.* **57**, 112 (1968).

buffer solution to make 100 ml. This stock solution is used to prepare solutions of lower thiamine concentrations, for preparation of calibration curves, by diluting with pH 6.6 buffer solution. All thiamine solutions should be prepared fresh daily since thiamine is relatively unstable at pH 6.6.

Procedures. An exact quantity of assay sample is placed in an appropriate volumetric flask and diluted with pH 6.6 buffer solution so that the final solution contains between 4 and 10 μg/ml of thiamine. A 10.0-ml quantity of this solution is placed in a 125-ml separatory funnel. A 10.0-ml quantity of dye solution in chloroform ($1 \times 10^{-4} M$) is added to the aqueous solution, and the separatory funnel is shaken vigorously by hand for 1 minute. The nonpolar and polar phases are allowed to separate, and the chloroform layer is collected. The aqueous phase is extracted with two additional 10.0-ml quantities of chloroform, and the three chloroform extracts are combined. The chloroform solution is evaporated in a 50-ml beaker using low heat (60°) to a volume less than 10 ml. After the solution has cooled to room temperature, chloroform is added to make 10.0 ml of solution. A portion of this solution is centrifuged for 5 minutes at 2500 rpm, and the absorbance is read at 420 mμ (slit width 0.03 mm) using a Beckman DU spectrophotometer. A blank is prepared by substituting 10.0 ml of plain buffer solution for thiamine solution in the above procedure.

Preparation of Calibration Curves. Using the previously described procedures, pH 6.6 buffer solutions containing various concentrations of thiamine hydrochloride or thiamine mononitrate are assayed for thiamine dye salt concentrations. The absorbance readings are used to prepare calibration curves. Beer's law is followed within the range of 4–10 μg of thiamine per milliliter.[3]

Discussion

Optimum Conditions for Assay. A method based on the reaction between bromothymol blue and thiamine in a pH 6.6 buffer solution can be used for the analysis of thiamine, and the best dye concentration is $1 \times 10^{-4} M$.[3] The optimum pH value of 6.6 is not in agreement with a 7.4 value reported by Schill.[4]

Beer's law is followed within the range of 4–10 μg of thiamine per milliliter at 420 mμ. The best solvent for extraction is chloroform, and the thiamine dye salt is very stable in this solvent.[3] The acid dye method is at least as accurate and precise as the USP method. The reproducibility of results based on 12 samples (4 μg/ml) was 4.00 ± 0.04 versus 4.00 ± 0.05 by the USP method.[3]

[4] G. Schill, *Acta Pharm. Suecica* **2**, 13 (1965).

Effect of Diverse Substances. No interference in the analytical procedure was observed from other vitamins, minerals, hormones (ethynylestradiol and progesterone), and excipients except for a slight interference from ethyl alcohol.[3] Lack of interference was expected since none of the substances tested was as strong a base as thiamine. The interference from alcohol can be eliminated if the alcohol present is evaporated using gentle heat before assaying.

Interference from the decomposition products of thiamine and other vitamins is also negligible.[3]

[13] Identification and Determination of Thiamine by Picrolonic Acid and Lithium Picrolonate

By P. Vassileva-Alexandrova and A. Alexandrov

Determination of the Molar Ratio by the Reaction between Thiamine and Picrolonic Acid

Thiamine can be precipitated with picrolonic acid as a yellow precipitate that is insoluble in water.[1] The precipitate is soluble in 2 N HNO$_3$, concentrated H$_2$SO$_4$, and concentrated HCl. Some organic solvents, e.g., alcohols and methyl ethyl ketone, dissolve the precipitate. The presence of the functional groups —NO$_2$ and —OH in the compound formed explains its solubility in alcohols. The molar ratio of the reaction between picrolonic acid and thiamine was determined in order to establish the constitution of the compound formed.[2]

Gravimetric Investigation by the Job-Ostromyslenski Method[3,4]

Isomolar solutions of thiamine and picrolonic acid were prepared. These solutions were used for preparing a series with molar ratios between the components from 1:9 to 9:1. The Job-Ostromyslenski method requires that the precipitate formed in each sample be extracted and the absorbance of the extract measured. The extracts of the precipitates have a very low molar extinction coefficient; therefore each sample was filtered, and the precipitate was weighed. The weights were used for drawing up the curve showing the constitution of the compound formed (Fig. 1).

[1] S. Ohdake, *Proc. Imp. Acad. Tokyo*, **10**, 95 (1934).
[2] P. Vassileva-Alexandrova and A. Alexandrov, *Mikrochim. Acta* p. 572 (1966).
[3] P. Job, *Ann. Chim.* **9**, 113 (1928).
[4] I. Ostromyslenski, *Ber. Deut. Chem. Ges.* **44**, 268 (1911).

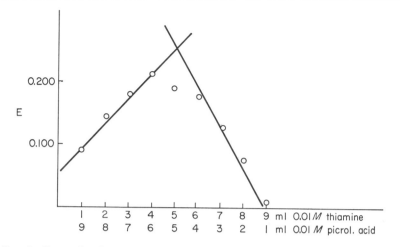

Fig. 1. Curve showing constitution of the compound formed by the Job-Ostromyslenski method. *Mikrochim. Acta* p. 572 (1966).

The maximum of the curve indicates a molar ratio of thiamine : picrolonic acid near 2 : 3. But if one compares the position of the points in the two parts of the curve, it is easy to establish that the maximum is moved to the left. An excess of thiamine increases the solubility of the precipitate. Thus, making a suitable extrapolation, one can conclude that thiamine and picrolonic acid react in a true molar ratio of 1 : 1.

Analysis of Thiamine Picrolonate

An analysis of the complex was carried out in order to confirm the molar ratio established. The quantity of the picrolonic acid in the complex was determined by reaction with triphenyltetrazolium chloride. The determination is based on the fact that the solubility product of triphenyltetrazolium picrolonate is less than that of thiamine picrolonate. Triphenyltetrazolium picrolonate is soluble in chloroform where it has a large molar extinction coefficient. Thus, thiamine picrolonate is converted into triphenyltetrazolium picrolonate in the presence of triphenyltetrazolium chloride; the latter is extracted with chloroform, and its extinction is measured. Analysis shows that the initial compounds react in a ratio of 1 : 1 with actual values of 0.8 to 1.2 moles of picrolonate per mole of thiamine.

The thiamine picrolonate was analyzed for chloride also, but the result was negative.

Taking into consideration the investigation described above as well as the well-known properties of the two initial compounds, one can conclude that the reaction proceeds in agreement with the equation:

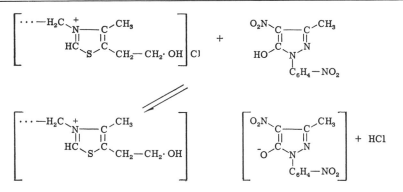

Our results exclude the allegation of other authors that thiamine reacts with picrolonic acid in a ratio of 1:2.[1]

Microcrystalloscopic Identification of Thiamine with Picrolonic Acid

The reaction for identification of thiamine with picrolonic acid is carried out on an object glass followed by an examination through a microscope at a magnification of ×400.

Other B vitamins, i.e., B_2, B_6, B_{12}, and niacin, which often appear together with vitamin B_1, do not react with picrolonic acid. Some inorganic ions do react, giving crystals that are different from those of thiamine picrolonate.

The sensitivity of the reaction is 0.0062 $\mu g/\mu l$. The limiting dilution is 1:150,000.

Procedure

Reagents

Thiamine
Picrolonic acid, saturated water solution

Samples (1 μl each) of thiamine and of picrolonic acid are mixed on an object glass. The mixture is allowed to stand for 3–4 minutes and then is examined under ×400 magnification. In the presence of thiamine, fanlike crystals are formed which appear as whorls (Fig. 2).

Conclusion

The reaction for identification of thiamine by means of picrolonic acid is very sensitive and specific because of the peculiar form of the crystals.

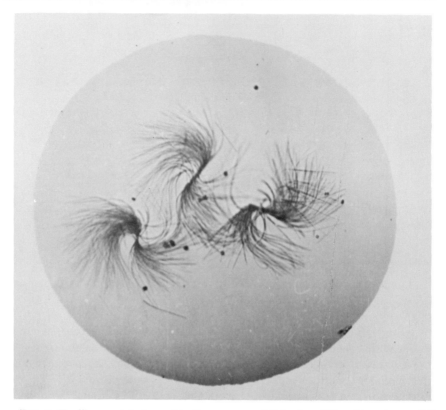

FIG. 2. Fanlike crystals formed in the presence of thiamine. *Mikrochim. Acta* p. 572 (1966).

Extractive Titration of Thiamine with Lithium Picrolonate

The reaction of thiamine with picrolonic acid can be used for quantitative determination of thiamine.[5]

Optimal Conditions

Thiamine can be precipitated quantitatively under special conditions. The first condition is that the solubility of the precipitate be lowered. This can be done by evaporating the water from a thiamine solution and adding the reagent as a saturated solution.

A very important condition relates to the acidity of the solution. It must be neutral or slightly alkaline. It is preferable to carry out the precipitation with lithium picrolonate because picrolonic acid makes the solution acid. The solubility of lithium picrolonate satisfies the requirements for a complete precipitation of thiamine.

[5] P. Vassileva-Alexandrova and A. Alexandrov, *Mikrochim. Acta* p. 277 (1967).

The temperature of the solution in which thiamine is precipitated must not be higher than 20°. The precipitate continues to form over about 2 hours.

The vitamins from the B group (B_2, B_6, B_{12}, and nicotinic acid), which often appear in combination with thiamine, do not react with lithium picrolonate, but do delay the crystallization of thiamine picrolonate. In the presence of these vitamins, the crystallization continues for not less than 7–8 hours.

After the precipitation of thiamine, the excess lithium picrolonate is titrated with an aqueous solution of methylene blue.[6,7] A blank assay must be titrated to determine the quantity of the picrolonate compounded with the thiamine. The blank assay contains only lithium picrolonate. Chloroform is added to the solution before titration. The chloroform dissolves the methylene blue picrolonate formed during the titration. Thus the chloroform layer is colored blue while the water layer remains colorless. At the end of the titration the chloroform layer is replaced by a fresh one. At the end point, the water layer is colored blue by the excess of methylene blue.

The concentration of the solution of methylene blue can by determined with a standard solution of thiamine. The titration must be carried out after separation of the precipitate of thiamine picrolonate by filtration with a glass filter G-4.

Procedure

   Reagents

      Thiamine chloride, 1 mg/ml
      Lithium picrolonate, saturated solution
      Methylene blue, 5×10^{-4} M
      Chloroform

Preparation of the Stock Solution of Methylene Blue. Methylene blue (5×10^{-4} M) is dissolved in water, and the solution is diluted to 1 liter. The concentration of this solution is determined by means of a solution of thiamine chloride (100 mg/100 ml).

Three parallel samples (each 0.5 ml) of the solution of thiamine chloride are put into 1-ml microglasses and left at 60° until evaporation is complete. The glasses are cooled, and then 0.4 ml of the lithium picrolonate solution is added to each. The samples are left for 2 hours for complete precipitation. To avoid any evaporation, the glasses must be stoppered. Each solution is

[6] D. Nonova, *Mikrochim. Acta* p. 111 (1958).

[7] N. Gantchev, P. Vassileva-Alexandrova, and A. Alexandrov, *Trav. Sci. École Normale Supérieure, Plovdiv, Bulgaria* **4**, 107 (1966).

filtered with a glass microfilter. Aliquots of 0.2 ml are taken and pipetted into three separatory funnels, each containing 1 ml of distilled water and 10 ml of chloroform. These samples are titrated in portions with methylene blue with shaking after every portion. The blue compound formed is extracted by the chloroform layer and the water layer remains colorless. When the water layer begins to become blue, the chloroform must be replaced. After shaking, the water layer becomes colorless, but after adding a new portion of methylene blue it turns blue again. The titration is continued until, with replacement of the chloroform layer, the water layer does not become colorless but remains blue. Three samples of lithium picrolonate, each 0.2 ml, are titrated in the same way. The calculation of the concentration of the solution of methylene blue is done in the following way:

$$\text{where} \qquad T_{mb} = T_t V_t M_{mb}/M_t 2(V_{1\,mb} - V_{2\,mb}) \qquad (1)$$

T_{mb} is the concentration of methylene blue in g/ml.
T_t is the concentration of thiamine chloride in g/ml.
V_t is the volume of the sample of thiamine chloride.
M_{mb} is the molecular weight of methylene blue.
M_t is the molecular weight of thiamine.
$V_{1\,mb}$ is the volume of the methylene blue solution taken for titration in the blank assay of thiamine picrolonate.
$V_{2\,mb}$ is the volume of the solution of methylene blue taken for the titration of the sample analyzed.

The stock solution of methylene blue prepared as described can be used for indirect titration of thiamine.

Determination of Thiamine. The thiamine solution must be prepared so that its concentration becomes about 1 mg/ml.

Three samples of 0.5 ml each of this solution are put into microglasses provided with stoppers.

The samples are then treated in the way described for determination of the concentration of the methylene blue solution. The quantity of thiamine is calculated as follows:

$$\% \text{ thiamine} = T_{mb} 2(V_{1\,mb} - V_{2\,mb}) M_t V_b 100/M_{mb} V_t w \qquad (2)$$

where

T_{mb}, $V_{2\,mb}$, V_t, M_t, and M_{mb} are as used in Eq. (1).
$V_{1\,mb}$ is the volume of the solution of methylene blue, taken for the titration of the blank assay of lithium picrolonate.
V_b is the volume of the measuring flask in which the solution of the sample was prepared.
w is the weight of the solid sample dissolved in the measuring flask V_b.

The mean relative error of the method is 3.25% (see table). The ions of some heavy metals as well as amines interfere with the determination, but do not occur in usual mixtures with thiamine.

DETERMINATION OF ERROR

Thiamine		Error	
Taken (mg)	Found (mg)	Absolute (mg)	Relative (%)
0.200	0.196	−0.004	2.00
0.400	0.427	+0.027	6.75
0.600	0.623	+0.023	3.84
0.800	0.807	+0.007	0.87
1.000	0.972	−0.028	2.80

Conclusion

The method for determination of thiamine by means of lithium picrolonate has a satisfactory accuracy and is very simple to carry out. As little as 0.1 mg of thiamine can be determined.

[14] Simultaneous Determination of Thiamine and Pyrithiamine[1]

By R. L. AIRTH and G. ELIZABETH FOERSTER

Thiamine + $K_3Fe(CN)_6 \rightarrow$ "thiochrome" ($\lambda_{excit} = 385$ mμ; $\lambda_{emit} = 440$ mμ)
Pyrithiamine + $K_3Fe(CN)_6 \rightarrow$ "pyrichrome" ($\lambda_{excit} = 410$ mμ; $\lambda_{emit} = 480$ mμ)

Assay Method

Principle. A spectrophotofluorometric assay of a mixture of thiamine and pyrithiamine based upon the difference in excitation and emission characteristics of their respective oxidation products, thiochrome and pyrichrome, is presented.

Reagents and Apparatus

Trichloroacetic acid, 1.53 M
Potassium acetate, 4 M

[1] R. L. Airth and G. E. Foerster, *Anal. Biochem.* **3**, 383 (1962).

NaOH, 7.5 M

$K_3Fe(CN)_6$, 0.059 M; 0.65 ml of the 0.059 M solution is diluted to 10.0 ml with 7.5 M NaOH. These solutions are made within 1 hour of use.

Hydrogen peroxide, 30%

Isoamyl alcohol, redistilled over activated carbon (128–129° fraction) and stored saturated with distilled water

Thiamine hydrochloride. A stock solution of 1 mg/ml in 0.1 N HCl is prepared, and appropriate dilutions are made of this standard.

Pyrithiamine hydrobromide. A stock solution of 1 mg/ml in distilled water is prepared, and appropriate dilutions are made of this standard.

Quinine sulfate. A stock solution of 10.0 mg/100 ml in 0.1 N H_2SO_4 is prepared; from this a working standard of 250 mμg/ml of 0.1 N H_2SO_4 is used to calibrate the spectrophotofluorometer.

Glass-redistilled water is used throughout. All compounds are analytical reagent grade unless specified otherwise. All glassware must be cleaned in nitric acid, well rinsed with glass redistilled water, and completely dry prior to each use. A major, but easily avoidable, source of error in this procedure is improperly prepared glassware.

Any spectrophotofluorometer capable of exciting at 385 mμ and 410 mμ and measuring fluorescence at 435 and 480 mμ is suitable. The instrument utilized for the initial development of this procedure was the Aminco-Bowman spectrophotofluorometer with slits 1, 3, and 4 having $\frac{1}{8}$ inch diameter and the remainder $\frac{1}{16}$ inch. The values reported have not been corrected for photocell sensitivity.

Procedure. The oxidation of vitamin B_1 or its analog was ostensibly accomplished by combining the methods recommended by Burch et al.[2] and Mickelsen and Yamamoto.[3]

To a reaction volume of 3.0 ml, 0.24 ml of 1.53 M trichloroacetic acid is added. Any protein precipitating at this stage is removed by centrifugation. To a 2.0-ml aliquot of this mixture is added 3.40 ml of 4 M potassium acetate and then 0.20 ml of alkaline ferricyanide. The contents of the tube are mixed and left standing for 5 minutes at room temperature; then 0.10 ml of 30% H_2O_2 is added. The reaction is then extracted into 10.0 ml of isoamyl alcohol by mixing vigorously. The isoamyl alcohol phase is siphoned off and dried with approximately 6 g of anhydrous granular sodium sulfate. An appropriate series of thiamine and pyrithiamine standards are prepared and processed simultaneously with each batch of unknown

[2] H. B. Burch, D. A. Bessey, R. A. Love, and O. H. Lowry, *J. Biol. Chem.* **198,** 477 (1952).
[3] O. Mickelsen and R. S. Yamamoto, *Methods Biochem. Anal.* **6,** 191 (1958).

reactions. After the oxidation and extraction steps, care must be taken to avoid undue exposure of the extracts to light and contamination with dust.

A 2.0-ml aliquot of the isoamyl alcohol extract is used for the fluorometric measurements. Appropriate reagent blanks are simultaneously prepared. The emission of the isoamyl alochol extract is measured at 435 and 480 mμ with an excitation wavelength of 385 mμ. The emission is also measured at the same wavelengths with an excitation light of 410 mμ.

Calculation of the Results

The emission spectra of thiochrome ($\lambda_{excit} = 385$ mμ) and pyrichrome ($\lambda_{excit} = 410$ mμ) overlap. Thus equations reflecting this fact are utilized. Table I gives the relative molar fluorescence of thiochrome and pyrichrome as determined on the Aminco-Bowman spectrophotofluorometer. The values given are the average for three separate determinations.

TABLE I

RELATIVE MOLAR FLUORESCENCE OF THIOCHROME AND PYRICHROME AT
DIFFERENT EXCITATION AND EMISSION WAVELENGTHS

Conditions	$\lambda_{excit} = 385$ mμ		$\lambda_{excit} = 410$ mμ	
	λ_{emit} = 435 mμ $\times 10^4$	λ_{emit} = 480 mμ $\times 10^4$	λ_{emit} = 435 mμ $\times 10^4$	λ_{emit} = 480 mμ $\times 10^4$
Thiochrome, from thiamine reactions	26339	9685	6419	2816
Pyrichrome, from pyrithiamine reaction	210	1022	252	1709
Thiochrome	47200	17620	7720	2679
Thiochrome extracted with reagent blank	41300	16050	7720	3040

It is evident that thiochrome is about 15 times more fluorescent than pyrichrome under optimal conditions. Sealock and White,[4] using the ferricyanide oxidation method and measuring the resulting fluorescence of the aqueous phase, found thiamine to be 5.7 times as fluorescent as pyrithiamine. The inadequacy of measuring the fluorescence of the aqueous phase for quantitative determinations has been reviewed by Mickelsen and Yamamoto[3] and reconfirmed in this laboratory. Values are also presented for the relative molar fluorescence of thiochrome as determined by using a known concentration of this compound dissolved in isoamyl alcohol. An estimate of the partition coefficient of thiochrome between isoamyl

[4] R. R. Sealock and H. White, *J. Biol. Chem.* **181**, 393 (1949).

alcohol and the reagent blank may be obtained from the final values presented. In this case a 10.0-ml isoamyl alcohol–thiochrome solution is extracted with the aqueous phase of a reagent blank, and the fluorescence of the alcohol phase is then determined. It is readily apparent that there is a reduction in fluorescence, although not sufficient to account for the fluorescence of the oxidation products. Hence, the assumption of complete extraction of thiochrome—under the experimental conditions—is unjustified. This line of reasoning assumes that no quenching agent is extracted from the aqueous phase of the reagent blank. However, with standard amounts of thiamine or pyrithiamine, repeated determinations give sufficiently reproducible results to justify the method. Whether pyrichrome has a corresponding partition coefficient between isoamyl alcohol and the aqueous phase of the reagent blank is not known.

The values of Table I were used to generate Eqs. (1) and (2) to calculate thiamine and pyrithiamine concentration.

$$T = 4.12 \times 10^{-2} E_{435/385} - 0.85 \times 10^{-2} E_{480/385}$$
$$= \text{millimicromoles thiamine/assay} \quad (1)$$

$$P = 1.06 E_{480/385} - 0.39 E_{435/385} = \text{millimicromoles pyrithiamine/assay} \quad (2)$$

Where E = emission, T = thiamine concentration, and P = pyrithiamine concentration. The subscripts describe the conditions under which the emission is measured, the numerator representing the emission wavelength and the denominator the excitation wavelength. It should be noted that these equations have been formulated on the basis of the volumes outlined in the procedure. One of the chief sources of error is the presence of isoamyl alcohol-soluble fluorescent contaminants in the sample. If the contamination is slight, blank corrections will suffice. If this approach proves inadequate, the adsorption of thiamine and pyrithiamine on such adsorbents

TABLE II

RECOVERIES OF VARYING AMOUNTS OF PYRITHIAMINE AND A CONSTANT
AMOUNT OF THIAMINE (1.48 MILLIMICROMOLES)

Pyrithiamine present (mμmoles)	Molar ratio, pyrithiamine: thiamine	Pyrithiamine recovered		Thiamine recovered	
		mμmoles	%	mμmoles	%
5.95	4.0	4.91	83	1.46	99
11.9	8.0	11.4	96	1.46	99
23.8	16	22.3	94	1.49	101
34.5	23	32.7	95	1.46	99
47.6	32	45.0	95	1.41	95
59.5	40	53.8	90	1.36	92
83.9	57	94.1	105	1.17	80

TABLE III
RECOVERIES OF VARYING AMOUNTS OF THIAMINE AND A CONSTANT
AMOUNT OF PYRITHIAMINE (35.7 MILLIMICROMOLES)

Thiamine present (mμmoles)	Molar ratio, pyrithiamine: thiamine	Thiamine recovered		Pyrithiamine recovered	
		mμmoles	%	mμmoles	%
0.296	124	0.128	43	36.1	101
0.741	55	0.540	73	39.2	110
1.48	25	1.38	93	39.4	110
2.22	17	1.99	90	40.1	112
2.96	12	2.88	97	40.5	113
7.41	5.0	6.83	92	44.8	126
11.1	3.2	11.4	102	50.4	149

as kieselguhr, Decalso, or Amberlite IRC-50 may be employed. The efficacy of this latter approach has been reviewed.[3]

The limits of detection and reliability of the method are indicated in Tables II–IV.

TABLE IV
RECOVERIES OF VARYING AMOUNTS OF THIAMINE AND A CONSTANT
AMOUNT OF PYRITHIAMINE (11.9 MILLIMICROMOLES)

Thiamine present (mμmoles)	Molar ratio, pyrithiamine: thiamine	Thiamine recovered		Pyrithiamine recovered	
		mμmoles	%	mμmoles	%
0.296	40	0.270	91	13.0	109
0.741	16	0.702	95	13.0	109
1.48	8	1.35	92	13.9	117
2.22	5.4	2.01	91	13.4	113
2.96	4.0	2.80	95	13.5	113
7.41	1.6	6.96	94	13.6	114
11.1	1.1	8.24	75	15.0	126

It may be seen that the sensitivity and reliability of the method is determined *both* by the absolute amount of thiamine and pyrithiamine and by their molar ratio. In practice this difficulty may be circumvented, in some instances, by carrying out the assay on a dilution series of the sample to be measured. One shortcoming of these equations is the fact that the constants employed are dependent upon such factors as the spectrophotofluorometer used, purity of the standards (thiamine, pyrithiamine, and quinine sulfate), and, probably most important, the presence of fluorescent materials of biological origin which are isoamyl alcohol soluble. This difficulty may be circumvented in part, by using a generalized solution for Eqs. (1) and (2), i.e.:

$$T = [d/(ad - bc)]E_{435/385} - [b/(ad - bc)]E_{480/385}$$
$$= m\mu moles\ thiamine/assay$$

$$P = [a/(ad - bc)]E_{480/385} - [c/(ad - bc)]E_{435/385}$$
$$= m\mu moles\ pyrithiamine/assay$$

where a and b equal the relative molar fluorescence of thiochrome and pyrichrome when excited at 385 mμ and emissions are measured at 435 mμ. Constants c and d equal the respective relative molar fluorescence of thiochrome and pyrichrome when excited at the same wavelength, but emissions are measured at 480 mμ. Values for a, b, c, and d are measured by separately determining the fluorescence of known concentrations of pyrithiamine and thiamine with each group of assays carried out, thus permitting the ready calculation of the constants to be applied for each set of assays.

[15] Thin-Layer Chromatography for the Separation of Thiamine, N'-Methylnicotinamide, and Related Compounds

By Z. Z. ZIPORIN and P. P. WARING

By the use of labeled and nonlabeled thiamine, it has been possible to demonstrate that products of thiamine metabolism in rat and man appear in the urine. Approximately 20–22 different labeled spots have been separated from the urine of rats and man given pyrimidine- or thiazole-labeled thiamine.[1,2] Quantitatively, many of these fractions represent insignificant amounts of the ingested thiamine, leaving relatively few substances which must be considered in the investigation of urinary excretion of thiamine metabolites. These are: pyramin[3] (a term used by Mickelsen to represent a "pyrimidine-like component of the thiamine molecule which is excreted in the urine," and which enhances CO_2 production in the yeast fermentation assay for thiamine); thiamine disulfide, thiochrome, and thiazole[4]; and 2-methyl-4-amino-pyrimidinecarboxylic acid.[5] However, when studies are conducted requiring isolation and identification of metabolites, in addition to the metabolites listed above, it is necessary to consider other compounds: (a) pyrimidinesulfonic acid; (b) α-hydroxyethylthiamine (HET); (c) N'-methylnicotinamide (NMN). The pyrimidinesulfonic acid is the

[1] R. A. Neal and W. N. Pearson, *J. Nutr.* **83**, 343 (1964).
[2] M. Balaghi and W. N. Pearson, *J. Nutr.* **91**, 9 (1967).
[3] O. Mickelsen, W. O. Caster, and A. Keys, *J. Biol. Chem.* **168**, 415 (1947).
[4] J. M. Iacono and B. C. Johnson, *J. Am. Chem. Soc.* **79**, 6321 (1957).
[5] R. A. Neal and W. N. Pearson, *J. Nutr.* **83**, 351 (1964).

pyrimidine product of a bisulfite cleavage of thiamine,[6] while α-hydroxy-ethylthiamine may be found in extracts of bacteria and yeasts,[7] and is thus likely to occur after the incubation required in the yeast resynthesis method.[8] N'-Methylnicotinamide is a substance found in urine which remains with thiamine during Decalso column chromatography. When thiamine is assayed by the thiochrome method, N'-methylnicotinamide, when present in high enough concentrations, will provide a nonthiamine fluorescent substance in the blank treated with alkali and will also provide a fluorescent substance in the aliquot oxidized with alkaline ferricyanide.[9] In the latter, the fluorescence is diminished by approximately 80% as compared with the alkali-treated aliquot so that the arithmetic subtraction of the blank may provide negative values leading to erroneous results. It may, therefore, be important to ascertain the identity and concentration of N'-methylnicotinamide products that are present.

The separation of these substances has been accomplished by the use of paper chromatography,[4] column chromatography followed by paper chromatography,[1] and thin-layer chromatography.[10–12] By the latter means it has been possible to separate thiamine, thiazole, pyrimidinesulfonic acid, α-hydroxyethylthiamine, and thiochrome. Others using thin-layer chromatography have been able to separate thiamine from its mono- and diphosphate esters, as well as thiamine derivatives.[13–15] The method of Johnson et al.[10] has been difficult to reproduce, since one component of the solvent system, pyridine, absorbs short-wavelength ultraviolet light very strongly, thereby masking the components which are not converted to fluorescent products when sprayed with alkaline ferricyanide.

Thin-Layer Chromatography

Reagents

Thiamine hydrochloride
2-Methyl-5-ethoxymethyl-6-aminopyrimidine, purchased from Merck
and Company, Rahway, New Jersey

[6] G. A. Goldstein and G. M. Brown, Arch. Biochem. Biophys. 103, 449 (1963).
[7] G. L. Carlson and G. M. Brown, J. Biol. Chem. 236, 2099 (1961).
[8] Z. Z. Ziporin, E. Beier, D. C. Holland, and E. L. Bierman, Anal. Biochem. 3, 1 (1962).
[9] V. A. Najjar and R. W. Wood, Proc. Soc. Exptl. Biol. Med. 44, 386 (1940).
[10] D. B. Johnson, D. J. Howells, and T. W. Goodwin, Biochem. J. 98, 30 (1966).
[11] R. A. Neal, J. Biol. Chem. 243, 4634 (1968).
[12] P. P. Waring, W. C. Goad, and Z. Z. Ziporin, Anal. Biochem. 24, 185 (1968).
[13] C. Levorato and L. Cima, J. Chromatog. 32, 771 (1968).
[14] T. E. Dobbs and W. N. Pearson, Intern. Z. Vitaminforsch. 37, 468 (1967).
[15] K. S. Yang, R. T. Wang, and P. C. Chiang, Chung Kuo Nung Yeh Hua Hsueh Hui Chih 5, 53 (1967); (Chem. Abstr. 67, 111485q).

4-Methyl-5-β-hydroxyethyl thiazole, purchased from Merck and Company, Rahway, New Jersey

Pyrimidinesulfonic acid, prepared according to the procedure of Goldstein and Brown[6]

Pyrimidinecarboxylic acid, prepared according to the procedure of Neal and Pearson[5]

N'-Methylnicotinamide methiodide ($K_3Fe(CN)_6$-treated). About 100 mg of N'-methylnicotinamide methiodide in 5 ml of acidic solution (pH $\cong$ 2) saturated with NaCl or KCl, treated with 3 ml of oxidizing solution (4 ml of 1% $K_3Fe(CN)_6$ to 100 ml with 15% NaOH) and extracted into 13 ml of isobutyl alcohol.

N'-Methylnicotinamide methiodide, prepared according to the procedure of Huff and Perlzweig[16]

α-Hydroxyethylthiamine, prepared according to the procedure of Miller[17]

Thiochrome, purchased from Nutritional Biochemicals Corporation

Solvents. All solvents to be used in chromatography must be redistilled. Ethyl alcohol is used as absolute alcohol. Water is distilled, demineralized.

Eastman Chromagram sheets, type K301R (silica gel with fluorescent indicator)

Ultraviolet viewing box, BLE Spectroline, model C-3F, Spectronics Corp., Westbury, L.I., New York, equipped with short wavelength (2537 Å) and long wavelength (3660 Å) lamps

Procedure. Standards are prepared in aqueous solutions at concentrations of 0.5–4 mg/ml, of which 0.2–2 μl may be spotted on the chromagram sheets for one- or two-dimensional ascending chromatography. Solvent mixtures should equilibrate for 15 minutes in covered chromatography jars before the chromagram sheets are placed in the tanks for development. After the solvent front has moved 16–18 cm (80 minutes), the plates are removed and air-dried rapidly with a hair dryer.

The short wavelength absorbing spots may be detected in the ultraviolet viewing box using the 2537 Å line, and the 3660 Å wavelength can be used to detect the fluorescent compounds. The latter are derived from thiamine, α-hydroxyethylthiamine, and N'-methylnicotinamide when sprayed with alkaline ferricyanide.

Factors Affecting Thin-Layer Chromatography. As may be seen in the table, the more acidic solvent systems resulted in greater movement of

[16] J. W. Huff and W. A. Perlzweig, *Science* **97**, 538 (1943).

[17] Personal communication from Dr. James M. Sprague, Merck Sharp & Dohme, West Point, Pennsylvania, describing the method of synthesis by Dr. Charles S. Miller.

EFFECT OF pH ON R_f OF THIAMINE AND RELATED COMPOUNDS AS WELL AS
N'-METHYLNICOTINAMIDE (NMN) AND RELATED COMPOUNDS[a]

Standard	Solvent pH			Ultraviolet	
	2.54	4.03	7.85	2537 Å	3660 Å
Thiamine	0.16	0.04	0.03	A[b]	—[c]
NMN	0.31	0.06	0.05	A	—
HET[e]	0.23	0.09	0.06	A	—
Pyrimidinecarboxylic acid	0.42	0.21	0.26	F[d]	—
Thiochrome	0.31	0.28	0.33	—	F
Pyrimidinesulfonic acid	0.48	0.39	0.46	A	—
Pyrimidine	0.64	0.65	0.68	A	—
Thiazole	0.85	0.79	0.81	A	—
NMN-NaOH	—	0.29	—	A	—
NMN-NaOH	—	0.79	—	A	—
NMN-K$_3$Fe(CN)$_6$, abs.	0.71	0.65	0.67	A	—
NMN-K$_3$Fe(CN)$_6$, flu.	0.79	0.75	0.75	F	—

[a] Eastman chromagram with silica gel, K301R; acetonitrile–H$_2$O–formic acid, 40:10:to indicated pH.
[b] A = absorbing.
[c] — = neither absorbing nor fluorescing.
[d] F = fluorescent.
[e] α-Hydroxyethylthiamine.

those substances with R_f values of 0.16–0.42, while those with R_f values >0.48 were not markedly affected by changes in pH. Thus, while N'-methylnicotinamide and thiochrome could not be separated in single-phase chromatography at pH 2.54, a second phase at right angles in pH 4.03 solvent separated these quite clearly without affecting the resolution of the other compounds (Fig. 1).

The preparation of a urine sample for thin-layer chromatography of thiamine metabolites would require the separation of these metabolites from interfering substances such as inorganic salts and possibly other compounds, the effect of which has not yet been ascertained in the chromatography. The removal of salts from urine is reported by Iacono and Johnson[4] as well as Neal and Pearson.[1] We have had no experience as to the effectiveness of these procedures in preparing a sample of urine for thin-layer chromatography.

General Comments

N'-Methylnicotinamide can be distinguished from thiochrome by the use of long and short wavelength ultraviolet light. The former absorbs 2537 Å ultraviolet light with no fluorescence whereas the thiochrome neither absorbs nor fluoresces at this wavelength. With long wavelength ultraviolet

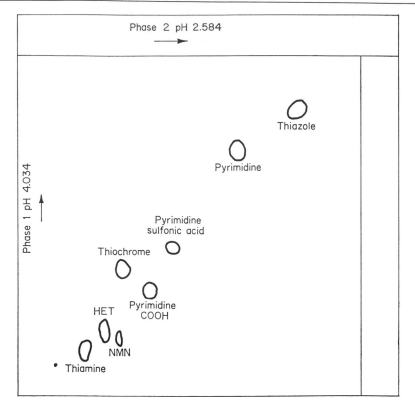

FIG. 1. Two-phase separation of thiamine, N'-methylnicotinamide, and related compounds.

light (3660 Å), N'-methylnicotinamide neither absorbs nor fluoresces, whereas thiochrome fluoresces intensely with a blue color.

As may be seen in the table, treatment of N'-methylnicotinamide with alkali or alkaline ferricyanide results in different substances being formed. Still other substances may be formed by the above treatments of N'-methylnicotinamide, but only those extracted into isobutyl alcohol at an alkaline pH have been chromatographed. These are the compounds one would encounter in the assay for thiamine where N'-methylnicotinamide is present. The use of thin-layer chromatography offers a means for separating these products.

[16] Electrophoretic Separation and Fluorometric Determination of Thiamine and Its Phosphate Esters

By Y. ITOKAWA and J. R. COOPER

Assay Method

Principle. This procedure is a modification of ·existing methods for electrophoretic separation[1] and subsequent fluorometric determination[2] of thiamine compounds. The advantage of the method to be described lies both in its speed of separation of the compounds (30 minutes) and its sensitivity (5 mμg).

Reagents

Acetate buffer, 0.05 M, pH 3.8

Cellulose polyacetate (Sepraphore[3] II or III) strips or Munktell paper strips No. S-311

Solution A. To 76 ml of water add 1 ml of 2% potassium ferricyanide and 15 ml of 15% NaOH. This reagent should be prepared just prior to use.

Hydrogen peroxide, 30%

Procedure. In addition to the sample, aliquots (total volume, 5 μl) of a standard solution containing thiamine, TMP, TPP, and TTP are spotted in the middle of strips that have been soaked in the acetate buffer and blotted. The strips are then subjected to electrophoresis for 30 minutes at a constant current of 8 mA. The strips are removed, air dried, and sprayed lightly with solution A. Fluorescent spots are visualized with a longwave UV light,[4] cut out from the strip along with a blank area of corresponding size, and placed in Turner[5] cuvettes containing 3.5 ml of solution A. After 2 minutes, 0.01 ml of 30% H_2O_2 is added to destroy the reagent and elution of the fluorophore is allowed to proceed for 20 minutes. The paper is removed, and the fluorescence of the sample is determined in a Turner fluorometer. The excitation filter is 7-60 and the emission filters are 2A and 47B; in a spectrophotofluorometer the excitation wavelength is 365 mμ and the emission maximum is 430 mμ.

Comments. When the thiamine compounds are to be determined in the

[1] D. Siliprandi and N. Siliprandi, *Biochim. Biophys. Acta* **14**, 52 (1954).

[2] L. M. Lewen and R. Wei, *Anal. Biochem.* **16**, 29 (1966).

[3] Gelman Instrument Company, Ann Arbor, Michigan.

[4] UVL-21. Ultra-violet Products Inc., San Gabriel, California.

[5] G. K. Turner Associates, Palo Alto, California.

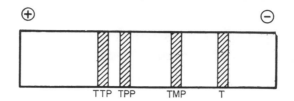

FIG. 1. Separation of thiamine, TMP, TPP, and TTP on Sepraphore II.

presence of tissue extracts, TCA is recommended as the protein precipitant with subsequent removal of the TCA by extraction with ether.

A variety of filter papers were tested for the electrophoretic separation of the thiamine phosphates; Munktell paper S-311 proved to yield the best separation and lowest blank. However, when maximum sensitivity was desired, Sepraphore was the support that was chosen.

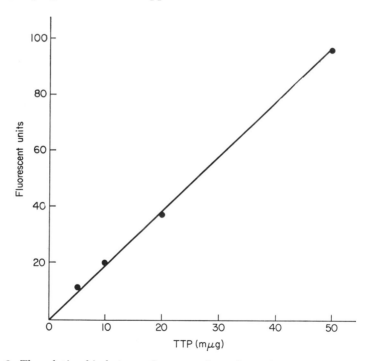

FIG. 2. The relationship between fluorescent intensity and concentration of TTP.

Figure 1 shows the separation of thiamine and its phosphate esters on Sepraphore II, and Fig. 2 illustrates the relationship between fluorescent units and concentration of TTP. This linear relationship continues up to a concentration of at least 200 mμg.

[17] Catalytic Polarography in the Study of the Reactions of Thiamine and Thiamine Derivatives

By SIGMUNDUR GUDBJARNASON

The reduction of hydrogen ions at the dropping mercury electrode (DME) in a cobaltous ammoniacal buffer can be catalyzed by a number of proteins and low molecular weight substances. The formation of a Co^{2+} complex with these substances results in lowering of the electrode potential at which the reduction of hydrogen ions can take place.[1] Several hypotheses have been suggested in order to explain the mechanism of the Brdicka catalytic hydrogen wave: (a) reaction of SH groups[1]; (b) absorption of the surface-active substances on the surface of the mercury electrode and subsequent desorption, permitting the depolarizer outside the layer to be electrolyzed[2]; (c) formation of different kinds of catalytic active complexes which alternatively appear and disappear, time after time, in accordance with the shift of the applied potential.[2] Extensive studies on the catalytic polarographic activity of a number of sulfhydryl and disulfide compounds indicate that the nature and proximity of neighboring groups, such as carboxyl groups, markedly affect the catalytic polarographic activity.[3] Recent studies indicate that the catalytic polarographic activity is not limited to sulfur-containing compounds, but extends to the reduction of protons of protonated amino and imidazole groups as well.[4] The electron to proton transfer appears to be catalyzed by complex formation of the protonated group with divalent cobalt. The catalytic polarographic activity reaches a maximum in the pH range of the proton dissociation, i.e., in the proximity of the pK value of the protonated group. In this pH range the proton and metal ion compete for the free electron pair of the base, and this competition appears to enhance the electrolytic reduction of the proton, i.e., lower the energy level or electrode potential required for electron transfer.[4]

The catalytic polarographic activity of thiamine and thiamine derivatives can be ascribed to the reduction of the proton of carbon-2 from the thiazole ring and the protonated nitrogen of the pyrimidine ring.[5] The catalytic polarographic activity of thiamine and its derivatives offers an

[1] R. Brdicka, *Collection Czech. Chem. Commun.* **5**, 112, 148 (1933).

[2] M. Shinagawa, H. Nezu, H. Sunahara, F. Nakashima, H. Okashita, and T. Yamada, *Advan. Polarog. (Proc. 2nd, Intern. Congr. Cambridge, England)* **3**, 1142 (1960).

[3] W. Lamprecht, S. Gudbjarnason, and H. Katzlmeier, *Z. Physiol. Chem.* **322**, 52 (1960).

[4] S. Gudbjarnason, *Biochim. Biophys. Acta* **177**, 303 (1969).

[5] S. Gudbjarnason, *Biochim. Biophys. Acta* **148**, 22 (1967).

opportunity to examine the electrochemical behavior of thiamine derivatives and to study the reactivity and reaction mechanism of thiamine derivatives.[5]

Experimental Procedure

The working electrode in polarography is the dropping mercury electrode (DME) which consists of a length of narrow-bore glass capillary tube attached to a mercury reservoir by means of rubber or plastic tubing. The polarograph that has been used for this type of study was Sargent Model XV; h (height of the mercury reservoir) = 80 cm, t (drop time) = 2.3 sec, m (rate of flow of mercury) = 3.5 mg/sec. The polarographic measurements are made in cobaltous (1 mM $CoCl_2$), buffered solutions ranging in pH from 4.0 to 9.8. The following buffers can be used: acetate buffer pH 4.0 and pH 5.0; phosphate buffer pH 6.0 and pH 7.0; Tris-buffer pH 7.0, pH 8.0, and pH 9.0; NH_3/NH_4OH buffer, pH 7.0, pH 8.0, pH 9.0, and pH 9.8.

The final volume of solution in the polarographic cell is 20 ml with (19.8 − x) ml of buffer placed in the cell. The oxygen is removed with a stream of nitrogen and during bubbling of nitrogen through the buffer, 0.2 ml of 0.1 M $CoCl_2$ is pipetted into the cell. Finally, x ml of 0.1 M thiamine or thiamine-derivative solution is added to the cobaltous buffer and mixed by bubbling nitrogen through the solution.

The polarographic activity is measured against a reference electrode, either an external electrode, i.e., the saturated calomel electrode (SCE) or an internal electrode, i.e., the mercury pool electrode (MPE). The choice of reference electrode has a significant effect upon the voltage of the various polarographic waves. The SCE gives an accurate measurement of a half-wave potential, whereas the MPE can alter its potential because of changes in adsorption on the electrode surface. The most widely used cell with an external reference electrode, and the one recommended here, is the H cell. It consists of two compartments, one containing the solution being studied and the other containing the reference electrode. To prevent polarization of the reference electrode, the compartment containing it should be made from tubing of at least 20 mm i.d., but the dimensions of the solution compartment can be varied widely to accommodate any desired volume of solution. In the Lingane-Laitinen cell,[6] these compartments are separated by a cross-member filled with a 4% agar-saturated potassium chloride gel, which is held in place by a medium-porosity, sintered-Pyrex disk. To facilitate rapid and complete deaeration of the solution, the disk should be placed as near to the solution compartment as possible; and for

[6] I. I. Lingane and H. A. Laitinen, *Ind. Eng. Chem. Anal. Ed.* **11**, 504 (1939).

the same reason the side tube through which the inert gas is passed through the solution should be as near to the bottom of the cell as possible.

The agar gel is prepared by warming a small flask containing 4 g of agar and 90 ml of water in a large beaker filled with boiling water, or on a steam bath, until solution is complete, then adding 30 g of potassium chloride and stirring thoroughly. When the salt has dissolved, the gel is pipetted into the cross-member until it is almost completely filled; then the cell is allowed to stand undisturbed, with the cross-member vertical, until the gel has solidified.

After the gel has solidified, the cell is turned upright and enough pure mercury is added to the reference-electrode compartment to give a layer 1–2 cm deep. This is covered with an equally thick layer of a paste made by stirring equal weights of mercurous and potassium chlorides with a little saturated potassium chloride, and the compartment is filled with saturated potassium chloride solution containing a large excess of the solid salt. Electrical connection to the mercury is made by means of a glass tube through which a small platinum wire is sealed so as to project into the mercury for a few millimeters; this tube is filled with mercury and inserted into a rubber stopper which serves to seal the reference-electrode compartment tightly. The wire leading to the reference-electrode terminal of the polarograph is simply dipped into the mercury in this tube. Creeping of the potassium chloride can be prevented by a coat of silicone grease on each of the stopper-glass boundaries.

The solution compartment of the cell is filled with saturated potassium chloride solution, and 2 days should be allowed for the SCE to reach its equilibrium potential.

The current sensitivity can be varied from 0.100 μA/mm to 0.300 μA/mm depending upon the concentration of thiamine. The polarogram is recorded from -0.8 to -2.0 V with damping in position 2.

Catalytic Polarographic Activity

This method offers an opportunity to examine the reactions and reactivity of specific protonated groups in the molecule; the influence of buffers, substrates, metal ions, etc. A few examples will be illustrated.

Catalytic Polarographic Activity of Thiamine and Thiamine Derivatives during Molecular Transformations in Alkali

In order to understand the changes in catalytic polarographic activity of thiamine in the cobaltous ammoniacal buffer at pH 9.8, it is necessary to review briefly the reactions of thiamine in alkali, which have been studied extensively in the past.[7]

[7] G. D. Maier and D. E. Metzler, *J. Am. Chem. Soc.* **79**, 4386 (1957).

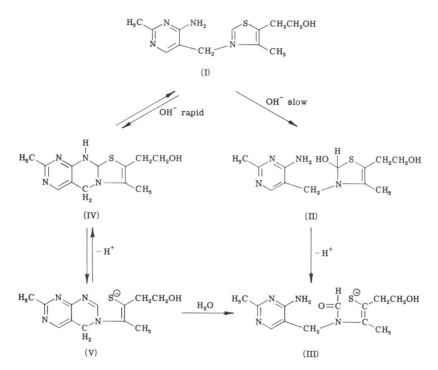

Fig. 1. Reactions of thiamine in alkaline solution.

When thiamine reacts slowly with alkali it forms the pseudo-base (II) (Fig. 1). This form is not present in high concentrations in thiamine solutions because the thiazole ring is easily opened yielding the colorless form (III). Oxidation of this thiol form gives the corresponding dimer, thiamine disulfide.

If thiamine is added to media at pH 11 or higher, it is converted to a yellow thiol form (V). The color of this compound fades rapidly due to the conversion of the yellow form (V) to the colorless thiol form (III). The reaction of thiamine with alkali giving the yellow thiol (V) is completely and rapidly reversible and requires the participation of two protons.

A form of thiamine intermediate between the cation (I) and the yellow thiol form (V) has also been isolated from methanol solutions of thiamine hydrochloride treated with two equivalents of sodium ethoxide. This form, isolated as a white crystalline compound, has been characterized as the cyclic compound (IV). When this cyclic form is dissolved in water, it disproportionates completely, but does so only partially in methanol, giving a mixture of the cation (I) and yellow form (V) (Fig. 1).

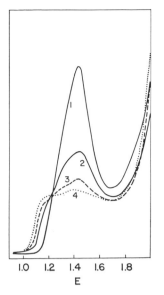

Fig. 2. Polarographic activity of thiamine, $5 \times 10^{-4} M$, in cobaltous ammoniacal buffer, pH 9.8, recorded against the SCE. Polarogram *1*, 1 minute after adding thiamine to the buffer; polarogram *2*, 10 minutes after mixing; polarogram *3*, 20 minutes after mixing; polarogram *4*, 30 minutes after mixing.

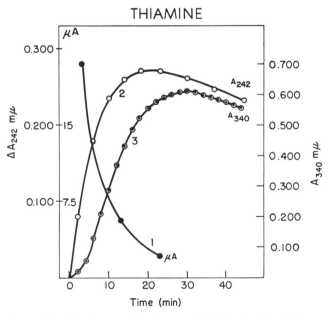

Fig. 3. Spectral and polarographic changes for thiamine in cobaltous ammoniacal buffer. Curve *1*, diminution in catalytic polarographic activity (μA) of thiamine ($5 \times 10^{-4} M$) in cobaltous buffer, pH 9.8. Curve *2*, difference spectrum at 242 mμ of thiamine ($5 \times 10^{-5} M$) in 0.1 M ammoniacal buffer (pH 9.8) minus thiamine ($5 \times 10^{-5} M$) in 0.05 M Tris-buffer (pH 7.4). Curve *3*, changes in absorption at 340 mμ of thiamine ($10^{-3} M$) in cobaltous buffer (pH 9.8).

When thiamine is converted to the tricyclic form, the absorption band around 240 mμ increases markedly and the absorption maximum at 265 mμ is shifted to 280 mμ. These spectral changes can then be used to identify the presence of the tricyclic form.

Thiamine ($5 \times 10^{-4} M$) possesses in cobaltous ammoniacal buffer at pH 9.8 a catalytic wave at -1.43 V SCE that decreases rapidly in the alkaline medium (Fig. 2). The diminution in catalytic polarographic activity parallels spectral changes in the ultraviolet range, which indicate conversion of thiamine to the tricyclic, dihydrothiachromine form (IV).[5] The increase in $\Delta A_{242\ m\mu}$ (Fig. 3) illustrates the formation of the tricyclic form (IV), and the accompanying decrease in catalytic polarographic activity at -1.43 V demonstrates that the tricyclic form does not possess catalytic polarographic activity in this medium. The conversion of the tricyclic form (IV) to the yeollw thiol form (V) is illustrated by the increase in absorbance at 340 mμ (Fig. 3).[5]

The rapid disappearance of the catalytic polarographic wave of thiamine parallel to the formation of the tricyclic form, $\Delta A_{242\ m\mu}$, designates either the pyrimidine N or C-2 as the proton donor of the catalytic wave at -1.43 V SCE, since the reaction of the amino group with C-2 and the resulting formation of the tricyclic form abolishes the polarographic activity.[5]

The reactions of thiamine pyrophosphate and thiamine monophosphate at pH 9.8 are similar to those of thiamine but proceed at different rates. The half-life of the polarographic waves at -1.3 to -1.4 V (the carbon-2 proton) is 6 minutes for thiamine, 19 minutes for thiamine pyrophosphate, and 35 minutes for thiamine monophosphate (Fig. 4). The thiamine phosphates appear to be more stable in the tricyclic dihydrothiachromine form, whereas thiamine is converted more readily to the yellow thiol form.[5]

Oxythiamine possesses a large wave at -1.43 V at pH 9.8. This wave is relatively stable with a half-life of 188 minutes.[5] Due to the absence of the amino group, a conversion of oxythiamine to the tricyclic intermediate (IV) is not possible, instead a slow conversion to the pseudo-base (II) and subsequent opening of the thiazole ring, yielding the colorless thiol form (III), appears to take place.[5]

Effect of Concentration, pH, and Buffer

Thiamine. Thiamine possesses in cobaltous buffer one, two, or three catalytic waves depending upon concentration of thiamine and pH of the buffer. In ammoniacal buffer at pH 7.0, thiamine in low concentrations ($5 \times 10^{-5} M$) possesses three polarographic waves at -1.30, -1.38, and -1.74 V (Fig. 5). At higher concentrations ($2 \times 10^{-4} M$) wave I at -1.30 V dominates over wave II, but wave III has become slightly more negative with a maximum at -1.76 V (Fig. 5). At pH 9.8 (Fig. 2), wave III does not appear in a fresh thiamine solution.

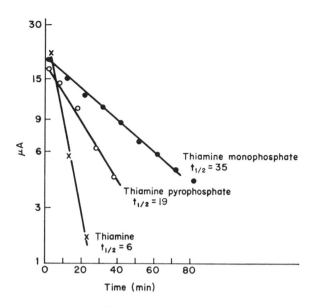

FIG. 4. Half-life of polarographic maxima in cobaltous ammoniacal buffer (pH 9.8). Thiamine, $t_{\frac{1}{2}} = 6$ minutes. Thiamine pyrophosphate, $t_{\frac{1}{2}} = 19$ minutes. Thiamine monophosphate, $t_{\frac{1}{4}} = 35$ minutes.

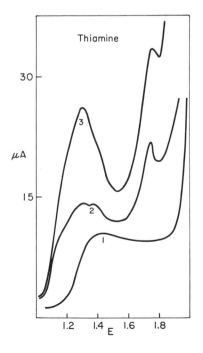

FIG. 5. Polarographic activity of thiamine in cobaltous ammoniacal buffer, pH 7.0. *1,* Polarogram of Co^{2+}, 1×10^{-3} *M*. *2,* Polarogram of thiamine, 5×10^{-5} *M*. *3,* Polarogram of thiamine, 2×10^{-4} *M*.

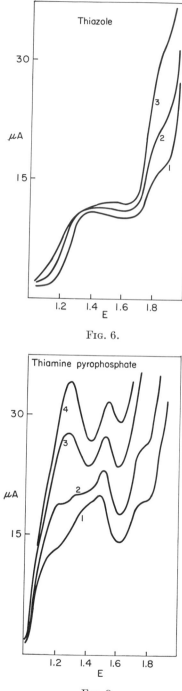

FIG. 6.

FIG. 8.

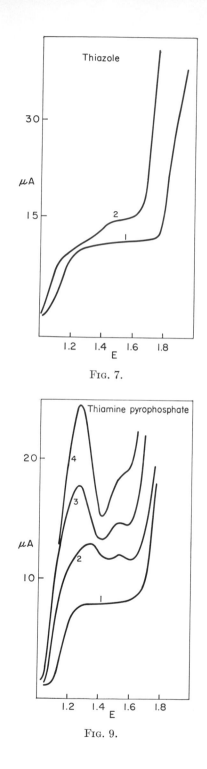

FIG. 7.

FIG. 9.

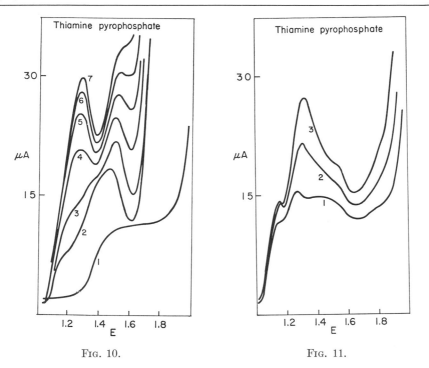

FIG. 10.   FIG. 11.

FIG. 10. (Left) Polarographic activity of thiamine pyrophosphate in cobaltous phosphate-buffer, pH 7.0. *1*, Polarogram of Co^{2+}, $1 \times 10^{-3} M$. *2*, Thiamine pyrophosphate, $5 \times 10^{-5} M$. *3*, Thiamine pyrophosphate, $1 \times 10^{-4} M$. *4*, Thiamine pyrophosphate, $2 \times 10^{-4} M$. *5*, Thiamine pyrophosphate, $3 \times 10^{-4} M$. *6*, Thiamine pyrophosphate, $4 \times 10^{-4} M$. *7*, Thiamine pyrophosphate, $5 \times 10^{-4} M$.

FIG. 11. (Right) Polarographic activity of thiamine pyrophosphate in cobaltous ammoniacal buffer, pH 9.0. *1*, Thiamine pyrophosphate, $1 \times 10^{-4} M$. *2*, Thiamine pyrophosphate, $2 \times 10^{-4} M$. *3*, Thiamine pyrophosphate, $3 \times 10^{-4} M$.

FIG. 6. Polarographic activity of thiazole in cobaltous ammoniacal buffer, pH 7.0. *1*, Thiazole, $0.5 \times 10^{-5} M$. *2*, Thiazole, $1 \times 10^{-5} M$. *3*, Thiazole, $2 \times 10^{-5} M$.

FIG. 7. Polarographic activity of thiazole in cobaltous ammoniacal buffer. *1*, Thiazole $5 \times 10^{-5} M$, at pH 9.0. *2*, Thiazole, $5 \times 10^{-5} M$, at pH 7.0.

FIG. 8. Polarographic activity of thiamine pyrophosphate in cobaltous ammoniacal buffer, pH 7.0. *1*, Thiamine pyrophosphate, $5 \times 10^{-5} M$. *2*, Thiamine pyrophosphate, $1 \times 10^{-4} M$. *3*, Thiamine pyrophosphate, $2 \times 10^{-4} M$. *4*, Thiamine pyrophosphate, $3 \times 10^{-4} M$.

FIG. 9. Polarographic activity of thiamine pyrophosphate in cobaltous Tris-buffer. pH 7.0. *1*, Polarogram of Co^{2+}, $1 \times 10^{-3} M$. *2*, Thiamine pyrophosphate, $5 \times 10^{-5} M$, *3*, Thiamine pyrophosphate, $1 \times 10^{-4} M$. *4*, Thiamine pyrophosphate, $2 \times 10^{-4} M$.

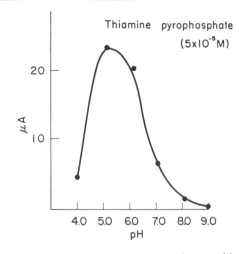

FIG. 12. The effect of pH upon the catalytic polarographic activity of thiamine pyrophosphate, 5×10^{-5} M. The following buffers were used, ionic strength 0.1: acetate-buffer (pH 4.0, pH 5.0); phosphate-buffer (pH 6.0, pH 7.0); Tris-buffer (pH 7.0, pH 8.0, pH 9.0).

Wave I represents the proton of carbon-2 of the thiazole ring,[5] wave II represents the protonated nitrogen of the pyrimidine ring,[5] but wave III has not been identified with a specific group (appears to be related to the sulfur of the thiazole group).[5]

Thiazole. Thiazole possesses polarographic activity at pH 7.0 in the range from -1.7 to -1.9 V (Fig. 6). At pH 9.0 this catalytic polarographic activity has disappeared (Fig. 7).

Thiamine Pyrophosphate. Thiamine pyrophosphate possesses two or three catalytic waves depending upon concentration, pH, and buffer. At pH 7.0 in an ammoniacal buffer, thiamine pyrophosphate (TPP) demonstrates waves II (-1.5 V) and III (-1.7 V) at low concentrations (5×10^{-5} M). With increasing concentration of TPP wave I (-1.27 V) increases at a faster rate than wave II (Fig. 8). In Tris buffer, at comparable concentrations, wave II of TPP is much smaller than wave I (Fig. 9), whereas in phosphate buffer, wave II is considerably larger than wave I (Fig. 10).

The catalytic polarographic activity of TPP is strongly pH dependent, particularly waves II and III. At pH 9.0 (Fig. 11), waves II and III have diminished or disappeared. The catalytic activity of wave II reaches a maximum at pH 5.0, the pK of one of the pyrimidinoid nitrogens (Fig. 12).

Thiamine Monophosphate. Thiamine monophosphate (TMP) possesses at low concentrations only wave II (-1.42 V) and possibly wave III

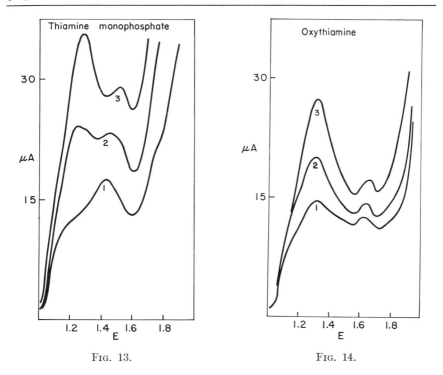

FIG. 13.   FIG. 14.

FIG. 13. Polarographic activity of thiamine monophosphate in cobaltous ammoniacal buffer, pH 7.0. *1*, Thiamine monophosphate, $5 \times 10^{-5} M$. *2*, Thiamine monophosphate, $1 \times 10^{-4} M$. *3*, Thiamine monophosphate, $3 \times 10^{-4} M$.

FIG. 14. Polarographic activity of oxythiamine in cobaltous ammoniacal buffer, pH 7.0. *1*, Oxythiamine, $5 \times 10^{-5} M$. *2*, Oxythiamine, $1 \times 10^{-4} M$. *3*, Oxythiamine, $2 \times 10^{-4} M$.

(-1.7 V). At higher concentrations of TMP, wave I (1.28 V) appears and increases more rapidly than wave II (Fig. 13).

Oxythiamine. Oxythiamine possesses one major catalytic wave at -1.32 V (wave I) and a small wave at -1.64 V (wave III) at pH 7.0 (Fig. 14). Wave II corresponding to the protonated pyrimidinoid nitrogen is absent. The absence of wave II is also evident from Figs. 15 and 16, which illustrate the catalytic activity in a phosphate buffer at pH 6.0. In this medium thiamine pyrophosphate demonstrates at low concentration ($5 \times 10^{-5} M$) one major wave (wave II; protonated nitrogen, whereas oxythiamine demonstrates wave I (Fig. 15). At higher concentration ($4 \times 10^{-4} M$), thiamine pyrophosphate demonstrates waves I and II, but oxythiamine only wave I (Fig. 16).

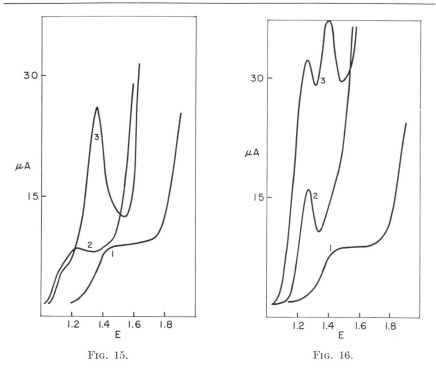

FIG. 15. Polarographic activity of oxythiamine and thiamine pyrophosphate in cobaltous phosphate buffer, pH 6.0. *1*, Polarogram of Co^{2+}, 1×10^{-3} *M*. *2*, Oxythiamine, 5×10^{-5} *M*. *3*, Thiamine pyrophosphate, 5×10^{-5} *M*.

FIG. 16. Polarographic activity of oxythiamine and thiamine pyrophosphate in cobaltous phosphate buffer, pH 6.0. *1*, Polarogram of Co^{2+}, 1×10^{-3} *M*. *2*, Oxythiamine, 4×10^{-4} *M*. *3*, Thiamine pyrophosphate, 4×10^{-4} *M*.

Conclusion

The catalytic polarographic waves represent the energy levels at which electrons can be transferred from the electrode to the acceptor proton (proton-polarography).

The involvement of the protonated group in formation of metal complexes (cobalt complexes) lowers the electrode potential at which the electron transfer can take place (catalytic effect). The polarographic waves illustrate the existence of distinct and specific energy levels for protons of specific protonated groups in these metal complexes, at which the protons can accept electrons. The catalytic proton-reduction reaches a maximum in the pH range of the p*K* of the protonated group.

The catalytic polarographic activity of thiamine and thiamine derivatives indicates that carbon-2 of the thiazole ring and the nitrogen of the pyrimidine ring react with Co^{2+}, forming a polarographically active complex. The effect of pH, buffer, and concentration illustrates the effect of the environment upon the reactivity of the individual groups.

[18] Thiamine and Its Phosphoric Acid Esters: Estimation by an Ion-Exchange Method

By Hiroyuki Koike and Takashi Yusa

Since the discovery of thiamine triphosphate (TTP) in rat liver,[1] attempts have been made by many investigators to estimate thiamine and its phosphates in biological materials.

Paper chromatography[2-4] and paper electrophoresis,[5,6] which have been widely employed, usually allow only a qualitative assay, although Kiessling[7] quantitatively determined thiamine diphosphate (TDP) and TTP in rat tissues using paper chromatography.

Quantitative estimations have been made by column chromatography using Dowex I X-8,[8] Dowex 50, and Amberlite IRC 50,[9] DEAE-Sephadex,[10] by starch-block electrophoresis,[11] and by a method based on barium salt formation.[12] The method originated by Rindi *et al.*[8] seems to have a great advantage, since it involves a purification procedure of biological samples by charcoal treatment prior to application to the ion-exchange column. Herein will be described a modification of this method in which Dowex I X-4 resin is employed instead of Dowex I X-8 resin with some technical alterations.[13]

Principle. The method consists of the treatment of a biological sample with charcoal to remove interfering substances, such as inorganic salts and

[1] A. Rossi-Fanelli, *Science* **116**, 711 (1952).
[2] K. H. Kiessling, *Nature* **172**, 1187 (1953).
[3] W. Bartley, *Biochem. J.* **56**, 379 (1954).
[4] T. Yusa, *Plant Cell Physiol. (Tokyo)* **2**, 471 (1961).
[5] H. P. Gurtner, *Helv. Physiol. Acta* **15**, C66 (1957).
[6] O. Wiss and G. Brubacher, *Ann. N.Y. Acad. Sci.* **98**, 508 (1962).
[7] K. H. Kiessling, *Biochim. Biophys. Acta* **46**, 603 (1961).
[8] G. Rindi and L. Giuseppe, *Biochem. J.* **78**, 602 (1961).
[9] A. Rossi-Fanelli, P. L. Ipata, and P. Fasella, *Biochem. Biophys. Res. Commun.* **4**, 23 (1961).
[10] A. Nakamura, K. Sanada, and E. Katsura, *Vitamins* **37**, 1 (1968).
[11] Yu. V. Khemelevskii, *Federation Proc.* **22**, T542 (1963).
[12] A. Ya. Rosanow, *Biokhimiya* **31**, 815 (1966).
[13] H. Koike, T. Wada, and H. Minakami, *J. Biochem.* **62**, 492 (1967).

proteins, with subsequent separation of thiamine compounds by column chromatography on Dowex I X-4 resin.

Reagents

Cellulose powder.[8] Ash-free cellulose powder (Toyo filter paper, 100–200 mesh) is freed from "fines" by repeated decantation from water and is washed with 50% (v/v) ethanol on a Büchner funnel. It is thoroughly washed with water until all the ethanol is removed, and dried in hot air.

Charcoal treated with cholesteryl stearate.[8] Ten grams of activated carbon (3SL, produced by Italian C.E.C.A.) is added with continuous stirring to a mixture of 300 ml of ethanol and 300 ml of ethyl ether containing 1 g of cholesteryl stearate. After 48 hours the mixture is diluted with 10 liters of water and the charcoal filtered off by suction on a Büchner funnel, followed by washing with water to completely remove ethanol and ether. The charcoal is dried in hot air and stored in an amber-colored bottle.

n-Propanol, 10% (v/v)

Dowex I X-4 resin (acetate form, 200–400 mesh)

Sodium acetate buffer, $0.1 M$, pH 4.5

Sodium acetate buffer, $1.0 M$, pH 4.5

Takadiastase solution, 3%. To 10 ml of 0.5 M acetate buffer are added 300 mg of Takadiastase and 300 mg of acid clay. This mixture is shaken for 2 minutes, and acid clay is removed by centrifugation.

Preparation of Charcoal Column.[8] The column used for charcoal chromatography is equipped with a glass filter and stopcock. The cellulose powder is settled to a height of 1.5 cm on the glass filter by pouring in a suitable amount of cellulose suspension. A homogeneous suspension in water of cellulose powder and charcoal (1200 mg:400 mg) is gently poured onto the settled cellulose layer with the stopcock opened. Another cellulose layer is placed on it to a height of 1.5 cm in the same manner as described above. During these procedures the surface of the column always must be covered with water so that no channeling occurs. A separatory funnel for the application of tissue extracts is attached to the charcoal column by means of a glass joint.

Preparation of Ion-Exchange Resin and Column. Dowex I X-4 resin (Cl⁻ type, 200–400 mesh, Dow Chemical Co.) is freed from "fines" by repeated decantation from water and washed by the method of Cohn.[14] The purified resin is transferred into a large tube and converted into an acetate form by passing 1 M sodium acetate through the column until the

[14] W. Cohn, Vol. III [107], p. 732.

effluent gives a negative test for chloride. Excess ions are removed by washing with water, then the resin is stored in 0.5 M acetic acid.

For the preparation of the column, the resin is slurried into a chromatographic tube (inner diameter 6 mm). The resin is allowed to settle to a height of 190 mm, and excess resin is removed by suction. The column is thoroughly washed with water and allowed to equilibrate at 10°.

Procedure. The extraction and purification of thiamine compounds from tissue are carried out as follows according to Rindi et al.[8]

The quickly excised tissue is weighed and homogenized in 3 volumes of cold 5% trichloroacetic acid solution with a Potter homogenizer. The homogenate is centrifuged (5000 g) at 0° for 5 minutes, and the supernatant fluid is collected. The precipitate is again extracted with 2 volumes of the trichloroacetic acid solution. Both extracts are combined, and the pH is adjusted to 5.5–6.5 with 40% sodium hydroxide. The extract thus prepared is passed through a column of charcoal, with gentle pressure if necessary. The flow rate is 40–60 ml/hour during all steps of charcoal treatment. The column is washed with 25 ml of cold water; then thiamine compounds, which are absorbed on the charcoal, are eluted with 70 ml of 10% (v/v) n-propanol. The eluate is concentrated to 5 ml at about 20° in a rotary evaporator under vacuum.

The purified extract is adjusted to pH 3–4.5 with 0.1 N hydrochloric acid and applied to a column of Dowex I X-4 resin (190 × 6 mm). All ion-exchange chromatographic procedures are carried out at 10°, and flow rates are 10 ml/hour during all steps of elution and washing. Since thiamine and thiamine monophosphate (TMP) are washed out of the column with 14 ml of water, eluate and washings are collected together in a volumetric flask and made up to 25 ml with 0.5 M acetate buffer, pH 4.5. This is divided into two equal portions in order to determine thiamine and TMP separately. To one portion is added 0.25 ml of 3% Takadiastase solution for the determination of thiamine and TMP; to the other is added 0.25 ml of 0.5 M acetate buffer, pH 4.5, for the determination of thiamine. After TDP is eluted with 24 ml of 0.1 M acetate buffer, pH 4.5, TTP is eluted with 24 ml of 1.0 M acetate buffer, pH 4.5. To each fraction is added 0.5 ml of Takadiastase solution, and the volumes are made up to 25 ml with 0.5 M acetate buffer, pH 4.5.

These four fractions are incubated at 38° overnight, and then thiamine contents are determined fluorometrically after oxidation by alkaline potassium ferricyanide.[15]

Comments. As shown in Fig. 1, separation of thiamine and its phosphates

[15] "Method of Vitamin Assay," 3rd. Ed. (Association of Vitamin Chemists, ed.), p. 123. Wiley (Interscience), New York, 1966.

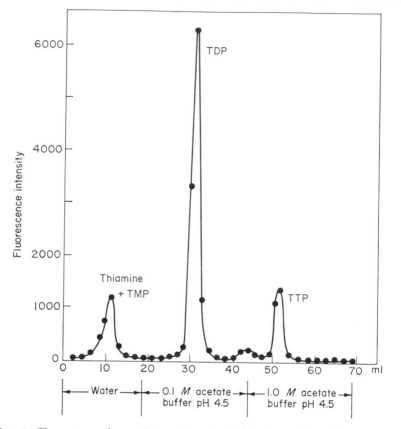

Fig. 1. Chromatography on Dowex I resin of thiamine and its phosphates from biological material. Sample: Rat liver 2.2 g. Resin: Dowex-I X-4, acetate form, 200–400 mesh. Column: 190 mm long, 6 mm in diameter. Flow rate: 10 ml/hour. Fraction volume: 2 ml. Temperature: 10°.

is complete, although thiamine and TMP are superimposed on each other. The recovery of each thiamine compound, when added separately to tissue extract, is 81–91%.[13]

For application of the method, the total thiamine content of tissue to be analyzed must exceed approximately 10 μg. When the thiamine content is extremely low, as in the case of thiamine deficiency, the quantity of thiamine in every 2-ml fraction of effluent is determined as illustrated in Fig. 1.

[19] Yeast Pyruvate Decarboxylase (2-Oxoacid Carboxy-lyase, EC 4.1.1.1) Assay of Thiamine Pyrophosphate

By JOHANNES ULLRICH

Principle

Yeast pyruvate decarboxylase (PDC) is usually isolated as a holoenzyme which is a ternary complex of the apoenzyme with thiamine pyrophosphate (TPP) and Mg^{2+}, exhibiting a rather high—"quasi-irreversible" —stability.[1-3] It can be resolved from TPP and Mg^{2+} by alkaline ammonium sulfate precipitation[4-6] or dialysis. The apoenzyme has no decarboxylating activity. The holoenzyme is fully reconstituted by the addition of large amounts of TPP and Mg^{2+} to freshly prepared apoenzyme. Both cofactors form labile salts (showing normal Michaelis–Menten characteristics) with the apoenzyme at different sites of the protein.[2] By a structural change of the protein, loaded with both cofactors at pH = 6.8, the stable ternary complex is again formed.[2] This reconstitution of the holoenzyme occurs even with very low (limiting) quantities of TPP, when a large excess of apoenzyme and Mg^{2+} is applied, and can thus be used for an assay of TPP,[7] the pyruvate-decarboxylating activity being a measure for the amount of TPP recombined:

$$\text{TPP} + \text{PDC-Mg} \rightleftharpoons \text{TPP-PDC-Mg} \xrightarrow{\text{pH 6.8}} \begin{matrix} \text{TPP-Mg} \\ \diagdown \\ \text{PDC} \end{matrix} \text{(Holo-PDC)} \qquad (1)$$
$$\text{(excess)}$$

$$\text{CH}_3\text{—CO—COO}^- + \text{H}_2\text{O} \xrightarrow[\substack{\text{Holo-PDC} \\ \text{(limiting)}}]{\text{pH 6.0–6.2}} \text{CH}_3\text{—CHO} + \text{OH}^- + \text{CO}_2 \qquad (2)$$

$$\text{CH}_3\text{—CHO} + \text{NADH} + \text{H}^+ \xrightarrow[\substack{\text{Alcohol dehydrogenase} \\ \text{(large excess)}}]{} \text{CH}_3\text{—CH}_2\text{OH} + \text{NAD}^+ \qquad (3)$$

The most convenient method for measuring the reaction is by enzymatic reduction of the acetaldehyde produced to ethanol with NADH (Eq.

[1] E. P. Steyn-Parvé and H. G. K. Westenbrink, *Z. Vitaminforsch.* **15**, 1 (1944).

[2] A. Schellenberger, *Angew. Chem.* **79**, 1050 (1967) (survey).

[3] A. V. Morey and E. Juni, *J. Biol. Chem.* **243**, 3009 (1968).

[4] H. Holzer, G. Schultz, C. Villar-Palasi, and J. Jüntgen-Sell, *Biochem. Z.* **327**, 331 (1956).

[5] H. Holzer and H. W. Goedde, *Biochem. Z.* **329**, 192 (1957).

[6] H. Holzer and K. Beaucamp, *Biochim. Biophys. Acta* **46**, 225 (1961).

[7] E. Holzer, H. D. Söling, H. W. Goedde, and H. Holzer, *in* "Methoden der enzymatischen Analyse" (H. U. Bergmeyer, ed.), p. 602. Verlag Chemie, Weinheim/Bergstr., Germany, 1962.

3), the disappearance of which is measured spectrophotometrically at 340 nm or 365–366 nm.[7,8] The reaction can also be followed by manometric measurement of the CO_2 evolved (Eq. 2) or by automatic titration of the OH^- ions left from the water consumed in Eq. (2).[9]

Reagents

1. Sodium phosphate buffer, 50 mM, pH 6.8
2. Ammonium sulfate solutions, 2.7, 2.2, and 1.8 M
3. Ammonium sulfate, 2.7 M, adjusted to pH 8.8
4. Magnesium sulfate (or chloride), 20 mM
5. Thiamine pyrophosphate, 20 mM (stored frozen)
6. Thiamine pyrophosphate, 10 μM, freshly prepared every day from TPP free of thiamine and thiamine mono- and triphosphates: 5 mg of TPP is dissolved in 100 ml of water, the concentration is adjusted to give an absorbance of 0.740 at 272.5 nm ($d = 1$ cm), and the solution is diluted 10-fold exactly.
7. Sodium pyruvate, 1 M (stored frozen)
8. NADHNa$_2$, 5 mM (stored frozen)
9. Alcohol dehydrogenase, commercial preparation diluted to 0.5 mg/ml with buffer 1 (stored frozen)
10. Apopyruvate-decarboxylase solution, 1–3 mg/ml in buffer 1, freshly prepared every day from the ammonium sulfate precipitate (see following section).
11. Biuret reagent: To 100 ml of 2 N NaOH is added 9 g of potassium-sodium tartrate, 3 g of copper sulfate ($CuSO_4 \cdot 5H_2O$) 5 g of potassium iodide, and water to give a total volume of 1 liter. The solution should be stored in a polyethylene bottle.
12. Citrate buffer, 0.3 M, pH 6.0
13. Commercial chemicals: Glycerol, disodium phosphate, acetone, ammonium sulfate, ethylenediaminetetraacetate (disodium salt), Sephadex G-200, 2 N sodium hydroxide, 25–27% aqueous ammonia, 70% (11.7 M) perchloric acid ($D = 1.67$), potassium bicarbonate.

Preparation of Pyruvate Decarboxylase[9a,10]

Dried Brewers' Yeast. Fresh yeast slurry (*Saccharomyces carlsbergensis*) from a local brewery is centrifuged and washed twice with water. The most

[8] G. Beisenherz, H. J. Boltze, T. Bücher, R. Czok, K. H. Garbade, E. Meyer-Arendt, and G. Pfleiderer, *Z. Naturforsch.* **8b,** 555 (1953).

[9] A. Schellenberger, G. Hübner, and H. Lehmann, *Angew. Chem.* **80,** 907 (1968).

[9a] Yeast Pyruvate Decarboxylase is now commercially available from C. F. Boehringer, D-68 Mannheim.

[10] J. Ullrich, J. H. Wittorf, and C. J. Gubler, *Biochim. Biophys. Acta* **113,** 595 (1966).

efficient device for handling large quantities is a laundry centrifuge lined with a sack of very thick linen. It produces yeast of about the same consistency as commercial bakers' yeast, which is very good for drying. The yeast lumps, 1–2 cm in size, are spread on several layers of filter paper at ca. 25° and low humidity (in damp summer air, drying is usually too slow). When broken at the right time, the lumps give a fine powder which is turned over from time to time during 2–3 days until it is dry. This material can be stored at 0° in air-tight containers for years without significant loss of pyruvate decarboxylase activity.

Crude Extract. Dried yeast (600 g) is stirred into a solution of 100 ml of glycerol, 1 g of EDTANa$_2$, and 6 g of ammonium sulfate in 2.5 liters of water. The slurry is left overnight at room temperature, stirred again, and centrifuged at more than 10,000 g.

Acetone Fractionation. The extract is vigorously stirred (foaming prevented) at 0°, while 0.6 volume of acetone, precooled to −15°, is added within 10 minutes. After standing at 0° for 1–2 hours, the mixture is centrifuged at more than 10,000 g. To the clear supernatant solution, 0.15 volume of precooled acetone is added within 5 minutes, followed by immediate centrifugation at 2000–3000 g and suspension of the sediment in 600 ml of buffer 1. After addition of 5 ml of 20 mM TPP and MgSO$_4$, the mixture is homogenized intermittently for 2 hours at 0° and centrifuged at over 10,000 g.

Ammonium Sulfate Fractionation. Ammonium sulfate (40 g) is added slowly to each 100 ml of the above supernatant solution with stirring or plunging (foaming prevented). The precipitated protein is centrifuged off at 2000–3000 g, washed with 300 ml of 2.7 M ammonium sulfate, and extracted for 2 hours with 300 ml of 2.2 M ammonium sulfate and with 300 ml of 1.8 M ammonium sulfate for another 2 hours. The 1.8 M ammonium sulfate extract contains the bulk of the enzyme, which is precipitated by the addition of 8 g of ammonium sulfate per 100 ml of solution and centrifuged off at over 10,000 g as a yellowish, amorphous paste. It can be stored at −20° for many months when kept wet, without appreciable loss of activity.

Gel Filtration. A column (100 × 2 cm) made from 12 g of Sephadex G-200, equilibrated in excess buffer 1, is thoroughly washed with this buffer at 0°. Two to three grams of ammonium sulfate paste, dissolved in 4 ml of buffer is applied to the column and eluted with buffer (20–40 ml/hour). Ten-milliliter fractions are collected and assayed soon for protein content and pyruvate decarboxylase activity. An example of a good run is shown in the table. All fractions containing 20 or more units per milligram of protein are good for the TPP assay. They are stabilized by the addition of 0.5 ml of 20 mM TPP and MgSO$_4$ and precipitated by

PURIFICATION OF PYRUVATE DECARBOXYLASE

Step	Volume (ml)	Protein (mg/ml)	Activity (units/ml)	Specific activity (units/mg)
Crude extract	1800	45–55	60–125	1.5–2.5
Acetone precipitate extract	650	20–30	150–250	6–10
Ammonium sulfate paste	(10–20 g)	(150–250 mg/g paste)		20–40
Sephadex G-200 eluate[a]				
Fraction no. 13	10	0.4	72	18
14	10	2.4	173	72
15	10	5.6	448	80
16	10	8.8	457	52
17	10	10.1	383	38
18	10	6.2	37	6
19	10	5.1	11	2
20	10	4.8	5.7	1.2

[a] From 3 g of ammonium sulfate paste containing 22% protein of 43 units/mg.

ammonium sulfate (4 g/10 ml). The purified enzyme is as stable as the crude paste of the previous step.

Resolution from TPP and Mg^{2+}. A solution of 0.3 g of glycine, 3.6 g of $Na_2HPO_4 \cdot 12 H_2O$, and 2 mg of $EDTANa_2$ in 45 ml of water is adjusted to pH 8.9 with 2 N NaOH at 0°, diluted to 50 ml, and added to 0.2–0.5 g of pyruvate decarboxylase paste (pure or crude), dissolved in 1 ml of water. After 30 minutes at 0°, 23 g of ammonium sulfate is stirred into the solution within 15 minutes, while the pH is carefully maintained at 8.6–8.8 by dropwise addition of 12–15% aqueous ammonia (main portion at the beginning). The precipitated apoenzyme is centrifuged off at 2000–3000 g, washed twice with alkaline 2.7 M ammonium sulfate (reagent 3) for 10 minutes each time, twice with ordinary 2.7 M ammonium sulfate (reagent 2) for a shorter period, and finally centrifuged at high speed (30,000 g or more). There is almost no loss of protein during the procedure, but the specific activity of the apoenzyme is usually somewhat lower than that of the starting material. It can be kept at −20° only for a few days and at 0° only for a couple of hours without serious loss of activity.[10a]

Assay of Enzyme (Apo- or Holopyruvate Decarboxylase)

Preincubation. Eight volumes of enzyme[10b] solution in buffer 1 are mixed with 1 volume of 20 mM TPP (reagent 5) and 1 volume of 20 mM

[10a] Morey and Juni[3] were able to obtain a very crude apopyruvate decarboxylase, stable for a long period, by a modified resolution procedure starting from unfractionated yeast extract.

[10b] Holopyruvate decarboxylase has to be treated in the same manner as apopyruvate decarboxylase, since about one-fourth of its TPP is lost during the isolation.

MgSO$_4$ (reagent 4). After 20–30 minutes at room temperature recombination is complete, and the solution can be applied to the activity test. It may be diluted with buffer 1 to a convenient concentration just prior to use.

Photometric Activity Test. The following solutions are pipetted into a cuvette of 1-cm light path, thermostatted at 30°:

Citrate buffer, 0.3 M pH 6.0 (reagent 12)	1.00 ml
Sodium pyruvate, 1 M (reagent 7)	0.10 ml
NADHNa$_2$, 5 mM (reagent 8)	0.10 ml
Alcohol dehydrogenase, 0.5 mg/ml (reagent 9)	0.10 ml
Water	1.69 ml
Preincubation mixture, diluted if necessary, for start	0.01 ml
Total volume	3.00 ml

The absorbance at 340 or 366 nm is read in 1-minute intervals or recorded during a few minutes before and after the start of the reaction.

Calculation. The activity units are defined as micromoles of pyruvate decarboxylated per minute. For calculating the units per milliliter in the original enzyme solution, the following equation applies:

$$\text{Activity (units/ml)} = \frac{10}{8} 300f \frac{\Delta A_\lambda/\text{min}}{\epsilon_\lambda} = 375f \frac{\Delta A_\lambda/\text{min}}{\epsilon_\lambda} \qquad (4)$$

where f = dilution factor of preincubation mixture before use; $\Delta A_\lambda/\text{min}$ = absorbance change per minute at λnm; and ϵ_λ = millimolar extinction coefficient at λnm for NADH (6.02 cm^2/micromole at 340 nm; 3.30 cm^2/micromole at 366 nm).

Protein Determination. For calculating the specific activity, the protein content of the enzyme solution must be known. For more or less purified pyruvate decarboxylase preparations, it can be obtained from the absorbances at 280 and 260 nm by the Warburg–Kalckar method,[11,12] simplified by multiplication of the A_{280} with 1.01 for apopyruvate decarboxylase and with 0.985 for holopyruvate decarboxylase, individual factors found by calibration of the method with freeze-dried purified enzyme to give the milligrams per milliliter.

For crude extracts, only the biuret method[8] can be applied. The sample, containing 1–5 mg of protein, and a blank (buffer) are diluted to 5 ml and mixed with 5 ml of biuret reagent. After 30 minutes the absorbance is read at 546 nm and a 1-cm light path. The difference between sample and blank, multiplied by 36, an individual factor for pyruvate decarboxylase found by calibration with freeze-dried enzyme, gives the total protein content (milligrams) in the sample.

[11] O. Warburg and W. Christian, *Biochem. Z.* **310**, 384 (1942).
[12] H. M. Kalckar, *J. Biol. Chem.* **167**, 461 (1947).

Thiamine Pyrophosphate Assay

Treatment of Samples. In order to prevent changes in TPP content, samples have to be fresh. If necessary, they are homogenized and deproteinized by the addition of a 0.1 volume of 11.7 M perchloric acid (70% w/w, $D = 1.67$) and centrifugation. Per milliliter of supernatant solution, 125 mg of solid $KHCO_3$ is added at 0°, followed by another centrifugation 1 hour later and, if necessary, exact neutralization. At this stage, the samples may be stored at $-20°$ for a few days or, if the TPP content is very low, concentrated by freeze-drying to a convenient concentration of 0.2–10 μM TPP.

Deproteinization is also achieved by treatment of the sample with 4–5 volumes of boiling methanol or ethanol, centrifugation, extraction of the precipitate with 80% alcohol, and evaporation of the combined supernatant solutions.

Preincubation. Fifty microliters of sample, containing 10–500 picomoles of TPP, 40 μl of apopyruvate decarboxylase solution (reagent 10), and 10 μl of 20 mM $MgSO_4$ (reagent 4) are mixed in cuvettes of 1-cm light path and left at approximately 25° for 30 minutes.

Activity Test. To the above preincubation mixture (0.10 ml), left in the same cuvette, the following additions are made at 30°:

Water	1.60 ml
Citrate buffer, 0.3 M, pH 6.0 (reagent 12)	1.00 ml
NADH-Na$_2$, 5 mM (reagent 8)	0.10 ml
Alcohol dehydrogenase, 0.5 mg/ml (reagent 9)	0.10 ml
Sodium pyruvate, 1 M for starting (reagent 7)	0.10 ml
Total volume	3.00 ml

The absorbance at 340 nm or at 366 nm is checked for a few minutes before the last addition and read every minute or recorded for a number of minutes after the start of the reaction.

Calibration Curve. A number of artificial "samples" of known TPP content (0, 10, . . . , 500 picomoles) are prepared from 10 μM TPP (reagent 6) and water and run together with the samples to be analyzed. Their reaction velocity (arbitrary units), plotted against the TPP content (picomoles), gives a straight line, from which the TPP content of the samples can be read directly. If necessary, the values found have to be corrected for volume changes during the "treatment of samples" prior to the test.

Controls. Considering the rather rapid deterioration of apopyruvate decarboxylase, it is recommended to run an assay of its specific activity at the same time with each set of samples or at least every day. This indicates the total capacity of the enzyme left. The TPP assay should use less than one-fourth of this total capacity, unless a bent calibration curve is accepted.

The efficiency of cofactor removal is indicated by the reaction velocity of the blank without added TPP. It should be less than 0.002/minute (at 366 nm) for pure apopyruvate decarboxylase and less than 0.005/minute for apopyruvate decarboxylase made from ammonium sulfate paste of crude holoenzyme.

Bent velocity curves indicate either incomplete recombination during the preincubation period or denaturation of the enzyme during the activity test. In any case, the steepest part of the curve should be used for calculating the enzymatic activity.

Specificity. The assay described is specific for only one naturally occurring substance: TPP. Thiamine and its mono- (TMP) and triphosphates (TTP) cannot react as coenzyme. By the action of phosphatases, which are very widespread in nature and thus likely to be contained in samples to be analyzed, TPP may be produced from TTP or the former degraded to TMP prior to deproteinization. Thus special care must be given to these possibilities. For coenzymatic activity of artificial homologs and analogs of TPP see the survey of Schellenberger.[2] The presence of large amounts of oxythiamine pyrophosphate may inhibit the assay by competition with TPP for the active sites of the apoenzyme during the recombination. Oxythiamine arises from thiamine when samples are "maltreated," mainly by heat under acidic conditions. The same reaction is likely to occur to the phosphorylated compound. But under normal conditions, the content of oxythiamine pyrophosphate should never be so high as to interfere with the TPP assay described.

Alternate Methods

Apotransketolase may be used for a similar TPP assay of approximately the same sensitivity and reliability (see this volume [22]).

If the presence of thiamine, TMP, and TTP can be excluded with certainty, the author prefers the simpler TPP → thiochrome pyrophosphate fluorescence assay, essentially described by Bessey *et al.*,[13] over all enzymatic procedures.

[13] O. A. Bessey, O. H. Lowry, and E. B. Davis, *J. Biol. Chem.* **195**, 453 (1952).

[20] Preparation of Apopyruvate Decarboxylase (2-Oxoacid Carboxy-lyase, EC 4.1.1.1)

By C. J. Gubler and J. H. Wittorf

Pyruvate decarboxylase ⇌ apopyruvate decarboxylase + thiamine diphosphate

Principle. The preparation of apopyruvate decarboxylase depends on the ability of a small-pore molecular sieve to remove thiamine pyrophosphate (TPP) from the apoprotein after separation of TPP from the holoprotein at alkaline pH.

Earlier methods used alkaline washing of yeast[1,2] or alkaline treatment of pyruvate decarboxylase followed by repeated ammonium sulfate precipitation[3] to obtain the apoenzyme. The method described here is simpler, less harsh on the enzyme, more effective, and less time-consuming.

Procedure. Dissolve 65 mg of purified pyruvate decarboxylase (last ammonium sulfate precipitate) in 15 ml of 0.1 M Tris buffer, pH 8.5–8.6 which is $10^{-3} M$ in EDTA. Readjust the pH to approximately 8.5 at 0° with cold, concentrated Tris solution. Prepare a column of Sephadex G-75 (25 × 2 cm) and equilibrate with the above Tris buffer at 0–5°. Place the enzyme solution on the column. Elution is accomplished with the same buffer, and the ultraviolet absorption is monitored manually in appropriate fractions or with a suitable ultraviolet monitor (250–280 mμ). The apoenzyme appears in the first 40–45 ml of eluate. If much larger amounts of the enzyme (i.e., 100–125 mg) are placed on such a column, some overlap with the TPP, which elutes next, may occur at the tail end of the protein peak. In this case, those fractions showing transmittances of 90–100% are discarded. The protein-containing fractions are combined and made 0.75 saturated with solid ammonium sulfate, allowed to stand 20–30 minutes, and centrifuged. Resolution and reprecipitation as above may be repeated several times, if desired, in order to remove all traces of the buffer. The final precipitate can be stored at −20° and used as required.

[1] H. Weil-Malherbe, *Biochem. J.* **33,** 1997 (1939).
[2] E. P. Steyn-Parvé, *Biochim. Biophys. Acta* **8,** 310 (1952).
[3] H. Holzer and H. W. Goedde, *Biochem. Z.* **329,** 192 (1957).

[21] Use of Coenzyme Analogs to Study Thiamine Diphosphate (Cocarboxylase) Binding in Yeast Pyruvate Decarboxylase (2-Oxoacid Carboxy-lyase, EC 4.1.1.1)

By C. J. Gubler and J. H. Wittorf

Apopyruvate decarboxylase + thiamine diphosphate + Mg(II) $\rightleftharpoons$
pyruvate decarboxylase

Preparation

Principle. Schellenberger and co-workers[1,2] have recently shown that thiamine diphosphate and Mg(II) ions are bound independently to different sites on apopyruvate decarboxylase, but that all three components must react together to form the stable holoenzyme complex. With Mg(II) in excess, the K_m and K_i values for the coenzyme and its analogs, respectively, may be considered to represent dissociation constants for the apoenzyme–coenzyme and apoenzyme–analog complexes, i.e., relative affinities of the coenzyme and its analogs for the apoenzyme. Since pyruvate decarboxylase activity depends upon the amount of the holoenzyme present, reconstitution of the holoenzyme in the presence of the coenzyme and its analog will indicate the extent of interference of the analog with coenzyme binding. The relative affinities of the coenzyme and its analogs for the apoenzyme may then be correlated with the structural features of the molecules involved. The apoenzyme is thus preincubated with appropriate concentrations of coenzyme and analog prior to determination of the enzyme activity.

Reagents

Sodium *N*-tris(hydroxymethyl)methyl-2-aminoethane sulfonic acid[3] buffer (TES), 0.1 M, pH 6.7

Magnesium sulfate, 0.5 M

Thiamine diphosphate (cocarboxylase, ThDP), purified[4] and neutralized to pH 6.5 with NaOH

[1] A. Schellenberger, *Angew. Chem. Intern. Ed. Engl.* **6**, 1024 (1967).

[2] A. Schellenberger, K. Winter, G. Hübner, R. Schwaiberger, D. Helbig, S. Schumacher, R. Thieme, G. Bouillon, and K.-P. Rädler, *Z. Physiol. Chem.* **346**, 123 (1966).

[3] N. E. Good, G. D. Winget, W. Winter, T. N. Connolly, S. Izawa, and R. M. M. Singh, *Biochemistry* **5**, 467 (1966). TES is commercially available from Calbiochem, Los Angeles, California.

ULTRAVIOLET ABSORPTION DATA FOR THIAMINE ANALOGS

Compound	pH	λ_{max} (mμ)	ϵ (cm^{-1} M^{-1})
Thiamine	7.0	266–267	8,550
Oxythiamine	7.0	266	8,900
Pyrithiamine	7.0	272	10,400
Tetrahydrothiamine	7.0	269	6,080
2′-Demethylthiamine	7.0	264	7,300
2′-n-Butylthiamine	7.0	268	8,640
2′-Methoxythiamine[a]	1.0	262	13,100
2′-Hydroxythiamine	1.0	273	11,600
2′-Ethylthiamine[b]	7.0	273	9,450
Thiochrome[c]	7.0	367	20,600

[a] This value was determined by H. C. Koppel, R. H. Springer, R. K. Robins, and C. C. Cheng, *J. Org. Chem.* **27,** 3614 (1966).

[b] This value is for the isobestic point, not for λ_{max}, it was determined by A. Schellenberger, K. Winter, G. Hubner, R. Schwaiberger, D. Helbig, S. Schumacher, R. Thieme, G. Bouillon, and K.-P. Rädler, *Z. Physiol. Chem.* **346,** 123 (1966).

[c] This value was determined by W. Hamanaka, *J. Vitaminol.* **12,** 231 (1966). All other values were determined by the writers at room temperature (25–27°). The pH 7.0 buffer used was 0.02 M sodium phosphate. The pH 1.0 values were determined in 0.1 N HCl.

Thiamine diphosphate analog, purified[4] and neutralized to pH 6.5 with NaOH

Apopyruvate decarboxylase,[4] 6–8 mg/ml, in 0.05 M TES buffer, pH 6.7

Sodium succinate buffer, 0.1 M, pH 6.0

Sodium pyruvate, 0.20 M, pH 6.0

DPNH, 0.003 M in 0.01 M disodium hydrogen phosphate solution

Yeast alcohol dehydrogenase,[5] 10 mg/ml, in 0.01 M sodium phosphate buffer, pH 7.0

Procedure. Preincubation is carried out by mixing 1.5 ml of 0.1 M TES buffer, pH 6.7, 0.45 ml of 0.5 M magnesium sulfate, appropriate amounts of thiamine diphosphate and analog, and water to 2.95 ml in a test tube. These are brought to 30°, and 0.3–0.4 mg of apopyruvate decarboxylase in 0.05 ml of 0.05 M TES buffer is added. This is carefully mixed to avoid foaming and kept at 30° for 2 minutes. An aliquot of the preincubation

[4] C. J. Gubler, L. R. Johnson, and J. H. Wittorf, this volume [22]. Purification of ThDP is necessary since it usually contains ThMP.

[5] Yeast alcohol dehydrogenase should be tested for extraneous pyruvate decarboxylase activity. A good preparation of the enzyme is available (Sigma Chemical Co., St. Louis, Missouri) as a dry preparation containing ammonium sulfate.

mixture is then added to the enzyme assay mixture and the reaction rate is measured as indicated below.

The concentration of thiamine diphosphate and its analogs are determined spectrophotometrically from the molar extinction coefficients given in the table. Values for additional analogs may be found in publications by Schellenberger et al.[2,6]

Time studies of recombination of the holoenzyme, in the presence of excess Mg(II) ions ($7.5 \times 10^{-3} M$), and in the presence of saturating concentrations of thiamine diphosphate ($5 \times 10^{-3} M$) and concentrations of thiamine diphosphate in the region of the K_m (2 to $3 \times 10^{-5} M$), indicate that recombination is maximal after 2 minutes. After 3.5 minutes, a decline in enzyme activity is observed. This continues with time and is probably to be attributed to partial denaturation of the enzyme.

Assay Method

Principle. The pyruvate decarboxylase reaction is coupled to that of yeast alcohol dehydrogenase. Enzymatic activity is measured spectrophotometrically by following the rate of oxidation of reduced diphosphopyridine nucleotide at 340 mμ. Under the conditions of the assay, this rate is proportional to the amount of acetaldehyde produced by the decarboxylation of pyruvate.

$$CH_3COCO_2^- + H^+ \rightarrow CH_3CHO + CO_2$$
$$CH_3CHO + DPNH + H^+ \rightleftharpoons C_2H_5OH + DPN^+$$

Procedure. Into a 3-ml optical cuvette having a 1-cm light path are placed 2.10 ml of sodium succinate buffer, pH 6.0, 0.20 ml of 0.20 M sodium pyruvate solution, 0.15–0.20 ml of 0.003 M DPNH, and sufficient water to bring the volume to 2.70 ml. The reagents are mixed and brought to 30°. The amount of DPNH to be added should be sufficient to bring the beginning optical density to between 1.0 and 1.2. Just prior to the initiation of the reaction, 0.20 ml of the yeast alcohol dehydrogenase solution is added. The reaction is initiated by addition of an aliquot of pyruvate decarboxylase from the preincubation mixture in 0.10 ml. The amount of enzyme added should be controlled so as to avoid reaction rates in excess of about 0.350 optical density unit per minute. Where less enzyme is used, water is substituted to make the volume of added enzyme 0.10 ml. The final reaction mixture volume is 3.00 ml, and the reaction temperature is 30°. Where desired, protein concentration is determined by the biuret method of Gornall et al.,[7] using a factor of 32.0 mg per 10 ml of biuret reaction mixture per unit of optical density at 540 mμ.

---

[6] A. Schellenberger and G. Hubner, *Z. Physiol. Chem.* **343**, 189 (1965).

[7] A. G. Gornall, C. J. Bardawill, and M. M. David, *J. Biol. Chem.* **177**, 751 (1949).

Where sufficient coenzyme analog is available, the inhibitor concentration may be varied at two different coenzyme concentrations and plotted after the graphical method of Dixon.[8,9] The advantage to this type of plot is that K_i may be determined directly from the graph as the intersection point of two lines. Otherwise, a Lineweaver–Burk[9,10] double reciprocal plot using the reciprocal of the coenzyme concentration as the abscissa is sufficient for the calculation of K_i values. In this case, a value of $2.3 \times 10^{-5} M$[11] may be used for the K_m for thiamine diphosphate.

[8] M. Dixon, *Biochem. J.* **55**, 170 (1953).
[9] J. L. Webb, "Enzymes and Metabolic Inhibitors," Vol. 1, pp. 149–180. Academic Press, New York, 1963.
[10] H. Lineweaver and D. Burk, *J. Am. Chem. Soc.* **56**, 658 (1934).
[11] J. H. Wittorf and C. J. Gubler, *in preparation*.

[22] Yeast Transketolase (Sedoheptulose-7-phosphate: D-Glyceraldehyde-3-phosphate Dihydroxyacetonetransferase, EC 2.2.1.1) Assay of Thiamine Diphosphate

By C. J. GUBLER, L. R. JOHNSON, and J. H. WITTORF

Principle. This yeast transketolase assay of thiamine diphosphate is a modification of a method suggested by Datta and Racker[1] in which they used yeast apotransketolase. In our hands, the apotransketolase method is preferable to methods using apopyruvic decarboxylase from yeast[2] because of the greater stability of the apotransketolase. The method consists of a coupling of the transketolase reaction with the triose isomerase and α-glycerophosphate dehydrogenase–DPNH system as follows[2a]:

$$\text{Apotransketolase}^1 + \text{TDP} \rightarrow \text{transketolase (active)} \qquad (1)$$

$$\text{Ribose 5-phosphate} + \text{ribulose 5-phosphate} \xrightarrow{\text{transketolase}}$$
$$\text{sedoheptulose 7-phosphate} + \text{glyceraldehyde 3-phosphate} \qquad (2)$$

$$\text{Glyceraldehyde 3-phosphate} \underset{\text{isomerase}}{\overset{\text{triose}}{\rightleftharpoons}} \text{dihydroxyacetone phosphate} \qquad (3)$$

$$\text{Dihydroxyacetone phosphate} \xrightarrow[\text{DPNH} \rightarrow \text{DPN}^+ + \text{H}^+]{\alpha\text{-Glycero-PO}_4\text{dehydrogenase}} \alpha\text{-glycerophosphate} \qquad (4)$$

[1] A. G. Datta and E. Racker, *J. Biol. Chem.* **236**, 624 (1961).
[2] H. Holzer, G. Schultz, C. Villar-Palasi, and J. Jentgen-Sell, *Biochem. Z.* **327**, 331 (1956).

The reaction is measured by following the oxidation of DPNH spectrophotometrically at 340 (or 366) mμ.

Reagents

Glycylglycine buffer, 0.1 M, pH 7.3
DPNH, 0.015 M
MgCl$_2$, 0.3 M
α-Glycerophosphate dehydrogenase–triose phosphate isomerase mixture[2b] from rabbit muscle containing about 1 mg/ml
Pentose phosphate mixture. A mixture of the enzymes ribose phosphate isomerase and xylulose phosphate epimerase is prepared from bovine spleen acetone powder according to the procedure of Ashwell and Hickman[3] through their heat-treatment step with two minor modifications: (1) four times the amount of acetone powder is extracted with 4 volumes of water, and (2) the (NH$_4$)$_2$SO$_4$ fraction is dissolved in 2 volumes of water prior to heat treatment. The final extract contains around 12 units/ml when assayed by the transketolase system described below, with an excess of all other reagents so that the isomerase–epimerase mixture is the limiting factor. This enzyme mixture from the crude extract, 8.35 ml, is added to a mixture containing 9.74 ml of water, 0.37 ml of 0.5 M glycylglycine buffer, pH 7.3, and 3.13 millimoles of ribose 5-phosphate 2Na$^+$. After incubation of this mixture for 30 minutes at 40°, the reaction is stopped by cooling to 0° in a cold bath and adding 2.78 ml of 50% trichloroacetic acid. The precipitate is removed by centrifugation at 10,000 g for 5 minutes at 0°. The pH of the supernatant is adjusted to 6 by adding 0.9 ml of 10 M NaOH with stirring followed by addition of 4.65 ml of 1.0 M barium acetate. The resulting precipitate is removed by centrifugation at 10,000 g at 0°, and four volumes (104 ml) of 95% ethanol is added to the supernatant solution at 0°. The precipitated material is collected by centrifugation at 8000 g for 20 minutes at 0°. These barium salts of the pentose phosphates are washed with 100 ml of 95% ethanol, dried, and stored over P$_2$O$_5$ in a vacuum desiccator. The yield is about 1.0 g. For use, 0.642 g of the dried mixed barium salts is dissolved in 8.9 ml of 0.1 M HCl. The Ba^{2+} is removed by the addition of 1.0 ml of 1.0 M ammonium sulfate followed by centrifugation and adjust-

---

[2a] Abbreviations used: TDP, thiamine diphosphate (cocarboxylase); TPK, thiamine pyrophosphokinase; TMP, thiamine monophosphate; TTP, thiamine triphosphate.

[2b] Obtainable commercially from Sigma Chemical Co., St. Louis, Missouri, or Calbiochem., Los Angeles, California.

[3] G. Ashwell and J. Hickman, *J. Biol. Chem.* **226,** 65 (1957).

ment of the supernatant solution with 0.8 ml of 1.0 N NaOH to pH 6.8. The resultant equilibrium mixture of pentose phosphates is stored at $-20°$ until used.

Yeast Apotransketolase. A modification of the method of Racker *et al.*[4] (Vol. I [54]) for transketolase is used. Extract 300 g of dried brewers' or bakers' yeast with 900 ml of 0.066 M Na_2HPO_4 for 3 hours at room temperature. The extract is centrifuged at 13,000 g for 30 minutes at 0°. After adjusting the pH of the supernatant extract to 6.5, add 0.5 volume of acetone, precooled to $-2°$, and centrifuge at 23,000 g for 20 minutes at $-2°$. Discard the precipitate, add another 0.5 volume of acetone to the supernatant solution, and centrifuge as above. Dissolve the precipitate in 100–150 ml of cold water, and dialyze for an hour (or overnight) at 5° against 25–50 volumes of water. Discard any precipitate formed, and adjust the pH to 6.5–6.6 with 1 N acetic acid. Distribute the solution into 10–12-ml aliquots in test tubes, and heat in a water bath for 15 minutes at 55°. Cool quickly and centrifuge at 23,000 g for 20 minutes at 0°. Wash the precipitate once with a small volume of water and add the washings to the first supernatant solution. Chill the combined supernatant solutions in a cold bath at $-12°$ (avoid freezing) and add 0.2 volume of precooled 95% ethanol slowly with constant stirring. The temperature should never exceed 0°, and should be -5 to $-6°$ at the end. Centrifuge at 8000 g for 20 minutes at $-5°$. Discard the precipitate, and add 0.9 volume of 95% ethanol as above to the supernatant solution. Centrifuge, discard the supernatant solution, and extract the precipitate for 10 minutes with 60 ml of double-distilled water. Remove any precipitate remaining by centrifugation. A volume of the extract containing not more than 500 mg of protein is run through a 5.5 × 7.0 cm column of DEAE-cellulose at 0–5° which has been pretreated with 0.005 M potassium phosphate buffer, pH 7.7. The enzyme is held up on the column and can be eluted with an additional 50 ml of the same buffer. Combine the fractions of eluate containing the enzyme, and fractionate with $(NH_4)_2SO_4$. The active fraction precipitates between 50 and 70% saturation. Readjust the pH to 7.6 with KOH as needed. Redissolve the 70% $(NH_4)_2SO_4$ precipitate in 1.0 ml of water at room temperature. Adjust the pH to 7.8 with KOH, and add either saturated $(NH_4)_2SO_4$ solution or solid $(NH_4)_2SO_4$ to 50–60% saturation or until slight turbidity appears, and allow the protein to crystallize slowly. The crystalline precipitate is redissolved in water and tested at this stage for the presence of thiamine diphosphate (TDP) by measuring for transketolase activity before and after incubation with TDP. It is also tested for

[4] P. Srere, J. R. Cooper, M. Tabachnick, and E. Racker, *Arch. Biochem. Biophys.* **74,** 295 (1958).

thiamine pyrophosphokinase (TPK) activity. Both are usually absent, and the preparation can be stored frozen as an $(NH_4)_2SO_4$ paste and used as needed. It can be stored even at 0–5° for several months without appreciable loss of activity. If TPK activity is present, it can be further removed by dialysis for 2 hours against 4 liters of 0.01 M Tris-HCl buffer, pH 8.0, and for 2 hours against 2 liters of 0.01 M potassium phosphate buffer, pH 6.0, followed by chromatography at 0–5° on a DEAE cellulose column (1.8 × 26 cm) which has been equilibrated with 0.01 M potassium phosphate, pH 6.0. The enzyme (apotransketolase) can be eluted with the same buffer. The most active fractions are combined and made 70–80% saturated with $(NH_4)_2SO_4$, and the resulting precipitate is stored as above. The TPK activity can be more readily removed if brewers' yeast is used as the source than with bakers' yeast.

Thiamine Diphosphate (TDP). Commercially available TDP usually contains some TMP and occasionally free thiamine, both of which increase upon storage. A number of effective methods have been developed for the separation of the free and phosphorylated forms of thiamine and some of its analogs. Siliprandi and Siliprandi[5] have described chromatographic methods utilizing starch columns, Amberlite IRC-50 ion-exchange resin, and electrophoretic methods with starch and cellulose powder columns. More recently Schellenberger and Huebner[6] have used Dowex 2 × 8 ion-exchange resin for quantitative separation of thiamine analog phosphate esters. We have found Bio-Rex 70 cation-exchange resin[7] (equivalent to Amberlite IRC-50), 200–400 mesh, hydrogen form, to be very effective for this purification of cocarboxylase. About 150 mg of commercial TDP can be easily purified on a 1.5 × 25 cm column. The sample is dissolved in a small volume of deionized water and placed on the column. Elution is accomplished with deionized water.

Horizontal starch gel electrophoresis of thiamine and thiamine analog phosphate esters has also proved to be an effective method for quantitative separation on a preparative scale. Starch gel is prepared as described earlier in this series [I. H. Fine and L. A. Costello, Vol. VI, p. 962].

Use 13 g of hydrolyzed starch per 100 ml of 0.05 M triethylammonium acetate buffer,[8] pH 5.4, and cast it in plastic trays 360 × 30 (or 20) × 6.5 mm inside diameter. Two or three pieces of Whatman No. 3 filter paper, cut to the proper size, are inserted in the center of the cooled gel. Apply about 75–100 mg of the phosphate ester mixture in 0.4 ml of deionized

---

[5] D. Siliprandi and N. Siliprandi, *Biochim. Biophys. Acta* **14**, 52 (1954).

[6] A. Schellenberger and G. Hübner, *Z. Physiol. Chem.* **343**, 189 (1965).

[7] Bio-Rad Analytical Grade Cation Exchange Resin, Bio-Rex 70, sodium form, 200–400 mesh, is distributed by Calbiochem, Los Angeles, California. To prepare the hydrogen form, 3 N HCl is used.

[8] J. Porath, *Nature* **175**, 478 (1955).

water, adjusted to about pH 5.5, to the filter paper strips by means of a Pasteur or other suitable pipette. A thickness of Saran Wrap,[9] or other thin sheet cellophane, covered with a thin film of mineral oil, is wrapped around the gel and tray to prevent evaporation. Lengths of Whatman No. 3 filter paper, cut to the width of the gel, are used as wicks to connect the gel with the buffer reservoir. These are also covered with Saran Wrap, but without the mineral oil coat. Electrophoresis is continued for 5–16 hours at 0–5° with a constant current of 10–18 mA. After completion of electrophoresis, the gel is frozen at −15 to −20°. Thorough freezing facilitates subsequent elution of phosphates from the gel. The frozen gel surface is scraped clean with a razor blade or knife and examined under ultraviolet light (about 250 mμ) to determine the extent of separation. As a general rule, the free thiamine analog migrates slightly toward the cathode, and the di- and triphosphates migrate toward the anode, the latter migrating the farthest in this direction. The gel is allowed to thaw to a sufficient extent to permit cutting into strips 5–15 mm in width along the length of the gel with a thin knife or razor blade. Care should be exercised so that thawing is not permitted to proceed too far before cutting into strips, as mixing of phosphate esters may occur. Each strip is macerated or dispersed in about 10 ml of deionized water, and the starch suspension is removed by centrifugation at 15,000–20,000 g. The eluates are evaporated to dryness *in vacuo* at room temperature, and taken up in about 1 ml of deionized water. Each fraction is then checked for homogeneity by means of paper electrophoresis.

Paper electrophoresis is carried out at room temperature for 1.5–3 hours with 0.05 M sodium acetate buffer, pH 5.4, at 16 mA constant current. After drying, the strips are analyzed by ultraviolet light (about 250 mμ), or sprayed with alkaline ferricyanide in ethanol[5] to form fluorescent thiochrome-like compounds, when possible. Phosphate analyses may be performed by the method of King,[10] using perchloric acid for the hydrolysis of the esters.

Procedure. Mix 25 micromoles of glycylglycine buffer, pH 7.3, 0.15 micromole of NADH, 3 micromoles of MgCl$_2$, at least 0.3 unit of apotransketolase and sufficient unknown to contain 0.01–0.1 nanomole of TDP in 0.25 ml final volume in a covered cuvette, and incubate for 30–40 minutes at 37°. Then add 20 μg of α-glycerophosphate dehydrogenase triose phosphate isomerase mixture, and increase the volume to 0.95 ml. Follow the A_{340} in a suitable spectrophotometer for 1 minute, then initiate the reaction by adding 1.0–1.3 micromole of pentose phosphate equilibrium mixture in 0.05 ml and follow the A_{340}. The rate (ΔA_{340}/min) is proportional to the

---

[9] Saran Wrap cellophane is manufactured by the Dow Chemical Co., Midland, Michigan it is generally available at grocery stores and supermarkets.

[10] E. J. King, *Biochem. J.* **26**, 292 (1932).

amount of TDP available to combine with the apoenzyme and produce transketolase, as can be shown when the rate is plotted vs. quantities of pure TDP added to the reaction. As little as 10 picomoles of TDP can be determined by this method. Under the above conditions, it requires 20–30 minutes for maximum recombination of TDP with apotransketolase. The reaction is inhibited by oxythiamine diphosphate, and to a lesser extent by pyrithiamine diphosphate, but not by the free antagonists.

Other Methods for Thiamine Diphosphate. (a) Recombination of TDP with yeast apopyruvic decarboxylase and coupling of the decarboxylation of pyruvate with alcohol dehydrogenase and NADH so that the rate of oxidation of NADH to NAD^+ can be followed at 340 mμ (see Ullrich, this volume [19]).

(b) Since thiochrome phosphates are not soluble in organic solvents, the thiamine phosphate ester content of a mixture or tissue extract can be determined by the usual thiochrome method before and after hydrolysis of the phosphates with taka-diastase, using isobutanol or a similar solvent for the extraction of thiochrome for fluorometric assay. This procedure is not specific for TDP as it will also measure the mono- and triphosphates.

[23] Transketolase (Sedoheptulose-7-phosphate: D-Glyceraldehyde-3-phosphate Dihydroxyacetonetransferase, EC 2.2.1.1) and the TPP Effect in Assessing Thiamine Adequacy

By MYRON BRIN

The transketolase enzyme is of interest to the nutritionist because thiamine pyrophosphate, the functional form of vitamin B_1, is its coenzyme. It is one of a series of enzymes in the pentose phosphate pathway (otherwise called the glucose oxidative pathway, the pentose shunt, etc.), and it is widely distributed in tissues of plants, microorganisms, and animals.[1,2] Following the demonstration that erythrocyte transketolase activity (RBC-TK) is progressively depressed as thiamine deficiency in rats becomes more severe, and that the addition of thiamine pyrophosphate (TPP) to the hemolyzate in the test tube results in the restoration of enzyme activity toward normal,[3,4] other workers subsequently applied this or other trans-

---

[1] B. L. Horecker and P. Z. Smyrniotis, *J. Am. Chem. Soc.* **75**, 1009 (1953).

[2] E. Racker, G. de la Haba, and I. G. Leder, *J. Am. Chem. Soc.* **75**, 1010 (1953).

[3] M. Brin, S. S. Shohet, and C. S. Davidson, *Federation Proc.* **15**, 224 (1956).

[4] M. Brin, S. S. Shohet, and C. S. Davidson, *J. Biol. Chem.* **230**, 319 (1958).

ketolase assay procedures to the study of thiamine deficiency in animals and man.[5–10]

This functional evaluation for vitamin B₁ adequacy by the assay of RBC-TK and the associated TPP effect is of particular importance as a diagnostic test in man for the following reasons: (1) It is specific for thiamine depletion. (2) It is unaffected by changes in plasma enzymes resulting from trauma or other disease. (3) The assay reveals two measurements, one for enzyme activity, and one for the TPP effect. From these values, one can differentiate between a depletion of apoenzyme and a simple coenzyme deficiency. (4) Since the assay data are both objective and specifically thiamine-sensitive, their use permits the elimination of false positive and negative diagnoses of thiamine inadequacy in man. (The latter may result either because clinical findings in thiamine deficiency are often similar to those of other nonthiamine-related clinical situations, or because a marginal biochemical thiamine deficiency may precede the appearance of clinical signs.[11,12])

Although absolute levels of transketolase activity may be obtained at all times in all tissues from all species, the restoration of enzyme activity by the addition of TPP *in vitro*, otherwise called the TPP effect, may not always be elicited. The extent of the TPP effect varies among different strains of rats, and may depend upon the completeness of the purified rat diet. While it is highly variable in rat erythrocytes and is nonevident in rat tissues,[13] it is readily elicited in duck erythrocytes, but not duck tissues.[14] However, a full TPP effect is obtained in both erythrocytes and tissues of thiamine-deficient dogs.[15]

Following the initial report on the relationship between transketolase activity and thiamine deficiency,[3,4] a large variety of alternate assays have appeared.[10] Some are suitable for use with erythrocytes, and those that are dependent upon ultraviolet absorption are appropriate primarily for tissues. Of these one is a microassay.[9]

[5] S. J. Wolfe, M. Brin, and C. S. Davidson, *J. Clin. Invest.* **37**, 1476 (1958).

[6] F. H. Bruns, E. Dunwald, and E. Noltmann, *Biochem. Z.* **330**, 497 (1958).

[7] M. Brin, M. Tai, A. S. Ostashever, and H. Kalinsky, *J. Nutr.* **71**, 273 (1960).

[8] M. Brin, *Ann. N.Y. Acad. Sci.* **98**, 528 (1962).

[9] P. M. Dreyfus, *New Engl. J. Med.* **267**, 596 (1962) [as discussed, *New Engl. J. Med.* **267**, 1265 (1962)].

[10] M. Brin, *Ciba Found. Study Group No.* **28**, pp. 82–84 (1967). Boston, Massachusetts, 1967.

[11] J. Gilroy, J. S. Meyer, R. B. Bauer, M. Vulpe, and D. Greenwood, *Am. J. Med.* **40**, 368 (1966).

[12] D. I. Webb, *Arch. Internal Med.* **120**, 494 (1967).

[13] M. Brin, *Israel J. Med. Sci.* **3**, 792 (1967).

[14] M. Brin, *Federation Proc.* **23**, 851 (1964).

[15] M. Brin and W. A. Vincent, *Abstr. 102nd Am. Vet. Med. Assoc. Mtg. Portland, Oregon, July, 1965.*

The assay to be described below has been employed with success in a number of species, both in the study of biological interrelationships of thiamine deficiency[13,16] and in the assessment of thiamine adequacy in man.[17,18] A similar semimicroassay has been described.[19] The macroassay has been adapted for use also under automated conditions.[20]

Assay for Transketolase Activity and the TPP Effect

Preparation of Materials for Assay

Hemolyzate. Five milliliters of heparinized blood is drawn (1 ml of blood is sufficient for the semimicro method) before administering any oral or parenteral therapy. The RBC are packed by centrifugation and, after the plasma and buffy coat have been removed by suction, transferred by a pipette to a preweighed vial. A volume of distilled water equal to the weight of packed cells is added, and the cells are resuspended by stirring with a wood applicator stick. The mixture is frozen with the vial tipped to prevent breakage. To prepare for immediate assay, freeze (in dry ice) and thaw the mixture three times to accelerate hemolysis. The hemolyzed sample is stable for a number of hours at room temperature, for 2–3 days in the refrigerator, and for at least 3 months when frozen.

Reagents

a. "B" Buffer. Combine the following solutions in the proportions shown, and adjust the final pH to 7.4:
0.9% NaCl:40 ml (9 g/1000 ml)
1.15% KCl:1030 ml (11.5 g/1000 ml)
1.75% K_2HPO_4:200 ml (17.5 g + 18–20 ml of 1 N HCl per 1000 ml, to adjust to pH 7.4)
3.82% $MgSO_4 \cdot 7 H_2O$:10 ml

b. Thiamine pyrophosphate (TPP)
TPP stock solution: 1 mg TPP per milliliter in "B" buffer. Keep frozen.
TPP working solution: 1 part of thawed stock TPP solution plus 8 parts "B" buffer. Distribute 1–2 ml into small vials and freeze.

c. Substrate ribose 5-phosphate: (R-5-P). Place 3.24 g of barium R-5-P in a 100-ml beaker and add 8.5 ml of 1 N HCl. Stir until dissolved. Add 45 ml of distilled water and distribute into four 40–

[16] M. Brin, *in* "Newer Methods of Nutritional Biochemistry," (A. A. Albanese, ed.), Vol. III, 407. Academic Press, New York, 1967.
[17] E. H. Morse, S. B. Merrow, and R. F. Clarke, *Am. J. Clin. Nutr.* **17,** 211 (1965).
[18] E. N. Haro, M. Brin, and W. W. Faloon, *Arch. Internal Med.* **117,** 175 (1966).
[19] M. Brin, M. Dibble, A. Peel, E. McMullen, A. Bourquin, and N. Chen, *Am. J. Clin. Nutr.* **17,** 240 (1965).
[20] C. O. Stevens, H. E. Sauberlich, and J. L. Long, *Technicon Symp.* **1,** 533 (1968).

50-ml centrifuge tubes. Rinse the beaker and add washings to the tubes. To each of the 4 tubes, add 2 ml of saturated Na_2SO_4. Stir well, and allow to precipitate in the cold room for 15 minutes. When settled, add Na_2SO_4 dropwise to determine whether precipitation is complete. If so, centrifuge the tubes for 5 minutes at high speed. Decant supernatant solutions of the 4 tubes into a 250-ml Erlenmeyer flask. Wash the $BaSO_4$ residues in each tube 3 times with 5 ml of H_2O. Centrifuge after each washing and combine the supernatant solutions in the Erlenmeyer flask. Adjust with $5 N$ KOH to pH 7.4. Filter if turbid. The volume will be approximately 125 ml; however, the actual volume must be determined. Assay for the concentration of ribose in this solution by the orcinol method as described below, and adjust the total volume so that the final concentration is 7.0 mg of ribose per milliliter of solution. (R-5-P gives a different color equivalent per mole than ribose alone, with orcinol.) Distribute approximately 10 ml into a series of small vials and freeze. (When sodium ribose 5-phosphate is available, dissolve it in distilled water, and standardize the solution to a final concentration of 7.0 mg ribose per milliliter as described.)

    d. Trichloroacetic acid (TCA), 7.5%. Dissolve 226 g in 3 liters with distilled water. Refrigerate.

    e. Hexose standard solutions:

        Glucose stock solution (D-glucose): Make a stock solution of 1 mg of D-glucose per milliliter in distilled water. Refrigerate.

        Hexose working standard: Dilute 10 ml of the glucose stock solution to 100 ml with 7.5% TCA to yield a solution of 100 μg/ml.

    f. Anthrone Reagent. Dissolve 0.5 g of anthrone and 10.0 g of thiourea in 66% H_2SO_4 by heating to 60–70° (do not exceed 90°). Cool to room temperature, and add additional 66% H_2SO_4 to make 1 liter. Refrigerate.

    g. Pentose Standard Solutions:

        D-Ribose stock solution: Make a stock solution of 1 mg of D-ribose per milliliter in distilled water. Refrigerate.

        Pentose working standard: Dilute 1 ml of the ribose stock solution to 100 ml with distilled water to yield a solution of 10 μg/ml. Refrigerate.

    h. Orcinol reagent. Dilute 4.0 g of orcinol and 0.2 g of $FeCl_3$ to 100 ml with distilled water. Add 30% HCl to a volume of 2 liters.

Procedure 1. Macromethod

Incubation Procedure. A unit of 4 test tubes is set up for each sample; the tubes are labeled A, B, D, and R, respectively, as indicated in Table I.

TABLE I
INCUBATION PROCEDURE

Tube	Hemolyzate (ml)	"B" Buffer (ml)	TPP working solution[a] (ml)	Incubate at 38° (min)	R-5-P[a] (ml)	Incubation at 38° (min)	7.5% TCA[a] (ml)
A	0.5	0.45	—	30	0.2	60	6.0
B	0.5	—	0.45	30	0.2	60	6.0
D	0.5	0.65	—	—	—	—	6.0
R	0.5 saline	0.45	—	—	0.2	—	6.0

[a] Addition is followed by mixing on a Vortex.

The final volume for each test tube is 7.15 ml. Tube A contains hemolyzate without TPP. Tube B contains hemolyzate with TPP. Tube D serves as a blank in order to determine the amount of hexose and/or pentose endogenous to the tissue sample. Tube R contains substrate only and is used in the determination of the amount of ribose utilized in the reaction. Tubes A and B are initially incubated at 38° for 30 minutes (as shown in Table I) to allow the enzyme to combine with the coenzyme, TPP, before the substrate is added. The substrate R-5-P is then added with shaking, and the time and order of each addition is noted. Exactly 60 minutes later, for each tube in turn, 6.0 ml of TCA is added with shaking, in the same tube order and time interval as for the substrate addition. The tubes are centrifuged at a high setting for 10 minutes. The protein-free filtrates are used for the determination (1) of the hexose formed by the anthrone method, and (2) of the pentose utilized by the orcinol method. The filtrates are stable for up to 5 days when refrigerated.

Determination of the Hexose Formed in the Reaction (Table II). Hexose

TABLE II
PROCEDURE FOR DETERMINATION OF HEXOSE

Tube	Filtrate (ml)	Hexose[a] (ml)	7.5% TCA (ml)	Anthrone, cold[b] (ml)	Boiling water bath[c] (min)	Cold water bath (min)	Dark (min)
All incubation tubes	1.0	—	—	10.0	10	5	20
50-μg standard	—	0.5	0.5	10.0	10	5	20
100-μg standard	—	1.0	—	10.0	10	5	20
Blank	—	—	1.0	10.0	10	5	20

[a] 100 μg/ml.
[b] Addition is followed by mixing on a Vortex.
[c] Permit to return to boiling before starting timing period.

is determined in the filtrates of tubes A, B, and D. Standards are run in duplicate. Read the solutions against the blank set at 0 optical density (OD), in a spectrophotometer or colorimeter, at 620 mμ.

Determination of Pentose (Table III). Pentose is determined in the solutions of the A, B, D, and R tubes. Standards are run in duplicate. Read the solutions against the blank set at 0 OD, in a spectrophotometer or colorimeter, at 670 mμ.

TABLE III
PROCEDURE FOR DETERMINATION OF PENTOSE

Tube	Filtrate (ml)	Pentose[a] (ml)	H_2O (ml)	Orcinol[b] (ml)	Boiling water bath[c] (min)	Cold water bath (min)
A, B, and D tubes	0.2	—	1.3	4.5	20	5
R tubes	0.1	—	1.4	4.5	20	5
5-μg standard	—	0.5	1.0	4.5	20	5
10-μg standard	—	1.0	0.5	4.5	20	5
Blank	—	—	1.5	4.5	20	5

[a] 10 μg/ml.
[b] Addition is followed by mixing on a Vortex.
[c] Permit to return to boiling before starting timing period.

Calculations

Hexose. To calculate the amount of hexose formed per milliliter of hemolyzate per hour, a dilution factor for each sample must be determined:

$$\underset{\text{(ml hemolyzate used)}}{1/0.5} \times \underset{\substack{\text{(total ml per} \\ \text{incubation tube)}}}{7.15/1} \times \underset{\text{(ml filtrate used)}}{1/1.0} = 14.3$$

Calculate the average optical density (OD) per microgram of hexose standard, and call it SH. The dilution factor and OD per microgram of hexose are constant for each tube of the determination. Thus, 14.3/SH = a constant, KH, for all tubes of the determination (assuming that 1.0 ml of filtrate was used for the determination of hexose in each case). The OD readings obtained from the A, B, and D tubes for each hemolyzate are referred to as A, B, and D, respectively.

(i) (A − D) × KH equals the micrograms of hexose per milliliter of hemolyzate per hour formed during the incubation without TPP = TH_1

(ii) (B − D) × KH equals the micrograms of hexose per milliliter of hemolyzate per hour formed during the incubation with TPP = TH_2

(*iii*) TPP effect (%) = $(TH_2 - TH_1)/TH_1 \times 100$

Pentose. To calculate the amount of pentose utilized per milliliter of hemolyzate per hour, a dilution factor for each sample read must be determined:

$$\underset{\text{(ml hemolyzate used)}}{1/0.5} \quad \times \quad \underset{\substack{\text{(total ml per} \\ \text{incubation tube)}}}{7.15/1} \quad \times \quad \underset{\text{(ml filtrate used)}}{1/0.2} \quad = 71.5$$

Calculate the average OD per microgram of pentose standard and call it SP. The dilution factor and OD per microgram of pentose are constant for each tube of the determination. Thus 71.5/SP = a constant, KP, for all tubes of the determination. The OD readings which were obtained from the A, B, D, and R tubes are referred to as A, B, D, and R, respectively. 2R + D = the amount of pentose originally present in each tube before incubation (2R is used, since only 0.1 ml R-5-P was added to the R, tube whereas 0.2 ml R-5-P was added to the A, B, and D tubes).

(*i*) $[(2R + D) - A] \times KP$ equals micrograms of pentose utilized per milliliter of hemolyzate per hour for the incubation without TPP = TP_1

(*ii*) $[(2R + D) - B] \times KP$ equals micrograms of pentose utilized per milliliter of hemolyzate per hour during the incubation with TPP = TP_2

(*iii*) "TPP effect" (%) = $(TP_2 - TP_1)/TP_1 \times 100$

Procedure 2. Semimicromethod

This procedure is essentially the same as the macromethod (Procedure 1) except for the use of smaller volumes.[19]

Notes

It is important to standardize the substrate as described in order to stabilize data on transketolase activity over a period of time in the laboratory. This is because one cannot saturate the enzyme with substrate, and therefore the observed enzyme activity increases as the substrate concentration increases. Small variations in substrate concentration, however, will not markedly affect the TPP effect, as this is a calculated value.

Interpretation of the Assay

1. Transketolase is an enzyme within the erythrocyte, and as such it is independent of nonspecific changes in the extracellular plasma. As vitamin

B_1 deficiency becomes more severe, (a) thiamine becomes limiting in the body cells, (b) the availability of the coenzyme for metabolic work becomes depleted of coenzyme, and therefore (c) the transketolase activity diminishes. The "TPP effect" measures the extent of depletion of the transketolase enzyme for coenzyme.

2. Most normal individuals have an RBC-TK activity range of 850–1000 μg of hexose per milliliter of hemolyzate per hour. Duplicate determinations may vary by 5%, and repeat assays by 10%. No differences have been observed between males and females. The ranges of TPP-effect are tabulated in (Table IV).

TABLE IV
CRITERIA FOR THIAMINE ADEQUACY

Clinical thiamine condition	TPP effect (%)
Normal	0–15
Marginally deficient	15–25
Severely deficient (with clinical signs)	≥ 25

3. Hexose activity of less than 800 μg hexose formed per milliliter of hemolyzate per hour has been generally associated with a positive TPP effect, i.e., 15% or higher.

4. The TPP effect is generally eliminated within 2–4 hours after the parenteral administration of 50 mg of thiamine to the patient. It responds slowly to oral therapy. Occasionally, however, samples with very low transketolase activity may be encountered (i.e., 200–400 μg of hexose per milliliter of hemolyzate per hour) in which the TPP effect is effectively eliminated by the parenteral therapy, but the total enzyme activity may not return to normal values for 1–14 days. These are generally from very malnourished individuals, often with liver involvement, who may have an apotransketolase deficit in addition to a coenzyme depletion. Although this may appear to be a deterrent to the use of the assay, we feel that it is not, as the TPP-effect data are still applicable in that they reflect the proportion of the available transketolase apoenzyme which was depleted of thiamine pyrophosphate.[13]

5. In our hands, the hexose activity and the TPP effect$_{hexose}$ have been more reflective of clinical objective and subjective findings. Some laboratories have found the pentose data more useful.[20]

Comment

Erythrocyte transketolase activity and the "TPP effect" present a functional evaluation which is both sensitive to, and specific for, thiamine deficiency. It may be used to confirm clinical beriberi or other manifestations such as Wernicke's encephalopathy, to reveal a biochemical defect in marginal vitamin B_1 deficiency, or to differentiate B_1 deficiency from clinically similar diseases of other etiology.

[24] Isolation and Identification of Thiamine Catabolites in Mammalian Urine; Isolation and Identification of Some Products of Bacterial Catabolism of Thiamine

By R. A. NEAL

Thiamine is extensively metabolized by mammals. In a study of the metabolism in the rat of thiamine labeled in the pyrimidine moiety with ^{14}C, some 22 distinct breakdown products were detected.[1] In an analogous study using thiazole-^{14}-C labeled thiamine of a higher specific activity, some 29 different metabolites were detected.[2]

Section I describes some techniques that have been useful in the isolation of metabolites of pyrimidine-^{14}C- and thiazole-^{14}C-labeled thiamine from the urine of rats.

The metabolism of thiamine by a microorganism that will use thiamine as the sole source of carbon and nitrogen has also been investigated. The isolation and properties of some of the metabolites of thiamine by this organism is described in Section II.

I. Mammalian Metabolism

Principle

Thiamine and its metabolic products are isolated and separated from mammalian urine by first isolating the metabolites in a form free of salt followed by ion-exchange chromatography on Amberlite CG-50.[1] The peaks of radioactivity from the Amberlite CG-50 columns are further purified by chromatography on Sephadex G-10. The metabolites in the peaks of radioactivity off the Sephadex G-10 columns were then isolated by chromatography on thin layers of microcrystalline cellulose.

Materials and Apparatus

Ion-Exchange Chromatography. The cation-exchange resin, Amberlite CG-50, 200–400 mesh, H$^+$ form (Mallinckrodt Chem. Works), is prepared

[1] R. A. Neal and W. N. Pearson, *J. Nutr.* **83,** 343 (1964).
[2] M. Balaghi and W. N. Pearson, *J. Nutr.* **89,** 265 (1966).

for use by suspending 500 g in 2 liters of distilled water and stirring for 10 minutes. After settling for 15 minutes, the fine particles are removed by decanting the supernatant solution. More water is added, and the process is repeated until a clear supernatant is obtained with a settling period of 15 minutes. The resin is then suspended in 1 liter of 10% sodium hydroxide and stirred for 10 minutes. After the resin has settled, the supernatant solution is decanted. Then 2 liters of distilled water is added; the preparation is stirred for 10 minutes, the resin is permitted to settle, and the supernatant solution is decanted. Again, 1 liter of 10% sodium hydroxide is added; the preparation is stirred for 1 hour, the resin is permitted to settle, and the supernatant solution is decanted. The sodium form of the resin is then washed with 5 successive 2-liter portions of distilled water. The resin is converted back to the hydrogen form by stirring for 20 minutes with 4 successive 1-liter portions of 4 N hydrochloric acid. Between each portion of hydrochloric acid the resin is washed by stirring with 2 liters of distilled water for 5 minutes. A column, 2.5 × 40 cm, is prepared and washed with distilled water until the pH is above 4.0.

Charcoal. Acid-washed charcoal (Norite-A, Matheson, Coleman and Bell) is prepared by refluxing 100 g of charcoal for 3 hours in 1 liter of 2 N hydrochloric acid. The charcoal is allowed to settle, the hydrochloric acid is decanted, and the procedure is repeated. The charcoal is transferred to a Büchner funnel (25 cm in diameter) and washed with distilled water until the pH of the washings is 5 or greater (this usually requires 25–40 liters of water). The charcoal is next washed with 4 liters of ethanolic ammonia [ethanol–30% aqueous NH_3 (w/v)–water, 50:5:45 (v/v)] and with distilled water until neutral. The charcoal is air-dried and stored in a brown bottle.

Sephadex Chromatography. Sephadex G-10 (Pharmacia, Inc.) is equilibrated with distilled water for 24 hours at room temperature. The Sephadex suspension is then placed in a suction flask, and vacuum is applied for a few minutes. This step prevents the subsequent formation of air bubbles on the column. A column, 1.5 × 90 cm (total volume of gel, 150 ml) is prepared and washed overnight with distilled water.

Thin-Layer Chromatography. An electric homogenizer is used to blend 28 g of microcrystalline cellulose (Avicel, FMC Corp.) with 110 ml of 50% ethanol. The suspension is applied as a 0.5-mm layer to five 20 × 20 cm plates. The plates are air-dried overnight. Aqueous solutions of the sample are streaked in a narrow band 20 mm from the lower edge of the plate; margins of 25 mm are allowed. The length of the run is marked by drawing a line 150 mm from the point of application of the sample. The amount of material applied to the plate ranges from 0.1 to 5.0 mg of solid material. The plates are developed in nonsaturated chambers (7.5 × 27 × 24 cm) charged with 100–200 ml of solvent just before use. As many as 8 plates are loaded into the chamber at one time, using a metal frame that holds pairs of plates at 1-cm intervals.

Autoradiography. The metabolites of thiamine on the thin-layer chromatograms are located by autoradiography. Thus, the developed thin-layer chromatograms are suitably marked with radioactive ink and placed in an X-ray exposure holder, 20.3 × 25.4 cm (Eastman Kodak Co.). A piece of No-Screen X-ray film 20 × 25 cm (Eastman Kodak Co.) is next placed in contact with the coated side of the thin-layer plate this step being carried out in a dark room. The plates are stored in a dry, dark container. The length of exposure varies from 7 to 30 days depending on the amount of radioactivity spotted on the thin-layer chromatogram. As a guide for estimation of the required time of exposure, an accumulated emission of 10^6 disintegrations per square centimeter has been found to give good exposure of the film. After the required time of exposure, the films are developed by conventional techniques.

Procedure

Acid-washed charcoal is used to obtain the urinary metabolites of thiamine-^{14}C in a form free of inorganic salts. In the desalting procedure, the pH of the urine is adjusted to 6.0–6.5, 1 g of acid-washed charcoal is added for every 100 ml of urine, and the suspension is stirred for 1 hour at room temperature. The charcoal is removed by centrifugation (5000 g for 10 minutes) and suspended in 100 ml of pyridine–ethanol–water (10:45:45, v/v) for each gram of charcoal originally added to the urine. This mixture is shaken for 3 hours at 37° and centrifuged; and the supernatant solution containing the radioactive metabolites is decanted. Some 85–90% of the radioactivity in the urine can be obtained free of inorganic salts by this procedure.

For chromatography the charcoal eluates of 300–400 ml of urine are reduced in volume in a rotary evaporator under vacuum at 40° to approximately 5 ml. The flask is washed with 5 ml of distilled water, which is added to the original concentrate, and its pH is adjusted to 5.5. After centrifugation at 5000 g for 10 minutes, the supernatant solution is placed on the Amberlite CG-50 column. The column is eluted with 400 ml of water followed by 600 ml of pyridine–acetic acid–water (7.5:1.5:91.0, v/v). Ten milliliter fractions are collected at a flow rate of approximately 100 ml per hour. A 0.5-ml aliquot of each fraction is assayed for radioactivity by gas flow counting. Four major areas of radioactivity appear during elution of the column: I (fractions 8–20), II (fractions 28–47), III (fractions 59–66), and IV (fractions 68–78). The pattern of radioactivity is nearly the same for thiamine labeled with ^{14}C in carbon 2 of the thiazole moiety and carbon 2′ or carbon 4′ of the pyrimidine moiety. The only significant difference is that area I from the urine of rats injected with pyrimidine-labeled thiamine is centered at about fraction 11 whereas, with thiazole-labeled thiamine, I is centered at about fraction 16.

The major fractions of radioactivity from ion-exchange chromatography are each reduced in volume to approximately 5 ml and applied to a 1.5 × 90 cm column of Sephadex G-10 which has been equilibrated with distilled water. The column is eluated by downward flow, first with 100 ml of water followed by 400 ml of pyridine–acetic acid–water (2.4:1.8:95.8, v/v). Five milliliter fractions are collected at a flow rate of approximately 50 ml per hour. The radioactivity in the column fractions is again determined by gas flow counting. By this technique, I can be resolved into four radioactive components: Sephadex fractions 8–11 (IA), 14–17 (IB), 21–24 (IC), 55–65 (ID); II into three radioactive components: Sephadex fractions 13–16 (IIA), 22–24 (IIB), 40–53 (IIC); and III into four radioactive components: Sephadex fractions 8–9 (IIIA), 10–12 (IIIB), 16–19 (IIIC), 20–26 (IIID). IV contains almost exclusively unmetabolized thiamine and is, therefore, not further purified.

The additional separation of the metabolites in the ion-exchange peaks of radioactivity on Sephadex G-10 is probably due to a combination of exclusion by way of size of the molecule, exclusion because of charge, and interaction of the metabolites with aromatic character with the gel. The exclusion because of charge is demonstrable using the two metabolites 4-methylthiazole-5-acetic acid and 5(2-hydroxyethyl)-4-methylthiazole. The latter compound is retained to a greater extent by Sephadex G-10. One explanation for this phenomenon[3] is that the gel contains a small amount of carboxyl groups and these charged groups tend to exclude low molecular weight anions from the gel phase.

Sephadex chromatography is very effective not only in further separating individual thiamine metabolites, but also in the separation of the various metabolites from the large amounts of other compounds extracted from the urine by the charcoal.

The individual peaks of radioactivity from the Sephadex G-10 columns are pooled, lyophilized, and redissolved in a minimum amount of water; aliquots are chromatographed on thin layers of microcrystalline cellulose. The plates are developed with n-propanol–sodium acetate (1 M, pH 5.0)–water (70:10:20, v/v), (solvent system A). The metabolites of thiamine

---

[3] B. Gelotte, *J. Chromatog.* **3**, 330 (1960).

on the thin-layer chromatograms are visualized using autoradiography. The cellulose areas containing the radioactivity are each scraped from the thin-layer plates and packed into small glass columns; the metabolites are eluted using water (2.5 ml) followed by 50% ethanol (2.5 ml). The eluates are lyophilized and respotted on thin layers of cellulose, and the chromatograms are developed, again using solvent system A. By these procedures, some 28 metabolites of pyrimidine-2′-^{14}C-labeled thiamine and 33 metabolites of thiazole-2-^{14}C-labeled thiamine were detected. The increased number of metabolites using thiazole-labeled thiamine may not reflect an increased number of metabolic products of this labeled species of thiamine, but rather the fact that the thiazole-labeled thiamine used in this experiment has a higher specific activity than the pyrimidine-labeled thiamine, and the metabolites are therefore more easily detected.

These results probably indicate the minimum number of metabolites, since there has been as yet no systematic approach to rechromatography of all these metabolites on thin layers of cellulose using solvent systems that differ in polarity, acidity, or basicity from the solvent system A.

The following metabolites have been isolated from the urine of rats.

2-Methyl-4-amino-5-pyrimidinecarboxylic Acid. This compound is eluted from the Amberlite column in I and from the Sephadex G-10 column in IA. It migrates on thin layers of cellulose with an average R_f of 0.48 in solvent system A. The physical and chemical properties of this compound have been described.[1]

4-Methylthiazole-5-acetic Acid. This compound is also eluted from the Amberlite column in I, and the Sephadex G-10 column in IA. This compound migrates on thin layers of cellulose with an average R_f of 0.89 in solvent system A, 0.45 in chloroform–methanol–aqueous 10% (w/v) NH$_3$ (65:35:3.5, v/v) (solvent system B), and 0.29 in hexane–chloroform–methanol (70:20:10, v/v) (solvent system C). The physical and chemical properties of this compound have also been described.[4]

5-(2-Hydroxyethyl)-4-methylthiazole. This compound is eluted from the Amberlite column in III and from the Sephadex G-10 column in IIIC. This compound migrates on thin layers of cellulose with an average R_f of 0.98 in solvent system A, 0.98 in solvent system B, and 0.42 in solvent system C. This compound was first reported to be present in urine by Iacono and Johnson.[5]

II. Bacterial Catabolism

By means of the enrichment culture technique, an aerobic microorganism that can grow on thiamine as the sole source of carbon and

[4] Y. Imai, S. Ziro, and A. Kobata, *J. Biochem.* **48**, 341 (1960).
[5] J. M. Iacono and B. C. Johnson, *J. Am. Chem. Soc.* **79**, 6321 (1957).

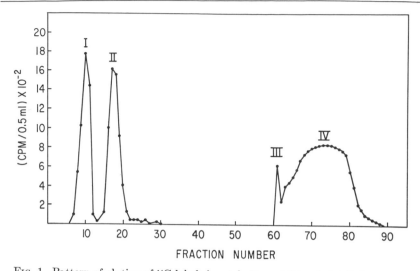

FIG. 1. Pattern of elution of ^{14}C-labeled metabolites of thiazole-2-^{14}C thiamine. Four liters of thiamine media containing 5 μCi labeled thiamine was inoculated and incubated until 75% of the thiamine had been metabolized (30 hours). The culture was centrifuged; the supernatant solution was reduced in volume to 20 ml and chromatographed on a Amberlite CG-50 column as described in Section I. Based on data of R. A. Neal, *J. Biol. Chem.* **243**, 4634 (1968).

nitrogen has been isolated.[6] This section describes the application of some of the techniques described in Section I to a study of the metabolism of thiamine by this organism.

Procedure

In a typical experiment, 4 liters of culture, which have been incubated until 50–75% of the ^{14}C-labeled thiamine (pyrimidine-2'-^{14}C or thiazole-2-^{14}C thiamine) has been metabolized, are centrifuged; the cells are discarded. The supernatant solution is reduced in volume, under vacuum, to 20 ml at a temperature of 40°. There is considerably less contaminating inorganic material in the bacterial supernatant than in rat urine, so it is not necessary to use charcoal to isolate the metabolites in a salt-free form. Therefore, the concentrated supernatant solution is applied directly to a column of Amberlite CG-50 and eluted as described previously. The pattern of radioactivity from the ion-exchange chromatography of the bacterial supernatant solution when the organism was grown on thiazole-2-^{14}C thiamine and on pyrimidine-2'-^{14}C thiamine is shown in Figs. 1 and 2,

[6] R. A. Neal, *J. Biol. Chem.* **243**, 4634 (1968).

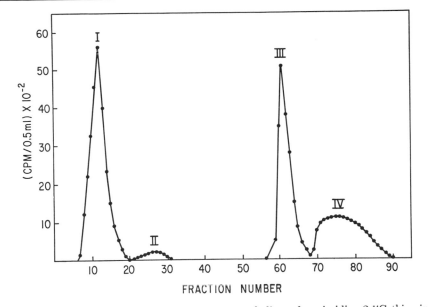

FIG. 2. Pattern of elution of ^{14}C-labeled metabolites of pyrimidine-2-^{14}C thiamine. Four liters of thiamine media containing 2 μCi of labeled thiamine was inoculated and incubated until about 50% of the thiamine had been metabolized (24 hours). The culture was centrifuged; the supernatant solution was reduced in volume to 20 ml and chromatographed on a Amberlite CG-50 column as described in Section I. Based on data of R. A. Neal, *J. Biol. Chem.* **243**, 4634 (1968).

respectively. The individual peaks of radioactivity from the Amberlite columns are lyophilized, a portion of the residue is dissolved with heating in a minimum amounts of water, and aliquots of these solutions are chromatographed on a series of thin layers of cellulose with the techniques described in Section I. The following solvent systems are used to develop the thin-layer chromatograms: solvent systems A, B, and C, *n*-propanol 0.1 *N* HCl (67:33, v/v) (solvent system D), and isopropanol-0.1 *N* HCl (85:15, v/v) (solvent system E). By means of these techniques, the following metabolites of thiamine are isolated in a chemically pure form: 2-methyl-4-amino-5-hydroxymethylpyrimidine, 2-methyl-4-hydroxy-5-hydroxymethylpyrimidine, 2-methyl-4-hydroxy-5-pyrimidinecarboxylic acid, 4-methylthiazole-5-acetic acid, 2-oxy-4-methylthiazole-5-acetic acid, and 5-(2-hydroxyethyl)-4-methylthiazole. The physical and chemical properties of these metabolites have been described previously.[6] The range of R_f values of thiamine and these metabolites in the various solvent systems, as well as the ion-exchange fraction in which these various metabolites appear, are shown in the table.

CHROMATOGRAPHIC PROPERTIES OF THIAMINE AND ITS METABOLITES[a]

Compound	R_f value					Amberlite fraction
	Solvent system A	Solvent system B	Solvent system C	Solvent system D	Solvent system E	
Thiamine	0.48–0.51	0.33–0.40	>0.05	0.32–0.38	0.07–0.09	1-IV, 2-IV
2-Methyl-4-amino-5-hydroxymethylpyrimidine	0.59–0.62	0.72–0.77	>0.05	0.50–0.56	0.32–0.34	2-III
2-Methyl-4-hydroxy-5-hydroxymethylpyrimidine	0.68–0.72	0.61–0.72	0.07–0.11	0.49–0.56	0.69–0.71	2-II
2-Methyl-4-hydroxy-5-pyrimidinecarboxylic acid	0.71–0.75	0.14–0.17	>0.05	0.70–0.75	0.69–0.72	2-I
4-Methylthiazole-5-acetic acid	0.87–0.91	0.41–0.48	0.26–0.33	0.90	0.90	1-II
2-Oxy-4-methylthiazole-5-acetic acid	0.87–0.91	0.14–0.22	0.11–0.13	0.90	0.99	1-I
5-(2-Hydroxyethyl)-4-methylthiazole	0.98	0.98	0.40–0.44	0.92	0.92	1-III

[a] Chromatograms were run on thin layers (0.5 mm) of microcrystalline cellulose by the ascending technique. The solvent systems used to develop the thin-layer plates are described in the text. Thiamine and its metabolites were visualized by autoradiography and by examining the chromatograms under ultraviolet light. The range of R_f values shown is, in each case, from five chromatograms. Based on data of R. A. Neal, J. Biol. Chem. **234**, 4634 (1968).

Discussion

The procedures described provide a means for isolating the metabolites of thiamine in a chemically pure form. The application of chromatography on Sephadex G-10 to the further purification the metabolites in the peaks of radioactivity off the Amberlite CG-50 column has provided a means of further separating the metabolites that are anionic from those that are neutral, and metabolites that are aromatic from those that are less aromatic in structure.

[25] Preparation of Thiamine Derivatives and Analogs

By TAIZO MATSUKAWA, HIROSHI HIRANO, and SHOJIRO YURUGI

I. Thiamine Derivatives

Thiamine Homologs and Analogs

Condensation Method (Fig. 1)

The condensation method, first adapted to the synthesis of thiamine (III) by Williams[1] and Andersag[2] in 1936, consists in the condensation of the pyrimidine with thiazole derivatives. For the preparation of the pyrimidine, 4-amino-5-bromomethyl-2-methylpyrimidine hydrobromide (I), the methods of Andersag[2] and Grewe[3] are excellent; and for the preparation of the thiazole, 4-methyl-5-(2-hydroxyethyl)thiazole (II), the method of Buchmann[4] is outstanding. These methods can be applied to the preparation of a number of thiamine homologs and analogs.

Pyrithiamine (IV), which exhibits the strongest activity as a thiamine antagonist, was synthesized first by Tracy[5] in 1940 by the condensation of pyrimidine (I) with 2-methyl-3-(2-hydroxyethyl)pyridine. 2-Methyl-3-(2-hydroxyethyl)pyridine also was synthesized later by Dornow,[6] Wilson,[7] and Schellenberger.[8]

[1] R. R. Williams, J. K. Cline, and J. Finkelstein, *J. Am. Chem. Soc.* **58**, 1504 (1936); *ibid.* **59**, 1052 (1937).

[2] H. Andersag and K. Westphal, *Ber.* **70**, 2035 (1937).

[3] R. Grewe, *Z. Physiol. Chem.* **242**, 89 (1936).

[4] E. R. Buchmann, *J. Am. Chem. Soc.* **58**, 1803 (1936).

[5] H. Tracy and R. C. Elderfield, *Science* **92**, 180 (1940); *J. Org. Chem.* **6**, 54 (1941).

[6] A. Dornow and W. Schacht, *Ber.* **82**, 117 (1949).

[7] A. N. Wilson and S. A. Harris, *J. Am. Chem. Soc.* **73**, 2388 (1951).

[8] A. Schellenberger, K. Winter, G. Hübner, R. Schwaiberger, D. Helbig, S. Schumacher, R. Thieme, G. Bouillon, and K-P. Rädler, *Z. Physiol. Chem.* **346**, 123 (1966).

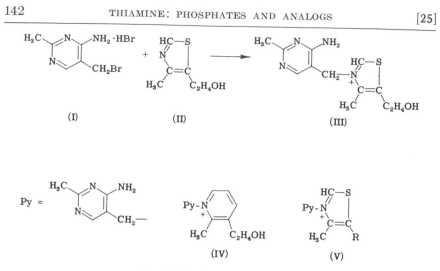

Fig. 1. Condensation method.

Also, thiamine sulfate (V: $R = C_2H_4—O—SO_3H$), 4-methyl-5-ethyl-thiazolium (V: $R = C_2H_5$), 4,5-dimethylthiazolium (V: $R = CH_3$), 4-methylthiazolium (V: $R = H$), and 4-methyl-5-(2-chloroethyl)thiazolium (V: $R = C_2H_4Cl$) have been prepared by a similar condensation method. The activities of these deoxythiamine homologs (V) toward microorganisms have been investigated in Japan.[9–11]

A practical method for the preparation of the 4,5-dimethyl compound (V: $R = CH_3$) as well as pyrithiamine is presented.

Pyrithiamine (IV).[5] A mixture of 600 mg of 2-methyl-3-(2-hydroxyethyl)pyridine dissolved in 3–4 ml of isopropanol and 240 mg of 4-amino-5-bromomethyl-2-methylpyrimidine HBr (I) is shaken to effect a solution. After filtration, the filtrate is allowed to stand overnight; during this time a product separates from the solution. It is centrifuged, washed with isopropanol–petroleum ether, and dried. The yield is 200 mg; m.p. 205–210° (decomp.).

3-(4-Amino-2-methylpyrimidinyl-5-methyl)-4,5-dimethylthiazolium (V: $R = CH_3$) *Bromide Hydrobromide.*[10] A mixture of 4.2 g of 4-amino-5-bromomethyl-2-methylpyrimidine HBr (I) and 2.5 g of 4,5-dimethylthiazole is heated at 100–110° with occasional stirring to form a solidified mass. The resulting mass is dissolved in water, decolorized, and evaporated to dryness

[9] C. Kawasaki, E. Hiraoka, I. Tomita, and H. Shimada, *Vitamins* **20,** 311, 425 (1960); *ibid.* **22,** 197, 200 (1961).
[10] T. Matsukawa and S. Yurugi, *Yakugaku Zasshi* **72,** 990 (1952); T. Ishii, Y. Takamatsu, S. Yurugi, and K. Masuda, Japanese Patent Publication No. 18113 (1966).
[11] G. Sunagawa, Y. Suzuki, T. Matsuzawa, T. Yusa, N. Kawakita, H. Katano, K. Kawada, and M. Nagawa, *Vitamins* **36,** 478 (1967).

under reduced pressure. To the residue is added 15 ml of hot EtOH to yield 4 g of crystals. For further purification, the crystals are recrystallized from EtOH to give colorless needles, m.p. 263–264° (decomp.).

Thiothiamine Method (Fig. 2)

The thiothiamine method was established by Matsukawa[12] in 1950. The intermediate, thiothiamine (IX: $R = CH_3$), which is formed by the condensation of 4-amino-5-aminomethyl-2-methylpyrimidine (VI: $R = CH_3$), CS_2, and 3-acetyl-3-chloro-1-propanol (VII), is a water-insoluble crystalline compound and is easily converted by oxidation into thiamine (III) in good yield.[13] This method is very useful not only in the laboratory but also in industry and can be applied to the preparation of thiamine homologs and analogs by allowing the various 5-aminomethylpyrimidines to react with various α-chloroketones.

The preparative methods[14] for ethylthiamine (X: $R = C_2H_5$) and butylthiamine (X: $R = C_4H_9$) are mentioned as examples. Both compounds have been prepared also by the condensation method.[15,16] It has been reported that the former analog has an activity equal to that of thiamine,[17] and the latter analog is a thiamine inhibitor.[18]

The above-mentioned deoxythiamine homologs (V) also can be prepared by this method in good yield.

Ethylthiamine (X: $R = C_2H_5$).[14] 4-Amino-5-aminomethyl-2-ethylpyrimidine (VI: $R = C_2H_5$)·2HCl (4.5 g) and 2.3 g of KOH are dissolved in 35 ml of 75% EtOH. To the solution, 3 ml of 20% NH_4OH, an aqueous solution of 3-acetyl-3-chloro-1-propanol (VII)[19] prepared by heating 3.9 g of its dimer (VII′) with 4 ml of H_2O, and 2 g of CS_2 are added in turn with stirring. The mixture is allowed to stand overnight to give a crystalline product, the 4-hydroxythiazolidine compound (VIII: $R = C_2H_5$); this is collected, washed with EtOH and H_2O, and dried. The yield is 2.5 g. Another 1.3 g of the crude product is isolated by treatment of the mother liquor. The 4-hydroxythiazolidine compound (3.8 g) is dissolved in 40 ml of 10% HCl, and the solution is warmed on a water bath for 15 minutes. After cooling,

12 T. Matsukawa and T. Iwatsu, *Yakugaku Zasshi* **70**, 28 (1950); *ibid.* **71**, 455, 667 (1951); *ibid.* **72**, 1203 (1952); H. Hirano, *ibid.* **74**, 56, 59 (1954); *ibid.* **75**, 244, 249, 252, 1184 (1955).
13 T. Matsukawa and T. Iwatsu, *Yakugaku Zasshi* **71**, 1215 (1951); T. Matsukawa and H. Hirano, *ibid.* **73**, 379 (1953).
14 T. Iwatsu, *Yakugaku Zasshi* **72**, 354, 366 (1952).
15 H. Andersag and K. Westphal, German Patent 685032.
16 R. T. Major, K. Folkers, O. H. Johnson, and P. L. Southwick, U.S. Patent 2,478,049.
17 F. Schultz, *Z. Physiol. Chem.* **265**, 113 (1940).
18 G. A. Emerson and P. L. Southwick, *J. Biol. Chem.* **160**, 169 (1945).
19 T. Matsukawa and T. Iwatsu, *Yakugaku Zasshi* **71**, 720 (1951).

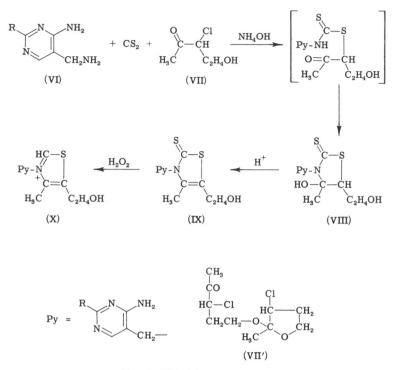

FIG. 2. Thiothiamine method.

the solution is neutralized with 10% NaOH to give a colorless crystalline mass. The collected thiazole-2-thione compound (IX: R = C_2H_5) is washed with H_2O and dried. The yield is 2.3 g.

To a solution of 3.5 g of the thiazole-2-thione compound dissolved in 20 ml of 10% HCl, 3.4 g of 30% H_2O_2 is added dropwise with stirring under cooling; the reaction is completed after 30 minutes. A water solution of $BaCl_2$ is added to the mixture until the precipitation of $BaSO_4$ is complete. The mixture, after removal of $BaSO_4$, is treated with charcoal and concentrated under reduced pressure. Hot EtOH is added to the residue to give a mass which crystallizes to yield 3.2 g of ethylthiamine HCl as needles, m.p. 233–234° (decomp.).

Butylthiamine (X: R = C_4H_9).[14] 4-Amino-5-aminomethyl-2-butylpyrimidine (VI: R = C_4H_9)·2 HCl (4.5 g) and 1.4 g of NaOH are dissolved in 50 ml of 90% EtOH. To the solution, 2 ml of 20% NH_4OH, an aqueous solution of 3-acetyl-3-chloro-1-propanol (VII)[19] prepared by heating 3 g of its dimer (VII') with 6 ml of H_2O, and 2 g of CS_2 are added with stirring. The resulting mixture is allowed to stand overnight, then acidified with

HCl and concentrated to remove EtOH under reduced pressure. The remaining aqueous solution is warmed on a water bath for 30 minutes, treated with charcoal, and made alkaline with an NaOH solution to separate an oil, which solidifies gradually. The solid product is washed with H_2O and recrystallized from EtOH to afford 3.5 g of the thiazole-2-thione compound (IX: R = C_4H_9) as colorless needles, m.p. 159°.

The thiazole-2-thione compound (4 g) is mixed with 40 ml of 0.15% HCl. The hydrochloride of the compound formed in dilute HCl is insoluble. To this mixture, 7.4 g of 30% H_2O_2 is added dropwise with stirring at a temperature below 40°. The resulting solution is warmed at 60° to complete the reaction. After cooling, a solution of $BaCl_2$ is added until the precipitation of $BaSO_4$ is complete. After removal of $BaSO_4$ by filtration, the solution is decolorized and concentrated under reduced pressure. The residue obtained is mixed with 40 ml of hot EtOH and allowed to stand at room temperature to give 2.3 g of butylthiamine HCl as colorless needles, m.p. 221° (decomp.).

Oxythiamine (Fig. 3)

Oxythiamine (XI), which was prepared first by Bergel,[20] has no thiamine activity. Soodak,[21] who prepared the compound by the diazotization of thiamine, found that the compound is an antagonist of thiamine.

This compound (XII), which can be made available also by the above mentioned thiothiamine method,[22] is most conveniently prepared by the treatment of thiamine with hydrochloric acid.[23]

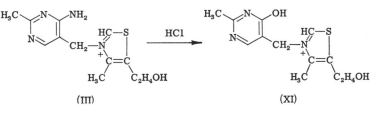

FIG. 3. Formation of oxythiamine.

Oxythiamine (XI) Hydrochloride.[23] A mixture of 10 g of thiamine HCl and 30 ml of 20% HCl is refluxed for 3 hours until the thiochrome test becomes negative. The resulting solution is concentrated under reduced

[20] F. Bergel and A. R. Todd, *J. Chem. Soc.* p. 1504 (1937).
[21] M. Soodak and L. R. Cerecedo, *J. Am. Chem. Soc.* **66**, 1988 (1944); *ibid.* **71**, 3566 (1949).
[22] T. Iwatsu, *Yakugaku Zasshi* **72**, 358 (1952).
[23] T. Matsukawa and S. Yurugi, *Yakugaku Zasshi* **71**, 827 (1951).

pressure, and hot EtOH is added to the residue. As the ethanolic solution cools, a crystalline product is separated; this is collected and recrystallized from 99% EtOH to give 8 g of oxythiamine HCl as colorless needles, m.p. 198–200° (decomp.).

Esters of Thiamine

Inorganic Esters (Fig. 4)

In 1937, the coenzyme of carboxylase was isolated from yeast by Lohmann,[24] and the structure was established to be thiamine diphosphate (TDP, XIII). This compound is prepared from thiamine by phosphorylation with phosphoryl chloride,[25] a mixture of sodium pyrophosphate and metaphosphoric acid[26] or methaphosphoric acid.[27,28] These procedures are always unavoidably accompanied by the formation of other phosphates, such as thiamine monophosphate (TMP, XII) and triphosphate (TTP, XIV), as by-products. Therefore, in the isolation of TDP (XIII), it is important to utilize chromatography to separate each phosphate.[28–30] TMP was prepared by the hydrolysis of TDP[24] and there are also several other reports[31,32] on preparative methods for TMP (XII).

The formation of thiamine triphosphate was recognized by Velluz[33] in the phosphorylation of thiamine with metaphosphoric acid, and TTP also exists in animal tissue. Rossi-Fanelli[34] reported that it is equal to TDP in coenzyme activity, but its physiological significance has not been clarified.

For the coenzyme action of diphosphoryl esters of thiamine analogs, the report of Schellenberger[35] is noteworthy.

Thiamine Diphosphate (TDP, XIII) Hydrochloride.[28] To polyphosphoric acid prepared from 9.4 g of 85% H_3PO_4 and 10.6 g of P_2O_5 is added 4 g of thiamine HCl, and the mixture is heated at 100° for 15 minutes. After cooling, 500 ml of absolute EtOH is added and the insoluble mass is separated, washed with EtOH and ether, and dried. The residue is dissolved in a little H_2O, and the aqueous solution is poured into 500 ml of absolute EtOH to give a precipitate; this treatment is repeated twice for removal of inorganic impurities. Two grams of the dried solid is dissolved in 500 ml

[24] K. Lohmann and P. Schuster, *Biochem. Z.* **294,** 188 (1937).

[25] K. G. Stern and J. W. Hofer, *Science* **85,** 483 (1937); *ibid., Enzymologia* **3,** 82 (1937).

[26] H. Tauber, *J. Am. Chem. Soc.* **60,** 730 (1938); J. Weijlard and H. Tauber, *ibid.* **60,** 2263 (1938); J. Weijlard, *ibid.* **63,** 1160 (1941).

[27] P. Karrer and M. Viscontini, *Helv. Chim. Acta* **29,** 711 (1946); M. Viscontini, G. Bonetti, P. Karrer, and C. Ebnöther, *ibid.* **32,** 1478 (1949); *ibid.* **34,** 1384 (1951).

[28] T. Tanaka, *Yakugaku Zasshi* **76,** 1314 (1956).

[29] Z. Suzuoki, M. Yoneda, and M. Hori, *J. Biochem.* **44,** 783 (1957).

[30] T. Yusa, *J. Biochem.* **46,** 391 (1959); Japanese Patent Publication No. 9497 (1962).

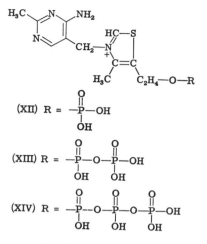

(XII) R = $-\overset{\overset{\displaystyle O}{\|}}{\underset{\underset{\displaystyle OH}{|}}{P}}-OH$

(XIII) R = $-\overset{\overset{\displaystyle O}{\|}}{\underset{\underset{\displaystyle OH}{|}}{P}}-O-\overset{\overset{\displaystyle O}{\|}}{\underset{\underset{\displaystyle OH}{|}}{P}}-OH$

(XIV) R = $-\overset{\overset{\displaystyle O}{\|}}{\underset{\underset{\displaystyle OH}{|}}{P}}-O-\overset{\overset{\displaystyle O}{\|}}{\underset{\underset{\displaystyle OH}{|}}{P}}-O-\overset{\overset{\displaystyle O}{\|}}{\underset{\underset{\displaystyle OH}{|}}{P}}-OH$

FIG. 4. Inorganic esters.

of H_2O, and the solution is adjusted to pH 7.0 with NH_4OH, and chromatographed through a column (Dowex 1 $\times$ 8 chloride type, 33.0 cm $\times$ 5.3 cm, 50–100 mesh) at a flow rate of 1 liter an hour. The column is washed with distilled water, then eluted with 6 liters of 0.005 N HCl. Two liters of the last effluent fractions, including TDP chloride, are collected, and concentrated under reduced pressure below 30°. The residue is mixed with EtOH to precipitate 1 g of thiamine diphosphate $HCl \cdot H_2O$ as a colorless crystalline powder, m.p. 215–216° (decomp.).

Thiamine Monophosphate (TMP, XII) Hydrochloride.[32] To a viscous liquid obtained by dissolving 5 g of P_2O_5 in 6.5 g of 85% H_3PO_4, 1 g of thiamine HCl is added with stirring, and the mixture is heated at 60° for 2 hours. The resulting mixture is dissolved in 50 ml of H_2O; the solution is heated on a water bath for 1 hour, then concentrated to dryness under reduced pressure. The residue is mixed with 200 ml of an ethanol–acetone mixture (1:1, v/v) to give a colorless precipitate. After cooling in an ice box overnight, the precipitate is separated by centrifugation and dried. A solution of the

[31] G. Leichssenring and J. Schmidt, *Ber.* **95**, 767 (1962).
[32] T. Kuroda and M. Masaki, *Vitamins* **29**, 109 (1964).
[33] M. L. Velluz, G. Amiard, and J. Bartos, *Bull. Soc. Chim. Fr.* **15**, 871 (1948).
[34] D. Siliprandi and N. Siliprandi, *Biochim. Biophys. Acta* **14**, 52 (1954); A. Rossi-Fanelli; N. Siliprandi, D. Siliprandi, and P. Ciccarone, *Arch. Biochem. Biophys.* **58**, 237 (1955).
[35] A. Schellenberger, A. Kolbe, and G. Hübner, *Z. Physiol. Chem.* **341**, 22 (1965); A. Schellenberger, V. Müller, K. Winter, and G. Hübner, *ibid.* **344**, 244 (1966); A. Schellenberger and G. Hübner, *ibid.* **348**, 491 (1967); A. Schellenberger, K. Wendler, P. Creutzburg, and G. Hübner, *ibid.* **348**, 501 (1967); A. Schellenberger, I. Heinroth, and G. Hübner, *ibid.* **348**, 506 (1967); A. Schellenberger, *Angew. Chem. (Intern.)* **6**, 1024 (1967).

powder (1.5 g) dissolved in 75 ml of water is adjusted to pH 7.4 with aqueous $Ba(OH)_2$ and cooled in a refrigerator. After removal of barium phosphate by filtration, the filtrate is adjusted to pH 5.4 with 10% H_2SO_4 to precipitate $BaSO_4$, which is removed. The solution is concentrated under reduced pressure, the residue is dissolved in a little dilute HCl, and the acidic solution is mixed with the ethanol–acetone mixture to give a colorless powder. For further purification, the powder is dissolved in a little H_2O and mixed again with the ethanol–acetone mixture to give TMP HCl·H_2O as a crystalline product. The yield is 0.85 g, m.p. 200° (decomp.).

Thiamine Triphosphate (TTP, XIV).[30] A mixture of 50 g of thiamine HCl and 50 g of P_2O_5 is added in several portions to a melting phosphoric acid mixture at 100–105°, which is prepared from 130 g of H_3PO_4 (84%) by preliminary heating at ca. 320°. The reaction is almost complete after the preparation has stood at 100–105° for a further 20 minutes with occasional stirring. After cooling, 80 ml of distilled H_2O is added and the resulting solution is centrifuged. The supernatant viscous liquid is slowly poured into 2 liters of cold ethanol–acetone mixture (1:1, v/v) under vigorous agitation. The white precipitate is collected, dissolved in minimal amount of H_2O, and reprecipitated as described for TMP. After such precipitations are repeated three times, the product is obtained as a white powder. The yield is 63 g. This product is a mixture of TTP including higher phosphorylated thiamine, TMP, and free phosphate.

Eight grams of the above product is dissolved in 10 ml of H_2O and chromatographed on an IRC-50 (H-form) column (5.5 cm × 83 cm) as described by Siliprandi.[34] Development is effected by washing with distilled H_2O at a flow rate of 50 ml/hour. Effluent fractions of pH 2.8–4.0 are collected (ca. 400 ml) and concentrated to ca. 10 ml under reduced pressure. Two such samples are combined (20 ml of concentrate) and chromatographed again, using a smaller column (2.5 cm × 38 cm). The rate of flow is adjusted to 10 ml/hour. Fractions of pH range 3.0–3.8 are collected (ca. 20 ml) and lyophilized to yield a snow-white powder, which shows a single spot (R_f = ca. 0.13) on an ascending paper chromatogram irrigated with n-propanol–H_2O–1 M sodium formate buffer at pH 5.0 (65:20:15, v/v/v) at 10°. The yield is 504 mg.

For further purification, 500 mg of the above substance is dissolved in 3 ml of H_2O, and 2.3 ml of absolute EtOH is slowly added so as to form moderate turbidity in the solution. On standing for several hours in an ice box (10°), the substance commences to separate completely in prismatic crystals. After recrystallizations are repeated twice, 346 mg of pure crystals are recovered. This corresponds to 2.2% of the original mixture. The substance is quite free from HCl and includes 2 moles of water of crystallization; m.p. 194–197° (decomp.).

Organic Esters (Fig. 5)

Organic esters of thiamine are prepared by the following three methods: (i) condensation of pyrimidine (I) with O-acylthiazole (XV),[38] (ii) oxidation of O-acylthiothiamine (XVII),[39, 40] and (iii) direct acylation of thiamine.[41, 42] Depending upon the sort of materials on hand, one can select the appropriate method. O-Acetyl- (XVIII: R = CH$_3$),[36] O-propionyl- (XVIII: R = C$_2$H$_5$),[37] and O-benzoylthiamine (XVIII: R = C$_6$H$_5$)[37] are selected as the representatives of O-acylthiamines.

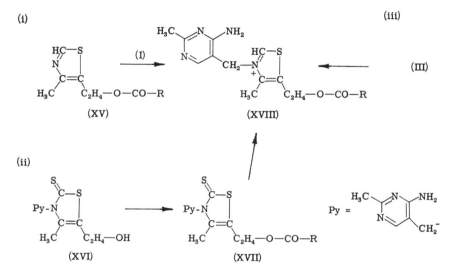

FIG. 5. Organic esters.

(i) *The Condensation Method.*[38] *O-Acetylthiamine (XVIII: R = CH$_3$) Hydrobromide.* 4-Amino-5-bromomethyl-2-methylpyrimidine HBr (I) (5 g) and 5 g of 5-(2-acetyloxyethyl)-4-methylthiazole (XV: R = CH$_3$) are mixed on heating at 110° for 30 minutes. After cooling, the resulting mixture is dissolved in H$_2$O, decolorized, and concentrated under reduced pressure. The residue is dissolved in hot absolute EtOH. Cooling of the EtOH solution gives a crude crystalline product, which is recrystallized from EtOH to yield 6.5 g of colorless plates, m.p. 241°.

[36] R. Kuhn, T. Wieland, and H. Huebschmann, *Z. Physiol. Chem.* **259**, 48 (1939).
[37] T. Sano, *Bull. Chem. Soc. Japan* **19**, 185 (1944).
[38] T. Matsukawa and S. Yurugi, *Yakugaku Zasshi* **71**, 69 (1951).
[39] S. Yoshida and M. Unoki, *Yakugaku Zasshi* **72**, 968 (1952); K. Uematsu, *Ann. Sankyo Res. Lab.* **11**, 42 (1959).
[40] T. Fujita, Y. Mushika, and K. Hagio, *Yakugaku Zasshi* **82**, 1452 (1962).

(ii) *The O-Acylthiothiamine Method.*[39,40] *O-Propionylthiamine (XVIII: R = C₂H₅) Hydrochloride.* A mixture of 5 g of thiothiamine (XVI) and 10 ml of propionic acid is refluxed for 3 hours. The reaction mixture is diluted with H_2O and neutralized by addition of dilute $NaHCO_3$ to separate a crystalline product, which is collected and washed with H_2O. The product (4.8 g) is recrystallized from dilute EtOH to give *O*-propionylthiothiamine as color-less prisms, m.p. 160–161°. These crystals are suspended in 40 ml of H_2O and added dropwise to 1.92 g of 30% H_2O_2. After 1 hour, a clear solution results, to which is added $BaCl_2$ solution until the precipitation of $BaSO_4$ is com-plete. After separation of $BaSO_4$, the filtrate is decolorized and evaporated under reduced pressure. The resulting residue is added to hot absolute EtOH. After standing, the crystals are collected and washed with EtOH to give 1.8 g of colorless needles, m.p. 215–216° (decomp.).

(iii) *The Direct Method.*[41,42] *O-Benzoylthiamine (XVIII: R = C₆H₅) Hydrochloride.* Thiamine HCl (20 g) is kneaded with 11 ml of 25% NH_4OH. After standing overnight, the crystals are collected, washed with cold MeOH, and dried at 70–75° to give 18.6 g of thiamine monochloride as colorless fine crystals. A mixture of 3 g of thiamine monochloride and 10.8 g of benzoyl chloride is shaken in an oil bath at 160° for 5 minutes to yield a product that solidifies. After heating 30 minutes more, it is added to 9 ml of EtOH and allowed to stand for a while. The separated crystals are collected and washed with EtOH to give 3.6 g of crude crystals, m.p. 223–224° (decomp.), which are recrystallized from 90% EtOH to give color-less prisms, m.p. 226–227° (decomp.).

Dihydrothiamine (**Fig. 6**)

Dihydrothiamine was synthesized in 1950 by Karrer[43] by the reduction of thiamine with lithium aluminium hydride. The compound can be prepared in another way, which was developed by Iwatsu,[44] by the condensation of 4-amino-5-aminomethyl-2-methylpyrimidine (XX) with formaldehyde and 3-acetyl-3-mercapto-1-propanol (XXI). Although Karrer ascribed the formula (XIX) to dihydrothiamine, the infrared and nuclear magnetic resonance spectra[45,46] strongly suggest that the structure of the compound

[41] H. Hagiwara, Y. Oka, Y. Hara, S. Yurugi, J. Suzuoki, K. Furuno, and T. Iida, *Ann. Rept. Takeda Res. Lab.* **22**, 1 (1963).

[42] M. Unoki and T. Miki, *Yakugaku Zasshi* **84**, 780 (1964).

[43] P. Karrer and H. Krishna, *Helv. Chim. Acta* **33**, 355 (1950); *ibid.* **35**, 459 (1952); *ibid.* **40**, 2476 (1957).

[44] T. Iwatsu, *Yakugaku Zasshi* **75**, 677 (1955); T. Matsukawa and T. Iwatsu, *J. Vitaminol.* **1**, 305 (1955).

[45] H. Hirano, T. Iwatsu, and S. Yurugi, *Yakugaku Zasshi* **76**, 1332 (1956); *ibid.* **77**, 241 (1957); T. Matsukawa, H. Hirano, T. Iwatsu, and S. Yurugi, *J. Vitaminol.* **3**, 213 (1957).

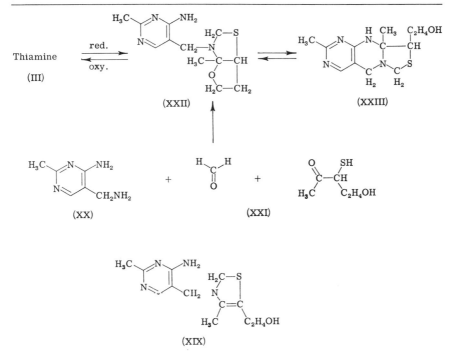

FIG. 6. Formation of dihydrothiamine.

should be in the perhydrofuro[2,3-*d*]thiazoline form (XXII). When an aqueous solution of (XXII) is heated, it is converted into its isomer, formula (XXIII).[47] Both dihydrothiamine and its isomer are oxidized to thiamine by many kinds of oxidizing agents, including air.[48] The biological activity of dihydrothiamine was reported by Karrer[43] and Iwatsu[44] to be one-tenth to one-fifteenth that of thiamine.

Dihydrothiamine (XXII, XXIII).[44] To an aqueous solution of 3-acetyl-3-chloro-1-propanol (VII)[19] prepared by heating 6 g of its dimer (VII') with 12 ml of H_2O at 100° for 15 minutes, 17.6 ml of 10% NaOH saturated with H_2S and a solution of 2 g of the 4-amino-5-aminomethyl-2-methylpyrimidine (XX) HCl are added. To this mixture is added 0.9 ml of 35% HCHO with stirring and cooling. An oily substance appears in a few minutes and solidifies gradually. The product is filtered off, washed with H_2O, and recrystallized from EtOH to give 7 g of colorless needles, m.p. 150°. This compound is identified as the tetrahydrofurothiazole form

---

[46] S. Yoshida and M. Kataoka, *Pharm. Bull. Japan* **5**, 176, 320 (1957).

[47] H. Hirano, T. Iwatsu, and S. Yurugi, *Yakugaku Zasshi* **77**, 244 (1957); T. Matsukawa, H. Hirano, T. Iwatsu, and S. Yurugi, *J. Vitaminol.* **3**, 218 (1957).

[48] H. Hirano, H. Yonemoto, and Y. Hara, *Chem. Pharm. Bull.* **7**, 545 (1959).

(XXII), which is converted to the condensed tricyclic form (XXIII) by isomerization.

Isomerization of Dihydrothiamine—Formation of the Compound (XXIII).[44] Dihydrothiamine (XXII) (5 g) is dissolved in 5 ml of hot H_2O on heating. After shaking with $CHCl_3$, the $CHCl_3$ layer is separated, dried and evaporated *in vacuo* to give a syrup. The residue is mixed with ether to give a crystalline product. Filtration and recrystallization from EtOH afford the condensed tricyclic form (XXIII), an isomer of dihydrothiamine, as colorless prisms, m.p. 175°.

Formation of Thiamine (III) by Oxidation of Dihydrothiamine (XXII, XXIII).[48] Activated carbon (1.3 g) is added to a solution of 3.3 g of dihydrothiamine (XXII), or its isomer (XXIII), in 70 ml of 2% HCl, and oxygen is introduced into the mixture at room temperature with shaking. When the theoretical amount of oxygen is absorbed in several hours, the carbon is filtered out and 99% EtOH is added to the concentrated filtrate, whereupon thiamine HCl separates out in crystalline form (3 g). Recrystallization from EtOH gives colorless needles, m.p. 250° (decomp.).

α-Hydroxyethylthiamine (Fig. 7)

It is now believed that the reaction of thiamine as a cofactor for an enzyme catalyzing the reversible decarboxylation of pyruvate involves electrophilic attack on the 2-position of the thiazole ring by the carbonyl carbon of pyruvate. This mechanism was proposed first by Breslow,[49] and lines of evidence confirming the mechanism have accumulated. One of these is the successful preparation of α-hydroxyethylthiamine (HET XXIV) by Krampitz[50] and Miller.[51] Miller prepared the compound in an excellent way by the condensation of thiamine with acetaldehyde. The same compound was also prepared by Oka[52] by treatment of one of the dihydrothiamine analogs (XXV).

The compound (HET, XXIV) thus prepared is of course a racemate at the α-carbon of α-hydroxyethyl group, and its optical isomers at the α-carbon were resolved by Murakami[53] using dibenzoyl-D-tartaric acid.

DL-HET (XXIV) is said to have 80% of the activity of thiamine toward some microorganisms,[50] and unexpectedly, both the D- and L-isomer were reported by Murakami[53] to have equal activity.

[49] R. Breslow, *J. Am. Chem. Soc.* **80**, 3719 (1958).

[50] L. O. Krampitz, G. Greull, C. S. Miller, J. B. Bicking, H. R. Skeggs, and J. M. Sprague, *J. Am. Chem. Soc.* **80**, 5893 (1958).

[51] C. S. Miller, J. M. Sprague, and L. O. Krampitz, *Ann. N.Y. Acad. Sci.* **98**, 401 (1962).

[52] Y. Oka, K. Yoshioka, and H. Hirano, *Chem. Pharm. Bull.* **15**, 119 (1967).

[53] M. Murakami, K. Takahashi, J. Matsumoto, Y. Hirata, K. Murase, Y. Shiobara, N. Sato, R. Hattori, and Y. Odani, *Vitamins* **32**, 165 (1965).

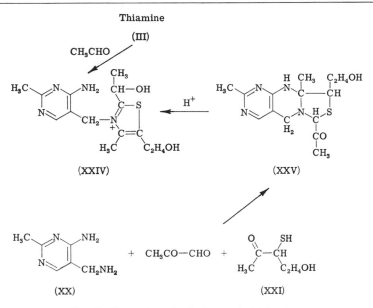

FIG. 7. Formation of α-hydroxyethylthiamine.

As the analogs of HET (XXIV), O-acyl derivatives such as the diphosphoryl ester of the primary alcohol[54–56] and the O,O'-diacetate,[57] and its thiol-form derivatives, such as alkyl disulfides[58] and S-acylates,[59] are known.

DL-α-*Hydroxyethylthiamine* (*HET XXIV*). *Step a.*[57] A solution of 110 g of thiamine HCl dissolved in 500 ml of H_2O is mixed with 620 ml of 80% aqueous CH_3CHO. After adjustment to pH 8.6 with ca. 24 ml of 10% aqueous NaOH, the solution is warmed at 40–50° on a water bath for 5 hours. The reaction mixture is acidified to pH 1.0 with 5% HCl and concentrated to dryness; the residue is mixed with hot MeOH. The solution, after removal of the precipitated NaCl, is allowed to stand to give a crystal-

[54] L. O. Krampitz and R. Votaw, Vol. IX, p. 65.

[55] C. L. Carlson and G. M. Brown, *J. Biol. Chem.* **235,** Pc 3 (1960); *ibid.* **236,** 2099 (1961).

[56] H. Holzer and K. Boaucamp, *Angew. Chem.* **71,** 776 (1959); *ibid. Biochim. Biophys. Acta* **46,** 225 (1961); H. W. Goedde, H. Inouye, and H. Holzer, *ibid.* **50,** 41 (1961); H. Holzer, H. W. Goedde, K. H. Göggel, and B. Ulrich, *Biochem. Biophys. Res. Commun.* **3,** 599 (1960); P. Scriba and H. Holzer, *Biochem. Z.* **334,** 473 (1961).

[57] Y. Oka and S. Yurugi, *Vitamins* **32,** 570 (1965).

[58] M. Nagawa, S. Miyazawa, T. Yoshioka, M. Kataoka, and T. Wada, Japanese Patent Publication No. 12150 (1968).

[59] M. Murakami, K. Takahashi, I. Tamazawa, and R. Kawai, *Vitamins* **32,** 164 (1965); M. Murakami, K. Takahashi, Y. Hirata, and H. Iwamoto, *ibid.* **34,** 71 (1966); *ibid.* Japanese Patent Publication No. 18471, 19595 (1967); Japanese Patent Publication No. 12152 (1968).

line product as the first crop. Recrystallization from a methanol–acetone mixture affords 26 g of α-hydroxyethylthiamine HCl as colorless needles, m.p. 217–219° (decomp.). A second crop is obtained by concentration of the mother liquors, and the product is purified by crystallization as above for 19 g. The total yield is 45 g.

Step b.[52] To an aqueous solution of 3-acetyl-3-chloro-1-propanol (VII)[19] prepared by heating 14 g of its dimer (VII′) with 50 ml of H_2O at 100° for 15 minutes, 40 ml of 10% NaOH saturated with H_2S and a solution of 21 g of 4-amino-5-aminomethyl-2-methylpyrimidine (XX) HCl are added. A 30% aqueous solution of 25 g of methylglyoxal is added to this mixture with stirring. An oily substance appears in a few minutes and solidifies gradually. The solid is filtered, washed with H_2O, and recrystallized from EtOH to give 9.2 g of 7-acetyl-2,9a-dimethyl-9-(2-hydroxyethyl)-5,9,9a,10-tetrahydro-7H-pyrimido[4,5-d]thiazolo[3,4-a]pyrimidine (XXV) as colorless needles, m.p. 198–200° (decomp.).

One gram of the compound (XXV) is heated at 80° with 3 ml of 20% ethanolic phosphoric acid for 1.5 hours. To the mixture is added 5 ml of H_2O and 2.2 ml of 10% HCl, and the solution is evaporated to dryness *in vacuo*. The residue dissolved in 3 ml of EtOH is allowed to stand overnight. The resulting crystals are filtered and recrystallized from methanol–acetone to give 0.6 g of α-hydroxyethylthiamine HCl as colorless needles, m.p. 217–219° (decomp.).

II. Thiol-Form Thiamine Derivatives

Esters of Thiamine Disulfide (Fig. 8)

In an aqueous solution of thiamine, its ammonium form (XXVI: R = H) and thiol form (XXVII: R = H) exist in an equilibrium.[60] Thiamine disulfide (XXVIII: R = H) was prepared first to prove the formation of the thiol form (XXVI: R = H),[61] and since then several reports[62–64] about its preparation have been presented. Thiamine disulfide (XXVIII: R = H) is usually produced from thiamine by oxidation with any oxidizing reagent when the reaction is carried out at pH 11–12,[64] and it is easily reduced to thiamine (XXVII: R = H) with cysteine or glutathione. Although such a redox system[65] was assumed to be concerned in the action of thiamine

[60] R. R. Williams, *Ergeb. Vitamin Hormonforsch.* **1**, 256 (1938).

[61] O. Zima and R. R. Williams, *Ber.* **73**, 941 (1940).

[62] P. Sykes and A. R. Todd, *J. Chem. Soc.* p. 534 (1951).

[63] G. D. Maier and D. E. Metzler, *J. Am. Chem. Soc.* **79**, 4386 (1957).

[64] C. Kawasaki, Japanese Patent Publication No. 14249 (1964); T. Horio, *Vitamins* **24**, 79 (1961); C. Kawasaki, T. Horio, and I. Daira, *ibid.* **26**, 298 (1962).

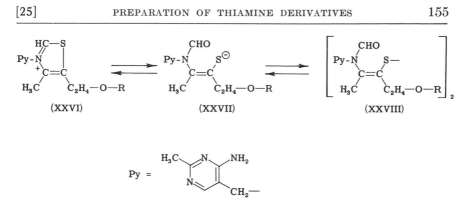

FIG. 8. Esters of thiamine disulfide.

in living systems, the disulfide of TDP (XXVII: P(O₂H)—O—PO₃H₂) has not yet been synthesized in a pure state, and its coenzyme action was reported to be feeble.[66]

Esters of thiamine disulfide are prepared from the thiamine esters (XXVI) by the oxidation of the thiol-form thiamine (XXVII), which is formed on the treatment of thiamine esters (XXVI) with a base such as sodium hydroxide, and from thiamine disulfide by acylation. Reagents used for the oxidation are iodine, chlorine, bromine, hydrogen peroxide, potassium ferricyanate, and sodium tetrathionate.[66–70]

The preparation of the disulfide of TMP (XXVIII: R = PO₃H₂) as well as *O*-acetate (XXVIII: R = COCH₃),[71] *O*-butyrate (XXVIII: COC₃H₇),[69] and *O*-benzoate (XXVIII: R = COC₆H₅)[67,72] of thiamine disulfide are described. The latter two are claimed to be absorbed from the digestive tract more easily than thiamine.

O-Acetylthiamine Disulfide (XXVIII: R = COCH₃).[71] A mixture of 0.5 g of thiamine disulfide and 5 ml of acetic anhydride is allowed to stand

[65] O. Zima, K. Ritsert, and Th. Moll, *Z. Physiol. Chem.* **267**, 210 (1941).

[66] O. Zima, G. Göttmann, A. Hoffmann, L. Hepding, and R. Hotovy, *E. Merck's Jahresber.* **67**, 1 (1953).

[67] T. Fujita, Y. Mushika, and K. Hagio, *Yakugaku Zasshi* **82**, 1452 (1962); *ibid.* **83**, 1056 (1963).

[68] T. Fujita and Y. Mushika, *Vitamins* **32**, 51 (1965).

[69] I. Utsumi, K. Harada, K. Kohno, and I. Daira, Japanese Patent Publication No. 17264 (1967).

[70] S. Yurugi, T. Fushimi, Y. Oka, and H. Asakawa, *Ann. Rept. Takeda Res. Lab.* **27**, 25 (1968).

[71] T. Matsukawa and T. Iwatsu, *Yakugaku Zasshi* **70**, 224 (1950).

[72] C. Kawasaki, Japanese Patent Publication No. 8228 (1964); C. Kawasaki, I. Utsumi, T. Fujita, and H. Kobayashi, *ibid.* No. 8229 (1964); C. Kawasaki, T. Horio, and I. Daira, *Vitamins* **26**, 298 (1962).

for 2 days at room temperature. The reaction mixture is evaporated to dryness under reduced pressure at a temperature which is as low as possible. The residue is recrystallized twice from EtOH to give 0.2 g of colorless granular crystals, m.p. 155° (decomp.).

O-Butyroylthiamine Disulfide (*XXVIII:* $R = COC_3H_7$).[69] *O*-Butyroyl-thiamine (XXVI: $R = COC_3H_7$) HCl (41 g) is dissolved in 100 ml of H_2O, mixed with 30 ml of 4% NaOH and adjusted to pH 11.5–11.9 by the addition of 10% NaOH. The mixture is allowed to stand for 15 minutes, then 180 ml of 20% $K_3Fe(CN)_6$ is added and the mixture is allowed to stand for 30 minutes. The separated yellowish oil is extracted with $CHCl_3$. The $CHCl_3$ solution is washed with water, dried, and evaporated to dryness *in vacuo;* the residue is dissolved in acetone. After addition of petroleum benzene and standing at room temperature, 10.6 g of colorless granular crystals are separated, m.p. 85–87°.

O-Benzoylthiamine Disulfide (*XXVIII:* $R = COC_6H_5$).[67,72] (a) Thiamine disulfide (20 g) is suspended in 200 ml of pyridine. To the solution, 10 ml of benzoylchloride is added dropwise with stirring; after 3 more hours of stirring, the reaction mixture is allowed to stand overnight. The resulting mixture is concentrated to dryness under reduced pressure. Water is added to the residue to separate a substance which is extracted with $CHCl_3$. The solution is washed with H_2O and dried, and evaporated under reduced pressure to give a solid. This solid is treated with 20 ml of benzene, and the separated crystals are collected. The product is recrystallized from 99% EtOH to give 20 g of colorless prisms, m.p. 147–148°.

(b) *O*-Benzoylthiamine (XXVI: $R = COC_6H_5$) HCl (10 g) is dissolved in 10 ml of H_2O. First, 15 ml of H_2O containing 2.7 g of NaOH is added to the solution with cooling, and then 40 ml of H_2O containing 2.9 g of I_2 and 1.5 g of KI is added dropwise. The separating colorless crystals are collected and washed with H_2O. The product is recrystallized from EtOH to give 8.5 g of colorless prisms, m.p. 143–146°.

O-Phosphonothiamine Disulfide (*XXVIII:* $R = PO_3H_2$).[66,68] Thiamine monophosphate (72.4 g) is dissolved in 250 ml of H_2O. To the solution, 50 ml of H_2O containing 26.4 g of NaOH is added at below 10° and chlorine gas is passed in gradually. When the pH of the reaction mixture reaches 10, the addition of chlorine ceases and the stirring is continued for 10 minutes.

The pH of the reaction mixture is adjusted to 3.9 by addition of 36% HCl (about 28 ml). After cooling with ice water for 1 hour, the crystals are filtered off. The filtrate, after decolorization, is concentrated to about 100 ml *in vacuo* and more crystals are obtained by adding 500 ml of MeOH and cooling. The combined crystals are recrystallized from H_2O to give 54.8 g of the product.

Thiamine Alkyl Disulfides (Fig. 9)

Allithiamine (thiamine allyl disulfide, XXX: R = CH_2—CH=CH_2) was discovered by Fujiwara[73] out of an admixture of thiamine with the juice of garlic (*Allium sativum* L.) in course of his search for a factor in plants which seems to decompose thiamine. The structure of allithiamine was established by Matsukawa[74] in 1952.

Thiamine allyl disulfide (XXX: R = CH_2—CH=CH_2) as well as its homologs (XXX: R = CH_3, $CH_2CH_2CH_3$) are also formed by mixing thiamine with the juice of many other plants of the *Allium* species.[75] Especially thiamine methyl disulfide (XXX: R = CH_3) is formed in fairly good yield by the treatment of thiamine with the juice, previously treated with allinase, of *Cruciferae*, such as cabbage, radish, and rape.[76] In the table, the reaction products of thiamine with various *Allium* plants are noted.[75]

Thiamine alkyl disulfides (XXX) can be synthesized by the reaction of thiol-form thiamine (XXIX) with alkyl alkanethiolsulfinate (R—SO—S—R),[77] alkyl alkanethiolsulfonate (R'—SO_2—S—R),[78,79] alkyl

THIAMINE ALKYL DISULFIDES (XXX) PRODUCED BY REACTION BETWEEN THIAMINE AND COMPONENTS OF PLANTS BELONGING TO *Allium* SPECIES

Plants	R		
	—C_3H_5	—CH_3	—C_3H_7
A. sativum (garlic)	++++	+	(+)
A. tuberosum Rottl. (Japanese leek)	+++	+++	−
A. victorialis L. var. *platyphyllum* Hultén (Japanese wild onion)	+++	++	−
A. bakeri Regal (garden shallot)	(+)	+++	(+)
A. grayi Regal (Japanese garlic)	+	++	(+)
A. thunbergii Don	++	++	(+)
A. togasii Hara	+	++	(+)
A. cepa L. (onion)	−	−	(+)

[73] M. Fujiwara and H. Watanabe, *Proc. Japan Acad.* **28**, 156 (1952).

[74] T. Matsukawa and S. Yurugi, *ibid.* **28**, 146 (1952); *Yakugaku Zasshi* **72**, 1602, 1616 (1952).

[75] T. Matsukawa, S. Yurugi, and T. Matsuoka, *Science* **118**, 325 (1953); S. Yurugi, *Yakugaku Zasshi* **74**, 502, 506, 511, 514, 519 (1954).

[76] M. Fujiwara, S. Tsuno, and M. Yoshimura, *Vitamins* **10**, 64, 506 (1955); S. Tsuno, *ibid.* **14**, 671 (1958).

[77] T. Matsukawa and H. Kawasaki, *Yakugaku Zasshi* **73**, 216 (1953).

[78] T. Matsukawa, T. Iwatsu, and H. Kawasaki, *Yakugaku Zasshi* **73**, 497 (1953).

[79] S. Yamada, T. Fujita, and T. Mizoguchi, *Yakugaku Zasshi* **74**, 963 (1954).

thiothiocyanate (R—S—SCN),[78] sodium alkylthiosulfate (NaO$_3$S$_2$R),[78,80] alkylsulfenechloride (R—SCl),[81] alkylmercaptan with oxidizing agents,[78] O,S-disubstituted thiosulfate (R—S—SO—O—R'),[82] S-alkylsulfenisothiourea (R—S—S—C(=NH)—NH$_2$),[83] and sodium alkylthiothiosulfate (NaO$_3$S$_3$R).[70]

In addition to these methods, the methods using the reactive thiamine derivatives and alkyl mercaptans are employed. As the reactive thiamine derivatives, S-cyanothiamine (XXXI: X = CN),[84] thiosulfate (XXXI: X = SO$_3$CH$_3$),[85] and thiosulfonate of thiol-form thiamine (XXXI: X = SO$_2$CH$_3$),[85] S-sulfothiamine (XXXI: X = SO$_3$H),[86] and the sulfoxide of thiamine disulfide (XXXII)[87,88] are applied.

These thiamine alkyl disulfides (XXX) are claimed to be excellent in their absorbability from the digestive tract and in the retention of thiamine produced from them in the tissues.

The synthetic methods for representative derivatives are described.

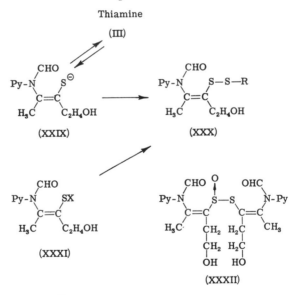

Fig. 9. Thiamine alkyl disulfides.

[80] T. Iwatsu, *Yakugaku Zasshi* **73**, 1115 (1953).
[81] H. Kawasaki and S. Noguchi, *Yakugaku Zasshi* **73**, 1307 (1953).
[82] H. Okuda, *Vitamins* **36**, 317 (1967).
[83] H. Hirano, K. Shirakawa, and O. Aki, Japanese Patent Publication No. 19549 (1968).
[84] H. Yonemoto, Japanese Patent Publication No. 13023 (1960).
[85] A. Takamizawa, K. Hirai, and Y. Sato, *Yakugaku Zasshi* **82**, 1202 (1962).
[86] H. Hirano, M. Hieda, Y. Oka, and K. Takiura, *Vitamins* **33**, 444 (1966).
[87] S. Yurugi, H. Kawasaki, and S. Noguchi, *Yakugaku Zasshi* **75**, 498 (1955).
[88] K. Kohno, G. Tsukamoto, Y. Kakie, and I. Utsumi, *Vitamins* **36**, 336 (1967).

Thiamine Propyl Disulfide $(XXX: R = C_3H_7)$.[78] Thiamine HCl (1 g) is dissolved in 10 ml of H_2O, 3.3 ml of 10% NaOH is added, and the solution is saturated with NaCl and allowed to stand for 30 minutes. One gram of crude sodium propylthiosulfate is mixed in with stirring, whereupon a turbidity occurs, a viscous oil separates, and gradually a solid forms. The product is filtered, washed with H_2O, and recrystallized from benzene to give 0.9 g of colorless prisms, m.p. 134–135° (decomp.).

Thiamine Tetrahydrofurfuryl Disulfide
$$(XXX: R = CH_2—CH—(CH_2)_2—CH_2).^{89}$$

$$\underset{O}{\underline{\qquad\qquad}}$$

Thiamine HCl (20.4 g) is dissolved in 30 ml of H_2O; 30 ml of H_2O containing 7.2 g of NaOH is added. The solution is allowed to stand for 30 minutes, then is added to 60 ml of $CHCl_3$ and mixed with 30 ml of aqueous solution containing 34 g of sodium tetrahydrothiosulfate with stirring for 15–20 minutes. The $CHCl_3$ layer is separated and extracted with 50 ml of 5% HCl by shaking, the acidic extract is neutralized with $NaHCO_3$ after decolorization to yield a viscous oil, which solidifies gradually. The crude product (16 g) is filtered and recrystallized from ethyl acetate to give colorless prisms, m.p. 132° (decomp.).

Thiamine [Methyl 6-Acetyl(dihydrothioctate)] 8-Disulfide $(XXX: R = CH_2—CH_2—CH—(SCOCH_3)—(CH_2)_4—COOCH_3)$.[90] Thiamine HCl (44 g) is dissolved in 140 ml of H_2O. To the solution is added 60 ml of H_2O containing 15.7 g of NaOH, and the solution is allowed to stand for 30 minutes. The solution is mixed with an aqueous solution of sodium alkylthiosulfate prepared from 50 g of methyl 6-acetylthio-8-iodooctanoate, 50 g of $Na_2S_2O_3 \cdot 5H_2O$, 400 ml of EtOH, and 100 ml of H_2O with stirring, and the separating oil is extracted with ethyl acetate, which is in turn extracted with 100 ml of dilute HCl by shaking. After decolorization, the extract is neutralized with $NaHCO_3$ to separate an oily substance which is extracted repeatedly with ethyl acetate. The organic layer is dried and concentrated *in vacuo* to give a syrup. The syrup is dissolved in as small volume of EtOH as possible, and 10% EtOH-HCl and a large amount of ether are added to produce the HCl salt of the product as crystals. The product is recrystallized from a mixture of EtOH and ether to give 51 g of crystals, m.p. 135–136°.

Thiamine Glutathione Disulfide $(XXX: R = CH_2—CH(CONHCH_2-COOH)—NHCO—CH_2CH_2—CH(NH_2)—COOH)$.[88] To a suspension of 2.3 g of thiamine disulfide monosulfoxide (XXXII)[91] in 40 ml of pyridine is

[89] S. Yurugi and T. Fushimi, *Yakugaku Zasshi* **78**, 602 (1958).
[90] Y. Deguchi and K. Miura, *Yakugaku Zasshi* **83**, 717 (1963).
[91] I. Utsumi, K. Harada, K. Kohno, and G. Tsukamoto, *Vitamins* **32**, 458 (1965).

added dropwise 100 ml of an aqueous solution containing 1.8 g of gluta-thione with stirring at room temperature. The reaction mixture is concen-trated *in vacuo*, and the residue is extracted with 300 ml of hot EtOH. After removal of insoluble oxidized glutathione by filtration, the filtrate is concentrated *in vacuo* and cooled to separate about 1 g of crude crystals. The product is recrystallized from EtOH to give a colorless amorphous powder, m.p. 179–181° (decomp.).

S-Acylthiamine Derivatives (Fig. 10)

S-Acylthiamines (XXXIII), prepared by Matsukawa[92] for the first time in 1953, can be produced from thiamine or thiamine esters by the acylation of their thiol forms with acid halides, acid anhydrides,[41,66,92–96] and any other acylating reagents.[97–99] In these reactions both SH and OH radicals of the thiol form (XXVII: R = H) could be acylated to yield the *O,S*-diacylthiamines (XXXIII). However, it may be possible to obtain only *S*-acylated thiamines under some condition.

A number of substances thus prepared have not only activity equal to that of thiamine, but also excellent absorbability from the digestive tract and good retention in the tissues.

Methods of synthesis for some important *S*-acylthiamines (XXXIII, XXXIV) are described.

O,S-Diacetylthiamine (*XXXIII: R' = CH₃, R = COCH₃*).[92] Thia-mine HCl (10 g) is dissolved in 25 ml of H_2O; 35 ml of 10% NaOH is added, and the solution is saturated with NaCl. The preparation is allowed to stand for 30 minutes, and 6 ml of acetic anhydride is added dropwise with stirring. During the reaction the mixture needs to be kept alkaline by the occasional addition of 10% NaOH to separate a viscous oil, which solidifies gradually.

---

[92] T. Matsukawa and H. Kawasaki, *Yakugaku Zasshi* **73**, 705 (1953).

[93] K. Shirakawa, *Yakugaku Zasshi* **74**, 367 (1954); H. Kawasaki, *ibid.* **74**, 588, 1189 (1954); *ibid.* **76**, 543 (1956).

[94] S. Yoshida, *ibid.* **74**, 993 (1954); M. Nagawa, M. Kataoka, T. Yoshioka, and Y. Baba, *Ann. Sankyo Res. Lab.* **13**, 20, 31 (1961); M. Kataoka and H. Itō, *ibid.* **13**, 24 (1961); T. Yoshioka and H. Itō, *ibid.* **13**, 35 (1961); N. Yoshida and Y. Nakamura, *ibid.* **13**, 37 (1961).

[95] A. Takamizawa, K. Hirai, and Y. Hamashima, *Chem. Pharm. Bull.* **10**, 1102, 1107 (1962).

[96] A. Takamizawa, K. Hirai, Y. Hamashima, H. Sato, M. Hata, and H. Itō, *Chem. Pharm. Bull.* **11**, 882, 1368 (1963); *ibid.* **12**, 558 (1964); *ibid.* **15**, 816 (1967); A. Takamizawa, Y. Hamashima, and H. Sato, *Vitamins* **29**, 26 (1964); A. Takamizawa, K. Hirai, T. Ishiba, and S. Hayakawa, *ibid.* **31**, 210 (1965); A. Takamizawa, K. Hirai, M. Ishihara, T. Ishiba, A. Takase, and S. Sumimoto, *Yakugaku Zasshi* **87**, 274 (1967).

[97] A. Itō, *Yakugaku Zasshi* **82**, 883 (1962).

[98] A. Takamizawa, K. Hirai, and K. Matsui, *Bull. Chem. Soc. Japan* **36**, 1214 (1963); A. Takamizawa and Y. Sato, *Chem. Pharm. Bull.* **12**, 398 (1964).

[99] M. Murakami, K. Takahashi, Y. Hirata, and H. Iwamoto, *Vitamins* **34**, 71 (1966).

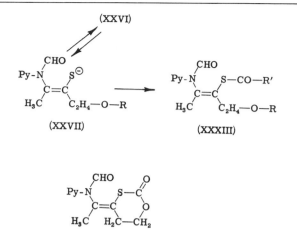

Fig. 10. *S*-Acylthiamine derivatives.

The product is filtered, washed with saturated NaCl, dried, and recrystallized from a mixture of benzene and petroleum benzene to give 7 g of colorless prisms, m.p. 122–123° (decomp.).

O,S-Dibenzoylthiamine (*XXXIII: R₆ = C₆H₅, R = COC₆H₅*).⁹² Thiamine HCl (10 g) is dissolved in 25 ml of H₂O, 35 ml of 10% NaOH is added, and the solution is saturated with NaCl and allowed to stand for 30 minutes. To the solution is added dropwise 10 ml of benzoyl chloride with stirring. During the reaction, the mixture needs to be kept alkaline by occasional addition of 10% NaOH to separate an oily substance, which solidifies gradually upon stirring with 100 ml of ethyl acetate. The product is collected and recrystallized from dilute EtOH to give 7 g of colorless prisms, m.p. 173–174° (decomp.).

O,S-Dicarboethoxythiamine (*XXXIII: R′ = OC₂H₅, R = COOC₂H₅*).⁹⁶ A solution of 17 g of thiamine HCl in 17 ml of H₂O is added to a solution of 3.5 g of Na in 200 ml of 99% EtOH. The preparation is allowed to stand for 30 minutes, then 5 g of triethylamine is added and then a further 10.8 g of ethyl chlorocarbonate in small portions. Stirring is continued for 2 hours at 45–48°. After filtration, the filtrate is concentrated under reduced pressure. The oily residue is dissolved in CHCl₃ and washed with dilute acetic acid and H₂O. After shaking with 15% HCl, the CHCl₃ layer is dried and evaporated. The oily residue solidifies gradually, yield 10.3 g, which is recrystallized from acetone to give colorless prisms, m.p. 121–123° (decomp.). The free base is recrystallized from ethyl acetate–petroleum ether to give colorless prisms, m.p. 113.5–114.5°.

S-Benzoylthiamine O-Monophosphate (*XXXIII: R′ = C₆H₅, R = PO₃H₂*).⁹⁷ Thiamine monophosphate HCl (4 g) is dissolved in 15 ml of

H$_2$O, adjusted to pH 10 by 30% NaOH under cooling with ice water, and mixed with 3.6 g of sodium benzoylthiosulfate[100] in small portions. The solution needs to be kept at pH 10 by the addition of 30% NaOH during the reaction. After filtration the solution is adjusted to pH 4 by 30% HCl to separate the crystals. The product is filtered, washed with H$_2$O, and dissolved repeatedly in alkaline solution. The solution is adjusted to pH 4 by HCl with cooling to separate 4 g of colorless plates, m.p. ca. 165° (decomp.).

Cyclocarbothiamine (*XXXIV*).[99] Thiamine HCl (13.5 g) is dissolved in 120 ml of 4% NaOH. After standing for 30 minutes, the solution is diluted with 100 ml of H$_2$O, and 14 g of COCl$_2$ is passed in with vigorous stirring within 30 minutes at 0–5°. During the reaction the solution needs to be kept at pH 9.5–10 by occasional addition of 20% NaOH. After further stirring for 15 minutes the mixture is extracted five times with ethylene chloride. The extract is washed with H$_2$O, dried, and concentrated *in vacuo* to separate colorless crystals. The product is filtered and washed with ethyl acetate to give 8.4 g of colorless crystals, m.p. 175–178° (decomp.). It is recrystallized from ethylene chloride to give pure colorless prisms, m.p. 181° (decomp.).

[100] A. Itō, *Yakugaku Zasshi* **82**, 866 (1962).

[26] Preparation of the Mono- and Pyrophosphate Esters of 2-Methyl-4-amino-5-hydroxymethylpyrimidine for Thiamine Biosynthesis

By GENE M. BROWN

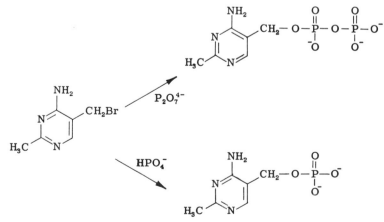

Both the mono- and pyrophosphate esters of hydroxymethylpyrimidine[1] are intermediates in the enzymatic synthesis of thiamine.[2] These compounds can be made enzymatically,[2] but if relatively large quantities are needed, it is more convenient to use the chemical method devised by Lewin and Brown[3] and described below. This method was devised as a result of the observation that bromomethylpyrimidine appears to hydrolyze spontaneously in aqueous solution to give hydroxymethylpyrimidine. In the presence of high concentrations of orthophosphate or pyrophosphate, the compound apparently is phosphorylyzed or pyrophosphorylyzed to yield the desired esters.

Preparation of the Pyrophosphate Ester. 2-Methyl-4-amino-5-bromomethylpyrimidine dihydrobromide (80 mg), obtained from Merck and Company, is dissolved in 1 ml of a saturated solution of sodium pyrophosphate and allowed to stand at room temperature for 4 hours. The pH of the solution is approximately 7.5. A saturated solution of barium hydroxide is then added until the pH has risen to 9.0. The resulting white precipitate of barium pyrophosphate is removed by centrifugation. The supernatant fluid can be shown by paper chromatography[2,4] to contain hydroxymethylpyrimidine and the pyrophosphate ester of this compound (as well as a somewhat smaller amount of the monophosphoester). The pyrophosphate ester is separated from the other products by chromatography on Dowex 1-formate. For this purpose, a column (2 × 28 cm) of Dowex 1-formate is prepared and the solution containing the pyrimidine compounds is applied to the column. Elution is effected with a linear gradient of 0 to 0.6 M ammonium formate at a rate of increase of 0.6 M per liter. The effluent liquid from the column is collected in 20-ml fractions, and each fraction is analyzed for ultraviolet light-absorbing material at 245 mμ. A large amount of such material is eluted with 0–0.4 M ammonium formate; another compound begins to be eluted with 0.55 M ammonium formate. In order to elute all of the latter compound, an extra 200 ml of 0.6 M ammonium formate is passed through the column.

The fast-eluting material (material that is eluted with 0–0.4 M ammonium formate) is a mixture of hydroxymethylpyrimidine and the monophosphoester of this compound, whereas the compound that elutes with 0.6 M ammonium formate can be shown to be the pyrophosphate ester of hydroxymethylpyrimidine by the following criteria. (a) Its absorption

---

[1] Abbreviations used: hydroxymethylpyrimidine for 2-methyl-4-amino-5-hydroxymethyl-pyrimidine; and bromomethylpyrimidine for 2-methyl-4-amino-5-bromomethyl-pyrimidine.

[2] L. M. Lewin and G. M. Brown, *J. Biol. Chem.* **236,** 2768 (1961).

[3] L. M. Lewin and G. M. Brown, *Arch. Biochem. Biophys.* **101,** 197 (1963).

[4] G. W. Camiener and G. M. Brown, *J. Biol. Chem.* **235,** 2404 (1960).

spectra at pH 3.0 and pH 9.0 are identical with those of hydroxymethyl-pyrimidine. (b) The compound can be shown to contain 2 phosphate residues (determined by the method of Fiske and SubbaRow[5]) per pyrimidine (pyrimidine is estimated spectrophotometrically at 245 mμ; molar extinction coefficient = 9.6 $\times$ 10^3). (c) Paper chromatography shows that its migration characteristics in isobutyric acid–NH$_4$OH–water (198:3:99, by volume) are identical (R_f = 0.25) with those of enzymatically formed pyrophosphoester. (d) Enzymatic evidence shows that the compound can be converted to thiamine monophosphate in the presence of thiazole phosphate, Mg^{2+}, and an enzyme preparation from yeast.[2]

The amount of pyrophosphoester recovered from the column varies with individual preparations from 15 to 25% of the total bromomethyl-pyrimidine used as reactant.

Preparation of the Monophosphoester. The monophosphoester of hydroxy-methylpyrimidine can be prepared by the same method as that described above for the pyrophosphate except that the bromomethylpyrimidine dihydrobromide is dissolved in a saturated solution of Na$_2$HPO$_4$. After removal of the excess phosphate as the barium salt, the remaining soluble materials are subjected to chromatography on a column (2 $\times$ 20 cm) of Dowex 1-formate in order to separate the monophosphoester from hydroxy-methylpyrimidine. The column is developed with a linear gradient of ammonium formate solution scaled to increase from 0 to 0.4 M at a rate of 0.4 M per 500 ml. The monophosphoester of hydroxymethylpyrimidine is eluted with 0.4 M ammonium formate, whereas hydroxymethylpyrimidine is eluted much earlier (from 0 to 0.2 M). The product analyzes for 1 phosphate group per pyrimidine residue and migrates (R_f = 0.45) on paper as enzymatically produced monophosphoester with isobutyric acid–NH$_4$OH–water as solvent.[4]

[5] C. H. Fiske and Y. SubbaRow, *J. Biol. Chem.* **66,** 375 (1925).

[27] Preparation of (2-Methyl-4-amino-5-pyrimidinyl)-methyl Pyrophosphate

By IRWIN G. LEDER

This procedure is based on the reaction described by Lewin and Brown,[1] who synthesized the pyrimidinyl pyrophosphate in milligram quantities, but did not isolate it. The compound has also been prepared by reaction of

[1] L. M. Lewin and G. M. Brown, *Arch. Biochem. Biophys.* **101,** 197 (1963).

the corresponding hydroxymethylpyrimidine with pyrophosphoric acid mixtures.[2,3]

Principle

(2-Methyl-4-amino-5-pyrimidinyl)methyl chloride dihydrochloride[4,5] is allowed to react with hot saturated aqueous sodium pyrophosphate. The pyrophosphate ester and the free hydroxymethyl pyrimidine are both formed in the reaction mixture. They are separated on a cation exchange column by elution of the pyrophosphate ester with water. The free ester is isolated and converted to the cyclohexylamine salt.

Procedure

Thirty-three grams of sodium pyrophosphate (75 millimoles) is dissolved in 60 ml of water at about 75° and 6 g of chloromethyl pyrimidine dihydrochloride (25 millimoles) is added within 1 minute. After 5 minutes the preparation is transferred to an ice bath and cooled to 5°; the pH is adjusted to 11 by slowly adding approximately 15 ml of 5 N NaOH. The suspension is chilled to 0°; after 1 hour it is filtered with suction, and the filter cake, mainly $Na_4P_2O_7$, is washed with a small quantity of cold water. The combined filtrate and wash (80 ml) are evaporated *in vacuo* using a 40° bath until the volume is approximately 20 ml. If a precipitate is present, it is removed by filtration and discarded.

The following column chromatography is carried out in the cold room. The brown solution (20–25 ml) is cooled to 0–4° and added directly to a column (4 × 27 cm) of Dowex 50 × 8 in the H^+ form. The compound is washed onto the column, then the elution is started with cold water, 8-ml fractions being collected at the rate of 1 ml/minute. Each tube is checked for ultraviolet-absorbing material by spotting on paper. The compound appears after about 175 ml of eluate have been collected. The contents of tubes containing ultraviolet-absorbing material are pooled (80 ml) and evaporated to dryness *in vacuo*. The white solid is collected with the aid of absolute ethanol, washed on a filter with ethanol and ether, and dried. The yield is 1.96 g, 6.6 millimoles. For recrystallization, the solid is dissolved in 21 ml of water at 65° and poured into 150 ml of ethanol followed by 200 ml of ether. After several hours at approximately 0°, the precipitate is

---

[2] T. Sato, *J. Japan. Biochem. Soc.* **32,** 822 (1960).

[3] I. G. Leder, *J. Biol. Chem.* **236,** 3066 (1961).

[4] Abbreviations used: hydroxymethylpyrimidine, (2-methyl-4-amino-5-pyrimidinyl)-methanol; chloromethylpyrimidine, (2-methyl-4-amino-5-pyrimidinyl)methyl chloride; hydroxymethylpyrimidine-P and hydroxymethylpyrimidine-PP, the corresponding monophosphate and pyrophosphate esters.

[5] I wish to thank Merck and Co., Elkton, Virginia, for providing generous amounts of this compound as well as β-(4-methyl-5-thiazolyl)ethanol.

collected, washed with ethanol and ether, and dried. The yield is 1.8 g, 6 millimoles. The product is essentially pure (2-methyl-4-amino-5-pyrimidinyl)methyl pyrophosphate. It contains no inorganic phosphate; on the basis of a molar extinction coefficient[2] of 12.16×10^3 cm²/mole in 0.1 M HCl, it contains 2 atoms of P per mole.

Conversion to the Cyclohexylammonium Salt. One gram of the free acid is suspended in 12 ml of water at 5°. The pH is adjusted to 9.5 with ice-cold 0.4 M cyclohexylamine, and the resulting solution is evaporated *in vacuo*, using a 40° bath, until the solid starts to separate at approximately 1 ml. Ten milliliters of ethanol is added. The solution is reduced once more almost to dryness, and the suspension is taken up in 6 ml of ethanol. Two volumes of ether are added, and after chilling in ice for several hours, the precipitate is collected, washed with ether, and dried. The yield is 1.7 g, 2.7 millimoles, of the crystalline cyclohexylammonium salt.

The product is recrystallized as follows: 1.1 g is dissolved in 10 ml of hot ethanol and filtered. The filtrate is evaporated to approximately 5 ml, and two volumes of ether are slowly added with shaking. The mixture is chilled for several hours in an ice bath and filtered with suction; the white crystalline solid is washed with alcohol and ether and dried *in vacuo* over P_2O_5. The yield is 0.9 g, 1.6 millimoles. The product is free of inorganic phosphate and contains 2 atoms of P per mole on the basis of a molar extinction coefficient[2] of 12.16×10^3 cm²/mole in 0.1 N HCl.

Properties

Both phosphate groups are acid labile. At 100° in 1 N sulfuric acid, the pyrophosphate bond is split first and both phosphate groups are hydrolyzed in 30 minutes. For purposes of identification by paper or thin-layer chromatography, the following R_f values are obtained with a solvent composed of a 1:1 mixture of acetone and 50% acetic acid: chloromethylpyrimidine, 0.9; hydroxymethylpyrimidine, 0.8; hydroxymethylpyrimidine-P, 0.45; hydroxymethylpyrimidine-PP, 0.23.

[28] Preparation of β-(4-Methyl-5-thiazolyl)-ethyl Phosphate

By IRWIN G. LEDER

β-(4-Methyl-5-thiazolyl)ethyl phosphate has been prepared by phosphorylation of the corresponding alcohol[1] and by the sulfite cleavage of

[1] J. Weijlard, *J. Am. Chem. Soc.* **64**, 2279 (1942).

thiamine monophosphate. The following procedure provides a simple isolation of the free thiazolyl monophosphate. The conditions for the bisulfite treatment are as described for thiamine pyrophosphate by Lohmann and Schuster.[2]

Two grams of thiamine monophosphate[3] is dissolved in 10 ml of water, and the pH is adjusted to 4.8 with 5 N NaOH. To this solution, 3.2 g of $Na_2S_2O_5$ is added. A white precipitate—(2-methyl-4-amino-5-pyrimidinyl)-methyl sulfonate—begins to form after approximately 10 minutes. The mixture is allowed to stand at room temperature for 3 days; the precipitate is filtered, washed with a small quantity of cold water, and discarded. To the combined filtrate and washings totaling 20 ml, 1.8 ml of 30% hydrogen peroxide is added in several portions to oxidize the residual sulfite to sulfate. The solution is concentrated *in vacuo* to approximately 8 ml.

The following column chromatography is carried out in the cold room. The 8 ml of concentrated solution is added to a column (1.5 × 15 cm) of Dowex 50 × 8 in H+ form. After the compound has been washed onto the column, connection is made to a reservoir of water, and 10-ml fractions are collected at the rate of 0.5 ml per minute. The fractions are checked for ultraviolet absorption by simply spotting on paper or measuring absorbancy in the spectrophotometer at 255 mμ. The thiazolylphosphate appears after about 6 bed volumes of eluate have been collected. The fractions containing ultraviolet-absorbing material are pooled (150 ml) and evaporated to dryness. The yield is 846 mg.

The product is recrystallized, as white needles, from approximately 25 ml of 80% ethanol to yield 740 mg. The compound has not been subjected to complete elementary analysis. It contains the theoretical percentage of phosphorus and no acid labile or inorganic phosphate.

[2] K. Lohmann and P. Schuster, *Biochem. Z.* **294**, 188 (1937).
[3] Commercially available.

[29] Oxidation Products of Thiamine Derivatives

By ISAMU UTSUMI, KIYOSHI HARADA, KEIICHI KOHNO, and
GORO TSUKAMOTO

A number of thiol-form thiamine derivatives have been synthesized mainly in Japan and are in current clinical use as long-acting vitamin B$_1$ derivatives which are easily absorbed from the gastrointestinal tract. However, relatively little is known about metabolic and oxidative products of these modified thiamine compounds.[1]

[1] C. Kawasaki, *Vitamins Hormones* **21**, 69 (1963).

The mild oxidation of thiamine hydrochloride (I) in an alkaline medium produces thiamine disulfide (III), which is reduced to thiamine by cysteine, gluthathione, and other reducing agents.

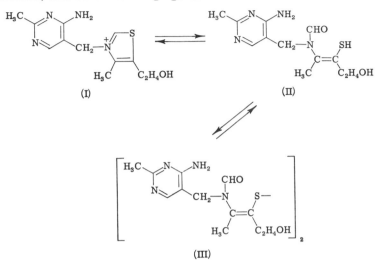

(I)   (II)

(III)

This oxidation-reduction system could be presumed to bear a close resemblance to that of cysteine ⇌ cystine *in vivo*. It is well known that cysteine sulfinic acid, cysteic acid, hypotaurine, and taurine appear in the course of metabolism of cysteine and cystine.[2]

Similar considerations have led us to suspect the presence of 2-(2-methyl-4-amino-5-pyrimidyl)methylformamido-5-hydroxy-2-pentene-3-sulfinic acid or sulfonic acid, which were named hypothiaminic acid or thiaminic acid on the basis of a structural resemblance to hypotaurine or cysteic acid, respectively.

The chemical oxidation of thiol-form thiamine derivatives made it clear that various derivatives, which are easily converted to thiamine *in vivo*, provide thiaminic acid as a final product.[3] It was confirmed that thiamine alkyl sulfides, which are hard to revert to thiamine *in vivo*, afford the corresponding sulfoxides and sulfones.[4] Hypothiaminic acid as an intermediate in the oxidation was synthesized from thiamine disulfide monosulfoxide, the formation of which is the first step in the oxidation of thiamine disulfide.[5] The monosulfoxide[6] and hypothiaminic ester[7] are biologi-

[2] L. Young and G. A. Maw, "Metabolism of Sulfur Compounds," Methuen Press, London, 1958.

[3] I. Utsumi, K. Harada, and G. Tsukamoto, *J. Vitaminol. (Kyoto)* **11**, 225 (1965).

[4] G. Tsukamoto, T. Watanabe, and I. Utsumi, *J. Vitaminol. (Kyoto)* **13**, 313 (1967).

[5] I. Utsumi, T. Watanabe, K. Harada, and G. Tsukamoto, *Chem. Pharm. Bull. (Tokyo)* **15**, 1485 (1967).

cally active. However, these hypothiaminic and thiaminic acids, the toxicity of which is extremely low, have neither thiamine nor antithiamine activity, but have a degree of analgesic and antiinflammatory effect similar to that of easily absorbable thiamine derivatives.[8]

Although no evidence has been obtained in support of the formation of these oxidation products as metabolites, it has been assumed that their formation may occur as with the metabolic reactions of cysteine and cystine.

I. Thiamine Disulfide Monosulfoxides

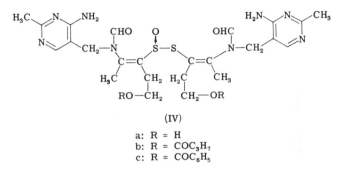

(IV)

a: R = H
b: R = COC₃H₇
c: R = COC₆H₅

b: R = COC_3H_7
c: R = COC_6H_5

Preparation

Principle. Oxidation of disulfides by hydrogen peroxide is generally known to produce the corresponding thiolsulfinates, i.e., monosulfoxides.[9] In the case of thiamine benzyl disulfide, one of the asymmetrical disulfides of thiamine, the corresponding monosulfoxide is obtained by oxidation with hydrogen peroxide.[10] However, the oxidation of symmetrical disulfides of thiamine with this reagent does not afford the desired products (IV).[6] The monosulfoxides (IV) are easily obtained in satisfactory yield by the oxidation of thiamine disulfides with perorganic acids, such as peracetic acid[11] and perbenzoic acid.[6]

Procedure

Thiamine Disulfide Monosulfoxide (IVa) and Its O-Benzoyl Ester (IVc).[11] While cooling, 0.12 mole of peracetic acid is added to a stirred solution of

[6] I. Utsumi, K. Harada, K. Kohno, and G. Tsukamoto, *J. Vitaminol. (Kyoto)* **13,** 26 (1967).

[7] G. Tsukamoto, T. Watanabe, and I. Utsumi, *Bull. Chem. Soc. (Japan),* **42,** 2686 (1969).

[8] I. Utsumi, K. Harada, K. Kohno, and G. Tsukamoto, *J. Vitaminol. (Kyoto)* **11,** 248 (1965).

[9] A. Stoll and E. Seebeck, *Experienta* **3,** 114 (1947).

[10] H. Kawasaki and H. Yonemoto, *Yakugaku Zasshi* **77,** 640 (1957).

[11] G. Tsukamoto, T. Watanabe, and I. Utsumi, *Bull. Chem. Soc. Japan,* **42,** 2566 (1969).

0.10 mole of thiamine disulfide (56.3 g) or O-benzoylthiamine disulfide [(IIIc), OH in structure (III) is $OCOC_6H_5$] (77.0 g) dissolved in 500 ml of 2 N H_2SO_4. The reaction mixture is allowed to stand overnight at room temperature. Adjustment of the reaction solution to pH 6–7 with saturated Na_2CO_3 or $NaHCO_3$ gives an 80–85% yield of the corresponding mono-sulfoxide (IVa, IVc), which is nearly free from other oxidation products, as pale yellow crystals.

Further purification of compound (IVa), m.p. 191–193° (decomp.), is difficult, since suitable solvents for the recrystallization are not known. Recrystallization of (IVc) from chloroform or chloroform–ether affords colorless crystals, m.p. 173–174° (decomp.).

O-Butyrylthiamine Disulfide Monosulfoxide (IVb).[6] A chloroform solution containing 1.6 g of perbenzoic acid is added with cooling to a solution of 7.4 g of O-butyrylthiamine disulfide [(IIIb), OH in structure (III) is $OCOC_3H_7$] dissolved in 5 ml of glacial acetic acid and 15 ml of chloroform. The mixture is allowed to stand overnight at room temperature; the mixture is then neutralized with 10% Na_2CO_3, and the chloroform layer is collected. The chloroform solution is washed with water and dried over Na_2SO_4. The solvent is evaporated under reduced pressure. The residue is dissolved in a small amount of ethanol, ether is added until the solution becomes turbid, and the mixture is allowed to stand at room temperature to afford 4.0 g of crystals. Recrystallization of the crude crystals from acetone gives pure material, m.p. 140–141° (decomp.).

Properties

Interestingly, the monosulfoxides (IV) show no change after several years at room temperatures. This stability may be due to the resonance of the $—S(O^-)=S^+—$ double bond with olefinic double bonds on both sides. Infrared absorption spectra of these monosulfoxides show the characteristic absorption band due to an S—O group near 1100 cm^{-1}.

Reaction with Hydrogen Iodide.[6] The monosulfoxides (IV) are allowed to react with potassium iodide in glacial acetic acid to smoothly liberate iodine quantitatively. Accordingly, the monosulfoxides (IV) can be analyzed by the determination of free iodine with sodium thiosulfate.

Reaction with Triphenylphosphine.[11] The monosulfoxide (IVc) readily reacts with triphenylphosphine at room temperature to afford the corresponding disulfide and triphenylphosphine oxide.

Reaction with Mercaptans.[7] These monosulfoxides react smoothly with mercaptans at room temperature to give asymmetrical disulfides within a few minutes. The reaction of the monosulfoxide with mercaptan is com-

petitive between the so-called mercaptan-thiolsulfinate reaction (1) and the
apparent oxidation-reduction reaction (2):

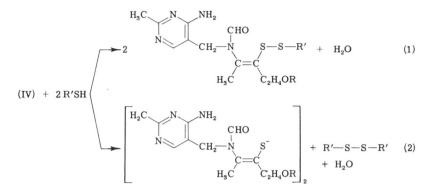

$$(IV) + 2 R'SH$$

(1)

(2)

Up to the present time, a number of asymmetrical disulfides of thiamine
have been prepared by a variety of methods,[12] but the method generally
used is presumed to be unavailable in order to prepare the mixed disulfide of
thiamine and *tert*-mercaptan.

On the other hand, the reaction of (IV) with *tert*-mercaptan offers an
excellent method for the preparation of the mixed disulfides, depending on
the acidity of mercaptan. In addition, this reaction is also suitable for
preparing unstable asymmetrical disulfides of thiamine such as thiamine-
glutathione disulfide[13] and thiamine pyridylmethyl disulfide derivatives.[14]

Biological Properties of O-Benzoylthiamine Disulfide Monosulfoxide
(IVc).[6] The effect of the monosulfoxide in the lovebird assay is the same
as for thiamine and O-benzoylthiamine disulfide (IIIc). When (IVc) is
orally or intravenously administered to rabbits, blood thiamine levels
remain somewhat higher than O-benzoylthiamine disulfide (IIIc).

Analysis

Separation by Chromatography.[6] The separation of thiamine disulfide
(III) and its monosulfoxide (IIIa) is possible by means of paper partition
chromatography (Toyo Roshi No. 51, butanol–acetic acid–H_2O (4:1:5),
ascending chromatography). The R_f value for (III) is 0.30; and for (IVa),
0.17. The convenient method for the separation of O-butyrylthiamine
disulfide (IIIb) or O-benzoylthiamine disulfide (IIIc) and the corresponding

[12] T. Matsukawa and S. Yurugi, *in* "Review of Japanese Literature on Beriberi and
Thiamine" (N. Shimazono and E. Katsura, eds.), p. 107–114. Vitamin B Research
Committee of Japan, Kyoto, 1965.
[13] K. Kohno, G. Tsukamoto, Y. Kakie, and I. Utsumi, *Vitamins (Kyoto)* **36**, 336 (1967).
[14] I. Utsumi, T. Watanabe, and G. Tsukamoto, *Vitamins (Kyoto)* **37**, 276 (1968).

monosulfoxides (IVb or IVc) involves thin-layer chromatography with alumina in benzene–acetic acid (9:1). The R_f values are: (IIIb) 0.4–0.6; (IVb) 0.2–0.3; (IIIc) 0.6–0.8; (IVc) 0.5–0.7.

Determination.[6] These monosulfoxides (IV), which are reduced with cysteine or sodium thiosulfate to two molecules of thiamine, can be determined under the same conditions as for the assay of *O*-benzoylthiamine disulfide.[15]

II. Hypothiaminic Acids

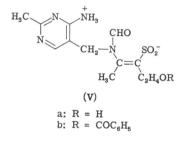

(V)

a: R = H
b: R = COC$_6$H$_5$

Preparation

Principle. Thiamine disulfide monosulfoxides are hydrolyzed to hypothiaminic acids and thiamines in alkaline solution, but they are decomposed in neutral solution to afford hypothiaminic acids and thiamine disulfides.[5] Hypothiaminic acid (Va) is prepared by alkaline hydrolysis of 2-(2-methyl-4-amino-5-pyrimidyl) methylformamido-5-benzoyloxy-2-pentene-3-sulfinic acid [(Vb), *O*-benzoylhypothiaminic acid] obtained by neutral hydrolysis of *O*-benzoylthiamine disulfide monosulfoxide.

Procedure

O-Benzoylhypothiaminic Acid (Vb).[5] A suspension of 20 g of *O*-benzoylthiamine disulfide monosulfoxide (IVc) in 200 ml of 80% aqueous ethanol is refluxed in a boiling water bath for 1.5 hours to afford an apparent solution. The solution is evaporated to dryness under reduced pressure. The residue is dissolved in 100 ml of ethanol, ether is added to turbidity, and the mixture is allowed to stand at room temperature to yield 4.5 g of colorless needles. The crystals thus obtained are refluxed in 100 ml of chloroform for 15 minutes, after which they are recrystallized from 90% aqueous ethanol and ether to afford 4 g of *O*-benzoylhypothiaminic acid as colorless needles, m.p. 204–205°. The ultraviolet absorption maximum at pH 2.5 is at 230 mμ (log ε = 4.35); at pH 7.3, maxima are at 230 mμ (log ε = 4.40) and 275 mμ (log ε = 3.81).

[15] I. Utsumi, K. Harada, Y. Kondo, and H. Hirao, *J. Vitaminol. (Kyoto)* **8**, 220 (1962).

Hypothiaminic Acid (Va).[5] A solution of 700 mg of *O*-benzoylhypothiaminic acid, in 30 ml of 80% aqueous ethanol, and 2.4 ml of 1 *N* NaOH is heated at 70° with adjustment of pH to 9–10 by the dropwise addition of 2.4 ml of 1 *N* NaOH for 3 hours. The reaction solution is condensed under reduced pressure, and the residue is dissolved in water. The solution is adjusted to pH 3.0 by the addition of dilute HCl, and separated crystals are extracted with ether. The aqueous solution is condensed under reduced pressure, the residue is extracted with hot ethanol to remove inorganic salts, and then the extract is evaporated to dryness under reduced pressure. The residue thus obtained is recrystallized from 95% aqueous ethanol to give 580 mg of hypothiaminic acid (Va) as colorless prisms, m.p. 121–122°. This compound (Va) has 2.5 moles of water as crystal solvent. The ultraviolet absorption maximum at pH 2.5 is at 247 mμ (log ϵ = 3.91); at pH 7.3 at 275 mμ (log ϵ = 3.67).

Properties

Hypothiaminic acids (V) are very stable in crystalline form and dilute alkaline solution. They are unstable at below pH 2.0. The pK_a values are the same as those of thiaminic acids (VI) described in Section III. These compounds can be converted by various methods[7] into hypothiaminic esters such as 2-(2-methyl-4-amino-5-pyrimidyl)methylformamido-5-benzoyloxy-3-alkoxysulfino-2-pentene.

Hypothiaminic acids cannot be reduced to thiamine with reducing agents such as sodium thiosulfate and sulfhydryl compounds. Accordingly, they have no thiamine activity like thiaminic acids (VI). This observation seems analogous to the biological irreversibility of cysteine sulfinic acid to cysteine. However, hypothiaminic esters which are reduced to thiamine have activity.[7] In addition, the toxicity of hypothiaminic acids, like that of thiaminic acids, is extremely low.

Analysis

Thin-Layer Chromatography. A convenient method for the detection of hypothiaminic acids involves thin-layer chromatography with Toyo chromatosheet (silica gel) in butanol–acetic acid–H$_2$O (4:1:5). The R_f for hypothiaminic acid (Va) is 0.15; for *O*-benzoylhypothiaminic acid,. the R_f is 0.38.

Hypothiaminic acids are not detected by means of Dragendorff's reagent, unlike thiamine and other derivatives. On the other hand, they can be visualized by the *tert*-butyl hypochlorite–iodine–starch procedure of R. H. Mazur *et al.*[16] or by the iodine vapor procedure.

[16] R. H. Mazur, B. W. Ellis, and P. S. Cammarata, *J. Biol. Chem.* **237,** 1619 (1962).

III. Thiaminic Acid

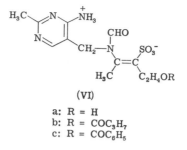

(VI)

a: R = H
b: R = COC$_3$H$_7$
c: R = COC$_6$H$_5$

Preparation

Principle. Thiaminic acids (VI) are prepared by the oxidation of thiol-form thiamine derivatives, such as thiamine disulfides, thiamine alkyl disulfides, S-acylated thiamines, S-carbalkoxythiamines, and cyanothiamine with hydrogen peroxide.[3,4] Especially, the oxidation of S-acylated thiamines and thiamine disulfide monosulfoxides gives thiaminic acids in good yield (80–90%). As thiaminic acid (VIa) is hard to crystallize when impure, a convenient method for the preparation of the acid is alkaline hydrolysis of O-benzoylthiaminic acid (VIc).

In the case of oxidizing O-benzoylthiamine disulfide (IIIc) or its monosulfoxide (IVc) with hydrogen peroxide, the geometric isomer (VIIa), which has the olefin-CH$_3$ and SO$_3^-$ in *cis*-configuration, can be obtained besides O-benzoylthiaminic acid (VIc).[17]

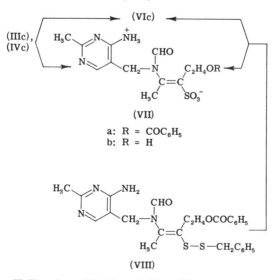

(VII)

a: R = COC$_6$H$_5$
b: R = H

(VIII)

[17] G. Tsukamoto, K. Harada, and I. Utsumi, *Chem. Pharm. Bull. (Tokyo)* **14,** 823 (1966).

The isomer (VIIa) is also obtained by oxidizing the geometric isomer (VIII) of O-benzoylthiamine benzyl disulfide,[18] which is prepared by benzoylation of the geometrical isomer of thiamine benzyl disulfide synthesized by the method of Murakami et al.[19]; however, considerable quantities of the *trans* isomer (VIc) also are formed. Oxidation of O-benzoylthiamine benzyl disulfide, however, gives only O-benzoylthiaminic acid (VIc), and does not afford the isomer (VIIa). The alkaline hydrolysis of the isomer (VIIa) affords the geometric isomer (VIIb) of thiaminic acid.

Procedure

O-Benzoylthiaminic Acid (VIc). To a solution of 6.3 g (1.4×10^{-2} mole) of O,S-dibenzoylthiamine dissolved in 60 ml of glacial acetic acid is added 8.4 ml (8.4×10^{-2} mole) of 30% hydrogen peroxide. The mixture is allowed to stand at room temperature. After 10 days, the mixture in which crystals appear is condensed to dryness under reduced pressure, and ethanol is added to the residue to yield 4.3 g (86%) of O-benzoylthiaminic acid (VIc) as colorless needles, m.p. 237–238°. Recrystallization of the crystals from 50% aqueous ethanol gives the hydrate of (VIc) ($C_{19}H_{22}O_6N_4 \cdot H_2O$) as colorless prisms, m.p. 234–235° (decomp.). The hydrate can be converted to the anhydride ($C_{19}H_{22}O_6N_4S$), m.p. 237–238° (decomp.), by being refluxed in ethanol or glacial acetic acid. The ultraviolet absorption maximum at pH 3.0 is at 232 mμ (log $\epsilon = 4.32$); at pH 7.4, maxima are at 232 mμ (log $\epsilon = 4.32$), and at 275 mμ (log $\epsilon = 3.74$).

Thiaminic Acid (VIa). A solution of 13.11 g of O-benzoylthiaminic acid (VIc) dissolved in 29 ml of 1 N NaOH and 100 ml of H_2O is heated at 80° and the pH maintained at 9–10 by the dropwise addition of 29 ml of 1 N NaOH for 3 hours. The solution is adjusted to pH 3.0 by the addition of dilute HCl, and separated crystals are extracted with ether. The aqueous layer is concentrated under reduced pressure. The residue is extracted with hot ethanol to remove inorganic salts, and then the extract is evaporated to dryness under reduced pressure. The residue is recrystallized from 90% aqueous ethanol to yield 9.5 g of thiaminic acid hydrate ($C_{12}H_{18}O_5N_4S \cdot H_2O$) as colorless needles, m.p. 213–214° (decomp.). This hydrate can be also converted to the anhydride ($C_{12}H_{18}O_5N_4S$), m.p. 215–216° (decomp.), by being refluxed in ethanol. The ultraviolet absorption maximum at pH 3.0 is at 242 mμ (log $\epsilon = 3.98$); at pH 7.4, maxima are at 232 mμ (log $\epsilon = 4.04$) and at 275 mμ (log $\epsilon = 3.69$).

O-Butyrylthiaminic Acid (VIb). To a solution of 3.7 g of O-butyryl-thiamine disulfide (IIIb) dissolved in 40 ml of glacial acetic acid is added

[18] Unpublished work.
[19] M. Murakami, K. Takahashi, M. Iwanami, and H. Iwamoto, *Yakugaku Zasshi* **85,** 752 (1965).

5 ml of 30% hydrogen peroxide, and the mixture is allowed to stand at room temperature for 3 days. The solvents of the mixture are removed under reduced pressure, and the residue is crystallized from ethanol to yield 1.8 g of crystals. Recrystallization of the crude crystals from ethanol gives O-butyrylthiaminic acid (VIb) as colorless needles, m.p. 212–214°. The ultraviolet absorption maximum in 0.1 N hydrochloric acid is at 243 mμ (log ϵ = 5.04); at pH 7.3, maxima are at 233 (log ϵ = 5.08) and 275 mμ (log ϵ = 4.72).

Isolation of the Geometric Isomer (VIIa) of O-Benzoylthiaminic Acid. To a solution of 6.0 g or O-benzoylthiamine disulfide (IIIa) in 60 ml of glacial acetic acid is added 5.3 ml of 30% hydrogen peroxide; the mixture is allowed to stand at room temperature. After a week, the mixture is condensed under reduced pressure, ethanol is added to the residue, and the separated crystals are recrystallized from 50% aqueous ethanol to give 4.5 g of O-benzoylthiaminic acid (VIc) as colorless prisms, m.p. 234–235° (decomp.).

The mother liquor, after filtration of (VIc), is again concentrated to dryness under reduced pressure, and the residue is dissolved in a little H$_2$O. The solution is allowed to stand at room temperature for a day to separate 0.4 g of crystals. The crystals thus obtained are recrystallized from water to give the geometric isomer (VIIa) of O-benzoylthiaminic acid, hydrated with 0.5 mole of water, as colorless needles, m.p. 225–226° (decomp.). This hydrate can also be converted to the anhydride (VIIa), m.p. 233–234° (decomp.), by being refluxed in ethanol. The ultraviolet absorption maximum at pH 3.0 is at 233 mμ (log ϵ = 4.34); at pH 7.4, maxima are at 232 mμ (log ϵ = 4.36) and 275 mμ (log ϵ = 3.77). Debenzoylation of this isomer (VIIa) yields the geometric isomer (VIIb) of thiaminic acid.

Properties

Solubility.[20] Thiaminic acids are almost insoluble in organic solvents. Thiaminic acid (VIa), its geometric isomer (VIIb), and O-butyrylthiaminic acid (VIb) are very soluble in water, as are the anionic forms of O-benzoylthiaminic acid (VIc) and its geometric isomer (VIIa).

The pK_1 and pK_2 for thiaminic acids and their isomers determined titrimetrically are shown in the table.[17]

DISSOCIATION CONSTANTS (pK_a) OF THIAMINIC ACIDS AND THEIR ISOMERS AT 25°

pK	Thiaminic acid	Isomer	O-Benzoylthiaminic acid	Isomer
pK_1	2.2	2.2	2.3	2.3
pK_2	6.2	5.9	6.1	5.8

[20] I. Utsumi, K. Harada, and G. Tsukamoto, *J. Vitaminol. (Kyoto)* **11**, 239 (1965).

Biological Properties.[8] Thiaminic acids (VI) have no thiamine activity and no antithiamine activity when assayed with lovebirds. The toxicity is very low among the compounds related to thiamine. The LD_{50} of thiaminic acid (VIa) for mice by intravenous injection is over 4 g/kg, and that of O-benzoylthiaminic acid (VIc) is 1.87 g/kg. No toxic effect is observed after either oral or intraperitoneal administration.

Analysis

Thin-Layer Chromatography. A convenient method for the detection of thiaminic acids involves thin-layer chromatography with Toyo chromato-sheet (silica gel) in butanol–acetic acid–H_2O (4:1:5). The R_f values are: thiaminic acid (VIa), 0.14; the isomer (VIIb), 0.15; O-butyrylthiaminic acid (VIb), 0.36; O-benzoylthiaminic acid (VIc), 0.35; the isomer (VIIa), 0.30. The detection of thiaminic acids on chromatography is carried out by the *tert*-butyl hypochlorite–iodine–starch procedure.

Paper Electrophoresis. Thiaminic acids (VI) of zwitterion structure can be separated from the corresponding thiamine compounds by paper electrophoresis at each pH level.[20]

IV. Miscellaneous Compounds

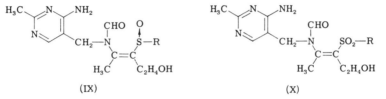

(IX)   (X)

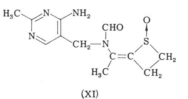

(XI)

Thiamine alkyl sulfides, which are difficult to convert to thiamine by sulfhydryl compounds, are oxidized with hydrogen peroxide and perbenzoic acid in glacial acetic acid to afford the corresponding sulfoxides (IX) and sulfones (X) in accordance with the amount of the peroxides.[4]

C. Kawasaki and Tomita have reported that thiamine anhydride, which has almost no thiamine activity, is metabolized to the corresponding sulfoxide (XI) in rats.[21, 22]

[21] C. Kawasaki and I. Tomita, *Vitamins (Kyoto)* **26,** 195 (1962).
[22] I. Tomita, *Vitamins (Kyoto)* **26,** 198 (1962).

[30] Thiamine—Methods for Studying the Biosynthesis of the Pyrimidine Moiety

By P. C. NEWELL and R. G. TUCKER

B-group vitamins are produced in microorganisms in extremely small amounts, and usually they are not formed in great excess over the organisms[1] current requirements. This presents problems, particularly of isolation, when studying vitamin biosynthesis. The procedure often adopted is to process large quantities of the microorganism and extract the small amount of vitamin present. While this method has been successfully used for studying thiamine biosynthesis in both yeast and bacteria,[1–4] it suffers from the disadvantage of requiring large amounts of expensive radioactive material when radioactive tracer studies are made. The study of the biosynthesis of the pyrimidine moiety of thiamine is normally further hampered because the usual method of cleavage of the thiamine after its isolation[5] produces the sulfonated derivative of the pyrimidine, which is inactive in microbiological assays.

In the methods described below for *Salmonella typhimurium*, use is made of a system that removes the normal control of thiamine over its own biosynthesis,[6,7] so that the organisms form excess thiamine or the pyrimidine moiety. The pyrimidine is excreted into the culture supernatant solution as 4-amino-5-hydroxymethyl-2-methyl pyrimidine, which is active in microbiological assays.

Derepression of Thiamine Biosynthesis Using Adenosine Inhibition

Adenosine inhibits the biosynthesis of the pyrimidine moiety of thiamine,[8] and growth of organisms in the presence of adenosine therefore lowers the cellular concentration of thiamine. The normal repressive control by thiamine is thereby lost, and the thiamine-synthesizing enzymes become derepressed. If the inhibiting adenosine is then removed by washing, the bacteria are able to form thiamine very rapidly, even in the absence of further growth.

[1] M. J. Pine and R. Guthrie, *J. Bacteriol.* **78,** 545 (1959).
[2] S. David and B. Estramareix, *Biochim. Biophys. Acta* **42,** 562 (1960).
[3] G. A. Goldstein and G. M. Brown, *Arch. Biochem. Biophys.* **103,** 449 (1963).
[4] D. B. Johnson, D. J. Howells, and T. W. Goodwin, *Biochem. J.* **98,** 30 (1966).
[5] R. R. Williams, R. E. Waterman, J. C. Keresztesy, and E. R. Buchman, *J. Am. Chem. Soc.* **57,** 536 (1935).
[6] P. C. Newell and R. G. Tucker, *Biochem. J.* **100,** 512 (1966).
[7] P. C. Newell and R. G. Tucker, *Biochem. J.* **100,** 517 (1966).
[8] H. S. Moyed, *J. Bacteriol.* **88,** 1024 (1964).

In the following method, the tryptophan-requiring mutant T of *S. typhimurium* LT2 is used in order that bacterial growth may be limited after derepression.

An overnight culture of mutant T, growth of which has been restricted by a limiting concentration of glucose, is diluted to 0.035 mg/ml, dry weight, in 200 ml of minimal medium[9] containing glucose (0.2%) and DL-tryptophan (100 μg/ml). Then 300 μg of adenosine is added per milliliter, and the culture is aerated on a rotary shaker in a 1-liter flask for precisely 2.5 hours at 37°. Cells are then chilled and immediately centrifuged and resuspended twice in water at 2°. After one more centrifugation, the cells are added to flasks containing minimal medium without added tryptophan. The flasks are then incubated for about 1.5 hours; during this time the thiamine concentration rises from the normal 40 mμg/mg, dry weight, to about 200 mμg/mg, dry weight. The thiamine is found almost completely in the form of thiamine pyrophosphate and is located intracellularly. This procedure may be used for studying factors that affect the biosynthesis of the pyrimidine moiety of thiamine in organisms that are prototrophic for thiamine biosynthesis.[6]

Depression of Pyrimidine Moiety Biosynthesis Using Thiamine-Requiring Mutants

The principle of depleting cells of thiamine so as to derepress thiamine biosynthesis may also be applied to mutants that form only the pyrimidine moiety of thiamine.[10, 11] Such mutants, which possess a block in the thiazole moiety biosynthetic pathway or in the reactions joining the two moieties of thiamine, freely excreted the pyrimidine moiety in great excess into the culture supernatant solution when derepressed. With these mutants, addition of adenosine to the growing culture is unnecessary as they cannot synthesize thiamine, and derepression may be achieved by supplying the growing organisms with suboptimal amounts of exogenous thiamine.

The pyrimidine-excreting organisms are inoculated into 1-liter flasks each with 180 ml of minimal medium containing glucose (0.015%), plus (per flask) 40 mg of DL-tryptophan and 1.5 μg of thiamine. The flasks are incubated on a rotary shaker at 37° overnight. Owing to the low concentration of glucose in the medium, the cultures cease growing at about 0.2 mg/ml, dry weight. In the morning 60 ml of 2% (w/v) glucose solution is added to each flask, and the organisms are allowed to grow for a further 2–2.5 hours so that they reach 0.6–0.7 mg/ml, dry weight. The cultures are then centrifuged, and the bacteria are washed twice with water at 2°

[9] B. D. Davis and E. S. Mingioli, *J. Bacteriol.* **60**, 17 (1950).
[10] P. C. Newell and R. G. Tucker, *Biochem. J.* **106**, 271 (1968).
[11] P. C. Newell and R. G. Tucker, *Biochem. J.* **106**, 279 (1968).

and resuspended in 1-liter flasks with 200 ml of minimal medium containing glucose (0.6%) so that the flasks contain 0.8 mg, dry weight, of bacteria per milliliter. During incubation at 37° for 1.5 hours, the cultures then excrete the pyrimidine into the culture supernatant solution. If the flasks are aerated by shaking, about 50 mμg of pyrimidine per milliliter of supernatant solution is obtained. However, if the flasks are left stagnant, about 150 mμg of the pyrimidine is excreted per milliliter (Newell and Tucker, unpublished data).

Isolation and Purification of the Pyrimidine Moiety of Thiamine from Culture Supernatant Solutions

The supernatant solutions from pyrimidine-excreting mutants obtained as just described are concentrated *in vacuo* from 200 ml to 20 ml in a rotary evaporator at a temperature below 70°. About 3 ml of 1 N KOH is then added to bring the pH to 8–9, and the alkaline solution is extracted eight times with 20-ml portions of isoamyl alcohol (3-methylbutan-1-ol). The combined extracts are concentrated *in vacuo* to 1–1.5 ml at a temperature below 80°. The precipitated salt is removed by centrifugation at minus 10°, and the solution is extracted three times with 5 ml of 0.001 N HCl. The combined extracts are then evaporated to a final volume of 0.1 ml, and any precipitate formed is removed by centrifugation. About 0.01 ml of the concentrated supernatant solution is then applied as a spot to a thin-layer chromatography plate coated with a 250 μ-thick layer of MN cellulose powder 300 (Macherey, Nagel and Co., Duren, Germany). The plate is developed in butanol–acetic acid–water (63:10:27, v/v) for 3 hours, dried, and then developed in the second dimension in methanol–*n*-amyl alcohol–benzene–ammonia–water (70:35:70:8:17, v/v) for 2.5 hours. The pyrimidine moiety can be localized on the plate by autoradiography (if radioactive precursors are added to the pyrimidine-excreting mutants) or by bioautography using a specific pyrimidine-requiring mutant. In microgram quantities, the pyrimidine can be seen as a characteristic purplish absorbing spot under ultraviolet illumination.

Bioassay of Thiamine Using *Lactobacillus fermenti* 36 (ATCC 9338)

This organism is suitable for thiamine assay since it shows no response to the pyrimidine and thiazole moieties even at 100 times the normal assay concentration in a 20-hour incubation. The inoculum for the assay is an 18-hour culture in medium B2,[12] with oxoid liver extract instead of fresh extracts. The organisms are washed three times with sterile water, and

---

[12] R. H. Nimmo-Smith, J. Lascelles, and D. D. Woods, *Brit. J. Exptl. Pathol.* **29**, 264 (1948).

0.1 ml of a 1:20 dilution is used to inoculate each assay tube. The assay is carried out in stagnant cultures of 6.5 ml total volume in upright sterile tubes (16 mm × 150 mm) using the medium of MaciasR.[13] For the standard curve, thiamine concentrations between 2 and 30 mμg per assay tube are suitable, and growth of the organism after a 20-hour incubation is measured colorimetrically using a neutral density filter. Before assay, thiamine is extracted from cells by boiling for 5 minutes in 0.05 M sodium acetate buffer, pH 5.0. This is followed by a 4-hour incubation with Takadiastase (1 mg/ml) at 45° to convert thiamine pyrophosphate to thiamine.[14]

Bioassay of the Pyrimidine Moiety of Thiamine

This is performed using a mutant of *S. typhimurium* LT2 that requires the pyrimidine of thiamine or thiamine itself for growth. The organisms are grown in 6.5 ml of minimal salts medium[9] in 16 mm × 150 mm, plugged, sterile tubes that are tilted and shaken on a reciprocating shaker at 37° for 16 hours. The cell density is then measured colorimetrically using a neutral density filter. Since mutants requiring the pyrimidine of thiamine for growth also respond to thiamine, a correction must be made for the thiamine content of the material being assayed for pyrimidine. This can be done by using standard curves for the pyrimidine and for thiamine so that the response due to thiamine (measure independently with the *L. fermenti* 36 assay) can be subtracted from that given by a mixture of the two substances.

Bioautography of Thiamine and the Pyrimidine Moiety Using Thin-Layer Chromatograms

The most positive method of locating a spot containing a few millimicrograms of thiamine or the pyrimidine moiety on a thin-layer chromatogram uses the technique of bioautography. Bioautography[15] uses the ability of auxotrophic organisms to grow in a layer of agar adjacent to a chromatogram only in the spots where the chromatogram contains the organisms' special growth requirements. Areas of growth in the agar are readily demonstrated by adding triphenyltetrazolium chloride, which is reduced to the red formazan.[16]

The procedure of placing a chromatogram on the surface of solid-inoculated agar, as used for bioautography with paper chromatograms, cannot be used with glass-backed thin-layer chromatography plates, since it is difficult to obtain a perfectly flat upper surface to the agar when it is

[13] F. M. MaciasR, *Appl. Microbiol.* **5**, 249 (1957).
[14] H. W. Kinnersley and R. A. Peters, *Biochem. J.* **32**, 1516 (1938).
[15] W. A. Winsten and E. Eigen, *Proc. Soc. Exptl. Biol. Med.* **67**, 513 (1948).
[16] E. Usdin, G. D. Shockman, and G. Toennies, *Appl. Microbiol.* **2**, 29 (1954).

poured at just a few degrees above its solidifying temperature. The alternative of pouring melted inoculated agar onto the thin-layer chromatogram is also unsatisfactory since it disturbs the position of the water-soluble active spots. A convenient procedure that overcomes these difficulties,[11] is to pour the agar layer onto a clean glass plate the same size as the chromatography plate, and then to transfer the solid agar to the chromatogram as a sheet. This sheet is flexible, and its bottom side is perfectly flat owing to contact with the glass plate. With this procedure 200 ml of agar-containing medium is required per thin-layer plate (200 mm × 200 mm). The medium is a minimal salts medium[9] with glucose (0.2% w/v) and triphenyltetrazolium chloride (0.015%); the glucose and triphenyltetrazolium chloride are autoclaved separately and added as sterile solutions to the melted agar. The hot sterile agar is cooled to just 50° in a water bath and then the bacteria (about 5 mg, dry weight, in 2 ml of water) are warmed to 50° and inoculated into the agar. (For detection of the pyrimidine of thiamine, mutants of *S. typhimurium* LT2 that require the pyrimidine or thiamine for growth can be used.) The agar is then rapidly poured onto a clean thin-layer glass plate. A strip of adhesive tape around the edge of the plate prevents loss of agar from the edge, and after the agar has set, removal of the tape allows the agar sheet to be slid neatly off the glass and transferred to the dried chromatography plate. Incubation for 16 hours at 37° then produces bioautograms with small red spots over the active spots on the chromatogram. The method is sensitive to very small quantities of pyrimidine moiety, 1–2 mμg being easily detectable.

[31] The Biosynthesis of the Pyrimidine and Thiazole Moieties of Thiamine[1]

By RAYMOND V. TOMLINSON, PAUL F. TORRENCE,
and HOWARD TIECKELMANN

Introduction

Although the mechanism of the enzymatic formation of thiamine (I) from preformed pyramin (2-methyl-4-amino-5-hydroxymethylpyrimidine) (II) and thiazole [4-methyl-5-(β-hydroxyethyl)thiazole] (III) has been known for some years,[2] relatively little has been discovered regarding the

[1] This investigation was supported by Public Health Service Research Grant No. CA-07793 from the National Cancer Institute.
[2] T. W. Goodwin, "The Biosynthesis of Vitamins and Related Compounds," Academic Press, London, 1963.

biosynthetic origins of these two unique compounds. In fact, even their basic precursors cannot be considered as being unequivocally established. This situation results partly from the lack of any cell-free system capable of the complete or partial synthesis of either ring, in contrast to the availability of several such systems that will effect their condensation. Thus studies on the *de novo* formation of pyramin and thiazole have been restricted to the use of intact microbial cells. With few exceptions,[3,4] the most popular and productive approach has been to incubate the cells with suspected radioactively labeled precursors, isolate the thiamine and determine the incorporation of label (or lack of it) into the molecule. Despite the obvious disadvantages of such a complex experimental system, such studies have established a number of compounds as probable direct precursors of thiamine, although discrepancies do exist. For example, in a few cases, the origin of some atoms is attributed to more than one precursor. The total body of data nevertheless does suggest strongly that the pathway of pyramin biosynthesis is not similar to that of the pyrimidines of nucleic acids, nor is thiazole formation comparable to that of the penicillins or bacitracins.

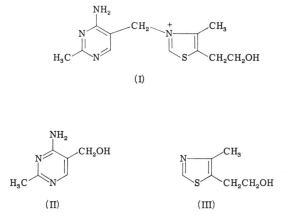

(I)

(II)   (III)

General Methodology

Experimental Technique

The microorganism is incubated in the appropriate minimal medium with the suspected thiamine precursor labeled with specific isotope (e.g., ^{14}C, ^{3}H, or ^{35}S). The cells are then harvested, washed, and lysed, usually by acid treatment. After enzymatic dephosphorylation of the thiamine

[3] P. C. Newell and R. G. Tucker, *Biochem. J.* **106**, 271 (1968).
[4] A. F. Dorio and L. M. Lewin, *J. Biol. Chem.* **15**, 3999, 4006 (1968).

esters, the labeled product (with or without "cold" carrier) is isolated by ion-exchange chromatography and then subsequently further purified by a combination of crystallization and chromatographic techniques. In many cases the thiamine is chemically cleaved into its component pyramin and thiazole moieties. Under these circumstances, extensive purification of the thiamine is omitted and rigorous purification is reserved for its cleavage products.

Criteria for Direct Incorporation

Specific Activities

When compounds containing a single label are used as substrates, two criteria have been employed to determine whether or not they are direct precursors of the thiamine molecule. One requires that the biosynthesized vitamin possess a specific activity comparable to that of the labeled precursor.[5] In the other, the total amount of isotope incorporated is compared among a number of candidate precursors to determine which is the most effective donor.[6] Neither is definitive, since such phenomena as metabolic randomization, endogenous synthesis, and transport properties may bias the results. It is, therefore, important that a variety of probes be utilized to ascertain whether incorporation from an isotopic precursor is truly indicative of direct participation in the biosynthesis.

Isotope Ratios

The presence of chemically and metabolically stable hydrogen and sulfur atoms in thiamine permits experiments testing incorporation from doubly-labeled substrates. The objectives in this approach are to show that both labels are incorporated and that the relationships of the isotope ratios on proceeding from precursor to product are of a nature consistent with the precursor's proposed role in a hypothetical biosynthetic scheme. Changes in $^{14}C:{}^{3}H$ ratios have been used to substantiate claims that acetate is directly involved in the biosynthesis of pyramin and that alanine and methionine are immediate precursors of thiazole.[6,7] The direct participation of methionine in thiazole biosynthesis has also been demonstrated by double labeling using ^{14}C and ^{35}S.[8]

Degradation

Degradation is a powerful tool for the study of a postulated scheme of biosynthesis. It cannot, however, be used to indicate directly whether or not an incorporation results from metabolic randomization. Moreover,

[5] P. Linnett and J. Walker, *J. Chem. Soc.*, Sec. *C* p. 796 (1967).

[6] R. V. Tomlinson, D. P. Kuhlman, P. F. Torrence, and H. Tieckelmann, *Biochim. Biophys. Acta* **148**, 1 (1967).

success requires the development of degradation schemes that will unequivocally isolate specific atoms of the thiamine. Such techniques have been described and employed to isolate various carbon atoms of pyramin,[8–10] as well as carbon 2 of thiazole.[5,11] Other procedures[12] for obtaining carbon 4 plus its methyl group, and the β-carbon of the 5-(β-hydroxyethyl) side chain of thiazole are described in this chapter. Degradation studies have found applicability in localizing the incorporation of carbon atoms from formate,[8–10] acetate,[6,7] and aspartate[6] in the biosynthesis of pyramin, as well as from formate,[5] glycine,[5] and methionine[11] in thiazole biosynthesis.

Specific Methodology

Incorporation of Radioactive Substrates into Thiamine by *Bacillus subtilis*[6]

A spore suspension (0.5 ml) of *Bacillus subtilis*, ATCC 6633, is incubated in 100 ml of modified Demain's medium[6,13] at 37° with shaking (140 rpm). This actively growing culture (0.5 ml) is used to inoculate a second 100 ml of medium, which is also incubated under the same conditions. If the experiment requires an extended exposure of the bacteria to an isotopic substrate, the culture is transferred to a reaction vessel at an absorbancy (660 mμ) of 0.1–0.2. In pulse-label experiments, the transfer is made when the absorbancy (660 mμ) is 0.55–0.70 ($\frac{2}{3}$ log phase growth). A 500-ml Erlenmeyer flask equipped with a $24/40$ ground glass joint is a suitable reaction vessel. The stopper is fitted with a large-bore stopcock to admit the labeled substrate, an inlet port extending to the bottom of the flask through which aeration is accomplished using compressed air, and an outlet port which is connected to a series of scrubbing towers to prevent contamination of the surrounding area by radioactive spray. The vessel is maintained at 37° in a constant-temperature water bath.

After several minutes' incubation under these conditions, an aqueous solution of the radioactive substrate is added. In extended labeling experiments, no "cold chaser" substrate is used, and the incubation is prolonged for 3–4 hours. However, in pulse labeling experiments, the addition of the radioactive substrate is followed in 10–20 minutes by the addition of "cold chaser" substrate (500- to 1000-fold excess); 20–30 minutes later the incubation is terminated.

[7] R. V. Tomlinson, *Biochim. Biophys. Acta* **115**, 526 (1966).

[8] D. B. Johnson, D. J. Howells, and T. W. Goodwin, *Biochem. J.* **98**, 30 (1966).

[9] H. Kumaoka and G. M. Brown, *Arch. Biochem. Biophys.* **122**, 378 (1967).

[10] S. David, B. Estramareix, and H. Hirshfeld, *Biochim. Biophys. Acta* **148**, 11 (1967).

[11] P. F. Torrence and H. Tieckelmann, *Biochim. Biophys. Acta* **158**, 183 (1968).

[12] P. F. Torrence and H. Tieckelmann, unpublished observations.

[13] A. L. Demain, *J. Bacteriol.* **75**, 517 (1958).

Isolation of Thiamine[6]

The bacteria are harvested by centrifugation in the cold, washed three times with cold 0.9% saline, resuspended in 5 ml of 0.1 N sulfuric acid, and autoclaved for 1 hour at 121°, 15 psi. The digest is cooled, its pH is adjusted to 4.5 with saturated barium hydroxide solution, and, after the addition of 100 mg of Takadiastase, the mixture is incubated at 50° for 3 hours.[14] At the end of the incubation period, further addition of saturated barium hydroxide solution is made to adjust the pH to 7.0, then the precipitate is removed by centrifugation. Carrier thiamine (30 mg) is added to the clear supernatant, the pH is readjusted to 7.0, and the solution is diluted to 15 ml. This solution is applied to a column (1.5 × 7.5 cm) of IRC-50 (H+) resin, and after the column has been washed with 15 ml of distilled water, the thiamine is eluted with 1 N hydrochloric acid. The thiamine-containing fractions are identified by their absorbancy at 267 mμ. The pooled eluate is concentrated *in vacuo* at room temperature, and if a solid product is desired, the concentrate is lyophilized in the presence of sodium hydroxide pellets.

Bisulfite Cleavage of Thiamine[15]

Thiamine (20–30 mg) is dissolved in 2.5 M sodium bisulfite solution (0.5 ml), and the pH of the solution is adjusted to 4.7 with 1 N sodium hydroxide. The mixture is heated in a boiling water bath for 30 minutes and then allowed to cool to room temperature. The pH is readjusted to 4.7 with 1 N sodium hydroxide if necessary, and the solution is permitted to stand for 24–48 hours at room temperature. Crystals of 2-methyl-4-aminopyrimidyl-5-methanesulfonic acid (IV) which form after several hours at 2° are recovered by centrifugation and then washed with 0.5 ml of cold water.

The washings and the supernatant are combined, the pH is adjusted to 10 with concentrated sodium hydroxide, and the resultant solution is extracted with nine 5-ml portions of chloroform. The chloroform extract is dried over solid anhydrous sodium sulfate, filtered, and then evaporated *in vacuo* at 40° leaving the thiazole as a viscous oil.

Purification Procedures for Thiamine, Pyramin, and Thiazole

A combination of the techniques of chromatography and crystallization can be used to purify thiamine and its component moieties. A selection of suitable chromatographic solvents is given in Table I.

[14] G. A. Goldstein and G. M. Brown, *Arch. Biochem. Biophys.* **103,** 449 (1963).
[15] R. R. Williams, R. E. Waterman, J. C. Keresztesy, and E. R. Buchman, *J. Am. Chem. Soc.* **57,** 536 (1935).

TABLE I
CHROMATOGRAPHIC DATA[a]

Solvent[b]	R_f thiamine	R_f pyrimidine–sulfonic acid	R_f thiazole
A	0.44	0.33[c]	0.74
B	0.23	0.10	0.88
C	0.83	0.91	0.85
D	0.21	0.14[c]	0.87
E	0.29[c]	0.12	0.87
F	0.20	0.13[c]	0.87
G	0.06	0.06	0.80
H	0.07	0.07	0.87
I	0.30	0.17	0.94
J	0.05	0.07	0.84
K	0.46	0.29	0.86
L	0.51	0.48	0.91

[a] Whatman No. 1 or 3 MM paper is used according to the amount of material available. All chromatograms are run by the descending technique and the compounds are visualized on the paper by ultraviolet light. Elution of thiamine, the pyrimidine–sulfonic acid and the thiazole is accomplished with water, 0.1 N ammonium hydroxide, and 0.1 N hydrochloric acid, respectively. Concentrations of the eluates are carried out by lyophilization using a Virtis Bio-dryer mounted on a Virtis Unitrap.

[b] Solvents: A, isopropanol–HCl–H_2O, 170:41:39 v/v[d]; B, n-butanol–3% NH_4OH, 4:1 v/v[e]; C, n-butanol–3% NH_4OH, 4:1 v/v (aq); D, n-butanol–acetic acid–water, 50:15:60 v/v (org)[f]; E, 95% ethanol–NH_4OH, 95.6:4.4 v/v[g]; F, n-butanol–acetic acid–water, 12:3:5 v/v[h]; G, n-butanol–2 N formic acid (org)[i]; H, n-butanol–water (org)[j]; I, n-propanol–acetic acid–water, 75:5:20 v/v[k]; J, isobutanol–formic acid–water, 100:20:45 v/v (org)[l]; K, n-butanol–ethanol–water, 8:2:1.3 v/v[m]; L, n-butanol–pyridine–water, v/v[n].

[c] Streaked.

[d] J. D. Smith and G. R. Wyatt, *Biochem. J.* **49**, 114 (1951).

[e] G. N. Kowkabany and H. G. Cassidy, *Anal. Chem.* **22**, 817 (1950).

[f] R. Robinson, *Nature*, **168**, 512 (1951).

[g] B. J. Stevens, I. Smith, and J. B. Jepson *in* "A Manual of Paper Chromatography and Electrophoresis" (R. J. Block, E. L. Durrum, and C. Zweig, eds.) 2nd ed., p. 93. Academic Press, New York, 1958.

[h] B. J. Stevens, I. Smith, and J. B. Jepson *in* "A Manual of Paper Chromatography and Electrophoresis" (R. J. Block, E. L. Durrum, and C. Zweig, eds.) 2nd ed., p. 319. Academic Press, New York, 1958.

[i] J. Gros and C. P. Leblond, *Endocrinology* **48**, 714 (1951).

[j] E. F. Mellon, A. H. Korn, and S. R. Hover, *J. Am. Chem. Soc.* **75**, 1675 (1953).

[k] J. R. S. Fincham, *Biochem. J.* **53**, 313 (1953).

[l] R. Allouf and R. Munier, *Bull. Soc. Chim. Biol.* **34**, 196 (1952).

[m] K. B. Augustinsson and M. Grahn, *Acta Chem. Scand.* **7**, 94 (1966).

[n] R. I. Morrison, *Biochem. J.* **53**, 474 (1953).

Thiamine, itself, may be crystallized most efficiently from absolute ethanol-concentrated hydrochloric acid; however, in the presence of gross impurities, trituration with absolute ether may be necessary to induce the crystallization.

The pyrimidinesulfonic acid, obtained as a product from the bisulfite cleavage of thiamine, crystallizes well from either a water or a water–acetone solution.

The thiazole is best purified by conversion to the crystalline picrate salt.[5] Thiazole (5–10 mg; obtained by evaporation of the chloroform extract of the bisulfite cleavage mixture) is dissolved in 2 ml of absolute ethanol, and to this is added an ethanolic solution of picric acid (50–60 mg, in excess). The solvent is evaporated *in vacuo* at 40°, then the yellow residue (thiazole picrate and unreacted picric acid) is scraped quantitatively from the flask and subsequently crystallized twice from ethyl acetate and twice from 1-butanol, thoroughly washing the product with anhydrous ether between each recrystallization. The picrate is now dissolved in 5 ml of absolute ethanol and the volume is diluted to 50 ml with ethyl acetate. This solution is applied to a column (1.0 × 15 cm) of alumina, and the eluted solvent is saved. Elution of the thiazole is continued with an additional 150 ml of ethyl acetate–ethanol (9:1). After evaporation of the total eluent *in vacuo* at 40°, the viscous residue of thiazole is dissolved in several drops of concentrated hydrochloric acid and then diluted to 2–3 ml with water. This acidic solution is lyophilized.

Incorporation of Label from Acetate-2-^{14}C and from Acetate-2-^{14}C, Me-^{3}H into the Methyl Group at Carbon 2 of Pyramin[6, 7, 16, 17] (Fig. 1, I → VII)

Two cultures of *B. subtilis* prepared as described above are labeled by exposing one to a 15-minute pulse of acetate-2-[14]C (0.5 mCi with S.A. of 29 mCi/millimole) and the other to a similar pulse of acetate-2-^{14}C, *Me*-[3]H (1.0 mCi with S.A. of 189 mCi/millimole). Following the procedure outlined earlier, 1 g of "cold" acetate is then added as a "chaser" and 30 minutes later the incubation is terminated. The [14]C- and ^{14}C, ^{3}H-labeled thiamine specimens are isolated and diluted with 100 mg and 200 mg, respectively, of "cold" carrier thiamine to minimize the loss of label in subsequent manipulation and purification.

After conversion of the thiamine-^{14}C to oxythiamine-[14]C (V), the vitamin is cleaved with α-mercaptoacetic acid, and the resultant pyrimidine thioether (VI) is purified. The specific activity (S.A.) of this product (VI)

[16] S. David and B. Estramareix, *Bull. Soc. Chim. Fr.* p. 2023 (1964).

[17] S. David, B. Estramareix, H. Hirshfeld, and P. Sinay, *Bull. Soc. Chim. Fr.* p. 936 (1964).

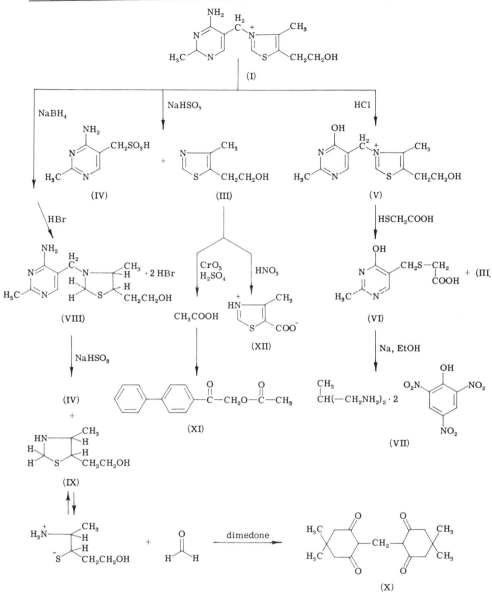

Fig. 1

is determined, and then it is degraded by reduction with sodium and alcohol to 1,3-diamino-2-methylpropane, which is isolated as the dipicrate salt (VII).[16,17] The radioactivity contents of both derivatives, (VI) and (VII), are determined by conventional procedures for beta counting using a low-background gas flow counter. The incorporation of ^{14}C into carbon-2 and the methyl group at carbon 2 of pyramin equals:

$$\frac{\text{S.A. (VI)} - \text{S.A. (VII)}}{\text{S.A. (VI)}} \times 100\%$$

The thiamine-^{14}C, ^3H is cleaved with bisulfite, and the pyrimidinesulfonic acid is purified by a combination of recrystallization and chromatography. The $^{14}C:^3H$ ratios of the acetate-2-^{14}C, Me-^3H precursor and the pyramin product (pyrimidinesulfonic acid) are determined after analyzing the radioactivity present by liquid scintillation spectroscopy.

Degradation of Thiazole

Carbon 2 of Thiazole[11,18] (*Fig. 1, I → X*)

Radioactive thiamine hydrochloride (20–30 mg, 60–90 micromoles), dissolved in 1.5 ml of water, is cooled to 2° in an ice bath and the pH of the solution is adjusted to 7.5–8.1 with sodium hydroxide (1 N and 0.1 N). This solution is now stirred continuously throughout all subsequent phases of the procedure. First, 0.5 ml of cold water containing a 2–3 M excess of sodium borohydride is added drop by drop. Then 30 minutes after this addition, the solution is removed from the ice bath and its temperature is allowed to rise to 25°. One hour later 0.4 ml of acetone is added to dissolve any tetrahydrothiamine which has precipitated and also to decompose the excess borohydride (which could interfere in subsequent procedures). Finally, 30 minutes after this addition, stirring is terminated and the mixture is lyophilized. If foaming occurs at any step, it is stopped by the addition of a drop of absolute ethanol.

The residue, consisting of tetrahydrothiamine and inorganic salts, is taken up in the minimum amount of hot (90–100°) distilled water. Upon cooling for 8 hours at 2°, crystals of tetrahydrothiamine form. These crystals are recovered by centrifugation, and the supernatant fluid (containing mostly inorganic salts) is removed and discarded. The moist crystals of product are dissolved in 1.0 ml of absolute ethanol, and the ultraviolet spectrum of an aliquot (5 μl) in ethanol is determined in order to check product yield and identity.[19] The ethanol solution is then evaporated, and the remaining water is removed by lyophilization.

[18] G. E. Bonvicino and D. J. Hennessy, *J. Am. Chem. Soc.* **79**, 6325 (1957).

[19] While both low-background gas flow Geiger-Müller and liquid scintillation counting has been employed in our work, we believe that except for highly colored substances, scintillation counting is the most advantageous, specifically with reference to the ease of sample preparation. All substances herein mentioned can be counted efficiently in a standard toluene fluor with the aid of an appropriate solubilizer (e.g., Nuclear Chicago NCS).

The tetrahydrothiamine is now taken up in the minimum quantity of concentrated hydrobromic acid (0.5–1.0 ml), and then 10–20 ml of absolute ethanol is added to the solution. Refrigeration of the mixture for 10 hours and trituration with absolute ether speeds the formation of crystals of tetrahydrothiamine dihydrobromide (VIII).

The tetrahydrothiamine dihydrobromide is recrystallized from absolute ethanol-ether, washed successively with absolute alcohol and ether, dried, and then redissolved in 1.0 ml of water. An aliquot of this solution is removed for determination of optical density and another for determination of radioactivity. The remaining aqueous solution is lyophilized. Crystallizations are repeated until the hydrobromide is at a constant ($\pm5\%$) specific activity (see Table II).

TABLE II
ULTRAVIOLET SPECTRAL DATA[a]

Compound	λ_{max}[b]	ϵ
Thiamine	267	9.0×10^3
2-Methyl-4-aminopyrimidyl-5-methane-sulfonic acid	267	7.2×10^3
Tetrahydrothiamine[c]	278	4.9×10^3
Tetrahydrothiamine dihydrobromide	242	1.02×10^4
Dimedone derivative of formaldehyde[c]	257	2.3×10^4
p-Phenylphenacylacetate[c]	285	2.06×10^4
4-Methylthiazole-5-carboxylic acid	258	5.9×10^3
4-Methyl-5-(β-hydroxyethyl)thiazole	253	4.0×10^3

[a] Since in most cases, the amount and condition of materials were unsatisfactory for accurate weighing, extensive use was made of ultraviolet absorption to determine quantities of material for various purification steps. This table lists some pertinent spectral data regarding thiamine, its component moieties, and various compounds in the degradation schemes.

[b] This need not represent the only absorption maximum of the indicated compound; it represents only that chosen for measurement. All spectra were run in phosphate buffer (pH 6.0) unless otherwise indicated.

[c] Spectra were determined in ethanol.

The purified tetrahydrothiamine dihydrobromide (15–30 mg) is dissolved in 2.5 M sodium bisulfite (0.3–0.5 ml), and the pH is adjusted to 4.7 with 0.1 N sodium hydroxide. After the mixture has been heated for 20 minutes at 80°, it is cooled to room temperature and allowed to stand for 10 hours. Subsequent refrigeration of the solution for 12 hours usually produces crystals of pyrimidinesulfonic acid. Occasionally, it is necessary to induce crystallization by scratching the inside of the test tube with a glass rod. The pyrimidinesulfonic acid crystals are centrifuged from the solution, washed once with cold water, and either purified for determination of radioactivity or discarded.

The wash water is combined with the supernatant solution containing the thiazole from the bisulfite cleavage reaction, and the pH of the mixture is adjusted to 8.0 with solid sodium carbonate. At this stage a pale pink color develops in the basic solution. The mixture is now extracted with eight 2-ml portions of ether. The ether is removed by evaporation with a stream of dry nitrogen, and the residue (IX) is dissolved in 1.0–1.5 ml of a solution of dimedone [60–80 mg in ethanol–water (1:1)]. After the addition of 1–2 drops of piperidine catalyst, the mixture is heated for 10 minutes at 80°, then water (5–6 drops) is added, and the solution is allowed to cool to room temperature. The crystals of the dimedone derivative (X) are centrifuged down and recrystallized to constant specific activity (ethanol–water, 1:1, followed by trituration with water). Based on thiamine, the yield of dimedone derivative is usually 10–20% (m.p. = 190–191°).

The percentage of ^{14}C at the 2-carbon position of the thiazole is calculated as

$$\frac{\text{S.A. (X)}}{\text{S.A. (VIII)} - \text{S.A. (IV)}} = \frac{\text{S.A. (X)}}{\text{S.A. (III)}} \times 100\%$$

Carbon-4 and the Methyl Group at Carbon 4
  of Thiazole[12] (Fig. 1, III → XI)

Labeled thiazole (23 mg, 161 micromoles), purified to constant specific activity, is dissolved in 10 ml of an ice-cold oxidizing solution,[20] prepared by the addition of 25 ml of concentrated sulfuric acid (specific gravity 1.84) to a solution of 16.8 g chromic anhydride (CrO_3) in 100 ml of distilled water. This solution is quickly transferred to a three-necked 25-ml round-bottom flask fitted with a 15-cm reflux condenser, a 10-ml capacity addition funnel, and a glass stopper in the third neck. The reaction mixture is heated at 145° for 1.5 hours in an oil bath and then is permitted to cool to 45°. The top of the reflux condenser is fitted with a water-cooled distillation condenser, after which the reflux condenser is drained of water and is used as a column for the steam distillation of acetic acid from the acidic reaction mixture.

Distilled water (10 ml) is added via the addition funnel, and 10 ml of water is distilled from the reaction mixture. This process is repeated until a total of 60 ml of distillate is collected.

The distillate is titrated against 0.0100 N sodium hydroxide delivered from a 25- or 50-ml buret to the end point at pH 8.0. A total of 0.15 meq

[20] E. J. Eisenbraun, S. M. McElvain, and B. F. Aycock, *J. Am. Chem. Soc.* **76,** 607 (1954).

(93%) of acetic acid is found in the distillate. A blank titration under identical conditions should also be performed.

The aqueous sodium acetate solution is then evaporated to dryness at 50° *in vacuo*, the residue is taken up in 2 ml of water, and the solution is lyophilized in a small test tube. The sodium acetate residue is dissolved in 0.4–0.5 ml of water, and the pH of the solution is adjusted to 7.0. After addition of 44 mg (160 micromoles) of *p*-phenylphenacyl bromide dissolved in 5 ml of absolute ethanol, the solution is refluxed for 1 hour. The mixture is then evaporated to dryness at 40° *in vacuo*, and the residue is extracted twice with 2-ml portions of chloroform–ethanol (3:1). The inorganic salts are removed by centrifugation, and the supernatant solution is applied to a 20 × 20 cm silica gel (with 2% phosphor) 1.1 mm-thick-layer plate. The chromatogram is developed in chloroform and dried. The *p*-phenyl-phenacyl acetate (XI) ($R_f = 0.55 \pm 0.05$) and unreacted *p*-phenylphenacyl bromide ($R_f = 0.85 \pm 0.05$) can be located by the use of ultraviolet light.[10]

The area containing the acetate is scraped off the plate, and the gel is extracted twice with 20-ml portions of chloroform and twice with 20-ml portions of hot ethanol. The combined chloroform and ethanol extracts are evaporated to dryness *in vacuo* at 40°, then the *p*-phenylphenacyl acetate is taken up in a minimum amount of chloroform and transferred to a small test tube. The chloroform is evaporated slowly by warming the test tube in a water bath while passing a stream of dry nitrogen over the surface of the liquid. The acetate derivative is recrystallized from hot ethanol–water to give 22 mg (93 micromoles, 58% based on thiazole) of colorless crystals (m.p. 109–111°,[21] mixture m.p. 109–111°). The percentage of ^{14}C in carbon 4 and the methyl group at carbon 4 of the thiazole may be expressed as

$$\text{S.A. (XI)/S.A. (III)} \times 100\%$$

The β-Carbon of the β-Hydroxyethyl Side Chain of Thiazole[12,22] *(Fig. 1, III → XII)*

Thiazole (20 mg, 139 micromoles), purified to constant specific activity, is dissolved in 2.0 ml of concentrated nitric acid, and the solution is heated at 40° for 48 hours. Nitrogen oxide vapors are given off during this time. The nitric acid is then removed *in vacuo* at 50°, the residue is taken up in water, and the solution is evaporated again. This procedure is repeated several times. The resulting semisolid residue is dissolved in the minimum

[21] F. Wild, "Characterisation of Organic Compounds," p. 172. Cambridge Univ. Press, London, England, 1962.

[22] E. R. Buchman, R. R. Williams, and J. C. Keresztesy, *J. Am. Chem. Soc.* **57,** 1849 (1935).

amount of water (1.0 ml) and the pH of the solution is slowly adjusted to approximately 3.0 with concentrated sodium hydroxide. Usually at this point, the product (XII) will precipitate from solution. The solution should be kept cool during the titration. The crystals are extracted from the solution with ether (6 times 5 ml). The ether extract is dried over anhydrous sodium sulfate, and the solvent is evaporated with the aid of a stream of dry nitrogen. The residue is recrystallized from the minimum amount of water until it has reached constant specific activity. The percentage of ^{14}C at the β-carbon of the side chain may be expressed as

$$[\text{S.A. (III)} - \text{S.A. (XII)}]/\text{S.A. (III)} \times 100\%$$

Conclusion

Despite the effort which has gone into attempts to elucidate the basic precursors for pyramin and thiazole biosynthesis, no unanimous agreement is yet apparent on their identities. Table III lists compounds which have

TABLE III
PROPOSED CARBON DONORS FOR THIAMINE BIOSYNTHESIS

Site	Source[a]
Pyramin	
Methyl group at carbon 2	Formate,[8] carbon 2 of acetate,[7] carbon 2 of glycine[3]
Carbon 2	Formate,[9] carbon 1 of glycine[3]
Carbon 4	Formate[10]
Hydroxymethyl group at carbon 5	Aspartate[6]
Carbons 5 and 6	Aspartate[6]
Thiazole	
Carbon 2	Methionine,[8,11] carbon 2 of glycine[5]
Carbon-4-methyl group	Alanine,[6,8] acetate[8]
Carbon 4	Alanine,[6,8] acetate[8]
Carbon-5-hydroxyethyl group	Methionine[6,8,12]
Carbon 5	Methionine[6,8,12]

[a] Superscript numbers refer to footnotes.

been proposed as carbon donors along with their sites of incorporation. Two of these, aspartic acid in pyramin biosynthesis and methionine in thiazole biosynthesis, would require novel transformation reactions to fulfill their roles. Definitive work to confirm the roles of these proposed precursors and the mechanisms of their transformations will probably require the development of systems that are less complex than those using whole organisms.

[32] The Use of Yeast to Assay the Separate Moieties of Thiamine

By Z. Z. ZIPORIN and P. P. WARING

Principle

Bakers' yeast (*Saccharomyces cerevisiae*) has the capability of joining the pyrimidine and thiazole moieties of thiamine to form vitamin B_1. The reaction is:

1 mole of pyrimidine (thiamine) + 1 mole of thiazole (thiamine) → 1 mole thiamine

The stoichiometry of the above equation establishes that the amount of thiamine formed is determined by the molarity of the two moieties. The one with the lesser molarity becomes the limiting factor with no more thiamine synthesized than is possible in the presence of this lesser molarity. The synthesized thiamine may then be assayed by the thiochrome or the microbiological method.

In applying this capability of yeast, it has been found possible to assay thiamine which has been altered by: (1) metabolism in the body (excreted in the urine)[1,2]; (2) chemical changes induced by irradiation[3]; (3) contact with alkaline media[4]; (4) deterioration of pharmaceutical products due to storage.[5] The use of the yeast resynthesis method has revealed information as to the events involved in the loss of thiamine under the conditions listed above. It has been possible to determine whether the vitamin has been destroyed by splitting the methylene bridge between the pyrimidine and thiazole moieties, or whether either of the entities has undergone further alteration.

Reagents, Equipment, and Incubation Medium

Reagents

Fleischmann's dry bakers' yeast. *Saccharomyces cerevisiae* sold as dried active bakers' yeast. Suspension is made by adding 6 g of yeast to 200 ml of water at 43°, and dispersing homogeneously throughout the medium.

[1] Z. Z. Ziporin, E. Beier, D. C. Holland, and E. L. Bierman, *Anal. Biochem.* **3**, 1 (1962).
[2] Z. Z. Ziporin, W. T. Nunes, R. C. Powell, P. P. Waring, and H. E. Sauberlich, *J. Nutr.* **85**, 287, 297 (1965).
[3] Z. Z. Ziporin, H. F. Kraybill, and H. J. Thach, *J. Nutr.* **63**, 201 (1957).
[4] Z. Z. Ziporin, unpublished data, 1952.
[5] Z. Z. Ziporin, Ph.D. Thesis, Georgetown University, Washington, D.C., 1953.

Pyrimidine.   2-Methyl-5-ethoxymethyl-6-amino-pyrimidine,   purchased from Merck, Sharp and Dohme Company, Rahway, New
Jersey. Weigh 495.8 mg of pyrimidine as equivalent to 1 g thiamine;
make up to 1000 ml with water made to pH 3 with dilute HCl.
Dilute an aliquot of the above solution with water acidified to pH 3
with HCl to obtain the desired concentration for assay.

Thiazole.   4-Methyl-5-β-hydroxyethylthiazole,   purchased   from
Merck, Sharp and Dohme Co., Rahway, New Jersey. Weigh
424.9 mg of thiazole as equivalent to 1 g of thiamine, make up to
1000 ml with water made to pH 3 with dilute HCl. Dilute an aliquot
of the above solution with water acidified to pH 3 with HCl to
obtain the desired concentration for assay.

Enzyme. Polidase S. Sold by Schwarz BioResearch, Inc., Orangeburg,
New York. Made up as 10% solution (w/v) in 2 N sodium acetate.

Decalso. Sold by Fisher Scientific Company as thiochrome Decalso
(Permutit T), 50–80 mesh. Prepared by pouring quantities of the
zeolite into hot water, stirring, allowing to settle, pouring off excess
water, and storing under water until used.

Isobutyl alcohol, boiling range 106.5–108°. Fluorometer reading
should not exceed 3 when the instrument is set to read 60–70 for
an isobutyl alcohol solution of thiochrome from 1 μg of oxidized
thiamine.

Equipment

Photofluorometer. Any of the standard instruments with input filters
having one transmission peak at 365–370 nm and output filters
with a peak at 435–445 nm.

Reaction vessel. Conical bottom, glass-stoppered centrifuge tube
having a 25–35-ml capacity.

Incubation Medium. Medium of Kline and Friedman[6] with 6 g of Difco
"vitamin-free casamino acids" per liter of solution A, in place of the
"acid-hydrolyzed casein solution." The composition of solutions A and
B is as follows (grams per liter):

SOLUTION A (1 liter)

$(NH_4)H_2PO_4$	180.0 g
$(NH_4)_2HPO_4$	72.0
Difco, vitamin-free casamino acids	6.0
Nicotinic acid	0.2
Pyridoxine HCl	0.004

[6] O. L. Kline and L. Friedman, *Biol. Symp.* **12**, 65 (1947).

SOLUTION B (1 liter)

Dextrose, anhydrous	200.0 g
$MgSO_4 \cdot 7H_2O$	7.0
KH_2PO_4	2.2
KCl	1.7
$CaCl_2 \cdot 2H_2O$	0.5
$FeCl_3 \cdot 6H_2O$	0.01
$MnSO_4 \cdot 4H_2O$	0.01

After preparation, the solutions are distributed into Erlenmeyer flasks, stoppered with a porous plug such as is used to prepare sterile solutions, and covered with an inverted beaker to prevent dust from contaminating the plug. The flasks are then autoclaved for 30 minutes at 15 psi on 3 successive days before use. After a flask has been opened for use, the remaining contents should be stored in the refrigerator.

Design of an Assay

Each assay should have: (1) "Reagent blank" containing solution A, solution B, and yeast equal to that contained in the assay flasks and water to provide the same final volume as the assay flasks; (2) a thiamine standard (20 μg) carried through the yeast resynthesis method to indicate the capability for recovering thiamine (95 $\pm$ 5% recovery expected); (3) a flask containing equimolar amounts of pyrimidine and thiazole (each equivalent to 20 μg of thiamine) also carried through the yeast resynthesis method, to demonstrate the yeast's capability for quantitative conversion of these moieties to thiamine (95 $\pm$ 5% recovery expected); (4) a flask containing the sample and components required for the incubation; (5) a flask containing the sample, an excess of pyrimidine (an amount equivalent to 100 μg of thiamine), as well as components required for the incubation; (6) a flask containing same as (5) above except that an excess of thiazole is added (an amount equivalent to 100 μg of thiamine). Flasks equivalent to (4), (5), and (6) are added for additional samples assayed within the same run. In calculating the concentrations of pyrimidine and thiazole moieties in the sample solutions, the amount of thiamine formed in the third flask (with standard pyrimidine and thiazole) is used as the basis for calculation, subtracting the thiamine found in flask 1 (reagent blank).

Procedure

An aliquot of the sample in a volume not exceeding 18 ml, estimated to contain between 5 and 40 μg of thiamine, or an equimolar amount of pyrimidine and/or thiazole, is added to each of three 125-ml Erlenmeyer flasks. After this, each flask will receive 2.5 ml of solution A, 7.5 ml of solution B, 10 ml of the yeast suspension (3% w/v), and sufficient water to bring the final volume to 40 ml. The suspension is swirled to distribute the contents homogeneously. The pH of this suspension has been found to vary between 5.7 and 5.9; this should be adjusted to 5.0 by use of HCl. The flasks are covered by inverting a beaker and are left to stand at room temperature with occasional swirling.

After 18 hours, the yeast is resuspended by rotation of the flask, and 8 drops of H_2SO_4–H_2O (1:1) are added along with 25 ml of 0.1 N H_2SO_4. The pH of this suspension should be 1–2. The flasks are steamed in an autoclave or heated on a steam bath for 30 minutes, then allowed to come to room temperature; 5 ml of a 10% solution (w/v) of Polidase S in 2 N sodium acetate is added, resulting in a pH of 4.0–4.5. Where necessary, additional drops of the sodium acetate are added to achieve the desired pH. The flasks are then incubated at 50° for 3 hours with frequent swirling. After they have been incubated and allowed to come to room temperature, the contents of each flask and water washings are transferred to 100-ml volumetric flasks and made up to volume with water. The latter flasks are shaken to provide a homogeneous dispersion of the contents and ca. 25–30 ml aliquots are withdrawn, placed in the 35-ml centrifuge tubes and centrifuged for 20 minutes at 2000 rpm. Measured aliquots from the supernatant solutions, estimated to contain 1–10 μg of thiamine are then added to a Decalso column, washed with water, eluted, and oxidized to thiochrome according to the procedure described by the AOAC.[7] By centrifuging the entire 100-ml volume after yeast incubation, it has been possible to chromatograph 95 ml of the supernatant. This method is employed with samples that are very low in thiamine.

Interpretation and Calculations

At the end of each assay, there are 3 values of thiamine content derived from the 3 flasks containing the assayed sample.

Flask 3, containing sample and incubation mixture, provides thiamine values which are limited by that moiety which exists in lesser concentration. Thiamine in the sample will also appear as thiamine in the values obtained from this flask. After subtracting the thiamine found in flask 1, the vitamin concentration determined by direct thiochrome assay on the sample must be subtracted from the yeast resynthesis method value to obtain concentrations of pyrimidine and thiazole in the sample.

Flask 4 containing sample, an excess of pyrimidine, and incubation mixture, provides a measure of the amount of thiazole in the sample. Again the thiamine content of the sample and the thiamine of flask 1 must be subtracted to obtain the concentrations of the vitamin moieties.

[7] Association of Official Agricultural Chemists, "Official and Tentative Methods of Analysis," 10th ed., pp. 758–760. Washington, D.C., 1965.

Flask 5 containing sample, an excess of thiazole and incubation mixture provides a measure of the amount of pyrimidine in the sample. As with the others, thiamine content of the sample and flask 1 must be subtracted.

Assuming three conditions, the following calculations may clarify the interpretation of the values obtained:

Condition 1. The thiamine has split into its separate moieties and there has been no destruction of pyrimidine or thiazole. Assuming 100 μg of thiamine at the start, the concentrations of pyrimidine and thiazole are equivalents of thiamine. Flask 4 has an excess of pyrimidine ($\cong$ to 100 μg thiamine) while flask 5 has an excess of thiazole added. The sample has been assayed by the yeast fermentation method.

Flask 3: 100 μg pyrimidine $+$ 100 μg thiazole $=$ 100 μg B_1
Flask 4: 200 μg pyrimidine $+$ 100 μg thiazole $=$ 100 μg B_1
Flask 5: 100 μg pyrimidine $+$ 200 μg thiazole $=$ 100 μg B_1

Condition 2. The thiamine (100 μg) has been split into its separate moieties and there has been a 50% destruction of pyrimidine.

Flask 3: 50 μg pyrimidine $+$ 100 μg thiazole $=$ 50 μg B_1
Flask 4: 150 μg pyrimidine $+$ 100 μg thiazole $=$ 100 μg B_1
Flask 5: 50 μg pyrimidine $+$ 200 μg thiazole $=$ 50 μg B_1

Condition 3. The thiamine (100 μg) has been split into its separate moieties and there has been 50% destruction of the thiazole.

Flask 3: 100 μg pyrimidine $+$ 50 μg thiazole $=$ 50 μg B_1
Flask 4: 200 μg pyrimidine $+$ 50 μg thiazole $=$ 50 μg B_1
Flask 5: 100 μg pyrimidine $+$ 150 μg thiazole $=$ 100 μg B_1

It is apparent that, in Condition 1, a sample containing equimolar quantities of pyrimidine and thiazole provides equal values in the 3 flasks assayed. Where there is destruction of one of the moieties this will be revealed in 2 of the 3 flasks having the same value while the third will be higher and will indicate the highest concentration of the moiety apparently not destroyed. Where thiamine is split and both moieties undergo degradation, the method cannot determine the extent of this degradation.

[33] Thiaminosuccinic Acid

By SABURO FUKUI and NOBUKO OHISHI

"Thiaminosuccinic acid" has been found as an intermediate in the transformation of oxythiamine to thiamine by a thiamineless mutant of

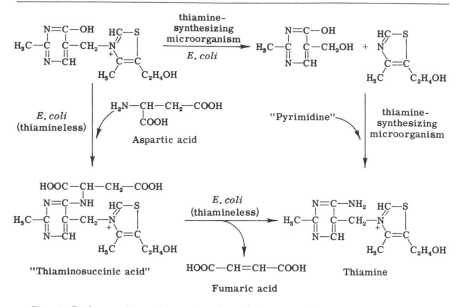

FIG. 1. Pathway of transformation of oxythiamine to thiamine by a thiamine-synthesizing microorganism and *Escherichia coli* thiamineless mutant.

Escherichia coli[1] (Fig. 1). It has the same activity as thiamine on the growth of such thiamine-requiring microorganisms as an *E. coli* thiamineless mutant and *Kloeckera brevis* (ATCC No. 9774).

Preparation Procedure

Chemical Synthesis. The diethyl ester of "thiaminosuccinic acid," 3-[2-methyl-4-(1,2-dicarboxyethylamino)-5-pyrimidylmethyl]-4-methyl-5-(2-hydroxyethyl) thiazole, is made according to the process shown in Fig. 2.

A solution of 5.2 g of 2-methyl-4-chloro-5-ethoxymethylpyrimidine (I), prepared according to Williams and Cline[2] and Cline *et al.*,[3] and 5.3 g of ethyl aspartate in 40 ml of benzene containing 4 ml of pyridine is heated in a sealed tube at 100–110° for 20–24 hours. The reaction mixture is washed with water, and the benzene phase is fractionally distilled under reduced pressure. 2-Methyl-4-(1,2-dicarboxyethylamino)-5-ethoxymethylpyrimidine (II) is distilled as a yellow oily substance at 145–160° and 150 mm Hg

[1] S. Fukui, N. Ohishi, S. Kishimoto, A. Takamizawa, and Y. Hamazima, *J. Biol. Chem.* **240**, 1315 (1965).

[2] R. R. Williams and J. K. Cline, *J. Am. Chem. Soc.* **58**, 1504 (1936).

[3] J. K. Cline, R. R. Williams, and J. J. Finkelstein, *J. Am. Chem. Soc.* **59**, 1052 (1937).

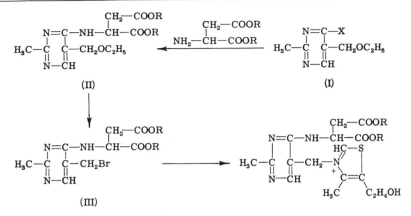

FIG. 2. Route of chemical synthesis of diethyl ester of "thiaminosuccinic acid."

with a yield of 5.8 g (62% of the theoretical amount). The R_f value of the compound during the paper chromatography developed with 1-butanol–acetic acid–water (40:10:50) is 0.75. Peaks in the infrared spectrum are observed at 3350, 1735, 1591, and 1190 cm^{-1}.

A solution of 4.2 g of compound (II) in 240 ml of acetic acid containing 10% hydrogen bromide is heated at 110–120° for 4 hours. After the reaction mixture is evaporated *in vacuo*, the resulting viscous residue of crude 5-bromoethyl derivative, compound (III), is washed with ether and allowed to react with the thiazole moiety of thiamine ("thiazole") without further purification. Compound (III), 6.7 g, is condensed with 20 g of "thiazole" by refluxing at 110° for 3 hours in 15 ml of isopropyl alcohol. The reaction mixture is evaporated under reduced pressure and a sticky residue is dissolved in ethanol. The alcohol solution, saturated with dry hydrogen chloride, is allowed to stand for 24 hours in an ice box and then evaporated to a viscous liquid *in vacuo*. Purification of the reaction product is carried out by cellulose column chromatography with 1-butanol–acetic acid–water (40:10:50) as a developing solvent.

The fraction containing the product is concentrated *in vacuo* and allowed to stand in an ice box after adding acetone until a slight turbidity results. A colorless crystalline product is obtained. On recrystallization from an ethanol–acetone mixture, the diethyl ester of 3-[2-methyl-4-(1,2-dicarboxyethylamino)-5-pyrimidylmethyl]-4-methyl-5-(2-hydroxyethyl) thiazolium bromide hydrobromide is obtained as colorless fine needles (m.p. 196°, dec.). The yield from compound (III) is about 20%.

Isolation from Cultured Broth of E. coli Thiamineless Mutant with Oxythiamine and Aspartic Acid. E. coli strain 70-23, originally obtained by Dr. B. D. Davis, is cultivated for 18 hours at 30° in 15 liters of Davis–

Mingioli's medium[4] containing 10 g of oxythiamine and 5 g of L-aspartic acid. The cultured broth is centrifuged at 5000 g for 10 minutes and purified by treatment with phenol according to Iacono and Johnson.[5] The concentrated aqueous phase contains thiamine, oxythiamine, aspartic acid, fumaric acid, small amounts of the pyrimidine and thiazole moieties of oxythiamine in addition to thiaminosuccinic acid. These compounds are separated by zone electrophoresis as follows. The concentrated aqueous phase is mixed with cellulose powder and placed in a slit cut in the support medium consisting of packed cellulose powder on a trough (33 $\times$ 1 $\times$ 5 cm). After electrophoresis for 3 hours at 12 V per centimeter in 0.1 M phosphate buffer, pH 5.7, each band of 1-cm width is extracted with distilled water and tested for ultraviolet absorption. The fraction containing thiaminosuccinic acid, positive to bromophenol blue reagent, ninhydrin reagent, and Dragendorff's reagent, stays near the origin. It is further purified by the phenol treatment and paper chromatography with a broad paper sheet [R_f 0.30 in 1-butanol–acetic acid–water (40:10:50)].The zone of thiaminosuccinic acid is extracted with 99% ethanol and concentrated *in vacuo*. About 100 mg of thiaminosuccinic acid is obtained as a hygroscopic, amorphous white powder by lyophilization of the concentrate.

The free acid thus obtained is dissolved in 50 ml of absolute ethyl alcohol previously saturated with dry hydrogen chloride. The solution is further treated with dry hydrogen chloride for 1 hour at 15° and then concentrated under reduced pressure to a viscous syrup. The syrup is streaked on a broad paper sheet and chromatographed by an ascending technique in 1-butanol–acetic acid–water (40:10:50). The zone of the diethyl ester (R_f 0.5–0.6) is extracted with absolute ethyl alcohol previously saturated with dry hydrogen chloride. The extract is concentrated to a viscous syrup under reduced pressure and allowed to stand in an ice box after addition of acetone until a slight turbidity occurs. A colorless crystalline product is obtained. On recrystallizing from an ethanol–acetone mixture, 22 mg of hydrochloride salt of diethyl ester of thiaminosuccinic acid is obtained as colorless fine needles (m.p. 196°, dec.).

Property

Ultraviolet absorption peaks of free thiaminosuccinic acid are 264 mμ at pH 3.0 and 267 mμ at pH 7.0, respectively. Diethyl ester of thiaminosuccinic acid shows an absorption maximum at 258 mμ at pH 4.0 and double peaks at near 240 mμ and 280 mμ in neutral and weak alkaline solutions. Their paper chromatographic behaviors are shown in the table.

[4] B. D. Davis and E. S. Mingioli, *J. Bacteriol.* **60,** 17 (1950).
[5] J. M. Iacono and B. C. Johnson, *J. Am. Chem. Soc.* **79,** 6321 (1957).

PAPER CHROMATOGRAPHIC BEHAVIOR OF THIAMINOSUCCINIC
ACID AND ITS DIETHYL ESTER

Solvent[a]	Thiamine (R_f)	Thiaminosuccinic acid (R_f)	Diethyl ester of thiaminosuccinic acid (R_f)
A	0.10	0.20	0.45
B	0.17	0.30	0.56
C	0.89	0.86	—
D	0.77	0.68	0.90
E	0.17	0.08	—

[a] Solvent A is water-saturated 1-butanol; solvent B, 1-butanol–acetic acid–water (40:10:50); solvent C, 95% ethanol–10% KCl aqueous solution (20:30); solvent D, 1-propanol–28% ammonia water–water (70:10:20); solvent E, ethyl acetate–pyridine–water (50:30:20). Ascending chromatography is for 15–17 hours.

[34] Formation of the Pyrophosphate Ester of 2-Methyl-4-amino-5-hydroxymethylpyrimidine by Enzymes from Brewers' Yeast in Thiamine Biosynthesis

By GENE M. BROWN

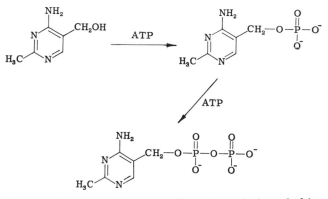

In order for the pyrimidine and thiazole moieties of thiamine to be converted enzymatically to the vitamin, these two compounds must first be "activated" by the formation of the pyrophosphate ester of hydroxymethylpyrimidine[1] and the monophosphoester of thiazole.[1] These two

[1] The following abbreviations are used: hydroxymethylpyrimidine, hydroxymethylpyrimidine-P, and hydroxymethylpyrimidine-PP for 2-methyl-4-amino-5-hydroxymethylpyrimidine, and the mono- and pyrophosphate esters of this compound, respectively; thiazole and thiazole-P for 4-methyl-5-(β-hydroxyethyl)thiazole and the phosphate ester of this compound; and thiamine-P for thiamine monophosphate.

"activated" compounds then can interact enzymatically with the displacement of inorganic pyrophosphate to yield thiamine monophosphate.[2-4] The enzymatic formation of the pyrophosphate ester of hydroxymethylpyrimidine proceeds in two steps with the intermediate formation of the monophosphoester of this pyrimidine.[5] The procedures for obtaining preparations of the two enzymes that are necessary to catalyze the formation of hydroxymethylpyrimidine-PP are described below.

Method of Detection

Principle. The method used to detect the formation of phosphate esters of hydroxymethylpyrimidine makes use of a mutant of *Aerobacter aerogenes* PD-1 that requires either thiamine or the pyrimidine moiety of thiamine for growth.[5] In order to separate hydroxymethylpyrimidine, hydroxymethylpyrimidine-P, and hydroxymethylpyrimidine-PP, samples containing these compounds are subjected to paper chromatography. The zones of migration on the chromatogram can then be located by bioautography with the *A. aerogenes* mutant as the test organism. Since this mutant cannot use the mono- and pyrophosphate esters of hydroxymethylpyrimidine, the developed chromatogram has to be sprayed with a phosphatase solution and incubated a short period of time to dephosphorylate the esters on the paper before the chromatogram is used for bioautography. The details of these procedures are described below.

Reagents

> Hydroxymethylpyrimidine (obtained from Merck and Co.), 0.58 mM
> MgCl$_2$, 100 mM
> ATP, 10 mM
> Phosphate buffer, 1 M, pH 7.0
> Purified alkaline phsophatase from *Escherichia coli* (Worthington),
>   2 μg/ml (0.1 M Tris, pH 8.5)

Procedure. To 0.1 ml of buffer are added 0.1 ml each of MgCl$_2$, ATP, and hydroxymethylpyrimidine solutions and enzyme preparation not to exceed 0.6 ml. Enough water is added to give a final reaction volume of 1.0 ml. Incubation is for 3 hours at 38°, after which a 3-μl sample is withdrawn and spotted 3 cm from the bottom edge of a sheet of Whatman No. 1 paper (20 cm wide and 30 cm long). The chromatogram is developed by the ascending technique with isobutyric acid–NH$_4$OH–water (198:3:99,

---

[2] G. W. Camiener and G. M. Brown, *J. Biol. Chem.* **235**, 2404 (1960).
[3] G. W. Camiener and G. M. Brown, *J. Biol. Chem.* **235**, 2411 (1960).
[4] I. G. Leder, *J. Biol. Chem.* **236**, 3066 (1961).
[5] L. M. Lewin and G. M. Brown, *J. Biol. Chem.* **236**, 2768 (1961).

by volume) as the solvent system. The developed chromatogram is dried and is then sprayed with a solution containing alkaline phosphatase (2 μg/ml) in 0.1 M Tris buffer, pH 8.5. The wet chromatogram is then incubated at 38° for 1 hour in a container saturated with water vapor. The chromatogram is then dried and used to prepare a bioautogram.

Bioautography. Bacterial growth medium is prepared to contain the following components (in amounts per liter): K_2HPO_4, 21 g; KH_2PO_4, 9 g; sodium citrate · $2H_2O$, 1.0 g; $MgSO_4$, 0.1 g; ammonium sulfate, 2 g; mannitol, 20 g; thiamine-free yeast extract (prepared as described by Niven and Smiley[6]), 10 g; calcium pantothenate, 1 mg; nicotinimide, 1.0 mg; riboflavin, 1.0 mg; p-aminobenzoic acid, 2.0 mg; pyridoxine, 4.0 mg; folic acid, 0.2 mg; and biotin; 0.02 mg. To a 200 ml-amount of this medium is added 4 g of Bacto agar (to give a 2% solution) and the medium is sterilized by autoclaving for 10 minutes. After the medium has cooled to 45–50°, it is seeded with the mutant of *A. aerogenes* by the addition of 1 ml of a washed suspension of cells that is only faintly turbid to the eye (absorbancy reading of 0.1). The cells are evenly distributed in the medium by gentle swirling, and the medium is then poured into a sterile 22 × 30 cm Pyrex baking dish to give a layer approximately 0.5 cm thick in the dish. The medium is allowed to cool to form a solid layer; after this the dried chromatogram, which has been treated with phosphatase, is carefully placed in contact with the surface of the solid layer of medium. The chromatogram is allowed to remain on the medium for approximately 5 minutes before it is removed. The dish is covered and the bioautogram is incubated overnight at 38°. Inspection of the solid medium shows that growth zones of turbid areas on the surface correspond to the zones of migration on the chromatogram of hydroxymethylpyrimidine (R_f of 0.75), hydroxymethylpyrimidine-P (R_f of 0.45) and hydroxymethylpyrimidine-PP (R_f of 0.25). Although this assay is not quantitative, some notion of the relative amounts of compounds present in the aliquot spotted on the chromatogram can be gained by the sizes of the growth zones.

The inoculum used to seed the medium for bioautography consists of a suspension of washed cells (washed with sterile 0.9% NaCl) taken from a 24-hour culture grown on a nutrient agar slant (10 ml). Stocks of this culture are carried in nutrient agar slants, and the organism should be transferred at weekly intervals to maintain vigorous cultures.

Preparation of Enzymes[3]

Cell-Free Extracts. The method is a modification of that devised originally by Meyerhof.[7] A mixture of 11 kg of brewers' yeast (Anheuser-Busch)

[6] C. F. Niven, Jr., and K. L. Smiley, *J. Biol. Chem.* **150**, 1 (1943).
[7] O. Meyerhof, *Biochem. Z.* **183**, 176 (1927).

and 1200 ml of toluene is placed in a 45° water bath and continuously stirred until it reaches a temperature of 38°. The suspension is then allowed to stand at room temperature (20–25°) for 2.5 hours with periodic stirring. The mixture is immersed in an ice bath to cool it to 5–8°; then 2400 ml of cold (4°) phosphate (K salts) buffer, pH 7.0, is added, and the suspension is adjusted to pH 7.0. This material is allowed to remain overnight at 4°; the aqueous (bottom) layer is then carefully siphoned off and centrifuged at 105,000 g for 2 hours in a Spinco ultracentrifuge. The resulting cloudy extract is clarified by filtration with the aid of Johns-Manville Hyflo-Supercel. For this purpose, the Supercel is added to the extract in an amount of 100 g per liter; after the suspension is thoroughly mixed, it is filtered on a Büchner funnel. The resulting clear solution (called the "crude extract") can be frozen and stored until needed, but should not be repeatedly thawed and refrozen, since this results in a loss of activity of the enzymes.

Fractionation with Ammonium Sulfate. The following operations are carried out at 4°. To a 210-ml portion of the crude extract is added (slowly, with stirring) enough of a saturated solution of ammonium sulfate to give a solution 58% saturated with the salt. The resulting precipitated protein is removed by centrifugation, and more ammonium sulfate solution is added to the supernatant solution until the solution is 65% saturated. The insoluble protein is recovered by centrifugation and dissolved in approximately 50 ml of 0.02 M phosphate buffer, pH 7.0. This solution is dialyzed against two successive (6 hours each) 2-liter portions of the same buffer which also contains 0.02 M cysteine to prevent enzyme inactivation. This 58–65% ammonium sulfate fraction contains most of the activity for formation of hydroxymethylpyrimidine-PP from hydroxymethylpyrimidine, but only 12% of the total protein of the crude extract.

Fractionation on DEAE-Cellulose. A DÉAE-cellulose column (2.2 × 35 cm) is prepared and equilibrated with a buffer solution (pH 7.0) containing 0.005 M each of phosphate, EDTA, and cysteine. The 58–65% ammonium sulfate fraction described above (50 ml, 1.5 g of protein) is dialyzed for 6 hours against 2 liters of the same buffer and then applied to the column. The column is washed by allowing 200 ml of the phosphate–EDTA–cysteine buffer to pass through at a rate of 2.5 ml per minute. These and all subsequent operations are carried out at 4°. The column is then developed by logarithmic gradient elution with a solution containing sodium chloride and cysteine. For this purpose, the mixing chamber initially contains 1000 ml of 0.005 M NaCl in pH 7.0 buffer (0.005 M each with respect to phosphate, EDTA, and cysteine). The reservoir initially contains 1500 ml of 0.13 M NaCl in pH 7.0 buffer (0.005 M phosphate, 0.005 M EDTA, and 0.04 M cysteine). After the solution in the reservoir is depleted, the column is washed finally with 100 ml of 0.12 M NaCl contained in the phosphate–

EDTA–cysteine (0.04 M) buffer. Fractions of 10 ml each are collected throughout the elution process.

The fractionation on DEAE-cellulose results in the separation of the enzymes necessary for the synthesis of thiamine monophosphate into two fractions. Fractions 40 through 100 from the column contain the enzymes that catalyze the conversion of hydroxymethylpyrimidine to hydroxymethylpyrimidine-P and the latter compound to hydroxymethylpyrimidine-PP. Fractions 120 through 180 contain the enzymes that catalyze the conversion of thiazole to thiazole-P and the formation of thiamine-P from thiazole-P and hydroxymethylpyrimidine-PP.

Properties

UTP, CTP, and GTP can all replace ATP as a phosphate donor for the enzymatic production of hydroxymethylpyrimidine-P from hydroxymethylpyrimidine, although only ATP can be used for the conversion of hydroxymethylpyrimidine-P to hydroxymethylpyrimidine-PP. The enzyme that catalyzes the formation of the phosphomonoester is stable to heating for 10 minutes at 55° and to treatment with 1 mM p-hydroxymercuribenzoate, whereas the enzyme that catalyzes the conversion of the monoester to hydroxymethylpyrimidine-PP is labile to both of these treatments. For the formation of the pyrophosphate ester either Mg^{2+} or Mn^{2+} must be supplied. Although the two active enzymes have not been separated, these observations provide convincing evidence that two enzymes are necessary for the formation of the pyrophosphoester of hydroxymethylpyrimidine.

[35] Thiamine Monophosphate Pyrophosphorylase (Crystalline) (2-Methyl-4-amino-5-hydroxymethyl-pyrimidine-pyrophosphate: 4-Methyl-5-(2′-phospho-ethyl)-thiazole-2-methyl-4-aminopyrimidine-5-methenyltransferase, EC 2.5.1.3)

By IRWIN G. LEDER

Hydroxymethylpyrimidine-PP[1] + thiazole-P $\rightleftharpoons$ thiamine-P + PP$_i$

A partial purification of this enzyme has been described.[2]

Assay

The thiamine phosphate formed is oxidized to thiochrome phosphate and determined fluorimetrically.[3]

[1] The following abbreviations are used: hydroxymethylpyrimidine-PP—(2-methyl-4

Reagents

ENZYME INCUBATION

Glycine buffer, 1 M, pH 9.2
$MgSO_4$, 0.01 M
2-Mercaptoethanol, 0.05 M
Hydroxymethylpyrimidine-PP, 0.01 M
Thiazole-P, 0.01 M
Enzyme, diluted with 0.005 M 2-mercaptoethanol

DETERMINATION OF THIAMINE-P

Potassium acetate, 4 M
Trichloroacetic acid, 10%
Potassium ferricyanide, 0.0038 M in sodium hydroxide, 7.0 M
H_2O_2, 0.06% in sodium dihydrogen phosphate, 5.5 M

Procedure. ENZYME INCUBATION. A mixture containing 50 μl of buffer, 25 μl of $MgSO_4$, 10 μl of 2-mercaptoethanol, 10 μl of hydroxymethylpyrimidine-PP, 10 μl of thiazole-P, and sufficient water to make the final volume (including enzyme) 0.5 ml, is equilibrated at 37°, and the reaction is started by the addition of enzyme. Where necessary, the enzyme is diluted with 0.005 M 2-mercaptoethanol. After 5 minutes, 0.5 ml of 10% trichloroacetic acid is added to stop the reaction. A corresponding mixture inactivated at zero time is included in each assay.

THIAMINE-P DETERMINATION. To a 0.1-ml aliquot of the acid filtrate, 0.2 ml of potassium acetate is added, followed by 0.1 ml of ferricyanide solution and, after mixing, 0.1 ml of hydrogen peroxide. The solution is diluted with 0.5 ml of water and the fluorescence is determined.

Definition of Unit and Specific Activity. A unit of enzyme activity is that amount of enzyme which produces 1 micromole of thiamine-P in 1 minute at 37°. Specific activity is defined in terms of units of enzyme per milligram of protein. Protein concentration is determined by the method of Lowry et al.[4]

---

amino-5-pyrimidinyl)methylpyrophosphate; thiazole-P—β-(4-methyl-5-thiazolyl) ethyl phosphate.

[2] I. G. Leder, *J. Biol. Chem.* **236,** 3066 (1961).

[3] H. B. Burch, O. A. Bessey, R. H. Love, and O. H. Lowry, *J. Biol. Chem.* **198,** 477 (1952).

[4] O. H. Lowry, N. J. Rosebrough, A. L. Farr, and R. J. Randall, *J. Biol. Chem.* **193,** 265 (1951).

Purification Procedure

Source of Yeast. Producers of bakers' yeast have, in recent years, made a practice of adding thiamine to the growth medium. This results in repressed levels of enzyme of less than 15% of that found when the vitamin is omitted. The yeast in these studies, cultivated without added thiamine, was obtained through the courtesy of Anheuser-Busch, Inc., St. Louis, Missouri. The fresh yeast is crumbled into liquid nitrogen and thawed for immediate use, or it may be stored in the deep freeze for at least 1 year with no significant loss in activity. Except where noted, all operations are carried out at 0–4°. Large volumes are centrifuged at approximately 11,000 *g* and small volumes at 35,000 *g*. Except where noted, all enzyme fractions are dissolved in 0.005 *M* 2-mercaptoethanol–0.005 *M* EDTA, pH 7 (ME-EDTA solution).

Step 1. Preparation of Crude Extract. The yeast, 1.5 kg, is allowed to thaw, whereupon it liquefies. To each kilogram of yeast, 160 mg of the protease inhibitor,[5] phenylmethylsulfonyl fluoride, dissolved in 10 ml of *n*-propanol, is added and the yeast is stirred at room temperature for 24–30 hours. The mixture is diluted with one-half volume of water and centrifuged for 30 minutes; the sediment is discarded. This simple procedure is readily applied to Anheuser-Busch yeast, but not necessarily to other commercial yeasts. Extracts may also be prepared by passing the thawed and diluted yeast, cooled to 2°, through a Gaulin press or by stirring in the cold for a minimum of 4 days. The purification procedure has been successfully applied to all three types of extracts.

Step 2. Heat Coagulation of Inactive Protein. The crude extract (1270 ml), is heated to 55° in 500-ml batches by stirring vigorously in a boiling water bath. The mixture is maintained at that temperature for 2.5–3 minutes then rapidly cooled to below 40° and placed in ice while further batches are treated. The combined heat-treated suspension is chilled to 5°, and the precipitate is removed by centrifugation.

Step 3. Removal of Nucleic Acid. The heat-treated supernatant solution is adjusted to pH 6.5 with 5 *N* NH₄OH and treated with 2.5% protamine sulfate (pH 6.5) until no further precipitate is formed. The volume required varies from 3 to 5% of the crude extract. A large excess of protamine should be avoided. The suspension is centrifuged, and the supernatant fluid is fractionated with ammonium sulfate.

Step 4. First Ammonium Sulfate Fractionation. Solid ammonium sulfate (29.1 g/100 ml) is added to bring the solution to 50% saturation. The addition should require no more than 30 minutes, and the suspension is stirred an additional 40 minutes before centrifuging. The precipitate is

[5] D. E. Fahrney and A. M. Gold, *J. Am. Chem. Soc.* **85,** 997 (1963).

discarded, and the supernatant solution is brought to 60% saturation by the addition of ammonium sulfate (6 g/100 ml) over a period of 1 hour. The suspension is stirred for an additional hour before centrifuging, or it may be kept overnight in the cold room. The precipitate is dissolved in 100 ml of ME-EDTA (ammonium sulfate I).

Step 5. Dialysis and Protamine Precipitation. The success of this step depends upon the thorough dialysis of the enzyme to establish a very low ionic strength. The enzyme is dialyzed with internal stirring against approximately 60 volumes of ME-EDTA. The dialyzate is changed three times and finally replaced by two changes of 0.005 M 2-mercaptoethanol. The dialyzed enzyme (200 ml) is treated serially with 0.4-ml portions of protamine sulfate (2.5%, pH 6.5). The temperature of the enzyme solution must be maintained at 0–2° throughout this operation. After each addition, the suspension is centrifuged and the supernatant solution is removed and assayed. Usually no enzyme is removed by the first two additions, and less than 5% of the total activity remains in the supernatant solution after the addition of 1.5–2% of the volume of the dialyzed enzyme. The precipitates containing enzymatic activity are individually extracted with two 10-ml portions of 0.2 M potassium phosphate buffer, pH 7.6, and the active fractions are combined. The combined extracts are diluted with additional buffer to adjust the protein content to approximately 10 mg/ml (protamine extract). The enzyme suffers no inactivation when stored overnight in the deep freeze at this point.

Step 6. Second Ammonium Sulfate Fractionation. Solid ammonium sulfate (31 g/100 ml) is added to bring the protamine extract to 52–53% saturation. The precipitate, collected by centrifugation, is dissolved in ME-EDTA to approximately 5 mg protein per milliliter and solid ammonium sulfate is added until, at approximately 42% saturation, the solution becomes definitely turbid. The suspension is stored overnight in the cold room. The next day the suspension is allowed to reach room temperature and then centrifuged for 20 minutes with the centrifuge temperature set for 25°. The temperature of the enzyme suspension may reach 30–38° during the centrifugation with no loss in activity. The enzyme is dissolved in ME-EDTA (ammonium sulfate II).

Step 7. DEAE-Sephadex Chromatography. The ammonium sulfate fraction from step 6 is dialyzed against 0.05 M K phosphate buffer, pH 6.5, containing 0.003 M 2-mercaptoethanol. The dialyzed enzyme is applied to a column (1.5 × 20 cm) of DEAE-Sephadex A50 previously equilibrated with the same phosphate–mercaptoethanol buffer. Elution is carried out at 4–5 drops per minute with 50 ml of buffer and then with a linear gradient consisting of 350 ml of buffer in the mixing flask and an equal volume of buffer containing 0.5 M KCl in the reservoir. Fractions of 5 ml are collected.

Two main 280-mμ absorbing peaks appear in the eluate. All the enzymatic activity is associated with the second and major peak, which appears at approximately 0.14 M KCl.

Step 8. Concentration and Crystallization. If the protein concentration is less than 0.5 mg/ml, it is concentrated by membrane ultrafiltration. To precipitate the enzyme, the combined DEAE-Sephadex fractions are brought to 65% saturation by the addition of solid ammonium sulfate (40 g/100 ml). The precipitate is dissolved in ME-EDTA at 5–10 mg of protein per milliliter, and sufficient solid ammonium sulfate is added (27 g/100 ml) to reach 47% saturation. Crystals may begin to appear at approximately 42% saturation. Crystallization is allowed to proceed overnight in the cold, and the crystals are collected by centrifugation the next day. If less than 90% of the activity is precipitated, the yield may be increased by warming the enzyme suspension to room temperature and centrifuging at 20°. Recrystallization is readily achieved from ME-EDTA solution by the addition of solid ammonium sulfate as described above. The crystals are thin needles measuring approximately 3–5 μ in the long axis. Overall yields of from 20–30% have been obtained. The purification procedure is summarized in the table.

PURIFICATION PROCEDURE

Fraction	Volume (ml)	Total units	Protein (mg/ml)	Specific activity ($\times 100$)	Recovery (%)
1. Crude extract	1270	54	50	0.08	100
2. Heated extract	1160	41	26.7	0.13	76
3. Ammonium sulfate I	104	40	96	0.4	74
4. Protamine extract	16	32	23.7	8.4	59
5. Ammonium sulfate II	4	25	6.5	96	46
6. DEAE-Sephadex A-50	80	17	0.21	100	31
7. First crystals	2	14	6.3	111	26
Second crystals	2	13	5.8	112	24

Properties

Specificity. The enzyme is specific for thiazole-P and hydroxymethyl-pyrimidine-PP. Neither the nonphosphorylated thiazole, nor its pyrophosphate can replace the monophosphate, nor can the corresponding hydroxymethylpyrimidine-monophosphate substitute for the pyrophosphate. The enzyme does not catalyze a pyrophosphate exchange reaction between inorganic pyrophosphate and the hydroxymethylpyrimidine-PP unless thiazole-P is added to complete the reaction mixture.

Activators and Inhibitors. The enzyme is stimulated by 0.001 M MgCl$_2$

and to a lesser extent by Mn^{2+}. Inorganic pyrophosphate (10^{-4} M) inhibits the reaction noncompetitively. p-Mercuribenzoate (10^{-5} M) causes complete and irreversible inactivation, but 10^{-3} M iodoacetic acid or iodoacetamide are without effect. Although dilutions prepared with 0.1% albumin or 0.005 M 2-mercaptoethanol are fully active, the enzyme is inactivated when diluted with quartz-distilled water.

Studies with relatively crude yeast extracts indicate[6] that the enzyme is inhibited in a partially competitive manner by pyridoxal phosphate.

Kinetic Properties and Equilibrium. The Michaelis constants, determined at 25° and pH 9.2, are 7×10^{-6} M for thiazole-P and 1×10^{-6} M for hydroxymethylpyrimidine-PP. A K_m for hydroxymethylpyrimidine-PP of 3.7×10^{-5} M, determined at pH 7.4 with a relatively crude enzyme preparation, has been reported.[6]

Stability and pH Optimum. The crystalline enzyme stored as a suspension in 60% saturated ammonium sulfate lost less than 10% of its activity in 2 months. Maximal activity is exhibited at pH 9.2 with half this rate at pH 8.

Other Properties. The enzyme migrates as a single, slow-moving component in disc gel electrophoresis. Preliminary studies in the ultracentrifuge indicate an $S_{20,w}$ of 11.3 and a molecular weight, determined by short-column sedimentation equilibrium and assuming a partial specific volume of 0.73 cm^3/g, of 337,000.

[6] L. M. Lewin and G. M. Brown, *Arch. Biochem. Biophys.* **101,** 197 (1963).

[36] Degradation of the Pyrimidine Moiety of Thiamine for the Use of Radioactive Determination of Individual Carbon Atoms

By S. DAVID, B. ESTRAMAREIX, H. HIRSHFELD, and P. SINAŸ

Bisulfite cleavage[1] of thiamine quantitatively yields 4-amino-2-methylpyrimidine-5-methylsulfonic acid. This is converted to the more easily purifiable 4-hydroxy derivative, which is then oxidized by the Kuhn-Roth method, giving C-2 and C-8 as acetic acid.

Thioglycolic acid cleavage[2] of oxythiamine yields 4-hydroxy-2-methylpyrimidine-5-methylthioacetic acid (I). After purification, this compound can be treated by two different methods:

[1] R. R. Williams, R. E. Waterman, J. C. Keresztesy, and E. R. Buchman, *J. Am. Chem. Soc.* **57,** 535 (1935).
[2] G. E. Bonvicino and D. J. Hennessy, *J. Org. Chem.* **24,** 451 (1959).

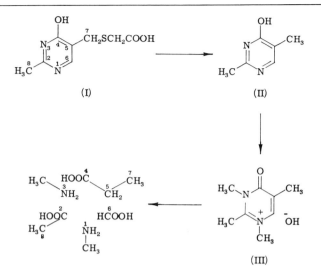

FIG. 1. Degradation of the pyrimidine moiety of thiamine.

1. Reduction by sodium[3] in alcohol gives 1,3-diamino-2-methylpropane, which can be oxidized by the Kuhn-Roth method[4] to give acetic acid from C-5 and C-7.

2. Raney nickel reduction[2] gives 4-hydroxy-2,5-dimethylpyrimidine (II), which can be exhaustively methylated[5] and the product (III) hydrolytically cleaved according to the scheme (Fig. 1) shown, thus giving information about all the carbon atoms.

In the case when 4-amino-5-hydroxymethyl-2-methylpyrimidine (pyramine) is directly available, it is converted to the 4-hydroxy analog which reacts with thioglycolic acid to give the same pyrimidine-thioacetic acid as above.

Isolation as Pyrimidine Sulfonic Acid

Reagents

Commercial sodium bisulfite solution (*D*, 1.24) brought to pH 5 by
5 *M* NaOH solution
NaOH, 1 *M*
HCl, 1 *M* and 6 *M*

[3] S. David, B. Estramareix, H. Hirshfeld, and P. Sinaÿ, *Bull. Soc. Chim. Fr.* p. 936 (1964).
[4] R. Kuhn and H. Roth, *Ber.* **66**, 1274 (1933).
[5] S. David and H. Hirshfeld, *Bull. Soc. Chim. Fr.* p. 527 (1966).

Acetic acid, 2 M
Ethanol
Anion-exchange resin, Dowex 2, 200–400 mesh, strongly basic

Procedure. Thiamine (100 mg) is dissolved in water (0.3 ml) in a 3-ml centrifuge tube. Sodium bisulfite solution (1.5 ml) is then added. The tube is stoppered and left 3 days at 50°. The 4-amino-2-methylpyrimidine-5-methylsulfonic acid is collected by centrifugation. The precipitate is washed and recrystallized by dissolving in NaOH (0.4 ml) and centrifuging, followed by removal of the supernatant solution and precipitation of the "aminopyrimidine sulfonic acid" from this liquid by addition of acetic acid (0.4 ml). The centrifuged crystals (54 mg) are washed with water (0.4 ml).

The above "aminopyrimidine sulfonic acid" is dissolved in 6 M HCl (2 ml) and heated under reflux for 6 hours. The HCl is evaporated in a rotary evaporator under reduced pressure, and the final traces of HCl are removed by redissolving the residue in water (2 × 2 ml) and evaporating each time.

This residue is dissolved in water (10 ml) and passed through a Dowex 2 (acetate) column (9 cm × 1.1 cm). The column is washed with water (50 ml) followed by acetic acid (200 ml). The "hydroxypyrimidine sulfonic acid" is eluted with 1 M HCl (50 ml). The HCl is evaporated. The residue is dissolved in water (1 ml); the product crystallizes when ethanol (4 ml) is added.

This "hydroxypyrimidine sulfonic acid" is suitable for radioactive determination and can easily be recrystallized to constant activity.

Kuhn-Roth Oxidation of the "Hydroxypyrimidine Sulfonic Acid"

Reagents

Garbers and Karrer oxidizing mixture[6]: analytical grade CrO_3 (22.33 g) dissolved in twice-distilled water (166 ml) and H_2SO_4 (D, 1.84; 42 ml)
NaOH, 0.01 M
Phenol red indicator: 0.1 g in 14.3 ml of 0.02 M NaOH, to 250 ml with water
p-Phenylphenacyl bromide

Procedure. The above "hydroxypyrimidine sulfonic acid" (20 mg) is treated with the Garbers and Karrer oxidizing mixture at 170° for 2 hours. The acetic acid is steam distilled for 100 ml of distillate, titrated with

---

[6] C. F. Garbers, H. Schmidt, and P. Karrer, *Helv. Chim. Acta* **37**, 1336 (1954).

NaOH to the phenol red end point and isolated as its p-phenylphenacyl ester, which is purified as described below for radioactive measurements.

Isolation as 4-Hydroxy-2-methylpyrimidine-5-methylthioacetic Acid (I)

Reagents

HCl, 6 M
Thioglycolic acid, 95%
NaOH, 5 M
Anion-exchange resin Dowex 1 (200–400 mesh), strongly basic
Acetic acid, 0.5 M and 1.0 M

Procedure. Thiamine (100 mg) is heated under reflux for 6 hours in HCl (10 ml). The HCl is evaporated off and the last traces are removed *in vacuo* over KOH. The residue is dissolved in water (1.5 ml), and thioglycolic acid (116 mg) is added. NaOH (0.36 ml) is cautiously added until a light purple color appears which persists for a few seconds. The pH is then approximately 4.5. The solution is heated under reflux for 8 hours (120–125°). The hot solution is transferred from the 5-ml round-bottom flask to a centrifuge tube and concentrated to about 0.6 ml *in vacuo* over KOH (24 hours). The crystals and the deposit on the walls are redissolved and then left in a refrigerator for 2 hours. The pale yellow crystals that form are centrifuged, washed with water (0.2 ml), and dried *in vacuo* (56 mg).

The crude product is redissolved in water (15 ml) and passed through a Dowex 1 (acetate, 23 cm × 1.1 cm) column with a flow rate of 10 ml/hour. The column is washed with water (15 ml) followed by 0.5 M acetic acid (210 ml). The product is eluted with 1 M acetic acid (130 ml). The eluate is evaporated to dryness, and the resultant white solid is crystallized from water (1 ml), giving 43 mg of crystals. These are suitable for radioactive measurements and can be recrystallized from water to constant activity.

Degradation into 1,3-Diamino-2-methylpropane

Reagents

Alcohol, freshly dried and distilled over magnesium ethylate
Sodium
HCl, 1 M
Picric acid solution, saturated
Anion-exchange Amberlite IRA 410 resin (100–200 mesh), strongly
    basic
Garbers and Karrer reaction mixture

Phenol red indicator
NaOH, 0.01 M
p-Phenylphenacyl bromide

Procedure. The above thioglycolic acid (60 mg) is placed in a 50-ml round-bottom flask, which directly receives the alcohol (7 ml) from its distillation. The condenser is rapidly put in its place and protected by a calcium chloride tube. The crystals are dissolved by heating under reflux for 45 minutes on an oil bath (110°). The heating is then stopped, and sodium (2 × 425 mg) is added. When the dissolution slows down, dry alcohol (4 × 0.8 ml) is added and the reaction mixture is slightly warmed. Once the solution is homogeneous, it is cooled and diluted with water (2.5 ml). A part of the alcohol (5 ml) is distilled off. The solution is then steam-distilled over 5 hours with addition of water (5 ml) for each 5 ml of distillate collected (total distillate, 80 ml). The distillate is acidified with HCl (0.8 ml) and evaporated to dryness. The residue is transferred to a centrifuge tube with water (1 ml), and this solution is treated with picric acid solution (10 ml). The crystals appear immediately; they are centrifuged, washed with water (3 ml), dried *in vacuo* over KOH, and recrystallized from hot water (5 ml). They are suitable for radioactive measurements and can be recrystallized to constant activity.

This picrate (76 mg) is dissolved in boiling water (20 ml). This solution is passed over an IRA 410 (SO_4^{2-}, 1 g) column. The column is washed with water (5 ml). The colorless effluent is concentrated almost to dryness and transferred to the Kuhn-Roth apparatus, where it is completely dried *in vacuo* over P_2O_5. When the product is dry, it is submitted to the Kuhn-Roth oxidation as above, and the acetic acid originating from C-5 and C-7 is counted as its p-phenylphenacyl ester.

Transformation to 2,5-Dimethyl-4-hydroxypyrimidine (II)

Reagents

Raney nickel W-2 prepared according to Mozingo[7]
NaCl
Chloroform

Procedure. The Raney nickel (2.5 ml under ethanol) is thoroughly washed with water (4 × 5 ml) so that no trace of ethanol is left. It is dispersed in water (5 ml) and added to the "pyrimidine thioacetic acid" (I) (100 mg) in a 25-ml conical flask. The reaction mixture is stirred under

[7] R. Mozingo, *Org. Syn.* **21**, 15 (1941).

reflux for 3 hours. The nickel suspension is centrifuged off while still hot and carefully washed with boiling water (6 × 4 ml).

The supernatant liquid and washings are concentrated to 5 ml on a steam bath and NaCl (500 mg) is added. A green flocculent precipitate forms, which is removed by filtration. The filtrate is concentrated to 3 ml, and the dimethylpyrimidine is extracted with chloroform (9 × 5 ml) by means of a microextractor.[8] The chloroform is evaporated; the yield is 42 mg. The product is sublimed at 150° under 18 mm pressure. The sublimed product (40 mg) is suitable for radioactive measurement and can be recrystallized from methyl ethyl ketone (5 ml) for a verification of its radioactive purity.

Methylation of 2,5-Dimethyl-4-hydroxypyrimidine (II) and Its Complete Degradation

Reagents

Methyl iodide
Na$_2$CO$_3$, 0.1 M
Amberlite IR 120 cation-exchange resin (100–200 mesh)
NH$_4$OH, 1.0 M
NaOH, 5.0 and 0.1 M
H$_2$SO$_4$, 6.0 and 1.0 M
Phenol red indicator
HCl, 0.1 M
Ethanol
p-Phenylphenacyl bromide
Fluorescent (254 mμ) thin-layer silica gel plates (20 cm × 20 cm), 0.5 mm thick
Dichloromethane

Procedure. The dimethylpyrimidine (II) is heated under reflux in a two-phase mixture of methyl iodide (5 ml) and Na$_2$CO$_3$ solution (5.6 ml) for 24 hours.

The aqueous layer is passed through an Amberlite IR 120 (H$^+$, 2 g) ion-exchange column. The methyl iodide is washed with water (4 × 5 ml), and the washings are passed through the column. The effluent is then neutral. The tetramethyl pyrimidine (III) is eluted with NH$_4$OH (100 ml), which is then evaporated to dryness.

The residue is transferred with 5.0 M NaOH (5 × 2 ml) to a pear-shaped flask which has two openings and is heated under reflux (130–140°)

[8] B. L. Browning, *Mikrochemie* **26**, 54 (1939).

for 24 hours. The liquid is then cooled to $-10°$, acidified by 6.0 M H_2SO_4 (6 ml) and steam distilled (125 ml). The total acid content of this distillate can be determined by titration against 0.1 M NaOH to a phenol red end point (0.96 meq).

The formic acid is selectively oxidized according to the method of Wood and Gest[9] to give barium carbonate originating from C-6.

The solution is then steam-distilled a second time (125 ml), and the distillate is neutralized by 0.1 M NaOH to the phenol red end point. The neutralized distillate is evaporated to dryness. The residue is dissolved in water (0.07 ml for each millimole present), neutralized with HCl, and heated under reflux with ethanol (0.7 ml for each millimole) and p-phenyl-phenacyl bromide (a stoichiometric quantity) for 45 minutes.

The solution is then deposited on the silica gel plates (1 millimole per plate) and chromatographed in dichloromethane. On visualization under UV light the second major dark band from the origin contains the acetate; the third major band contains the propionate. The acetate and propionate bands are eluted with boiling ethanol (4 $\times$ 10 ml per band). The eluates are evaporated to dryness, and the residues are crystallized from ethanol (1.3 ml per millimole) to which water is added (0.7 ml per millimole) while still hot. The p-phenylphenacyl propionate (m.p. 102.5–104°) and acetate (m.p. 110–112°) are suitable for radioactive measurements and can be further crystallized to constant activity. The propionate originates from C-4, C-5, C-7, and the acetate from C-2, C-8.

The free acids can be regenerated by heating the ester (0.25 meq) under reflux with 1 M H_2SO_4 (5 ml) for 19 hours and steam distilling. The free acids are determined by addition of 0.1 M NaOH to the phenol red end point and submitted to the Schmidt degradation according to Phares[10] as adapted by Greenberg and Rothstein.[11] This Schmidt degradation yields the information for all the remaining carbon atoms one by one.

Isolation of 4-Hydroxy-2-methylpyrimidine-5-methylthioacetic Acid from Pyramine

Reagents

HCl, 6 M and 2 M
Thioglycolic acid, 95%
NaOH, 5 M

Procedure. Pyramine (40 mg) is heated under reflux with 6 M HCl (50 ml) in a 250-ml round-bottom flask for 6 hours. The HCl is evaporated

---

[9] H. G. Wood and H. Gest, Vol. III, p. 286.
[10] E. F. Phares, *Arch. Biochem. Biophys.* **33,** 173 (1951).
[11] D. M. Greenberg and M. Rothstein, Vol. IV, p. 718.

in a rotatory evaporator. The dry residue is transferred to a 25-ml round-bottom flask by washing with water (3 × 3 ml). This solution is evaporated to dryness.

Thioglycolic acid (145 mg) and 2 M HCl (2.5 ml) are added to the residue, and the solution is heated under reflux (130°) for 48 hours. The solution is evaporated to dryness, and the residue is transferred to a centrifuge tube by washing with water (4 × 0.1 ml). This solution is adjusted to pH 4.5 by NaOH (about 0.15 ml) until a purple color persists for a few seconds. The solution is left in the refrigerator overnight. Crystals (51 mg) appear which are washed by cold water (0.25 ml).

[37] Thiamine Pyrophosphokinase
(ATP: Thiamine Pyrophosphotransferase, EC 2.7.6.2)

By C. J. GUBLER

$$\text{Thiamine} + \text{ATP} \underset{\text{Mg}^{2+}}{\overset{\substack{\text{thiamine} \\ \text{pyrophosphokinase}}}{\longrightarrow}} \text{TDP} + \text{AMP}$$

Assay Method

Principle. This method is based on a procedure suggested by Datta and Racker[1] and modified by Johnson and Gubler[2] (this volume [22]), and depends upon the recombination of apotransketolase with thiamine diphosphate (TDP) to form active transketolase. The activity of transketolase can then be measured spectrophotometrically and is proportional to the available TDP.

The reagents and assay procedure are described in detail (this volume [22]).

An alternative procedure has been used with good results; it involves the use of [14]C- or ^{35}S-labeled thiamine and measurement of the amount converted to the corresponding labeled TDP. This assay is more sensitive and more direct than the coupled spectrophotometric assay.

Deus (this volume [38]) has described a method whereby the radioactive TDP is separated from radioactive thiamine by paper chromatography. We have used a separation by electrophoresis on paper for 3 hours at pH 5.4 in acetate buffer followed by cutting out the appropriate strips and measuring the radioactivity with a liquid scintillation counter. Electrophoresis has one advantage in that it allows separation in a much shorter time than paper chromatography.

[1] A. G. Datta and E. Racker, *J. Biol. Chem.* **230,** 624 (1961).
[2] L. R. Johnson and C. J. Gubler, *Biochim. Biophys. Acta* **156,** 85 (1968).

Definition of Unit. A unit of enzyme activity is the amount of enzyme that will produce 1 micromole of TDP per hour.

Purification Procedure

This procedure is essentially that of Mano[3] for liver thiamine pyrophosphokinase.

Step 1. Preparation of Acetone Powder. Fresh rat liver is washed thoroughly with ice-cold physiological saline, blotted dry with filter paper, and homogenized in a Waring blendor with 10 volumes of acetone at $0°$. The precipitate is collected by centrifugation at 6000 g for 10 minutes at $0°$, rehomogenized in acetone, and recentrifuged as above. The resulting pellet is again homogenized in 10 volumes of acetone as above, and the final acetone powder is collected and dried on a Büchner funnel. The dried cake is powdered and stored in a desiccator over anhydrous calcium sulfate. It can be stored thus for several months at $-20°$ without loss of activity.

Step 2. Extraction. A convenient amount of the acetone powder is extracted with 6 volumes (w/v) of 0.05 M Tris buffer pH 7.8 by homogenizing with a Potter-Elvehjem homogenizer at $0°$ followed by centrifugation at 34,000 g for 15 minutes. The precipitate is extracted again in the same manner with 4 volumes of the buffer and centrifuged. The two supernatant solutions are combined.

Step 3. Heat Treatment. The combined supernatant solutions are adjusted to pH 5.8 with 0.5 N acetic acid, transferred to appropriate containers, heated for 5 minutes at $52°$ in a water bath with stirring, cooled immediately, and centrifuged at 34,000 g for 20 minutes.

Step 4. $(NH_4)_2SO_4$ Fractionation. To the clear orange supernatant solution, slowly add finely ground solid $(NH_4)_2SO_4$ to 52% saturation (32.8 g for each 100 ml of extract). Let stand 3–5 hours, centrifuge off, and discard the precipitate. To the supernatant solution, add more $(NH_4)_2SO_4$ to 68% saturation (an additional 10.3 g/100 ml). Let stand overnight at $0-5°$ and centrifuge. Dissolve the precipitate in a small volume of water, dialyze against distilled water for 6 hours at $0°$, and centrifuge off any precipitate formed. Dilute the supernatant solution to give 10 mg of protein per milliliter.

Step 5. Acid Treatment. Adjust the pH to 4.3 by the slow addition of 0.5 N acetic acid, and add 0.02 volume of 0.5 M sodium acetate buffer, pH 4.3. Incubate at $37°$ for 30 minutes, centrifuge, and discard the precipitate.

Step 6. Alumina C_γ Gel Chromatography. A column is prepared with alumina C_γ gel and Hyflo-Supercel mixed in the ratio of 1:5 by weight. The

[3] Y. Mano, *J. Biochem. (Tokyo)* **47,** 283 (1960).

ratio of gel to protein is $2:1$. The column is washed thoroughly with $0.01\ M$ acetate buffer, pH 4.3, and the charge of supernatant solution from step 5 is placed on the column. The enzyme is recovered by gradient elution, starting with $0.1\ M$ KH$_2$PO$_4$, pH 4.5, until the first inert protein peak disappears. Then a gradient is started, using $0.133\ M$ sodium phosphate buffer, pH 6.7, in the reservoir and 100 ml of $0.1\ M$ KH$_2$PO$_4$ in the mixer system. When the pH of the eluate reaches 5.2, the reservoir system is changed to $0.16\ M$ Na$_2$HPO$_4$. The fractions between pH 5.2 and 5.8 contain most of the TPK activity. These are combined and brought to saturation with (NH$_4$)$_2$SO$_4$. After centrifugation at high speed, the precipitate is redissolved in a minimum of water and may be stored at $-20°$ for several months without appreciable loss of activity.

Properties

The enzyme, as prepared, contains negligible activity of thiamine diphosphatase, adenylate kinase, or ATPase, is approximately 68% pure, and should have a specific activity of approximately 40–50 milliunits/mg with 10% recovery of original activity. Pyrithiamine is a potent inhibitor of thiamine pyrophosphokinase (K_i about 10^{-7}), and oxythiamine is a 1000-fold weaker inhibitor ($K_i = 10^{-4}$). The K_m for thiamine is of the order of $10^{-7}\ M$. The enzyme is quite specific for ATP, although UTP has about 40% of the activity of ATP. AMP and ADP are inactive unless an ATP-generating system is operating.

[38] Assay of Thiamine Pyrophosphokinase (ATP: Thiamine Pyrophosphotransferase, EC 2.7.6.2) using ^{14}C- or ^{35}S-Labeled Thiamine

By Bruno Deus

Labeled thiamine $+$ ATP $\rightarrow$ labeled thiamine pyrophosphate $+$ AMP

The rate of enzymatic conversion of thiamine to thiamine pyrophosphate (TPP) provides a convenient means of determining thiamine pyrophosphokinase activity. For the estimation of TPP, pyruvate decarboxylase (2-oxoacid carboxy-lyase, EC 4.1.1.1),[1,2] and transketolase (D-sedoheptulose-7-phosphate: D-glyceraldehyde-3-phosphate glycolaldehyde transferase, EC

[1] H. G. K. Westenbrink, Vol. II [107], p. 636.
[2] R. Näveke, H. W. Goedde, and H. Holzer, *Arch. Mikrobiol.* **44**, 93 (1962).

2.2.1.1),[3,4] which are completely dependent on their coenzyme, are used. A radiometric procedure,[5] which allows direct measurement of TPP, is described here in detail.

Principle. The amount of labeled TPP formed in the reaction from labeled thiamine and ATP is determined after separation from the reaction mixture by paper chromatography or by electrophoresis.

Reagents

    Glycylglycine buffer, 0.5 M, pH 7.3
    ATP, disodium salt, 0.1 M, pH 7.0
    $MgSO_4$, 0.1 M
    Thiamine hydrochloride,[6] 0.01 M labeled; specific activity 1 to 3 $\times$
    10^7 dpm per micromole

All reagents are stored at $-15°$.

Procedure. In a final volume of 0.5 ml the following solutions are pipetted into a 7 $\times$ 40-mm metal centrifuge tube: 0.1 ml glycylglycine buffer, 0.04 ml $MgSO_4$, 0.06 ml ATP, and the enzyme preparation. The first three ingredients may be added together. After equilibration, the reaction is started by addition of 0.01 ml ^{14}C- or ^{35}S-labeled thiamine. The stoppered tubes are incubated at 37° for 60 minutes with vigorous shaking. After boiling for 1 minute, the tubes are placed into an ice bath. Precipitated protein is removed by centrifugation. A 0.1-ml aliquot is applied as a narrow band 6 cm from one end of a 3.8-cm strip of chromatography paper.[7] The chromatograms are developed in the descending manner at 20° for 12–15 hours with *n*-propanol–1 M sodium acetate buffer, pH 5.0–water (7:1:2, v/v).[8] The solvent front runs about 40 cm under these conditions. Thiamine, thiamine monophosphate, and thiamine pyrophosphate give average R_f values of 0.46, 0.27, and 0.10, respectively, and are completely separated. The part of the air-dried chromatogram containing the TPP peak (usually within 12 cm from the start) is cut into 1 $\times$ 3.8 cm strips. These are analyzed for radioactivity in 10 ml of scintillation medium.[9] The counting

---

[3] A. G. Datta and E. Racker, *J. Biol. Chem.* **236**, 624 (1961).

[4] L. R. Johnson and C. J. Gubler, *Biochim. Biophys. Acta* **156**, 85 (1968).

[5] B. Deus, H. E. C. Blum, and H. Holzer, *Anal. Biochem.* **27**, 492 (1969).

[6] The Radiochemical Centre, Amersham, England, provides thiamine-(thiazole-2-^{14}C) hydrochloride (10–30 mCi/millimole) and thiamine-^{35}S hydrochloride (5–70 mCi/ millimole).

[7] Chromatography paper No. 2043b Mgl (used without prewashing), Schleicher & Schüll, 3354 Dassel, Germany.

[8] D. Siliprandi and N. Siliprandi, *Biochim. Biophys. Acta* **14**, 52 (1954).

[9] Distilled toluene containing 0.3% 2,5-diphenyloxazole (PPO) and 0.03% *p*-bis-2-(5-phenyloxazolyl)benzene (POPOP).

efficiency in the liquid scintillation spectrometer[10] is $71 \pm 0.7\%$ with a lower/upper discriminator setting of $80-\infty$ and a gain of 1000. A control is obtained by running an incubation without the enzyme.

Alternative Separation Procedure. In addition to paper chromatography, TPP may be separated from the reaction mixture by electrophoresis on paper.[8,11]

A 0.05-ml sample of the deproteinized incubation mixture is applied as a narrow band to the middle of a 3.8×40 cm strip of Whatman No. 3 MM paper which has been soaked with $0.05\,M$ sodium acetate buffer, pH 5.5, previously. TPP is isolated by electrophoresis[12] for 4 hours at $0°$. The voltage gradient is set to 10 V per centimeter at 3.5 mA. After drying, TPP is located and radioactivity is estimated as described above.

Definition of Units and Specific Activity. The amount of TPP produced enzymatically is calculated from the disintegrations per minute obtained in the TPP peak. It must be checked that the specific activities of thiamine and TPP are the same. One unit is defined as the amount of enzyme which catalyzes the formation of 1 micromole of TPP per minute under the conditions specified. Specific activity is expressed as units of enzyme per milligram of protein. The biuret method[13] is used for the determination of protein. Thiamine and TPP are measured by the thiochrome procedure.[14]

Comments on the Assay Procedure. The method described above has been applied to the purified enzyme from rat liver,[5] to rat liver crude extract,[5] and to rat hemolysate.[15]

Depending on the activities of interfering enzymes, the optimal concentrations of ATP or $MgSO_4$ in the assay mixture may vary considerably.[15] Therefore these conditions must be checked for each enzyme source. Interfering activities which may occur are: TPP degrading enzymes,[16] ATP degrading enzymes,[15] and ADP- or AMP-forming enzymes.

The reaction is strongly inhibited by high concentrations of ATP. In

[10] Packard Instrument Company, Inc., La Grange, Illinois, Tri-Carb Model 314 EX or 3375.

[11] E. Sandner and B. Gassmann, *Z. Ernährungswiss.* **8,** 222 (1967).

[12] Pherograph-Original-Frankfurt (Wieland and Pfleiderer), obtained from Hormuth, 69 Heidelberg, Germany.

[13] G. Beisenherz, H. J. Boltze, T. Bücher, R. Czok, K. H. Garbade, E. Meyer-Arendt, and G. Pfleiderer, *Z. Naturforsch.* **8b,** 555 (1953).

[14] G. Kohlhaw, B. Deus, and H. Holzer, *J. Biol. Chem.* **240,** 2135 (1965).

[15] H. E. C. Blum and B. Deus, reported at the Herbsttagung der Gesellschaft für Biologische Chemie, Münster, Germany, October 3–4, 1968; *Physiol. Chem.* **349,** 1240 (1968).

[16] Thiamine pyrophosphate degrading activity is measured under the test conditions by substituting labeled TPP for labeled thiamine in the above incubation mixture.[5] The decrease in the amount of TPP during the reaction time is determined. Labeled TPP is prepared enzymatically and purified by paper chromatography.[5]

the presence of $1.2 \times 10^{-2} M$ ATP, thiamine pyrophosphokinase from rat liver is inhibited about 70% by $3 \times 10^{-3} M$ ADP or AMP, and about 90% by $1 \times 10^{-2} M$ ADP.[5]

The sensitivity of the method depends on the specific activity of the available thiamine. With thiamine of 1 to 3×10^7 dpm/micromole, 200 picomoles are easily detected.

The coefficient of variation is less than ±5%. (The volume of the incubation mixture should not be diminished, as the results become less reproducible.)

The author prefers separation of TPP from the assay system by paper chromatography because it is more sensitive and more convenient than electrophoresis, especially when many samples have to be assayed simultaneously.

A Rapid Test for the Detection of Thiamine Pyrophosphokinase Activity

For determining the presence of thiamine pyrophosphokinase in effluents from chromatography or gel filtration columns, a simplified procedure, which eliminates the time-consuming paper chromatography, is described.[17]

Principle. The method is a modification of an assay devised for thymidine kinase.[18] TPP is adsorbed to anion-exchange paper disks,[19] and thiamine is removed by a washing procedure.

Procedure. The reaction mixture is as described above. If sufficiently high enzyme activities are to be expected, the incubation period may be reduced. After 15–60 minutes at 37° the reaction mixture is cooled to −5°. Samples of 20 μl are spotted on anion-exchange paper disks. The papers must not be dried, but are immediately transferred to a funnel with a fritted disk of 22-mm diameter. At 15 seconds after the application of the aliquot, the disks are washed 5 times with 20 ml of $1 \times 10^{-4} M$ sodium acetate. The time interval between the washings is 0.5 minute. The eluent is withdrawn by suction. The papers are dried in hot air, then the radioactivity is measured in 5 ml of toluene scintillation medium.[9]

Comments. More than 99% of the thiamine is removed from the papers by the washing procedure. Figure 1 demonstrates, that the percentage of TPP which is adsorbed to the anion-exchange paper is dependent on the absolute amount of TPP applied. In addition, it is dependent on the volume in which a constant amount of TPP is applied to the papers. Therefore, the results obtained by this procedure are qualitative rather than quantitative.

However, provided the conditions in the assay system are standardized,

[17] B. Deus and H. E. C. Blum, in preparation.
[18] T. R. Breitman, *Biochim. Biophys. Acta* **67**, 153 (1963).
[19] Whatman DE 81 (23 mm in diameter).

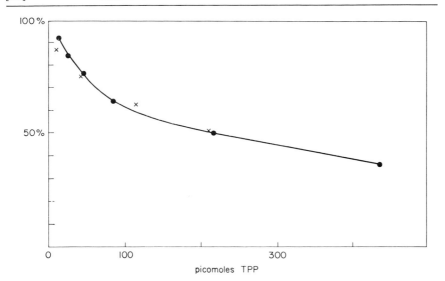

FIG. 1. Percentage of ^{14}C-labeled thiamine pyrophosphate remaining adsorbed to anion-exchange paper disks after the washing procedure as a function of the amount added. Crosses represent the results obtained in the presence of 335 micromoles of unlabeled thiamine.

this method is applicable to the detection of thiamine pyrophosphokinase activity in effluents from columns and to related problems.

Because of the highly friable condition of the wet paper disks, it is not advisable to wash the disks in a beaker all together.

Occurrence of the Enzyme

The enzyme has been partially purified from brewers' yeast[20] and liver.[20-22] It has been detected in several animal tissues[11,23] and in microorganisms.[24-26] In *Escherichia coli* it is located in the spheroplasts,[26] whereas it is found exclusively (more than 99%) in the soluble fraction of rat erythrocytes[15] and of rat liver cells.[17]

[20] Nguyen-Van-Thoai and L. Chevillard, *Bull. Soc. Chim. Biol.* **31**, 204 (1949).
[21] F. Leuthardt and H. Nielsen, *Helv. Chim. Acta* **35**, 1196 (1952).
[22] Y. Mano, *J. Biochem. (Tokyo)* **47**, 283 (1960).
[23] K. Lohmann and P. Schuster, *Biochem. Z.* **294**, 188 (1937).
[24] R. Suzue, *Bull. Inst. Chem. Res., Kyoto Univ.* **36**, 8, 13 (1958); *Chem. Abstr.* **52**, 18565b–g (1958).
[25] A. K. Sinha and G. C. Chatterjee, *Biochem. J.* **104**, 731 (1967).
[26] I. Miyata, T. Kawasaki, and Y. Nose, *Biochem. Biophys. Res. Commun.* **27**, 601 (1967).

[39] Thiamine Pyrophosphate (TPP)— ATP Phosphoryl Transferase

By Y. Itokawa and J. R. Cooper

$$\text{TPP} + \text{ATP} \overset{\text{Mg}^{2+}}{\rightleftharpoons} \text{TTP} + \text{ADP}$$

Assay Method

Principle. The enzyme may be assayed spectrophotometrically in either direction by coupling the adenine nucleotide to the appropriate enzymes and pyridine nucleotide.[1,2] Because of the difficulty in obtaining thiamine triphosphate (TTP), the reaction is usually followed in our laboratory from left to right.

Reagents

Sodium acetate buffer (1 M) pH 5.0
$MgCl_2$, 0.1 M
Phosphoenolpyruvate, 0.05 M
ATP, 0.01 M
TPP, 0.01 M
NADH, 1 mg/ml
Pyruvate kinase, 2 mg/ml (Boehringer Mannheim Corp.)
Lactate dehydrogenase, 5 mg/ml (Boehringer Mannheim Corp.)
Rat brain mitochondrial suspension, 10 mg protein/ml

To a 1-ml quartz cuvette are added 0.05 ml of the acetate buffer, 0.05 ml of $MgCl_2$, 0.05 ml of phosphoenolpyruvate, 0.02 ml of TPP, 0.1 ml of NADH, 0.01 ml of pyruvate kinase, 0.01 ml of lactate dehydrogenase, 0.03 ml of the rat brain mitochondrial suspension, and water to a final volume of 0.95 ml. The reaction is started by the addition of 0.05 ml ATP. TPP is omitted from a control cuvette. The latter is mandatory because of ATPase in the mitochondrial preparation. The disappearance of NADH is followed at 340 mμ for 10–15 minutes.

Preparation of the Enzyme. Rat brain is homogenized in 9 volumes of 0.25 M sucrose containing 1 mM EDTA and 10 mM glycylglycine buffer pH 7.4. After centrifugation at 700 g for 15 minutes to remove nuclei, mitochondria are isolated by centrifugation of the supernatant solution at 7000 g for 15 minutes. The mitochondrial suspension is washed 3 times

[1] T. Yusa, *Plant Cell Physiol. (Tokyo)* **3,** 95 (1962).
[2] Y. Itokawa and J. R. Cooper, *Biochim. Biophys. Acta* **158,** 180 (1968).

with sucrose medium before it is finally suspended in sucrose to a concentration of approximately 10 mg of protein per milliliter.

Rabbit brain acetone powder may also be used as a source of the enzyme. The powder is extracted with 10 volumes of 0.01 M phosphate buffer, pH 7.4, at room temperature for 20 minutes and centrifuged at 45,000 g for 15 minutes. The supernatant fluid is decanted and the residue reextracted with 5 volumes of fresh buffer as above. The supernatant solutions are pooled. Stored at $-20°$ the enzyme activity lasts for about 10 days.

Because this preparation contains NADH oxidase it is necessary to perform the assay in two steps. The same reagents as above are used except that NADH and lactate dehydrogenase are omitted. Incubation with usually 0.1 to 0.3 ml of enzyme is carried out for 15 minutes at 37° and the reaction is terminated by the addition of 0.05 ml of 100% TCA. After centrifugation the supernatant solution is neutralized with 3 N NaOH and an aliquot is then assayed for pyruvate using lactate dehydrogenase, NADH, and Tris buffer, pH 7.4 (final concentration, 0.05 M).

Properties

First described in yeast,[1] the enzyme has been isolated from brain[2] and liver.[3] When kept at 0–4°, the preparation retains its activity for about 3 days; for a reason that is as yet unclear, rewashing the preparation at this stage often results in the return of full activity. This phosphoryl transferase is of particular interest in that it has been found that patients with subacute necrotizing encephalomyelopathy contain a factor in their blood, urine, and spinal fluid that inhibits the brain mitochondrial preparation.[4]

[3] J. R. Cooper, unpublished data, 1969.
[4] J. R. Cooper, Y. Itokawa, and J. H. Pincus, *Science* **164,** 72 (1969).

[40] Thiamine Pyrophosphatase from Rabbit Brain

By J. R. COOPER

$$TPP \xrightarrow{\text{Mg}^{2+}} TMP + P$$

Thiamine pyrophosphatase has been purified from liver[1] as well as brain.[2] In both tissues it appears to have an absolute specificity for thiamine pyrophosphate (TPP) as substrate among the various thiamine phosphate esters, but in addition it also catalyzes the hydrolysis of certain nucleoside diphosphates.[1,2]

[1] M. Yamazaki and O. Hayaishi, *J. Biol. Chem.* **243,** 2934 (1968).
[2] J. R. Cooper and M. M. Kini, in preparation.

Assay

Reagents

Tris buffer, 0.5 M, pH 7.8
MgCl$_2$, 0.1 M
TPP, 0.05 M

Procedure. To a test tube containing 0.1 ml of Tris buffer is added 0.05 ml of MgCl$_2$, 0.1 ml of TPP, enzyme (0.2–0.8 mg of protein), and water to a final volume of 1 ml. A zero-time control tube kept in ice is also carried through the procedure. The contents of the experimental tube are incubated at 37° for 60 minutes, and the reaction is terminated by the addition of 1 ml of 10% trichloroacetic acid. After centrifugation an aliquot of the supernatant solution is assayed for inorganic phosphate (cf. Vol. III, [115]). Alternatively, the enzyme may be assayed by measuring the formation of TMP using column chromatography to separate TPP from TMP, followed by fluorometric analysis.[1]

Purification Procedure

A rabbit brain is homogenized with 9 volumes of 0.32 M sucrose. This and subsequent operations are carried out at 0–4°. The homogenate is centrifuged at 8000 g for 15 minutes, and the supernatant solution is decanted. The deposit is resuspended in 3 volumes of sucrose and recentrifuged at 8000 g for 15 minutes. The supernatant solution is combined with the first supernatant solution and centrifuged at 48,000 g for 45 minutes. The supernatant solution is discarded, and the deposit is resuspended in 40 ml of H$_2$O. The suspension is placed in a Rosette cell immersed in an ice-water bath and sonicated intermittently (1 minute on, 1 minute off) for a total sonication time of 5 minutes. The material is then centrifuged at 48,000 g for 45 minutes. The supernatant solution is decanted, and the deposit is discarded. About 15–20 ml of the supernatant solution is applied to a 2% agarose[3] column (3 × 30 cm) that is equilibrated with 0.01 M Tris-maleate buffer, pH 7.4. The enzyme is eluted off the column with this buffer and usually appears between 100 and 140 ml of eluent. The enzyme activity in these fractions disappears rapidly unless bovine serum albumen at a concentration of 3 mg/ml is added. The overall purification of the enzyme with this procedure is about 30-fold; the yield is 10%.

---

[3] Among the various preparations of agarose that were tested, Gelarose from the Litex Corp. in Denmark (supplied by the Aldrich Chemical Company) was the only material that was satisfactory.

Comments

For both the liver and brain enzymes, ATP has an allosteric effect and lowers the K_m of the substrate regardless of whether it is a nucleoside diphosphate or TPP. The nucleoside diphosphates that serve as substrate are GDP, UDP, and CDP; ADP and d-TDP are inactive. Among the thiamine phosphate esters that were tested, only TPP is active as substrate; no hydrolysis was observed with TMP or TTP.

With the liver enzyme, the K_m for TPP without ATP is $2 \times 10^{-2}\ M$; in the presence of ATP it is $3.5 \times 10^{-3}\ M$.[1] These values have not been determined for the enzyme as prepared from brain.

[41] Thiaminase I (Thiamine: Base 2-Methyl-4-aminopyrimidine-5-methenyltransferase, EC 2.5.1.2)[1]

By JAMES L. WITTLIFF and R. L. AIRTH

Thiaminase I catalyzes the decomposition of thiamine by a base-exchange reaction involving a nucleophilic displacement on the methylene group of the pyrimidine moiety.

Assay Method

Principle. The enzyme activity is determined by a modification[1a] of the method of Douthit and Airth,[2] which employs aniline as the base in the exchange reaction. The formation of the product, N-(4-amino-2-methyl-pyrimidin-5-ylmethyl)aniline, is measured spectrophotometrically by an increase in optical density at 248 mμ.

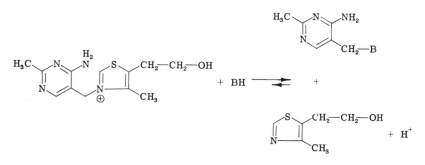

[1] Manuscript prepared in The Cell Research Institute, The University of Texas.
[1a] J. L. Wittliff and R. L. Airth, *Biochemistry* **7**, 736 (1968).
[2] H. A. Douthit and R. L. Airth, *Arch. Biochem. Biophys.* **113**, 331 (1966).

Reagents

Aniline, freshly redistilled, 5.75×10^{-3} M dissolved in 0.5 M sodium phosphate buffer, pH 5.8

Thiamine · HCl, 3×10^{-4} M

Enzyme: filtrate from a 24-hour culture of *Bacillus thiaminolyticus* grown in brain–heart infusion[2]

Procedure. To 0.6 ml of buffered aniline solution in a 3-ml cuvette (1-cm light path), add glass-distilled water so that the volume, including the subsequent enzyme addition, will be 2.6 ml. The diluted aniline and buffer should be allowed to stand for 10–12 minutes (25°) before further additions are made. The enzyme solution is then added to the experimental cuvette, and the reaction is initiated by adding 0.4 ml of the thiamine solution. The blank is a mixture of only the aniline, buffer, and water. Controls are performed with all the reactants and heated enzyme (10 minutes, 100°) or with only the reactants. The change in absorbancy at 248 mμ is measured for 6 minutes at 25°. The initial rate of absorbancy increase (corrected for a heated control) is used to calculate the enzyme activity according to:

$$\text{Enzyme unit (U)} = \frac{(\Delta A_{248})(0.269)}{(\text{min})}$$

where ΔA_{248} is the absorbancy change at 248 mμ and 0.269 is a constant which converts absorbancy to micromoles for a 3-ml reaction mixture. The assay method is very reproducible and is satisfactory for measuring the thiaminase activity of crude or purified enzyme preparations.

Definition of Unit and Specific Activity. One unit of enzyme is defined as that amount which catalyzes the formation of 1 micromole of product in 1 minute at 25° under the above conditions. Specific activity is expressed as units per milligram of protein. The method of Lowry *et al.*[3] is used to determine protein concentration.

Alternate Methods. Thiaminase I activity has also been measured by the procedure of Fujita *et al.*,[4] in which the thiamine remaining in the reaction mixture is determined fluorometrically by the thiochrome method.[5] Kenten[6] has proposed a method for the measurement of heteropyrithiamine, *N*-(4-amino-2-methylpyrimin-5-ylmethyl) pyridine, formed from thiamine and pyridine in the presence of thiaminase I from bracken ferns.

---

[3] O. H. Lowry, N. J. Rosebrough, A. L. Farr, and R. J. Randall, *J. Biol. Chem.* **193**, 265 (1951).

[4] A. Fujita, Y. Nose, S. Kozuka, T. Tashiro, K. Ueda, and S. Sakamoto, *J. Biol. Chem.* **196**, 289 (1952); see also Vol. II [103].

[5] R. L. Airth and G. E. Foerster, this volume [14].

[6] R. H. Kenten, *Biochem. J.* **67**, 25 (1957).

Source of Enzyme

In addition to *Bacillus thiaminolyticus*, another bacterium, *Clostridium thiaminolyticum*, produces thiaminase I.[7] This enzyme has also been found in the viscera of freshwater fish and shellfish, as well as in bracken ferns.[8]

Purification Procedure

Cultivation of Bacteria. The organism used as a source of thiaminase I is *Bacillus thiaminolyticus*, strain M. It is grown aerobically in the defined medium of Douthit and Airth[2] which consists of: glycerol, $2 \times 10^{-1} M$; monosodium glutamate, $1 \times 10^{-1} M$; sodium citrate $\cdot$ 2H$_2$O, $1.9 \times 10^{-2} M$; KH$_2$PO$_4$, $7.3 \times 10^{-3} M$; Na$_2$HPO$_4$, $5.6 \times 10^{-3} M$; NaCl, $1.7 \times 10^{-2} M$; MgSO$_4$ $\cdot$ 7H$_2$O, $2.8 \times 10^{-3} M$; FeSO$_4$ $\cdot$ 7H$_2$O, $1.1 \times 10^{-4} M$; (NH$_2$)$_2$CO, $5 \times 10^{-2} M$; and thiamine $\cdot$ HCl, $3 \times 10^{-7} M$. Spores from a 24-hour agar slant culture are used to inoculate 20 ml of brain–heart infusion which, after 24 hours, is used to inoculate 1 liter of synthetic medium contained in a 2-liter flask. The inoculated medium is incubated at 37° for 28–30 hours in a shaker. Cells are removed from the culture by centrifugation in a continuous-flow centrifuge (30,000 g, 4°); the thiaminase I, which is an extracellular enzyme,[2] is purified from the supernatant solution according to the method of Wittliff and Airth.[1] Twelve liters of the culture filtrate are sufficient starting material for the purification of 1.5–2.0 mg of purified enzyme. *Bacillus thiaminolyticus* has also been grown successfully in mass culture (350 liters) in defined medium.[9]

Step 1. First Ammonium Sulfate Precipitation. Unless otherwise noted, all operations were performed at 0–4°. Solid ammonium sulfate is added slowly with stirring to the clear, yellow supernatant solution to a concentration of 516 g/liter. The mixture is left unstirred 14–16 hours. The precipitate is removed by centrifugation in a continuous-flow centrifuge (30,000 g) and stored frozen.

Step 2. Sephadex G-25 Chromatography. The precipitate is suspended in cold, glass-redistilled water (1.5 ml/g wet weight) using a TenBroeck homogenizer and centrifuged at 30,000 g, for 15 minutes. The reddish tan supernatant solution is removed and desalted by passing through a Sephadex G-25 column (2.5 $\times$ 38 cm). Fractions containing thiaminase I activity are eluted with glass-redistilled water and pooled.

Step 3. Reprecipitation with Ammonium Sulfate. Solid ammonium sulfate to a concentration of 313 g/liter is added slowly with stirring to the

[7] R. Kimura, *in* "Review of Japanese Literature on Beriberi and Thiamine," p. 255. Vitamin B Research Committee of Japan, Tokyo, 1965.

[8] K. Murata, *in* "Review of Japanese Literature on Beriberi and Thiamine," p. 220. Vitamin B Research Committee of Japan, Tokyo, 1965.

[9] J. L. Wittliff, Ph.D. Dissertation, The University of Texas, Austin, Texas, 1967.

Sephadex G-25 fractions. After stirring for 1 hour, the suspension is centrifuged for 15 minutes at 30,000 g and the precipitate is discarded. The supernatant solution is brought to a final concentration of 472 g/liter with the further addition of solid ammonium sulfate, stirred for 1 hour, then centrifuged at 30,000 g for 15 minutes. The supernatant solution is discarded, and the precipitate is frozen at $-20°$.

Step 4. Sephadex G-100 Gel Filtration. The precipitate is dissolved in a minimal amount of 0.05 M sodium phosphate buffer (pH 7.0) and dialyzed overnight against two 4-liter portions of the same buffer. After dialysis, the fraction is centrifuged at 30,000 g for 15 minutes to remove a small amount of insoluble material. The supernatant solution is applied to a Sephadex G-100 column (2.5 × 38 cm) equilibrated with 0.05 M sodium phosphate buffer (pH 7.0). Thiaminase I, eluted with the same buffer, emerges as an early peak; fractions containing the enzyme are pooled.

Step 5. DEAE-Sephadex Ion-Exchange Chromatography. The enzyme from the Sephadex G-100 step is dialyzed overnight against two 4-liter portions of 0.1 M Tris·HCl buffer (pH 8.2). The dialyzed enzyme is centrifuged at 30,000 g for 15 minutes to remove any insoluble material that might have formed during this step. The supernatant solution is applied to a DEAE-Sephadex A-50 column (2.5 × 38 cm) previously equilibrated with the Tris·HCl buffer. When all the material is loaded onto the column, the gel is washed with two bed volumes of the Tris·HCl buffer to remove unbound proteins. The remaining proteins bound to the gel are eluted using 0.1 M Tris·HCl buffer (pH 8.2) and a gradient of 0.0–1.0 M sodium chloride. The fractions containing thiaminase I activity are pooled and frozen at $-20°$.

The enzyme resolves as a single peak after ion-exchange chromatography and can be frozen at $-20°$ for 6 months without an appreciable loss in activity. A representative purification is presented in the table.

PURIFICATION OF EXTRACELLULAR THIAMINASE I

Step	Total volume (ml)	Protein (mg/ml)	Total activity (mU)	Specific activity (mU/mg of protein)	Recovery (%)
Supernatant, 30 kg	11655.0	0.30	114,000	32.8	100.0
(NH₄)₂SO₄, 0–75%	21.4	6.35	35,540	261.3	31.2
Sephadex G-25	63.0	1.28	30,500	378.0	26.8
(NH₄)₂SO₄, 50–70%	9.1	4.25	15,850	410.0	13.9
Sephadex G-100	22.0	0.43	12,250	1287.0	10.8
DEAE-Sephadex	11.6	0.15	11,200	6007.0	9.8

Properties

Ultraviolet Absorption Spectrum. The spectrum of bacterial thiaminase I is that of a simple protein with an absorption maximum at 277 mμ and a minimum at 252 mμ.[1a]

Molecular Weight. The enzyme purified from *Bacillus thiaminolyticus* by Wittliff and Airth[1a] was homogeneous by three criteria of purity: ultracentrifugation, polyacrylamide gel electrophoresis, and immunodiffusion in agar gel. Thiaminase I has a sedimentation velocity coefficient, $S_{20,w}$, of 3.1 S and a diffusion coefficient of 6.6 $\times$ 10^{-7} cm^2/sec suggesting a molecular weight of 44,000.[1a] Ebata and Murata[10] reported a molecular weight of 40,000, calculated from a sedimentation velocity coefficient of 2.0 S and a diffusion coefficient of 4.7 $\times$ 10^{-7} cm^2/sec.

Substrate Specificity. The bacterial and shellfish enzymes are specific for thiamine, thiamine pyrophosphate, and other thiamine derivatives with the 4-amino group of the pyrimidine moiety intact.[11] Wang, Wilkins, and Airth[12] have recently shown that the bacterial enzyme will utilize pyrithiamine, but not oxythiamine, as a substrate in the base-exchange reaction.

Activators and Inhibitors. The enzyme from bacteria, shellfish, fish, and ferns is markedly activated by many aromatic amines, heterocyclic amines, and sulfhydryl compounds at concentrations of 10^{-3} M.[8,11] The bacterial enzyme is unaffected by FeSO$_4$, Fe$_2$(SO$_4$)$_3$, and CuSO$_4$ at concentrations of 10^{-3} M; however, these metal ions are slightly inhibitory to the enzyme from fish, shellfish, and ferns.[8] The activity of thiaminase I from all sources is inhibited slightly by MnSO$_4$ (10^{-3} M).[8] Both p-chloromercuribenzoate (10^{-4} M) and monoiodoacetic acid (10^{-4} M) are potent inhibitors of bacterial thiaminase I activity.[8]

Effect of pH. Thiaminase I from *Bacillus thiaminolyticus* has a broad pH optimum of 5.8–6.8.[1] However, Fujita *et al.*[4] reported that the enzyme from *B. thiaminolyticus* and certain shellfish (*Meretrix meretrix*) has a pH optimum of 5.0 and that the enzyme isolated from the viscera of a fish (*Carassius carassius*) has a pH optimum of 6.0. Kenten[6] found a pH optimum of 7.8 for the thiaminase I purified from bracken ferns. The pH optima of the thiaminase I isolated from other sources are reported elsewhere.[8,11]

Effect of Temperature. Wittliff and Airth[1] reported a temperature optimum of 37° for the bacterial enzyme; the temperature coefficient Q$_{10}$ (10–20°) was 1.93 and Q$_{10}$ (20–30°) was 1.33. These workers[1] also reported that thiaminase I has two activation energies, 9800 cal/mole and 2700

[10] J. Ebata and K. Murata, *J. Vitaminol. (Kyoto)* **7**, 115 (1961).
[11] A. Fujita, Vol. II [103].
[12] L. Wang, J. H. Wilkins, and R. L. Airth, *Can. J. Microbiol.* **14**, 1143 (1968).

cal/mole, and they interpret these data as indicating that the enzyme is reversibly inactivated at temperatures above 45° but not exceeding 65°. Mazrimas et al.,[13] working with thiaminase I isolated from carp viscera, calculated an energy of activation of 9600 cal/mole. This enzyme also had a temperature-activity optimum of 40°. The temperature optima of thiaminase I isolated from other sources are reported by Murata[8] and Fujita.[11]

Thermostability Studies. More than 90% of the activity of bacterial thiaminase I remains after heating 20 minutes at 60° in 0.5 M sodium phosphate buffer (pH 5.8).[1a] However, heating for 20 minutes at 65° completely destroys the thiaminase I activity. The temperature inactivation coefficient (T_i, temperature at which 50% of the original activity remains after heating 20 minutes) of the bacterial enzyme is 63.5°.[1]

Effect of Substrate Concentration. The K_m values of bacterial thiaminase I at pH 5.8 and 25° are 8.7×10^{-6} M for thiamine and 2.9×10^{-3} M for aniline.[1a] Ebata and Murata[10] reported that the K_m value for thiamine was 0.9×10^{-3} M and for pyridine was 1.0×10^{-3} M at pH 6.5 and 30° using enzyme purified from *Bacillus thiaminolyticus.* Wittliff and Airth[1] reported that the maximum velocity of the thiaminase I reaction at pH 5.8 and 25° is 6.4 mg of thiamine decomposed per minute per milligram of protein.

Serological Properties. Recently, Wittliff et al.[14] obtained a rabbit antiserum to bacterial thiaminase I. The antiserum contained approximately 0.93 mg of antienzyme antibody per milliliter of serum; it inhibited thiaminase I activity in a noncompetitive manner with respect to thiamine and was neither competitive nor noncompetitive with respect to aniline.

[13] J. A. Mazrimas, P.-S. Song, L. L. Ingraham, and R. D. Draper, *Arch. Biochem. Biophys.* **100,** 409 (1963).
[14] J. L. Wittliff, W. J. Mandy, and R. L. Airth, *Biochemistry* **7,** 2380 (1968).

[42] Thiaminase II (Thiamine Hydrolase, EC 3.5.99.2)[1]

By James L. Wittliff and R. L. Airth

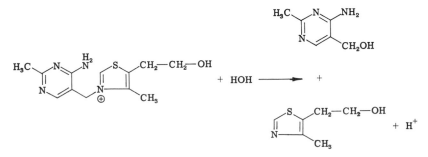

Thiaminase II, unlike the base-transfer enzyme (thiaminase I), catalyzes the hydrolysis of thiamine.

Assay Method

Principle. The enzyme activity is measured by determining the amount of thiamine decomposed in a reaction mixture of thiamine, buffer, and enzyme preparation after 10-minute incubation at 37°.[1a] The remaining thiamine is measured fluorometrically by the thiochrome method.[2]

Reagents

Sodium phosphate buffer, 0.067 M, pH 7.0
Thiamine·HCl, 3 × 10⁻⁶ M
Ethylenediaminetetraacetic acid (EDTA), 10⁻³ M
Metaphosphoric acid, 10% solution
Enzyme: filtrate from an 8-day culture of *Bacillus aneurinolyticus* grown in Ikehata's synthetic medium[1a] (see section on cultivation of bacteria)

Procedure. To 1.0 ml of distilled H_2O in a test tube, add 1.0 ml each of the phosphate buffer, EDTA solution, and enzyme preparation. After a 5-minute equilibration at 37° in a water bath, the reaction is initiated by the addition of 1.0 ml of the thiamine solution. After 10 minutes the reaction is stopped by the addition of 1.0 ml of 10% metaphosphoric acid. The residual thiamine is determined by the thiochrome method[2] with aliquots of the reaction mixture. A blank containing water, phosphate buffer, EDTA solution, and enzyme and a control containing all the reactants and heated enzyme (10 minutes, 100°) are performed simultaneously with the experimental mixture. Thiaminase II activity is expressed by the first-order reaction constant:

$$K = \frac{1}{t} \log(a/a - x) \tag{1}$$

where t is time in minutes, a is the initial amount of thiamine, and x is the amount of thiamine decomposed in time t.[1a]

Specific Activity. Specific activity is expressed as thiaminase II activity (K) per milligram of protein. The method of Lowry *et al.*[3] is used to determine protein concentration.

---

[1] Manuscript prepared in The Cell Research Institute, The University of Texas.
[1a] H. Ikehata, *J. Gen. Appl. Microbiol.* **6,** 30 (1960).
[2] See Vol. II [103]; R. L. Airth and G. E. Foerster, this volume [14].
[3] O. H. Lowry, N. J. Rosebrough, A. L. Farr, and R. J. Randall, *J. Biol. Chem.* **193,** 265 (1951).

Source of Enzyme

In addition to *Bacillus aneurinolyticus*, thiaminase II has been reported from various bacteria and yeastlike fungi; among those listed are *Micrococcus pyogenes*, *Escherichia coli*, *Candida aneurinolytica*, and *Trichosporon aneurinolyticum*.[4]

Purification Procedure

Cultivation of Bacteria. The organism used as a source of thiaminase II is *Bacillus aneurinolyticus*. It is grown aerobically in Ikehata's synthetic medium,[1] which consists of: glycerol, $2.18 \times 10^{-1} M$; monosodium glutamate, $0.53 \times 10^{-1} M$; sodium citrate·$2H_2O$, $1.7 \times 10^{-2} M$; KH_2PO_4, $3.35 \times 10^{-3} M$; K_2HPO_4, $3.35 \times 10^{-3} M$; NaCl, $1.7 \times 10^{-2} M$; $MgSO_4$·$7H_2O$, $2.8 \times 10^{-3} M$; $FeSO_4$·$7H_2O$, $1.1 \times 10^{-4} M$; and thiamine·HCl, $3.0 \times 10^{-7} M$. The pH of the medium is brought to 6.8 with dilute NaOH solution. According to Ikehata,[1] the bacteria are cultured at 37° for 8 days before the cells are removed, and the supernatant solution is used for the purification of thiaminase II. The enzyme presumably is located intracellularly but is released into the medium during the late spore-forming stage.[5] Wittliff[6] has suggested a shorter method of cultivation using the media and inoculation procedures given for *Bacillus thiaminolyticus*.[7] According to the method of Ikehata,[1a] thiaminase II is purified from the filtrate of an 8-day culture.

Step 1. First Ammonium Sulfate Precipitation. Unless otherwise noted, all operations are performed at 0–4°. Solid ammonium sulfate is added slowly with stirring to 5 liters of filtrate to a final concentration of 400 g/liter. The pH of the solution is adjusted occasionally to 7.0 with 20% NH_4OH. The precipitate is removed by filtration or centrifugation (30,000 g) and dissolved in 500 ml of distilled water.

Step 2. Reprecipitation with Ammonium Sulfate. To 500 ml of crude enzyme solution from the first salt precipitation, 330 ml of saturated ammonium sulfate solution is added with stirring at pH 7.0. The precipitate is removed by centrifugation (30,000 g), and an additional 160 ml of saturated ammonium sulfate solution is added with stirring to the 800 ml of supernatant solution at pH 7.0. The second precipitate is removed by centrifugation (30,000 g) and dissolved in 50 ml of distilled water.

Step 3. Calcium Phosphate Gel Treatment. Calcium phosphate gel is

[4] K. Murata, *in* "Review of Japanese Literature on Beriberi and Thiamine," p. 220. Vitamin B Research Committee of Japan, Tokyo, 1965.

[5] R. Kimura, *in* "Review of Japanese Literature on Beriberi and Thiamine," p. 255. Vitamin B Research Committee of Japan, Tokyo, 1965.

[6] J. L. Wittliff, Ph.D. Dissertation, The University of Texas, Austin, Texas, 1967.

[7] J. L. Wittliff and R. L. Airth, this volume [41].

prepared by mixing 100 ml of 0.3 M CaCl$_2$ solution with 100 ml of 0.2 M Na$_3$PO$_4$ solution.[8] After the pH is corrected to 8.4 with dilute NH$_4$OH, the gel is washed with distilled water until the removal of ammonia is complete.

The enzyme solution is dialyzed overnight against distilled water, and the pH is adjusted to 6.0. Calcium phosphate gel (6.0 g dry weight) is added with stirring to the dialyzed preparation. After the suspension is centrifuged, the precipitate is washed with distilled water and eluted twice with minimal amounts of 0.067 M sodium phosphate buffer, pH 7.0.

Step 4. Starch Zone Electrophoresis. The enzyme solution from the calcium phosphate gel step is dialyzed for 2 days against distilled water and concentrated to about 1.0 ml in a vacuum desiccator over solid NaOH. The concentrated enzyme solution is dialyzed overnight against sodium phosphate buffer, pH 7.7 (ionic strength, $\mu = 0.05$) and mixed with small amounts of washed starch previously equilibrated with the same buffer. Electrophoresis is carried out for 15 hours at 4° in a trough of starch (33 × 6 × 1.5 cm) in phosphate buffer ($\mu = 0.05$) pH 7.7. The potential applied is 200 V (6 V/cm); the current is approximately 7 mA (0.8 mA/cm^2). After electrophoresis, bands of starch (each 1 cm wide) are cut, removed, and eluted with 10 ml of distilled water. Thiaminase II migrates as a single band toward the anode. A representative purification of the enzyme is presented in the table. By using this procedure, Ikehata[1a] has obtained crystals of thiaminase II.

PURIFICATION OF THIAMINASE II[a]

Step	Total protein (mg)	Total activity (K)	Specific activity (K/mg protein)	Recovery (%)
Culture filtrate	11200	5565	3.1	100.0
(NH$_4$)$_2$SO$_4$ fractionation	845	2043	15.1	37.0
Calcium phosphate gel treatment	92	1552	52.0	28.0
Starch zone electrophoresis	29	460	77.0	8.0

[a] H. Ikehata, *J. Gen. Appl. Microbiol.* **6**, 30 (1960).

Properties

Ultraviolet Spectrum. The spectrum of purified thiaminase II is that of a simple protein with an absorption maximum at 276 mμ and a minimum at 252 mμ.[9]

[8] See Also Vol. I [11].

[9] R. Kimura, E. Sakakibara, and M. Katsumata, *J. Vitaminol.* (*Kyoto*) **4**, 199 (1958).

Molecular Weight. The purified enzyme from *Bacillus aneurinolyticus*, which migrated as a single peak in the ultracentrifuge and zone electrophoresis, has a sedimentation velocity coefficient (uncorrected) of 4.5 S and an apparent diffusion coefficient of 4.3×10^{-7} cm^2/sec. By the use of these values, the molecular weight was calculated as 100,000.[1a]

Substrate Specificity. The bacterial enzyme will hydrolyze thiamine and thiamine derivatives with the side chain intact at the 5-position of the thiazole moiety.[4] Thiaminase II will also hydrolyze thiothiamine, but not thiamine pyrophosphate.[4]

Activators and Inhibitors. Thiaminase II is activated by cysteine and EDTA in a concentration range of 2×10^{-5} M to 10^{-1} M.[1a,9] The enzyme is markedly inhibited by many aromatic and heterocyclic amines at concentrations of 10^{-3} M,[4] and heavy metal salts such as $CuSO_4$, $ZnSO_4$, and $FeSO_4$ at concentrations of 10^{-5} M.[1a] The activity of purified thiaminase II is almost totally inhibited by 10^{-5} M *p*-chloromercuribenzoate.[1a]

Effect of pH. The optimum pH of purified thiaminase II from *Bacillus aneurinolyticus* is 8.6;[1a] the enzyme isolated from *Trichosporon aneurinolyticum*, *Candida aneurinolytica*, and *Oospora lactis* has a pH optimum of 7.0.[4]

Effect of Temperature. The enzyme purified from *B. aneurinolyticus* by Kimura *et al.*[9] has a temperature optimum of 60° as does the enzyme from *O. lactis*.[4] The enzyme from both of these sources is inactivated by heating at 80° for 20 minutes. However, thiaminase II isolated from *T. aneurinolyticum* exhibits a temperature optimum at 40–45°, whereas the enzyme from *C. aneurinolytica* has the lowest temperature optimum at 30–40°.[4] Heating for 10 minutes at 60° will completely inactivate the enzyme from these two sources.[4]

Effect of Substrate Concentration. The K_m value of thiaminase II purified from *B. aneurinolyticus* is 3.0×10^{-6} M for thiamine at pH 8.6 and 37° as reported by Ikehata.[1a] Ikehata also calculated the maximum velocity of the thiaminase II reaction under the same conditions as 1.3 mg of thiamine decomposed per minute per milligram of protein.

[43] Resolution, Reconstitution, and Other Methods for the Study of Binding of Thiamine Pyrophosphate to Enzymes

By A. V. Morey and Elliot Juni

Since the discovery in 1911 of the first thiamine pyrophosphate enzyme, yeast pyruvate decarboxylase (2-oxoacid carboxy-lyase, EC 4.1.1.1),[1] the

---

[1] C. Neuberg and L. Karczag, *Biochem. Z.* **36**, 96 (1911).

number of TPP-enzymes studied has steadily increased. A basic feature of the catalytic action of TPP-enzymes is the splitting of carbon–carbon bonds in compounds with the general structure, R_1—CO—CO—R_2 and R_3—CO—C(H)(OH)—R_4. In the decarboxylation of pyruvate the splitting of C—C results in the immediate release of carbon dioxide and the formation of enzyme-bound α-hydroxyethyl-TPP from carbons 2 and 3 of the substrate.[2] The α-hydroxyethyl moiety is decomposed to acetaldehyde in simple decarboxylation reactions. In more complex reactions such as the synthesis of acetoin, α-acetolactate, diacetylmethylcarbinol, sedoheptulose-7-phosphate, and tartronic semialdehyde the thiazolium adduct (α-hydroxyethyl-TPP, α,β-dihydroxyethyl-TPP, or hydroxymethyl-TPP) reacts with another molecule of substrate, or with a molecule of one of the products, before being released as part of the acyloin. In such cases, where condensation reactions take place, firm binding of TPP to apoenzymes would be most efficient, since free dissociation of the coenzyme would then require saturating concentrations of the "active aldehyde" intermediates to accumulate. "Active acetaldehyde" bound to the pyruvate decarboxylase component of the pyruvate dehydrogenase complex,[3] may, however, be required to dissociate so that the second enzyme of the complex, lipoic reductase–transacetylase, could act on its presumed substrate, α-hydroxyethyl-TPP. It might be expected, therefore, that the pyruvate decarboxylase component of the pyruvate dehydrogenase complex would not strongly bind the cofactor. Experimental findings, however, are not consistent with such expectations in all cases.

Pyruvate decarboxylase from yeast,[4] transketolase from spinach,[5] yeast,[6] and rat liver,[5] α-ketoglutarate dehydrogenase from pig heart muscle,[7] and the enzyme systems catalyzing the phosphoroclastic split of pyruvate from *Clostridium butyricum*[8] and *Clostridium kluyveri*,[9] all seem to bind TPP in a nondissociating manner. α-Acetolactate synthetase from *Aerobacter aerogenes* and glyoxylate carboligase from *Escherichia coli* appear to be heterogeneous in that only part of each enzyme binds TPP irreversibly.[4] Cofactors are bound in a reversible manner by pyruvate decarboxylase from wheat germ,[10] pyruvate dehydrogenase from *E. coli*,[4] diacetylmethyl-

[2] L. O. Krampitz, I. Suzuki, and G. Greull, *Federation Proc.* **20**, 971 (1961).

[3] M. Koike, L. J. Reed, and W. R. Carroll, *J. Biol. Chem.* **238**, 30 (1963).

[4] A. V. Morey and E. Juni, *J. Biol. Chem.* **243**, 3009 (1968).

[5] B. L. Horecker, P. Z. Smyrniotis, and H. Klenow, *J. Biol. Chem.* **205**, 661 (1953).

[6] G. de la Haba, I. G. Leder, and E. Racker, *J. Biol. Chem.* **214**, 409 (1955).

[7] T. Hayakawa, H. Muta, M. Hirashima, S. Ide, K. Okabe, and M. Koike, *Biochem. Biophys. Res. Commun.* **17**, 51 (1964).

[8] R. S. Wolfe and D. J. O'Kane, *J. Biol. Chem.* **215**, 637 (1955).

[9] C. W. Shuster and F. Lynen, *Biochem. Biophys. Res. Commun.* **3**, 350 (1960).

[10] T. P. Singer and J. Pensky, *J. Biol. Chem.* **196**, 375 (1952).

carbinol synthetase from *Micrococcus ureae*,[4] pyruvate oxidase from *E. coli*,[11] and pyruvate oxidase from *Proteus vulgaris*,[4] and they are lost during purification of these enzymes.

The different methods that have been used for cofactor resolution of TPP-enzymes from several sources are described below.

Procedures for Resolving TPP-Enzymes

Alkaline Washing of Yeast Cells[12]

One gram of dried yeast is suspended in distilled water to a final volume of 20.0 ml. The suspension is mixed with 20.0 ml of 0.2 M dibasic sodium phosphate solution whose pH has been made alkaline by adding sufficient concentrated NaOH so that the final pH is 8.0 upon mixing the yeast suspension with the phosphate solution. The preparation is stirred for 5 minutes, then the yeast cells are separated by centrifugation and suspended in 20 ml of water. The suspension is immediately centrifuged and washed with water two more times. The final sediment is suspended in 0.1 M acetate buffer, pH 5.6. The pyruvate decarboxylase in washed yeast is resolved for TPP. About 45% of the pyruvate decarboxylase activity is recovered in the yeast cells.

Precipitation of Yeast Pyruvate Decarboxylase with Alkaline Ammonium Sulfate[13]

Reagents

Citrate buffer, 0.5 M, pH 6.0
Ammonium sulfate, saturated solution
Ammoniacal ammonium sulfate solution, 9 parts of saturated ammonium sulfate mixed with one part of concentrated ammonia

Procedure. The enzyme solution is diluted with 2 volumes of water, and 4 volumes of ammoniacal ammonium sulfate are added to the diluted mixture. The precipitate is centrifuged, dissolved in water, and the process is repeated three more times. The final protein sediment is suspended in a solution consisting of equal parts of saturated ammonium sulfate solution and 0.5 M citrate buffer, pH 6.0. The suspension is stable at room temperature for several weeks. Phosphate buffer, pH 6.5, may be used in place of citrate buffer.

[11] R. Williams and L. P. Hager, *J. Biol. Chem.* **236**, PC 36 (1961).
[12] E. P. Steyn-Parve and H. G. K. Westenbrink, *Z. Vitaminforsch.* **15**, 1 (1944).
[13] D. E. Green, D. Herbert, and V. Subrahmanyan, *J. Biol. Chem.* **138**, 327 (1941).

Precipitation of Liver and Spinach Transketolase with Acidic Ammonium Sulfate[5]

Reagents

Ammonium sulfate, saturated solution
Sodium acetate, 4.0 M
Sulfuric acid solution, 2.2 N
Glycylglycine buffer, 0.25 M, pH 7.4
Ammonium hydroxide, 2.0 N

Procedure. Saturated ammonium sulfate solution is added to a preparation of transketolase to bring the ammonium sulfate concentration to 45%. An amount of 4.0 M sodium acetate equal to 1.3% of the volume of enzyme–ammonium sulfate mixture is also added to buffer the solution. The solution is then placed in an ice bath, and 2.2 N sulfuric acid is added until the pH becomes acid to methyl orange (pH $\sim$ 3.5). After 5 minutes at 0° the precipitate is centrifuged in the cold, dissolved in 0.25 M glycylglycine buffer, pH 7.4, and neutralized with 2.0 N ammonium hydroxide. The yield of resolved enzyme is 70–80% with no loss in specific activity.

Phosphoketolase from *Leuconostoc mesenteroides* has also been resolved for TPP by precipitation from acidic ammonium sulfate.[14]

Differential Precipitation of Metal Ions and Apoenzyme at pH 8.5, Yeast Pyruvate Decarboxylase, and Other Enzymes[4]

Reagents

Ammonium sulfate, saturated solution
Tris, 2.0 M
NaOH, 0.1 M in 2.0 M Tris
Sodium phosphate, 0.1 M and 0.5 M, pH 6.0

Procedure. Ammonium sulfate is added with stirring to ice-cold, enzyme solution to bring the final concentration to 0.4 M. The pH of the mechanically stirred mixture is then adjusted to 8.5 by dropwise addition of 2.0 M Tris. Stirring is continued for 20 minutes; then the enzyme solution is centrifuged at 20,000 g for 20 minutes at 5°. The sediment, which consists of phosphates of magnesium, calcium, and other naturally occurring di- and trivalent cations and some precipitated protein, is discarded. Ammonium sulfate is added to the mechanically stirred supernatant fraction to bring the final concentration to 70%. The pH of the suspension is again

[14] R. Votaw, W. J. Williamson, L. O. Krampitz, and W. A. Wood, *Biochem. Z.* **338,** 756 (1963).

adjusted to 8.5 by dropwise addition of 0.1 N NaOH in 2.0 M Tris, and this is followed by centrifugation at 20,000 g for 15 minutes at 4°. The supernatant fraction, which should contain TPP, is discarded. The sedimented protein is suspended in 0.1 M sodium phosphate, pH 6.0, and the pH of the suspension is adjusted to 8.5 by dropwise addition of 2.0 M Tris. The insoluble material from the suspension is removed by centrifugation at 20,000 g for 20 minutes at 5°. Ammonium sulfate is added to the supernatant solution after the above centrifugation, to a concentration of 70%, at which time the pH is adjusted to 8.5 by dropwise addition of 0.1 N NaOH in 2.0 M Tris. The precipitated protein is sedimented by centrifugation at 20,000 g for 15 minutes at 5° and suspended in a solution made by mixing equal volumes of 0.5 M sodium phosphate, pH 6.0, and saturated ammonium sulfate. This suspension, containing resolved pyruvate decarboxylase, can be stored at −18° for periods longer than 6 months without significant loss of activity. All operations are performed at 0–5°.

The recovery of resolved pyruvate decarboxylase prepared by this procedure ranges from 75 to 80%. The yields of resolved enzyme obtained by the procedure of Green et al.,[13] described above under procedure for precipitation of yeast pyruvate decarboxylase with alkaline ammonium sulfate, are much lower. Moreover, it is important to note that resolved pyruvate decarboxylase prepared by the procedure of Green et al.[13] contains small amounts of naturally occurring divalent cations, since these cations form insoluble phosphates at pH 8.0 or higher, the pH at which enzyme resolved for TPP is precipitated.

Gel Filtration at Alkaline pH[4]

Reagents

Sephadex, G-25, fine
Tris-phosphate, 0.1 M, pH 8.5
Acetic acid, 4.0 M
Tris, 2.0 M

Procedure. A Sephadex G-25 column of appropriate size is prepared in a cold room (∼5°) and equilibrated with 0.1 M Tris-phosphate, pH 8.5. The pH of the solution containing the TPP-enzyme to be resolved is adjusted to 8.5 by slow addition of 2.0 M Tris. The alkaline solution is stirred for 15 minutes at 5°, then filtered through the Sephadex column in the cold. The filtrate, collected after passage of the column void volume, contains the resolved enzyme and is adjusted to pH 6.2 by addition of 4.0 M acetic acid solution.

Pyruvate decarboxylases from *Zymomonas mobilis* and yeast are resolved by gel filtration at alkaline pH. It should be possible to resolve phospho-ketolase from *L. mesenteroides* and transketolase from spinach and rat liver by gel filtration at acid pH (~ 3.5). Thiamine pyrophosphate enzymes that bind Mg^{2+} and TPP reversibly can be separated from these cofactors more conveniently by gel filtration at the usual pH of the enzyme solutions rather than by dialysis.

Reconstitution of Enzymes Resolved for TPP and Mg^{2+}

Formation of holoenzyme from pyruvate decarboxylase apoenzyme and cofactors, TPP and Mg^{2+}, is a relatively slow process.[4] The rate of reconstitution is a function of the temperature and the concentrations of TPP and divalent cations. The extent of reconstitution depends upon the pH of the reaction mixture and the concentrations of cofactors when they are presented below the optimum concentrations. The rate of reconstitution is determined from the amount of holoenzyme formed at any time during reconstitution, which in turn is determined either directly by arresting the progress of reconstitution, or indirectly from kinetic parameters.[15]

Rate of Cofactor Reconstitution of Yeast Pyruvate Decarboxylase by Direct Determination of Holoenzyme Formed[4]

Procedure. The rate of reconstitution is determined by incubating re-solved enzyme in 0.1 M sodium phosphate, pH 6, containing 0.01 M $MgSO_4$ at a constant temperature for various lengths of time after addition of a given amount of TPP. Reconstitution is stopped by diluting aliquots of reaction mixture into 0.2 M sodium phosphate, pH 6.0 containing 0.046 M EDTA. Control experiments have shown that this procedure prevents further reconstitution and is without effect on that part of the enzyme that is already saturated with cofactors.[4] The rate of reconstitution is a function of TPP concentration. In order to achieve complete reconstitution of apoenzyme in about 1 hour at 23° in the presence of excess Mg^{2+} (0.01 M), the concentration of TPP must be 6.25×10^{-5} M or greater. With TPP concentrations of 2×10^{-4} M and higher, the rates of reconstitution of yeast pyruvate decarboxylase are too fast to determine at temperatures above 20° with this procedure. The rate of cofactor reconstitution increases as the temperature is raised.

It has been shown recently by Schellenberger and Hübner[15] that the substrate, pyruvate, also influences the rate of cofactor reconstitution of resolved yeast pyruvate decarboxylase. At a pyruvate concentration of 3.43×10^{-2} M, the rate of cofactor reconstitution is 10 times that observed in its absence.

---

[15] A. Schellenberger and G. Hübner, *Z. Physiol. Chem.* **348**, 491 (1967).

It is clear from the study of the rate of cofactor reconstitution of yeast pyruvate decarboxylase[4] that precautions must be taken while assaying thiamine pyrophosphate by the enzymatic method[16] to be certain that sufficient incubation time for maximum enzyme activation is used.

Functional Groups of TPP Involved in Binding to Enzymes

The nature of the binding of TPP to enzymes depends on the individual interactions among the functional groups of the apoenzyme and TPP. With the possible exception of rat liver transketolase,[5] and benzoylformate decarboxylase,[17] divalent cations have a significant positive affect on TPP binding. Information concerning primary, secondary, and tertiary structures of TPP-enzymes is lacking at the present time.

Use of TPP Analogs

It is known that thiamine, thiamine monophosphate, deaminothiamine, 4-hydroxy-4-deaminothiamine, and pyrithiamine do not serve as cofactors for pyruvate decarboxylase,[18, 19] nor do they act as inhibitors of cofactor reconstitution when added to apoenzyme together with TPP. The pyrophosphate derivatives of deaminothiamine, pyrithiamine, N-methylthiamine, and 4'-hydroxy-4'-deaminothiamine, however, are inactive in replacing TPP as coenzyme, but they are inhibitory to the activity of pyruvate decarboxylase if added before or simultaneously with TPP to reconstitution mixtures.[20] It appears that the presence of a pyrophosphate group permits the analogs of TPP to bind to the cofactor site on the enzyme. "Thiazole pyrophosphate" (4-methyl-5-(pyrophosphoryl-2-hydroxyethyl)thiazole) inhibits cofactor reconstitution of pyruvate decarboxylase when added simultaneously with TPP.[4,21] However, addition of thiazole pyrophosphate after reconstitution of the apoenzyme with TPP and Mg^{2+} results in negligible inhibition. Since thiazole pyrophosphate binds reversibly to the enzyme, its presence does not affect the degree of reconstitution if TPP is present in sufficiently high concentration even though cofactor reconstitution proceeds at a slower rate in the presence of the analog.

The TPP analogs 4-nor-TPP and 6'-methyl-4-nor-TPP are active as coenzymes to some extent.[22] The analogs 6'-methyl-4'-hydroxy-4'-deamino-TPP and 6'-methyl-TPP, with methyl groups in both the 6' and 4 positions

---

[16] J. Ullrich, this volume [19].

[17] R. Y. Stanier, Vol. II [37].

[18] S. Eich and L. R. Cerecedo, *J. Biol. Chem.* **207**, 295 (1954).

[19] S. Eich and L. R. Cerecedo, *Arch. Biochem. Biophys.* **57**, 285 (1955).

[20] A. Schellenberger, W. Rodel and H. Rodel, *Z. Physiol. Chem.* **339**, 122 (1964).

[21] E. R. Buchman, E. Heelgoard, and J. Bonner, *Proc. Natl. Acad. Sci. U.S.* **26**, 561 (1940).

[22] A. Schellenberger, I. Heinroth, and G. Hübner, *Z. Physiol. Chem.* **348**, 506 (1967).

of the pyrimidine and the thiazole rings, respectively, are totally inactive as coenzymes and have low affinity for the apoenzyme as determined by lack of binding during gel filtration.[22] The affinity of the analogs with one methyl group in either the 4 or 6' positions, namely, 4'-hydroxy-4'-deamino-TPP and 6'-methyl-4-nor-TPP, as determined by binding during gel filtration, seems to be comparable to that of TPP.[22] The analog with no methyl group in either the 6' or 4 positions, 4-nor-TPP, does not bind firmly to the apoenzyme.[22]

[44] Thiamine Antagonists

By EDWARD F. ROGERS

A thiamine antagonist may be defined as a compound that can compete with thiamine, or thiamine precursors or derivatives, in enzyme reactions. Most of the known antagonists inhibit either the synthesis of thiamine pyrophosphate (TPP, cocarboxylase) or thiamine pyrophosphate-mediated reactions. Well-known antagonists are pyrithiamine (II),[1] oxythiamine (III),[2] and amprolium (IV)[3]; their structures are compared with that of thiamine (I, note numbering) in Fig. 1. Because such agents have been useful in the elucidation of thiamine biochemistry, treatment in this text seems justified. The reader may also consult other reviews on the subject.[4–7]

Standard techniques for measurement of the effectiveness of antagonists are illustrated in the references cited for the following types of tests: enzyme inhibition,[4,8,9] microbial growth inhibition,[10–12] animal growth or

[1] A. H. Tracy and R. C. Elderfield, *J. Org. Chem.* **6**, 54 (1941); A. W. Wilson and S. A. Harris, *J. Am. Chem. Soc.* **71**, 2231 (1949).

[2] F. Bergel and A. R. Todd, *J. Chem. Soc.* p. 1504 (1937).

[3] E. F. Rogers, R. L. Clark, A. A. Pessolano, H. J. Becker, W. J. Leanza, L. H. Sarett, A. C. Cuckler, E. McManus, M. Garzillo, C. Malanga, W. H. Ott, A. M. Dickinson, and A. Van Iderstine, *J. Am. Chem. Soc.* **82**, 2974 (1960).

[4] L. R. Cerecedo, *Am. J. Clin. Nutr.* **3**, 273 (1955).

[5] E. P. Steyn-Parve, *in* "Thiamine-Deficiency: Biochemical Lesions and Their Clinical Significance" (G. E. W. Wolstenholme and M. O'Connor, eds.), p. 26. Little, Brown, Boston, Massachusetts, 1967.

[6] G. Rindi, *Bull. Chim. Farm.* **102**, 363 (1963).

[7] E. F. Rogers, *Ann. N.Y. Acad. Sci.* **98**, 412 (1962).

[8] L. R. Johnson and C. J. Gubler, *Federation Proc.* **24**, 481 (1965); *Biochim. Biophys. Acta* **156**, 85 (1968).

[9] J. C. Koedam and E. P. Steyn-Parve, *Proc. Koninkl. Ned. Acad. Wetenschap. Ser.* **C63**, 318 (1960).

[10] T. Sakuguri, *Arch. Biochem. Biophys.* **74**, 362 (1958).

[11] D. W. Woolley and A. G. C. White, *J. Exptl. Med.* **78**, 489 (1943).

[12] M. C. Goldschmidt and R. P. Williams, *J. Bacteriol.* **96**, 609 (1968).

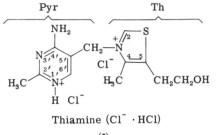

Thiamine (Cl⁻ · HCl)

(I)

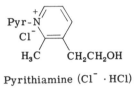

Pyrithiamine (Cl⁻ · HCl)

(II)

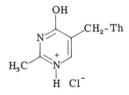

Oxythiamine (Cl⁻ · HCl)

(III)

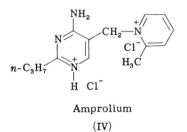

Amprolium

(IV)

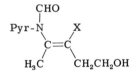

X

(V) SH
(V-a) SSCH₂CH=CH₂
(V-b) OH (or keto form)

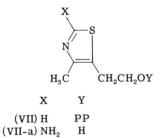

X	Y
(VII) H	PP
(VII-a) NH₂	H

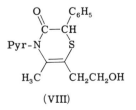

(VIII)

Fig. 1

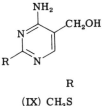

R

(IX) CH₃S
(IX-a) CH₃O

(X) NH₂CH₂CH₂
(X-a) 2-NH₂—C₆H₄CH₂

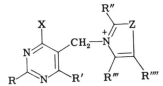

	R	X	R'	Z	R''	R'''	R''''
VI	CH₃	NH₂	CH₃	S	H	H	CH₂CH₂OPP
VIa	CH₃	NH₂	CH₃	S	H	CH₃	CH₂CH₂OPP
VIb	CH₃	NH₂	H	S	CH₃	CH₃	CH₂CH₂OH
VIc	CH₃	NH₂	H	N-CH₃	H	CH₃	CH₂CH₂OH
VId	CH₃	NH₂	H	N-n-C₄H₉	H	CH₃	CH₂CH₂OH
VIe	CH₃	NH₂	H	N-CH₂C₆H₅	H	CH₃	CH₂CH₂OH
VIf	CF₃	NH₂	H	S	H	CH₃	CH₂CH₂OH
VIg	n-C₄H₉	NH₂	H	S	H	CH₃	CH₂CH₂OH
VIh	CH₃S	NH₂	H	S	H	CH₃	CH₂CH₂OH
VIi	CH₃	H	H	S	H	CH₃	CH₂CH₂OH
VIj	CH₃	NH₂	H	S	H	CH₃	CH₃
VIk	CH₃	NH₂	H	S	H	CH₃	C₂H₅
VIl	CH₃	NH₂	H	O	H	CH₃	CH₂CH₂OH
VIm	CH₃O	NH₂	H	S	H	CH₃	CH₂CH₂OH
VIn	CH₃	NH₂	NH₂	S	H	CH₃	CH₂CH₂OH
VIo	CH₃	NH₂	H	S	H	CH₃	CH₂CHOHCH₂OH

Fɪɢ. 1. Structures of thiamine antagonists compared with that of thiamine.

weight maintenance.[13, 14] A novel enzyme approach, which has provided fresh insight on structure–activity relationships of thiamine, deserves special mention. This method permits determination of the binding capacity

[13] W. H. Ott, A. M. Dickinson, and A. Van Iderstine, *J. Poultry Sci.* **44**, 920 (1965).
[14] L. R. Cerecedo, *Intern. J. Vitaminol. Res.* **37**, 189 (1967).

of a TPP analog in reconstituted yeast pyruvate decarboxylase holoenzyme, as well as the enzyme activity of the analog system so formed.[15]

After a consideration of structural aspects of thiamine and its antagonists, the inhibition data will be discussed.

Structural Aspects

As the design, or structural features, of antithiamines necessarily relate to thiamine, some aspects of thiamine chemistry deserve first attention.

The only permissible, metabolically irreversible changes in the thiamine molecule, in terms of vitamin activity, involve the 2'-position. The 2'-ethyl and 2'-n-propyl homologs are, respectively, slightly superior to and about one-third as effective as thiamine.[16] All other structural modifications result in complete loss of activity. The requirements for coenzyme activity are much less stringent, as will be indicated later.

The term "metabolically irreversible" was introduced above to exclude a large class of irrelevant "modified thiamines,"[17] typified by allithiamine (Va, Fig. 1). These are derivatives of the ring-opened form of thiamine (V), which is obtained when thiamine chloride hydrochloride is treated with two moles of alkali. On metabolism of the modified thiamines, generally disulfides like allithiamine or thiol esters, the open form (V) of thiamine is regenerated. In the physiological pH range, this rapidly cyclizes to the thiazolium structure (I). 2,3-Dihydrothiamine[18] is another example of a variation of the basic thiamine structure, convertible *in vivo* to thiamine.

In thiamine pyrophosphate, the coenzyme form of the vitamin, the catalytic function is associated with the carbanion generated at the 2-position of the thiazolium moiety.[19a–19c, 20] The mechanism, as it applies to pyruvate decarboxylase, is illustrated in Fig. 2. More generally, the TPP carbanion is visualized as first adding to the carbonyl group of an α-keto acid or an aldehyde, then a molecule, either carbon dioxide or an aldehyde is ejected, leaving a new carbanion, which reacts further in one of three

[15] A. Schellenberger, *Angew. Chem. Intern. Ed.* **6**, 1024 (1967).

[15b] A. V. Morey and E. Juni, *J. Biol. Chem.* **243**, 3009 (1968).

[16] F. Schultz, *Z. Physiol. Chem.* **265**, 113 (1946), **272**, 29 (1941); H. Andersag and K. Westphal, *Ber.* **70**, 2035 (1937); G. A. Stein, W. L. Sampson, J. K. Cline, and J. R. Stevens, *J. Am. Chem. Soc.* **63**, 2059 (1941).

[17] C. Kawasaki, *Vitamins Hormones* **21**, 69 (1963); T. Matsukawa, S. Yurugi, and Y. Oka, *Ann. N.Y. Acad. Sci.* **98**, 430 (1962).

[18] P. Karrer and H. Krishna, *Helv. Chim. Acta* **33**, 555 (1950).

[19a] R. Breslow, *J. Am. Chem. Soc.* **79**, 1762 (1957).

[19b] R. Breslow, *J. Am. Chem. Soc.* **80**, 3719 (1958).

[19c] R. Breslow and E. McNelis, *J. Am. Chem. Soc.* **81**, 3080 (1959).

[20] T. C. Bruice and S. Benkovic, "Bioorganic Mechanisms," Vol. 2, p. 181. Benjamin, New York, 1966.

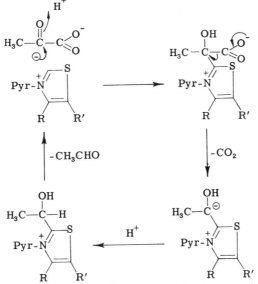

FIG. 2. Mechanism of catalytic function in thiamine pyrophosphate as it applies to pyruvate decarboxylase. R = CH₃; R' = CH₂CH₂OPP.

ways: (a) by protonation, to give a labile aldehyde adduct of TPP (decarboxylases); (b) by direct oxidation, to give a high-energy 2-acyl TPP or, alternatively, by reaction with lipoate to form an acyldihydrolipoate (oxidases); (c) by addition to an aldehyde carbonyl, producing a new ketol (transketolase, phosphoketolase, acetolacetate synthetase, carboligases).

The fact that the coenzyme form of thiamine is the pyrophosphate mplies that a thiamine antagonist, to compete at the coenzyme level, must be an acceptable substrate for thiamine kinase. The thiamine transport mechanism, to be considered later, also bears on this requirement. Obviously nonclassical designs would be unsuitable. "Nonclassical," a medicinal chemist's term,[21] connotes a major structural deviation in some part of an analog molecule—for example, replacement of the 2'-methyl of thiamine by diphenylmethyl; an implied consequence in this case is rejection of the analog as an alternate substrate by thiamine kinase. Since antagonists employed at nonphosphorylated substrate levels, such as thiamine kinase or thiaminase inhibitors, are not thus limited, greater structural freedom may be found in these types.

Acetoin formation, a typical TPP-catalyzed reaction, is promoted nonenzymatically by a compound as simple as N-methylthiazolium chloride.

[21] B. R. Baker, "Design of Active-Site-Directed Irreversible Enzyme Inhibition." Wiley, New York, 1967.

However, since benzyl quaternaries afford better yields than simple alkyl quaternaries in the acetoin test, there has been speculation on the possibility of a catalytic role for the 4'-amino-2'-methyl-5'-pyrimidinylmethyl moiety of thiamine.[19b] The 4'-amino group has attracted special attention,[22,23] partly because of its involvement with the important 2-carbon in the formation of the tricyclic thiochrome and related products produced by alkali treatment of thiamine,[24] although these compounds have no biochemical significance. X-Ray diffraction studies of thiamine pyrophosphate hydrochloride indicate bonding by the 2-hydrogen to the 4'-nitrogen and suggest that proton transfer in enzyme reactions may be facilitated by the amino group.[25]

If interaction does occur between the 4'-amino and the 2-hydrogen of thiamine, or adducts on the thiamine 2-carbon, then the 4- and 6'-positions should be found opposite, on the other side of the methylene bridge. A close 4–6' fit is deduced from the fact that 6'-methyl-4-norTPP (VI, Fig. 1) has coenzyme activity in the reconstituted pyruvate decarboxylase holoenzyme assay, while 6'-methylTPP (VIa) has not. Apparently only one methyl, either at 6' or at 4, as in TPP, can be tolerated. Steric hindrance between the 4- and 6'-methyls must produce a skewed configuration in 6'-methylTPP. Coenzyme activity is absent either because the molecule cannot locate properly on the apoenzyme surface or because access of the 4'-amino to 2-carbon adducts is impossible. The latter alternative is favored and is supported by additional data on oxythiamine pyrophosphate.[15]

In holoenzyme studies with oxythiamine pyrophosphate, it was found that the analog is as acceptable as TPP at the bonding site, furthermore, that pyruvate reacts with the holoenzyme produced. Decarboxylation is effected, but without acetaldehyde release. These results imply that the function of the missing 4'-amino group is purely operational, to facilitate acetaldehyde liberation, a hitherto unrecognized second step in the enzymatic reaction. Oxythiamine emerges as an established "antioperational"[7] antagonist.

Because of the catalytic function of the 2-thiazolium carbanion, one would expect analogs with a blocked 2-position, such as 2-methylthiamine (VIb) to be antioperational also. Actually, the binding of 2-methylTPP by pyruvate decarboxylase is negligible so, for this enzyme at least, the classification is meaningless. Pyrithiamine (II) and similar analogs containing thiazolium ring replacements, such as the N-methyl, N-n-butyl-,

[22] G. E. Risinger and M. F. Dove, *Chem. Ind. (London)* p. 510 (1965).
[23] C. D. May and P. Sykes, *J. Chem. Soc.* p. 649 (1966).
[24] G. D. Maier and D. E. Metzler, *J. Am. Chem. Soc.* **79**, 4386 (1947).
[25] J. Pletcher and M. Sax, *Science* **154**, 1331 (1966).

and N-benzylimidazolium analogs (VIc, VId, VIe),[26,27] represent another type of potential antioperational thiamine inhibitor.

Other studies with the pyruvate decarboxylase holoenzyme system suggest that the 1′-nitrogen and 4-methyl of thiamine pyrophosphate are required for coenzyme binding. Probably the 2′-methyl is also essential, as the 2′-H analog is inactive.[16] Only variants at the 2′-position have been checked to any extent for antivitamin activity. It is believed that 2′-trifluoromethylthiamine (VIf), 2′-butylthiamine (VIg), and 2′-methylthio-thiamine (VIh), all effective antithiamines, function as "antipositional" antagonists,[7] the 2′-substituent causing improper positioning in the holo-enzyme, with consequent malfunction. The 2′-trifluoromethyl group may act indirectly by reducing the basicity of the 1′-nitrogen and so affecting its bonding capacity.

Antagonists containing a thiazolium ring may be designed in alternate modified thiamine structures, probably convertible *in vivo* to the parent quaternary. Thus, the 2′-butyl analog of allithiamine may be expected to metabolize to give 2′-butylthiamine, just as allithiamine (Va), yields thiamine. This approach has been employed in the design of antithiamine coccidiostats.[28]

Further comments on structure–activity relationships are reserved for the next section.

Thiamine Antagonists: Inhibition Data

Thiamine antagonists are classified below in three major categories and a final fringe group. There appears to be a reasonable correspondence between structural types and the inhibition patterns observed at enzyme, microbial, and animal levels.

Oxythiamine Type: Thiazole Ring Intact

Oxythiamine (III) stands alone as a broadly active and potent inhibitor of TPP-mediated enzymes, after conversion to the pyrophosphate. Typical enzymes so inhibited are wheat germ carboxylase and acetoin synthetase.[9] The analog is a good substrate and a poor inhibitor of thiamine kinase.[8,9,29] The TPP-competitive binding of oxythiamine pyrophosphate by yeast pyruvate decarboxylase holoenzyme was mentioned earlier. Even more striking is the observation that the pyrophosphate has an affinity for yeast

[26] G. Kurata, T. Sakai, and T. Miyahara, *Bitamin* **36**, 388 (1967).
[27] H. A. Staab and G. A. Schwalbach, *Ann.* **715**, 128 (1968).
[28] E. F. Rogers and R. L. Clark, U.S. Patent 3,268,403 (1966).
[29] S. Eich and L. R. Cerecedo, *J. Biol. Chem.* **207**, 295 (1954).

transketolase several orders greater than TPP: the enzyme is 60% inactivated by $3.6 \times 10^{-8} M$ analog.[30]

Loss of weight results from administration of oxythiamine at 30–200 times thiamine levels to the chick, pigeon, mouse, rat, and dog. To cite one experiment: on subcutaneous injection of 50 µg of oxythiamine, using 10–14 g mice which were made slightly deficient, then maintained on 1 µg of thiamine per day, the animal survival time was 13–21 days and the incidence of polyneuritis was zero.[4] The reader is referred to reviews[4–6] for further details and references to animal studies. The significance of the biochemical lesions in thiamine deficiency is a complex problem. Transketolase may be most critical,[31] particularly in the deficiency induced by oxythiamine.

The growth of the thiamine-requiring bacterial species *Kloeckera brevis* and *Lactobacillus fermenti* is affected only at high concentrations of oxythiamine.[10] Surprisingly, oxythiamine is much more toxic to *Staphylococcus aureus* than is pyrithiamine,[32] perhaps because of unusually high absorption. The analog inhibits growth of the protozoan *Tetrahymena geleii*[33] and, at 100 times the level of vitamin used, reduces thiamine-induced production of prodigiosin by *Serratia marcescens*.

Deaminothiamine (VIi) pyrophosphate is reported to be a strong carboxylase inhibitor,[34] but enzymatic pyrophosphorylation of this analog, which is necessary for activity *in vivo*, has not been studied. Similarly, the pyrophosphate of the pyridine analog of thiamine, in which the 3′-nitrogen is replaced by methynyl, has cocarboxylase activity, while the pyrophosphate of the analog with 1′-nitrogen replacement has not.[35] These facts argue that the 1′-nitrogen alone is essential for binding. The significance of the 3′-nitrogen is not clear. As there is evidence that adenosine may have regulatory roles in thiamine biosynthesis,[36] it may be important that thiamine and adenosine resemble each other in several respects, one being

[30] G. A. Datta and E. Racker, *J. Biol. Chem.* **236**, 617 (1961).

[31] M. Brin, *in* "Thiamine Deficiency: Biochemical Lesions and Their Clinical Significance" (G. E. W. Wolstenholme and M. O'Connor, eds.), p. 112. Little, Brown, Boston, Massachusetts, 1967.

[32] T. L. V. Ulbricht and J. S. Gots, *Nature* **178**, 913 (1956).

[33] S. Suzuki and T. Yishida, *J. Vitaminol. (Kyoto)* **2**, 53 (1956); M. Mizunoya and F. Wada, *Kyoto Furitsu Ika Daigaku Zasshi* **69**, 1046 (1961).

[34] A. Schellenberger and W. Rodel, *Angew. Chem. Intern. Ed.* **3**, 227 (1964); A. Schellenberger, W. Rodel, and H. Rodel, *Z. Physiol. Chem.* **339**, 122 (1964); A. Schellenberger, V. Müller, K. Winter, and G. Hubner, *ibid.* **344**, 244 (1966).

[35] A. Dornow and A. Hargesheimer, *Ber.* **86**, 461 (1953); A. Schellenberger, K. Wendler, P. Creutzburg, and G. Hubner, *Z. Physiol. Chem.* **348**, 501 (1967).

[36] H. S. Moyed, *J. Bacteriol.* **88**, 1024 (1964); P. C. Newell and R. G. Tucker, *Biochem. J.* **100**, 512, 517 (1966).

the common feature of a 4-aminopyrimidine grouping. Pursuing this thought further, it was recently disclosed that a relatively nonspecific nucleoside diphosphatase functions as a thiamine diphosphatase.[37] Now, since this enzyme will operate upon ribose-5-pyrophosphate also, the TPP thiazolium moiety must bear some resemblance to ribose diphosphate.

The first recognized coenzyme inhibitor was the thiamine thiazole pyrophosphate (VII).[38] Unlike TPP and oxythiamine pyrophosphate, the "half-molecule" combines reversibly with yeast pyruvate decarboxylase, proving the importance of the missing pyrimidine half for binding.[15b]

Amprolium Type

Amprolium (IV) is the prototype of a large class of antithiamines with well-defined structural features and biochemical properties. In these compounds, the pyrimidine moiety of thiamine or 2'-homologs, 4'-amino-2'-loweralkyl-5'-pyrimidinylmethyl, is joined as a quaternizing group to a heteroaryl base, such as pyridine or thiazole, usually alkylated. Amprolium resembles pyrithiamine (II), but differs in one important respect, the hydroxyethyl group is missing. Lacking this, amprolium cannot be converted to a pyrophosphate ester; hence, it is exempt from direct involvement in TPP-mediated reactions.

The discovery that amprolium is a useful poultry coccidiostat[3,7] first focused attention upon this type. The anticoccidial effect is reversed by thiamine.[3,39] Studies on the antithiamine action of amprolium in the chick,[13] led to experiments in thiamine absorption across the intestinal wall.[40] It was demonstrated that amprolium blocks thiamine absorption. This proof of thiamine absorption inhibition established definitely, for the first time, the existence of an inhibitable active transport mechanism. A number of investigators[41-44] soon established that, in various preparations, the transport mechanism employs thiamine kinases. Kinase has been

[37] M. Yumazaki and O. Hayaishi, *J. Biol. Chem.* **243**, 2034 (1968).

[38] E. R. Buchman, E. Hergard, and J. Bonner, *Proc. Natl. Acad. Sci. U.S.* **26**, 561 (1940).

[39] A. C. Cuckler, M. Garzillo, C. Malanga, and E. C. McManus, *Poultry Sci.* **39**, 1241 (1960).

[40] D. Polin, E. R. Wynosky, and C. C. Porter, *Poultry Sci.* **41**, 1673 (1962); *ibid.* **42**, 1057 (1963); *Federation Proc.* **21**, 261 (1962); *Proc. Soc. Exptl. Biol. Med.* **114**, 273 (1963).

[41] S. K. Sharma and J. H. Quastel, *Biochem. J.* **94**, 790 (1965); J. A. Menon and J. H. Quastel, *ibid.* **99**, 766 (1966).

[42] U. Ventura and G. Rindi, *Experientia* **21**, 645 (1965); G. Rindi, V. Ventura, L. de Guiseppe, and G. Sciorelli, *ibid.* **22**, 473 (1966).

[43] H. Y. Neujahr, *Acta Chem. Scand.* **20**, 786 (1966).

[44] T. Bauchop and L. King, *Appl. Microbiol.* **16**, 961 (1968).

found in the membrane fraction of *Escherichia coli*.[45] The inhibition of the enzyme by amprolium and by pyrithiamine correlates satisfactorily with transport phenomena.

Essentially similar results are observed with dimethallium (VIj)[46] and deoxythiamine (VIk).[47]

Besides analogs with pyridinium and thiazolium ring systems, quaternaries derived from *N*-alkylimidazoles, pyridazines, pyrimidines, pyrazines, and other heterocyclics are known as coccidiostats. It is of some interest that the quaternary group is not essential for anticoccidial and antithiamine activity; analogous 1-(4-amino-2-loweralkyl-5-pyrimidinylmethyl)-4-loweralkylpiperazines have similar properties.[48]

The antithiamine index of amprolium is 503 compared with 14 for pyrithiamine.[13] Dimethallium is said to be about equally potent.[46] While compounds of this class tend to be fairly weak B_1-antagonists, most effective orally,[49] a wide range of activity exists and indices as low as 90 and as high as 6500 have been reported.[7]

Pyrithiamine Type; Pyrimidine Ring Intact

The principal enzyme target of pyrithiamine is thiamine kinase,[8, 29, 50-52] the K_i of rat brain kinase, for example, being 1.3×10^{-7} M. Pyrithiamine, however, is also a kinase substrate, although a poor one.[8] This situation seems to account for its considerable differences from amprolium and oxythiamine. While pyrithiamine pyrophosphate inhibits TPP-mediated enzymes, such as the pyruvate decarboxylases from wheat germ[4] and yeast, and acetoin synthetase,[9] it is less effective than oxythiamine pyrophosphate. More important is the systemic action of pyrithiamine, particularly the penetration of nervous tissue, which may be explained by an ability simultaneously to disrupt thiamine-TPP interconversion with consequent loss of thiamine from tissue and to utilize the kinase system for transport. The turnover of thiamine through the diphosphate is reported to be continual and rapid.[5,53]

Pyrithiamine precipitates the neurological symptoms of thiamine

[45] I. Miyata, T. Kawasaki, and Y. Wose, *Bitamin* **38**, 55 (1968).
[46] Z. Suzuoki, K. Furuno, K. Murakami, T. Fujita, T. Matsuoka, and K. Takeda *J. Nutr.* **94**, 427 (1968).
[47] H. Kishi and E. Hiraoka, *Bitamin* **35**, 146; *ibid.* **36**, 326 (1967).
[48] R. L. Clark and E. F. Rogers, U.S. Patents 3,060,183–4 (1962); 3,141,820 (1964).
[49] G. Rindi, G. Ferrari, U. Ventura, and A. Trotta, *J. Nutr.* **89**, 197 (1966).
[50] Y. Kajiro, *J. Biochem. (Tokyo)* **46**, 1523 (1959).
[51] Y. Mano and R. Tanaka, *J. Biochem. (Tokyo)* **47**, 401 (1960).
[52] M. Morita, T. Kanaya, and T. Mineshita, *J. Vitaminol.* **14**, 223 (1968).
[53] K. H. Kiessling, *Arkiv Kemi* **11**, 451 (1957); J. E. Vincent, *Rec. Trav. Chim.* **76**, 779 (1957).

deficiency and has marked effects upon central and peripheral nervous tissue. Accumulation in brain has been noted.[54] The action potential of nonmyelinated nerve fiber is increased within minutes after immersion in a 2.5 mM pyrithiamine bath.[55] This response is not observed with either amprolium or oxythiamine and does not seem to be associated with the levels of TPP-enzymes utilized in nerve metabolism.[56] Ca^{2+}-stimulated receptors in frog tongue are inhibited by pyrithiamine at 10^{-4} M.[57]

It has been proposed that thiamine may function in a previously unrecognized way in nervous tissue.[58] A transport mechanism is suggested by evidence[59] of malfunction of the blood-brain barrier in thiamine deficiency. However, one transport mechanism, the Na–K-activated ATPase system, is not affected by pyrithiamine.[60] Variously established inhibition indices for pyrithiamine in animals have been reported: mouse, 40,[61] 50 times oxythiamine[32, 62]; rat, 20[63]; chick, 4,[64] 14[13]; pigeon and dog, 5–7.[61–66] I_{50} values for microbial systems are as follows: *Ceratostomella* spp. 7, 10; *Phytophora cinnamoni*, 12; *Mucor ramanianus*, 800[11]; *L. fermenti*, 50[67]; *L. acidophilus*, 1900[68]; *E. coli*, >2,000,000,[11] 20,000.[69] In addition, pyrithiamine inhibits growth of *Phycomyces blakesleeanus*[70] and anerobically grown (high thiamine requirement) *Mucor rauxii*.[71] At a 10-fold thiamine level, pyrithiamine almost completely inhibits the thiamine-induced production of prodigiosin

[54] G. Rindi, V. Perri, and L. DeCaro, *Experientia* **12,** 546 (1961); *Biochem. J.* **80,** 214 (1961).
[55] C. J. Armett and J. R. Cooper, *J. Pharmacol.* **148,** 137 (1965).
[56] J. R. Cooper, *Biochim. Biophys. Acta* **156,** 368 (1968).
[57] V. Perri, G. Rapuzzi, and L. Chiesa, *Boll. Soc. Ital. Biol. Sper.* **43,** 1466, 1470 (1967).
[58] A. von Muralt, *Ann. N.Y. Acad. Sci.* **98,** 499 (1962).
[59] L. G. Warnock and V. J. Burkhalter, *J. Nutr.* **94,** 256 (1968).
[60] J. R. Cooper and J. H. Pincus, *in* "Thiamine Deficiency: Biochemical Lesions and Their Clinical Significance" (G. E. W. Wolstenholme and M. O'Connor, eds.), p. 112. Little, Brown, Boston, Massachusetts, 1967.
[61] D. W. Woolley and A. G. C. White, *J. Biol. Chem.* **149,** 285 (1943).
[62] A. J. Eusebi and L. R. Cerecedo, *Science* **110,** 162 (1949).
[63] G. A. Emerson, *Abstr. Papers, 111th Am. Chem. Soc. Meeting* p. 42B (1947); L. De Caro, G. Rindi, V. Perri, and G. Farrari, *Experientia* **12,** 300 (1956); *Intern. J. Vit. Res.* **26,** 343 (1956); **28,** 252 (1958).
[64] E. C. Naber, W. W. Cravens, C. A. Baumann, and H. H. Bird, *J. Nutr.* **64,** 579 (1954).
[65] J. E. Koedam, *Biochim. Biophys. Acta* **29,** 333 (1958).
[66] H. R. Hulpieu, W. C. Clark, and H. Pon Onyett, *Quart. J. Studies Alcohol* **15,** 189 (1954).
[67] H. P. Sarett and V. H. Cheldelin, *J. Biol. Chem.* **156,** 91 (1944).
[68] S. Dreizen, E. Scholz, and T. D. Spies, *Proc. Soc. Exptl. Biol. Med.* **68,** 620 (1948).
[69] O. Wyss, *J. Bacteriol.* **46,** 483 (1943).
[70] W. J. Robbins, *Proc. Natl. Acad. Sci. U.S.* **27,** 19 (1941).
[71] S. Burtnick-Garcia and W. T. Wickerson, *J. Bacteriol.* **82,** 142 (1961).

by *S. marcescens*.[12] The algal species *Chlamydomonas eugametos*, is sensitive to one part per billion (mμg/ml) of pyrithiamine.[72]

The 2'-ethyl homolog of pyrithiamine has a chick antithiamine index of 11, compared with 14 for pyrithiamine.[73] It produces action potential increases in nonmyelinated fibers of rabbit vagus nerve at levels appreciably lower than pyrithiamine.[74]

Benzyl (VIe) and butyl (VId) imidazolium analogs of thiamine prevent absorption of the vitamin by *Kloeckera apiculata*, the former more effectively than pyrithiamine.[26] The I_{50} index for *L. fermenti* of the corresponding methyl analog (VIc) is low, 5–10.[75]

Although the oxazole analog (VI l) has been synthesized, no inhibition data are available.[76] Dethiothiamine (Vb), which is actually the open form of the oxazole, inhibits growth of *L. fermenti* and *K. apiculata*.[77]

An unexpected variation in the B$_1$ thiazolium moiety is found in the nonquaternary phenylthiazinothiamine (VIII). When this compound is fed to rats at 200 times the level of maintenance thiamine, neuromuscular symptoms of thiamine deficiency are produced.[78]

2'-Substituted thiamines, such as 2'-*n*-butyl (VIg), 2'-methoxy (VIm),[79] 2'-methylthio (VIh), and 2'-trifluoromethyl (VIf), may be tentatively placed in the pyrithiamine group; for a firmer classification, enzyme inhibition data would be helpful. Some of these antagonists are quite active. The S-methyl compound is comparable with pyrithiamine as a growth inhibitor for *K. brevis* and *L. fermenti*[10] and also arrests growth of a thiamine-requiring *E. coli* mutant.[80]

Bacillus subtilis growth is affected by the 2'-trifluoromethyl analog. However, the inhibition produced, in contrast to that of pyrithiamine and oxythiamine, could not be reversed with either the thiamine pyrimidine or thiazole components. Nervous system deterioration followed administration of the antagonist to mice fed a thiamine-deficient diet.[81] The 2'-butyl analog[82–84] appears similar, but more potent. Antithiamine indices of 25,

[72] J. C. McBride and C. S. Gowan, *Genetics* **56,** 405 (1967).

[73] W. H. Ott, A. A. Pessolano, and E. F. Rogers, unpublished results.

[74] C. J. Armett and J. R. Cooper, *Experientia* **21,** 605 (1965).

[75] H. A. Staab and G. A. Schwalbach, *Ann.* **715,** 128 (1968).

[76] A. Dornow and H. Hell, *Ber.* **94,** 1248 (1961).

[77] G. Kurata, T. Sakai, T. Miyahara, and H. Yokoyama, *Bitamin* **35,** 136 (1967).

[78] M. Morita, T. Kanaga, and T. Mineshita, *J. Vitaminol.* (*Kyoto*) **14,** 223 (1968).

[79] H. C. Koppel, R. H. Springer, R. K. Robins, and C. C. Cheng, *J. Org. Chem.* **27,** 3614 (1962).

[80] T. L. V. Ulbricht and J. S. Gots, *Nature* **178,** 913 (1956).

[81] J. A. Barone, H. Tieckelman, R. Guthrie, and J. F. Holland, *J. Org. Chem.* **25,** 211 (1960).

[82] G. A. Emerson and P. L. Southwick, *J. Biol. Chem.* **160,** 169 (1945).

[83] L. R. Cerecedo, *Intern. J. Vit. Res.* **37,** 189 (1967).

[84] H. Haenel, *Vitamine Hormone* **7,** 113 (1956).

40, and 200 are found with pigeon, rat, and mouse, while the I_{50} value for *P. blakesleeanus* is 30. Again, nervous system effects are observed.

With other 2′-substituents also, such as methoxymethyl, 2-ethoxyethyl, 2-phenoxyethyl, and 2-bromoethyl groups, moderately effective antithiamines are obtained. These are toxic to *L. fermenti* and *K. brevis*.[85] The broad activity supports the postulated antipositional role of 2′-substituents.

6′-Aminothiamine (VIn)[23] is a competitive inhibitor of thiamine kinase and thiamine absorption by *L. fermenti; I_{50}* is 450–470.[86]

In the early vitamin synthesis research,[16] a number of thiamine analogs were made in which the 5-(2-hydroxyethyl) group is replaced by other hydroxyalkyls. Unfortunately, the compounds were tested for vitamin activity only, there being at that time little interest in antithiamines. It is evident now that these analogs, with variations at or near the phosphorylation site, are attractive candidates for kinase inhibition studies. The one compound of this class on which there is any information is homothiamine glycol (VIo),[87] an unpromising double-variant, which nevertheless retains one-sixteenth of pyrithiamine toxicity to *L. fermenti*.[88]

Inhibitors of Thiamine Biosynthesis and Metabolism

Inhibitors in this interesting category have not received much attention.

4-Amino-2-methylthio-5-pyrimidylcarbinol (methioprim, IX) is toxic to an *E. coli* mutant which requires the B₁-pyrimidine for growth,[89] and the corresponding 2-methoxy compound (IXa) is the antibiotic bacimethrin, also lethal to *E. coli* and other bacteria and yeasts.[79,90] The antibacterial activity of bacimethrin is reversed, not only by the thiamine pyrimidine, but also by pyridoxine. This point illustrates one complication of biosynthesis inhibition studies *in vivo*, an antagonism between 5-hydroxymethylpyrimidines and B₆-compounds.

E. coli toxicity is also observed with 2-amino-5-(2-hydroxyethyl)-4-methylthiazole (VIIa), which is an inhibitor of the kinase involved in synthesis of the B₁-thiazole phosphate,[91] a thiamine phosphate biosynthesis intermediate.[92]

While bacterial thiaminases have been the subject of considerable

[85] D. Mucke, *Zentr. Bacteriol. Parasitenk. Abt. II* **113**, 470 (1960).

[86] A. Iwashima, T. Uematsu, and T. Masuda, *Bitamin* **34**, 490 (1966).

[87] P. Karrer and M. Schoeller, *Helv. Chim. Acta* **34**, 826 (1951).

[88] H. Schopper, M. L. Bein, G. Besson, and R. Eichin, *Z. Vitaminforsch.* **23**, 47 (1951).

[89] T. L. V. Ulbricht and C. C. Price, *J. Org. Chem.* **21**, 567 (1956).

[90] F. Tanaka, N. Tanaka, H. Yonehara, and H. Umezawa, *J. Antibiotics* **15**, 191 (1962).

[91] Y. Nose and A. Iwashima, *in* "Current Aspects of Biochemical Energetics" (N. O. Kaplan, ed.), p. 243. Academic Press, New York, 1966; *J. Biochem.* *(Tokyo)* **62**, 537 (1967).

[92] I. G. Leder, *J. Biol. Chem.* **236**, 3066 (1961).

research,[93] surprisingly little interest in inhibition has been manifested. Pyrithiamine and oxythiamine do not affect thiaminase activity.[94]

Other sources of thiaminases are carp viscera and the fern *Pteridium aquilinum* and plausible antagonists of thiamine, the 2'-aminoethyl and *o*-aminobenzyl quaternaries of 4-methylthiazole (X, Xa), are claimed to be effective inhibitors of these enzymes.[95] Recently, however, the isolation of nonenzymatic thiamine-destroying agents from carp and fern has been reported. The agent from fern has been identified as caffeic acid (3,4-dihydroxycinnamic acid),[96] and that from carp as a conjugate of hemin or a related compound.[97] In the latter case, the unconjugated heminlike substance is fully active. The mechanisms for thiamine inactivation by these compounds and their relationships to the alleged inhibitors are not clear.

Conclusion

With improved understanding of mechanisms of action, the major thiamine antagonists can be employed more confidently as research tools. The investigator should be aware, however, that the most available or best-studied inhibitors are not necessarily the most suitable for every purpose. There should be kept in mind the possibility that more definitive results may be obtained with a congener of the better-known inhibitors, or perhaps some quite novel type of antagonist.

[93] A. Fujita, *Advan. Enzymol.* **15**, 389 (1954).
[94] H. A. Douthit and R. L. Airth, *Arch. Biochem. Biophys.* **113**, 331 (1966).
[95] R. R. Sealock and R. L. Goodland, *J. Am. Chem. Soc.* **66**, 507 (1944); R. H. Kenton, *Biochem. J.* **69**, 439 (1958).
[96] J. Beruter and J. C. Somogyi, *Experientia* **23**, 996 (1967).
[97] H. Kundig and J. C. Somogyi, *Intern. Z. Vitaminforsch.* **37**, 476 (1967).

[45] Enzymatic Preparation, Isolation, and Identification of 2-α-Hydroxyalkylthiamine Pyrophosphates[1]

By BRUNO DEUS, JOHANNES ULLRICH, and HELMUT HOLZER

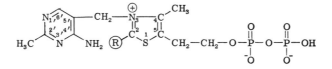

TPP[2] and its 2-α-hydroxyalkyl derivatives:

R = H   TPP
R = CH₂OH   2-Hydroxymethyl-TPP (TPP-linked formaldehyde)
R = CH₃CHOH   2-α-Hydroxyethyl-TPP (TPP-linked acetaldehyde[3])
R = CH₂OH·CHOH   2-α,β-Dihydroxyethyl-TPP (TPP-linked
                                 glycolaldehyde)
R = COOH·CH₂·CH₂·CHOH   2-α-Hydroxy-γ-carboxypropyl-TPP (TPP-linked
                                 succinic semialdehyde[4])

Introduction

2-Hydroxymethyl-TPP,[5] 2-α-hydroxyethyl-TPP,[6] 2-α,β-dihydroxyethyl-TPP,[7-9] and 2-α-hydroxy-γ-carboxypropyl-TPP,[4,10] have been isolated from incubation mixtures of TPP-dependent enzymes. These compounds

---

[1] The chemical syntheses of 2-α-hydroxyethyl-TPP and of 2-α,β-dihydroxyethyl-TPP are described in Vol. IX, p. 65.

[2] Abbreviations used: TPP, thiamine pyrophosphate; HMTPP, 2-hydroxymethyl-thiamine pyrophosphate; HETPP, 2-α-hydroxyethylthiamine pyrophosphate; DETPP, 2-α,β-dihydroxyethylthiamine pyrophosphate; HCPTPP, 2-α-hydroxy-γ-carboxypropylthiamine pyrophosphate.

[3] 2-α-Hydroxy-α-carboxyethylthiamine pyrophosphate (TPP-linked pyruvate) has been identified as the unstable precursor of 2-α-hydroxyethylthiamine pyrophosphate (see reference cited in Holzer and Beaucamp).[6]

[4] The structure of this substituent is inferred to be α-hydroxy-γ-carboxypropyl by analogy with the other TPP-linked aldehydes. The yield of HCPTPP obtained by the procedure described below is rather variable (see Fig. 1).

[5] G. Kohlhaw, B. Deus, and H. Holzer, *J. Biol. Chem.* **240**, 2135 (1965).

[6] H. Holzer and K. Beaucamp, *Angew. Chem.* **71**, 776 (1959); *Biochim. Biophys. Acta* **46**, 225 (1961).

[7] H. Holzer, F. Da Fonseca-Wollheim, G. Kohlhaw, and Ch. W. Woenckhaus, *Ann. N.Y. Acad. Sci.* **98**, 453 (1962).

[8] F. Da Fonseca-Wollheim, K. W. Bock, and H. Holzer, *Biochem. Biophys. Res. Commun.* **9**, 466 (1962).

[9] F. Pohlandt, G. Kohlhaw, and H. Holzer, *Z. Naturforsch.* **22b**, 407 (1967).

[10] B. Deus and H. Holzer, in preparation.

were considered to represent the TPP-activated aldehyde intermediates of the enzymatic reactions.[7,11]

In the case of 2-α-hydroxyethyl-TPP, however, there is substantial evidence[12-14] that the nucleophilic α-carbanion, which arises from the decarboxylation step in the yeast pyruvate decarboxylase reaction, is protected from protonization by the lipophilic environment of the active site. Thus, the α-carbanion of 2-α-hydroxyethyl-TPP is considered to be the real activated intermediate. In the absence of oxidized lipoic acid, the reaction is blocked at this point. Under these conditions, the α-carbanion dissociates from the protein and adds a proton at the same time. The resulting saturated compound has lost most of its reactivity. It might be concluded that the same conditions apply to the other TPP-catalyzed reactions.

Principle. Crude preparations of mitochondrial α-oxoacid dehydrogenase complexes (EC 1.2.4.1 and EC 1.2.4.2) are prepared almost free of NAD[+] and CoA. Such preparations still decarboxylate the α-oxoacids, but are unable to oxidize the TPP-linked decarboxylation products. The α-carbanions of the 2-α-hydroxyalkylthiamine pyrophosphates add a proton and are replaced at the active site of the enzyme by free TPP. Because of their stability, the protonated compounds can be isolated by the procedure described below.

Reagents

> Phosphate buffer, 20 mM, pH 7.0
> Ammonium sulfate
> Potassium dihydrogen phosphate, 1.0 M
> Sodium fluoride, 1.0 M
> Magnesium sulfate, 1.0 M
> Thiamine pyrophosphate, 0.1 M
> Sodium bicarbonate
> Sulfuric acid, ca. 2 N
> Alkali salts of glyoxylate, or pyruvate, or hydroxypyruvate,[15] or α-oxoglutarate, 1.0 M
> Dowex 2 × 8, 200–400 mesh, analytical grade[16]
> Hydrochloric acid, ca. 2 N
> Sodium acetate, ca. 2 N
> Acetic acid, ca. 2 N
> Acetic acid, 16.7 mM (dilute 1 ml of glacial acetic acid to 1 liter)

[11] H. Holzer, *Angew. Chem.* **73,** 721 (1961).
[12] A. Schellenberger, *Angew. Chem.* **79,** 1050 (1967).
[13] J. Ullrich, B. Deus, and H. Holzer, *Intern. Z. Vitaminforsch.* **38,** 273 (1968).
[14] J. Ullrich, *Angew. Chem.* **81,** 87 (1969).
[15] Prepared by the method of F. Dickens and D. H. Williamson, *Biochem. J.* **68,** 74 (1958).
[16] Obtained from Bio-Rad Laboratories, Richmond, California.

Preparation of Crude α-Oxoacid Dehydrogenase Complexes[17]

All operations are carried out at 4°, unless otherwise stated. Hearts of young pigs are collected in ice immediately after slaughter. Fat and connective tissue are carefully removed. The muscle is cut to pieces and passed through a mincer with approximately 0.25-cm diameter holes. The minced tissue is washed 4–6 times with about 3 volumes of water, until the supernatant is opalescent. Two hundred-gram portions of the sediment are wrapped in polyethylene and stored below −18° without considerable loss of activity for 4–6 weeks.

Two hundred grams of the mince is homogenized with 700 ml of cold 20 mM phosphate buffer, pH 7.0, in a refrigerated Waring blendor. To avoid temperatures above 8°, the homogenization is performed in 5 periods of 1 minute each. After the preparation has stood for 30 minutes, the sediment is collected by centrifugation and extracted again with 250 ml of buffer as described above. The combined extracts are filtered through cheesecloth or glasswool to remove traces of lipids. To 100 ml of the extract is added 24.3 g of solid ammonium sulfate (40% saturation). The suspension is stirred for 2 hours. The precipitate obtained after centrifugation at 12,500 g for 30 minutes is suspended by use of a Potter-Elvehjem homogenizer in 25 ml of phosphate buffer. The suspension is frozen for at least 12 hours below −18°, then undissolved material is removed by centrifugation at 35,000 g for 1 hour. The supernatant contains the α-oxoacid dehydrogenase complexes (ca. 35–40 mg of protein per milliliter). It may be stored in the deep-freeze for 2–4 weeks without appreciable decrease in activity.

Determination of Activity. The activity of the pyruvate dehydrogenase complex can be measured by a modification[17] of the arylamine acetyltransferase (EC 2.3.1.5) assay.[18] For convenience, the initial rates of CO_2 evolution from the α-oxoacids are taken as a measure of enzymatic activity.[10] The main compartment of a Warburg vessel contains in a final volume of 3.0 ml: 300 micromoles of phosphate buffer, pH 6.0, 10 micromoles of $MgSO_4$, 10 micromoles of TPP, 3 micromoles of NAD^+, 1 micromole of CoA, 5 micromoles of cysteine-HCl, and the enzyme. After equilibration at 37°, 10 micromoles of the substrate are tipped in from the side arm. The rate of CO_2 evolution is estimated from the first 10 minutes of the reaction.

The relative decarboxylation rates of the above enzyme preparation are 1.0 for glyoxylate, 1.5 for pyruvate, 2.0 for hydroxypyruvate, and 5.0 for α-oxoglutarate.

---

[17] P. Scriba and H. Holzer, *Biochem. Z.* **334**, 473 (1961). This method is based on an earlier procedure by S. Korkes, A. Del Campillo, and S. Ochoa, *J. Biol. Chem.* **195**, 541 (1952).

[18] H. Tabor, A. H. Mehler, and E. R. Stadtman, *J. Biol. Chem.* **204**, 127 (1953).

Preparation of 2-α-Hydroxyalkylthiamine Pyrophosphates

The composition of the incubation mixtures is given in Table I.

The pH is adjusted with 2 N H_2SO_4 or solid $NaHCO_3$. The volume is measured (10–15 ml), then the mixture is transferred to a Warburg vessel of 40 to 70-ml volume. After equilibration at 37°, the neutralized substrate is added from the side arm, and the evolution of CO_2 is followed manometrically. The approximate incubation times are listed in Table I. The reaction is stopped by addition of 10 volumes of boiling methanol (65°). Precipitated protein is removed by centrifugation, and the supernatant is evaporated to dryness in a vacuum evaporator at ca. 35°. The residue is redissolved in 3–4 ml of water. After removal of undissolved material by centrifugation at 35,000 g for 30 minutes, the supernatant is ready for anion-exchange column chromatography.

Isolation of 2-α-Hydroxyalkylthiamine Pyrophosphates

Preparation of the Column. All operations are conducted at room temperature. Dowex 2 resin suspended in 2 N HCl is allowed to settle to a

TABLE I
INCUBATION CONDITIONS FOR PREPARATION OF TPP-LINKED ALDEHYDES[a]

Conditions	HMTPP[b]	HETPP[c]	DETPP[d]	HCPTPP[e]
Substrate	Glyoxylate, 25	Pyruvate, 300	Hydroxy-pyruvate, 75	α-Oxoglutarate, 30
KH_2PO_4	300	300	1000	300
NaF	—	500[f]	—	500[f]
$MgSO_4$	10	240	10	240
TPP	20	50	50	50
α-Oxoacid dehydrogenase (mg protein)	350	400	100	320
pH	6.0	5.6	5.6	5.8
Incubation at 37° (hours)	1	3–8	3–6	1–2
CO_2 liberated	16	80–90	50	30
Yield of product (% of added TPP)	50–80	50–70	50–70	1–4[g]

[a] Numbers represent micromoles, unless otherwise stated in column 1.
[b] G. Kohlhaw, B. Deus, and H. Holzer, *J. Biol. Chem.* **240**, 2135 (1965).
[c] J. Ullrich and A. Mannschreck, *European J. Biochem.* **1**, 110 (1967).
[d] F. Pohlandt, G. Kohlhaw, and H. Holzer, *Z. Naturforsch.* **22b**, 407 (1967).
[e] B. Deus and H. Holzer, in preparation (see also footnote 4).
[f] Added in order to minimize the action of phosphatases.
[g] So far no better yield has been obtained, because NAD^+ and CoA are rather firmly bound to the α-oxoglutarate dehydrogenase complex.

height of ca. 35 cm in a column of 2-cm diameter. The column is washed successively with 1 liter of 2 N HCl and 1 liter of water. The resin is suspended in 2 M sodium acetate and allowed to swell. Then it is washed with 2 M sodium acetate, until the eluate is free of chloride (upon addition of a few drops of HNO_3 and $AgNO_3$ no precipitate should form). One liter of water is passed through the column before the resin is resuspended in 2 N acetic acid. The column is washed with water until the eluate reacts neutrally. At the end of the procedure the resin is packed approximately 40 cm high.

Regeneration of the column is achieved by repeating the last steps: resuspension and washing with 500 ml of 2 N acetic acid and elution with water until the eluate is neutral. Only if the separation qualities of the resin decrease considerably, the whole regeneration procedure must be repeated.

Chromatography. The extract of the incubation mixture is applied to the column, which is subsequently washed with ca. 100 ml of water. Thiamine derivatives are eluted from the resin with an exponential gradient of acetic acid at a flow rate of 100–150 ml per hour. The mixing chamber contains 300 ml of water and the reservoir vessel is filled with 16.7 mM acetic acid. Fractions of 20 ml are collected. Thiamine derivatives are detected by measuring the absorbance at 272.5 nm (isosbestic points of thiamine and its derivatives). An automatic recording ultraviolet analyzer for flow streams greatly facilitates the localization of the substances. Typical elution patterns are shown in Fig. 1. The fractions containing pure 2-α-hydroxyalkyl-TPP are combined and lyophilized. During freeze-drying the use of particle traps minimizes losses of the products. Further purification is achieved by precipitation with ethanol or acetone from concentrated aqueous solutions. The resultant white powder is hygroscopic. It can be stored in the presence of a desiccant below $-18°$ for several weeks.

Chemically prepared 2-α-hydroxyalkylthiamine pyrophosphates[1] may be isolated by the same chromatographic procedure.

Identification of 2-α-Hydroxyalkylthiamine Pyrophosphates[19]

Position in the Eluate. As seen in Fig. 1, the 2-α-hydroxyalkylthiamine pyrophosphates are eluted from the resin immediately in front of TPP,[20] which is identified by its ultraviolet absorption spectrum (see below). If the enzyme preparation is incubated with α-oxoacids labeled with ^{14}C in

---

[19] Paper chromatography and paper electrophoresis using ^{14}C-labeled materials and other isotopic methods, which led to the identification of the nature of the substituents in the 2-position of TPP, are described in references cited in Holzer *et al.*[5–9,11]

[20] Although the separation qualities of different batches of the resin vary considerably, separation from TPP is usually good for HMTPP, very good for HETPP, poor for DETPP, and good for HCPTPP (see Fig. 1).

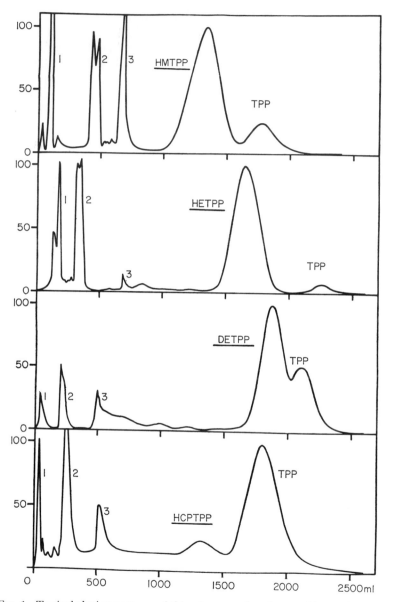

Fig. 1. Typical elution patterns of thiamine pyrophosphate and its 2-α-hydroxyalkyl derivatives.

TABLE II
Spectral Characteristics of TPP-Linked Aldehydes

Characteristic	pH	TPP[a]	HMTPP[b]	HETPP[a]	DETPP[c]	HCPTPP[a]
λ_{max} (thiazolium)[e]	8.0	267	268	269	270	268
Molar extinction coefficient[f]	—	7.8	7.7	7.5	7.5	—
λ_{max} (aminopyrimidine)[e]	1–3	247	248	247	247	246
Molar extinction coefficient[f]	—	13.0	10.4	9.6	9.4	—
λ_{max} (aminopyrimidine)[e]	8.0	233	230	229	230	227
Molar extinction coefficient[f]	—	10.8	9.6	9.1	8.8	—
λ (isosbestic point)[e]	1–8	272.5	272.5	272.5	272.5	272.5
Molar extinction coefficient[f]	—	7.4	7.4	7.4	7.4	—
λ (isosbestic point)[e]	—	237	236	237	237	234
Molar extinction coefficient[f]	—	10.2	8.5	7.6	7.8	—

[a] J. Ullrich and A. Mannschreck, *European J. Biochem.* **1**, 110 (1967).

[b] G. Kohlhaw, B. Deus, and H. Holzer, *J. Biol. Chem.* **240**, 2135 (1965).

[c] F. Pohlandt, G. Kohlhaw, and H. Holzer, *Z. Naturforsch.* **22b**, 407 (1967).

[d] B. Deus and H. Holzer, in preparation.

[e] Wavelengths are given in nanometers.

[f] As centimeters $M^{-1} \times 10^3$.

any position except the carboxyl group, the 2-α-hydroxyalkyl derivatives of TPP can easily be detected by their radioactivity.

Ultraviolet Absorption Spectra. The 2-α-hydroxyalkylthiamine pyrophosphates are distinguished from TPP by their ultraviolet absorption spectra. Table II gives the spectral characteristics of the substances. The rather small differences between the spectra of the 2-α-hydroxyalkyl derivatives do not suffice to distinguish one TPP derivative from another. However, depending on the conditions of the experiment, in most cases only one of them becomes of practical importance.

Nuclear Magnetic Resonance Spectra (NMR). An absolute method to distinguish between the individual 2-α-hydroxyalkylthiamine pyrophosphates is to compare the NMR spectra of the substances with that of TPP in D_2O.[21] About 50–100 micromoles of the compound is needed for a 60 MHz NMR spectrum, and 5–10 micromoles for a 100 MHz NMR spectrum.

Quantitative Determination of 2-α-Hydroxyalkylthiamine Pyrophosphates

Ultraviolet Absorption. The molar extinction coefficients of TPP and its 2-α-hydroxyalkyl derivatives are 7.4×10^3 at the isosbestic points of 272.5 nm (see Table II). On this basis the substances are estimated by their ultraviolet absorption.

Thiochrome Assay.[22] In the thiochrome test, equimolar amounts of HMTPP,[5] HETPP,[10] and DETPP[9] yield 15, 100, and 55% thiochrome fluorescence, respectively, as compared to TPP. The low thiochrome fluorescence yields of HMTPP and DETPP may be used to calculate the percentage of the TPP derivative in a mixture with TPP, if no other compound absorbing at 272.5 nm interferes.[5,9]

---

[21] J. Ullrich and A. Mannschreck, *European J. Biochem.* **1,** 110 (1967).
[22] B. Deus, H. E. C. Blum, and H. Holzer, *Anal. Biochem.* **27,** 492 (1969).

Section III

Lipoic Acid and Lipoamide

[46] Turbidimetric and Polarographic Assays for Lipoic Acid using Mutants of *Escherichia coli*

By A. A. Herbert and J. R. Guest

Several turbidimetric and manometric methods have been described for the assay of lipoic acid using organisms which require lipoic acid as a growth factor.[1,2] More recently, mutants of *Escherichia coli* that are unable to synthesize lipoic acid have been characterized.[3,4] Their lipoic acid requirement is approximately 0.5 ng/ml for half-maximal aerobic growth, but in glucose minimal medium this requirement can be spared or replaced by acetate plus succinate or lysine plus methionine. However, with succinate as carbon and energy source, the requirement for lipoic acid is absolute, and acetate, glucose, lysine, and methionine are without effect. The organisms respond to α- and to β-lipoic acid, but not to lipoamide, lipoylglycinamide, or conjugated derivatives present in biological material unless they are first hydrolyzed. In addition, suspensions of organisms grown in the absence of lipoic acid cannot oxidize pyruvate or α-ketoglutarate unless this cofactor is supplied. Consequently these mutant organisms can be used for assaying lipoic acid by methods that are analogous to, but more convenient than, the procedures described previously.

Turbidimetric Method

Principle. Lipoic acid can be determined by assessing turbidimetrically the response of mutant W1485*lip*2[4] to graded amounts of lipoic acid over the range 0.2–2.0 ng using a growth medium containing succinate as principal substrate.

Reagents

Basal growth medium (see table)
Sodium succinate, 1 M, pH 7.0
DL-α-lipoic acid, crystalline
Potassium phosphate buffer, 0.2 M, pH 7.0
NaCl, 0.9%
Inoculum of lipoic acid mutant of *E. coli*

[1] E. L. R. Stokstad, G. R. Seaman, R. J. Davis, and S. H. Hutner, *Methods Biochem. Anal.* **3**, 23 (1956).

[2] I. C. Gunsalus and W. E. Razzell, see Vol. III [138].

[3] A. B. Vise and J. Lascelles, *J. Gen. Microbiol.* **48**, 87 (1967).

[4] A. A. Herbert and J. R. Guest, *J. Gen. Microbiol.* **53**, 363 (1968). Mutant W1485*lip*2, A.T.C.C. No. 25645, has been submitted to the National Collection of Industrial Bacteria, U.K.

BASAL GROWTH MEDIUM FOR TURBIDIMETRIC ASSAY

Medium[a]	Grams/liter
Acid-hydrolyzed casein (vitamin free)	4
K_2HPO_4	14
KH_2PO_4	6
Na_3 citrate·$3H_2O$	1
$MgSO_4$·$7H_2O$	0.2
$(NH_4)_2SO_4$	2
L-Asparagine	8
L-Arginine	0.20
L-Glutamate	0.20
Glycine	0.20
L-Histidine	0.20
L-Proline	0.20
L-Tryptophan	0.36
L-Cysteine	0.16
Na thioglycolate	0.20

[a] Adjust pH to 6.8; autoclave at 121° for 15 minutes for storage.

Procedure. STANDARD SOLUTION OF LIPOIC ACID. Prepare a solution of lipoic acid (1 mg/ml) by dissolving a known amount of crystalline lipoic acid in sterile potassium phosphate buffer (0.2 M, pH 7.0). Dilute this in sterile water to contain 10 μg/ml and 100 ng/ml. These solutions may be stored for several months at 0° without any quantitative decrease in the response given by the lipoic acid mutant. Further dilutions are prepared from these solutions when necessary.

LIPOIC ACID FROM NATURAL MATERIAL. Lipoic acid is released from natural material by autoclaving in 6 N H_2SO_4 at 120° for 2 hours. The material is then adjusted to pH 7.0 with 4 N NaOH, made up to a known volume, and filtered to remove any insoluble material. The solution can then be stored at −20° until required for assay. Addition of known amounts of lipoic acid to natural material prior to hydrolysis gives recoveries of 90–110%.

STANDARD CURVE. Transfer the following reagents to a series of 150 × 19-mm test tubes: 1.0 ml of basal growth medium, an appropriate volume of lipoic acid solution, and sufficient water to bring the volume to 1.8 ml. Sterilize the tubes by autoclaving at 115° for 10 minutes and cool. Then add aseptically 0.1 ml of sodium succinate solution (1 M) and 0.1 ml of mutant inoculum (see below). Incubate the tubes vertically on a gyratory shaker (shaking speed 210 rpm) for 40 hours at 37°. The growth response in each tube can then be measured turbidimetrically with a suitable color-imeter. A typical standard curve relating the extent of growth to the amount of added DL-α-lipoic acid is shown in Fig. 1A. Unknown samples containing

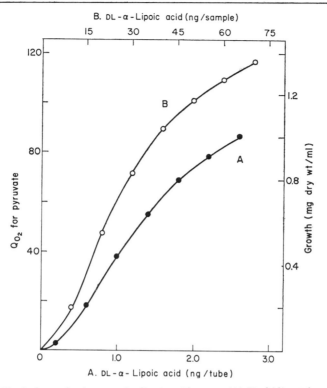

FIG. 1. Typical standard curves for lipoic acid assay. (A) Turbidimetric method (●) after conversion of turbidity to milligrams dry weight of cells per milliliter. (B) Polarographic method (○) with oxygen uptake for pyruvate expressed as microliters of O_2 per milligram dry weight per hour.

lipoic acid are assayed at four or five levels by the same procedure, and their lipoic acid content can be calculated by comparison with the standard curve which should be set up in duplicate with each assay.

PREPARATION OF THE INOCULUM. Prepare the inoculum by transferring a single colony of the mutant organism to 5 ml of minimal salts medium E^5 containing succinate (50 mM) as carbon and energy source and supplemented with lipoic acid (1 ng/ml). Incubate with shaking for 24 hours, then harvest, and wash twice in sterile saline (0.9% NaCl) by centrifuging. Resuspend the organisms in saline, dilute the suspension to 10 μg dry weight of cells per milliliter, and use 0.1 ml of this to inoculate each assay tube.

Polarographic Method

Principle. Lipoic acid can be assayed at levels of 5–50 ng by measuring the Q_{O_2} for pyruvate of lipoic acid-deficient mutant organisms which have been preincubated with solutions containing lipoic acid. Oxygen consumption is measured polarographically with a simple oxygen electrode, a reaction vessel with 1 ml capacity, and a recorder, as described by Estabrook,[6] rather than manometrically as in the analogous assay using *Streptococcus faecalis* 10C1 pyruvate apodehydrogenase.[2]

Reagents

Potassium phosphate buffer, 0.04 M, pH 7.0
Sodium pyruvate, 1 M, pH 7.0
DL-α-lipoic acid, crystalline
Suspension of lipoic acid mutant of *E. coli*

Procedure. The standard solution of lipoic acid and the release of lipoic acid from natural material are as described for the turbidimetric method (see above).

STANDARD CURVE. Mix the following components in $\frac{1}{4}$ oz screw-capped bottles and incubate with shaking for 1 hour at 37°: 1.0 ml of suspension of mutant organisms (see below), an appropriate volume of lipoic acid solution, and 0.04 M phosphate buffer to give a final volume of 2.0 ml. After preincubation, aerate the suspension at 25° and introduce it into a suitable closed reaction vessel for measuring oxygen consumption.[6] Record the endogenous oxygen consumption by the suspension, and when the rate is constant (in about 5 minutes), add 10 μl of 1 M sodium pyruvate, and record the new rate. A typical suspension of W1485*lip*2 has an endogenous Q_{O_2} value of 10 μl of O_2 per milligram dry weight per hour at 25° in the absence of added lipoic acid, and the rate does not change when pyruvate is added. After preincubation with lipoic acid, the response to sodium pyruvate is immediate, and oxygen consumption occurs at a constant rate. A curve relating the rate of oxygen consumption (Q_{O_2}) for pyruvate to the amount of lipoic acid added during preincubation serves as the standard curve (Fig. 1B). Unknown samples of lipoic acid are tested at three or four levels by the same procedure and assayed by reference to a standard curve obtained with the same batch of cells.

PREPARATION OF SUSPENSION OF ORGANISMS. Grow the mutant overnight in nutrient broth from a single colony. Inoculate 0.2 ml of the broth culture into 500 ml of minimal medium E containing glucose (0.2%) and acetate plus succinate (each 4 mM). Incubate with shaking for 18 hours at 37°, harvest, wash twice in 0.04 M phosphate buffer, pH 7.0, and finally resuspend at 2 mg of dry weight of organisms per milliliter in the same buffer.

[5] H. Vogel and D. M. Bonner, *Microbial Genet. Bull.* **13,** 43 (1956).
[6] R. W. Estabrook, see Vol. X [7].

[47] Chemical Syntheses of ^{14}C-α-Lipoic Acid

By ALAN B. PRITCHARD, DONALD B. McCORMICK,
and LEMUEL D. WRIGHT

General Techniques and Materials

Lipoic acid can be synthesized with ^{14}C in positions 1 and 6 beginning with the commercially available 1,6-^{14}C-labeled adipic acid or in positions 2 and 5 or 3 and 4 after first synthesizing 2,5- or 3,4-^{14}C-labeled adipic acid from the commercially available 1,4- or 2,3-^{14}C-labeled succinic acid.

Adipic Acid-2,5- or 3,4-^{14}C

Principle. Succinic acid-1,4- or 2,3-^{14}C is esterified with methanol, the dimethyl ester is reduced by lithium borohydride to 1,4-butanediol, which is treated with hydrobromic acid to form 1,4-dibromobutane. This, in turn, is allowed to react with sodium cyanide to form adiponitrile, and the latter is hydrolyzed in acid to yield adipic acid-2,5- or 3,4-^{14}C.

Procedure. Succinic acid-1,4- or 2,3-^{14}C is converted to its dimethyl ester by refluxing for 1.5 hours 2.4 g (20 millimoles) of the ^{14}C-labeled acid in 200 ml of methanol containing 1.2 ml of concentrated sulfuric acid. The solution is cooled, neutralized by stirring in 5 g of calcium carbonate, and filtered through Celite. The filtrate is evaporated to dryness with warming under partial vacuum to obtain 2.4 g (80% yield) of dimethyl succinate, m.p. 19°.

A mixture of 2.3 g (16 millimoles) of the dimethyl succinate-^{14}C and 1.2 g of lithium borohydride in 50 ml of diethyl ether is stirred under reflux for 1.75 hours. Excess lithium borohydride is decomposed by the careful addition of water in small portions, and ether is removed under partial vacuum at room temperature. The residue is extracted twice with 25-ml portions of acetone and the solvent is removed to yield a crude syrup of 1,4-butanediol.

Without further purification, the ^{14}C-labeled 1,4-butanediol is dissolved in 60 ml of 32% hydrogen bromide in glacial acetic acid, and the solution is heated on a steam bath for 3 hours. This treatment is repeated with another 60 ml of hydrogen bromide–acetic acid, which is then removed by warming under partial vacuum. The syrup is dissolved in 15 ml of water and extracted with 30 ml of diethyl ether; the ether layer is washed twice with 10-ml portions of water, and the solvent is evaporated to obtain 1.43 g (46% yield) of 1,4-dibromobutane, b.p. 190°.

The ^{14}C-labeled 1,4-dibromobutane, 1.43 g (6.6 millimoles), is added over a 30-minute period to 1.06 g of sodium cyanide in 5 ml of dimethyl sulfoxide at 60°. The mixture is then heated at 90° for 15 minutes, cooled, diluted with 10 ml of water, and extracted 3 times with 50-ml portions of diethyl ether. The combined ether extracts are washed with 10 ml of 6 N hydrochloric acid followed by water, and the solvent is evaporated to obtain 0.65 g (90% yield) of adiponitrile, b.p. 153° (6 mm Hg).

A solution of 0.65 g (6 millimoles) of adiponitrile-2,5- or 3,4-^{14}C in 18 ml of 6 N hydrochloric acid is refluxed for 3 hours and evaporated to dryness with warming under partial vacuum. The residue is dissolved in the minimum volume of hot ethyl acetate to obtain 0.76 g (87% yield) of crystalline adipic acid-2,5- or 3,4-^{14}C, m.p. 152–153°.

α-Lipoic Acid-^{14}C

Principle. Adipic acid-^{14}C is refluxed with its diethyl ester to obtain the monoethyl adipate, which is allowed to react with thionyl chloride to form ethyl adipoyl chloride as the precursor to α-lipoate-^{14}C.

Procedure. Adipic acid-^{14}C is converted to its ^{14}C-labeled monoethyl ester, essentially as described by Brown *et al.*[1] for the unlabeled compound, by refluxing 11.6 g (79 millimoles) of the ^{14}C-labeled acid and 9.4 g of its diethyl ester in 6.2 ml of absolute ethanol, 4.0 ml of dibutyl ether, and 2.0 ml of concentrated hydrochloric acid for 4.5 hours. The solvent is removed by warming under partial vacuum, 100 ml of benzene is stirred into the warm residue, and the mixture is stored at 5° overnight. The precipitate of unreacted adipic acid is recovered by filtration and washed with benzene. The combined benzene solutions are slowly added with continuous stirring to 8.5 g of sodium bicarbonate in 60 ml of water. Unreacted diethyl adipate is recovered by evaporation of the upper, benzene layer. The lower, aqueous phase, is stirred with 10 ml of benzene and acidified with 9 ml of concentrated hydrochloric acid. The benzene layer is concentrated by warming under partial vacuum and the monoethyl adipate distilled over at 135–140° (1 mm Hg) to yield 9.2 g (67% based on initial free acid).

The monoethyl adipate-^{14}C is converted to its acid chloride, as described by Durham *et al.*[2] for the unlabeled compound, by allowing a solution of 2.5 g (14 millimoles) of the ^{14}C-labeled ester in 3.75 g (2.3 ml) of thionyl chloride to stand overnight. Excess thionyl chloride is removed with warming under partial vacuum and the monoethyl adipoyl chloride is distilled at 112–115° (1 mm Hg) to yield 2.5 g (84%).

[1] G. S. Brown, M. D. Armstrong, A. W. Moyer, W. P. Anslow, M. V. Guerry, S. Bernstein, and S. R. Safir, *J. Org. Chem.* **12**, 160 (1947).

[2] L. Durham, D. J. McLeod, and J. Cason, *Org. Syn. Coll. Vol.* **4**, 556 (1963).

The monoethyl adipoyl chloride-^{14}C is converted to DL-α-lipoate-^{14}C through a series of reactions first described by Reed and Niu,[3] but improved by Acker and Wayne.[4] The ^{14}C-labeled D-α-isomer (natural isomer) can be separated from the racemate.[4] The biologically active dihydro and 6-acetyl derivatives can be made as described by Gunsalus and Razzell.[5]

Chromatographic Assay

A simple and rapid technique to follow progress of the reactions and purity of the intermediates involves application to thin-layer sheets with development of the chromatograms in benzene–methanol–acetic acid.[6] The mobilities of compounds formed in the overall synthesis of α-lipoic acid-^{14}C in this solvent are given in the table.

MOBILITIES ON THIN-LAYER CHROMATOGRAMS^a OF INTERMEDIATES IN THE
CHEMICAL SYNTHESIS OF ^{14}C-α-LIPOIC ACID

Compound	R_f value^b in benzene–methanol–acetic acid (45:8:4, v/v/v)
Succinic acid	0.28
Dimethyl succinate	0.72
1,4-Butanediol	0.05
1,4-Dibromobutane	0.63
Adiponitrile	0.34
Adipic acid	0.44
Diethyl adipate	0.75
Monoethyl adipate	0.64
Ethyl adipoyl chloride	0.58
Ethyl 8-chloro-6-ketooctanoate	0.72
Ethyl 8-chloro-6-hydroxyoctanoate	0.89
Ethyl DL-6,8-dichlorooctanoate	0.81
DL-α-Lipoic acid	0.56

^{a} Thin-layer sheets were MN-Polygram SIL GEL N-HR.

^{b} Locations of compounds were visualized after exposure of the chromatograms to iodine vapor.

[3] L. Reed and C. Niu, *J. Am. Chem. Soc.* **77**, 416 (1955).

[4] D. S. Acker and W. J. Wayne, *J. Am. Chem. Soc.* **79**, 6483 (1957).

[5] I. C. Gunsalus and W. R. Razzell, see Vol. III [138].

[6] E. Stahl, "Thin-Layer Chromatography," p. 357. Academic Press, New York, 1965.

[48] Lipoic Acid Transport

By FRANKLIN R. LEACH

Reaction Catalyzed

Definitions[1]

Because of the varying meanings of the word "transport" as used by different investigators, definitions will be given for the terms used in this article.

Movement of a substance across a permeability barrier (that is, a phase change) in which the substance is in the same chemical form (same state) on either side of the permeability barrier is defined as transport.

Active transport occurs when the substance is concentrated in one phase against an electrochemical gradient.

The internal phase will be called pool, and that lipoic acid which is in a different state will be called incorporated (such as protein bound). The total incorporation and pool will be known as uptake and operationally defined by separation of the cells from the external medium which originally contained the lipoic acid.

Scheme of the Reaction

$$\text{Lipoic acid}_{\text{external}} \rightleftharpoons \text{cell membrane} \rightleftharpoons \text{lipoic acid}_{\text{internal}}$$

VITAMIN UPTAKE STUDIES IN MICROORGANISMS

Vitamin	Organism	Reference[a]
B_{12}	*Escherichia coli*	1
B_{12}	*Lactobacillus delbrueckii*	2
Biotin	*Lactobacillus arabinosus*	3
Folic acid	*Streptococcus faecalis*	4
Thiamine	*Lactobacillus fermenti*	5
Niacinamide	*Streptococcus faecalis*	6
Lipoic acid	*Streptococcus faecalis*	7

[a] References: (1) E. L. Oginsky, *Arch. Biochem. Biophys.* **36**, 72 (1952); (2) K. Kitahara and T. Sasaki, *J. Gen. Appl. Microbiol.* **9**, 213 (1963); (3) H. C. Lichstein *et al.*, *J. Biol. Chem.* **233**, 243 (1958); *J. Bacteriol.* **81**, 65 (1961); (4) R. C. Wood and G. H. Hitchings, *J. Biol. Chem.* **234**, 2381 (1959); (5) H. Y. Neujahr, *Acta Chem. Scand.* **17**, 1902 (1963); (6) H. Y. Neujahr, *6th Intern. Congr. Biochem.* **VIII-75**, 661 (1964); (7) D. C. Sanders and F. R. Leach, *Biochim. Biophys. Acta* **62**, 604 (1962).

[1] H. N. Christensen, "Biological Transport." Benjamin, New York, 1962.

Occurrence of Vitamin Transport and Uptake

The table lists studies on vitamin transport and uptake. Thus, the transport of vitamins has been studied in several systems and this article will present methods generally applicable to uptake studies.

Sanders and Leach[2,3] have demonstrated an energy-requiring, temperature-dependent system capable of concentrating lipoic acid 100-fold in the cell pool of *Streptococcus faecalis*. The lipoic acid in the cell pool was in the free form. This system is easily saturated with lipoic acid and is constitutive. The system has also been demonstrated in *Escherichia coli*, *Staphylococcus aureus*, and *Aerobacter aerogenes*.

Assay Method

Preparation of Labeled Lipoic Acid

Equations of Reactions

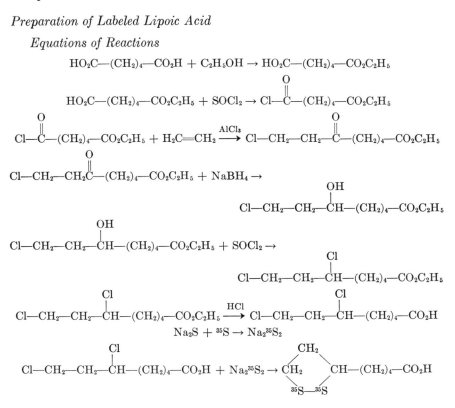

Procedure. Adipic acid monoethylester was prepared by the method of Fichter and Laurie[4] in 57% yield. This compound was allowed to react with

[2] D. C. Sanders and F. R. Leach, *Biochim. Biophys. Acta* **62,** 604 (1962).

[3] D. C. Sanders and F. R. Leach, *Biochim. Biophys. Acta* **62,** 41 (1964).

[4] F. Fichter and S. Laurie, *Helv. Chim. Acta* **16,** 887 (1933).

thionyl chloride as described by Berg[5] to produce ethyl δ-chloroformyl-valerate in an overall yield of 54%. Ethyl-DL-6,8-dichlorooctanoate was prepared by allowing ethyl δ-chloroformylvalerate to react with ethylene in a Friedel-Crafts type reaction followed by reduction with sodium borohydride and halogenation via thionyl chloride as described by Reed and Niu.[6] Overall yield at this stage was 12%. The ester was hydrolyzed by refluxing for 18 hours in 12 N HCl as described by Acker and Wayne[7] producing DL-6,8-dichlorooctanoic acid in a 6% overall yield.

Detailed instructions are given below for the incorporation of radio-active elemental sulfur into lipoic acid. This procedure was adapted from that of Acker and Wayne[7] using the distillation techniques of Thomas and Reed[8] to isolate crystalline lipoic acid-^{35}S.

One hundred milligrams of carrier S is placed into a round-bottom flask. The ^{35}S (50 mCi) in benzene is transferred to the flask using benzene for washing. The benzene is distilled off, and the flask is allowed to cool. A suspension of 770 mg of Na_2S in 8 ml of 95% ethanol is added and the mixture is refluxed until a brown solution is obtained. The system is con-tinually flushed with a stream of N_2 during the refluxing procedure. A solution containing 98 mg of NaOH and 532 mg of dichlorooctanoic acid in 8 ml of ethanol is added over a 3-hour period to the refluxing solution. The mixture is refluxed 30 minutes after the last addition, and then about one-half of the ethanol is removed by distillation. A solution of 100 mg of NaOH in 2 ml of H_2O is added and the system flushed with N_2 and refluxed for 30 minutes. After cooling, the mixture is acidified with 8 ml of 5% HCl and extracted 3 times with benzene. The benzene is then washed 3 times with a saturated aqueous solution of NaCl. The benzene is dried overnight over Na_2SO_4. The benzene solution is filtered into a round bottom flask and then the benzene is removed with a stream of N_2. The residue is taken up in a small amount of benzene and transferred to a tube with a side arm. The benzene is evaporated in a stream of N_2. The tube is connected to a line under 0.01 mm vacuum and allowed to pump for 2 hours. The tube is heated in a 160–170° oil bath and the lipoic acid is distilled up to a cold finger filled with crushed dry-ice. The lipoic acid from the cold finger is scraped into a bottle and weighed.

The overall yield for the entire synthesis is 2% based on adipic acid. The final product has an uncorrected melting point of 57–58°. Paper chromatography in two solvent systems reveals a single spot corresponding

[5] H. Berg, *Ber.* **67B**, 1622 (1934).

[6] L. J. Reed and C. Niu, *J. Am. Chem. Soc.* **77**, 1416 (1955).

[7] D. S. Acker and W. J. Wayne, *J. Am. Chem. Soc.* **79**, 6483 (1957).

[8] R. C. Thomas and L. J. Reed, *J. Am. Chem. Soc.* **78**, 5446 (1956).

to authentic lipoic acid which contains 99% of the radioactivity while 1% of the radioactivity remains at the origin. The specific activity is 46 μCi/mg.

Other Methods. Thomas and Reed[8] allowed benzyl mercaptan-^{35}S to react with DL-ethyl-6,8-dibromooctanoate to yield lipoic acid with a specific radioactivity of 68 μCi/mg. Adams[9] also used benzyl mercaptan-^{35}S and ethyl-6,8-dibromooctanoate to obtain lipoic acid with a specific activity of 1 μCi/mg. Attempts to isolate crystalline material which would have a specific activity of 200 μCi/mg failed. Acker and Wayne,[7] using essentially the procedure described in detail above, obtained lipoic acid-^{35}S with a specific activity of 4 μCi/mg. Attempts in our laboratory to effect tritium exchange into lipoic acid by the procedure of Wilzbach[10] resulted in production of a lipoic acid polymer.[11] Mitra, Mandal, and Burma[12] have developed a biosynthetic method for producing labeled lipoic acid using *Azotobacter vinelandii.* The specific activity of the product was not given.

Methods of Stopping the Reaction

Because of the metabolism of many of the compounds for which transport measurements are required and the desirability of measuring initial rates of transport, techniques for accurate measurement of small quantities and for stopping the reaction after short intervals are essential. Use of radioactive substances is the general method adopted for accurate measurement of the small quantities of substance transported and for distinguishing the transported molecules from those already present in the cell. Rapid separation of the cells from the suspending medium has been obtained by two procedures: filtration and centrifugation. Membrane filtration[13,14] allows separations to be obtained in 10–30 seconds and trains of multiple filters may be used for kinetic measurements. High concentrations of cells (2 mg/ml or greater) and certain organisms with capsules, such as *Leuconostoc mesenteroides,* lengthen the time of filtration and obviate rapid sampling. The filtration procedure is not suitable for use with lipoic acid-containing solutions because of the retention of approximately 10% of the lipoic acid by the filters in the absence of bacterial cells. This background retention is linear with increasing concentrations of lipoic acid and cannot be reduced by treatment of the filters with nonradioactive lipoic acid or by exhaustively washing the membrane filters.

[9] P. T. Adams, *J. Am. Chem. Soc.* **77,** 5357 (1955).

[10] K. E. Wilzbach, *J. Am. Chem. Soc.* **79,** 1013 (1957).

[11] R. C. Thomas and L. J. Reed, *J. Am. Chem. Soc.* **78,** 6148 (1956).

[12] S. K. Mitra, R. K. Mandal, and D. P. Burma, *Biochim. Biophys. Acta* **107,** 131 (1965).

[13] R. J. Britten, R. B. Roberts, and E. F. French, *Proc. Natl. Acad. Sci. U.S.* **41,** 803 (1955).

[14] F. R. Leach and E. E. Snell, *Biochim. Biophys. Acta* **34,** 292 (1959).

The centrifugation technique requires a means of stopping the uptake reaction prior to separation of the cells and the external medium, that is, a rapid transfer of the cells to conditions where uptake does not occur without damage to the permeability integrity of the cells. The following technique has been developed for use with lipoic acid as the substrate, but is generally applicable for use with other substrates. Samples are taken with a Cornwall syringe fitted with an automatic stop and are squirted into a cold centrifuge tube containing finely chopped frozen medium (the same medium as used for the incubation). This procedure results in rapid cooling (5 seconds) of the sample to 0°, which stops the transport reaction.

Measurement Procedure

Cells of *S. faecalis* strain 10C1 are grown for 8–10 hours at 37° in the lipoic acid-free synthetic medium described by Gunsalus and Razzell,[15] harvested by centrifugation, washed twice with cold salt solution, and then suspended in the neutral salts solution described by Leach and Snell.[14] The cells are used within 2 hours of preparation and the suspension is kept in an ice bath until the experiment is started. *E. coli* cells are grown on medium M-9[16] and treated similarly.

The concentration of cells is determined using a Spectronic 20 spectrophotometer by comparing the absorbance reading with a standard dry weight of cells versus absorbance curve. A quantity of cell suspension sufficient to give a final concentration of 0.3 mg/ml dry weight is added to a tube containing the neutral salts solution (M-9 medium with *E. coli*). The cells are allowed to equilibrate at the temperature of the experiment for 15 minutes. Glucose is added to a final concentration of 1 mg/ml to ensure an adequate energy source, and the incubation is continued for 15 minutes longer. At this time, labeled lipoic acid is added using a micropipette, and aliquots are removed at appropriate intervals using a Cornwall syringe set for 0.5 ml. The aliquot and a 0.5-ml syringe wash are rapidly ejected into finely chopped frozen medium. The cells are removed by centrifugation at 7000 rpm for 10 minutes in a Sorvall RC-2 centrifuge and washed twice with cold neutral salts solution. The two 1-ml washings remove essentially all the lipoic acid which is not accumulated. After the final washing, the cell pellet is suspended in 0.5 ml of distilled water and counted using either a thin-window gas flow counter or a liquid scintillation spectrometer. All aliquots are counted for 3000 total counts, which gives approximately a 2% counting error.

[15] I. C. Gunsalus and W. D. Razzell, see Vol. III p. 941.
[16] E. H. Anderson, *Proc. Natl. Acad. Sci. U.S.* **32,** 120 (1946).

Characterization of the Reaction

Specificity. Octanoic acid competes with lipoic acid for transport, there is slight inhibition with 8-methylthioctic acid, and no effect of 1,2-dithiolane-3-caproic or 1,2-dithiolane-3-butyric acids (the C-7 and C-9 analogs of lipoic acid) upon uptake of radioactive lipoic acid.

Effect of Temperature. The amount of lipoic acid uptake is proportional to the temperature of incubation. There is little uptake at 0° and a slightly increased uptake at 10°, while significant uptake occurs at both 20 and 37°. As stated before, a temperature of 20° is used for most experiments.

Energy Requirement. 2,4-Dinitrophenol inhibits uptake even in the presence of glucose. With incubated cells which have exhausted endogenous energy sources, glucose is required.

Occurrence in Protoplasts. Removal of the cell wall by lysozyme treatment does not modify uptake.

Intracellular Location of Enzymes. The reactions catalyzed by the lipoic acid-activating system are similar to those expected for a transport system. However, both in *E. coli* and *S. faecalis* the lipoic acid activating system as well as the pyruvate dehydrogenase are in the soluble fraction rather than membrane bound.[17] This finding makes unlikely the function of the lipoic acid-activating system in the transport of lipoic acid.

Binding of Lipoic Acid to Cell Membranes.[18] A glucose-dependent binding of radioactive lipoic acid to cell membranes prepared from *E. coli* has been demonstrated. The uptake of lipoic acid is reduced by osmotic shocking and the concentrated shock fluid binds lipoic acid. These findings suggest that the lipoic acid transport system is similar to those for the amino acids glycine and proline.

[17] Y. K. Oh and F. R. Leach, *Can. J. Microbiol.* **15**, 183 (1969).
[18] Y. K. Oh and F. R. Leach, unpublished observations.

[49] Lipoic Acid Activation [Biosynthesis of Lipoamide (Enzyme)]

By FRANKLIN R. LEACH

The Reaction

$$E_1 + ATP + lipoic\ acid \rightleftharpoons E_1—lipoyl—AMP + PP \tag{1}$$

$$E_1—lipoyl—AMP + E_2 \rightleftharpoons E_2\text{-}lipoyl + AMP + E_1 \tag{2}$$

$$E_2\text{-}lipoyl + A\text{-}PDC\text{-}NH_2 \rightarrow PDC\text{-}NH\text{-}lipoyl + E_2 \tag{3}$$

A-PDC = apopyruvate dehydrogenase complex; PDC-NH-lipoyl = holopyruvate dehydrogenase complex containing lipoic acid attached through the ϵ-NH_2 group of lysine[1,2]; E_1 = lipoic acid-activating enzyme (fraction PS-2A); E_2 = lipoyl transfer enzyme (fraction PS-2B).

Occurrence

Streptococcus faecalis

Reed and co-workers[3,4] observed that incubation of cell-free extracts of *S. faecalis* with lipoic acid resulted in activation of the apopyruvate dehydrogenase which was present in these extracts. By protamine sulfate fractionation, the *S. faecalis* extract was separated into two fractions, both of which had to be incubated simultaneously with lipoic acid to obtain a holopyruvate dehydrogenase. The protamine sulfate supernatant fraction (PS) was required only with apopyruvate dehydrogenase and was not required if the extract had been "activated" by prior incubation with lipoic acid. Thus the PS fraction is called the lipoic acid activating system. The PS fraction was further separated into PS-1, which contains lipo-amidase, an enzyme that removes bound lipoic acid from the holoenzyme, and PS-2, which contains the activating system. Alkaline $(NH_4)_2SO_4$ fractionation of PS-2 gave two fractions, PS-2A and PS-2B, both of which are required during the incubation of the apopyruvate dehydrogenase with lipoic acid.

In Other Organisms

Reed *et al.*[4] showed the existence of a lipoic acid-activating system in *Escherichia coli*. Mitra and Burma[5] have demonstrated the formation of

[1] H. Nawa, W. T. Brady, M. Koike, and L. J. Reed, *J. Am. Chem. Soc.* **82**, 896 (1960).
[2] K. Daigo and L. J. Reed, *J. Am. Chem. Soc.* **84**, 659 (1962).
[3] F. R. Leach, K. Yasunobu, and L. J. Reed, *Biochim. Biophys. Acta* **18**, 297 (1955).

radioactive ATP from $^{32}PP_i$ which was dependent upon added lipoyl adenylate (the reverse of reaction 1) in spinach and mung bean seedling extracts. Tsunoda and Yasunobu[6] have demonstrated the existence of a lipoic acid-activating enzyme in dog and beef liver. They were unable to detect any activity in beef heart extracts.

Assay Methods

Activation of Apopyruvate Dehydrogenase Complex[4]

Principle. Growth of *S. faecalis* strain 10C1 on a lipoic acid-free synthetic medium results in production of an apopyruvate dehydrogenase complex which requires activation by the lipoic acid-activating system plus lipoic acid before pyruvate dismutation can be observed. With limiting amounts of one of the fractions of the lipoic acid-activating system, the amount of pyruvate dismutation occurring in a second incubation is a measure of the amount of holopyruvate dehydrogenase formed during the first incubation.

Treatment of *E. coli* pyruvate dehydrogenase with "lipoamidase" results in the production of an apopyruvate dehydrogenase which can be used in the same way as the *S. faecalis* 10C1 apopyruvate dehydrogenase.

STEP 1

Reagents

> ATP, dipotassium, 6 mg per 10 ml of water
> Lipoic acid, 1 mg per 10 ml of 95% ethanol
> Potassium phosphate buffer, 1 M, pH 7.0
> Cysteine, 12 mg of free base per 1 ml of water
> TPP, 9.6 mg per 10 ml of water
> MgSO$_4$, 96 mg per 10 ml of water
> PP-1, prepared from *S. faecalis* 10C1 (see below)

Procedure. The complete reaction mixture contains: 0.005 micromole of *dl*-lipoic acid, 0.01 ml; 0.04 micromole of TPP, 0.02 ml; 0.02 micromole of ATP, 0.02 ml; 0.8 micromole of MgSO$_4$, 0.01 ml; 6 micromoles of potassium phosphate buffer (pH 7.0), 0.06 ml; 1 micromole of cysteine, 0.01 ml; PP-1, 5 units; and varying amounts of the lipoic acid-activating system, in a final volume of 0.25 ml. This mixture is incubated for 1 hour at 30° in 13 × 100-mm test tubes.

---

[4] L. J. Reed, F. R. Leach, and M. Koike, *J. Biol. Chem.* **232**, 123 (1958).

[5] S. K. Mitra and D. P. Burma, *J. Biol. Chem.* **240**, 4072 (1965).

[6] J. N. Tsunoda and K. T. Yasunobu, *Arch. Biochem. Biophys.* **118**, 395 (1967).

Step 2

Reagents

Potassium phosphate buffer, 1 M, pH 7.0

Potassium pyruvate or α-ketobutyrate, 1 M, pH 7.0, prepared as described by Korkes *et al.*[7]

Supplement consisting of 4.8 mg of CoA, 9.6 mg of NAD+, 46.5 mg of cysteine (free base), dissolved in 6 ml of water and used for 3 days after preparation

Phosphotransacetylase, cell extract prepared from *Clostridium kluyveri* by the method of Stadtman[8]

Lactate dehydrogenase, commercial crystalline, is determined as described by Kornberg[9] except the units of Mehler *et al.*[10] are used (A_{340} of 0.01 per minute = 1 unit)

Procedure. The reaction mixtures from step 1 are completed by adding 100 micromoles of potassium phosphate buffer, pH 7.0, 0.1 ml; 50 micromoles of potassium pyruvate or α-ketobutyrate, 0.05 ml; supplement, 0.1 ml giving 0.1 micromole of CoA, 0.23 micromole of NAD+, and 6.4 micromoles of L-cysteine; 12 units of phosphotransacetylase; and 2000 units of lactic dehydrogenase, in a final volume of 1 ml. The mixture is incubated for 30 minutes at 30° and then assayed for acetyl or propionyl phosphate by the method of Lipmann and Tuttle[11] with the modifications that citrate buffer is used in place of acetate buffer and the color is measured at 500 mμ in a Bausch and Lomb Spectronic 20 spectrophotometer.

Units. One unit of the lipoic acid activating system results in the activation of that amount of apopyruvate dehydrogenase which forms 1 micromole of acetyl or propionyl phosphate under the above conditions.

Lipohydroxamate Trapping[4]

Principle. In the presence of hydroxylamine, the activated lipoyl moiety formed in reaction 1 can be trapped as lipohydroxamate and determined colorimetrically.

Reagents

Potassium lipoate, 0.1 M

Na$_2$ATP, 0.1 M

[7] S. Korkes, A. Del Campillo, I. C. Gunsalus, and S. Ochoa, *J. Biol. Chem.* **193,** 721 (1951).

[8] E. R. Stadtman, see Vol. I, p. 595.

[9] A. Kornberg, see Vol. I, p. 441.

[10] A. H. Mehler, A. Kornberg, S. Grisolia, and S. Ochoa, *J. Biol. Chem.* **174,** 961 (1948).

[11] F. Lipmann and L. C. Tuttle, *J. Biol. Chem.* **159,** 21 (1945).

MgCl$_2$, 0.1 M
Tris buffer, 1 M, pH 7.4
Salt-free hydroxylamine, 5 M, pH 7.2[12]
KF, 0.1 M

Procedure. The complete system contains 10 micromoles of potassium lipoate, 0.1 ml; 10 micromoles of Na$_2$ATP, 0.1 ml; 5 micromoles of MgCl$_2$, 0.05 ml; 2500 micromoles of salt-free hydroxylamine (pH 7.2), 0.5 ml; and enzyme, in a final volume of 1.1 ml. The reaction mixture is incubated 1 hour at 30°, and the lipohydroxamic acid is determined as described by Berg[12] using a Beckman DU spectrophotometer. The precipitate is removed by centrifugation and filtration.

Units. The amount of enzyme forming 1 micromole of lipohydroxamate under the above conditions is a unit of activity.

Binding of ^{35}S-Lipoic Acid[4]

Principle. When ^{35}S-labeled lipoic acid is used as the substrate in lipoic acid-activation assays, the amount of radioactivity bound to protein is a measure of the activation of lipoic acid.

Reagents

[35]S-Lipoic acid: see this volume [48], Lipoic Acid Transport, for method of preparation
Potassium phosphate buffer, 0.02 M, pH 7.0

Procedure. Incubations are done as described above in section on activation of apopyruvate dehydrogenase complex (Step 1), with the use of 10 μg of ^{35}S-lipoic acid. The solution is dialyzed for 12 hours at 4° against three changes of 0.02 M potassium phosphate buffer, pH 7.0.

Units. That amount of protein which catalyzes the binding of 1 μg of lipoic acid to the apopyruvate dehydrogenase is a unit of activity.

[32]PP-ATP Exchange[6]

Principle. Tsunoda and Yasunobu[6] have found that [32]PP-ATP exchange, the reverse of reaction (1), can be used to determine the activity of the lipoic acid-activating enzyme.

Reagents

[32]PP neutralized with Tris buffer to pH 7.2 with 10[5] cpm and 178 micromoles per milliliter
Tris buffer, 1 M, pH 7.2
Potassium lipoate, 0.1 M

---

[12] P. Berg, *J. Biol. Chem.* **222,** 991 (1956).

MgCl$_2$, 0.1 M
KF, 1 M
Na$_2$ATP, 0.1 M

Procedure. The reaction mixture contains: 17.8 micromoles of ^{32}PP, 0.1 ml; 100 micromoles of Tris buffer, pH 7.2, 0.1 ml; 10 micromoles of potassium lipoate, 0.1 ml; 5 micromoles of MgCl$_2$, 0.05 ml; 50 micromoles of KF, 0.05 ml; 10 micromoles of Na$_2$ATP, 0.1 ml; and enzyme to a total of 1.5 ml. The incubation is at 31° for varying time intervals. At selected time intervals, aliquots are removed, and the reaction is terminated by the addition of 2 N trichloroacetic acid and 6 N HCl. The solution is centrifuged, and the ATP in the supernatant solution is adsorbed on charcoal as described by Crane and Lipmann.[13]

Pyrophosphate-Dependent Disappearance of Lipoyl Adenylate[5]

Principle. Synthetic lipoyl adenylate can be hydrolyzed in a pyrophosphate-dependent reversal of reaction (1). The remaining lipoyl adenylate is determined by hydroxamate color.

Reagents

Lipoyl adenylate,[4] 15 mM, prepared as described below
Tris buffer, 1 M, pH 7.0
KF, 0.1 M
MgCl$_2$, 0.1 M
Potassium pyrophosphate, 0.1 M
Neutralized hydroxylamine, 2 M
FeCl$_3$ reagent composed of equal volumes of 1:3 HCl, 12% trichloroacetic acid, and 5% FeCl$_3$ in HCl

Lipoyl adenylate is prepared by a modification of the method of Avison[14] with lipoic anhydride and AMP in aqueous pyridine. Lipoic anhydride is prepared by reaction of lipoic acid with N,N'-dicyclohexylcarbodiimide.[15]

To an ice-cold solution of 288 mg (1.4 millimoles) of *dl*-lipoic acid in 1.5 ml of acetonitrile is added dropwise, with stirring, a solution of 144 mg (0.7 millimole) of N,N'-dicyclohexylcarbodiimide in 0.5 ml of acetonitrile. The reaction mixture is stirred for 30 minutes at room temperature and then filtered. The yellow filtrate contains 0.68 millimole of lipoic anhydride, as measured by the hydroxamic acid method.[11] The instability of the anhydride necessitates immediate conversion to the desired product.

[13] R. K. Crane and F. Lipmann, *J. Biol. Chem.* **201**, 235 (1953).
[14] A. W. D. Avison, *J. Chem. Soc.* p. 732 (1955).
[15] F. Zetzsche and A. Fredrich, *Ber.* **72**, 1477 (1939).

AMP·2H$_2$O (202 mg, 0.525 millimole) is dissolved in 5.25 ml of 32% aqueous pyridine. The solution is cooled in an ice–salt bath, and a solution of lipoic anhydride (0.68 millimole) in acetonitrile (2.5 ml) is added with vigorous stirring during a period of 10 minutes. The mixture is stirred for an additional 50 minutes, and then 5 ml of ice-cold water is added. The mixture is extracted with two 10-ml portions of cold peroxide-free ether and then filtered. The filtrate is extracted in a centrifuge-type separatory funnel with one 15-ml and two 10-ml portions of cold chloroform. During the last extraction an emulsion forms. The mixture is centrifuged, as much liquid as possible is removed, and the light yellow precipitate is collected on a sintered-glass funnel. The precipitate is washed consecutively with two 5-ml portions of ice-cold water, two 5-ml portions of absolute ether, and then dried *in vacuo* over calcium chloride. The yield of product is 111 mg (39%, based on the amount of AMP used).

Lipoyl adenylate migrates as a single substance on paper chromatograms as determined by ultraviolet "quenching" and by means of spray reagents for organic phosphate,[16] and disulfide[17] and anhydride linkages.[18] The R_f values (at 25°) are 0.68 and 0.44, respectively, in isobutyric acid–concentrated NH$_4$OH–water (66:1:33) and isopropyl alcohol–water (70:30).[19] The ultraviolet absorption spectrum of the synthetic material at pH 7 shows maxima at 259 and 332 mμ. From the absorbance at these wavelengths, and the extinction coefficients of 15.4×10^3 and 154 for AMP and lipoic acid, respectively, the synthetic material is calculated to contain 1.81 micromoles of AMP and 1.80 micromoles of lipoic acid per milligram. The material contains 1.80 micromoles of labile lipoyl groups per milligram, measured as lipohydroxamic acid. These values are in reasonable agreement with the theoretical value of 1.86 micromoles per milligram.

Procedure. The reaction mixture contains 100 micromoles of Tris buffer, pH 7.0, 0.1 ml; 5 micromoles of MgCl$_2$, 0.05 ml; 10 micromoles of potassium pyrophosphate, 0.1 ml; 1.5 micromoles of lipoyl adenylate, 0.1 ml; 10 micromoles KF, 0.1 ml; and extract in a total volume of 1 ml. The incubation is at 35° for 20 minutes. Then 1 ml of 2 M neutralized hydroxylamine is added and the incubation is continued for 10 minutes at room temperature. Then 3 ml of the FeCl$_3$ reagent are added, and the volume is made up to 6 ml. The protein is removed by centrifugation, and hydroxamic acid is determined as described above.

Units. A unit of enzyme activity causes the disappearance of 1 micromole of lipoyl adenylate under the above conditions.

[16] R. S. Bandurski and B. Axelrod, *J. Biol. Chem.* **193,** 405 (1951).

[17] L. J. Reed and B. G. DeBusk, *J. Biol. Chem.* **199,** 881 (1952).

[18] E. R. Stadtman and H. A. Barker, *J. Biol. Chem.* **184,** 769 (1950).

[19] P. T. Talbert and F. M. Huennekens, *J. Am. Chem. Soc.* **78,** 4671 (1956).

Purification Procedures

From S. faecalis[4]

Growth of Cells. S. *faecalis* strain 10C1 is grown on lipoic acid-deficient medium as described by Gunsalus and Razzell[20] except that thiamine and ascorbic acid are omitted.

Preparation of Cell-Free Extract.[3] After 12–14 hours of incubation at 37°, the cells are harvested by means of a Sharples centrifuge and washed once with distilled water. About 25 g (wet weight) of cells is obtained from each carboy, which contains 10 liters of medium.

The cell paste (25 g) is suspended in sufficient 0.02 M potassium phosphate buffer (pH 7.0) to give a final volume of 50 ml. The suspension is subjected to sonic vibration for 45 minutes with a Raytheon 10 kc oscillator and then centrifuged for 40 minutes at the top speed of a Sorvall SS1 centrifuge at 4°. The cell-free extract contains approximately 80 mg of protein per milliliter.

Protamine Sulfate Fractionation.[4] The fractionation is carried out at 4° as follows. The extract is diluted with 0.02 M potassium phosphate buffer (pH 6.0) to a protein concentration of 20 mg/ml. To the diluted extract are added stepwise, with stirring, 0.12, 0.02, 0.01, and 0.01 volumes of 2% protamine sulfate solution (pH 5.0). After each addition of protamine sulfate, the mixture is centrifuged for 45 minutes at 54,000 g in a Spinco preparative ultracentrifuge, and aliquots (0.1 and 0.2 ml) of the supernatant fluid are assayed by the two-step dismutation procedure described above. If the supernatant fluid shows no loss in dismutation activity, the corresponding precipitate is discarded. Usually the first two precipitates are discarded. A complete loss of dismutation activity usually results from the third and fourth additions of protamine sulfate. The corresponding precipitates are combined and suspended, by means of a glass homogenizer, in a volume of 1 M KCl equal to 0.1 the volume of the diluted extract. The suspension is centrifuged at 95,000 g for 30 minutes, and the insoluble material is discarded. The clear solution is designated fraction PP. The supernatant fluid from the protamine sulfate fractionation is designated fraction PS.

Further Purification of Fraction PP. When fraction PP is dialyzed, a precipitate appears. The supernatant fluid exhibits little or no activity. The precipitate is soluble in 1 M KCl, and the resulting solution is active, suggesting that the precipitate is a protamine-protein complex. This difficulty in working with fraction PP is overcome by fractionation with ammonium sulfate. Fraction PP is brought to 0.6 saturation by adding a

---

[20] I. C. Gunsalus and W. E. Razzell, see Vol. III, p. 941.

saturated solution of ammonium sulfate slowly with mechanical stirring. The precipitate, designated fraction PP-1, is collected by centrifugation, dissolved in 0.02 M potassium phosphate buffer (pH 7.0), and then dialyzed for 4 hours with stirring against the same buffer.

Further Purification of Fraction PS. Fraction PS is brought to 0.6 saturation by addition of solid ammonium sulfate. The precipitate, designated fraction PS-1, is subsequently found to contain an enzyme, lipoamidase, which releases lipoic acid from the protein-bound form.[21] The supernatant fluid from the ammonium sulfate treatment is brought to 1.0 saturation by addition of solid ammonium sulfate. The precipitate, designated PS-2, is treated in the same manner as fraction PP-1.

Separation of Fraction PS-2. When fraction PS-2 is fractionated with ammonium sulfate at an alkaline pH, it is found to consist of two essential components. In a typical fractionation, 116 ml of a 4-molal solution of ammonium sulfate, adjusted to pH 8 (Beckman glass electrode) with concentrated NH_4OH, is added slowly with stirring to 50 ml of fraction PS-2. Then 17.4 g of solid ammonium sulfate is added slowly to the mixture. The precipitate, designated fraction PS-2A, is collected by centrifugation, and the supernatant fluid is then brought to saturation with solid ammonium sulfate. The precipitate obtained by this latter treatment is designated fraction PS-2B. The two precipitates are dissolved separately in 25 ml of 0.02 M potassium phosphate buffer (pH 7.0) and the solutions are dialyzed with stirring for 4 hours against the same buffer.

Purification of Fraction PS-2B. Fraction PS-2B is heated for 5–10 minutes in boiling water. The denatured protein is removed by centrifugation and is discarded. Heated fraction PS-2B is adjusted to pH 2.0 with 1 N hydrochloric acid. The precipitate is removed by centrifugation and discarded. To the supernatant solution (25 ml) is added slowly with stirring 2.5 ml of 50% trichloroacetic acid. The precipitate is collected by centrifugation and suspended in 7.5 ml of deionized water. The suspension is adjusted to pH 4.7 with 0.1 N potassium hydroxide and then centrifuged. The supernatant fluid is dialyzed for 4 hours with stirring against deionized water. The recovery of fraction PS-2B activity is 34% (8.4 mg of protein, specific activity 96).

Summary of Purification.[4] Table I summarizes the results of a typical purification.

From Beef Liver[6]

Procedure. Fresh beef liver (1000–3000 g) is chopped and suspended in 0.05 M Tris buffer, pH 7.2, in 1:1.5 ratio. The mixture is homogenized for

---

[21] L. J. Reed, M. Koike, M. E. Levitch, and F. R. Leach, *J. Biol. Chem.* **232**, 143 (1958).

TABLE I

TYPICAL FRACTIONATION OF *Streptococcus faecalis* EXTRACT[a]

Fraction	Volume (ml)	Protein (mg)	Activity (units per mg of protein)	Recovery (%)
Diluted extract	500	10,250	0.85[b]	100
PP	40	440	18.0	92
PS	550	3,900	2.4	106
PP-1	30	111	21.0	27[c]
PS-2	63	227	18.5	49
PS-2A	31	139	27.3	44
PS-2B	31	130	28.0	42
PS-2B (heated)	31	63	57.0	40

[a] See Reed *et al.*[4]

[b] This value varied from 0.65 to 1.7 in different preparations.

[c] The recovery in some instances is as high as 50%.

4 minutes in a large Waring blendor. The homogenate is passed through cheesecloth and centrifuged at 8000 g in a Sorvall centrifuge. The supernatant solution is brought to 0.35 saturation by the addition of solid ammonium sulfate, centrifuged, and the precipitate discarded. The supernatant is then brought to 0.55 saturation and centrifuged for 15 minutes at 8000 g. The precipitate is dissolved in a minimum amount of 0.05 M Tris buffer, pH 7.2, and dialyzed overnight against 0.005 M Tris buffer pH 7.2. The pH of the extract is adjusted to pH 5.8 with cold 1 N acetic acid. The solution is treated with calcium phosphate gel (16 mg gel/milliliter) at protein : gel ratios of 1:10, 1:8, and 1:4 in succession with centrifugation steps in between the adsorption steps. The specific activity of the purified preparations are 50 to 70-fold greater than that of the homogenate.

Summary. A typical result of the purification procedure is summarized in Table II.

Properties

Enzymatic Activities. Fraction PS-2A catalyzes the formation of lipoyl hydroxamate and the hydrolysis of exogenously added lipoyl adenylate. The activation of lipoic acid is dependent upon ATP and Mg^{2+}. Fraction PS-2B has been shown to be protein in nature by proteolytic enzyme treatments, but no activity other than its essentiality for the activation of the apopyruvate dehydrogenase complex has been detected.

Requirements and Specificity. Lipoic acid (10 μg/ml, 84 mg of protein) gives maximum binding and activation. 5-Methyl-1,2-dithiolane-3-valeric

TABLE II

PURIFICATION OF THE LIPOIC ACID ACTIVATING ENZYME[a]

Procedure	Volume (ml)	Total units[b]	Total protein (g)	Specific activity (μmoles/g)	Yield (%)
1. Crude homogenate	3480	1744	581.2	0.003	100.0
2. Supernatant	2210	690	344.8	0.002	35.8
3. 0.0–0.30 $(NH_4)_2SO_4$ supernatant	1710	297	47.9	0.006	18.4
4. 0.30–0.60 $(NH_4)_2SO_4$ precipitate	540	149	74.5	0.002	8.6
5. 1:30 Gel:protein $Ca_3(PO_4)_2$ supernatant	653	242	48.3	0.005	13.3
6. 1:10 Gel:protein supernatant	810	248	27.5	0.009	14.5
7. 1:8 Gel:protein supernatant	925	363	25.9	0.014	21.6
8. 1:4 Gel:protein supernatant	1380	1175	23.5	0.050	70.3
9. 1:4 Gel:protein supernatant (2nd treatment)	2300	2768	21.2	0.126	166.0

[a] See Tsunoda and Yasunobu.[6]

[b] One unit is defined as the amount of enzyme required to form 1 micromole of lipoylhydroxamate in 1 hour at 31° under standard assay conditions.

acid (8-methylthioctic acid) is a competitive inhibitor; fatty acids inhibit lipoic acid activation with increasing effectiveness as the length of the carbon chain increases. Lipoamide, lipoanilide, and N,N-diethyllipoamide fail to inhibit lipoic acid activation. Chemically synthesized lipoyl adenylate replaces both lipoic acid and ATP in the incubation mixture but does not replace any of the enzymes, suggesting that lipoyl adenylate is an enzyme-bound intermediate. The K_m values are 2.5×10^{-6} M for lipoic acid and 3.6×10^{-6} M for lipoyl adenylate. Mg^{2+} is required for activation, and binding with 1 mM $MgSO_4$ giving maximum activity. The magnesium ion can be replaced by manganous, cobaltous, or zinc ions, but not by calcium, ferrous, or ferric ions. ATP is required with 5×10^{-6} M giving half-maximal activation. Because of the presence of adenylate kinase in fraction PS-2, ADP is also effective; CTP, GTP, ITP, and UTP are, respectively, 40, 25, 15, and 50% as active as ATP. With aged fractions, 4 mM of a thiol, e.g., cysteine, glutathione, or 2-mercaptoethylamine, is required for activation. No requirement of TPP for binding of radioactive lipoic acid is demonstrable, but greater dismutation activity is obtained when TPP is included in the first incubation, presumably due to a time lag in attachment to the apoenzyme.[22] Arsenite (0.1 mM) does not inhibit the binding of lipoic acid, suggesting that the dithiol form of lipoic acid is not an intermediate in the binding of lipoic acid to protein. A pH optimum of 7.2 was shown by Tsunoda and Yasunobu.[6] A similar enzyme activates biotin,[23,24] but Tsunoda and Yasunobu[6] could not demonstrate biotin activation by their beef liver preparation.

Stability. The *S. faecalis* fractions can be stored in the deep freeze for several months. After 3 months 45% of the PS-2A and 80% of the PS-2B activity remains. Tsunoda and Yasunobu[6] report that their enzyme is "very unstable," losing all activity after 7 days of storage.

[22] D. E. Green, D. Herbert, and V. Subrahmanyan, *J. Biol. Chem.* **138,** 327 (1941).
[23] M. D. Lane, D. L. Young, and F. Lynen, *J. Biol. Chem.* **239,** 2858 (1964).
[24] M. D. Lane, K. L. Ronenger, D. L. Young, and F. Lynen, *J. Biol. Chem.* **239,** 2865 (1964).

[50] Purification and Properties of Lipoamidase

By MASAHIKO KOIKE and KANTARO SUZUKI

The lipoyl moieties in the pyruvate and 2-oxoglutarate dehydrogenase complexes from *Escherichia coli* (Crookes strain) are bound in amide linkage to the ε-amino group of lysine residues.[1] An enzyme in *Streptococcus faecalis* (strain 10Cl), which releases lipoic acid from the protein-bound form, has been purified approximately 100-fold.[2-4] In addition to releasing lipoic acid from the protein-bound form, the enzyme hydrolyzes methyl lipoate, lipoamide, and several N-lipoyl amino acids and peptides, including ε-N-lipoyl-L-lysine. Lipoamidase activity was not detectable in homogenates of bovine tissues or in *E. coli* (Crookes strain) extracts. The presence of an apparently similar enzyme in bakers' yeast extract has been reported.[5]

Assay Methods

Lipoamidase Assay

Principle. The assay is based on the loss of pyruvate dismutation activity (Eq. 1) accompanying the release of protein-bound lipoic acid

[1] H. Nawa, W. T. Brady, M. Koike, and L. J. Reed, *J. Am. Chem. Soc.* **82,** 896 (1960).
[2] L. J. Reed, M. Koike, M. E. Levitch, and F. R. Leach, *J. Biol. Chem.* **232,** 143 (1958).
[3] M. Koike and L. J. Reed, *J. Biol. Chem.* **235,** 1931 (1960).
[4] K. Suzuki and L. J. Reed, *J. Biol. Chem.* **238,** 4021 (1963).
[5] G. R. Seaman, *J. Biol. Chem.* **234,** 161 (1959).

$$2 \text{ Pyruvate} + P_i \rightarrow \text{acetyl-P} + CO_2 + \text{lactate} \qquad (1)$$

from a purified preparation of the *E. coli* pyruvate dehydrogenase complex. The dismutation assay[6,7] is based on measurement of acetyl phosphate [Eq. (1)] by means of the hydroxamic acid method.[8]

Reagents

Potassium phosphate buffer, 1 M and 0.02 M, pH 7.0

Potassium pyruvate, 0.5 M

Magnesium sulfate, 0.2 M

Thiamine pyrophosphate, 0.01 M

CoA, 0.001 M, containing 0.64 M cysteine, prepared before use

NAD, 0.0023 M

Sodium citrate buffer, 0.1 M, pH 5.4

Hydroxylamine solution, 2 M, pH 6.4. This is freshly prepared by mixing an equal volume of 28% hydroxylamine hydrochloride (4 M) and 14% sodium hydroxide (3.5 M).

Ferric chloride solution: a 5% solution of ferric chloride·$6H_2O$, in 0.1 N hydrochloric acid

Trichloroacetic acid, 12%

E. coli pyruvate dehydrogenase complex.[7,9] The complex is diluted with 0.02 M phosphate buffer, pH 7, to give 5 units per 0.01 to 0.05 ml.

Phosphotransacetylase. A cell-free extract prepared from *Clostridium kluyveri* dried cells[10] is suitable.

Crystalline lactate dehydrogenase

Enzyme. The enzyme is diluted with 0.02 M phosphate buffer, pH 7.0, and the aliquot assayed is such that the loss of pyruvate dismutation activity is less than 4 units.

Procedure. To a 13 × 100 mm Pyrex test tube are added 0.01 ml (5 units) of *E. coli* pyruvate dehydrogenase complex, the enzyme to be assayed, and 0.02 M potassium phosphate buffer to make a final volume of 0.25 ml. One control tube contains the complex, and another the phosphate buffer. The mixture is incubated for 1 hour at 30°, then cooled in ice during

[6] S. Korkes, A. del Campillo, I. C. Gunsalus, and S. Ochoa, *J. Biol. Chem.* **193,** 721 (1951).

[7] L. J. Reed and C. R. Willms, Vol. IX [50]; see also L. J. Reed, F. R. Leach, and M. Koike, *J. Biol. Chem.* **232,** 123 (1958).

[8] F. Lipmann and L. C. Tuttle, *J. Biol. Chem.* **159,** 21 (1945); see also E. R. Stadtman, Vol. III [39].

[9] M. Koike, L. J. Reed, and W. R. Carroll, *J. Biol. Chem.* **235,** 1924 (1960).

[10] E. R. Stadtman, Vol. I [84].

addition of components necessary to complete the pyruvate dismutation assay system. These components include 0.1 ml of phosphate buffer (1 M), 0.02 ml of thiamine pyrophosphate, 0.02 ml of magnesium sulfate, 0.1 ml of CoA-cysteine solution, 0.1 ml of NAD, 0.1 ml of potassium pyruvate, 2000 units[11] of lactate dehydrogenase, 10 units[12] of phosphotransacetylase and water in a final volume of 1.0 ml. The mixture is incubated for 30 minutes at 30°. One milliliter of citrate buffer and 1 ml of hydroxylamine solution are added, and the mixture is allowed to stand for 10 minutes at room temperature. One milliliter of trichloroacetic acid solution, hydrochloric acid solution, and ferric chloride solution are added successively, and the mixture is centrifuged for 10 minutes in a clinical centrifuge. After the supernatant fluid has stood for 20 minutes, the optical density is determined at 540 mμ against the control with a suitable spectrophotometer. A standard curve is prepared with synthetic acetylhydroxamate.[8,13]

Definition of Unit and Specific Activity. One unit of lipoamidase is defined as the amount of enzyme which produces a loss of one unit of pyruvate dismutation activity. One unit of pyruvate dismutation activity corresponds to the production of 1 micromole of acetyl phosphate per hour. Specific activity is expressed as units per milligram of protein. Protein is determined by the method of Lowry *et al.*,[14] crystalline bovine serum albumin being used as standard.

Application of Lipoamidase in the Determination of Protein-Bound Lipoic Acid

Principle. Lipoic acid released from lipoyl derivatives by lipoamidase is reduced with sodium borohydride, and the dithiol produced is determined colorimetrically.[15]

Reagents

> Potassium phosphate buffer, 1 M, pH 7.0
> Lipoic acid derivative
> Hydrochloric acid, concentrated
> Hydrochloric acid, 0.03 M
> Benzene
> Nitrogen

[11] A. Kornberg, Vol. I [67]; see also A. H. Mehler, A. Kornberg, S. Grisolia, and S. Ochoa, *J. Biol. Chem.* **174,** 961 (1948).

[12] E. R. Stadtman, Vol. I [98].

[13] W. R. Jencks, *J. Am. Chem. Soc.* **80,** 4584 (1958).

[14] O. H. Lowry, N. J. Rosebrough, A. L. Farr, and R. J. Randall, *J. Biol. Chem.* **193,** 265 (1951).

[15] G. L. Ellman, *Arch. Biochem. Biophys.* **82,** 70 (1959).

Potassium phosphate buffer, 0.04 M, pH 8.0

Potassium phosphate buffer, 0.04 M, pH 7.0

Potassium phosphate buffer, 0.02 M, pH 7.0

Sodium borohydride solution, 0.2%

5,5'-Dithiobis(2-nitrobenzoic acid) solution, 0.396%. This solution is prepared by dissolving 39.6 mg of DTNB in 10 ml of 0.04 M phosphate buffer, pH 7.0

Enzyme (purified lipoamidase). The lipoamidase is diluted with 0.02 M phosphate buffer to give 204 or 520 units per 0.05 ml.

Procedure. An amount of protein containing approximately 10 millimicromoles of protein-bound lipoic acid is incubated with 400 units of lipoamidase in 0.5 ml of 0.02 M phosphate buffer, pH 7.0, for 1 hour at 30°. The incubation mixture is successively cooled in ice, acidified carefully with 0.05 ml of concentrated hydrochloric acid, and extracted with two 1.5-ml portions of benzene. The benzene extracts are combined and evaporated by means of a stream of nitrogen. The residue is dissolved in a small volume of 0.04 M phosphate buffer, pH 8.0, to give a solution containing 5–25 millimicromoles of lipoic acid per 0.05 ml. To the 0.05-ml aliquot is added 0.05 ml of a 0.2% aqueous solution of sodium borohydride. The solution is allowed to stand at 30° for 15 minutes and then is adjusted to pH 7.0 with 0.03 N hydrochloric acid (approximately 0.07 ml) to destroy unreacted sodium borohydride. To the solution is added 0.9 ml of 0.04 M phosphate buffer, pH 8.0, and 0.01 ml of a 0.39% solution of 5,5'-dithiobis(2-nitrobenzoic acid) in 0.04 M phosphate buffer, pH 7.0. The reference cuvette contains the latter reagent and buffer. The absorbance is measured at 412 mμ with a suitable spectrophotometer, and the content of sulfhydryl groups is calculated using a molar extinction coefficient of 13,600 M^{-1} cm^{-1}.[15] Lipoic acid (5–25 millimicromoles) is used as an internal standard.

Purification Procedure

Reagents

Protamine solution, 2%, pH 5.0, prepared before use and kept at room temperature. Protamine sulfate (salmine) is suspended in water, the pH is adjusted to 5 with 10% KOH, and the mixture is centrifuged to remove insoluble material.

Potassium phosphate buffer, 0.02 M, pH 7.0

Acetic acid, 1 N

Growth of Cells. S. *faecalis*, strain 10C1 (ATCC 11,700), is grown according to the procedure of Gunsalus[16] with some modification. S. *faecalis*

[16] I. C. Gunsalus and W. E. Razzell, Vol. III [138].

from a stock agar (containing AC broth[17]) stab culture is inoculated into 10 ml of AC broth and incubated at 37° for 24 hours; 10 ml is transferred to 1 liter of the synthetic medium. After 15 hours' incubation at 37° the culture is transferred to 9 liters of the same medium. After an 8-hour incubation the cells are harvested in a Sharples centrifuge; yield, 1.7–2.3 g of wet cells per liter. For large-scale preparation, the cells are grown in 30 liters of lipoic acid-deficient medium in a Biogen continuous culture apparatus at 37° for 5 hours with vigorous aeration.[4] The cells used for inoculum are grown in 3 liters of synthetic medium for 10–12 hours at 37° with the use of a reciprocal shaker. The yield of wet cells is 75 to 100 g. They are kept frozen until used.

Step 1. Extraction. The cells are thawed and suspended in cold 0.02 M phosphate buffer, pH 7.0, at a ratio (w/v) of 30 g to 100 ml. The suspension is divided into 50-ml portions, and each is treated for 30 minutes at maximal power in a 10-kc sonic oscillator (Raytheon). Cell debris is removed by centrifugation at 53,700 g (20,000 rpm) for 30 minutes in the 21 rotor of a Beckman Model L ultracentrifuge.

Step 2. Protamine Sulfate Fractionation. The cell-free extract is diluted with 0.02 M phosphate buffer to a protein concentration of 20 mg/ml, as determined by the turbidimetric method,[18] and the pH is adjusted to 6.0 with 1 N acetic acid. To 415 ml of diluted extract is added slowly with stirring 33 ml (0.08 volume) of a 2% protamine solution. The mixture is stirred for an additional 15 minutes and then centrifuged at maximum speed for 30 minutes in the No. 850a head of an International Model PR-2 centrifuge. The precipitate is discarded, and protamine solution is added until the apopyruvate dehydrogenase complex is completely precipitated.[19] Usually, a total of 0.1–0.12 volume of protamine solution is required.

Step 3. First Ammonium Sulfate Fractionation. The supernatant fluid from the preceding step is brought to 0.5 ammonium sulfate saturation by adding slowly with stirring 0.35 g of the salt per milliliter. The mixture is stirred for an additional 15 minutes and then centrifuged for 30 minutes at 3,180 g. The precipitate is dissolved in 0.02 M phosphate buffer, and the solution is dialyzed for 8 hours against 2 liters of the same buffer. The yellow solution after dialysis (36.5 ml) contains 18.5 mg of protein per milliliter.

Step 4. Second Ammonium Sulfate Fractionation. To the dialyzed solution from the preceding step is added slowly 29.8 ml of a saturated solution of ammonium sulfate, adjusted to pH 7.0 with concentrated ammonium hydroxide. The mixture is centrifuged for 20 minutes at 10,000 g, and the

[17] I. C. Gunsalus, M. I. Dolin, and L. Struglia, *J. Biol. Chem.* **194**, 849 (1952).
[18] E. R. Stadtman, G. D. Novelli, and F. Lipmann, *J. Biol. Chem.* **191**, 365 (1951).
[19] L. J. Reed, F. R. Leach, and M. Koike, *J. Biol. Chem.* **232**, 123 (1958).

precipitate is discarded. To the supernatant fluid is added 25.2 ml of saturated salt solution. The precipitate is collected by centrifugation and is dissolved in a minimal volume of 0.02 M phosphate buffer. The solution is dialyzed for 8 hours against two changes of the same buffer. The dialyzed solution (13.2 ml) contains 26.2 mg of protein per milliliter.

Step 5. DEAE-Cellulose Column Chromatography. The enzyme solution from the preceding step is applied to a DEAE-cellulose column (2.5 × 15 cm) preequilibrated with 0.02 M phosphate buffer, and the column is washed with about 80 ml of the same buffer. Protein is eluted stepwise from the column with solutions containing 0.1 M and 0.15 M potassium chloride in 0.02 M phosphate buffer. Five-milliliter fractions are collected at a flow rate of 30 ml/hour. Lipoamidase is eluted with 0.15 M potassium chloride in 0.02 M phosphate buffer. The active fractions are combined and dialyzed for 7 hours against 2 liters of 0.02 M phosphate buffer. The dialyzed solution (35 ml) contains 0.78 mg of protein per milliliter. Dilute solutions of the enzyme can be concentrated approximately 10-fold (with essentially quantitative recovery of activity) by adsorption on a small column (0.5 × 1.5 cm) of DEAE-Sephadex, followed by elution with a solution of 0.4 M potassium chloride in 0.02 M phosphate buffer. A summary of the purification procedure is given in the table.

PURIFICATION OF LIPOAMIDASE

Step	Volume (ml)	Protein (mg)	Specific activity	Recovery (%)
1. Diluted extract[a]	415	3650	32[b]	100
2. Protamine sulfate fractionation	420	1930	60	102
3. First $(NH_4)_2SO_4$ fractionation	36.5	675	140	82
4. Second $(NH_4)_2SO_4$ fractionation	13.2	346	268	80
5. DEAE-cellulose fractionation	35	27.3	1960	46

[a] From 100 g of cells.

[b] The specific activity of extracts obtained from different batches of cells varies from 16 to 32.

Properties

Substrate Specificity. The natural substrate of lipoamidase is the lipoyl moiety, which is covalently bound to the ε-amino group of a lysine residue in lipoate acetyltransferase. However, the enzyme also hydrolyzes lipoic acid derivatives, including methyl lipoate, etc. At a concentration of the lipoyl moiety of 4×10^{-5} M, ε-N-lipoyl-L-lysine is hydrolyzed at approximately 29% of the rate at which lipoic acid is released from the pyruvate dehydrogenase complex. This enzyme shows rather high specificity for lipoyl derivatives such as DL-lipoamide, DL-methyl lipoate, ε-N-(DL-lipoyl)-L-lysine, and its derivatives.

Stability and pH Optimum. Solutions of the purified enzyme have been stored at −15° for 6 months without significant loss of activity. Optimal activity is observed at pH 7.8.

Inhibitor. Lipoamidase activity is inhibited slightly (20%) by $10^{-4}\,M$ p-chloromercuribenzoate. It is not affected by EDTA at $10^{-3}\,M$.

[51] Purification and Properties of Lipoamide Dehydrogenases from Pig Heart α-Keto Acid Dehydrogenase Complexes

By MASAHIKO KOIKE and TARO HAYAKAWA

$$\text{Lipoamide} + \text{NADH} + \text{H}^+ \rightleftharpoons \text{dihydrolipoamide} + \text{NAD}^+ \tag{1}$$

Lipoamide dehydrogenase is an essential component of the pig heart α-keto acid dehydrogenase complexes, which catalyze a CoA- and NAD-linked oxidative decarboxylation of pyruvate[1,2] and 2-oxoglutarate.[3,4] This enzyme is a flavoprotein and catalyzes the oxidation of protein-bound dihydrolipoic acid.

Assay Methods

NADH-Lipoamide Oxidoreductase Assay[5,6]

Principle. The assay is based on spectrophotometric determination of the rate of NADH oxidation (at 340 mμ) in the presence of lipoamide.

Reagents

Potassium phosphate buffer, 1 M, pH 6.5
NADH, $2 \times 10^{-3}\,M$, in 0.002 M phosphate buffer, pH 7.6
NAD, $2 \times 10^{-3}\,M$
EDTA, $2.5 \times 10^{-2}\,M$
(±)-Lipoamide, $1.6 \times 10^{-2}\,M$, in 95% ethanol, prepared just before use

[1] T. Hayakawa, M. Hirashima, S. Ide, M. Hamada, K. Okabe, and M. Koike, *J. Biol. Chem.* **241**, 4694 (1966).

[2] T. Hayakawa and M. Koike, *J. Biol. Chem.* **242**, 1356 (1967).

[3] V. Massey, *Biochim. Biophys. Acta* **38**, 447 (1960).

[4] M. Hirashima, T. Hayakawa, and M. Koike, *J. Biol. Chem.* **242**, 902 (1967).

[5] S. Ide, T. Hayakawa, K. Okabe, and M. Koike, *J. Biol. Chem.* **242**, 54 (1967).

[6] V. Massey, Vol. IX [52].

Procedure. To a cuvette with a 1-cm light path are added 0.1 ml of phosphate buffer, 0.1 ml of NADH, 0.1 ml of NAD, 0.05 ml of lipoamide, 0.1 ml of EDTA, water, and enzyme to make a total volume of 2.0 ml. The reaction is initiated at 25° by the addition of enzyme. The blank cell contains 0.1 ml of phosphate buffer, 0.1 ml of NAD, 0.05 ml of lipoamide, 0.1 ml of EDTA, and 1.75 ml of water. The decrease in absorbance at 340 mμ is followed with a recording spectrophotometer. A decrease in absorbance of 0.1–0.25 during the initial phase of the reaction (about 1 minute) is a linear function of protein concentration.

Definition of Unit and Specific Activity. One unit is defined as the amount of enzyme that catalyzes an initial rate of oxidation of 1 micromole of NADH per minute. Specific activity is expressed as units of enzyme per milligram of protein.

Dihydrolipoamide-NAD Oxidoreductase Assay[5]

Principle. The assay is based on spectrophotometric determination of the rate of NADH formation in the presence of lipoamide [Eq. (1), right to left].

Reagents

Potassium phosphate buffer, 1 M, pH 8.5
NAD, 2 × 10⁻³ M
EDTA, 2.5 × 10⁻² M
(±)-Dihydrolipoamide, 8 × 10⁻³ M, in an equal volume of water and 95% ethanol

Procedure. To a cuvette with a 1-cm light path are added 0.1 ml of phosphate buffer, 0.1 ml of NAD, 0.1 ml of EDTA, 0.1 ml of dihydrolipoamide, water, and enzyme to make a total volume of 2.0 ml. The reaction is started at 25° by the addition of enzyme. The blank cell contains 0.1 ml of phosphate buffer, 0.1 ml of EDTA, 0.1 ml of dihydrolipoamide, and 1.7 ml of water. The increase in absorbance at 340 mμ is followed with a recording spectrophotometer.

Definition of Unit and Specific Activity. One unit is defined as the amount of enzyme that catalyzes an initial rate of reduction of 1 micromole of NAD per minute. Specific activity is expressed as units of enzyme per milligram of protein.

Diaphorase Assay[5–7]

Principle. The assay is based on spectrophotometric determination of the rate of reduction of 2,6-dichlorophenolindophenol (at 600 mμ) in the presence of NADH.

[7] N. Savage, *Biochem. J.* **67**, 146 (1957).

Reagents

> Potassium phosphate buffer, 1 M, pH 7.2
> NADH, 2×10^{-3} M, in 0.002 M potassium phosphate buffer, pH 7.2
> 2,6-Dichlorophenolindophenol, 6.7×10^{-4} M
> EDTA, 2.5×10^{-2} M

Procedure. To a cuvette with a 1-cm light path are added 0.1 ml of phosphate buffer, 0.1 ml of NADH, 0.1 ml of 2,6-dichlorophenolindophenol, 0.1 ml of EDTA, water, and enzyme to make a total volume of 2.0 ml. The reaction is initiated at 25° by the addition of enzyme. The blank cell contains phosphate buffer, NADH, EDTA, and 1.7 ml of water. The decrease in absorbance at 600 mμ is followed with a recording spectrophotometer. A molecular extinction coefficient of 21×10^3 M^{-1} cm^{-1} is used to calculate the amount of reduced dye.

Definition of Unit and Specific Activity. One unit is defined as the amount of enzyme that catalyzes an initial rate of reduction of 1 micromole of 2,6-dichlorophenolindophenol per minute. Specific activity is expressed as units of enzyme per milligram of protein.

Transhydrogenase Assay[5,8]

Principle. The assay is based on spectrophotometric determination (at 395 mμ) of the rate of reduction of thionicotinamide adenine dinucleotide (TNAD) by NADH.

Reagents

> Potassium phosphate buffer, 1 M, pH 7.5
> NADH, 2×10^{-3} M, in 0.002 M phosphate buffer, pH 7.6
> TNAD, 4×10^{-3} M
> EDTA, 2.5×10^{-2} M

Procedure. To a cuvette with a 1-cm light path are added 0.1 ml of buffer, 0.1 ml of NADH, 0.1 ml of TNAD, 0.1 ml of EDTA, water, and enzyme to make a total volume of 2.0 ml. The reaction is initiated at 25° by the addition of enzyme. The blank cell contains all components except enzyme. The increase in absorbance at 395 mμ is followed with a recording spectrophotometer. A molecular extinction coefficient of 11.3×10^3 M^{-1} cm^{-1} for TNADH is used.

Definition of Unit and Specific Activity. A unit of enzyme activity is defined as the amount of enzyme which catalyzes an initial rate of reduction

---

[8] M. M. Weber and N. O. Kaplan, *J. Biol. Chem.* **225,** 909 (1957).

of 1 micromole of TNADH per minute. Specific activity is expressed as units of enzyme per milligram of protein.

Purification Procedure

From the Pig Heart Pyruvate Dehydrogenase Complex[1,2]

Reagents

Urea, 4 M, and 1% ammonium sulfate in 0.1 M potassium phosphate buffer, pH 7.5. Ultrapure-grade urea is recrystallized from water or 95% ethanol.

Ammonium sulfate, 6%, in 0.1 M potassium phosphate buffer, pH 7.5

Ammonium sulfate, 1%, in 0.1 M potassium phosphate buffer, pH 7.5

Potassium phosphate buffer, 0.05 M, pH 7.0

Calcium phosphate gel suspended on Whatman cellulose powder (CF-II) is prepared as described by Price and Greenfield,[9] and 15–20 g of the same powder per batch is added to obtain the appropriate flow rate of the column.

Pig heart pyruvate dehydrogenase complex[1]

Procedure. A slurry of gel-cellulose is added in portions to the chromatographic tube (3 × 27 cm). The gel-cellulose is allowed to pack by gravity to give a column approximately 4.5 cm in height, comprising about 26 ml of gel-cellulose. The column is cooled to about 5°. A flow rate of about 30 ml per hour is maintained throughout the resolution procedure.

A solution of the pyruvate dehydrogenase complex containing 30 mg of protein in 1.4 ml of 0.05 M phosphate buffer (pH 7.0) is applied to the column, which has been previously washed with a solution of 1% ammonium sulfate in 0.1 M phosphate buffer. After adsorption of the protein, the column is washed with 18 ml of a solution of 4 M urea and 1% ammonium sulfate in 0.1 M phosphate buffer. The column is then washed with about 45 ml of a solution of 1% ammonium sulfate in 0.1 M phosphate buffer. A colorless protein is eluted, leaving a broad yellow, fluorescent band on the column. The protein contents of the fractions are estimated by ultraviolet absorption.[10] Immediately after elution, the colorless fraction, comprising 26.8 mg of protein in a volume of 21 ml, is brought to 0.55 ammonium sulfate saturation by adding 0.375 g of the salt per milliliter. The mixture is stirred for an additional 15 minutes. The precipitate is collected by centrifugation for 20 minutes at 16,000 g and dissolved in a minimum

---

[9] V. E. Price and R. E. Greenfield, *J. Biol. Chem.* **209**, 363 (1954); see also L. J. Reed and C. R. Willms, Vol. IX [50].

[10] E. Layne, Vol. III [73].

volume of 0.05 M phosphate buffer. The solution is dialyzed against the same buffer. The yellow band is eluted from the column with a solution of 6% ammonium sulfate in 0.1 M phosphate buffer. The eluate, comprising 1.7 mg of protein in a volume of 7 ml, is fractionated with solid ammonium sulfate. The precipitate obtained between 0.50 and 0.90 saturation is collected by centrifugation for 30 minutes at 16,000 g and dissolved in a minimum volume of 0.05 M phosphate buffer. The solution is dialyzed overnight against the same buffer. The recovery of lipoamide dehydrogenase in this fraction is 1.5 mg (58% yield); specific activity is 71 units/mg.

From the Pig Heart 2-Oxoglutarate Dehydrogenase Complex[3,11]

Reagents

Urea, 8 M, and 2% ammonium sulfate in 0.1 M potassium phosphate buffer, pH 7.5

Urea, 4 M, and 1% ammonium sulfate in 0.1 M potassium phosphate buffer, pH 7.5

Ammonium sulfate, 4%, in 0.1 M potassium phosphate buffer, pH 7.5

Ammonium sulfate, 1%, in 0.1 M potassium phosphate buffer, pH 7.5

Potassium phosphate buffer, 0.1 M, pH 7.5

Potassium phosphate buffer, 0.05 M, pH 7.0

Calcium phosphate gel suspended on Whatman cellulose powder (CF-II). This gel is useful for 4 weeks after preparation.

Pig heart 2-oxoglutarate dehydrogenase complex[4]

Procedure. The gel-cellulose column (4 × 8 cm) is prepared as described above. The column is cooled to about 5°. A flow rate of about 60–70 ml per hour is maintained.

A solution of the 2-oxoglutarate dehydrogenase complex containing 200 mg of protein in a volume of 12 ml of 0.05 M phosphate buffer is mixed with an equal volume of 8 M urea and 2% ammonium sulfate in 0.1 M phosphate buffer. The mixture is applied to the column, which has been previously washed successively with 200 ml of a solution of 0.1 M phosphate buffer, pH 7.5, and 15 ml of a solution of 1% ammonium sulfate in 0.1 M phosphate buffer, pH 7.5. After adsorption of the protein, the column is washed successively with 26 ml of a solution of 4 M urea and 1% ammonium sulfate in 0.1 M phosphate buffer, and 250 ml of a solution of 1% ammonium sulfate in 0.1 M phosphate buffer. A colorless protein is eluted, leaving a broad yellow, fluorescent band on the column. The protein content is

[11] Y. Fukuyoshi, Y. Sakurai, and M. Koike, unpublished data.

estimated by ultraviolet absorption.[10] In the colorless fractions a total of 147 mg of protein in a volume of 100 ml of effluent is recovered. The yellow band is eluted from the column with 60 ml of a solution of 4% ammonium sulfate in 0.1 M phosphate buffer. The fractions exhibiting fluorescence are combined and fractionated with solid ammonium sulfate. The precipitate obtained between 0.50 and 0.90 saturation is collected by centrifugation and dissolved in a minimum volume of 0.05 M phosphate buffer. The solution is dialyzed overnight against the same buffer. The recovery of lipoamide dehydrogenase in this fraction is 27 mg (79% yield); specific activity is 70 units/mg.

Properties

Physical and Chemical Constants. Lipoamide dehydrogenases from pig heart pyruvate and 2-oxoglutarate dehydrogenase complexes are flavoproteins, containing FAD as prosthetic group (2 moles of FAD per mole of the enzyme). These enzymes show a single boundary in the ultracentrifuge, with sedimentation coefficients ($s_{20,w}$) of 5.3 to 5.4 S. The diffusion constants ($D_{20,w}$) of the two lipoamide dehydrogenases are 4.71×10^{-7} cm^2 sec^{-1} (at 5.6 mg/ml) and 4.80×10^{-7} cm^2 sec^{-1} (at 5.9 mg/ml), respectively. From these data average molecular weights are calculated to be 115,000 and 114,000, respectively. Recent studies indicate[12,13] that these two enzymes migrate toward the anode as a single band. The enzyme from the pyruvate dehydrogenase complex migrates farther than that from the 2-oxoglutarate dehydrogenase complex. Absorption spectra of these two enzymes are essentially identical, with maxima at 273, 356, and 456 mμ, a shoulder at 438 and 480 mμ, and a minimum at 396 mμ. The optical rotatory dispersion constants of the two enzymes are virtually identical with $[m']_{233} = -4,241$ and $-4,140$, respectively. The circular dichroism spectra of the two enzymes show positive bands between 300 mμ and 400 mμ. The molecular ellipticity of these two enzymes are $[\theta]_{374} = +51,700$ and $[\theta]_{369} = +28,600$, respectively.

Enzymatic Activities. The catalytic activities of the two enzymes are summarized in Table I. The two enzymes differ only with respect to their diaphorase activity.

Specificity for Lipoic Acid and Its Derivatives. The natural substrate of the two mammalian lipoamide dehydrogenases is the dihydrolipoyl moiety which is covalently bound to lipoate acyltransferase, presumably to the

---

[12] T. Hayakawa, Y. Sakurai, T. Aikawa, Y. Fukuyoshi, and M. Koike, *in* "Proceedings of the Conference on Flavins and Flavin Enzymes" (K. Yagi, ed.), p. 99. Tokyo Univ. Press, Tokyo, Japan, 1968.

[13] Y. Sakurai, T. Hayakawa, Y. Fukuyoshi, and M. Koike, *J. Biochem. (Tokyo)* **65,** 313 (1969).

TABLE I

CATALYTIC ACTIVITIES OF TWO PIG HEART LIPOAMIDE DEHYDROGENASES

Compound	Specific activities[a]				Activity ratios (×10²)		
	NADH-lipoamide	Dihydrolipo-amide-NAD	Transhydro-genase	Diaphorase	Dihydro-lipoamide-NAD:NADH-lipoamide	Transhydro-genase:NADH-lipoamide	Diaphorase:NADH-lipoamide
Lipoamide dehydrogenase from the pyruvate dehydrogenase complex	71.8	73.7	59.3	1.24	103	82.5	1.7
Lipoamide dehydrogenase from the 2-oxoglutarate dehydrogenase complex	71.8	73.0	61.4	0.68	101	85.5	0.97

[a] Units per minute per milligram of protein.

ε-amino group of a lysine residue. However, the enzyme also catalyzes reaction [Eq. (1)] with lipoic acid and lipoic acid derivatives. Both optical isomers of lipoic acid undergo reaction. A comparison of substrate specificities, turnover numbers, pH optima, and K_m values of oxidized and reduced forms of lipoic acid and its derivatives[14],[15] is presented in Table II. It is apparent that the reactivity of the enzyme is greatly influenced by the nature of the lipoyl derivative. With the derivatives possessing a charged group (lipoic acid, lipoylglycine, lipoyl-β-alanine, lipoylglycylglycine, and ε-N-lipoyllysine), the turnover number increases the farther the charged carboxyl group is separated from the dithiolane ring, and is even larger when the charged carboxyl group is effectively neutralized in a zwitterion structure as in ε-N-lipoyllysine. In keeping with this trend, three structures that are nonionized in the oxidized form show the highest activity of all (lipoanilide, carboethoxylipoanilide, lipoamide). This variation in activity suggests that an anionic charge on the substrate hinders approach to the enzyme. The reactions of lipoamide dehydrogenases from beef liver mitochondria[16] and human liver particles[5] with lipoic acid and its derivatives are quite similar to the two pig heart enzymes.

Effect of NAD on the Rate of Oxidation of NADH. A pronounced lag period is observed in the oxidation of NADH by lipoamide. This lag period is abolished by the addition of NAD (in equal amount to NADH). As shown in Table II, pH optima of two enzymes for various catalytic reactions are the same. The pH optimum of the reaction in the presence of added NAD is approximately 6.5 with lipoamide. The pH optima for the reaction with lipoic acid and lipoyllysine in the presence of NAD are approximately 5.7 and 6.1, respectively.

Specificity for NADP and NAD Analogs. NADP and α-NAD are inactive in the NADH-lipoamide oxidoreductase assay with two enzymes. TNAD shows a similar reaction rate to NAD with dihydrolipoamide as substrate. 3-Acetylpyridine adenine dinucleotide and pyridine 3-aldehyde dinucleotide react, respectively, only at rates one-tenth and one-thirtieth of the rates of reaction with NAD. Pyridine 3-aldehyde adenine dinucleotide at 0.1 mM shows complete inhibition of the oxidation of NADH by lipoamide and about 10% inhibition of the oxidation of dihydrolipoamide by NAD. 3-Acetylpyridine hypoxanthine dinucleotide, pyridine 3-aldehyde hypoxanthine dinucleotide, and nicotinamide hypoxanthine dinucleotide,

[14] V. Massey, *Biochim. Biophys. Acta* **37,** 314 (1960); see also V. Massey, *in* "The Enzymes" (P. D. Boyer, H. Lardy, and K. Myrbäck, eds.), Vol. 7, p. 275. Academic Press, New York, 1963.

[15] Y. Sakurai, T. Hayakawa, Y. Fukuyoshi, and M. Koike, unpublished data.

[16] C. J. Lusty, *J. Biol. Chem.* **238,** 3443 (1963).

TABLE II

REACTION OF PIG HEART LIPOAMIDE DEHYDROGENASES WITH LIPOIC ACID AND ITS DERIVATIVES[a]

Compound	pH optimum		K_m (mM)			Presumed maximum turnover number[c]		
	OGDC-Fp and PDC-Fp[b]	Massey's preparation[e]	OGDC-Fp	PDC-Fp	Massey's preparation[e]	OGDC-Fp	PDC-Fp	Massey's preparation[e]
(±)-Lipoic acid	5.7	5.9	5.25	5.00	2.00	8,700	8,300	1,000
(±)-Lipoamide	6.5	6.5	1.33	1.25	5.00	37,400	35,800	80,000
(±)-ε-N-Lipoyllysine	6.1	6.2	1.18	1.21	1.20	52,100	53,600	50,000
(±)-Lipoylglycine	—	6.3	—	—	1.00	—	—	1,700
(±)-Lipoyl-β-alanine	—	d	—	—	1.70	—	—	25,000
(±)-Lipoylglycylglycine	—	6.1	—	—	1.60	—	—	28,000
(±)-Carboethoxylipoamilide	—	d	—	—	About 0.70	—	—	About 30,000
(±)-Dihydrolipoic acid	6.4	—	0.80	0.83	—	420	500	—
(±)-Dihydrolipoamide	7.9	7.9	0.16	0.16	0.14	14,800	14,100	9,400
(±)-ε-N-Dihydrolipoyllysine	8.0	—	1.82	2.00	—	19,700	20,700	—
(±)-Dihydrolipoylglycine	—	7.0	—	—	0.70	—	—	2,500

[a] All activities were determined in 0.06 M or 0.05 M phosphate buffer at 25°.
[b] OGDC-Fp and PDC-Fp, lipoamide dehydrogenases from the 2-oxoglutarate and pyruvate dehydrogenase complexes, respectively.
[c] With 1×10^{-4} M NAD or NADH, extrapolated V_{max} obtained with varying concentrations of lipoyl derivatives.
[d] pH optimum not determined; values reported at pH 6.3.
[e] See V. Massey.[14]

at 0.1 mM, show almost no effect on the oxidation of both NADH and dihydrolipoamide. Similar results are obtained with human liver enzyme.[5]

Effect of Dithiol Inhibitors on Catalytic Activities. The NADH-lipoamide oxidoreductase, dihydrolipoamide-NAD oxidoreductase, and transhydrogenase activities of these two enzymes are strongly inhibited by DTNB (4×10^{-4} M), arsenite (5×10^{-5} M), cadmium ion (2.5×10^{-4} M), cupric ion (2.5×10^{-4} M), and p-chloromercuric benzoate (5×10^{-6} M). In contrast, the diaphorase activity of lipoamide dehydrogenase from the 2-oxoglutarate dehydrogenase complex is increased about 2-fold by preincubation of the enzyme with NADH and arsenite, and the diaphorase activity of the enzyme from the pyruvate dehydrogenase complex is increased 7- to 9-fold. The inhibitory effect of arsenite is prevented by L-cysteine or BAL.

Immunochemical Properties. The two enzyme preparations show complete identity in Ouchterlony double-diffusion experiments.[13,17] No significant antigenic differences are observed between these two enzymes with the microcompliment fixation technique.[13,18]

[17] T. Hayakawa, T. Aikawa, K.-I. Otsuka, and M. Koike, *J. Biochem.* **62,** 396 (1967).
[18] M. Reichlin, M. Hay, and L. Levine, *Immunochemistry* **1,** 21 (1964); see also L. Levine and H. V. Vunakis, Vol. XI [92].

Section IV

Pantothenic Acid, Coenzyme A, and Derivatives

[52] Gas Chromatography of Pantothenic Acid

By A. J. Sheppard and A. R. Prosser

Principle

Pantothenic acid is widely distributed throughout nature. Pantothenyl alcohol is a synthetic analog that has pantothenic activity in man and rat. The alcohol analog is used in the preparation of injectable and other liquid pharmaceutical preparations. The calcium or sodium salt of pantothenic acid is a constituent of numerous pharmaceutical multivitamin and B-vitamin preparations.

Two reports[1,2] on the gas chromatographic behavior of the acetate derivatives of pantothenic acid and pantothenyl alcohol have been made, both from this laboratory. No studies are available that indicate any application of gas chromatography (GLC) for assaying the vitamin in biological materials or pharmaceutical products. The techniques in this chapter are limited to the GLC operating parameters and the preparation of acetates of pantothenic acid ethyl ester and pantothenyl alcohol. Application of the GLC method for assaying the vitamin has not yet been established. The techniques are described in the hope that they may be of help to other investigators. All solvents used are ACS reagent grade unless otherwise specified.

Preparation of Packed Columns for Gas Chromatography

The most satisfactory column used by the authors to measure the acetates of pantothenic acid and pantothenyl alcohol is a Pyrex column, 244 × 0.4 cm i.d., packed with 2% neopentyl glycol sebacate (NPGSeb) immobile phase, coated on 110/120-mesh Anakrom ABS inert support as follows: 0.500 g NPGSeb is dissolved in 50 ml of chloroform, with heat if necessary. The NPGSeb:chloroform solution is transferred to a Morton flask containing 24.5 g of the silane-treated inert support, and enough additional chloroform to form a slurry. The mixture is allowed to stand for 1 hour with occasional shaking. The solvent is evaporated *in vacuo* with warming on a flash evaporator, and the prepared packing is dried at 80° for 1 hour. The finished product is stored in a dust-free container.

The chromatographic column is filled with the prepared packing in the manner described (this volume [235]) for gas chromatography of vitamin K₃.

[1] A. R. Prosser and A. J. Sheppard, *Federation Proc.* **27**, 256 (1968) (abstr. 231).
[2] A. R. Prosser and A. J. Sheppard, *J. Pharm. Sci.* **58**, 718 (1969).

Instrumentation

The GLC instrumentation requirements are identical to those given (this volume [101]) for gas chromatography of niacin.

The operating parameters for the acetates of pantothenic acid and pantothenyl alcohol are as follows: column, 230°; detector, 280°; injection area, 280°; carrier gas flow, 60 ml per minute. A cell voltage of 900 V, direct current, is used with the β-argon ionization detector.

The column is preconditioned for 24 hours at 235° with a carrier gas flow rate of 60 ml per minute. Then the column outlet is connected to the detector, and the instrument is set to operate at the previously mentioned GLC operating parameters. The acetates of pantothenic acid and pantothenyl alcohol appear at 8 and 11 minutes, respectively.

Preparation of Super-Dry Ethanol

Super-dry ethanol is prepared from commercial absolute ethanol by treatment with magnesium turnings activated by iodine, as follows: 5 g of clean, dry magnesium and 0.5 g of resublimed iodine are placed in a 2-liter round-bottom Pyrex flask in a heating mantle. An Allihn condenser is attached, and 50–75 ml of absolute ethanol is added through the condenser, with warming if necessary, until the iodine disappears. Hydrogen will be vigorously evolved, and it may be necessary to reduce or remove the heat source from the flask. It usually is not necessary to apply heat, as the reaction is almost violently exothermic. If hydrogen is not evolved, another 0.5 g of iodine is added and the mixture is heated until all the magnesium is converted to ethoxide. Then 900 ml of absolute ethanol is added and the mixture is boiled under reflux for 30 minutes. The product is then distilled with the exclusion of moisture in a closed system. The first 25 ml of the distillate is discarded. The super-dry ethanol is stored in a stoppered flask until used.

Preparation of 2.5% HCl in Super-Dry Ethanol

A 2-liter filtering flask with a filled drying tube attached to the hose connection is weighed, and 1000–1500 ml of super-dry ethanol is added. The filtering flask is stoppered and the total weight is obtained. The weight of the empty flask is subtracted from the total weight to obtain the ethanol weight. A bubbling tube is inserted into the flask stopper fitting, and two drying towers connected in series are attached to the tube. Enough concentrated H_2SO_4 is added to each drying tower to cover the bubbler. The last drying tower is attached to an anhydrous HCl-gas cylinder, and the HCl is permitted to bubble slowly through the drying towers into the super-dry ethanol until 2.5% HCl by weight is added to the ethanol. The bubbler is then removed from the filtering flask and a standard taper stopper is inserted before storage.

Esterification

The following typical preparation of the ethyl ester from calcium pantothenate or its equivalent of sodium pantothenate is provided as a guideline: 5.44 mg of calcium pantothenate is weighed into a 50-ml round-bottom flask, 5 ml of 2.5% ethanol-HCl reagent is added, and a magnetic stirring bar is inserted. The mixture is stirred on a magnetic mixer for 1.5 hours at room temperature. The ethanol-HCl solution is removed *in vacuo* on a flash evaporator.

Acetylation

Pantothenic acid ethyl ester and pantothenol are converted quantitatively to their acetate derivatives by reaction with a 1:1 mixture of acetic anhydride and pyridine. The reaction mixture can be injected directly onto the GLC column after the reaction is completed. The acetate derivatives are prepared as follows: Using the flask (from the esterification step above) containing the ethyl ester or a round-bottom flask containing 5.0 mg of pantothenol, 3 ml of a 1:1 mixture of acetic anhydride and pyridine is added and the mixture is stirred on the magnetic mixer for 1 hour at room temperature. The acetylating mixture is then removed *in vacuo* on a flash evaporator at a bath temperature of 40–50°. The acetate product is dissolved in spectro-grade chloroform. The final volume of the chloroform solution is dependent on the sensitivity of the instrument used. A final concentration of 0.5 mg/ml has been found to be satisfactory.

Standards and Calibration

The pantothenic acid ethyl ester diacetate and pantothenol triacetate are prepared by proportionately increasing the quantities in the above acetylation and esterification steps. More than 50 mg of the derivative needed should be prepared for calibration.

Two primary standard solutions are prepared:

1. Pantothenic acid ethyl ester diacetate: 50 mg is weighed into a 25-ml volumetric flask and diluted to the mark with spectro-grade chloroform.

2. Pantothenol triacetate: 50 mg is weighed into a 25-ml volumetric flask and diluted to the mark with spectro-grade chloroform.

Working standards are prepared by diluting aliquots of each primary standard to 0.5, 1.0, and 2.0 mg/ml.

A calibration plot covering 1–4 µg appears to be suitable for most GLC units equipped with hydrogen flame ionization detectors. Since instruments vary considerably in response, the calibration range must be adjusted to cover individual situations. No data are available regarding the sensitivity with these derivatives for the β-argon ionization detector. A 2-μl injection of each working standard will provide three calibration points of 1, 2, and 4 μg each. The response in square centimeters for each amount is plotted on standard graph paper with the peak area in square centimeters as the ordinate and the quantity injected as the abscissa. A straight line is fitted to the calibration data of individual compounds. Figure 2 in article [82] on gas chromatography of vitamin B_6 provides an example of how to treat the calibration data.

Injection Technique

This technique is described (this volume [235]) for gas chromatography of vitamin K_3.

Application Possibilities

This technique should have widespread application in the pharmaceutical field. There are fewer problems in sample cleanup with pharmaceutical products than with biological materials. Any sample preparation must provide for the removal of those compounds that will be competitive with pantothenic acid or pantothenyl alcohol for the acetylating reagents.

[53] Pantothenic Acid and Coenzyme A: Determination of CoA by Phosphotransacetylase from *Escherichia coli* B[1]

By Yasushi Abiko

Assay Method

This method is a modification of the method originally reported by Stadtman and Kornberg.[2]

Principle. The method is based on the fact that the arsenolytic decomposition of acetyl phosphate is CoA dependent in the presence of phosphotransacetylase (acetyl-CoA:orthophosphate acetyltransferase, EC 2.3.1.8).

$$\text{Acetyl phosphate} + \text{CoA} \underset{\longleftarrow}{\overset{\text{phosphotransacetylase}}{\rightleftharpoons}} \text{acetyl-CoA} + \text{phosphate}$$

$$\text{Acetyl-CoA} + \text{arsenate} \underset{\longleftarrow}{\overset{\text{phosphotransacetylase}}{\rightleftharpoons}} \text{acetylarsenate} + \text{CoA}$$

$$\text{Acetylarsenate} + \text{H}_2\text{O} \longrightarrow \text{acetate} + \text{arsenate}$$

[1] Y. Abiko, T. Suzuki, and M. Shimizu, *J. Biochem. (Tokyo)* **61,** 10 (1967).
[2] E. R. Stadtman and A. Kornberg, *J. Biol. Chem.* **203,** 47 (1953); see also Vol. I [98] and Vol. III [132].

The residual amount of acetyl phosphate is determined by the hydroxamic acid method.[3]

Reagents

Dilithium acetyl phosphate, 0.06 M
Potassium arsenate, 0.25 M, adjusted to pH 8.0 with HCl
L-Cysteine, 0.1 M, freshly prepared
Tris-HCl buffer, 0.2 M, pH 8.0
Hydroxylamine hydrochloride, 28%
NaOH, 14%
HCl, diluted to 1:3 (v/v) with water
Acetate buffer, 0.1 M, pH 5.4
Trichloroacetic acid, 12%
Ferric chloride, 5% in 0.1 N HCl
Hydroxylamine–acetate buffer, freshly prepared before use by mixing hydroxylamine hydrochloride with NaOH and acetate buffer (1:1:2, v/v)
Phosphotransacetylase

Preparation of Phosphotransacetylase

Culture Conditions. *Escherichia coli* B (ATCC 11303) is grown with shaking or aeration in a medium containing 1% glucose, 0.4% powdered yeast extract, 0.4% peptone, and 0.8% K_2HPO_4 at 37° for 6 hours. This incubation time (the early stationary phase) is preferable for the preparation of phosphotransacetylase. The specific activity of the enzyme in the extract from *E. coli* B increases in parallel with the bacterial growth during 7 hours' culture, but the enzyme activity falls considerably in a preparation incubated overnight. The cells are harvested by centrifugation and washed twice with water. The washed-cell suspension in water is lyophilized, and the dried cells are stored in a cold room as a source of phosphotransacetylase.

Extraction of the Enzyme. One gram of the dried cells is resuspended in 20 ml of 0.02 M $KHCO_3$ and ultrasonicated at 20 kc for 20 minutes under chilling with ice-water. The sonicated cell suspension is centrifuged at 21,000 g for 30 minutes. The resultant supernatant solution is brought to 20% saturation with ammonium sulfate by adding a saturated ammonium sulfate solution at 4°, and then to 30% saturation by adding a saturated acid ammonium sulfate solution (prepared by mixing 3 ml of sulfuric acid with 1 liter of the saturated ammonium sulfate solution). The precipitate formed is removed by centrifugation, and the resultant supernatant solution is brought to 35% saturation by the further addition of the saturated acid ammonium sulfate solution. The precipitate is col-

---

[3] F. Lipmann and L. C. Tuttle, *J. Biol. Chem.* **159**, 21 (1945).

lected by centrifugation and dissolved in 4 ml of Tris buffer. The enzyme solution thus prepared can be stored in a frozen state or in a lyophilized form for 6 months or longer. The lyophilized preparation is stable at room temperature (about 25°) for at least 3 months in a desiccator.

Assay Procedure

The stock enzyme is thawed at 25° (or the lyophilized enzyme is dissolved in water to the original volume) and diluted 10–20 times with Tris buffer before use. The activity of the diluted enzyme should be approximately 30 (20–40) units[4] per milliliter.

To a 15-ml tube are added 0.1 ml of Tris buffer, 0.1 ml of acetylphosphate, 0.1 ml of a test sample, 0.1 ml of cysteine, 0.1 ml of the diluted enzyme solution, and sufficient water to give a volume of 0.8 ml. After the mixture has stood for 5 minutes at 25°, 0.2 ml of potassium arsenate is added, and the mixture is incubated at 25° for exactly 15 minutes. Then 2.0 ml of the hydroxylamine–acetate buffer is added, and the mixture is allowed to stand at room temperature for 10 minutes. Finally, 1.0 ml each of HCl, trichloroacetic acid, and $FeCl_3$ are successively added to the mixture. The precipitate is removed by centrifugation. The remaining amount of acetyl phosphate is determined by colorimetry of the supernatant solution at 510 mμ within 30 minutes after the addition of $FeCl_3$.

The CoA content in the sample is calculated by a calibration curve obtained with known amounts of standard CoA.

Calibration Curve of CoA

Plots of hydrolyzed amounts of acetyl phosphate versus CoA give a straight line with a fixed amount of the enzyme (Fig. 1).

Specificity of the Method

The present method is highly specific for CoA. The determination of CoA by this method is not affected by various known precursors of CoA.

Effects of Inorganic Salts on the Phosphotransacetylase Reaction

K_2SO_4, $(NH_4)_2SO_4$, $MgCl_2$, and $CaCl_2$ cause 30–40% inhibition of the enzyme activity at $10^{-2} M$ and about 50% inhibition at $2 \times 10^{-2} M$, although K_2SO_4 and $(NH_4)_2SO_4$ have a marked stabilizing effect on the enzyme. NaCl and LiCl cause about 15% inhibition at $10^{-2} M$ and about 25% inhibition at $2 \times 10^{-2} M$. The rate of inhibition by KCl and NH_4Cl

---

[4] One unit of phosphotransacetylase is the amount of the enzyme required to catalyze the decomposition of 1 micromole of acetyl phosphate in the presence of 5 Lipmann units of CoA at pH 8.0 and 25° for 15 minutes. E. R. Stadtman, G. D. Novelli, and F. Lipmann, *J. Biol. Chem.* **191,** 365 (1951); see also Vol. I [98].

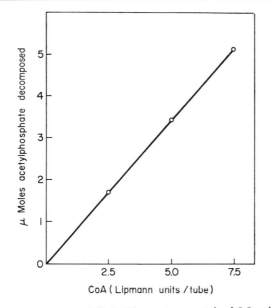

FIG. 1. Calibration curve of CoA. The system contained 3.3 units of phosphotransacetylase.

is below 10% at $2 \times 10^{-2} M$. Phosphate inhibits the enzyme activity by about 13% at $2 \times 10^{-3} M$.

Other Methods

The phosphotransacetylase method using the enzyme from *Clostridium kluyveri* was described by Stadtman and Kornberg.[2] Bergmeyer *et al.*[5] reported an assay method for phosphotransacetylase which was based on the measurement of the increase in the optical density at 233 mμ caused by the formation of acetyl-CoA. This method seems to be applicable also to the determination of CoA by phosphotransacetylase. Several other suitable methods for the determination of CoA have been reported. These include the oxoglutarate dehydrogenase (EC 1.2.4.2) method of Korff,[6] the citrate cleavage enzyme (EC 4.1.3.8) method of Srere,[7] and the sorbyl-CoA method of Wakil and Hubsher.[8] These methods were described to be specific for CoA, and the last one is the only method measuring directly an absolute amount of CoA. The sulfanilamide acetylation method of Kaplan and

[5] H. U. Bergmeyer, G. Holz, H. Klotzsch, and G. Lang, *Biochem. Z.* **338,** 114 (1963).
[6] R. W. Korff, *J. Biol. Chem.* **200,** 401 (1953); see also Vol. I [120] and Vol. III [132].
[7] P. A. Srere, *J. Biol. Chem.* **236,** 50 (1961); see also Vol. V [88].
[8] S. J. Wakil and G. Hubsher, *J. Biol. Chem.* **235,** 1554 (1960).

Lipmann[9] is a classic method for determination of CoA, but it has a very low specificity for CoA[1] although it is useful for standardization.

[9] N. O. Kaplan and F. Lipmann, *J. Biol. Chem.* **174,** 37 (1948); see also Vol. III [132].

[54] Preparation and Assay of CoA-SS-Glutathione

By RAÚL N. ONDARZA

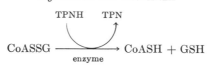

$$CoASSG \xrightarrow[\text{enzyme}]{} CoASH + GSH$$

Assay Method[1,2]

Principle. The estimation of CoASSG[3] can be carried out with the TPNH-dependent CoASSG-reductase which is specific for this substrate. The change of TPNH to the oxidized form can be measured spectrophotometrically and becomes directly dependent on the amount of substrate degraded. This enzymatic activity is present in commercial preparations of yeast GSSG-reductase[4] (EC 1.6.4.2).

Reagents

Phosphate buffer, 0.05 M, pH 5.5
EDTA, 1×10^{-3} M
TPNH, 1.2×10^{-4} M
Yeast GSSG-reductase, 5 mg/ml, used as a source of CoASSG-reductase (from C. F. Boehringer and Soehne, 80 U per milligram of protein. It contains a CoASSG-reducing activity equivalent to 2.0 U per milligram of protein).
CoASSG (see preparation below)

Procedure. In a quartz cuvette with 1-cm light path, containing 0.05 phosphate buffer, pH 5.5, are added to a final volume of 1 ml 10^{-3} M

[1] R. N. Ondarza and J. Martinez, *Biochim. Biophys. Acta* **113,** 409 (1966).
[2] R. N. Ondarza, *Natl. Cancer Inst. Monogr.* **27,** 81 (1967).
[3] CoASSG is a mixed disulfide compound formed by coenzyme A and glutathione moieties linked through an —SS— bridge (molecular weight, 1072.7).
[4] We have also obtained CoASSG-reductase from rat liver and yeast cells by following the procedure as for GSSG-reductase according to Racker[5] but omitting the heating step at 55° [R. N. Ondarza, R. Abney, and A. M. López-Colomé, *Biochim. Biophys. Acta* **191,** 239 (1969)].

EDTA, 10 μl of TPNH (approximately 120 millimicromoles), and 5 μl of yeast GSSG-reductase. The decrease in the optical density at 340 mμ is measured (control reading) for five minutes against water as a blank. A sample containing 50–100 millimicromoles of CoASSG is added in a small volume (10–20 μl), and the optical change is followed until the decrease in the optical density attains the slope of the control (final reading). The reaction should go to completion (approximately 94%) within 20 minutes. The difference between the control and final reading divided by the molecular extinction coefficient of TPNH (6200 at 340 mμ) is equal to the amount of substrate in millimoles per milliliter of incubation mixture.

Preparation of CoASSG

Chemical Synthesis.[1] The CoASSG can be prepared from CoASH and excess of GSSG in the presence of O_2, according to the following reaction:

$$CoASH + GSSG \rightleftharpoons CoASSG + GSH$$

$$GSH \xrightarrow{O_2} GSSG$$

Similar results are obtained with an alternate procedure designed by Eriksson[6]:

$$CoASH + GSSO_2G \rightarrow CoASSG + GSO_2H$$

Procedure. About 65 micromoles of CoASH[7] are mixed with 260 micromoles of GSSG (previously adjusted to pH 7.0 with 1 N KOH) in 1 ml of 0.05 M phosphate buffer pH 7.5 for 3 hours at 35° with constant bubbling of O_2.

The above mixture is made up to a volume of 50 ml with water, adjusted in the cold to pH 7.6–8.0 with a few drops of 10% NH_4OH, and resolved by chromatography on a Dowex 1 chloride column, 200–400 mesh (1.2 × 20 cm) according to Cohn.[9] The charged column is washed with approximately 1 liter of water followed by 1.5 liters of 0.01 N HCl–0.01 M NaCl solution. The CoASSG can now be eluted from the column with 0.01 N HCl–0.05 M NaCl. The eluate (about 1.5 liters) is adsorbed on activated charcoal[10] for 10 minutes, centrifuged, and recovered by desorp-

---

[5] E. Racker, *J. Biol. Chem.* **217**, 855 (1955).

[6] B. Eriksson, *Acta Chem. Scand.* **20**, 1178 (1966).

[7] The CoASH from C. F. Boehringer and Soehne, when assayed in our laboratory with β-hydroxyacyl-CoA-dehydrogenase[8] was 57–84.5% pure.

[8] G. Michal and H. U. Bergmeyer, *in* "Methods in Enzymatic Analysis" (H. U. Bergmeyer, ed.), p. 512. Academic Press, New York, 1963.

[9] W. E. Cohn, Vol. III [107], p. 724.

[10] Approximately 30 mg of activated charcoal/micromole of nucleotide with a molecular extinction coefficient of 15,000.

tion with 100 ml of ethanol–water (1:1, v/v) at pH 7.5 for 3 hours in the cold. Sometimes a small amount of charcoal remains as contaminant in the ethanol–water extract, which must be eliminated by filtration under vacuum, using a filter paper disk with a layer of talc. Be sure the pH is 7.5 when doing this.

The ethanol–water extract is concentrated in a flash evaporator at 35° to a small volume and applied to a piece of Whatman No. 3 MM filter paper for separation by high-voltage electrophoresis for 1.30 hours (3.3 mA/cm with a buffer of formic acid–acetic acid–water, 1:4:45, v/v, pH 2.0). The amount of material applied to the starting-line position must be of the order of 1 micromole of nucleotide per lineal centimeter. When a high-voltage apparatus is not available, the extract can be separated by low-voltage paper electrophoresis in a 0.1 M citrate buffer pH 3.2, with a current of 0.5 mA/cm (375 V) for 5 hours at 4–6°. Further purification of the component comprises the elution from the ionopherogram of the material showing both ultraviolet-absorbing and ninhydrin-positive properties and separation by paper chromatography in isopropanol–water (75:30, v/v) for 48 hours.

The yield of CoASSG averages 35%, depending on the purity of the commercial CoASH used and the critical charcoal step.

Estimation of CoASSG from Biological Sources

It has been shown that CoASSG is present in acid-soluble extracts from rat liver[11] and bovine liver.[12] As a matter of fact, the procedure reported by Kornberg and Stadtman[13] for the isolation of CoASH from yeast revealed that their main product was largely the mixed disulfide derivative of glutathione; the amount recovered from 3 kg of dried yeast was about 429 micromoles. We have found in rat liver about 0.2 micromole/100 g of fresh tissue.

Procedure.[11] About 55 g of fresh rat liver from 5–6 animals (Wistar rats of 200–300 g, fed on stock laboratory diet) is extracted in a Waring Blendor for 1 minute with two volumes of cold 10% trichloroacetic acid and centrifuged; the supernatant fraction is washed 5–6 times with ether. The glycogen is precipitated from the extract with 1 volume of 95% ethanol in the cold for 1 hour and centrifuged; the supernatant solution is reduced to a small volume in a flash evaporator at 35°. The extract is made up to a volume of 100 ml with water, adjusted to pH 7.6–8.0, and applied to a column of Dowex 1 chloride, according to Cohn.[9] After this, the column is washed with 1 liter of water followed by 1.5 liters of 0.01 N HCl–0.01 M

[11] R. N. Ondarza, *Biochim. Biophys. Acta* **107**, 112 (1965).
[12] S. H. Chang and D. R. Wilken, *J. Biol. Chem.* **240**, 3136 (1965).
[13] A. Kornberg and E. R. Stadtman, Vol. III [131], p. 907.

NaCl. Now the first 1.5 liters of 0.01 N HCl–0.05 M NaCl collected in bulk is treated with activated charcoal for 10 minutes, as indicated above, and centrifuged; the charcoal is eluted with 100 ml of ethanol–water (1:1, v/v), for 3 hours in the cold at pH 7.5. The solvent is evaporated in a flash evaporator and chromatographed on paper with the isopropanol–water solvent for 48 hours.

The ultraviolet-absorbing area of the chromatogram (with an R_f value of $R_{AMP} = 0.55$, $R_{ATP} = 0.97$), which also gives positive ninhydrin reaction, is eluted by descending technique with 0.05 M phosphate buffer, pH 5.5, and estimated by the specific enzymatic reaction for CoASSG as described. Up to this step, the material isolated is still impure but suitable for enzymatic estimation; if so desired, it can be further purified by high or low voltage electrophoresis followed by paper chromatography in isopropanol–water for 60 hours.

Properties of CoASSG

The pure mixed disulfide (CoASSG) obtained from synthesis or from biological sources, gives $CoASO_3H$ (see table, component 1) and GSO_3H (see table, component 2), after performic acid treatment.[14] Some characteristics of these compounds, CoASSG, and standards are reported in the table.

CHROMATOGRAPHIC AND ELECTROPHORETIC CHARACTERISTICS OF CoASSG, GSH, COENZYME A, AND THEIR CORRESPONDING OXIDIZED PRODUCTS[a]

	Method of separation		
	Paper chromatography in isopropanol–water for 48 hours		Paper ionophoresis in 0.1 M citrate buffer (pH 3.2) 0.5 mA/cm for 4 hours at 4–6°
Compound	R_{AMP}	R_{ATP}	M_{ATP}
CoASSG	0.55	0.97	0.71–0.73
CoASSG incubated with performic acid			
Component 1	0.86	1.75	1.25
Component 2	1.07	2.4	0.70
GSH	1.48	3.25	0.19
GSSG	0.61	1.2	0.19
GSO_3H	1.07	2.4	0.70
CoASH	0.73	1.4	0.91
$CoASO_3H$	0.86	1.75	1.25

[a] *Biochim. Biophys. Acta* **107**, 112 (1965).

[14] C. H. W. Hirs, *J. Biol. Chem.* **219**, 611 (1956).

CoASSG migrates toward the anode behind CoASH and ATP, in high-voltage electrophoresis at pH 2.0. In low-voltage electrophoresis (0.1 M citrate buffer, pH 3.2), it has a mobility of $M_{ATP} = 0.73$ (GSH and GSSG move toward the cathode).

The characteristic ultraviolet-absorbing spectrum shows at pH 2.0 a maximum at 257 mμ; min = 232 mμ; 250/260 = 0.916; and 280/260 = 0.325. The molecular extinction coefficient at 257 mμ is 15,900 at pH 2.0.

A 24-hour hydrolyzate of 108 millimicromoles of purified CoASSG prepared from rat liver,[11] when subjected to amino acid analysis on the Spinco analyzer showed the following composition, expressed in millimicromoles recovered: cysteine (cysteic acid), 79; glycine, 107; glutamic acid, 91; taurine, 83; β-alanine, 76. A small amount (10 millimicromoles) of a substance coming off at "the serine position" from the column is also found. This unknown ninhydrin-positive component is probably a product of incomplete hydrolysis of the peptide.

A distilled water solution of CoASSG can be stored in the freezer at least for a month without noticeable degradation.

[55] Pantothenic Acid and Coenzyme A: Preparation of CoA Analogs

By Masao Shimizu

Introduction

CoA analogs reported to date are the compounds modified on the cysteamine moiety, except for guano-CoA, as shown in Table I. The principles of the methods used for the chemical synthesis of CoA have been also applied for preparing the analogs, except in the case of desulfo-CoA (Table I).

In reference to the methods reported on the chemical synthesis of CoA, the following two points seem to be essential for constituting the structure of CoA: (1) the formation of the pyrophosphate bond between the pantetheine and adenosine moieties, (2) the attachment of the phosphomonoester group in the 3'-position of the adenosine moiety. Moffatt and Khorana[1] succeeded in the first point by condensing an adenosine 5'-phosphoromorpholidate derivative with pantetheine 4'-phosphate, Michelson[2] developed

[1] J. G. Moffatt and H. G. Khorana, *J. Am. Chem. Soc.* **81**, 1265 (1959); **83**, 663 (1961).
[2] A. M. Michelson, *Biochim. Biophys. Acta* **50**, 605 (1961); **93**, 71 (1964).

the procedure involving anhydride-anion exchange, and Gruber and Lynen[3] reported the concurrent condensation of pyrophosphoryl tetrachloride with a pantetheine derivative and adenosine 3'-phosphate. The second point may be overcome by selecting adenosine 2',3'-cyclic phosphate 5'-phosphoromorpholidate as the starting material and by opening the 2',3'-cyclic phosphate of the resulting product according to Moffatt and Khorana. The fission with an acidic agent employed by them, nevertheless, brought about inevitably the concurrent formation of 2'-phosphate. Michelson remedied this defect by use of the specific ability of ribonuclease T_2 from Takadiastase to afford exclusively 3'-phosphate from the 2',3'-cyclic phosphate of adenosine. The modification of the route of Moffatt and Khorana by the use of the enzymatic method of Michelson, was reported by Shimizu et al.[4] to improve the quality as well as the yield of CoA. Meanwhile, Shimizu et al. developed the thiazoline method involving the condensation step of the appropriate nitrile derivatives with cysteamine and the acidic fission step of the resulting thiazoline intermediates to prepare the metabolites[5-7] of pantothenic acid containing the cysteamine moiety inclusive of CoA[4] itself as shown in the following general scheme.

$$-NH-CH_2-CH_2-CN \; + \quad \begin{array}{c} NH_2-CH_2 \\ | \\ SH-CH_2 \end{array} \quad \rightarrow \quad -NH-CH_2-CH_2-C \overset{N-CH_2}{\underset{S-CH_2}{\diagdown}} \quad + \; NH_3 \overset{H^+}{\rightarrow}$$

$$-NH-CH_2-CH_2-CO \overset{NH-CH_2}{\underset{HS-CH_2}{\diagup}}$$

This thiazoline method is convenient for preparing CoA analogs modified on the cysteamine moiety, since the key nitrile derivative is the same as that in the synthesis of CoA and can be stored without any decomposition. The method has been applied for the preparation of α-methyl-, β-methyl-, and α-carboxy-CoA.[8,9]

The original method by Moffatt and Khorana has also been applied for the preparation of seleno-CoA (diselenide form alone) by Günther and

---

[3] W. Gruber and F. Lynen, Ann. **659**, 139 (1962).

[4] M. Shimizu, O. Nagase, S. Okada, Y. Hosokawa, H. Tagawa, Y. Abiko, and T. Suzuki, Chem. Pharm. Bull. (Tokyo) **13**, 1142 (1965); **15**, 655 (1967).

[5] M. Shimizu, G. Ohta, O. Nagase, S. Okada, and Y. Hosokawa, Chem. Pharm. Bull. (Tokyo) **13**, 180 (1965).

[6] G. Ohta, O. Nagase, Y. Hosokawa, H. Tagawa, and M. Shimizu, Chem. Pharm. Bull. (Tokyo) **15**, 644 (1967).

[7] O. Nagase, Chem. Pharm. Bull. (Tokyo) **15**, 648 (1967).

[8] M. Shimizu, O. Nagase, Y. Hosokawa, H. Tagawa, Y. Abiko, and T. Suzuki, Chem. Pharm. Bull. (Tokyo) **14**, 681 (1966).

[9] M. Shimizu, O. Nagase, Y. Hosokawa, and H. Tagawa, Tetrahedron **24**, 5241 (1968).

Mautner[10] and of oxy-CoA by Stewart et al.[11, 12] Homo-CoA also has been synthesized by a similar method modified by enzymatic fission.[9] Guano-CoA is the sole analog modified on the purine base moiety and has been prepared[13] by the same principle as above.

Direct desulfurization of CoA with Raney nickel yields desulfo-CoA.[14] 3'-Dephospho-CoA has been employed as a model compound prior to the total synthesis of CoA.[1–4] Therefore, 3'-dephospho-CoA analogs corresponding to seleno-, oxy-, and α-carboxy-CoA also have been prepared by the respective methods.

Many of these analogs have been shown to be competitive inhibitors of CoA in the phosphotransacetylase reaction with the respective K_i values in Table I.

Method A. Chemical Synthesis by the Thiazoline Method

Step 1. D-*Pantothenonitrile*[5]

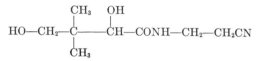

Reagents

D-Pantolactone,[15] m.p. 89–91.5°, $[\alpha]_D^{23} = -51.1°$ (c = 2.0, water)
3-Aminopropionitrile,[16] b.p. 84–85°/20 mm, prepared from acrylonitrile and ammonia

Procedure. A mixture of D-pantolactone (40 g, 0.31 mole) and 3-aminopropionitrile (21.5 g, 0.31 mole) is heated at 50° for 4 hours. The above reaction mixture is crystallized from ethyl acetate (15 ml) to yield D-pantothenonitrile as prisms, m.p. 82–84° (41 g). From the mother liquor, a second (13.7 g) and third crop (2.2 g) with the same melting point are obtained. The total yield is 92.4% (56.9 g). Molecular formula = $C_9H_{16}O_3N_2$; $[\alpha]_D^{20} + 31.5°$ (c = 1.75, water).

[10] W. H. H. Günther and H. G. Mautner, *J. Am. Chem. Soc.* **87**, 2708 (1965).
[11] C. J. Stewart and T. L. Miller, *Biochem. Biophys. Res. Commun.* **20**, 433 (1965).
[12] T. L. Miller, G. L. Rowley, and C. J. Stewart, *J. Am. Chem. Soc.* **88**, 2299 (1966).
[13] M. Shimizu, O. Nagase, S. Okada, Y. Abiko, and T. Suzuki, *Chem. Pharm. Bull.* (*Tokyo*) **14**, 683 (1966).
[14] J. F. A. Chase, B. Middleton, and P. K. Tubbs, *Biochem. Biophys. Res. Commun.* **23**, 208 (1966).
[15] E. T. Stiller, S. A. Harris, J. Finkelstein, J. C. Keresztesy, and K. Folkers, *J. Am. Chem. Soc.* **62**, 1785 (1940).
[16] S. R. Buc, *Org. Syn. Coll. Vol.* **3**, 93 (1955).

TABLE I

PARTIAL STRUCTURE OF CoA ANALOGS AND THEIR K_i VALUES IN THE PHOSPHOTRANSACETYLASE REACTION

CoA analogs	Purine moiety	Cysteamine moiety	K_i — Enzyme from E. coli B (M)[a]	K_i — Enzyme from Cl. kluyveri (M)[b]	
CoA	A	$-NH-CH_2-CH_2-SH$	2.9×10^{-4} c	1.3×10^{-4} c	5.6×10^{-4} d
Seleno-CoA	A	$-NH-CH_2-CH_2-Se_2$	—	—	2.0×10^{-4} e
Oxy-CoA	A	$-NH-CH_2-CH_2-OH$	—	—	3.5×10^{-7} f
Desulfo-CoA	A	$-NH-CH_2-CH_3$	3.4×10^{-6}	4.0×10^{-6}	3.5×10^{-6} g
α-Methyl-CoA	A	$-NH-CH(CH_3)-CH_2-SH$	2.5×10^{-4}	3.1×10^{-4}	—
β-Methyl-CoA	A	$-NH-CH_2-CH(COOH)-SH$	1.4×10^{-4}	2.5×10^{-4}	—
α-Carboxy-CoA	A	$-NH-CH-CH_2-SH$	3.0×10^{-4}	3.1×10^{-4}	—
Homo-CoA	A	$-NH-CH_2-CH_2-CH_2-SH$	4.8×10^{-4}	1.0×10^{-3}	—
Guano-CoA	G	$-NH-CH_2-CH_2-SH$	—	—	—

[a] Reaction velocity was determined by the arsenolytic decomposition of acetylphosphate (M. Shimizu, Y. Abiko, T. Suzuki, and K. Kameda, unpublished data).

[b] The reaction rate was measured from the formation of acetyl-CoA.

[c] K_m for CoA (M. Shimizu, T. Suzuki, and Y. Abiko, unpublished data).

[d] K_m for CoA [H. U. Bergmeyer, G. Holz, H. Klotzsch, and G. Lang, Biochem. Z. **338**, 114 (1963); also see C. J. Stewart and T. L. Miller, Biochem. Biophys. Res. Commun. **20**, 433 (1965)].

[e] K_m for CoA [J. F. A. Chase, B. Middleton, and R. K. Tubbs, Biochem. Biophys. Res. Commun. **23**, 208 (1966)].

[f] K_m for CoA [T. L. Miller, G. L. Rowley, and C. J. Stewart, J. Am. Chem. Soc. **88**, 2299 (1966).

[g] J. F. A. Chase, B. Middleton, and R. K. Tubbs, Biochem. Biophys. Res. Commun. **23**, 208 (1966).

Step 2. D-*Pantothenonitrile 4'-Dibenzyl Phosphate*[7]

$$(C_6H_5-CH_2-O-)_2-P-O-CH_2-\overset{\overset{\displaystyle CH_3}{|}}{\underset{\underset{\displaystyle CH_3}{|}}{C}}-\overset{\overset{\displaystyle OH}{|}}{CH}-CONH-CH_2-CH_2CN$$

Reagents

Dibenzyl phosphorochloridate.[17] Prepare immediately before use. To a solution of dibenzyl phosphite[18] (b.p. 110–120°/3 mm) (13.1 g, 0.05 mole) in benzene (80 ml) is added N-chlorosuccinimide[19] (6.75 g, 0.05 mole) which dissolves with evolution of heat. After the succinimide is filtered off, the benzene solution is instantly available for the following reaction.

Procedure. D-Pantothenonitrile (5.0 g, 0.025 mole) in anhydrous pyridine (300 ml) is cooled to −40° in a dry-ice–acetone bath. The solution of dibenzyl phosphorochloridate in benzene prepared above is then added under stirring over a period of 20 minutes. The mixture is allowed to stand in the bath for 20 hours; during this time the temperature rises to −10°. Water (45 ml) is added, and after the mixture has stood at room temperature for 15 minutes, the solvent is distilled off *in vacuo.* The residual pyridine is removed by several evaporations with methanol, and the residue is dissolved in ethyl acetate (200 ml) and washed three times successively with 2 N H_2SO_4, 10% $NaHCO_3$ and saturated sodium sulfate (each 100 ml). The organic layer is dried over anhydrous sodium sulfate and evaporated *in vacuo* to give the desired product (8.6 g, 73.2%) as a thick syrup. For further purification, a 90% methanolic solution of the syrup (0.2 g) is passed through a column of a mixture of Amberlite IR 120 (H+) and IRA 410 (OH−) (1:1, 1 ml) and the column is washed with 90% methanol. The effluent is evaporated to dryness *in vacuo*, and the residue is dissolved in ether (10 ml) and stirred with petroleum ether (10 ml) to precipitate an oil, which is collected and dried *in vacuo* over P_2O_5. Molecular formula = $C_{23}H_{29}O_6N_2P\cdot\frac{1}{2}H_2O$; $[\alpha]_D^{23} + 19.1°$ ($c = 1.1$, ethanol).

Step 3. D-*Pantothenonitrile 4'-Phosphate*[7]

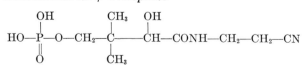

$$HO-P-O-CH_2-\overset{\overset{\displaystyle CH_3}{|}}{\underset{\underset{\displaystyle CH_3}{|}}{C}}-\overset{\overset{\displaystyle OH}{|}}{CH}-CONH-CH_2-CH_2-CN$$

[17] G. W. Kenner, A. R. Todd, and F. J. Weymouth, *J. Chem. Soc.* p. 3675 (1952).
[18] O. M. Friedmann, D. L. Klass, and A. M. Seligman, *J. Am. Chem. Soc.* **76,** 916 (1954).
[19] C. A. Grob and H. J. Schmid, *Helv. Chim. Acta* **36,** 1763 (1953).

Procedure. A solution of D-pantothenonitrile 4'-dibenzyl phosphate (4.7 g, 0.01 mole) in 90% methanol (50 ml) is hydrogenated at room temperature and atmospheric pressure over a 10% palladium–charcoal catalyst (1 g). Hydrogen is absorbed quite rapidly for the initial 5 minutes and then slowly. After absorption of hydrogen (520 ml) during 10 minutes, the catalyst is filtered off and the solvent is evaporated *in vacuo*. A solution of the residue in water (50 ml) is passed through a column of Amberlite IR 120 (H⁺) (6 ml), and the column is washed with water. The combined eluate is adjusted to pH 7.3 with 0.1 N Ba(OH)₂ and evaporated to dryness *in vacuo* after removal of a small amount of insoluble material. The residue is dissolved in methanol (20 ml) and the filtered solution is concentrated *in vacuo* to a volume of 15 ml. Addition of ether (180 ml) with stirring affords a precipitate which is dried over P_2O_5 *in vacuo* giving the barium salt (3.11 g, 68.9%). Molecular formula = $C_9H_{15}O_6N_2PBa \cdot 2H_2O$; $[\alpha]_D^{25} +$ 11.4° ($c = 2.0$, water). The lithium salt is obtained by passage of the aqueous solution of the barium salt through IR 120 (H⁺) and neutralization of the eluate with LiOH followed by precipitation from the methanolic solution with ether. Molecular formula = $C_9H_{15}O_6N_2PLi_2 \cdot \frac{1}{2}H_2O$; $[\alpha]_D^{20} +$ 18.2° ($c = 2.0$, water).

Step 4. P¹-Adenosine 3'-Phosphate 5'-P²-D-Pantothenonitrile 4'-Pyrophosphate[4]

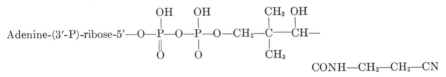

Reagents

Adenosine 2'(3'),5'-diphosphate.[20] The barium salt of 2-cyanoethyl phosphate[21] (1.94 g, 6 millimoles) is dissolved in a mixture of water (10 ml) and Amberlite IR 120 (H⁺) (6 ml). The mixture is applied to the top of a column (1.1 × 5 cm) of IR 120 (H⁺), and the column is washed with water (80 ml). After addition of pyridine (3 ml), the eluate is concentrated to dryness *in vacuo*, the residue is rendered anhydrous by repeated evaporation *in vacuo* with added dry pyridine, and the residue is dissolved in 10 ml of pyridine. Adenosine (267 mg, 1 millimole, dried at 110°, 1 mm over P_2O_5) is dissolved in boiling anhydrous pyridine (15 ml) and after cooling, this solution is added to the solution of cyanoethyl phosphate. The

[20] As modified from L. M. Fogarty and M. R. Rees, *Nature* **193**, 1180 (1962).
[21] G. M. Tener, *J. Am. Chem. Soc.* **83**, 159 (1961).

mixture is concentrated to dryness and the residue is dissolved in a mixture of anhydrous pyridine and dimethyl formamide (each 7.5 ml). Dicyclohexylcarbodiimide (DCC) (3.71 g, 18 millimoles) is added, and the mixture is stirred for 5 hours at room temperature. Water (5 ml) is added, and after 1 hour at room temperature, dicyclohexylurea is filtered off. The filtrate is concentrated to dryness *in vacuo*, and a solution of the residue in 25% acetic acid (20 ml) is heated in a steam bath for 30 minutes. An insoluble material is filtered and the filtrate concentrated *in vacuo*. The residue is dissolved in 10 N NH$_4$OH (12 ml), and the solution is kept at 80° for 1 hour. After cooling, the solution is filtered and the filtrate is concentrated to dryness *in vacuo*. The residue is dissolved in water and applied to a column (1.6 × 12 cm) of Dowex 1-X4 (Cl$^-$, 200–400 mesh), and the column is washed with water and then with 0.01 M NaCl in 0.003 N HCl (2 liters) at a flow rate of 20 ml per 15 minutes, with which adenosine monophosphate is eluted. Adenosine 2′(3′),5′-diphosphate is eluted with 0.075 M NaCl in 0.003 N HCl, and the appropriate eluate is passed through a charcoal column (1.2 × 10 cm). The column is washed with water and then with 50% aqueous methanol (1 liter) containing concentrated NH$_4$OH (6 ml). The fractions containing nucleotide are concentrated *in vacuo* to a volume of approximately 20 ml and passed through a column (1 × 5 cm) of Dowex 50-X4 (H$^+$); the column is washed with water (20 ml). The effluent is concentrated *in vacuo* to 1 ml, and ethanol–acetone (1:1, 20 ml) is added with stirring to separate a gum, which solidifies and powders gradually. It is collected by filtration and dried over P$_2$O$_5$ *in vacuo* to give the adenosine 2′(3′),5′-diphosphate as a powder (304 mg) in a yield of 64.3%. Molecular formula = C$_{10}$H$_{15}$O$_{10}$N$_5$P$_2$·C$_2$H$_5$OH.

Adenosine 2′,3′-cyclic phosphate 5′-phosphoromorpholidate. A solution of DCC (8.25 g, 40 millimoles) in *tert*-butanol (100 ml) is added to a solution of adenosine 2′(3′),5′-diphosphate (1.90 g, 4 millimoles) in water (10 ml) containing morpholine (2.09 g, 24 millimoles). The mixture is refluxed with stirring for 4 hours; during refluxing it soon becomes homogeneous. After cooling, the separated dicyclohexylurea is filtered off, and the filtrate is evaporated *in vacuo*. The residue is dissolved in a mixture of water (80 ml) and ether (80 ml), and insoluble dicyclohexylurea is filtered off. The aqueous layer of the filtrate is separated, extracted twice with ether (each 40 ml), and evaporated *in vacuo*. The residue is dissolved in methanol (5 ml), and dry ether (40 ml) is added with shaking to separate a gum. The solvent is decanted, and the resulting gum is stirred with dry ether (40 ml) until it changes to a powder. It is filtered and washed with ether and dried *in vacuo* to give bis(4-morpholine N,N'-dicyclohexylcarboxamidinium) adenosine 2′,3′-cyclic phosphate 5′-phosphoromorpholidate (4.1 g) in a yield of 93.2%. Molecular formula = $C_{48}H_{82}O_{11}N_{12}P_2 \cdot 2H_2O$.

Ribonuclease T₂.[22] Ribonuclease T₂ (RNase T₂) is prepared from Takadiastase according to the method of Uchida and Egami, which consists of water extraction of Takadiastase powder, batchwise treatment of the extracts with DEAE-cellulose, heat treatment at 80° for 2 minutes, ammonium sulfate fractionation, the first DEAE-cellulose chromatography, alcohol fractionation, and the second DEAE-cellulose chromatography. For the present purpose, two final steps of the above purification steps may be omitted. Activity of RNase T₂ is measured according to the method of Takahashi.[23] RNase T₂ thus obtained has a specific activity of 22–28, and the ratio of the activity at pH 4.5 to that at pH 7.5 is 1.3–2.3. The preparation is free of phosphomonoesterase[24] and nucleotide pyrophosphatase activities,[25] contamination of which must be avoided for the purpose of the present experiment.

Procedure. An aqueous solution of the barium salt of D-pantotheno-nitrile 4′-phosphate (271 mg, 0.6 millimole) is passed through a column of Amberlite IR 120 (H⁺) (0.9 × 4 cm), and the column is washed with water (30 ml). Pyridine (1 ml) is added to the total eluates, the solution is evaporated to dryness *in vacuo*, and the residue is dried by three evaporations with anhydrous pyridine (10 ml) and finally redissolved in pyridine (5 ml). Separately, 4-morpholine N,N'-dicyclohexylcarboxamidinium adenosine 2′,3′-cyclic phosphate 5′-phosphoromorpholidate (217 mg, 0.2 millimole) is dried by three evaporations with anhydrous pyridine (5 ml) and finally redissolved in pyridine (5 ml). Both solutions are combined and then evaporated *in vacuo* two times with pyridine (10 ml); the solution is kept at room temperature overnight. The major part of solvent is removed *in vacuo* and the persistent pyridine is removed by several evaporations

[22] T. Uchida and F. Egami, see Vol. XII, Part A [30b].

[23] K. Takahashi, *J. Biochem. (Tokyo)* **49**, 1 (1961).

[24] Phosphomonoesterase activity is assayed with *p*-nitrophenyl phosphate as the substrate at pH 4.5 in the presence of $2 \times 10^{-3} M$ EDTA, essentially according to the method described in Vol. IV, p. 371.

[25] Nucleotide pyrophosphatase activity is assayed in the presence of $1.6 \times 10^{-3} M$ EDTA at pH 4.5, essentially by the method of A. Kornberg, Vol. II [112]. For the purpose of the present experiment, the contamination of undesirable activities can be checked by examining the stability of CoA during incubation with the RNase T₂ preparation.

with methanol. The residue is dissolved in water (4 ml) and the pH is brought to 4.6 with 2% NH_4OH. After addition of 0.16 M EDTA (0.15 ml, 10^{-3} M final concentration) and an aqueous solution (1 ml) of 1.3 mg of ribonuclease T_2 (activity, 373 units/ml), the mixture is incubated at 37° for 2.5 hours. The pH is readjusted to 4.6 with diluted HCl. Enzyme solution (1.3 mg) is again added, and the solution is incubated for 1 hour. The solution, adjusted to pH 6.0 with 2% NH_4OH, is chromatographed on a column of DEAE-cellulose (Cl⁻) (2.8 × 30 cm) using a linear salt gradient with 0.003 N HCl (1.5 liters) and 0.15 N LiCl in 0.003 N HCl (1.5 liters) at a flow rate of 1.5 ml per minute. Following two small fractions (adenosine 3′,5′-diphosphate and an unidentified substance), the third fraction containing the desired nitrile derivative is adjusted to pH 4.5 with 0.1 N LiOH and evaporated to dryness. The residue is dried *in vacuo* over P_2O_5 and dissolved in methanol (5 ml). Acetone (50 ml) is added to the solution to give a white powder, which is again purified by treatment with methanol (5 ml) and acetone (50 ml). The product dried *in vacuo* over P_2O_5 is obtained in a yield of 61.8% (101 mg) as a chromatographically homogeneous white powder. Molecular formula = $C_{19}H_{27}O_{15}N_7P_3Li_3 \cdot 6H_2O$. This substance can be stored in the cold for about 1 year without any decomposition.

Step 5. CoA⁴ and Its Analogs⁹

$$\text{Adenine-(3'-P)-ribose-(5')} - O - \underset{\underset{O}{\|}}{P} - O - \underset{\underset{O}{\|}}{P} - O - CH_2 - \underset{\underset{CH_3}{|}}{\overset{\overset{CH_3}{|}}{C}} - \overset{\overset{OH}{|}}{CH} -$$

$$CONH - CH_2 - CH_2 - CONH - \overset{\overset{R_1}{|}}{CH} - \overset{\overset{R_2}{|}}{CH} - SH$$

Reagents

Cysteamine (2-aminoethanethiol),[26] m.p. 96–99°; hydrochloride, m.p. 70°

α-Methylcysteamine (*dl*-2-amino-1-propanethiol),[27] m.p. 67–69°; hydrochloride, m.p. 94–97°

β-Methylcysteamine (*dl*-1-amino-2-propanethiol),[28] m.p. 59–62°; hydrochloride, m.p. 84–88°

Monosodium salt of L-cysteine. This is prepared by treatment of L-cysteine hydrochloride monohydrate with two molar equivalents of sodium methoxide in methanol and used as the methanolic solution for the present preparation.

[26] E. E. Snell and E. L. Wittle, see Vol. III [133].

[27] M. Böse, *Ber.* **53**, 2000 (1920).

[28] S. Gabriel, *Ber.* **49**, 1110 (1916).

Procedure. FORMATION OF THE THIAZOLINE INTERMEDIATE. To a methanolic solution (5 ml) of the lithium salt of the nitrile derivative (57–80 mg, 0.07–0.1 millimole) obtained in step 4, add either cysteamine, methylcysteamine, or monosodium salt of cysteine (5–6 equivalents to the nitrile derivative) in methanol. The mixture is refluxed in nitrogen media for 6–9 hours during which evolution of ammonia is observed. Evaporation of the solvent gives the crude thiazoline intermediate as a pale yellow powder which does not exhibit the absorption band of the C≡N group at 2250 cm^{-1} in its infrared spectrum. In the use of cysteamine, paper chromatography of this compound with ethanol–water (7:3) gives two spots detectable with ultraviolet light. The main spot has an R_f of 0.42 and the minor spot, an R_f of 0.34, both distinguishable from the R_f of 0.48 of a synthetic sample of CoA. The material at R_f 0.42 is the thiazoline intermediate, since it does not react with nitroprusside-KCN reagent but reacts as a disulfide compound when exposed in air for 2 days. The spot at R_f 0.34 corresponds to adenosine 2'(3'),5'-diphosphate.

FISSION OF THE THIAZOLINE INTERMEDIATE. The crude substance is dissolved in water (5 ml) and adjusted to pH 4.7–5.0 with 0.1 N HCl. The mixture is heated in nitrogen media at 55–60° for 2–3.5 hours, during which the pH of the solution is changed to 4.2. After concentration *in vacuo* to about 1 ml and addition of 2-mercaptoethanol (1 ml), the mixture is kept at room temperature overnight, diluted with water (30 ml), and passed through a column of Dowex 50 (H$^+$) (2 ml). The column is washed with water (150 ml). The combined effluent is adjusted to pH 4.5 with 0.1 N LiOH and concentrated to dryness *in vacuo*. The residue is dissolved in methanol (1 ml) and acetone (20 ml) is added. The precipitate formed is filtered off and dried *in vacuo*. The desired CoA or its analog is obtained as the crude lithium salt.

Purification

The crude salt is dissolved in water (1 ml), 2-mercaptoethanol (1 ml) is added, and the solution is kept at room temperature overnight. The solution, diluted with water (20 ml) and adjusted to pH 6.0 with diluted NH$_4$OH, is applied to a column of DEAE-cellulose (Cl$^-$) (2.0 × 28 cm) using a linear salt gradient with 0.003 N HCl (0.75 liter) and 0.225 N LiCl in 0.003 N HCl (0.75 liter). Each 5-ml portion is collected at a flow rate of 0.5 ml per minute until ultraviolet absorption disappears. The first fraction is adenosine 2'(3'),5'-diphosphate. The second fraction, containing the desired compound (SH form), is adjusted to pH 4.5 with 0.1 N LiOH and evaporated to dryness. The residue is mixed with methanol (1 ml), and acetone (20 ml) is added to give a white powder that is purified by repeated precipitations as above until the supernatant solution becomes free of chloride ion. The compound is dried over P_2O_5 *in vacuo* to yield the purified salt. The third fraction containing the SS-form is worked up similarly after treatment with 50% aqueous 2-mercaptoethanol to yield an additional crop of the SH form.

TABLE II

R_f VALUES OF CoA ANALOGS AND THEIR INTERMEDIATES[a]

Compound	R_f in solvent	
	I	II
CoA (SH)	0.28	0.12
CoA (SS)	0.08	0.02
α-Me-CoA (SH)	0.30	0.12
α-Me-CoA (SS)	0.09	0.02
β-Me-CoA (SH)	0.30	0.12
β-Me-CoA (SS)	0.09	0.02
α-Carboxy-CoA (SH)	0.26	0.05
α-Carboxy-CoA (SS)	0.06	0.01
Homo-CoA (SH)	0.28	0.13
Homo-CoA (SS)	0.11	0.03
Guano-CoA (SH)	0.21	0.07
Guano-CoA (SS)	0.02	0.0
Desulfo-CoA	0.30	0.13
Dephospho-CoA (SH)	0.38	0.38
Dephospho-CoA (SS)	0.10	0.09
α-Carboxy-dephospho-CoA (SH)	0.32	0.23
α-Carboxy-dephospho-CoA (SS)	0.04	0.04
Pantetheine 4'-phosphate (SH)	0.63	0.47
Pantethine 4',4''-diphosphate (SS)	0.57	0.22
Pantothenonitrile 4'-phosphate	0.60	0.47
Homopantetheine 4'-phosphate (SH)	0.63	0.47
Homopantethine 4',4''-diphosphate (SS)	0.57	0.24
Adenosine 2'(3'),5'-diphosphate	0.16	0.02
Adenosine 2',3'-cyclic phosphate 5'-phosphoromorpholidate	0.34	0.35
P^1-Adenosine 3'-phosphate 5'-P^2-pantothenonitrile 4'-pyrophosphate	0.23	0.09
Adenosine 5'-phosphate	0.27	0.11
Adenosine 5'-phosphoromorpholidate	0.42	0.48
P^1-Adenosine 5'-P^2-pantothenonitrile 4'-pyrophosphate	0.35	0.32
Guanosine 2'(3'),5'-diphosphate	0.10	0.01
Guanosine 2',3'-cyclic phosphate 5'-phosphoromorpholidate	0.21	0.21

[a] Paper chromatography is carried out by the ascending technique on Toyo Roshi No. 50 paper: solvent I, ethanol–0.5 M ammonium acetate buffer, pH 3.8 (5:2) solvent II, ethanol-1 M ammonium acetate buffer, pH 7.5 (5:2).

Characterization

The results obtained by the present method are as follows: CoA (molecular formula = $C_{21}H_{35}O_{16}N_7P_3SLi_3 \cdot 8H_2O$, adenosine:P = 1:3) in a yield of 29.7%; α-methyl-CoA (molecular formula = $C_{22}H_{35}O_{16}N_7P_3SLi_3 \cdot 10H_2O$, adenosine:P = 1:3) in a yield of 32.3%; β-methyl-CoA (molecular formula = $C_{22}H_{35}O_{16}N_7P_3SLi_3 \cdot 7H_2O$, adenosine:P = 1:3.02) in a yield of 37.2%; and α-carboxy-CoA (molecular formula = $C_{22}H_{32}O_{18}N_7P_3SLi_4 \cdot 20H_2O$, adenosine:P = 1:3.06) in a yield of 28.9%. Alkaline hydrolysis (1 N NaOH, 100°, 20 minutes) of CoA and the analogs yields adenosine 3',5'-diphosphate, which is detected by paper chromatography. Acid hydrolysis (1 N HCl, 100°, 5 minutes) gives pantetheine 4'-phosphate from CoA and pantothenoylcysteine 4'-phosphate from α-carboxy-CoA. R_f values of CoA analogs and their intermediates are shown in Table II. Assuming that CoA has 316 units per micromole,[29] CoA activities assayed by the phosphotransacetylase method[30] are as follows: the crude preparation = 252 units per micromole (adenosine) purity 80%, 242 units per micromole (ribose) purity 77%; the purified preparation = 338 units per micromole (adenosine) purity 106%, 323 units per micromole (ribose) purity 103%.

Method B. Chemical Synthesis by the Method of Moffatt and Khorana Modified with the Enzymatic Fission of 2',3'-Cyclic Phosphate

1. CoA[4]

Reagents

   D-Pantetheine 4'-phosphate.[7] A solution of the barium salt of D-pantothenonitrile 4'-phosphate (1.25 g, 2.77 millimoles) and cysteamine (0.255 g, 3.3 millimoles) in methanol (5 ml) is refluxed under nitrogen for 6 hours. The reaction mixture is concentrated to dryness *in vacuo* to give a crude thiazoline intermediate as a white powder (1.3 g). A solution of the above powder in water (10 ml) is adjusted to pH 5.0 with 1 N oxalic acid and heated in nitrogen media at 60° for 2 hours. After removal of the white precipitate, the reaction mixture is passed through a column of Amberlite IR 120 (H+) (8 ml), which is then washed with water. The combined effluent is adjusted to pH 7.2 with 0.1 M Ba(OH)₂ and then evapo-

[29] F. Lipmann, *J. Am. Chem. Soc.* **74**, 4017 (1952).
[30] This volume [53]; Y. Abiko, T. Suzuki, and M. Shimizu, *J. Biochem.* (*Tokyo*) **61**, 10 (1967).

rated to dryness *in vacuo*. A solution of the residue dissolved in methanol (10 ml) is separated from the trace of insoluble material and concentrated *in vacuo* to a volume of 5 ml. The addition of dry, peroxide-free ether gives a white precipitate, which is filtered and dried *in vacuo* over P_2O_5 giving the barium salt of D-pantetheine 4′-phosphate (1.20 g) in a yield of 82%. Molecular formula = $C_{11}H_{21}O_7N_2PSBa\cdot2H_2O$; $[\alpha]_D^{25} + 12.3°$ ($c = 2.3$, water).

D-Pantethine.[5] To a solution of sodium isopropoxide prepared from sodium (3.8 g, 0.165 mole) and isopropanol (60 ml) is added a solution of cysteamine hydrobromide (26.1 g, 0.165 mole) in isopropanol (24 ml). The mixture is stirred thoroughly to liberate the amine, and D-pantothenonitrile (30 g, 0.15 mole) is then added. The mixture is refluxed for 7 hours. Removal of NaBr by filtration followed by evaporation of the solvent *in vacuo* affords the crude thiazoline intermediate (46.2 g), which is dissolved in water (90 ml) and adjusted to pH 5.1 with added 1 N oxalic acid solution (25.5 ml). The solution is heated under nitrogen at 60° for 3 hours. The solution is then adjusted to pH 8.3 with 10% NH_4OH solution, and ferrous sulfate ($FeSO_4\cdot7H_2O$) (10 ml) is added. Aqueous 3.5% H_2O_2 solution is added dropwise to the stirred and cooled solution (0–10°) until the reaction mixture gives no further color with sodium nitroprusside reagent. The reaction mixture is brought to room temperature and passed through a column of a mixture of Amberlite IR 120 (H^+) (90 ml) and Amberlite IRA 410 (OH^-) (90 ml) over a period of 1 hour. The column is washed with water (1.8 liters), the eluted solution is evaporated *in vacuo* below 60°, and the residue is dried *in vacuo* at 55° for 8 hours to give D-pantethine (32.2 g) in a yield of 77.2%. Molecular formula = $C_{22}H_{42}O_8N_4S_2$; $[\alpha]_D^{23} + 17.1°$ ($c = 3.2$, water).

D-Pantethine 4′,4″-diphosphate.[9] A mixture of D-pantethine (277 mg, 0.5 millimole), the pyridinium salt of cyanoethyl phosphate (1.2 millimoles), and DCC (248 mg, 1.2 millimoles) in dry pyridine (5 ml) is kept for 22 hours at room temperature. One hour after addition of water (5 ml) to the reaction mixture, dicyclohexylurea is filtered off and the filtrate is evaporated to dryness. An aqueous solution of the residue is neutralized with 0.2 M $Ba(OH)_2$ and concentrated to a volume of 3 ml. Addition of ethanol (12 ml) gives a precipitate which is removed by filtration. The filtrate is evaporated to dryness. The residue is treated with 2 N NaOH (6 ml) at 0° for 30 minutes, and an aqueous slurry of Amberlite IR 120 (H^+) resin is then added to acidify the solution. The mixture is applied to the top of a column of IR 120 (H^+) resin (10 ml), the column

being washed thoroughly with water. The eluate is neutralized
with $Ca(OH)_2$ and evaporated to dryness. The crude product is
dissolved in water (3 ml). Addition of methanol (6 ml) gives a
precipitate which is repeatedly treated with water and methanol
and dried *in vacuo* at room temperature to give the calcium salt
of D-pantethine 4′,4″-diphosphate (162 mg) in a yield of 38.3%.
Molecular formula = $C_{22}H_{40}O_{14}N_4P_2S_2Ca_2 \cdot 3H_2O$; $[\alpha]_D^{25} + 11.5°$ ($c =$
2.5, water).

Procedure. The barium salt of D-pantetheine 4′-phosphate (318 mg,
0.6 millimole) in water (10 ml) is passed through a column of Dowex 50
(pyridinium form) (3 ml), and the solution is evaporated to dryness *in
vacuo*. The residue is dried three times by evaporation with anhydrous
pyridine (each 10 ml), dissolved in anhydrous pyridine (5 ml) and added to
a solution in anhydrous pyridine (10 ml) of the salt of adenosine 2′,3′-cyclic
phosphate 5′-phosphoromorpholidate (217 mg, 0.2 millimole) which is
dried beforehand in the same way as above. The reaction mixture is further
dehydrated by evaporation three times with anhydrous pyridine, dissolved
in pyridine (10 ml), and kept at room temperature overnight. The solvent
is evaporated *in vacuo*, and the trace of remaining pyridine is thoroughly
removed by several evaporations with water (each 10 ml). The solution
of the residue in water (3 ml) is adjusted to pH 4.6 with 10% NH_4OH and
submitted to enzymatic fission with ribonuclease T_2 as done in the case of
P^1-adenosine 3′-phosphate 5′-P^2-pantothenonitrile 4′-pyrophosphate. The
reaction mixture is adjusted to pH 6 with 10% NH_4OH and kept at room
temperature overnight after addition of 2-mercaptoethanol (2 ml). The
subsequent treatments for obtaining the purified CoA are the same as
described in method A. The total yield is 37.8% (65 mg as trilithium salt).

2. Homo-CoA[9]

Reagents

D-Homopantethine.[31] To a solution of D-pantothenonitrile (7.0 g,
35 millimoles) and homocysteamine[32] (3-amino-1-propanethiol,
b.p. 154–156°, m.p. 98–105°) (3.85 g, 42 millimoles) in isopropanol
(30 ml) is added 35% HCl (0.35 ml), and the mixture is refluxed
under nitrogen for 6 hours. Evaporation of the solvent *in vacuo*
gives the oily substance [2-(2-D-pantamidoethyl)-5,6-dihydro-4*H*-
1,3-thiazine] (11.0 g). A solution of this crude thiazine in water

[31] O. Nagase, H. Tagawa, and M. Shimizu, *Chem. Pharm. Bull. (Tokyo)* **16**, 977 (1968).
[32] S. D. Turk, R. P. Louthon, R. L. Cobb, and C. R. Bresson, *J. Org. Chem.* **27**, 2846 (1962).

(40 ml) is adjusted to pH 7 with 1 N oxalic acid and heated at 60° for 15 hours. The solution is cooled with ice-water, adjusted to pH 8.2 with 10% NH_4OH, and oxidized with 3% H_2O_2 until it no longer colors sodium nitroprusside reagent. The reaction mixture is passed through a column of a mixture of Amberlite IR 120 (H^+) and IRA 410 (OH^-) (each 30 ml). The eluate together with washings is evaporated *in vacuo* to give 5.8 g of D-homopantethine in a yield of 56.8%. Molecular formula = $C_{24}H_{46}O_8N_4S_2 \cdot \frac{1}{2}H_2O$; $[\alpha]_D^{22}$ + 27.9° (c = 1.0, methanol).

Homopantethine 4',4"-diphosphate. A mixture of D-homopantethine (1.66 g, 2.8 millimoles), the pyridinium salt of cyanoethyl phosphate (7.84 millimoles), and DCC (1.45 g, 7.02 millimoles) in anhydrous pyridine (15 ml) is kept at room temperature for 40 hours. The work-up, done in the same manner as described for D-pantethine 4',4"-diphosphate, gives the calcium salt of the desired compound (635 mg), which is precipitated several times from aqueous solution with methanol to give a pure sample. Molecular formula = $C_{24}H_{44}O_{14}N_4P_2S_2Ca_2 \cdot 8H_2O$; $[\alpha]_D^{24}$ + 7.9° (c = 1.0, water).

Procedure. A mixture of the pyridinium salt of homopantethine 4',4"-diphosphate (0.3 millimole) and the salt of adenosine 2',3'-cyclic phosphate 5'-phosphoromorpholidate (220 mg, 0.2 millimole) is rendered anhydrous by evaporation from pyridine; it is dissolved in anhydrous pyridine (10 ml) and left overnight at room temperature. The subsequent treatments may be performed in a way similar to those used in the preparation of CoA by method B. The yield of homo-CoA as reduced form is 35%. Molecular formula = $C_{22}H_{35}O_{16}N_7P_3SLi_3 \cdot 8H_2O$.

3. Guano-CoA[8]

Reagents

Guanosine 2'(3'),5'-diphosphate. Guanosine (283 mg, 1 millimole, dried at 110°, 1 mm over P_2O_5) is dissolved in boiling dimethylformamide (30 ml), and the solution is cooled. A solution of pyridinium 2-cyanoethyl phosphate (6 millimoles) in anhydrous pyridine (30 ml) is added. DCC (3.71 g, 18 millimoles) is added, and the mixture is allowed to stand for 4 days. Water (15 ml) is added and the mixture is worked up in the same manner as with the adenosine derivative. The product is treated with acetic acid and dissolved in 2 N LiOH (15 ml); the mixture is heated at 80° for 1 hour and then stored in a refrigerator. Insoluble material is filtered off, and the filtrate is passed through a column (1.6 × 18 cm) of Dowex 50 (H+). The column is washed with water (350 ml), and the effluent adjusted with 10% NH_4OH to pH 7.5. The solution is applied to a column (1 × 16 cm) of Dowex 2-X8 (Cl−), and the column is washed with water and then with 0.01 M LiCl in 0.01 N HCl to elute guanosine monophosphate. Further elution with 0.1 M LiCl in 0.01 N HCl removes the guanosine $2'(3'),5'$-diphosphate. The latter eluate, which contains the diphosphate, is adjusted with 0.1 N LiOH to pH 4.5 and concentrated to dryness *in vacuo*. The residue is thoroughly stirred with methanol (4 ml), and acetone (50 ml) is added. The solid is repeatedly treated with methanol and acetone until the supernatant solution is free of chloride ion and dried over P_2O_5 *in vacuo* to yield the lithium salt of guanosine $2'(3'),5'$-diphosphate (240 mg) in a yield of 48%. Molecular formula = $C_{10}H_{13}O_{11}N_5P_2Li_2 \cdot 3H_2O$.

Guanosine $2',3'$-cyclic phosphate $5'$-phosphoromorpholidate. An aqueous solution of the lithium salt of guanosine $2'(3')$, $5'$-diphosphate (235 mg, 0.471 millimole) is passed through a column (1 × 5 cm) of Dowex 50 (morpholinium form), and the column is washed with water. The effluent is evaporated to dryness, and the residue is dissolved in a mixture of water (5 ml), *tert*-butanol (5 ml), and morpholine (328 mg, 3.77 millimoles). A solution of DCC (972 mg, 4.71 millimoles) in *tert*-butanol (7 ml) is added dropwise to the refluxing solution during 4 hours, and the mixture is refluxed for another 4 hours. A work-up in the same manner as with the adenosine derivative gives bis(4-morpholine-N,N'-dicyclohexylcarboxamidinium)guanosine $2',3'$-cyclic phosphate $5'$-phosphoromorpholidate (467 mg) in a yield of 90.3%. Molecular formula = $C_{48}H_{82}O_{12}N_{12}P_2 \cdot H_2O$.

Procedure. An aqueous solution of the barium salt of D-pantetheine $4'$-phosphate (318 mg, 0.6 millimole) is passed through a column (0.9 × 6 cm) of Dowex 50 (pyridinium form), and after evaporation of the total effluent, the residue is dried by three evaporations with anhydrous pyridine (10 ml) and dissolved in anhydrous pyridine (5 ml). Bis(4-morpholine-N,N'-dicyclohexylcarboxamidinium)guanosine $2',3'$-cyclic phosphate $5'$-phosphoromorpholidate (220 mg, 0.2 millimole) is dried by three evaporations from a mixture of dimethylformamide (3 ml) and pyridine (6 ml), and dissolved in dimethylformamide (2.5 ml) and pyridine (5 ml). After 24 hours at room temperature, the solvent is evaporated *in vacuo* and the residue is dissolved in water (10 ml). The pH is adjusted to 7.5 with 10%

NH$_4$OH, and 0.2 M EDTA (0.2 ml), an aqueous solution of ribonuclease T$_1$[33] (5 ml, 16 + 10^4 units),[34] and 2-mercaptoethanol (2 ml) are added. The mixture is incubated at 37° for 13 hours, adjusted to pH 6.0 with 10% HCl, and concentrated to a volume of 3 ml. After addition of 2-mercaptoethanol (2 ml), the mixture is allowed to stand at room temperature overnight. The reaction mixture is diluted with water (100 ml) and applied to a column (2.8 × 33 cm) of DEAE-cellulose (Cl$^-$). The column is washed with water (600 ml), and elution is carried out using a linear salt gradient with 0.003 N HCl (3 liters) and 0.34 M LiCl in 0.003 N HCl (3 liters). Fractions of 15 ml are collected at a flow rate of 1.5 ml per minute to obtain 8 separated materials. The eighth component eluted in about 0.25 M LiCl concentration (the disulfide form of guano-CoA) is brought to pH 4.5 with 0.1 N LiOH and evaporated *in vacuo* to dryness. The residue is stirred with methanol (5 ml), and acetone (50 ml) is added. The solid is collected and dissolved in 50% 2-mercaptoethanol (2 ml). After being left at room temperature overnight, the solution is evaporated to dryness *in vacuo* and the residue is repeatedly treated with methanol and acetone to give the lithium salt of guano-CoA as a white powder (51.7 mg) in a yield of 23.6%. Molecular formula = C$_{21}$H$_{33}$O$_{17}$N$_7$P$_3$SLi$_3$·17H$_2$O.

[33] T. Uchida and F. Egami, see Vol. XII, Part A [30a].
[34] Ribonuclease T$_2$ (3300 units) also can be used instead of ribonuclease T$_1$. In this case a reaction time of 11 hours is necessary for completion.

[56] Synthesis of Coenzyme A Analogs

By HENRY G. MAUTNER

Coenzyme A plays a crucial role in group transfer and energy transfer reactions.[1,2] However, while it is well understood what functions are carried out by this coenzyme, the molecular basis of these functions remains poorly defined. The role of the single sulfur atom of coenzyme A in transferring acyl groups or in activating the methylene group adjacent to the thioacyl carbon of esters of coenzyme A for condensation reactions, can be visualized fairly well in terms of organic reaction mechanisms. We lack information, however, about the roles, in biological reactions, of the adenine, the ribose, the phosphate, or the pantothenic acid portions of this complex molecule.

[1] L. Jaenicke and F. Lynen, *in* "The Enzymes" (P. D. Boyer, H. Lardy, and K. Myrbäck, eds.), Vol. 3, Part B, p. 3. Academic Press, New York, 1960.
[2] F. Lynen, *Angew. Chem.* **77**, 929 (1965).

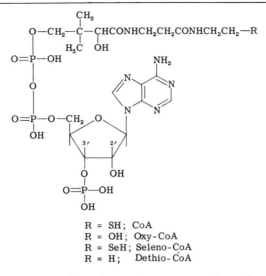

R = SH; CoA
R = OH; Oxy-CoA
R = SeH; Seleno-CoA
R = H;   Dethio-CoA

In recent years several analogs of coenzyme A and of some of its component building blocks have been synthesized. Presumably, antimetabolites of this type should be capable of providing information about the relative importance and the roles of the various parts of coenzyme A in the biological reactions in which it participates.

This review will be concerned exclusively with analogs involving the replacement of the sulfur of pantetheine, 4'-phosphopantetheine, and coenzyme A with other atoms. The synthesis of analogs, e.g., "guanocoenzyme A," in which the adenine of coenzyme A was replaced by guanine,[3] will be discussed elsewhere.

Analogs of Pantetheine and 4'-Phosphopantetheine

Pantetheine represents a major portion of coenzyme A. It is required for the growth of certain microorganisms such as *Lactobacillus helveticus*[4] and will replace coenzyme A, at least partially, in certain systems such as the ATP-dependent acetylation of 4-aminoazobenzene by an extract of pigeon liver acetone powder.[5]

4'-Phosphopantetheine can fully replace coenzyme A in some enzymatic preparations, the pigeon liver system being an example. This compound promotes the growth of *Acetobacter suboxydans*. New interest was centered on phosphopantetheine recently when it was shown that this compound is

[3] M. Shimizu, O. Nagase, S. Okada, Y. Abiko, and T. Suzuki, *Chem. Pharm. Bull.* **14**, 683 (1966).

[4] W. L. Williams, *J. Biol. Chem.* **177**, 933 (1949).

[5] O. Brenner-Holzach and F. Leuthardt, *Helv. Chim. Acta* **39**, 1796 (1956).

the prosthetic group of the acyl carrier protein carrying substrates, in thiolester linkage, during fatty acid synthesis in *Escherichia coli*.[6,7] Similarly, 4'-phosphopantetheine was found to be the prosthetic group of fatty acid synthetase in yeast[8] and in pigeon liver.[9]

The syntheses of pantetheine analogs are based on the various syntheses of pantethine.[10] These involve the condensation of activated derivatives of pantothenic acid with isologs or derivatives of cysteamine or cystamine or the condensation of aletheine derivatives with D-pantolactone.[11]

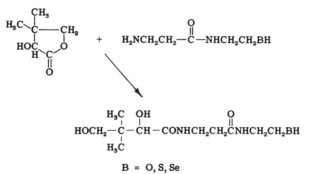

—COOR' = ester, azide, mixed anhydride
—B = O, S, Se

The Synthesis of Pantethine Analogs Using Analogs of Cysteamine

B = O, S, Se

The Synthesis of Pantethine Analogs Using Analogs of Aletheine

Recently a new synthesis of pantethine was introduced, the initial step of which involves the condensation of D-pantolactone with 3-aminopropio-

[6] P. W. Majerus, A. W. Alberts, and P. R. Vagelos, *Proc. Natl. Acad. Sci. U.S.* **53**, 410 (1965).

[7] E. L. Pugh and S. J. Wakil, *J. Biol. Chem.* **240**, 4727 (1965).

[8] W. W. Wells, J. Schultz, and F. Lynen, *Biochem. Z.* **346**, 474 (1967).

[9] C. J. Chesterton, P. H. Butterworth, A. S. Abramovitz, E. J. Jacob, and J. W. Porter, *Arch. Biochem. Biophys.* **124**, 386 (1968).

[10] E. E. Snell and G. M. Brown, *Advan. Enzymol.* **14**, 49 (1953); J. Baddiley, *ibid.* **16**, 1 (1955); E. E. Snell and E. L. Wittle, see Vol. III [133].

[11] C. J. Stewart, V. H. Cheldelin, and T. E. King, *J. Biol. Chem.* **215**, 319 (1955).

nitrile.[12] Treatment of D-pantothenonitrile with cysteamine produces the thiazoline analog of pantetheine, hydrolysis of which, in weak acid, yields pantetheine.

This approach has been utilized for the synthesis of α-methyl coenzyme A, β-methyl coenzyme A,[13] homocoenzyme A,[13a] and other coenzyme A analogs.[13a]

Syntheses of 4′-phosphopantethine analogs, like the synthesis of the parent compound,[14] involve the phosphorylation of a suitably blocked pantetheine analog, dibenzylphosphorochloridate being the preferred phosphorylating agent.[15] Pantothenonitrile has been found to be a useful intermediate for the synthesis of 4′-phosphopantethine[16] and will undoubtedly be useful for the synthesis of phosphopantethine analogs.

Syntheses Utilizing Aletheine Analogs

The following synthesis[17] is cited as an example of the condensation of D-pantolactone with an analog of an aletheine derivative. The reaction yields the benzyloxy derivative of oxypantetheine; this compound is then phosphorylated with dibenzylphosphorochloridate.

Benzyloxyaletheine Hydrochloride. N-(2-Benzyloxyethyl)-3-phthalimidopropionamide (28.2 g) is synthesized by the reaction of 2-benzyloxyethylamine hydrochloride (18.8 g) with phthalimidopropionyl chloride (23.8 g); this reaction is carried out in 200 ml of ice-cold N,N-dimethylformamide in the presence of 16.6 g of pyridine. The product (28.2 g) is dissolved in 200 ml of methanol. The solution is treated with 2.7 ml of 95% hydrazine. The mixture is refluxed for 1.5 hours with constant stirring. On cooling to room temperature, the mixture is filtered, and the filtrate is evaporated under vacuum. The residue is dissolved in absolute ethanol. Insoluble material is removed and discarded; the filtrate is evaporated to dryness. The residue is dried by two more evaporations with absolute ethanol, dissolved in 50 ml of absolute ethanol, and precipitated with ether. The hygroscopic product is collected by centrifugation and dried in a vacuum desiccator. Benzyloxyaletheine hydrochloride (15.8 g, 77% yield) is obtained as a white powder.

The product is purified by recrystallization from 100 ml of anhydrous

[12] M. Shimizu, G. Ohta, O. Nagase, S. Okada, and Y. Hosokawa, *Chem. Pharm. Bull. (Tokyo)* **13**, 180 (1965).

[13] M. Shimizu, O. Nagase, Y. Hosokawa, H. Tagawa, Y. Abiko, and T. Suzuki, *Chem. Pharm. Bull. (Tokyo)* **14**, 681 (1966).

[13a] M. Shimizu, O. Nagase, Y. Hosokawa, and H. Tagawa, *Tetrahedron* **24**, 5241 (1968).

[14] J. Baddiley and E. M. Thain, *J. Chem. Soc.* p. 1610 (1953).

[15] J. G. Moffatt and H. G. Khorana, *J. Am. Chem. Soc.* **83**, 633 (1961).

[16] O. Nagase, *Chem. Pharm. Bull. (Tokyo)* **15**, 648 (1967).

[17] T. L. Miller, G. L. Rowley, and C. J. Stewart, *J. Am. Chem. Soc.* **88**, 2299 (1966).

isopropanol. The white crystals are collected in the absence of moisture and washed with anhydrous diethyl ether. A yield of 12.1 g is obtained, m.p. 125–127°. An additional 2.8 g of crude product can be recovered by adding more ether to the filtrate.

D-*Benzyloxypantetheine.* To a solution of 5.2 g of benzyloxyaletheine hydrochloride in 60 ml of absolute ethanol, 17.4 ml of 1.15 N methanolic sodium methoxide is added dropwise with swirling. The mixture is allowed to stand for 20 minutes at room temperature, it is then filtered and the filtrate is evaporated under vacuum to yield an oily residue. After two evaporations with absolute ethanol, dry benzyloxyaletheine is obtained. Ethanol-insoluble material is discarded. The oil is treated with 2.6 g of D-pantoyllactone in the presence of one drop of sodium methoxide. The mixture is heated to 55° for 1 hour and to 65° for 12 hours. The contents are dissolved in 75 ml of water and passed through a Dowex 1-X4 (OH form) column. The effluent is concentrated to a volume of 75 ml and passed through an Amberlite IR-120 (H⁺ form) column. The effluent is concentrated under vacuum and evaporated twice with absolute ethanol. A yield of 6.0 g of D-benzyloxypantetheine (85%) is obtained as a colorless, viscous liquid.

The parent compound of the above product, oxypantetheine, has also been synthesized[11,18]; it has been reported to be an active, competitive inhibitor of pantetheine in the *Lactobacillus helveticus* system.[11]

*4'-Phospho-*D-*oxypantetheine.* D-Benzyloxypantetheine (2.9 g) is evaporated twice with anhydrous pyridine, dissolved in 50 ml of anhydrous pyridine, and frozen in an acetone–dry-ice bath. A solution of dibenzyl-phosphorochloridate[15] is prepared by treating a solution of 3.3 g of N-chloro-succinimide in 50 ml of warm, dry benzene with the dropwise addition of 6.5 g of dibenzyl phosphite. After standing at room temperature for 2 hours, the phosphorochloridate solution is decanted into the frozen pyridine solution; the mixture is thawed, quickly refrozen, and then permitted to stand for 20 hours at −18°. After addition of 28 ml of water and mixing, the solution is left to stand for 20 minutes at room temperature and then evaporated to dryness under vacuum at a temperature below 35°. The residue is shaken with a mixture of 25 ml of ethyl acetate and 25 ml of 2 N sulfuric acid. The organic layer is extracted three times each with 2 N sulfuric acid (25 ml), 10% sodium bicarbonate (25 ml), and saturated sodium sulfate (25 ml). The ethyl acetate layer is dried over sodium sulfate, filtered, and evaporated to dryness under vacuum. The residual syrup is dissolved in a mixture of 40 ml of isopropanol and 10 ml of water and treated with 2 g of Adams catalyst. Hydrogenolysis is permitted to proceed at room

[18] J. Baddiley and A. P. Mathias, *J. Chem. Soc.* p. 2803 (1954).

temperature and atmospheric pressure until no more hydrogen is taken up (6.5 hours). The catalyst is removed by centrifugation, washed with the isopropanol–water mixture, and recentrifuged. The supernatant solution is combined with the wash and evaporated to dryness at room temperature. The residual pale yellow syrup is dissolved in 50 ml of water, and the pH is adjusted to 7.5 with 1 N barium hydroxide. After centrifugation at 10,000 rpm, the clear supernatant solution is evaporated to dryness under vacuum to yield a clear glass. The product is dissolved in 15 ml of methanol, and the slightly turbid solution is centrifuged at 18,000 rpm. To the clear supernatant solution, 200 ml of acetone is added. The white powdery product is collected by centrifugation and dried under vacuum. A yield of 2.34 g (59%) of the crude barium salt of 4'-phospho-D-oxypantetheine is obtained.

For purification, 0.315 g of the crude salt is dissolved in 10 ml of water and applied to a 2.7 × 50 cm DEAE-cellulose (Cl⁻) column. The column is washed with water, then eluted with a linear gradient using 2 liters of 0.5 N lithium chloride in the reservoir and 2 liters of water in the mixing vessel. The peptide-containing peak can be located either by a positive biuret reaction[19] or by the application of phosphate reagent[20] sprayed on aliquots of the eluate spotted on paper. The peptide- and phosphate-containing fractions are pooled, and the solvent is evaporated off. The white residue is treated repeatedly with 40 ml of a 1:15 methanol–acetone mixture until a negative chloride test is obtained. On drying of the precipitate, a yield of 0.18 g of the dilithium salt of 4'-phospho-D-oxypantetheine is obtained. The white powder exhibits a specific rotation of $[\alpha]_D^{26} + 16.7°$.

It has recently been shown that 4'-phospho-D-oxypantetheine can be converted to oxycoenzyme A[21] by the beef liver coenzyme A synthesizing system of Hoagland and Novelli.[22]

Syntheses Utilizing Pantothenic Acid Derivatives

The following synthesis is presented as an example of the condensation of an activated ester of pantothenic acid with an analog of cysteamine. The preparation of selenopantethine[23] is based on the amide synthesis of Boissonas.[24] The condensation of the carbonic ester anhydride of pantothenic acid[25] with a cysteamine analog can also be applied to the synthesis

[19] E. Layne, see Vol. III, p. 450.
[20] R. S. Bandurski and B. Axelrod, *J. Biol. Chem.* **193**, 405 (1951).
[21] C. J. Stewart and W. J. Ball, *Biochemistry* **5**, 3883 (1966).
[22] M. B. Hoagland and G. D. Novelli, *J. Biol. Chem.* **207**, 757 (1954).
[23] W. H. H. Günther and H. G. Mautner, *J. Am. Chem. Soc.* **82**, 2762 (1960).
[24] R. A. Boissonas, *Helv. Chim. Acta* **34**, 874 (1951).
[25] R. Schwyzer, *Helv. Chim. Acta* **35**, 1903 (1952).

of the sulfide analog of pantethine and to the synthesis of N-pantothenoyl-1,2-dithio-5-azepane, a bifunctional analog of pantetheine.[23]

Selenopantethine. To a solution of 10.0 g of calcium pantothenate in 25 ml of water is added 10 ml of triethylamine. Calcium is precipitated by the addition of 1.9 g of oxalic acid in 25 ml of water. The precipitate is centrifuged and washed; the combined supernatant layer and washings are evaporated under vacuum. The residual oil is dried by azeotropic distillation under vacuum with ethanol and benzene and then dried to constant weight in a vacuum desiccator. The oily triethylamine salt (12.5 g) is dissolved in 50 ml of dry dimethylformamide and cooled to $-5°$ in a flask fitted with stirrer, dropping funnel, and drying tube. A solution of 4.1 g of ethyl chloroformate in 25 ml of ethyl acetate is added dropwise. Stirring and cooling of the resulting solution of mixed ester anhydride is continued for 30 minutes before 4.9 g of selenocystamine[22,26] in 25 ml of ethyl acetate and 10 ml of triethylamine are added through the dropping funnel.

The reaction mixture is stirred for 1 hour at room temperature and filtered; the filtrate is evaporated to dryness under vacuum at a temperature below 50°. The light brown oily residue is dissolved in 20 ml of methanol followed by the addition of acetone until a slight cloudiness persists (ca. 110 ml). The solution is passed through a column of activated alumina ("Woelm," neutral, Brockmann grade III) and eluted with a 1.5:8.5 methanol–acetone mixture. The first 150 ml of eluate is discarded, and the following fraction of 250 ml is evaporated to dryness under vacuum. On drying, 4.0 g (35%) of essentially pure selenopantethine is obtained as a highly hygroscopic, yellow glass.

This compound can also be obtained by the condensation of benzylselenoalletheine with benzyloxypantolactone, followed by debenzylation with sodium in liquid ammonia.[27] While selenopantethine is a hygroscopic oil, it can be converted readily to Se-benzoylselenopantethine, a crystalline compound.[27] Like other selenolesters, the selenobenzoyl derivative is relatively stable to hydrolysis but extremely susceptible to aminolysis,[28] permitting ready recovery of selenopantethine from its ester.

Selenopantethine is fully functional in the *Lactobacillus helveticus* system[4] and will replace pantethine on a mole for mole basis.[29] The ultraviolet spectrum of selenobenzoylselenopantethine, with its typical peaks at 242 and 286 mμ, provides another method of assaying selenopantethine.

4′-Phospho-D-selenopantethine.[27] This compound can be prepared in 20% overall yield by the phosphorylation of D-benzylselenopantethine[27] with

[26] V. Coblentz, *Ber.* **24**, 2131 (1891).
[27] W. H. H. Günther and H. G. Mautner, *J. Am. Chem. Soc.* **87**, 2708 (1965).
[28] S. H. Chu and H. G. Mautner, *J. Org. Chem.* **31**, 308 (1966).
[29] H. G. Mautner and W. H. H. Günther, *Biochim. Biophys. Acta* **36**, 561 (1959).

dibenzyl phosphorochloridate, in the usual fashion,[15] followed by debenzylation with sodium in liquid ammonia. Attempts to phosphorylate D-selenopantethine directly were unsuccessful. The barium salt of 4'-phosphoselenopantethine is a pale yellow powder exhibiting a specific rotation of $[\alpha]_D^{29}$ 8.4° ($c = 2$, H_2O).

The acetylselenol ester of 4'-phosphoselenopantethine can be prepared by the reduction of the diselenide with hypophosphorous acid,[30] followed by acetylation with acetic anhydride in the presence of potassium bicarbonate.[31] Se-Malonyl-4'-phosphoselenopantetheine can be prepared by the reaction of the selenol with S-malonyl-N-caproylcysteamine.[32] The equilibrium of this reaction can be shifted to prepare a selenolester from the thiolester by having the aqueous reaction mixture extracted with ether in a liquid–liquid extractor.[31] N-Caproylcystamine is extracted into the ether with the desired selenolester remaining in the aqueous layer.

Analogs of Coenzyme A and Dephosphocoenzyme A

Several syntheses of coenzyme A have been developed.[15,33–35] The first three methods involve the condensation of pantetheine derivatives with derivatives of adenosine or its phosphates. The syntheses of selenocoenzyme A and of oxycoenzyme A have been based on the procedure of Moffatt and Khorana,[15] which gives comparatively good yields. Condensation of analogs of 4'-phosphopantetheine with adenosine 5'-phosphoromorpholidate in anhydrous pyridine results in the formation of analogs of dephosphocoenzyme A, while the condensation of isologs of phosphopantetheine with 4-morpholine N,N'-dicyclohexylcarboxamidinium adenosine-2',3'-cyclic phosphate 5'-phosphoromorpholidate results in the formation of a mixture of the corresponding isolog of coenzyme A and of isocoenzyme A (the 2'-phospho-3'-hydroxy isomer of coenzyme A), following the mild acid hydrolysis of the 2',3'-cyclic phosphate of the condensation product. Thus, the synthesis of isologs of coenzyme A is reduced to the synthesis of analogs of 4'-phosphopantetheine or 4'-phosphopantethine analogs followed by condensation with the suitable adenosine derivative. In the case of "guano-coenzyme A," the adenine of the coenzyme was replaced by guanine.[3]

The recent synthesis of coenzyme A[35] using the condensation of D-pantothenonitrile with adenosine 2',3'-cyclic phosphate 5'-phosphoromor-

[30] W. H. H. Günther, *J. Org. Chem.* **31**, 1202 (1966).

[31] H. G. Mautner, unpublished data.

[32] H. Eggerer and F. Lynen, *Biochem. Z.* **335**, 540 (1962).

[33] A. M. Michelson, *Biochim. Biophys. Acta* **50**, 605 (1961); **93**, 71 (1964).

[34] W. Gruber and F. Lynen, *Ann.* **659**, 1 (1962).

[35] M. Shimizu, O. Nagase, S. Okada, Y. Hosokawa, H. Tagawa, Y. Abiko, and T. Suzuki, *Chem. Pharm. Bull. (Tokyo)* **15**, 655 (1967).

pholidate, avoids the problem of the separation of coenzyme A and isocoenzyme A by the use of ribonuclease T$_2$,[36] an approach originally introduced
in the coenzyme A synthesis of Michelson.[33] The resulting 3'-phosphoadeno-
sine 5'-D-pantothenonitrile-4'-pyrophosphate is then condensed with
cysteamine to yield the corresponding thiazoline derivative, acid hydrolysis
of which yields coenzyme A. It seems likely that replacement of cysteamine
with 2-hydroxyethylamine or with selenocysteamine should yield the
oxazoline and selenazoline analogs of oxycoenzyme A and of selenocoenzyme
A, respectively. For labeling purposes this approach is likely to be superior
to the syntheses used in the past.

As already noted, oxycoenzyme A has been prepared from 4'-phospho-
oxypantetheine using a beef liver enzyme system.[21] It has not proved
possible to convert phosphoselenopantethine to selenocoenzyme A enzymatically, perhaps because of the ready oxidizability of the selenium compound.

An interesting analog of coenzyme A, dethiocoenzyme A, has been
prepared by the desulfurization of coenzyme A with Raney nickel,[37] however, this compound does not appear to have been isolated in analytically
pure form by means of this approach. Recently, 4'-phosphodethiopante-
theine was synthesized by the condensation of D-pantolactone with N-
(carbobenzoxy-β-alanyl)aminoethane, followed by phosphorylation. Use
of the CoA-synthesizing enzyme system of Hoagland and Novelli[22] yielded
dethiocoenzyme A.[37a]

An alkylating derivative of coenzyme A, bromoacetylcoenzyme A, has
been reported recently. This compound is claimed to be a potent inhibitor
of carnitine acetyltransferase, β-ketothiolase, and choline acetylase.[38]

The synthesis of selenocoenzyme A is presented as a representative
example of the synthesis of a coenzyme A analog. The main problem that
presented itself was the separation of selenocoenzyme A from isoselenocoenzyme A. The ECTEOLA-cellulose column used by Moffatt and
Khorana[15] to separate coenzyme A from its 2'-phospho isomer becomes
incapable of resolving these compounds when they are in the disulfide
rather than in the thiol form. Because of the very great susceptibility of
selenols to oxidation to diselenides, it proved necessary to block the
selenium atom during the separation. Since the benzoyl esters of selenocoenzyme A and isoselenocoenzyme A, like other selenolesters,[28] proved fairly

[36] M. Naoi-Tada, K. Sato-Asano, and F. Egami, *J. Biochem. (Tokyo)* **46,** 757 (1959);
T. Uchida and F. Egami, *Progr. Ribonucleic Acid Res.* **3,** 59 (1964).
[37] J. F. A. Chase, B. Middleton, and P. K. Tubbs, *Biochem. Biophys. Res. Commun.*
23, 208 (1966).
[37a] C. J. Stewart, J. A. Thomas, W. J. Ball, and A. R. Aguirre, *J. Am. Chem. Soc.* **90,**
5000 (1968).
[38] J. F. A. Chase and P. K. Tubbs, *Biochem. J.* **100,** 47P (1966).

stable to hydrolysis but highly susceptible to aminolysis, they could be separated on the ECTEOLA column, while deacylation of the esters with n-butylamine yielded the resolved isomers.

Selenocoenzyme A.[27] A solution of 0.81 g of the barium salt of 4'-phosphoselenopantethine in 20 ml of pyridine is added to a dry solution of 0.44 g of bis(4-morpholine-N,N'-dicyclohexylcarboxamidinium) adenosine 2',3'-cyclic phosphate 5'-phosphoromorpholidate in 10 ml of pyridine. The mixture is dried by two evaporations with anhydrous pyridine. The residue is dissolved in 20 ml of pyridine and left to stand overnight. The solvent is evaporated off; several evaporations with water follow. The residue is taken up in 20 ml of 0.1 N HCl, kept at room temperature for 1 hour, evaporated to dryness again, and dissolved in 100 ml of water. The solution is applied to a DEAE (Cl⁻) column (25 × 400 mm), which is washed with 0.003 N HCl until the effluent has a pH of 3 and no more material absorbing at 257 mμ can be eluted. Linear gradient elution, with 4 liters of 0.003 N HCl in the mixing flask and 4 liters of 0.4 M LiCl in 0.003 N HCl in the reservoir flask, is used to separate the products. The drop rate is maintained at 1.5 ml/min; ultraviolet density of the effluent is monitored continuously at 257 mμ, or at 285 mμ if the optical density at the adenine peak becomes too great. The isomeric mixture of selenocoenzyme A and isoselenocoenzyme A is eluted as the last fraction of the chromatogram at 0.2 M LiCl concentration. After neutralization with lithium hydroxide, the combined fractions are freeze-dried. The product is washed with a 1:10 methanol–acetone mixture until the supernatant fraction is chloride free. A yield of 0.084 g of the isomeric mixture is obtained in the form of a white powder.

A 0.11-g sample of the selenocoenzyme–isoselenocoenzyme A lithium salt mixture is dissolved in 50 ml of water and treated with 0.3 g of 2-dimethylaminoethylselenolbenzoate,[39] a very potent acylating agent, and 0.2 g of sodium bicarbonate. Saturated sodium carbonate is added until the solution is slightly turbid (pH 8.5). The reaction is allowed to proceed for 1 hour at room temperature; sodium carbonate is added dropwise to maintain turbidity. The solution is then acidified (pH 2) by the addition of Amberlite IR-120 resin. The resin is removed by filtration and washed several times with water. The combined filtrates and washings are diluted to 350 ml and applied to an ECTEOLA-cellulose (Cl⁻) column (15 × 600 mm) which has been packed with a 200-cm liquid head and washed in turn with 1 N LiCl in 0.003 N HCl and with 0.003 N HCl. Application to the column of 1 liter of 0.05 N LiCl in 0.003 N HCl results in the elution of benzoic acid. This is followed by linear gradient elution with 2 liters of 0.05 N LiCl in 0.003 N HCl in the mixing flask and 2 liters of 0.09 N LiCl

[39] W. H. H. Günther and H. G. Mautner, *J. Med. Chem.* **7**, 229 (1964).

in 0.003 N HCl in the reservoir flask. The effluent is monitored continuously at 250 mμ. Benzoylselenocoenzyme A and benzoylisoselenocoenzyme A are obtained as two clearly separated fractions. Further elution with 0.25 N LiCl yields a crop of nonacylated selenocoenzyme A–isomer mixture. Total recovery is 94%.

Aminolysis of both benzoyl isomer fractions is carried out with n-butyl amine. The identity of selenocoenzyme A and of isoselenocoenzyme A is assigned on the basis of incubation at 37° with crude rattlesnake (*Crotalus adamanteus*) venom (1 mg/ml) in 0.2 N Tris buffer, 0.004 N in magnesium ion at a pH of 9.0. The first fraction yields adenosine 2′,5′-diphosphate; the second fraction yields adenosine 3′,5′-diphosphate exclusively. These are identified by comparison on paper chromatograms with authentic material.[15]

The various selenium compounds related to coenzyme A can be separated and identified by means of thin-layer chromatography. Slurries prepared from cellulose powder (MN 300, Macherey, Nagel and Co., Dueren, Switzerland) and distilled water are applied to glass plates with a variable thickness applicator set to 0.35 mm and run in the usual fashion. The table shows the comparative R_f's of the various sulfur and selenium isologs.

Compounds containing sulfur and adenine are detected by a nitroprusside spray and by observation under ultraviolet light, respectively. Selenium compounds are detected by a spray reagent (freshly prepared) containing soluble starch (1 g), sodium bicarbonate (1 g), and 0.1 N iodine solution (4 ml) in 100 ml of water. Application of this spray shows selenols, diselenides, monoselenides, but not selenoacyl esters, as white spots against a blue background. This method is capable of detecting as little as 10 millimicromoles of selenium. To differentiate sulfur from selenium compounds, the dried plates are sprayed with dilute nonoxidizing acids, such as phosphoric acid. Then only the white spots due to the presence of selenium compounds return to the blue color of the background, owing to the reduction of selenoxides and seleninic acids by hydrogen iodide. This method can be adapted to the quantitative determination of selenium compounds.[27]

Dethiocoenzyme A.[37] Raney nickel W-2 is prepared from nickel-aluminum alloy by the method of Mozingo[40]; it is stored under ethanol at 4°. Before use it is washed with distilled water. To 0.056 g of coenzyme A, 0.5 g of the catalyst is added (1 ml of slurry), in the presence of 4 ml of 0.1 M ammonium acetate (pH 5.1) and 0.5 ml of saturated disodium EDTA. The mixture is shaken continuously at room temperature for 65 minutes and then centrifuged. The pale blue supernatant layer is frozen overnight. The

[40] R. Mozingo, *Org. Syntheses, Coll. Vol.* **3,** 181 (1955).

COMPARATIVE R_f'S OF VARIOUS SULFUR AND SELENIUM ISOLOGS

A. Solvent: 1-Butanol–acetic acid–water, 5:2:1

	Sulfur compound R_f		Selenium compound R_f	
Pantetheine	SH	0.63	SeH	0.61
Pantethine	S_2	0.72	Se_2	0.86
4′-Phosphopantetheine	SH	0.67	SeH	0.72
4′-Phosphopantethine	S_2	0.50	Se_2	0.50

B. Solvent: Ethanol–0.5 N ammonium acetate, 3:2

	Sulfur compound R_f			Selenium compound R_f		
		pH 4	pH 8		pH 4	pH 8
Dephosphocoenzyme A	SH	0.48	—	SeH	—	—
	S_2	0.23	0.34	Se_2	0.25	0.36
Coenzyme A	SH	0.36	0.29	SeH	(0.34^a)	—
(or isocoenzyme A)	S_2	0.10	0.06	Se_2	0.10	0.06
Benzoylcoenzyme A		0.63^b	—		0.59^b	—
(or isobenzoyl CoA)						

[a] The value quoted for selenocoenzyme A(SeH) is based on the spot produced when Se-benzoylselenocoenzyme A (or Se-benzoylisoselenocoenzyme A) was chromatographed in the ammonium acetate solvent. Benzoylcoenzyme A was also subject to aminolysis in this solvent and yielded a single spot of coenzyme A(SH).

[b] The compounds were chromatographed in a solvent containing sodium acetate instead of ammonium acetate; even so there was some hydrolysis of the selenol ester resulting in a trail.

solution is applied to a DEAE-cellulose column equilibrated with 0.003 N HCl and eluted with a KCl gradient. The mixing vessel contains initially 750 ml of 0.003 N HCl, and the reservoir 0.2 M KCl in 0.003 N HCl. Most of the desulfocoenzyme A is obtained in a well-defined fraction.

The product is identified by means of ultraviolet spectroscopy, phosphate analysis, and amino acid analysis, which indicate the presence of adenine, β-alanine, ethylamine, and phosphate in the expected proportions. Unfortunately, no attempt has been made to carry out an elemental analysis of the product, the yield of which, based on spectroscopic measurements, was 50%. As already noted, this compound was obtained recently by a different method.[37a]

Methods of Assaying Coenzyme A Analogs and Derivatives

Numerous methods exist for the determination of coenzyme A; these differ widely in their specificity, however, ranging from methods requiring coenzyme A in the thiol form to methods in which even pantetheine is fully active. Several summaries of these methods have been published.[1,41,42]

No attempt will be made here to discuss in any detail the literature dealing with the enzymatic actions of coenzyme A analogs. Selenocoenzyme A is a "partial antagonist" of coenzyme A in acetocoenzyme A synthetase.[43] These isologs exhibit the same K_m for this enzyme, whereas the V_{max} for the seleno compound is only one-third that of its thio analog; however, in the presence of the seleno compound the catalytic activity of the thio isolog is reduced. Oxycoenzyme A and dethiocoenzyme A are reported to be competitive antagonists of coenzyme A in a variety of enzyme systems.[17,37] The observation that analogs of coenzyme A, in which the mercapto group has been replaced either by a hydrogen or by a hydroxy group, are potent antagonists of coenzyme A, implies that the sulfur of coenzyme A not only carries acyl groups, but also affects the ability of the coenzyme molecule to be bound to the respective apoenzymes.

[41] G. Michal and H. U. Bergmeyer, in "Methoden der enzymatishen Analyse" (H. U. Bergmeyer, ed.), pp. 512–528. Verlag Chemie, Weinheim/Bergstr., 1962.
[42] G. Michal and H. U. Bergmeyer, Biochim. Biophys. Acta 67, 599 (1963).
[43] W. H. H. Günther and H. G. Mautner, unpublished data.

[57] Pantothenic Acid and Coenzyme A: Phosphopantothenoylcysteine Synthetase from Rat Liver[1,2] (Pantothenate 4′-Phosphate: L-Cysteine Ligase, EC 6.3.2.5)

By Yasushi Abiko

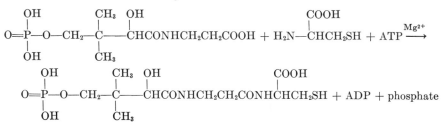

Assay Method

Principle. The method is based on the measurement of decrease in the amount of the substrate, 4′-phosphopantothenic acid, which is determined microbiologically with *Lactobacillus arabinosus* 17-5 after phosphatase digestion of a sample. *L. arabinosus* 17-5 requires pantothenic acid for growth,

[1] Y. Abiko, J. Biochem. (Tokyo) 61, 290 (1967).
[2] Y. Abiko, M. Tomikawa, and M. Shimizu, J. Biochem. (Tokyo) 64, 115 (1968).

but 4'-phosphopantothenic acid,[3-5] pantothenoyl-L-cysteine,[1] and its phosphate[1] do not support the growth of this organism.

Reagents

Phosphopantothenate-ATP mixture, $1.1 \times 10^{-4}\ M$ potassium 4'-phosphopantothenate,[3,6] $2 \times 10^{-3}\ M$ ATP and $2 \times 10^{-3}\ M$ MgCl$_2$ in $0.02\ M$ Tris-HCl buffer, pH 7.5

L-Cysteine, $10^{-3}\ M$, freshly prepared

Intestinal alkaline phosphatase, approximately 16 units[7]/ml of $0.1\ M$ bicarbonate buffer, pH 8.0

Procedure. Fifty microliters of the phosphopantothenate–ATP mixture and 50 microliters of cysteine are mixed, and the reaction is started by adding 0.1 or 0.2 ml of an enzyme solution to the mixture. After incubation at 37° for 15–30 minutes, the reaction mixture is heated in a boiling water bath for 1 minute, diluted with water to a volume of 5 ml, and centrifuged. Two milliliters of the supernatant solution is incubated with 0.2 ml of phosphatase at 37° for 1 hour and then heated in a boiling water bath for 3 minutes. The phosphatase-treated solution is assayed for pantothenic acid with *L. arabinosus* 17-5,[8] and the amount of pantothenic acid not recovered by the phosphatase digestion is calculated from the comparison with a value before the reaction.

Definition of Unit and Specific Activity. One enzyme unit is defined as the amount of the enzyme that catalyzes the formation of 1 micromole of pantothenate not recovered by phosphatase digestion per minute. Specific activity is in the units of activity per milligram of protein.

Purification Procedure

All operations are carried out in the cold.

Step 1. Extraction. Fresh livers from rats are homogenized with 3 volumes of $0.02\ M$ phosphate buffer (pH 7.2). The homogenate is centrifuged at 25,600 g for 30 minutes. The supernatant is again centrifuged at 105,000 g for 1 hour. The resultant supernatant solution is used as the starting material (the crude extract).

[3] T. E. King and F. M. Strong, *J. Biol. Chem.* **191**, 515 (1951).

[4] W. S. Pierpoint, D. E. Hughes, J. Baddiley, and A. P. Mathias, *Biochem. J.* **61**, 368 (1955).

[5] Y. Abiko and M. Shimizu, *Chem. Pharm. Bull.* (*Tokyo*) **15**, 884 (1967).

[6] Vol. III [134]; see also J. Baddiley and E. M. Thain, *J. Chem. Soc.* p. 246 (1951); S. Okada, O. Nagase, and M. Shimizu, *Chem. Pharm. Bull.* (*Tokyo*) **15**, 713 (1967).

[7] Units are in micromoles of *p*-nitrophenol liberated from *p*-nitrophenylphosphate per minute at 37° and pH 10.5.

[8] H. R. Skeggs and L. D. Wright, *J. Biol. Chem.* **156**, 21 (1944).

Step 2. Protamine Treatment. To the crude extract, 0.065 volume of 2% protamine sulfate solution is added dropwise with stirring, and the precipitate formed is removed by centrifugation. Total units of the synthetase activity largely increases after the treatment of this step. The reason for this excess yield of activity is unknown.

Step 3. Ammonium Sulfate Fractionation. The supernatant solution of step 2 is brought to 40% saturation with solid ammonium sulfate at pH 7. The precipitate is collected by centrifugation, dissolved in 0.01 M phosphate buffer (pH 7.0), and dialyzed overnight against the same buffer. The dialyzed solution is centrifuged to remove precipitate formed during dialysis.

Step 4. Calcium Phosphate Gel Treatment. About one-third volume of calcium phosphate gel[9] is added dropwise with stirring to the supernatant solution of step 3, and the gel is removed by centrifugation. The synthetase activity is not adsorbed on the gel, but is recovered in the supernatant solution in good yield. The preparation is free from pantothenate kinase (EC 2.7.1.33) or pantetheine kinase (EC 2.7.1.34).[10]

Step 5. CM-Sephadex C-50 Chromatography. The supernatant solution of step 4 is adjusted to pH 6.3 with 1 N HCl and centrifuged to remove precipitate. The resultant supernatant solution is applied on a CM-Sephadex C-50 column, equilibrated with 0.01 M phosphate buffer (pH 6.3); the column is washed with the same buffer, then the synthetase activity is eluted with a linear gradient of NaCl $(0 \rightarrow 0.2\ M)$ in 0.01 M phosphate buffer (pH 7.5). Seven grams of CM-Sephadex C-50 is sufficient for the chromatography of about 1 g of the calcium phosphate gel-treated enzyme. Active fractions are combined and rechromatographed on CM-Sephadex C-50 after dialysis against 0.01 M phosphate buffer (pH 6.3). The synthetase preparation thus obtained is free from 4'-phosphopantothenoyl-L-cysteine decarboxylase, phosphatase, and inorganic pyrophosphatase activities.

Step 6. Sephadex G-75 Gel Filtration. The active fractions obtained from the rechromatography of step 5 are combined, dialyzed against 0.02 M Tris-HCl buffer (pH 7.5), and concentrated to a small volume by the use of a "Collodion bag."[11] The concentrated enzyme solution is then subjected

---

[9] Preparation of calcium phosphate gel: 500 ml of 0.5 M CaCl$_2$ is added to 500 ml of 0.33 M Na$_3$PO$_4$ with stirring. The mixture is washed with water by decantation or centrifugation. The washed precipitate is resuspended in water and stored at 2–4°; see also T. P. Singer and E. B. Kearney, *Arch. Biochem. Biophys.* **29**, 190 (1950).

[10] Pantetheine kinase is identical with pantothenate kinase [see Abiko[1]; G. B. Ward, G. M. Brown, and E. E. Snell, *J. Biol. Chem.* **213**, 869 (1955); G. M. Brown and J. J. Reynold, *Ann. Rev. Biochem.* **32**, 419 (1963)].

[11] A product of Membrane Filter Co., West Germany.

to Sephadex G-75 gel filtration employing the 0.02 M Tris-HCl (pH 7.5) buffer. The synthetase activity is eluted at the V_e/V_o value[12] of about 1.33. The active fractions are combined and concentrated with the Collodion bag. The final product is free from adenylate kinase (EC 2.7.4.3).

The purification of the synthetase is summarized in the table.

PURIFICATION OF PHOSPHOPANTOTHENOYL-L-CYSTEINE SYNTHETASE FROM RAT LIVER

Step	Volume (ml)	Protein (mg/ml)	Specific activity, $\times 10^3$	Total units, $\times 10^3$
1. Crude extract[a]	140	39.2	0.014	76.8
2. Protamine treatment	144	28.0	0.054	216.0
3. $(NH_4)_2SO_4$ fractionation	21.6	50.0	0.147	159.0
4. Calcium phosphate gel	23.5	42.0	0.164	164.0
5. CM-Sephadex C-50 Chromatography	96.0	0.9	0.385	33.3
CM-Sephadex C-50 Rechromatography	27.0	0.44	1.018	11.9
6. Sephadex G-75 Gel filtration	3.7	1.34	1.64	8.1

[a] Sixty grams of livers from 5 rats was used for the purification of the enzyme.

Properties

Substrate Specificity. This enzyme catalyzes the condensation of 4′-phosphopantothenate with L-cysteine, but not of pantothenate with L-cysteine. Pantothenate does not affect the rate of the condensation reaction. β-Mercaptoethylamine, its disulfide and α-methylcysteine, all can be substrates of the synthetase.[13]

Nucleotide Requirement. ATP is required for the synthesis of 4′-phosphopantothenoyl-L-cysteine by this enzyme. ATP plays a role in amide formation that is analogous to the synthesis of glutamine[14] or glutathione.[15] A study of the stoichiometry for the synthetase reaction has shown the formation of equimolar amounts of ADP and phosphate accompanied by a decrease in the amount of 4′-phosphopantothenate. Phosphopantothenoyl-L-cysteine synthetase from *Proteus morganii* requires CTP in place of ATP.[13]

Activators and Inhibitors. The synthetase requires Mg^{2+} for action, and it is inhibited completely by EDTA (10^{-3} M) and phosphate (0.08 M) and by about 50% by KCl and NaCl (0.05 M).

Other Properties. The molecular weight of the synthetase is about

---

[12] Relative elution volume where V_e is elution volume for the synthetase activity; V_o, the void volume of the Sephadex column.

[13] G. M. Brown, *J. Biol. Chem.* **234,** 370 (1959).

[14] Vol. II [44].

[15] Vol. II [45].

37,000.[16] The optimal pH of the reaction is about 7.5. The K_m value of the synthetase is 7.1 to 8.3 $\times$ 10^{-5} M for 4'-phosphopantothenate at pH 7.5.

[16] Measured by the Sephadex gel filtration technique [J. R. Whitaker, *Anal. Chem.* **35,** 1950 (1963); P. Andrews, *Biochem. J.* **91,** 222 (1964)].

[58] Pantothenic Acid and Coenzyme A: Phosphopantothenoylcysteine Decarboxylase from Rat Liver[1]
[4'-Phospho-*N*-(D-pantothenoyl)-L-cysteine Carboxy-lyase, EC 4.1.1.36]

By YASUSHI ABIKO

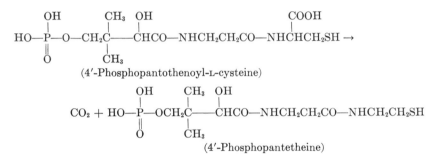

(4'-Phosphopantothenoyl-L-cysteine)

(4'-Phosphopantetheine)

Assay Method

Principle. The method is based on the microbiological measurement of pantetheine which is liberated by phosphatase treatment of 4'-phosphopantetheine, one of the products of the decarboxylase reaction.[2]

Reagents

> 4'-Phosphopantothenoyl-L-cysteine,[3] 1 $\times$ 10^{-4} M in Tris-HCl buffer
> Tris-HCl buffer, 0.1 M, pH 8.0
> L-Cysteine, 0.04 M, freshly prepared
> Intestinal alkaline phosphatase, approximately 16 units[4] per milliliter of 0.1 M bicarbonate buffer, pH 8.0

[1] Y. Abiko, *J. Biochem. (Tokyo)* **61,** 300 (1967).
[2] G. M. Brown, *J. Biol. Chem.* **234,** 370 (1959).
[3] J. Baddiley and A. P. Mathias, *J. Chem. Soc.* p. 2803 (1954). O. Nagase, *Chem. Pharm. Bull. (Tokyo)* **15,** 648 (1967).
[4] Units are micromoles of *p*-nitrophenol liberated from *p*-nitrophenylphosphate per minute at 37° and pH 10.5.

Procedure. 4'-Phosphopantothenoyl-L-cysteine (0.2 ml) and cysteine (0.05 ml) are mixed, and the reaction is started by adding 0.25 ml of an enzyme solution to the mixture. After incubation at 37° for 20 minutes, the reaction mixture is heated in a boiling water bath for 1 minute to stop the reaction, diluted twice with water and centrifuged, if necessary, to obtain a clear supernatant solution. To 0.2 ml of the supernatant solution is added 0.2 ml of intestinal alkaline phosphatase, and the mixture is incubated at 37° for 1 hour. After heating in a boiling water bath for 1 minute, the phosphatase-treated solution is microbiologically assayed for pantetheine.[1,5]

Definition of Unit and Specific Activity. One enzyme unit is defined as the amount of the enzyme which catalyzes the formation of 1 micromole of pantetheine per minute under the above conditions. The specific activity is represented as the units of activity per milligram of protein.

Purification Procedure

All operations are carried out in the cold.

Step 1. Extraction. Fresh rat livers are homogenized with 3 volumes of 0.01 M phosphate buffer (pH 7) and centrifuged at 21,000 g for 30 minutes. The supernatant solution is again centrifuged at 105,000 g for 1 hour. The soluble fraction is used as the starting material (the crude extract, specific activity, 0.24–0.26 $\times$ 10^{-3}).

Step 2. Ammonium Sulfate Fractionation. The crude extract is brought to 40% saturation with solid ammonium sulfate, and the pH is adjusted to 7 with diluted ammonium hydroxide. The precipitate is collected by centrifugation, dissolved in 0.01 M phosphate buffer (pH 7), and dialyzed against 0.01 M phosphate buffer (pH 6.3). The precipitate formed during dialysis is removed by centrifugation. The treatment of this step somewhat lowers the specific activity of the enzyme, although no activity is found in the fractions which precipitate at higher concentrations of ammonium sulfate (specific activity, 0.18–0.21 $\times$ 10^{-3}; recovery, 30–35%).

Step 3. Calcium Phosphate Gel Treatment. One-third volume of calcium phosphate gel[6] (sedimented fraction of the gel suspension at 2000 rpm for 3 minutes) is added dropwise with stirring to the clear supernatant solution of step 2, and the mixture is centrifuged at 21,000 g for 15 minutes. The decarboxylase activity is not adsorbed on the gel, but is recovered in the supernatant fraction without any loss of activity. Reproducible results are

---

[5] E. E. Snell and E. L. Wittle; Vol. III [133].

[6] Preparation of calcium phosphate gel: 500 ml of 0.5 M CaCl$_2$ is added to 500 ml of 0.33 M Na$_3$PO$_4$ with stirring. The mixture is neutralized with 1 N HCl, and the precipitate is washed with water by decantation or centrifugation until a AgNO$_3$ test of the supernatant is negative. The washed precipitate is resuspended in water and stored at 2–4°; see also T. P. Singer and E. B. Kearney, *Arch. Biochem.* **29**, 190 (1950).

obtained when the protein content of the enzyme solution is about 50 mg/ml, and 6 ml of thick gel is added to the enzyme solution per gram of protein (specific activity, 0.39–0.52 × 10^{-3}; overall recovery, 36–42%).

Step 4. DEAE-Cellulose Chromatography after Aging. The gel-treated enzyme solution of step 3 is allowed to stand at 3° for 3 weeks and then chromatographed on a DEAE-cellulose column which had been previously equilibrated with 0.01 M phosphate buffer (pH 6.2–6.3). One gram of DEAE-cellulose (0.84 meq/g; Brown Co.) is sufficient for the chromatography of 300 mg of protein of the gel-treated enzyme. After washing the loaded column with 0.01 M phosphate buffer containing 0.05 M NaCl (pH 6.2–6.3), adsorbed proteins are eluted with 0.01 M phosphate buffer containing 0.1 M NaCl (pH 6.2–6.3). The decarboxylase is associated with the second protein fraction (specific activity, 25 to 30 × 10^{-3}; overall recovery, 11–15%). Aging of the enzyme prior to chromatography is an important process for success in the chromatography. If the enzyme is chromatographed immediately after the treatment with calcium phosphate gel, the activity is eluted together with a large amount of inactive protein which can be removed by chromatography after aging.

The purification of the decarboxylase is summarized in Table I. The contamination with phosphatase activity is also shown in the table, because its elimination from the decarboxylase preparation is desirable for kinetic studies on the enzyme with 4′-phosphopantothenoyl-L-cysteine as the substrate. The final product is free from the phosphatase activity.

Properties

Substrate Specificity. This enzyme decarboxylates 4′-phosphopanto-thenoyl-L-cysteine, but not pantothenoyl-L-cysteine. Pantothenoyl-L-cysteine does not affect the rate of decarboxylation of 4′-phospho-pantothenoyl-L-cysteine. The substrate is active in the SH-form. Dephospho-α-carboxy-CoA[7] and α-carboxy-CoA[7] are decarboxylated by the enzyme, although the rate of the decarboxylation of these CoA analogs is lower than that of 4′-phosphopantothenoyl-L-cysteine.[8]

Stoichiometry of the Reaction. The formation of CO_2 during the decarboxylase reaction is manometrically observable with the concomitant formation of an equimolar amount of 4′-phosphopantetheine.

Activators and Inhibitors. The purified decarboxylase is activated by cysteine or reduced glutathione. Mg^{2+} has no effect. The reaction product, 4′-phosphopantetheine, inhibits the enzyme strongly and competitively with a K_i value of 4.3 × 10^{-4} M at pH 8.0. ATP markedly inhibits the

[7] This volume [55]; M. Shimizu, O. Nagase, Y. Hosokawa, and H. Tagawa, *Tetrahedron* **24**, 5241 (1968).

[8] Y. Abiko, T. Suzuki, and M. Shimizu, *J. Biochem. (Tokyo)* **61**, 309 (1967).

TABLE I

PURIFICATION OF PHOSPHOPANTOTHENOYL-L-CYSTEINE DECARBOXYLASE FROM RAT LIVER

Step	Volume (ml)	Protein (mg/ml)	Decarboxylase			Phosphatase		
			Specific activity, $\times 10^3$	Total units, $\times 10^3$	Recovery (%)	Specific activity,[a] $\times 10^3$	Total units,[b] $\times 10^3$	Recovery (%)
1. Crude extract	100	41.8	0.267	1115	100	7.10	29700	100
2. (NH₄)₂SO₄ fraction	37	54.8	0.181	368	32.9	5.95	12090	40.7
3. Gel treatment	42.8	21.2	0.521	473	42.4	2.45	2295	7.5
4. DEAE-cellulose	46	0.12	30.0	165.6	14.8	0	0	0

[a] The specific activity of phosphatase is presented as units of activity per milligram of protein.

[b] One phosphatase unit is defined as the amount of the enzyme that catalyzes the liberation of 1 micromole of p-nitrophenol from p-nitrophenylphosphate per minute at pH 8.0 and 37°.

reaction, whereas ADP and AMP have no effect. UTP is also inhibitory. Pyridoxal phosphate and pyridoxal cause a considerable inhibition, which is reversed by equimolar addition of semicarbazide. Rates of inhibition by these compounds are shown in Table II.

TABLE II
INHIBITION BY VARIOUS COMPOUNDS OF
4'-PHOSPHOPANTOTHENOYL-L-CYSTEINE DECARBOXYLASE

Compound, 10^{-3} M	Inhibition rate (%)
ATP	39.1
ADP	3.2
AMP	1.0
UTP	21.8
Pyridoxal phosphate	51.3
Pyridoxal	18.8

Kinetic Properties. The optimal pH of the decarboxylase reaction is about 8. The K_m for 4'-phosphopantothenoyl-L-cysteine is 1.33–1.50 × 10^{-4} M at pH 8.0.

[59] Pantothenic Acid and Coenzyme A:
Dephospho-CoA Pyrophosphorylase and Dephospho-CoA Kinase as a Possible Bifunctional Enzyme Complex[1] (ATP: Pantetheine-4'-phosphate Adenyltransferase, EC 2.7.7.3 and ATP: Dephospho-CoA 3'-phosphotransferase, EC 2.7.1.24)

By YASUSHI ABIKO

These enzymes catalyze the conversion of 4'-phosphopantetheine to CoA via dephospho-CoA which is the final step in CoA biosynthesis.[2]

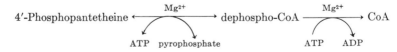

Dephospho-CoA pyrophosphorylase and dephospho-CoA kinase of rat liver cannot be separated from each other by the procedures described here,

[1] T. Suzuki, Y. Abiko, and M. Shimizu, *J. Biochem.* (*Tokyo*) **62,** 642 (1967).
[2] M. B. Hoagland and G. D. Novelli, *J. Biol. Chem.* **207,** 767 (1954); see also Vol. II [110] and [115].

and studies on the properties of these enzymes suggest the existence of a bifunctional enzyme complex composed of these two enzymes.

Assay Method

Principle. The assay method for dephospho-CoA pyrophosphorylase is based on the measurement of inorganic pyrophosphate liberated from ATP. The crude enzyme is assayed by determining CoA formed from 4'-phosphopantetheine and excess ATP.[2] The assay method for dephospho-CoA kinase is based on the measurement of CoA formed from dephospho-CoA and ATP.[2]

Reagents

    4'-Phosphopantetheine,[3,4] $2 \times 10^{-3}\ M$

    3'-Dephospho-CoA,[4,5] $2 \times 10^{-3}\ M$

    ATP–MgCl₂ mixture, $0.02\ M$ ATP and $0.01\ M$ MgCl₂ in Tris-HCl buffer for assay of the pyrophosphorylase; $0.01\ M$ ATP and $0.01\ M$ MgCl₂ in Tris-HCl buffer for assay of the kinase

    Tris-HCl buffer, $0.25\ M$, pH 8.0

    L-Cysteine, $0.05\ M$, freshly prepared

Procedure 1. Dephospho-CoA Pyrophosphorylase. One-tenth milliliter of 4'-phosphopantetheine, 0.1 ml of ATP-MgCl₂, 0.1 ml of cysteine, and 0.2 ml of an enzyme solution are mixed, and the mixture is incubated at 37° for 20–40 minutes. The reaction is stopped by heating in a boiling water bath for 1 minute, and the mixture is centrifuged at 650 *g* for 5 minutes. Two-tenths milliliter of the supernatant solution is assayed for CoA by the phosphotransacetylase method.[6] With purified preparations, the activity is assayed by measuring inorganic pyrophosphate liberated during the reaction.[7]

Procedure 2. Dephospho-CoA Kinase. Conditions and reactants are the same as for Procedure 1 except that dephospho-CoA is used with the appropriate ATP-MgCl₂ solution. The mixture is heated in a boiling water

---

[3] J. Baddiley and E. M. Thain, *J. Chem. Soc.* p. 1610 (1953); see also Vol. III [134]; O. Nagase, *Chem. Pharm. Bull. (Tokyo)* **15**, 648 (1967).

[4] J. G. Moffat and H. G. Khorana, *J. Am. Chem. Soc.* **81**, 1265 (1959); **83**, 663 (1961).

[5] M. Shimizu, O. Nagase, S. Okada, Y. Hosokawa, H. Tagawa, Y. Abiko, and T. Suzuki, *Chem. Pharm. Bull. (Tokyo)* **15**, 655 (1967).

[6] This volume [53].

[7] It is advisable to determine inorganic pyrophosphate by measuring inorganic phosphate liberated after digestion of a sample with yeast inorganic pyrophosphatase (EC 3.6.1.1), (Vol. II [91]). Microdetermination of inorganic phosphate in the presence of organic phosphates can be satisfactorily performed by the method of T. Yanagita, *J. Biochem. (Tokyo)* **55**, 260 (1964).

bath for 1 minute, then centrifuged. Two-tenths milliliter of the supernatant solution is assayed for CoA formed during the reaction.

Definition of Unit and Specific Activity. One unit of dephospho-CoA pyrophosphorylase is defined as the amount of the enzyme that catalyzes the liberation of 1 micromole of inorganic pyrophosphate per minute under the above conditions. For the crude enzyme preparations, a unit of the pyrophosphorylase is defined as the amount of the enzyme required to synthesize 1 micromole of CoA per minute under the above conditions. One unit of dephospho-CoA kinase is defined also as the amount of the enzyme which catalyzes the formation of 1 micromole of CoA per minute under the above conditions. The specific activity is represented as units of activity per milligram of protein.

Purification Procedure

All operations are performed in a cold room.

Step 1. Extraction. Fresh livers from 20 rats are homogenized with 3 volumes of 0.1 M KCl containing 0.02 M KHCO₃, and the homogenate is centrifuged at 3000 g for 30 minutes. The resultant supernatant solution is further centrifuged at 59,000 g for 2 hours. This supernatant solution is used as the starting material (the crude extract).

Step 2. Protamine Sulfate Treatment. To 870 ml of the crude extract is added, dropwise, 0.06 volume of 2% protamine sulfate solution. The precipitate is removed by centrifugation at 3000 g for 30 minutes.

Step 3. Ammonium Sulfate Fractionation. The supernatant solution of step 2 (900 ml) is adjusted to pH 8.0 with 1 M Tris solution and brought to 25% saturation with solid ammonium sulfate (171 g). The precipitate formed is removed by centrifugation at 3000 g for 30 minutes, and the supernatant solution is brought to 40% saturation by the addition of 97 g of ammonium sulfate. The precipitate is collected by centrifugation and dissolved in distilled water to a volume of 115 ml.

Step 4. Calcium Phosphate Gel Treatment. An equal volume of calcium phosphate gel[8] is added with stirring to the enzyme solution of step 3. The mixture is centrifuged at 15,000 g for 30 minutes to remove the gel.

Step 5. CM-Cellulose Chromatography. The supernatant solution of step 4 (170 ml) is divided into two parts, each of which is chromatographed on CM-cellulose. A column of CM-cellulose (2.6 × 19 cm) is prepared and washed with 2 liters of 0.01 M phosphate buffer (pH 6.4) through which nitrogen has been bubbled for several hours before use. The solution of step 4 (85 ml) is dialyzed for 16 hours against 6 liters of the N₂-bubbled

---

[8] See footnote 6 of article [58] in this volume.

phosphate buffer and centrifuged at 15,000 g for 20 minutes. The supernatant solution is passed through the column, and the column is washed with the same buffer until the E_{280} of the effluent becomes less than 0.4. The adsorbed protein is then eluted with N_2-bubbled 0.01 M phosphate buffer (pH 6.4) containing 0.05 M KCl at a flow rate of 30 ml per hour. Fractions of 7 ml each are collected. Protein is eluted in 2 peaks in this system. The first protein material appears in the effluent at the bed volume of the column, and the second at 3–4 times the bed volume. The latter fraction which contains the pyrophosphorylase and the kinase is pooled and dialyzed against 2 liters of N_2-bubbled 0.01 M Tris-HCl buffer (pH 8.8). The bubbling of N_2 through the buffers used is effective for protection of the enzymes from inactivation, and the N_2-bubbling cannot be replaced by the addition of mercaptoethanol or EDTA.

Step 6. DEAE-Cellulose Chromatography. The pooled effluent of step 5 (258 ml) is chromatographed on a DEAE-cellulose column (1.8 × 20 cm) equilibrated with N_2-bubbled 0.01 M Tris-HCl buffer (pH 8.8). The column is washed with the same buffer, then a linear gradient elution is performed with 500 ml of 0.01 M Tris-HCl buffer (pH 8.8) in a mixing vessel and 500 ml of 0.01 M Tris-HCl buffer containing 0.2 M KCl in a second container (pH 8.8). Nitrogen is bubbled into the mixing vessel during the chromatography. Fractions of 5 ml each are collected at a flow rate of 10 ml per hour. The pyrophosphorylase and the kinase are eluted at between 0.06 M and 0.08 M of KCl, completely superimposing each other. The active fractions are combined (105 ml) and lyophilized.

Step 7. Sephadex G-200 Gel Filtration. The lyophilized preparation of step 6 is dissolved in a small volume of distilled water (5 ml), and the solution is dialyzed against 1 liter of N_2-bubbled 0.01 M Tris-HCl buffer containing 0.05 M $(NH_4)_2SO_4$ (pH 8.8). The dialyzed solution is passed through a column of Sephadex G-200 gel (1.7 × 57 cm) equilibrated with the $(NH_4)_2SO_4$-containing Tris buffer, and protein is eluted with the same buffer at a flow rate of about 4 ml per hour. The pyrophosphorylase and the kinase are associated with the second protein fraction with a K_D value of about 0.36. These two activities are completely superimposed in the effluent of this gel filtration. The active fractions are combined (10 ml).

The results of the purification of the enzymes are summarized in Table I.

Properties

Substrate Specificity.[9] The pyrophosphorylase and the kinase are specific for 4′-phosphopantetheine and 3′-dephospho-CoA, respectively, as the sub-

---

[9] Y. Abiko, T. Suzuki, and M. Shimizu, *J. Biochem. (Tokyo)* **61,** 309 (1967).

TABLE I

PURIFICATION OF DEPHOSPHO-CoA PYROPHOSPHORYLASE AND DEPHOSPHO-CoA KINASE FROM RAT LIVER

Step	Volume (ml)	Protein (mg/ml)	Dephospho-CoA kinase			Dephospho-CoA pyrophosphorylase					Ratio of specific activities, kinase:pyrophosphorylase	
			Specific activity $\times 10^3$	Total units	Recovery (%)	Specific activity		Total units			From CoA	From PP_i
						From CoA,[a] $\times 10^3$	From PP,[b] $\times 10^3$	From CoA	From PP_i	Recovery (%)		
1. Crude extract	870	36.0	1.32	41.5	100	0.68	—	21.3	—	100	1.95	—
2. Protamine treatment	900	27.0	1.43	34.7	83.7	0.78	—	18.9	—	88.9	1.85	—
3. $(NH_4)_2SO_4$ fraction	115	88.0	2.82	28.5	68.7	1.55	—	15.6	—	73.5	1.82	—
4. Gel treatment	170	33.0	4.03	22.8	54.9	2.15	—	11.9	—	55.8	1.88	—
5. CM-Cellulose	258	1.05	29.5	8.0	19.3	14.5	27.8	3.92	7.59	18.4	2.03	1.06
6. DEAE-Cellulose	105	0.105	136	1.49	3.6	62.5	128	0.69	1.42	3.2	2.17	1.06
7. Sephadex G-200 gel	10	0.13	346	0.45	1.1	168	300	0.22	0.39	1.0	2.06	1.15

[a] The pyrophosphorylase activity was assayed by measuring the formation of CoA.
[b] The pyrophosphorylase activity was assayed by measuring the liberation of pyrophosphate (PP_i).

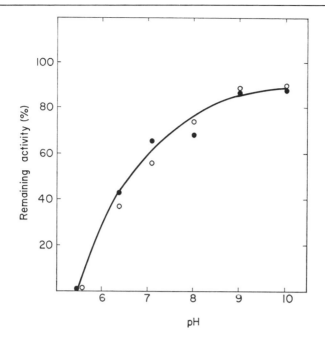

FIG. 1. pH-Stability curve of dephospho-CoA pyrophosphorylase and dephospho-CoA kinase. The purified enzyme preparation (180 µg/ml) was heated at 46° in the following buffers for 2 minutes: pH 5.5–6.4, 0.1 M acetate; pH 7.1–9.0, 0.1 M Tris-HCl; pH 9.1–10.0, 0.1 M glycine-KOH. After heating, the remaining activities of dephospho-CoA pyrophosphorylase (●) and dephospho-CoA kinase (○) were assayed.

strates. 4'-Phosphopantothenoyl-L-cysteine and 3'-dephospho-α-carboxy-CoA[10] cannot be the substrates of the pyrophosphorylase and the kinase, respectively.

Activators. The pyrophosphorylase and the kinase are activated by Mg^{2+}. The optimal concentration of Mg^{2+} is $5 \times 10^{-4} M$ for both the enzymes. These enzymes require cysteine for their actions.

Stability. The pyrophosphorylase and the kinase are partially inactivated to the same extent by heat treatment above 40° for 2 minutes, and completely lose activity after being heated at 50° for 2 minutes at pH 8.0. These enzymes show the same pH-stability curve (Fig. 1).

Kinetic Properties. Table II shows pH optima and K_m values for the pyrophosphorylase and the kinase.

Nature as a Bifunctional Enzyme Complex. The pyrophosphorylase and the kinase are not separated by the above purification procedures, and the

[10] This volume [55]; M. Shimizu, O. Nagase, Y. Hosokawa, and H. Tagawa, *Tetrahedron* **24,** 5241 (1968).

TABLE II

pH Optima and K_m Values of Dephospho-CoA Pyrophosphorylase and Dephospho-CoA Kinase

	Pyrophosphorylase	Kinase
Optimal pH	8–10	10
$K_m{}^a$ for		
ATP	$1.0 \times 10^{-3}\,M$	$3.6 \times 10^{-4}\,M$
4′-Phosphopantetheine	$1.4 \times 10^{-4}\,M$	—
3′-Dephospho-CoA	—	$1.2 \times 10^{-4}\,M$

^{a} Determined at pH 8.0.

ratio of specific activities is constant throughout the purification steps (Table I). These two activities are eluted as a single peak in the chromatographies on CM-cellulose, DEAE-cellulose, and Sephadex G-200 gel. These two enzymes show parallel sensitivities to heat treatment at various temperatures and pH values, and to magnesium activation. Deoxycholate (0.01%) shows different effects on these two enzymes: stimulation of the pyrophosphorylase and inhibition of the kinase, although these enzymes are completely inactivated by deoxycholate at higher concentrations (0.2%). These facts suggest that these two enzymes behave as a bifunctional enzyme complex.

[60] Preparation of Pantothenate-^{14}C Labeled Pigeon Liver Fatty Acid Synthetase: Release of the 4′-Phosphopantetheine ^{14}C-Labeled Prosthetic Group

By P. H. W. Butterworth, C. J. Chesterton, and John W. Porter

Present experimental evidence[1-5] suggests that 4′-phosphopantetheine functions biologically both as a precursor in the biosynthesis of coenzyme A and as a prosthetic group in fatty acid-synthesizing systems. In the latter

[1] P. R. Vagelos, P. W. Majerus, A. W. Alberts, A. R. Larrabee, and G. P. Ailhaud, *Federation Proc.* **25,** 1485 (1966).

[2] R. D. Simoni, R. S. Criddle, and P. K. Stumpf, *J. Biol. Chem.* **242,** 573 (1967).

[3] W. W. Wells, J. Schultz, and F. Lynen, *Biochem. Z.* **346,** 474 (1967).

[4] A. R. Larrabee, E. G. McDaniel, H. A. Bakerman, and P. R. Vagelos, *Proc. Natl. Acad. Sci. U.S.* **54,** 267 (1965).

[5] C. J. Chesterton, P. H. W. Butterworth, A. S. Abramovitz, E. J. Jacob, and J. W. Porter, *Arch. Biochem. Biophys.* **124,** 386 (1968).

role it has been shown to be present in the acyl carrier proteins of bacteria[1] and plants,[2] and in the multienzyme fatty acid–synthetase complexes of yeast,[3] rat liver,[6] rat adipose tissue,[4] and pigeon liver.[5] Possibly 4'-phosphopantetheine acts in fatty acid synthesis by providing a long, flexible molecular arm on which the substrate is mounted via a thioester linkage.[7] By this means the substrate, held in the required orientation, could be transferred sequentially through a series of enzyme-active sites without being subject to free diffusion.

As a part of our studies on the pigeon liver fatty acid synthetase, we prepared enzyme labeled with ^{14}C in the 4'-phosphopantetheine group.[5] To secure these preparations, a recent finding was utilized. Starvation of pigeons has been shown to result in a decrease in the absolute level of hepatic fatty acid synthetase.[8] Rapid adjustment to the normal level occurs on refeeding the starved birds. Also, it was found in earlier work that maximal incorporation of pantothenate in adult pigeons occurs during the period of rapid resynthesis of liver synthetase. Hence, pigeons were starved for 3 days, refed, and then injected twice (at 6 and 12 hours after the start of refeeding) with pantothenate-1-^{14}C. The fatty acid synthetase complex was then purified from livers of birds killed 30–35 hours after refeeding was initiated.

In Vivo Synthesis of Enzyme

Materials. Pigeons, obtained locally, are fed with a mixture of grains. Sodium D-pantothenate-1-^{14}C, purchased from New England Nuclear Corporation, is made up in 0.9% saline (400–450 µg/ml).

Procedure.[5] Pigeons are maintained on water alone for 3 days and then refed in the early morning daylight hours (a pigeon's appetite seems to be keener at this time of day). Subsequently, two injections of ^{14}C-labeled pantothenate are administered, one 6 hours, and the other 12 hours, after the start of refeeding. Each injection, containing 40–45 µg (5.45 mCi/millimole) of pantothenate-1-^{14}C in 0.1 ml of 0.9% saline, is made via a wing vein. The birds are killed by decapitation, and the livers are excised between 30 and 35 hours after refeeding. Sixteen adult pigeons is a convenient number for this procedure, and they will yield 150–200 mg of purified fatty acid synthetase (100–150 g of liver) with a specific radioactivity of approximately 1000 dpm/mg of protein.

Comments. In early experiments a single injection was given 6 hours after the initiation of refeeding. However, with this procedure the rate of incorporation of pantothenate into a liver supernatant protein-bound form

[6] D. N. Burton, A. G. Haavik, and J. W. Porter, *Arch. Biochem. Biophys.* **126,** 141 (1968).

[7] F. Lynen, *Biochem. J.* **102,** 380 (1967).

[8] P. H. W. Butterworth, R. B. Guchhait, H. Baum, E. B. Olson, S. A. Margolis, and J. W. Porter, *Arch. Biochem. Biophys.* **116,** 453 (1966).

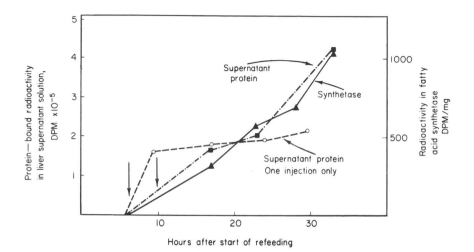

Fig. 1. Time study of the incorporation *in vivo* of pantothenate-1-¹⁴C into pigeon liver supernatant protein and pigeon liver fatty acid synthetase. Four pigeons were taken for each time point and treated as described in the text. The times are given as hours after the initiation of refeeding. The times after refeeding at which pantothenate-1-¹⁴C was injected were 6 and 12 hours. The symbols are as follows: ○ --- ○, only the 6-hour injection administered, incorporation into liver supernatant protein; ■ ··· ■, both injections, incorporation into liver supernatant protein; ▲ —— ▲, both injections, incorporation into purified fatty acid synthetase.

leveled off 10 hours after refeeding (Fig. 1). Later it was found that a second injection at 12 hours stimulated the rate of incorporation approximately 4-fold. This increase is reflected in the specific radioactivity of the fatty acid synthetase purified from the liver supernatant solution (Fig. 1). Longer periods of synthesis of the fatty acid synthetase *in vivo* were not examined. It may be, however, that higher specific radioactivities of the enzyme could be obtained.

Purification of Synthetase

Reagents

Homogenization buffer, pH 8.0, containing 70 ml of 1 M KHCO₃, 85 ml of 1 M K₂HPO₄, and 9 ml of 1 M KH₂PO₄ per liter.

Potassium phosphate buffers: (1) containing 1 mM EDTA and 1 mM 2-mercaptoethanol (added just before use): 0.005 M, pH 7.0; 0.04 M, pH 7.0; 0.25 M, pH 7.02; (2) containing 1 mM EDTA and 1 mM dithiothreitol: 0.2 M, pH 6.8; A volume of 250 ml of each buffer will suffice for a single preparation of fatty acid synthetase from 16 birds and 80 ml of liver supernatant solution, except for the 0.04 M phosphate buffer. One liter of this latter buffer is required.

Ammonium sulfate solution (saturated) containing 3 mM EDTA and 1 mM 2-mercaptoethanol (added just before use). This solution is adjusted to pH 7.0 with NH$_4$OH.

DEAE-cellulose, obtained from Mann Research Laboratories, is stirred in thick suspension with 1 M phosphate buffer, pH 7.0, for 2 days and then washed with water until the eluate contains no phosphate. It is stored at 4° in the 0.04 M phosphate buffer.

Calcium phosphate gel prepared as described in Vol. I (p. 98).

G-100 Sephadex is obtained from the Pharmacia Company.

Procedure.[8,9] The livers are washed, blotted dry, weighed, and homogenized in buffer (1.5 ml per gram of liver) in a Potter-Elvehjem all-glass homogenizer. When larger preparations are used, livers are homogenized for 30 seconds at maximum speed in a Waring blendor. The homogenate is filtered through 1 layer of cheesecloth and then centrifuged for 30 minutes at 3,000 g. The supernatant solution is filtered through 2 layers of cheesecloth and centrifuged again for 45 minutes at 100,000 g to remove microsomes. This preparation of soluble liver proteins may be stored for several months at $-15°$ under N$_2$ without loss of activity. Sixteen adult pigeons normally yield 200 ml of supernatant solution; the solution is stored in 40-ml batches.

The following purification steps are carried out at room temperature except where stated otherwise. Eighty milliliters of supernatant solution is used, and the purification procedure is completed in 8 hours.

Saturated ammonium sulfate is added dropwise under N$_2$ to the thawed supernatant solution, with stirring, to 25% saturation. After 15 minutes, the precipitated material is removed by centrifugation for 10 minutes at 18,000 g. This material is discarded, and the supernatant solution is brought to 40% saturation. The precipitated protein is again removed from solution by centrifugation. This is dissolved in 40 ml of 0.005 M buffer after the excess ammonium sulfate is wiped from the tube with a tissue. This fraction contains most of the original synthetase activity; it is stable for 1–2 hours.

The protein concentration is determined by the biuret assay.[10] Then the solution is diluted to 300 ml with the 0.005 M buffer in preparation for the calcium phosphate gel treatment. A solution containing a weight of gel equal to one-half that of the total protein is added and mixed well. The

[9] R. Y. Hsu, G. Wasson, and J. W. Porter, *J. Biol. Chem.* **240,** 3736 (1965).

[10] A. G. Gornall, C. J. Bardawill, and M. M. David, *J. Biol. Chem.* **177,** 751 (1949).

suspension is immediately centrifuged for 3 minutes at 4000 g, and the gel precipitate is washed with 25 ml of 0.04 M buffer. Since the enzyme is unstable at low ionic strength, this and the next step must be carried out rapidly. The first supernatant solution is divided into two equal portions, and each is loaded onto separate 10 × 3.5 cm DEAE-cellulose columns previously equilibrated with 0.04 M buffer. The second supernatant solution is then used to wash the first solution onto the columns. Contaminating protein is eluted with 0.04 M buffer until the optical density of the eluate at 280 mμ is below 0.05. Fatty acid synthetase is then eluted with 0.25 M buffer. The enzyme fraction, which bears a slight yellow color, is collected in the minimal volume possible (usually 20 ml). The enzyme is stable for several hours in this buffer.

The eluate protein is fractionated with ammonium sulfate as before. The 26–32% ammonium sulfate precipitate is saved. After wiping away excess ammonium sulfate solution, the protein is dissolved over a 60-minute period in 1.0 ml of 0.2 M buffer at 0°. The enzyme solution is then desalted by passage through a G-100 Sephadex column (22 × 1.0 cm, bed volume 18 ml) previously equilibrated with the 0.2 M buffer. The enzyme is collected in the minimal volume of eluate (3–5 ml). This preparation is stable for several days at 4° but is normally used immediately.

Comments. This procedure consistently yields a pure protein as characterized by ultracentrifugation, moving boundary electrophoresis, starch gel electrophoresis, and ion-exchange chromatography.[9]

Assay of Synthetase

Reagents

Acetyl-CoA-^{14}C, synthesized by the method of Simon and Shemin[11]
Malonyl-CoA (nonradioactive), synthesized by the method of Trams and Brady[12]
NADPH, obtained from Sigma Chemical Company

Procedure.[9] Protein concentration is measured by the biuret method.[10] Enzyme solutions are diluted where necessary with 0.2 M phosphate buffer, pH 7.0, containing 1 mM EDTA and 1 mM 2-mercaptoethanol.

Fatty acid-synthesizing activity is measured as follows. Phosphate buffer, pH 7.0, 100 micromoles; 2-mercaptoethanol, 5 micromoles; EDTA, pH 7.0, 3 micromoles; malonyl-CoA, 39 millimicromoles; acetyl-CoA-1-^{14}C, 12 millimicromoles, and 25,000 cpm; NADPH, 0.3 micromole; enzyme; and water are added to a ground glass-stoppered extraction tube (1.5 ×

[11] E. J. Simon and D. Shemin, *J. Am. Chem. Soc.* **75,** 2520 (1953).
[12] E. G. Trams and R. O. Brady, *J. Am. Chem. Soc.* **82,** 2972 (1960).

15 cm) to give a final volume of 1.0 ml. Enzyme is added to start the reaction, and the mixture is incubated for 6 minutes at 38°. One drop of 60% perchloric acid is added to stop the reaction, followed by 1 ml of absolute alcohol. ^{14}C-labeled fatty acids are extracted 3 times with petroleum ether to give a combined extract of 10 ml. A 1-ml portion of the extract is mixed with 15 ml of toluene scintillator solution (3.0 g of 2,5-diphenyloxazole and 0.3 g of 1,4-di[2-(5-phenyloxazole)]benzene per liter of toluene) and then assayed in a liquid scintillation spectrometer. The micromoles of acetyl-CoA converted to fatty acids per minute per milligram of protein are then calculated.

Specific radioactivity of the enzyme (dpm per milligram of protein) is determined as follows: an aliquot (0.1 ml) of the relevant fraction is diluted to 0.5 ml, and the protein is precipitated with one drop of perchloric acid. The precipitate is spun down in a bench centrifuge, washed 3 times with 3.0-ml portions of 0.2 N acetic acid to remove nonprotein radioactivity and then digested in 0.5 ml of 1 N KOH at 37° for 15 minutes. Aliquots (0.1 ml) of this solution are dissolved in 15 ml of Bray's scintillator solvent.[13] Assays for radioactivity are made as reported previously.

Comments. Assays of the fractions obtained at each step of the purification procedure for protein, fatty acid synthetase activity, and acid-stable, protein-bound radioactivity demonstrated that these activities purified together (see table). This result indicates that all the supernatant protein-bound radioactivity derived from pantothenate-1-^{14}C is covalently bound to the fatty acid synthetase by an acid-stable linkage.

Release of 4′-Phosphopantetheine Covalently Bound to Protein

4′-Phosphopantetheine is thought to be bound to the fatty acid synthetase protein via a phosphodiester linkage between the phosphate moiety and the hydroxyl of a serine residue in the polypeptide chain.[1] Such a linkage can be split by mild alkaline hydrolysis.

Reagents

4′-Phosphopantetheine synthesized as reported by Moffatt and Khorana[14]

Coenzyme A, supplied by P-L Biochemicals.

Procedure.[5] The following procedure is slightly modified from that given by Larrabee *et al.*[4] for the release of 4′-phosphopantetheine from a partially purified rat adipose tissue fatty acid synthetase. Approximately 2 ml (30–40 mg of protein) of the purified ^{14}C-labeled synthetase preparation

[13] G. A. Bray, *Anal. Biochem.* **1,** 279 (1960).
[14] J. G. Moffatt and H. G. Khorana, *J. Am. Chem. Soc.* **83,** 663 (1961).

COPURIFICATION OF THE FATTY ACID SYNTHETASE AND THE PROTEIN-BOUND ^{14}C-LABELED MOIETY

Fraction	Protein (mg)	Fatty acid synthetase specific activity[a]	Specific radioactivity[b]	Synthetase purification factor	Radioactivity purification factor	Ratio[c]
Liver supernatant[d]	1850	5.5	48.6	1.0	1.0	8.8
25–40% $(NH_4)_2SO_4$	575	14.0	170	2.5	3.5	12.1
$Ca_3(PO_4)_2$ gel supernatant	339	21.5	248	3.9	5.1	11.5
DEAE-cellulose eluate	84	58.2	569	10.6	11.7	9.8
Sephadex eluate	46	90.3	705	16.4	14.5	7.8

[a] Specific activity as millimicromoles of acetyl-CoA incorporated into fatty acids per minute per milligram of protein.
[b] Specific radioactivity as dpm per milligram of protein.
[c] Ratio = specific radioactivity/specific activity of fatty acid synthetase.
[d] Supernatant solution (47 ml) from the livers (37 g) of 4 pigeons.

from the G-100 Sephadex column is dialyzed overnight at 4° against 1000 ml of 0.01 M Tris-HCl buffer, pH 7.4, containing 1 mM dithiothreitol. This step removes excess ionic material that would otherwise overload the DEAE-cellulose column used in a later purification step. The prosthetic group is then hydrolyzed from the protein with mild alkali. This step is achieved by adjusting the pH of the solution to 12 with 1 N KOH and heating at 70° for 60 minutes. It is convenient to use microelectrodes and a microstirrer to accurately determine the pH before heating. The pH is then adjusted to 9 by addition of Dowex 50 (hydrogen form). Carrier CoA (5–10 micromoles), 4'-phosphopantetheine (10–15 micromoles), and 15 mg of dithiothreitol are added. Incubation at 37° for 20 minutes serves to reduce any oxidized sulfhydryl groups. Protein is removed by centrifugation after the pH of the solution is reduced to 2 by the addition of Dowex 50. The supernatant solution is lyophilized, and the residue is dissolved in water prior to DEAE-cellulose column chromatography (see this volume [60]).

Comments. The procedure described above quantitatively liberates the bound 4'-phosphopantetheine from the fatty acid synthetase. Mild alkaline hydrolysis of 36.3 mg of ^{14}C-fatty acid synthetase containing 38,300 dpm gave a protein-free supernatant solution containing 37,000 dpm of ^{14}C.[5] The ^{14}C-labeled material bound to synthetase prepared as above was identified as 4'-phosphopantetheine by cochromatography with authentic material on DEAE-cellulose and by conversion to coenzyme A in the presence of ATP and the appropriate enzymes (see this volume [61]). The formation of ^{14}C-labeled coenzyme A was confirmed by preparation of the acetylated derivative and by coelectrophoresis on paper with synthetic acetyl-CoA-1-^{14}C.[5]

[61] Enzymatic Synthesis of ^{14}C-Labeled Coenzyme A

By C. J. CHESTERTON, P. H. W. BUTTERWORTH, and JOHN W. PORTER

The method described below for the enzymatic synthesis of coenzyme A labeled with ^{14}C in either the adenine or pantetheine moieties, was developed from the procedures reported by Novelli (Volume II, page 667) and Larrabee *et al.*[1] It was devised in order to identify the 4'-phosphopantetheine prosthetic group of pigeon liver fatty acid synthetase[2] and to prepare

---

[1] A. R. Larrabee, E. G. McDaniel, H. A. Bakerman, and P. R. Vagelos, *Proc. Natl. Acad. Sci. U.S.* **54**, 267 (1965).

[2] C. J. Chesterton, P. H. W. Butterworth, A. S. Abramovitz, E. J. Jacob, and J. W. Porter, *Arch. Biochem. Biophys.* **124**, 386 (1968).

372 PANTOTHENIC ACID, COENZYME A, AND DERIVATIVES [61]

labeled coenzyme A for the study of the partial reactions of fatty acid synthesis. 4'-Phosphopantetheine and ATP, either of which may carry the ^{14}C label, are converted to coenzyme A by the enzymes dephospho-CoA pyrophosphorylase and dephospho-CoA kinase, and the product of this reaction is purified by DEAE-cellulose column chromatography.

Reagents

Synthetic 4'-phosphopantetheine, prepared as described by Moffatt and Khorana[3]

^{14}C-labeled ATP, obtained from New England Nuclear Corporation

Coenzyme A, purchased from P-L Biochemicals

Enzymes: pyruvate kinase and lactic dehydrogenase, obtained from Sigma Chemical Co.

NADH and phosphoenolpyruvate, obtained from Sigma Chemical Co.

DEAE-cellulose (Mannex-DEAE from Mann Research Laboratories). This is washed successively with 5 volumes of 1 N HCl, 10 volumes water, 5 volumes 1 N NaOH, 10 volumes water, 5 volumes 1 N HCl, and water until the eluate is neutral. Washing is carried out by decantation except for the acid washings. The latter are performed rapidly on resin held in a Büchner funnel.

Chromatographic Purification of 4'-Phosphopantetheine and Coenzyme A

The method used is essentially that of Moffatt and Khorana[3] as modified by Larrabee *et al.*[1] The reagents listed above are dissolved in 1 ml of distilled deionized water containing 130 micromoles of dithiothreitol. Twenty minutes' incubation of this mixture at 37° reduces any oxidized sulfhydryl groups. The solution is then diluted with distilled deionized water to a conductivity value equal to that given by 6×10^{-4} M HCl (the volume at this stage should not be more than 70 ml) and added to the top of a 1×6 cm column of DEAE-cellulose. A 400-ml linear gradient from water to 0.075 M LiCl in 0.003 N HCl and containing 0.1 mg dithiothreitol per milliliter (to maintain sulfhydryl groups in the reduced state), is applied to the column. Three- or 6-ml fractions of eluate are collected. 4'-Phosphopantetheine in the eluate is detected by assay for total phosphorus.[4] Coenzyme A is determined by light absorption at 260 mμ or by a specific enzyme assay using the α-ketoglutarate dehydrogenase system.[5] Radioactivity in each fraction is measured by counting an aliquot in 15 ml of

[3] J. G. Moffatt and H. G. Khorana, *J. Am. Chem. Soc.* **83,** 663 (1961).
[4] B. N. Ames and D. T. Dubin, *J. Biol. Chem.* **235,** 769 (1960).
[5] P. B. Garland, *Biochem. J.* **92,** 10C (1964).

Bray's scintillator solvent[6] in a liquid scintillation spectrometer. Removal of LiCl from the column eluate is achieved by passing the appropriate fractions through a 7-ml Dowex 50 (hydrogen form) column. The eluate is then lyophilized.

Preparation and Assay of Dephospho-CoA Pyrophosphorylase (ATP: Pantetheine-4'-phosphate Adenylyltransferase, EC 2.7.7.3) and Dephospho-CoA Kinase (ATP: Dephospho-CoA 3'-Phosphotransferase, EC 2.7.1.24)

The enzymes are extracted from pig liver and then purified together according to the procedure of Novelli (Vol. II, p. 667). The final enzyme preparation, approximately 30 mg of protein per milliliter, is stored in 0.5-ml aliquots at $-20°$. This preparation is dialyzed overnight at $4°$ against 1000 ml of 0.1 M Tris-HCl, pH 8.0, before use. The activity of the enzymes synthesizing coenzyme A is assayed spectrophotometrically by coupling these reactions to the oxidation of NADH$^+$ via pyruvic kinase, lactic dehydrogenase, and phosphoenol pyruvate. The latter system utilizes the ADP liberated in the phosphorylation of dephospho-CoA to coenzyme A (Eqs. 1-4).

$$\text{4'-Phosphopantetheine + ATP} \xrightleftharpoons{\overset{\text{dephospho-CoA}}{\text{pyrophosphorylase}}} \text{dephospho-CoA + pyrophosphate} \quad (1)$$

$$\text{Dephospho-CoA + ATP} \xrightarrow{\overset{\text{dephospho-CoA}}{\text{kinase}}} \text{coenzyme-A + ADP} \quad (2)$$

$$\text{Phosphoenolpyruvate + ADP} \xrightleftharpoons{\overset{\text{pyruvate}}{\text{kinase}}} \text{pyruvate + ATP} \quad (3)$$

$$\text{Pyruvate + NADH}^+ \text{ + H}^+ \xrightleftharpoons{\overset{\text{lactic}}{\text{dehydrogenase}}} \text{lactate + NAD} \quad (4)$$

The incubation mixture for the above assay contains: 4'-phosphopantetheine, 0.675 micromole; ATP, 1 micromole; MgCl$_2$, 5 micromoles; dithiothreitol, 20 micromoles; Tris-HCl buffer, pH 8, 100 micromoles; phosphoenolpyruvate, 0.5 micromole; lactic dehydrogenase, 0.05 mg of protein; pyruvic kinase, 0.05 mg of protein; NADH$^+$, 0.165 micromole; and dephospho-CoA pyrophosphorylase and dephospho-CoA kinase, 0.025 ml, and 0.75 mg protein; in a final volume of 1 ml. It is convenient to carry out the reaction in a 1-ml quartz cuvette with a 1-cm light path at $37°$ and to follow the reaction (decrease in light absorption at 340 mμ) continuously in a recording spectrophotometer.

[6] G. A. Bray, *Anal. Biochem.* **1**, 279 (1960).

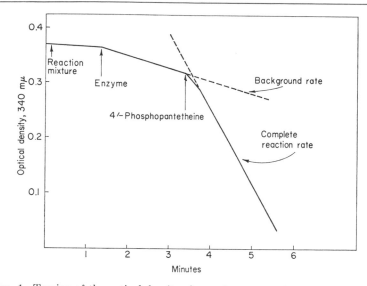

Fig. 1. Tracing of the optical density change in an assay for dephospho-CoA pyrophosphorylase and dephospho-CoA kinase activities. The formation of CoA and ADP from 4'-phosphopantetheine and ATP at 37° is coupled to NADH⁺ oxidation via pyruvate kinase, lactic dehydrogenase, and the auxiliary substrate, phosphoenolpyruvate. The incubation mixture, which contained 0.025 ml of enzyme preparation (0.78 mg of protein), contained the components reported in the text. Additions are made at the time points indicated.

The assay procedure is as follows: The complete incubation mixture, minus the enzyme preparation and 4'-phosphopantetheine, is placed in a cuvette. Essentially no change in optical density of the system with time should be detected. Enzyme is added, and the background rate of NADH⁺ oxidation is determined. Substrate is then added and the total rate of change in optical density is measured (see Fig. 1). The rate of coenzyme A formation is calculated by subtracting the blank from the total rate of change. The following equation (molar extinction coefficient of NADH⁺ = 6.22×10^3) is then used to calculate coenzyme A formation.

$$\text{Micromoles coenzyme A formed/min/mg protein} = \frac{\Delta\epsilon_{340 \, m\mu}/\text{min}}{\text{mg protein} \times 6.22}$$

In this assay system, the dialyzed enzyme preparation normally gives a specific activity of approximately 0.01 micromole of coenzyme-A formed per minute per milligram of protein.

Preparative Formation of Coenzyme A

The incubation mixture is made up as follows: 4'-phosphopantetheine, 10 micromoles; ATP, 20 micromoles; MgCl₂, 30 micromoles; dithiothreitol,

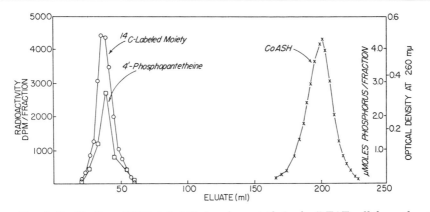

FIG. 2. Purification of ^{14}C-labeled 4'-phosphopantetheine by DEAE-cellulose column chromatography before conversion to coenzyme A. The radioactive compound was released from purified pigeon liver fatty acid synthetase by mild alkali treatment. ^{14}C-labeled 4'-phosphopantetheine (37,000 dpm), synthetic 4'-phosphopantetheine (13 micromoles), and coenzyme A (6.5 micromoles) were chromatographed and eluted from the column as described in the text. Values are given for radioactivity (○), phosphorus content (⊔), and optical density (×) at 260 mμ. Where not indicated, radioactivity levels were negligible.

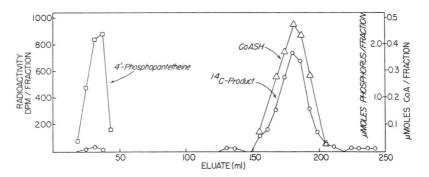

FIG. 3. Purification of ^{14}C-labeled coenzyme A, formed enzymatically from ^{14}C-labeled 4'-phosphopantetheine and ATP, by DEAE-cellulose column chromatography. ^{14}C-labeled coenzyme A (11,000 dpm) and synthetic 4'-phosphopantetheine were separated as described in the text. Values are given for radioactivity (○), phosphorus content (□), and coenzyme A content (△), determined enzymatically. Where not indicated, radioactivity levels were negligible.

60 micromoles; enzyme preparation, 6.2 mg protein; and Tris-HCl buffer, pH 8, 200 micromoles; in a total volume of 2 ml. This mixture is incubated for 4 hours at 37°. To prepare coenzyme A labeled with ^{14}C in the adenine or pantetheine residues, ATP-^{14}C or 4'-phosphopantetheine-^{14}C, respec-

tively, is included in the mixture. The 4'-phosphopantetheine-^{14}C used in our laboratory was prepared from pigeon liver fatty acid synthetase labeled with ^{14}C in the 4'-phosphopantetheine prosthetic group (see this volume [60] and purified on DEAE-cellulose as described above. In order to obtain maximal yields of labeled coenzyme A, it is suggested that during the last hour an excess of the unlabeled substrate be added (5 micromoles of 4'-phosphopantetheine or 10 micromoles of ATP).

After incubation, protein is precipitated by heating at 100° for 3 minutes, separated by centrifugation, and washed with 2 ml of water. The combined supernatant solution and washings are passed through a 5-ml column of Dowex 50 (hydrogen form) to remove Li$^+$, and the eluate is lyophilized to remove residual HCl prior to the purification of ^{14}C-labeled coenzyme A by DEAE-cellulose chromatography. The specific radioactivity of the product (dpm per micromole CoA) is determined by the assay methods described above.

The efficiency of this method is shown in Figs. 2 and 3. When 4'-phosphopantetheine-^{14}C is released from ^{14}C-labeled fatty acid synthetase and purified on DEAE-cellulose, only one fraction of radioactivity is seen coincident with carrier synthetic 4'-phosphopantetheine (Fig. 2). After incubation with dephospho-CoA pyrophosphorylase and dephospho-CoA kinase as described above, 98% of the ^{14}C recovered on rechromatography is eluted from the column with coenzyme A (Fig. 3). Proof that the ^{14}C material is, in fact, coenzyme A is obtained by conversion to the acetylated derivative and by comparison of the electrophoretic mobility of this product with that of synthetic 1-^{14}C-labeled acetyl-CoA.[2] Specific radioactivity of the coenzyme A may be determined by counting an aliquot, as described above, and by enzymatic assay of another aliquot.

Section V

Biotin and Derivatives

[62] *Amphidinium carterae* Assay for Biotin

By A. F. CARLUCCI

Principle. Radiocarbon ($^{14}CO_2$) uptakes by starved cells of the test alga, *Amphidinium carterae,* during a 2-hour exposure to the isotope after a preincubation under carefully controlled conditions are proportional to dissolved biotin concentrations in seawater.[1] If facilities for radiocarbon studies are not available, the preincubation is continued for 2–3 more days; after this time cell numbers are proportional to vitamin concentrations. Biotin concentrations are calculated using a similar equation for either the $^{14}CO_2$ uptake or cell number data. To assay freshwater systems it is necessary to add enough seawater salts to each sample to bring the salinity to about 3–3.5%. Alternatively, the assay could employ a biotin-requiring freshwater alga.

Capabilities

RANGE. From 0.2 to 5 ng of biotin per liter (higher amounts after dilution).

Precision at 3 ng per liter level: The correct value lies in the range

Mean of n determinations $\pm\ 0.28/n^{1/2}$ ng biotin per liter

Precision at 1 ng per liter level: The correct value lies in the range

Mean of n determinations $\pm\ 0.09/n^{1/2}$ ng biotin per liter

The precision of the method depends largely on the presence or absence of inhibitors in the seawater sample.

Special Apparatus and Equipment

Micro-Fernbach flasks, 50-ml, with deLong necks to accommodate stainless steel Morton enclosures

Erlenmeyer flasks, 125-ml, closed with cotton plugs wrapped in cheesecloth

All-glass Millipore filtration units with 1-liter capacity flasks, to take a 47-mm diameter filter

Millipore filter funnel to take a 25-mm diameter filter

Bacteriological fritted-glass filter, UF grade

Illuminated incubator with temperature controlled at 20 $\pm$ 2°. Lights should be cool-white fluorescent bulbs, preferably placed below the

[1] A. F. Carlucci and S. B. Silbernagel, *Can. J. Microbiol.* **13,** 979 (1967).

bioassay vessels. The light should be as uniform as possible with an intensity of about 0.05 cal per cm² per minute.

All glassware should be washed with detergent, treated with chromic-sulfuric acid for 12–24 hours, rinsed thoroughly with deionized water, and baked for 12 hours at approximately 260° in an oven.

Sample Collection and Storage. Samples should be filtered immediately after collection. Filter under aseptic conditions, if possible, through a PH Millipore filter (0.3-μ pore size). Store the filtered sample (50 ml is sufficient) in clean polypropylene bottles frozen at −20° in a deep-freezer.

Special Reagents and Their Preparation. VITAMIN-FREE WATER. In a bioassay of approximately 40 determinations (including external and internal standards), 500 ml of vitamin-free water is required. Prepare 1 liter at a time, using seawater with a salinity within 5% of the salinity of the samples to be analyzed. Use 10 g of Norit A (decolorizing carbon) for every liter of seawater. Pretreat the Norit by shaking it for 10 minutes with 500 ml of 5% w/v solution of analytical reagent-quality sodium chloride in distilled water. Filter the suspension through a Whatman No. 1 filter. Transfer the charcoal to a fresh sodium chloride solution and repeat the above operation twice more. Add the washed charcoal to 1 liter of the seawater and shake for 0.5 hour. Filter the seawater through a Whatman No. 1 paper and then aseptically through a 47-mm PH Millipore filter. Store aseptically.

NUTRIENT SOLUTIONS. (a) Chelated metals, nitrate, and phosphate. Dissolve 0.08 g of cobalt chloride, $CoCl_2 \cdot 6H_2O$, and 0.08 g of copper sulfate pentahydrate, $CuSO_4 \cdot 5H_2O$, in 100 ml of distilled water. Add 1 ml of this solution to approximately 800 ml of a solution containing 0.2 g ferric chloride, $FeCl_3 \cdot 6H_2O$; 0.06 g of zinc sulfate, $ZnSO_4 \cdot 7H_2O$; 0.12 g manganous sulfate, $MnSO_4 \cdot H_2O$; and 0.03 g of sodium molybdate, $Na_2MoO_4 \cdot 2H_2O$. Add 1.2 g of disodium ethylenediaminetetraacetate, and dilute to about 900 ml with distilled water. Adjust the pH of the solution with dilute sodium hydroxide until it is just over 7.5. Avoid any permanent precipitate formation that may arise from the addition of too much alkali. Add 10 g of potassium nitrate and 1.4 g of potassium dihydrogen phosphate. Dilute the solution to 1 liter, and autoclave it at 15 psi for 15 minutes. Store the solution in a glass bottle in the dark.

(b) Vitamin solution. Prepare 100 ml of a solution containing 1 mg of thiamine hydrochloride and 1 mg of crystalline vitamin B_{12}. The vitamin solution is sterilized by passing it through a fritted-glass, UF grade filter. Store in a sterilized glass container frozen at −20° in a deep-freezer.

RADIOACTIVE CARBONATE. Prepare sealed glass ampules each of which contains 1 ml of a sterile solution of radioactive carbonate ($Na_2{}^{14}CO_3$) in

5% w/v sodium chloride. The activity of each milliliter of solution should be about 1 μCi.

ALGAL INOCULUM (BIOTIN-FREE). Add 50 ml of vitamin-free seawater to each of three sterile 125-ml Erlenmeyer flasks. Plug each flask with a sterile cotton plug enclosed in cheesecloth. Add to each flask, 0.25 ml of nutrient solution (a) and 0.05 ml of solution (b). To one flask add 0.05 ml of the standard biotin solution prepared as described later. This flask contains the complete medium. Transfer 1 ml of an actively growing culture of *Amphidinium carterae* (culture should be visibly turbid with cells) to this flask. Incubate the transfer in the light incubator at 20° for 7 days, then transfer 0.5 ml of the resulting culture to one of the remaining two flasks containing biotin-free medium. Incubate this culture for 7 days and then add 5.5 ml of this second transfer to the remaining flask of biotin-free medium. After 6 days' further growth, the cells will be in the log phase of growth, almost stripped of biotin, and ready to be used as an inoculum.

The correct preparation of this inoculum is essential and the times of incubation given should be adhered to closely. Some trial experiments may be necessary to obtain a final suitable inoculum which must contain approximately 5.0×10^5 cells per milliliter, with no excess biotin, and the cells still physiologically active.

Procedure

1. Thaw the samples (see note a in next section) and add duplicate aliquots of between 5 and 20 ml (note b) aseptically to 50-ml micro-Fernbach flasks. Where necessary, bring the volume in each flask to 20 ml by the aseptic addition of vitamin-free seawater.

2. Add to each flask 0.1 ml of nutrient solution (a) and 0.02 ml of nutrient solution b (note c). To one of each duplicate add a known amount of biotin to serve as an internal standard as described below (note d).

3. To each bioassay flask add 0.5 ml of the inoculum of *Amphidinium carterae* prepared as described above. The concentration of cells in the inoculum should be approximately 5.0×10^5 cells/ml, making the initial concentration in each bioassay flask about 10^4 cells/ml. Should the inoculum cell concentration be notably lower or higher than this, add to each flask volumes proportionally greater or less than 0.5 ml. The addition should have a volume in the range 0.3–0.7 ml.

4. Allow the flasks to incubate in the light incubator at 20° for exactly 94 hours.

5. Add 1 ml of ^{14}C-labeled bicarbonate solution, containing 1 μCi of activity, to each flask, at 2-minute intervals; mix the contents well, and replace the flask in the incubator for exactly 2 more hours.

6. Filter the contents of each flask through a 25-mm diameter HA

Millipore filter (0.45 μ pore size), washing the sides of the flask with a police-man to detach any cells and rinsing with filtered seawater. Arrange the times for the addition of radioactive carbonate and filtration so that each sample receives exactly 2 hours (± 5 minutes) of incubation with the isotope. Count the activity of the cells on the filter with a suitable gas-flow Geiger counter.

7. Read the apparent concentration of biotin from the radioactive counts obtained from each sample from a calibration curve prepared with each batch of samples as described below. Let this concentration be A ng of biotin per liter. Read the apparent concentration of biotin in the flask containing sample plus internal standard which was taken through the analysis with each sample. Let this concentration be B ng biotin per liter. Calculate the concentration of biotin in the sample from the expression:

$$\text{ng biotin per liter} = A[1/(B - A)](20/v)$$

where v is the number of milliliters of sample originally taken for the anal-ysis. If $B - A$ is less than 0.65 ng biotin per liter, repeat the assay using only half the volume of sample and the rest vitamin-free seawater (note d).

Notes

a. Although the samples should be sterile as a result of aseptically filtering them soon after collection (see above), it may be necessary to sterilize them again by refiltering through PH Millipore filters immediately prior to the bioassay.

b. It is necessary to dilute the sample if its biotin content is greater than about 5 ng per liter or if it has greater than 35% inhibition as deter-mined by the recovery of internal standards (see note d).

c. To avoid adding nutrients by separate aliquots, suitable proportions of the nutrients may be mixed together just prior to use, and supplementa-tion can be made with one aliquot. The final concentration of nutrients in all flasks should be the same as obtained by adding the volumes separately.

d. Internal standardization is important because unknown substances in natural seawater can cause considerable inhibition to the growth of the alga. If the inhibition exceeds about 25–35%, calculations presented above become unreliable, warranting a repeat determination with a greater dilu-tion of the sample with vitamin-free seawater.

Calibration. STANDARD BIOTIN SOLUTION. Dissolve 10 mg of pure crys-talline biotin in 100 ml of distilled water. Sterilize the solution by passing it through a fritted-glass filter, UF grade, and store in 10-ml portions at $-20°$.

DILUTION OF BIOTIN SOLUTION. Dilute 1 ml of the concentrated stand-ard solution prepared above to 100 ml with distilled water. Take 1 ml of this solution and again dilute to 100 ml. From this second dilution, 5 ml is taken and diluted to 25 ml with distilled water: 1 ml = 2000 pg. This solution is to be referred to as solution A. Make further dilutions using vitamin-free seawater as follows:

> 6 ml of solution A to 10 ml (solution B)
> 5 ml of solution A to 10 ml (solution C)
> 4 ml of solution A to 10 ml (solution D)
> 3 ml of solution A to 10 ml (solution E)
> 2 ml of solution A to 10 ml (solution F)
> 1 ml of solution A to 10 ml (solution G)

INTERNAL STANDARDS. Add to 1 of each duplicate sample 0.1 ml of solution G. This addition is equivalent to 1 ng of biotin per liter. The sample with the internal standard is analyzed with the duplicate containing no added biotin and used for the calculation described above.

EXTERNAL STANDARDS. With each *series* of samples being assayed, prepare 7 micro-Fernbach flasks containing 20 ml of vitamin-free seawater enriched with the nutrient solutions. To 1 flask make no addition and to the other 6 flasks add 0.1 ml of solutions B–G, respectively. The concentrations of added biotin in the seawater of the external series will be 6, 5, 4, 3, 2, and 1 ng of biotin per liter, respectively. Inoculate and incubate these standards along with the samples being bioassayed. Prepare a calibration curve by plotting the counts per minute uptake of $^{14}CO_2$ per hour of exposure of each flask of the standard against the concentration of biotin in that standard. The biotin concentrations in the samples and the internals are read from the calibration curve.

[63] Colorimetric Determination of Biotin and Analogs

By DONALD B. McCORMICK and JEROME A. ROTH

Assay Method

Principle. Biotin and analogs are reacted under anhydrous conditions with *p*-dimethylaminocinnamaldehyde in sulfuric acid-ethanol to produce maximal absorbance at 533 mμ as described by Shimada *et al.*[1] This cyclic ureide-specific test is used to quantitate and differentiate microgram amounts of biotin analogs in alcoholic solutions and on paper or thin-layer chromatograms.[2]

[1] K. Shimada, Y. Nagase, and U. Matsumoto, *Yakugaku Zasshi* **89,** 436 (1969).
[2] D. B. McCormick and J. A. Roth, *Anal. Biochem.* **34,** 226 (1970).

Reagents

 p-Dimethylaminocinnamaldehyde, 0.2% in absolute ethanol
 Sulfuric acid-ethanol, 2% (v/v) of concentrated acid in absolute
   ethanol

Procedure. A. FOR SOLUTIONS: Ten to 100 μg of dry compound (solutions dried over silica gel in a vacuum desiccator) are dissolved in 1 ml of 2% sulfuric acid-ethanol, 1 ml of 0.2% *p*-dimethylaminocinnamaldehyde in ethanol is added, and the solution is diluted with 8 ml of absolute ethanol. After an hour at room temperature, the reddish-orange color is measured at 533 mμ against a reagent blank in a suitable spectrophotometer.

B. FOR CHROMATOGRAMS: A spray reagent comprised of equal volumes of 2% sulfuric acid-ethanol and 0.2% *p*-dimethylaminocinnamaldehyde is used to develop the chromatograms. Compounds are discerned as reddish spots on a nearly clear background. The amounts present may be quantitated by densitometric measurements with appropriate standards or by this procedure in conjunction with radioactively labeled compounds which can be additionally detected with a radiochromatogram scanner.

Specificity. The intensity of color development of biotin and analogs is shown in the table. A ureido carbonyl function is prerequisite for suitable color response, but alterations in the aliphatic side chain do not markedly influence the response except for the modest decreases seen with such

COLOR RESPONSE OF *d*-BIOTIN AND ANALOGS

Compound	$A_{533}/0.1$ μmole	Relative percent
Ureido-ring variations		
Biotin	0.176	100
Diaminocarboxylic acid sulfate[a]	0.030	17
2′-Thiobiotin	0.011	6
2′-Iminobiotin	0.010	6
Side chain variations		
Biotin methyl ester	0.177	100
Biotinol	0.177	100
Biotinyl ω-bromide	0.169	96
Biocytin	0.161	91
α-Dehydrobiotin	0.137	77
Tetranorbiotin	0.126	72
Thiophane-ring variations		
Dethiobiotin	0.175	99
Biotin *l*-sulfoxide	0.170	97
Biotin *d*-sulfoxide	0.044	25
Biotin sulfone	0.048	27

[a] The biotin derivative which lacks a ureido carbonyl.

catabolites as α-dehydrobiotin and tetranorbiotin. However, the stereo-specificity of color reaction for the sulfur end of biotin and such natural oxidized forms as the sulfoxides and sulfone is rather remarkable. The color response from d-sulfoxide and sulfone is only about one-fourth that of biotin and the l-sulfoxide. As seen by the comparison in Fig. 1 of radio-activity (quantity) with color response of the peroxide-treated products from tetranorbiotin, one can readily distinguish the initial thioether from both sulfoxide stereoisomers on the basis of intensity of color per relative amount of such compounds which are chromatographically separated.

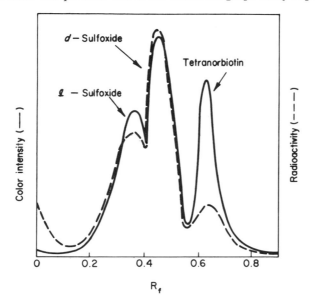

FIG. 1. Differentiation by color response and radioactivity of tetranorbiotin and the sulfoxides on a paper chromatogram. Approximately 0.5 μmole containing 10^3 dpm of material from reaction of carbonyl-labeled ^{14}C-tetranorbiotin with equimolar hydrogen peroxide in glacial acetic acid was chromatographed and assayed as described under "procedure."

Comments. The color reaction has been shown to proceed in a classical way for formation of the conjugate imine (Schiff's base) with biotin deriva-tives and other cyclic ureido compounds, i.e., substituted 2-imidazolidones, but is not interferred with by similar quantities of usual biochemicals.[2]

[64] Preparation of Biotin *l*-Sulfoxide

By HELMUT RUIS, DONALD B. McCORMICK, and LEMUEL D. WRIGHT

$$\text{Biotin} \xrightarrow{+H_2O_2} d\text{-} + l\text{-sulfoxide} \xrightarrow{+HCl} l\text{-sulfoxide}$$

Principle. Biotin *l*-sulfoxide has been demonstrated to occur in culture filtrates of various microorganisms.[1-4] It has been prepared by oxidation of biotin with 1 mole of hydrogen peroxide and by fractional crystallization of the resulting mixture of biotin *d*-sulfoxide and biotin *l*-sulfoxide.[5] It can be prepared more conveniently and in better yield by oxidation of biotin with 1 mole of hydrogen peroxide and treatment of the resulting mixture of sulfoxides with 1 *N* hydrochloric acid.[6] This method will be described here in detail.

Procedure. Biotin (1 g) is dissolved in 100 ml of warm glacial acetic acid. The solution is cooled to room temperature and 0.5 ml of 30% H_2O_2 is added. The solution is kept at room temperature for 24 hours and then the solvent is evaporated *in vacuo*. The residue is dissolved in 100 ml of 1 *N* HCl, and the solution is refluxed under N_2 for 10 hours. The solvent is then evaporated *in vacuo*, the residue is dissolved in water, and the solution is treated with Norit, filtered, and evaporated to dryness. The crude sulfoxide is recrystallized from water to yield 0.75 g (70%) of material melting at 239–243° (dec.).[7] $[\alpha]_D^{20}$, −39.4° (ca. 1.00, 0.1 *N* NaOH).

[1] L. D. Wright and E. L. Cresson, *J. Am. Chem. Soc.* **76**, 4156 (1954).
[2] L. D. Wright, E. L. Cresson, J. Valiant, D. E. Wolf, and K. Folkers, *J. Am. Chem. Soc.* **76**, 4160 (1954).
[3] L. D. Wright, E. L. Cresson, J. Valiant, D. E. Wolf, and K. Folkers, *J. Am. Chem. Soc.* **76**, **4163** (1954).
[4] L. F. Li, Ph.D. thesis, Cornell University, Ithaca, New York, 1964.
[5] D. B. Melville, *J. Biol. Chem.* **208**, 495 (1954).
[6] H. Ruis, D. B. McCormick, and L. D. Wright, *J. Org. Chem.* **32**, 2010 (1967).
[7] Melting point on Fisher-Johns block (corrected).

[65] Synthesis of *d*-Biotinyl 5′-Adenylate

By JAMES E. CHRISTNER and MINOR J. COON

Preparation

Principle. This procedure for the synthesis of *d*-biotinyl 5′-adenylate[1,2] follows the same principle as that used earlier in the synthesis of an aminoacyl adenylate.[3] The treatment of a solution of biotin and adenylic acid in pyridine with dicyclohexylcarbodiimide leads to the formation of biotinyl 5'-adenylate. The crude product is purified by chromatography on DEAE-cellulose. The preparation and purification of this compound have also been described by other workers.[4]

Reagents

Adenylic acid
d-Biotin
Pyridine and 75% aqueous pyridine
Dicyclohexylcarbodiimide
Acetone
Acetone–ethanol (3:2)
Diethyl ether
Potassium hydroxide, 0.5 N

Procedure. Adenylic acid (684 mg, 2 millimoles) and d-biotin (488 mg, 2 millimoles) are dissolved in 10 ml of 75% aqueous pyridine at 0°. A solution of 8 g of dicyclohexylcarbodiimide in 8 ml of pyridine precooled to 0° is added. The mixture is stirred in a stoppered flask in an ice bath by means of a magnetic stirrer for 20 hours. The yield of product at this point as judged by the amount of acyl phosphate formed is about 90%.

Acyl phosphate is determined on small samples (0.05 ml or less) of the reaction mixture as follows. A sufficient amount of 2 M hydroxylamine, pH 6.5, prepared just before use by mixing equal volumes of 4 M hydroxylamine hydrochloride and 3.5 M KOH, is added to bring the volume to 1.0 ml. After 5 minutes 3 ml of a solution consisting of 10% $FeCl_3 \cdot 6H_2O$ and 5% trichloroacetic acid in $\frac{2}{3}$ N HCl is added, and the absorbance is determined with a Klett photometer using filter No. 54. The molar absorbance of the ferric complex of biotin hydroxamic acid[2] is 5.26×10^5 Klett units.

The product is precipitated by the addition of 150 ml of acetone precooled to −15° and filtered on a Büchner funnel. The precipitate is washed on the filter with three 5-ml portions of acetone–ethanol (3:2) and then with three 5-ml portions of ether at 0°. It is dried overnight in an evacuated

[1] L. Siegel, J. L. Foote, J. E. Christner, and M. J. Coon, *Biochem. Biophys. Res. Commun.* **13,** 307 (1963).

[2] J. E. Christner, M. J. Schlesinger, and M. J. Coon, *J. Biol. Chem.* **239,** 3997 (1964).

[3] P. Berg, *J. Biol. Chem.* **233,** 608 (1958).

[4] M. D. Lane, K. L. Rominger, D. L. Young, and F. Lynen, *J. Biol. Chem.* **239,** 2865 (1964).

desiccator over P_2O_5 and paraffin shavings. The dry powder, which contains the desired product and dicyclohexylurea, is dispersed by means of a Teflon-glass homogenizer in 25 ml of cold water, and the suspension is adjusted to pH 6 by the addition of 0.5 N KOH. The insoluble dicyclohexylurea is collected on a filter and washed with 8 ml of cold water in two portions. The filtrate, which contains the product in about 50% yield, is concentrated to about 10 ml in a rotary evaporator at 40°. The crude product is stable for at least several months when stored at $-15°$. The molar ratio of adenine to acyl phosphate is about 1.5.

The crude material may be purified by paper chromatography. Approximately 80 micromoles of hydroxamate-forming material are applied to Whatman No. 3 MM filter paper as a streak about 8 inches long and subjected to chromatography in isobutyric acid–water–concentrated ammonium hydroxide (66:33:1). The paper is viewed under ultraviolet light, and the absorbing areas are marked. The product is then detected by spraying a portion of the paper with neutral hydroxylamine, followed by acidic ferric chloride in alcohol. The hydroxamate-positive area (R_f, 0.52), which corresponds to one of the ultraviolet light-absorbing areas, is cut out and eluted with 1 ml of water. The molar ratio of adenine to acyl phosphate is 1.0. The recovery is about 35%.

The crude product may also be purified by ion-exchange chromatography. DEAE-cellulose (Schleicher and Schuell No. 20, with a capacity of 0.83 meq/g) is washed with 20 volumes of each of the following solutions in the order given: 1 M NaCl, 0.5 M KOH, 0.1 M HCl in 95% ethanol, and 0.5 N KOH; each step is carried out by stirring the DEAE-cellulose in the solution for 15 minutes, filtering, and washing on the filter with water until the washings are free of solute before proceeding with the next step. A column of the washed DEAE-cellulose (1.5 × 15 cm) is packed by gravity and washed with 1 liter of 0.1 M potassium phosphate buffer, pH 7.4, and then with 1 liter of H_2O. Approximately 100 micromoles of biotinyl adenylate are applied to the column. Elution is carried out at room temperature by washing the column first with 20 ml of water, then with 20 ml of 0.01 M KCl, and finally with 0.05 M KCl. The product, closely preceded by biotin, is eluted in about 2 column volumes of the latter salt solution. By combining the fractions containing acyl phosphate starting with the peak tube, contamination by free biotin is eliminated. The combined fractions are concentrated to a volume of about 10 ml. The purified product, obtained from the column in about 80% yield, is stable for at least several months at $-15°$. When the compound is subjected to paper chromatography on Whatman No. 1 filter paper in isobutyric acid–water–concentrated ammonium hydroxide, a single ultraviolet light-absorbing spot is detected (R_f 0.59) which corresponds to the hydroxamate-positive area as detected by spraying with neutral hydroxylamine followed by acidic ferric chloride in 95% ethanol.

As a further test of the purity of the compound, the relative concentrations of acyl phosphate, adenine, vicinal hydroxyl groups, and organic phosphate may be determined. Acyl phosphate is determined by its reaction with neutral hydroxylamine to form the hydroxamic acid as indicated previously, adenine by the absorbance at 260 mμ, and organic phosphate by the colorimetric determination of P$_i$ after acid hydrolysis.[5] Vicinal hydroxyl groups are determined by spectrophotometric measurement of the uptake of periodate.[6] Since biotin reacts readily with an equimolar amount of periodate, presumably to form a sulfoxide, the value for vicinyl hydroxyls in biotinyl 5'-adenylate needs to be corrected accordingly. The results of typical analyses are shown in the table. The results eliminate the possibility

ANALYSIS OF BIOTINYL ADENYLATE

Functional group	Molar ratio relative to acyl phosphate	
	Expt. 1	Expt. 2
Acyl phosphate	1.00	1.00
Adenine	0.96	0.99
Organic phosphate	0.98	1.05
Vicinal hydroxyls[a]	0.93	1.01

[a] Corrected as described in the text.

of the presence of significant amounts of biotin, AMP, biotin anhydride, or adenylic anhydride. The presence of the biotinyl group in a 2'- or 3'-ester linkage is ruled out not only by the detection of the expected amount of vicinal hydroxyls, but also by the fact that such esters react slowly with neutral hydroxylamine, whereas with the acyl phosphate group of biotinyl adenylate the reaction is complete in less than 3 minutes.

Metabolic Role

Biotinyl 5'-adenylate is an intermediate in the enzymatic conversion of d-biotin to biotinyl-CoA[2] and has also been identified as an intermediate in the conversion of propionyl-CoA apocarboxylase to the biotin-containing holocarboxylase, catalyzed by a synthetase from liver.[1,7,8] Biotinyl adenylate is similarly involved in the synthesis of holotranscarboxylase,[4] β-meth-

[5] C. H. Fiske and Y. SubbaRow, *J. Biol. Chem.* **66**, 375 (1925).

[6] G. V. Marinetti and G. Rouser, *J. Am. Chem. Soc.* **77**, 5345 (1955).

[7] L. Siegel, J. L. Foote, and M. J. Coon, *J. Biol. Chem.* **240**, 1025 (1965).

[8] H. C. McAllister and M. J. Coon, *J. Biol. Chem.* **241**, 2855 (1966).

ylcrotonyl-CoA holocarboxylase,[9] and apparently of acetyl-CoA carboxylase.[10]

[9] T. Höpner and J. Knappe, *Biochem. Z.* **342**, 190 (1965).
[10] F. Lynen and K. L. Rominger, *Federation Proc.* **22**, 537 (1963).

[66] Microbial Synthesis of Dethiobiotin and Biotin

By KOICHI OGATA

Dethiobiotin[1]

Preparation

Principle. A large amount of biotin vitamers is accumulated by many microorganisms grown in the presence of pimelic acid. *Bacillus sphaericus* IFO 3525, selected by the monocell culture, accumulates the vitamers in especially large amounts (150–200 μg/ml) under optimal conditions. The dominant component of the biotin vitamers accumulated in the culture filtrates of these microorganisms has been assumed to be dethiobiotin by chromatographic methods.

"Total biotin" is quantitatively determined by microbiological assay with *Saccharomyces cerevisiae* according to the method of Snell *et al.*[2] "Total biotin" includes biotin, dethiobiotin, diaminobiotin, diaminopelargonic acid, biocytin, and other vitamers with activity for the growth of *Saccharomyces cerevisiae*. The microbiological assay of "true biotin" is carried out by the general assay procedure of Wright and Skeggs[3] using *Lactobacillus arabinosus*.

Reagents

Pimelic acid
Lead acetate, saturated solution
Dowex 1 X2, formate type
Formic acid, 0.012 M

Procedure. CULTURE METHOD. The medium employed for the production of biotin vitamers by the bacterium has the following composition: peptone, 10 g; casamino acid, 5 g; soybean meal, 100 g; glycerol, 20 g; K_2HPO_4, 1.0 g;

[1] S. Iwahara, Y. Emoto, T. Tochikura, and K. Ogata, *Agr. Biol. Chem. (Tokyo)* **30**, 64 (1966).
[2] E. E. Snell, R. E. Eakin, and R. J. Williams, *J. Am. Chem. Soc.* **62**, 175 (1940).
[3] L. D. Wright and H. R. Skeggs, *Proc. Soc. Exptl. Biol. Med.* **56**, 95 (1944).

KCl, 0.5 g; MgSO$_4$ · 7H$_2$O, 0.5 g; FeSO$_4$ · 7H$_2$O, 10 mg; MnSO$_4$ · 4~6H$_2$O. 10 mg; thiamine · HCl, 200 μg; pimelic acid, 1.0 g in 1000 ml of tap water. The pH is adjusted to 7.2. The medium (30 ml) is taken into 300-ml shaking flasks and sterilized. The cells of the bacterium from the slant culture are incubated in the medium for 4 to 5 days at 28° on a reciprocal shaker (140 reciprocations per minute). The cultures, after growth, are centrifuged to remove the cells and other insoluble materials. The clear supernatant solution contains about 100 μg of dethiobiotin per milliliter.

ISOLATION. Saturated lead acetate solution, 100 ml, is added to 3000 ml of the supernatant solution. The precipitates formed are removed by centrifugation, and then the residual lead is precipitated by the addition of a slight excess of sulfuric acid. The precipitates are removed by centrifugation. After the pH is adjusted to 3.0 with sodium hydroxide, 60 g of active carbon is added to the supernatant and stirred mechanically for 4 hours at room temperature. The active carbon is collected by filtration through a Büchner funnel and washed with about 500 ml of water. The biotin vitamers adsorbed on the active carbon are eluted with about 1000 ml of ethanol–ammonia mixture (prepared by mixing 18 volumes of 50% ethanol and 1 volume of 28% ammonia–water). The eluate is concentrated to a syrup *in vacuo* at room temperature. Two hundred milliliters of 99% ethanol is added to the syrup, the precipitates formed are removed by centrifugation, then the clear supernatant is again concentrated to about

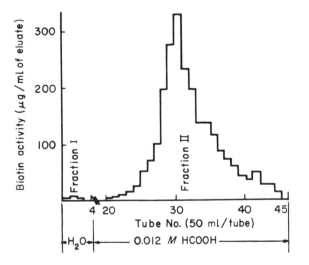

FIG. 1. Elution pattern of biotin vitamers on anion-exchange column. Exchanger: Dowex 1 X2-formate, 200–400 mesh; volume of resin: 7 cm^2 × 40 cm; flow rate: 3.0 ml/min.

50 ml *in vacuo* at room temperature. The pH is adjusted to 8.0, then the concentrate is taken onto a Dowex 1 X2-formate column (7 cm^2 × 40 cm), and the column is washed with 500 ml of deionized water. The biotin vitamers are eluted with 0.012 M formic acid. Fifty-milliliter fractions of eluate are collected. The biotin activity is determined by microbiological assay with *Saccharomyces cerevisiae*. A typical column chromatographic pattern is shown in Fig. 1. The fractions showing biotin activity corresponding to dethiobiotin (fraction II) are combined, and the combined solution is concentrated to about 10 ml. When the concentrate is kept overnight at 0°, colorless crystals are obtained. These are recrystallized three times from a minimal amount of hot water. About 30 mg of pure material (colorless, fine, needlelike crystals) is obtained.

Properties

The crystalline dethiobiotin melts at 158–160°. Found (%): C, 55.92; H, 8.17; N, 13.03. Calculated for $C_{10}H_{18}O_3N_2$ (%): C, 56.05; H, 8.47; N, 13.08%. The isolated dethiobiotin is twice as active as *dl*-dethiobiotin and is identified by paper chromatography with the natural isomer of dethiobiotin. The infrared spectrum also gives confirmation.

Biotin

Conversion of Dethiobiotin to Biotin[4]

Principle. A large number of microorganisms, such as molds, yeasts, bacteria, and *Streptomyces* convert dethiobiotin to biotin during the cultivation of the microorganisms.

Culture Method. The medium employed for the conversion of dethiobiotin to biotin by the mold and yeast has the following composition: glucose, 100 g; peptone, 10 g; casamino acid, 10 g; KH_2PO_4, 0.5 g; K_2HPO_4, 0.5 g; KCl, 0.5 g; $MgSO_4 \cdot 7H_2O$, 0.5 g; cystine · HCl, 0.2 g; DL-methionine, 0.2 g; taurine, 0.2 g; *dl*-dethiobiotin, 50 mg in 1000 ml of tap water. The

CONVERSION OF DETHIOBIOTIN TO BIOTIN

Microorganisms tested	Biotin produced (mμg/ml)	
	No addition	Dethiobiotin addition
Torulopsis calliculosa IFO 0381	Trace	100
Aspergillus oryzae AKU 3311	Trace	300
Aspergillus oryzae IFO 4117	29	300
Fusarium culmorum IFO 5902	Trace	230

[4] S. Iwahara, S. Takasawa, T. Tochikura, and K. Ogata, *Agr. Biol. Chem. (Tokyo)* **30,** 385 (1966).

pH is adjusted to 7.0. Five milliliters of medium is taken into each test
tube (25 mm × 200 mm) and sterilized. The microorganisms are cultivated
on the medium for 5 days at 28° with reciprocal shaking (340 reciprocations
per minute). After growth, the cultures are centrifuged to remove the cells.
The contents of biotin vitamers synthesized are determined. Typical data
are shown in the table.

Separation of Biotin and Biotin Vitamers[5]

Separation of each biotin vitamer is carried out by ion-exchange
column chromatography. Figure 2 shows a typical column chromatographic
pattern of known biotin vitamers. The chromatographic procedure employed is as follows. A few milliliters of the solution of biotin vitamers

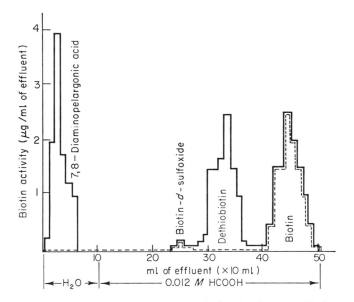

FIG. 2. Anion-exchange analysis of authentic biotin vitamers. Exchanger: Dowex
1 X2-formate, 200–400 mesh; volume of resin: 0.6 cm² × 19 cm. —, Assayed by
Saccharomyces cerevisiae; - - - -, assayed by *Lactobacillus arabinosus.*

(containing about 20–500 μg of total biotin, pH 7.0) is quantitatively
taken onto a Dowex 1 X2-formate column (0.6 cm² × 19 cm). The column
is washed with 100 ml of deionized water, which removes the unadsorbed
biotin vitamers, then the biotin vitamers (e.g., biotin, biotin sulfoxide, and
dethiobiotin) are eluted with 0.012 M formic acid, and 10-ml fractions

[5] K. Ogata, T. Tochikura, S. Iwahara, K. Ikushima, S. Takasawa, M. Kikuchi, and
A. Nishimura, *Agr. Biol. Chem. (Tokyo)* **29,** 895 (1965).

are collected. Biotin concentrations are quantitatively determined by microbiological assays with *Saccharomyces cerevisiae*[2] and *Lactobacillus arabinosus*.[3]

Biotin and biotin vitamers accumulated in the culture media may alternatively be adsorbed on active carbon before ion-exchange column chromatography. The pH of the culture medium of each microorganism is adjusted to about 2–3 with concentrated hydrochloric acid. Fifteen grams of active carbon is added to 1000 ml of the culture medium and stirred mechanically for 4 hours at room temperature. The active carbon is collected on a Büchner funnel and washed with about 200 ml of water. The active carbon cake is suspended in about 200 ml of ethanol–ammonia mixture (prepared by mixing 18 volumes of 50% ethanol and 1 volume of 28% ammonia water) and stirred for 4 hours at room temperature. The active carbon is filtered off on a Büchner funnel, and the active carbon cake is washed with ethanol–ammonia mixture three to four times. The filtrates are concentrated to dryness *in vacuo* at room temperature. The concentrate obtained is dissolved in about 40 ml of 90% ethanol, and insoluble materials are removed by centrifugation; the clear supernatant solution is again concentrated to a volume of about 5 ml *in vacuo* at room temperature. This concentrated solution is subjected to the ion-exchange chromatography.

[67] Determination of the Conversion of Dethiobiotin to Biotin in *Aspergillus niger*

By Judith P. Tepper, Heng-Chun Li, Donald B. McCormick, and Lemuel D. Wright

$$\text{Dethiobiotin} \rightarrow \text{biotin} \xrightarrow{\text{H}_2\text{O}_2} \text{biotin sulfone}$$

Assay Method

Principle. The mold is grown aerobically for several days on a liquid medium[1] to which a small amount of radioactive *d*-dethiobiotin is added. Mycelia are filtered off, and the filtrate, which contains *d*-biotin *l*-sulfoxide and the β-oxidative catabolites of *d*-dethiobiotin as principal metabolites,[2] is subjected to a series of treatments.[3,4] In this procedure, the radioactive biotin vitamers are concentrated by adsorption on and elution from charcoal, carrier biotin is added, biotin and its sulfoxide are oxidized to the

[1] L. D. Wright and E. L. Cresson, *J. Am. Chem. Soc.* **74**, 4156 (1954).

[2] H. C. Li, D. B. McCormick, and L. D. Wright, *J. Biol. Chem.* **243**, 4391 (1968).

[3] J. P. Tepper, D. B. McCormick, and L. D. Wright, *J. Biol. Chem.* **241**, 5734 (1966).

[4] H. C. Li, D. B. McCormick, and L. D. Wright, *J. Biol. Chem.* **243**, 6442 (1968).

sulfone by treatment with excess hydrogen peroxide, biotin sulfone is separated by anion-exchange chromatography on Dowex 1 (formate) with ammonium formate as eluent, ammonium ions are removed on Dowex 50 (H$^+$), and the sulfone is recrystallized from water to constant specific radioactivity.

Dethiobiotins. Both radioactive and nonradioactive d-dethiobiotin are prepared by desulfuration of the corresponding d-biotin with Raney nickel according to the method of Melville *et al.*[5] To ensure complete removal of any contaminating biotin from the radioactive dethiobiotin, the product is treated with a little hydrogen peroxide in acetic acid and chromatographed on 100- to 200-mesh Dowex 1 X8 (formate) from which it is eluted, following any trace of biotin sulfone, by formic acid.

Procedure. Aspergillus niger is grown on 500 ml of a Czapek-Dok medium as described by Wright and Cresson,[1] except that 1–2.5 mg of radioactive d-dethiobiotin is included.[3,4] Incubation is for several days, usually 7, at 30° with shaking at 200 rpm, after which time the mycelia are filtered off. From the filtrate, biotin vitamers are concentrated by adsorption on 1.5 g of Norit A and elution with two 75-ml portions of ethanol–water–ammonium hydroxide (10:10:1, v/v/v). The combined eluates are evaporated to dryness, and enough carrier biotin is added, e.g., 240 mg, to ensure sufficient sulfone for final crystallization. The total material is dissolved in 110 ml of glacial acetic acid and treated with 20 ml of 30% hydrogen peroxide to convert all biotin plus sulfoxide to biotin sulfone as described by Hofmann *et al.*[6] After 1 day, the remaining hydrogen peroxide is decomposed by gradual addition of 250 mg of palladium black with stirring. The catalyst is filtered off, the filtrate is evaporated to dryness under reduced pressure below 50°, and the residue is dissolved and neutralized in a small volume of dilute NaOH. The solution is poured over a 1.6 × 65 cm-column of 100- to 200-mesh Dowex 1 X8 (formate), and 10-ml fractions are collected during elution with a linear gradient from 1 liter of water to 1 liter of 1 M ammonium formate. Radioactivity is determined in appropriate aliquots, usually 0.5 ml, with 10 ml of Bray's solution[7] in a liquid scintillation spectrometer.

A typical elution pattern from column chromatography of peroxide-treated radioactive material from the culture filtrates of *A. niger* grown for 5 days with 1.7 mg of ^{3}H(random),^{14}C(carboxyl)-labeled d-dethiobiotin and with 240 mg of carrier d-biotin added to the charcoal-adsorbed material is shown in Fig. 1. The biotin sulfone fractions, which come after side-chain catabolites and before remaining dethiobiotin, are combined, evaporated to dryness, redissolved in a small volume of water, poured over a 3 × 25 cm-

[5] D. B. Melville, K. Dittmer, G. B. Brown, and V. du Vigneaud, *Science* **98**, 497 (1943).
[6] K. Hofmann, D. B. Melville, and V. du Vigneaud, *J. Biol. Chem.* **141**, 207 (1941).
[7] G. A. Bray, *Anal. Biochem.* **1**, 279 (1960).

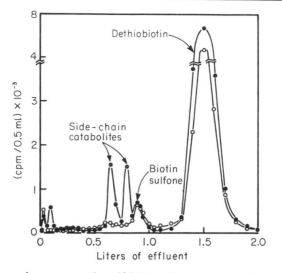

Fig. 1. Anion-exchange separation of biotin sulfone from dethiobiotin and its catabolites. ●, ³H; ○, ¹⁴C.

column of Dowex 50 W (H⁺) to remove ammonium ions, and washed through with 700 ml of water. The effluent is evaporated to dryness, and the powder is crystallized repeatedly from hot water to achieve constant specific radioactivity as determined by counting portions, usually 5 mg, dissolved in 0.5 ml of hydroxide of Hyamine and 10 ml of toluene plus scintillators.

Comments. The extent of conversion of dethiobiotin to biotin can be assessed by using dethiobiotin which is ³H(random)-, ¹⁴C(carbonyl)-, or ¹⁴C(carboxyl)-labeled. The first two offer some advantage in that one can also ascertain how much of the side chain-degraded catabolites of dethiobiotin are formed and how well they are chromatographically separated from biotin sulfone. The loss of tritium from the ³H(random)-dethiobiotin can be used to approximate the number of hydrogens abstracted during conversion to biotin. When combinations of the radioactive dethiobiotins are used, their ratios, e.g., ³H:¹⁴C, can indicate whether or not incorporation of the intact molecule has occurred.

Results

Extent of Incorporation. For *A. niger*, the amount of initial dethiobiotin converted to biotin has been found to be 1.32 and 1.35% for 5 days' growth on 1.3 mg of ¹⁴C-carbonyl-*d*-dethiobiotin and 1.7 mg of ¹⁴C-carboxyl-*d*-dethiobiotin, respectively.[4] After 7 days of growth on 1 mg of ¹⁴C-carbonyl-*d*-dethiobiotin, 2.4% incorporation was found, but only 0.8% when the level of substrate was 2.5 mg.[3] After 8 days of growth on 1.7 mg of ^{14}C-carboxyl-*d*-dethiobiotin, 1.87% was converted to biotin.[4] Thus, the amount of biotin formed from dethiobiotin increases during growth of the mold, but this level is not increased by large excesses of the precursor.

Nature of Incorporation. As 80–85% of the ^{3}H-radioactivity of dethiobiotin is incorporated into biotin sulfone finally isolated, little or no fragmentation and rearrangement can occur in *A. niger*.[4] Moreover, the 15–20% of tritium that is lost in this conversion indicates that 4 hydrogens may be removed in the overall process.

[68] Isolation and Identification of Dethiobiotin and Biotin Catabolites[1,2]

By KOICHI OGATA

Preparation

Principle. Dethiobiotin is converted to bisnordethiobiotin, and biotin to bisnorbiotin and bisnorbiotin sulfoxide by various molds and yeasts during cultivation.

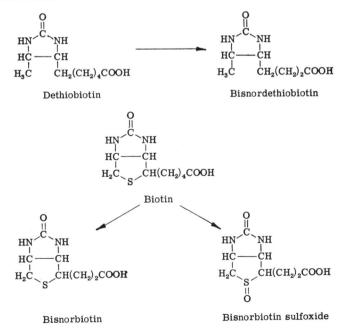

Dethiobiotin   Bisnordethiobiotin

Biotin

Bisnorbiotin   Bisnorbiotin sulfoxide

[1] S. Iwahara, S. Takasawa, T. Tochikura, and K. Ogata, *Agr. Biol. Chem. (Tokyo)* **30,** 1069 (1966).

These catabolites are quantitatively determined by microbiological assay with *Bacillus subtilis* which was isolated from "Natto" (fermented beans) and preserved in the Laboratory of Applied Microbiology, Kyoto University, Kyoto.

Reagents

dl-Dethiobiotin
d-Biotin
Dowex 50, H⁺ type
Dowex 1 X2, formate type
Formic acid, 0.01 M

Procedure for Bisnordethiobiotin. Culture method. *Aspergillus oryzae* IFO 3301 is inoculated into medium composed as follows: glucose, 50 g; $NaNO_3$, 9 g; KH_2PO_4, 0.5 g; K_2HPO_4, 0.5 g; NaCl, 0.5 g; $FeCl_3$, 1 mg; $ZnSO_4 \cdot 7H_2O$, 1 mg; $MnSO_4 \cdot 4 \sim 6H_2O$, 1 mg; $CuSO_4 \cdot 5H_2O$, 0.1 mg in 1000 ml of tap water (pH 6.8). Before inoculation, 2000 ml of the medium is supplemented with 1.5 g of *dl*-dethiobiotin. Cultivation is carried out for 7 days at 28° on a reciprocal shaker (140 reciprocations per minute).

Isolation. Isolation of the bisnordethiobiotin is carried out by the following procedure. The pH of the culture filtrate is adjusted to 2.0 with hydrochloric acid solution. Biotin vitamers are adsorbed on 40 g of active carbon at room temperature, and the active carbon is collected on a Büchner funnel and washed with about 100 ml of water. Biotin vitamers are eluted from the active carbon with about 500 ml of the ethanol–ammonia mixture (prepared by mixing 9 volumes of 50% ethanol and 1 volume of 28% ammonia-water). The eluate is concentrated to syrup *in vacuo* at 50°, the biotin vitamers are extracted with 100 ml of 75% ethanol, and the extract is again concentrated to about 30 ml *in vacuo*. The concentrate is taken onto a Dowex 50 (H⁺) column (3 × 15 cm). The biotin vitamers are washed out with 1000 ml of deionized water from the column. The washings are concentrated to about 30 ml *in vacuo* at 50°. The concentrate is placed onto a Dowex 1 X2-formate column (2.8 × 30 cm), and then the column is washed with 500 ml of water. The bisnordethiobiotin and other biotin vitamers are eluted with 0.01 M formic acid, and fractions of 50 ml are collected. The contents of biotin vitamers in each fraction are quantitatively determined by microbiological assays with *Lactobacillus arabinosus*,[3] *Bacillus subtilis*,[1] and *Saccharomyces cerevisiae*,[4] respectively. The eluate fractions that show biotin activity only for *Bacillus subtilis* are concentrated

[2] H. C. Yang, M. Kusumoto, S. Iwahara, T. Tochikura, and K. Ogata, *Agr. Biol. Chem. (Tokyo)* **32**, 399 (1968).
[3] L. D. Wright and H. R. Skeggs, *Proc. Soc. Exptl. Biol. Med.* **56**, 95 (1944).
[4] E. E. Snell, R. E. Eakin, and R. J. Williams, *J. Am. Chem. Soc.* **62**, 175 (1940).

to about 10 ml. The concentrate is allowed to stand overnight at 0°. About 500 mg of crystals is obtained. After recrystallization from 70% ethanol, 300 mg of pure crystals is obtained.

Procedure for Bisnorbiotin and Bisnorbiotin Sulfoxide.[2] CULTURE METHOD. *Rhodotorula rubra* IFO 889 is inoculated into a medium composed as follows: glucose, 75 g; K_2HPO_4, 3 g; $MgSO_4 \cdot 7H_2O$, 7.5 g; $Ca_3(PO_4)_2$, 2.4 g; yeast extract, 15 g; $(NH_4)_2SO_4$, 6 g in 3000 ml of tap water (pH 5.8). Before inoculation, 2 g of *d*-biotin is added to 3000 ml of the medium. Cultivation is carried out for 7 days at 28° on a reciprocal shaker (140 reciprocations per minute).

ISOLATION. The cells are removed by centrifugation. The supernatant solution (about 2.6 liters, pH 5.0) is concentrated *in vacuo* to 1.2 liters. The pH is adjusted to 2.3 with hydrochloric acid, then 48 g of active carbon is added to the supernatant solution and stirred mechanically for 10 hours at room temperature. The active carbon is collected by filtration through a Büchner funnel and washed with about 1000 ml of water. The biotin vitamers adsorbed on the active carbon are eluted with about 700 ml of ethanol–ammonia mixture (prepared by mixing 9 volumes of 50% ethanol

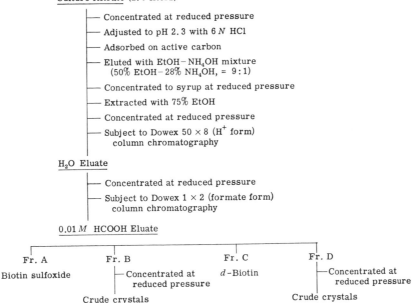

and volume of 28% ammonia-water). The eluate is concentrated to a syrup *in vacuo* at 50°. Then 300 ml of 75% ethanol is added to the syrup and the precipitates formed are removed by centrifugation. The clear supernatant solution is again concentrated to about 50 ml *in vacuo* at 50° and adjusted to pH 5.4. The concentrate is poured onto a Dowex 50 (H^+) column (6 × 75 cm). The biotin vitamers are washed out with 3800 ml of deionized water. The washings are concentrated to about 100 ml *in vacuo* at 50°. The concentrate is placed onto a Dowex 1 X2-formate column (3 × 100 cm), and the column is washed with 1000 ml of water. The biotin vitamers are eluted with 0.01 M formic acid, and fractions of 50 ml are collected. The content of biotin vitamers in each fraction is determined quantitatively by microbiological assays with *Saccharomyces cerevisiae*[4] and *Bacillus subtilis*,[1] respectively. As indicated in the scheme, four peaks (fractions A, B, C, and D) appear in the effluent of the column. Fractions A and C reveal the same biological activity for *Saccharomyces cerevisiae* and *Bacillus subtilis*. On the other hand, fraction B (about 4 liters) and fraction D (about 9 liters), which show biotin activity only for *B. subtilis*, are evaporated to dryness *in vacuo* at 50°. The crude crystals formed are washed with ethyl acetate plus ether (1:1). After recrystallization from hot water, 255 mg of bisnorbiotin and 508 mg of bisnorbiotin sulfoxide are obtained from fractions B and D, respectively.

Properties

Bisnordethiobiotin. The crystalline bisnordethiobiotin melts at 140–141°. Found (%): C, 51.57; H, 7.67; N, 14.85. Calculated for $C_8H_{14}O_3N_2$ (%): C, 51.60; H, 7.58; N, 15.05. The properties are identical to those of the chemically synthesized substance of Duschinsky and Dolan.[5] This substance has no optical rotation, suggesting that both *d-* and *l-*form of dethiobiotin are nonspecifically converted to bisnordethiobiotin.

Bisnorbiotin and Bisnorbiotin Sulfoxide. The crystalline bisnorbiotin is needlelike in shape and melts at 213–216°. Found (%): C, 44.33; H, 5.70; N, 12.70. Calculated for $C_8H_{12}O_3N_2S$ (%): C, 44.33, H, 5.59; N, 12.59. The crystalline bisnorbiotin sulfoxide is needlelike in shape and melts at 170–173°. Found (%): C, 40.16; H, 5.39; N, 11.45. Calculated for $C_8H_{12}O_4N_2S$ (%): C, 41.37; H, 5.21; N, 12.07.

[5] R. Duschinsky and L. A. Dolan, *J. Am. Chem. Soc.* **67**, 2079 (1945).

[69] Isolation and Characterization of Bisnordethiobiotin and Tetranordethiobiotin from Catabolism of Dethiobiotin in *Aspergillus niger*

By Heng-Chun Li, Donald B. McCormick, and Lemuel D. Wright

Dethiobiotin → bisnordethiobiotin → tetranordethiobiotin

Isolation

Principle. The mold is grown aerobically for 1–2 weeks on a liquid medium[1] to which ^{14}C(carbonyl)- or ^{3}H(random)-*d*-dethiobiotin is added. Mycelia are filtered off, remaining dethiobiotin and metabolites are concentrated by adsorption on and elution from charcoal, and the catabolites are separated by anion-exchange chromatography on Dowex 1 (formate) with formic acid as eluent. The two principal radioactive catabolites, bisnordethiobiotin (5-methyl-2-keto-4-imidazolebutyric acid) and tetranordethiobiotin (5-methyl-2-keto-4-imidazoleacetic acid), are separately evaporated to dryness and recrystallized from water and acetone, respectively.

Dethiobiotin. The radioactive *d*-dethiobiotin is prepared by desulfuration of radioactive *d*-biotin with Raney nickel[2] (see this volume [67]).

Procedure. Aspergillus niger is grown on 2 liters of a Czapek-Dok medium as described by Wright and Cresson,[1] except that 2 g of ^{14}C(carbonyl)- or ^{3}H(random)-*d*-dethiobiotin is included.[3] The medium is adjusted to pH 7.2 with dipotassium phosphate solution and divided into twenty 500-ml Erlenmeyer flasks. Incubation is for 12–14 days, at 30° with shaking at 250 rpm, after which time the mycelia are filtered off. The filtrate is adjusted to pH 2 with 1 N HCl and treated with 20 g of Norit A, which is filtered off and washed twice with deionized water. The adsorbed radioactive materials are eluted by stirring the Norit twice with 250-ml portions of ethanol–water–ammonium hydroxide (10:10:1, v/v/v), filtering off the charcoal each time, and evaporating the combined filtrates to dryness under reduced pressure with warming below 50°. The residue is largely dissolved in 75% aqueous ethanol and filtered; the filtrate is evaporated to a small volume. The solution is poured over a 1.6 × 95 cm column of Dowex 1-X8 (formate), and 15-ml fractions are collected during elution with a linear gradient from 2 liters of water to 2 liters of 0.03 M formic acid. Radioactiv-

[1] L. D. Wright and E. L. Cresson, *J. Am. Chem. Soc.* **74,** 4156 (1954).

[2] D. B. Melville, K. Dittmer, G. B. Brown, and V. du Vigneaud, *Science* **98,** 497 (1943).

[3] H. C. Li, D. B. McCormick, and L. D. Wright, *J. Biol. Chem.* **243,** 4391 (1968).

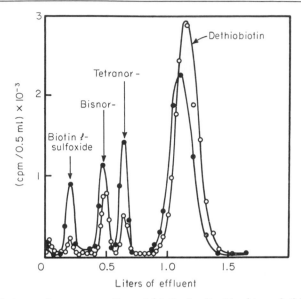

FIG. 1. Anion-exchange separation of biotin *l*-sulfoxide, bisnordethiobiotin, tetra-nordethiobiotin, and remaining dethiobiotin. ○, Relative amounts at 3 days; ●, at 8 days.

ity is determined in appropriate aliquots, usually 0.5-ml, with 10 ml of Bray's solution[4] in a liquid scintillation spectrometer.

Typical elution patterns from column chromatography of ^{14}C(carbonyl)-dethiobiotin and metabolites from 100-ml culture volumes of *A. niger* grown for 3 or 8 days with 5 μg of ^{14}C-dethiobiotin per milliliter are shown in Fig. 1.

The side chain-degraded catabolites of dethiobiotin, bisnor- and tetra-nordethiobiotin, are separately evaporated to dryness under reduced pressure below 50°. Bisnordethiobiotin is recrystallized from water, yielding about 0.5 g. Tetranordethiobiotin is recrystallized from acetone, yielding about 0.3 g.

Comments. Degradation of dethiobiotin to its 2-carbon shorter and 4-carbon shorter side-chain analogs by *A. niger* occurs extensively only after the mold has exhausted the supply of more readily metabolizable substrate, e.g. sucrose. With the growth conditions mentioned above, this has been estimated by loss of ^{14}CO$_2$ from ^{14}C(carboxyl)-*dl*-dethiobiotin to be more than a week.[3] It was also found that both *d*- and *l*-stereoisomers of dethiobiotin are degraded to approximately the same extent. Hence, the above procedure can be used to prepare the corresponding *l*-isomers of bisnor- and tetranordethiobiotin. ^{14}C(Carbonyl)- and ^{3}H(random)-labeled

[4] G. A. Bray, *Anal. Biochem.* **1,** 279 (1960).

catabolites with high specific radioactivities can be obtained with the correspondingly labeled dethiobiotin.

Characterization

The compounds present in each major fraction obtained from anion-exchange chromatography of dethiobiotin and its metabolites have been characterized by mobilities on column and paper chromatograms, melting points, specific radioactivities, infrared and mass spectra, and elemental compositions.[3]

Bisnordethiobiotin melts at 139–140°; tetranordethiobiotin at 157–159°.

The R_f values of dethiobiotin and its catabolites on paper chromatograms are compared in the table. These different mobilities are useful in ascertaining the purity as well as nature of the products.

The infrared spectra of dethiobiotin and its catabolites are compared in Fig. 2. Certain identification can be made on the basis of these spectra.

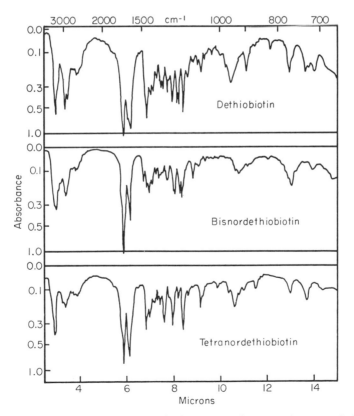

FIG. 2. Infrared spectra of dethiobiotin, bisnordethiobiotin, and tetranordethiobiotin. Approximately 1 mg of compound was compressed in 300 mg of KBr.

MOBILITIES OF DETHIOBIOTIN, BISNORDETHIOBIOTIN, AND
TETRANORDETHIOBIOTIN ON PAPER CHROMATOGRAMS

	R_f values[a] in solvent systems	
Compound	Butanol–acetic acid–water (2:1:1, v/v/v)	Butanol–methanol– benzene–water (2:1:1:1, v/v/v/v)
Dethiobiotin	0.78	0.80
Bisnordethiobiotin	0.78	0.71
Tetranordethiobiotin	0.70	0.55

[a] Locations of radioactive compounds were determined with a radiochromatogram strip scanner.

[70] Isolation and Characterization of Bisnorbiotin, α-Dehydrobisnorbiotin, and Tetranorbiotin from Catabolism of Biotin by *Pseudomonas* sp.

By SHOJIRO IWAHARA, DONALD B. McCORMICK, and LEMUEL D. WRIGHT

Biotin → bisnorbiotin → α-dehydrobisnorbiotin → tetranorbiotin

Isolation

Principle. A pseudomonad is grown aerobically for a week on a yeast extract–salts medium to which ^{14}C(carbonyl)-*d*-biotin is added.[1] The cells are centrifuged off, the supernatant solution is treated with Dowex 50 (H⁺) to remove cations, precipitates are removed after additions of ethanol, the solution is concentrated, and catabolites are separated by anion-exchange chromatography on Dowex 1 (formate) with formic acid as eluent. Bisnorbitin (*cis*-hexahydro-2-oxo-1*H*-thieno[3,4]imidazole-4-propionic acid) is purified on paper chromatograms developed with descending butanol–formic acid–water, dried, and recrystallized from water. α-Dehydrobisnorbiotin (*cis*-hexahydro-2-oxo-1*H*-thieno[3,4]imidazole 4-acrylic acid) is crystallized from methanol, as is tetranorbiotin (*cis*-hexahydro-2-oxo-1*H*-thieno[3,4]imidazole-4-carboxylic acid).

Procedure. Catabolites are isolated from 10 liters of liquid culture[1] which initially contains, per liter, 8 g of ^{14}C(carbonyl)-*d*-biotin, 2 g of yeast extract, and an inorganic salt mixture.[2] Incubation is for 7 days at 30° with continuous aeration. The bacterial culture is centrifuged for 20 minutes at

[1] S. Iwahara, D. B. McCormick, L. D. Wright, and H. C. Li, *J. Biol. Chem.* **244**, 1393 (1969).

16,000 g to remove cells. Three hundred grams of 100- to 200-mesh Dowex 50W-X8 (H⁺) is stirred into the total of about 9 liters of supernatant solution which is filtered through paper. The resin is washed with 6 liters of 70% ethanol and the washings are combined with the first filtrate. This combined solution is concentrated to about 600 ml, and the precipitate formed is removed by centrifugation. Six liters of 95% ethanol is then added to the supernatant solution, additional precipitate is removed, and the solution is concentrated to approximately 60 ml. After removal of the final precipitate, the solution (pH 7) is divided into equal portions and each portion is poured over a 1.5 × 70 cm column, 100- to 200-mesh, of Dowex 1-X2 (formate). Radioactive materials are eluted by successive treatment with water, 0.01 M and 0.1 M formic acid. Fractions of 15 ml are collected, and radioactivity is determined in suitable aliquots, usually 0.5 ml, with 10 ml of Bray's solution[3] in a liquid scintillation spectrometer.

A typical elution pattern from column chromatography of ¹⁴C(carbonyl)-d-biotin and catabolites from 20 ml of concentrated solution (equivalent to 3 liters of culture filtrate) is shown in Fig. 1.

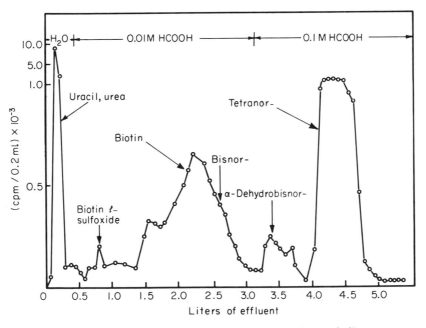

FIG. 1. Anion-exchange separation of biotin and its catabolites.

[2] R. N. Brady, L. F. Li, D. B. McCormick, and L. D. Wright, *Biochem. Biophys. Res. Commun.* **19,** 777 (1965).
[3] G. A. Bray, *Anal. Biochem.* **1,** 279 (1960).

The fraction from the column that contains biotin and bisnorbiotin is evaporated to dryness under reduced pressure below 50°, applied to preparative paper chromatograms, and developed in descending n-butyl alcohol–formic acid–water (4:1:1, v/v/v). After drying, the slower-moving bisnorbiotin is eluted with water and recrystallized from water to yield 5 mg. The fraction from the column that contains α-dehydrobisnorbiotin is evaporated to dryness, and the residue is dissolved in 10 ml of methanol and partially evaporated at room temperature to allow crystallization of 9 mg. The fraction from the column that contains tetranorbiotin is evaporated to dryness, the residue is dissolved in 15 ml of methanol, and an equal volume of ether is added to precipitate impurities that are removed by centrifugation. The supernatant solution is evaporated to dryness, the residue is dissolved in 5 ml of hot methanol, and the compound is allowed to crystallize overnight at 0°. The tetranorbiotin was collected and recrystallized 3 times from methanol to yield 500 mg.

Comments. The amounts of catabolites obtained can be changed by varying the conditions of bacterial growth. ^{14}C(Carbonyl)- and ^{3}H(random)-labeled catabolites with high specific radioactivities can be obtained with the correspondingly labeled biotin. Larger quantities of the bisnorbiotin can be isolated using similar techniques from the culture filtrates of *Penicillium oxalicum* as well.[4]

Characterization

The side-chain degraded catabolites of biotin as well as other metabolites, e.g., biotin d-sulfoxide, uracil, and urea, have been characterized by mobilities on column and paper chromatograms, melting points, specific

TABLE I

MOBILITIES OF BIOTIN AND CATABOLITES ON PAPER CHROMATOGRAMS

Compound	R_f values[a] in solvent systems	
	Butanol–formic acid–water (4:1:1, v/v/v)	Butanol–acetic acid–water (2:1:1, v/v/v)
Biotin	0.85	0.86
Bisnorbiotin	0.61	0.70
α-Dehydrobisnorbiotin	0.58	0.64
Tetranorbiotin	0.64	0.78

[a] Locations of radioactive compounds were determined with a radiochromatogram strip scanner.

[4] H. C. Li, Ph.D. thesis, Cornell University, Ithaca, New York, 1969.

radioactivities, elemental compositions, and mass, ultraviolet, infrared, and nuclear magnetic resonance spectra.[1]

Bisnorbiotin melts at 212–214°; α-dehydrobisnorbiotin at 235–236°; tetranorbiotin at 213–214°.

The R_f values of biotin and its catabolites on paper chromatograms are compared in Table I. These different mobilities are useful in ascertaining the purity as well as nature of the products.

The infrared spectra of biotin and its catabolites are compared in Fig. 2. Certain identification can be made on the basis of these.

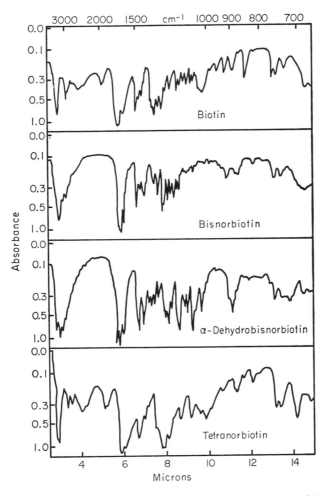

FIG. 2. Infrared spectra of biotin, bisnorbiotin, α-dehydrobisnorbiotin, and tetra-norbiotin. Approximately 1 mg of compound was compressed in 300 mg of KBr.

TABLE II

NUCLEAR MAGNETIC RESONANCE CHARACTERISTICS OF BIOTIN AND CATABOLITES[a]

Chemical shift for protons as δ (ppm)[b]

Compound	Methylene				Methine			Vinyl		Imino
	C-3'	C-4'	C-5'	C-2'	C-5	C-2	C-3,4	α	β	N-1,3
Biotin	←———	1.48	———→	2.28	2.93	3.24	4.38	—	—	6.53
Bisnorbiotin	1.91	—	—	2.37	2.98	3.73	4.38	—	—	6.57
α-Dehydrobisnorbiotin	—	—	—	—	2.89	3.13	4.43	6.01	6.53	6.53[c]
Tetranorbiotin	—	—	—	—	3.11	3.70	4.60	—	—	6.61

[a] All compounds were determined as 0.1 M in d$_6$-dimethylsulfoxide with tetramethylsilane as reference. The δ values are taken from the center of the signals even where multiplets occur.

[b] Positions in the acid side chains are numbered from the carboxyl as 1'; those in the thiophane ring are numbered counterclockwise from sulfur as 1.

[c] The chemical shift for these protons is the same as for the β-vinyl proton, and hence not readily distinguished.

The nuclear magnetic resonance characteristics of biotin and its catabolites are summarized in Table II. The presence or absence of particular protons in the side chains are indicative for each compound.

[71] Isolation and Characterization of Norbiotin and Trisnorbiotin from Catabolism of Homobiotin by *Pseudomonas* sp.

By HELMUT RUIS, ROBERT N. BRADY, DONALD B. MCCORMICK, and LEMUEL D. WRIGHT

Homobiotin → norbiotin → trisnorbiotin

Isolation

Principle. A pseudomonad is grown aerobically for a week on a biotin-salts medium.[1] The cells are ruptured by sonic oscillation[2] and incubated with ^{14}C(carbonyl)-*d*-homobiotin plus cofactors.[3] Particulate matter is removed by centrifugation, remaining homobiotin and catabolites are concentrated by adsorption on and elution from charcoal, the catabolites are separated by anion-exchange chromatography on Dowex 1 (formate) with ammonium formate as eluent, ammonium ions are removed on Dowex 50 (H$^+$), and the two principal radioactive catabolites, norbiotin (*cis*-hexahydro-2-oxo-1*H*-thieno[3,4]imidazole-4-butyric acid) and trisnorbiotin (*cis*-hexahydro-2-oxo-1*H*-thieno[3,4]imidazole-4-acetic acid), are separately evaporated to dryness and recrystallized from water.

Homobiotin. The *d*-homobiotin, from Hoffmann-La Roche, is labeled in the carbonyl carbon[4] by hydrolysis with barium hydroxide[5] and treatment of the resulting diaminocarboxylate with ^{14}COCl$_2$ essentially according to Melville *et al.*[6] A high-vacuum manifold similar to that described by Calvin *et al.*[7] is used for transferring the radioactive phosgene.

[1] L. F. Li, Ph.D. thesis, Cornell University, Ithaca, New York, 1964.

[2] R. N. Brady, L. F. Li, D. B. McCormick, and L. D. Wright, *Biochem. Biophys. Res. Commun.* **19**, 777 (1965).

[3] H. Ruis, R. N. Brady, D. B. McCormick, and L. D. Wright, *J. Biol. Chem.* **243**, 547 (1968).

[4] R. N. Brady, H. Ruis, D. B. McCormick, and L. D. Wright, *J. Biol. Chem.* **241**, 4717 (1969).

[5] K. Hoffman, D. B. Melville, and V. du Vigneaud, *J. Biol. Chem.* **141**, 207 (1941).

[6] D. B. Melville, J. G. Pierce, and C. W. H. Partridge, *J. Biol. Chem.* **180**, 299 (1949).

[7] M. Calvin, C. Heidelberger, J. C. Reid, B. M. Tolbert, and P. E. Yankwich, "Isotopic Carbon," p. 142. Wiley, New York, 1949.

Cells and Particulate Preparations. The soil pseudomonad, which grows on biotin as a sole source of carbon, nitrogen, and sulfur, is maintained on slants of 0.3% biotin agar.[1] Liquid medium at pH 7 is also prepared to contain 0.3% biotin, salts,[8] and trace elements.[9] The bacteria are grown at room temperature with aeration for a week or more. Cells are harvested by centrifugation, washed in isotonic saline, and lyophilized.[10] Suspensions of 10 mg of cells per milliliter are prepared in cold, 0.05 M potassium phosphate buffer at pH 6.2. The cells are ruptured by sonic oscillation for 5 minutes at 0° with a Branson sonifier model LS75. Rosett cooling vessels are used to maintain the temperature. The broken-cell preparations are centrifuged for 20 minutes at 12,100 g in a refrigerated centrifuge. The pellet obtained is resuspended in the above buffer with a hand homogenizer.

Procedure. Catabolites are isolated from 4320 ml of incubation mixture, 1.2×10^{-4} M in ^{14}C(carbonyl)-d-homobiotin.[3] The reactions are carried out in 18 four-liter flasks, each containing a total of 240 ml. Each reaction mixture contains 400 mg of broken cells and is 0.067 M in potassium phosphate buffer (pH 6.2) and 10^{-5} M in ATP, Mg^{2+}, NAD, and CoA. After 3 hours, an additional 400 mg of broken cells is suspended in 15 ml of 0.05 M potas-

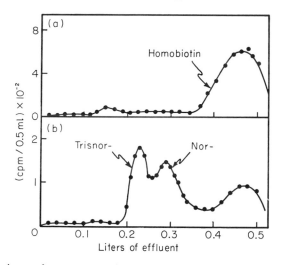

Fig. 1. Anion-exchange separation of homobiotin, norbiotin, and trisnorbiotin. Panel a shows data for the control after addition of boiled particulate preparation; panel b, after addition of active particulate preparation. Elution in these cases was with 1 M formic acid.

[8] V. M. Rodwell, B. E. Volcani, M. Ikawa, and E. E. Snell, *J. Biol. Chem.* **233,** 1548 (1958).

[9] D. B. Hoagland and W. C. Snyder, *Proc. Am. Soc. Hort. Sci.* **30,** 288 (1933).

[10] R. N. Brady, Ph.D. thesis, Cornell University, Ithaca, New York, 1967.

sium phosphate buffer, and the suspension is added to the incubation mixtures. The reactions are continued for an additional 5 hours. The mixtures are centrifuged for 20 minutes at 16,300 g, and the supernatant solutions are combined and concentrated under reduced pressure to 500 ml. Five grams of Norit is added, and the suspension is stirred for 1 hour. The Norit is removed by filtration and washed with water, and the ^{14}C-materials are eluted by stirring the Norit twice with 125 ml of ethanol–water–ammonia (50:45:5, v/v/v). After removal of the Norit by filtration, the combined solutions are evaporated under reduced pressure below 40°, and the residue is dissolved in 60 ml of water. The solution is poured over a 2.5 × 35 cm column of Dowex 1-X8, and 20-ml fractions are collected during elution with a linear gradient from 1 liter of water to 1 liter of 1 M ammonium formate. Radioactivity is determined in appropriate aliquots, usually 0.5 ml, with 10 ml of Bray's solution[11] in a liquid scintillation spectrometer.

Typical elution patterns from column chromatography of ^{14}C(carbonyl)-d-homobiotin and catabolites from 30-ml incubation volumes are shown in Fig. 1.

The side-chain degraded catabolites of homobiotin, i.e., nor- and trisnorbiotin, are separately evaporated to small volumes and poured over a 2.5 × 35 cm column of Dowex 50W-X8 (H$^+$) to remove ammonium ions; ^{14}C-materials are washed through with water and evaporated to dryness under reduced pressure. The residues were recrystallized from water for 7 mg of norbiotin and 12 mg of trisnorbiotin.

Comments. Nor- and trisnorbiotin accumulate during β-oxidative cleavage of the side chain of homobiotin, as the trisnorbiotin cannot be oxidized to lose one more methylene carbon and form tetranorbiotin which undergoes further catabolism.[12] Hence, broken-cell preparations are used in the present isolation of catabolites from homobiotin, since cultures of the growing pseudomonad require a source of N and S such as can be supplied by biotin but not homobiotin.

Characterization

Both catabolites have been characterized by mobilities on column and paper chromatograms, melting points, specific radioactivities, infrared spectra, and elemental compositions. In addition, the isolated norbiotin has been shown to be converted to trisnorbiotin upon incubation with the particulate preparation.[3]

Norbiotin melts at 259–263° (decomp.); trisnorbiotin at 226–230° (decomp.).

[11] C. A. Bray, *Anal. Biochem.* **1**, 279 (1960).
[12] S. Iwahara, D. B. McCormick, L. D. Wright, and H. C. Li, *J. Biol. Chem.* **244**, 1393 (1969).

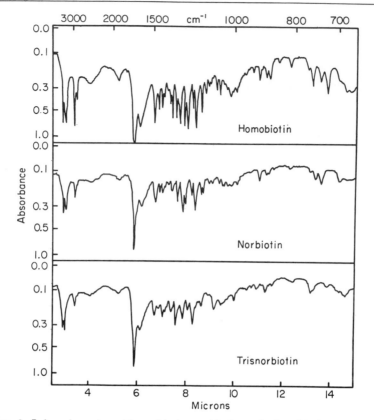

FIG. 2. Infrared spectra of homobiotin, norbiotin, and trisnorbiotin. Approximately 1 mg of compound was compressed in 300 mg of KBr.

MOBILITIES OF HOMOBIOTIN, NORBIOTIN, AND
TRISNORBIOTIN ON PAPER CHROMATOGRAMS

| | R_f values[a] in solvent systems | | |
Compound	Butanol–acetic acid–water (2:1:1, v/v/v)	Butanol–methanol–benzene–water (2:1:1:1, v/v/v/v)	Water-saturated butanol
Homobiotin	0.88	0.69	0.69
Norbiotin	0.81	0.48	0.40
Trisnorbiotin	0.70	0.32	0.17

[a] Locations of radioactive compounds were determined with a radiochromatogram strip scanner.

The R_f values of homobiotin and its catabolites on paper chromatograms are compared in the table. These different mobilities are useful in ascertaining the purity as well as nature of the products.

The infrared spectra of homobiotin and its catabolites are compared in Fig. 2. Certain identification can be made on the basis of these.

[72] Purification of Biocytin by Ion-Exchange Chromatography

By Donald B. McCormick and Werner Föry

Principle. A weakly acidic solution of biocytin contaminated by lysine and biotin is washed through Dowex 50 (NH$_4^+$) to remove lysine, the condensed effluent is neutralized and washed through Dowex 1 (chloride) to remove biotin, and biocytin is precipitated with acetone from a reduced volume at pII 6.

Procedure. Although synthesis of biocytin is readily accomplished by condensation of biotin methyl ester with excess lysine base, purification of the product by the countercurrent distribution method used earlier[1] is not very satisfactory. An easier method which is applicable in any scale to purification of biocytin from both synthetic and biological materials is the use of alternate column chromatography on a strong cation exchanger to remove lysine, other protonated bases, and metal ions and on a strong anion exchanger to remove biotin and other acid anions.

The material to be purified is dissolved in a minimal volume of water, the pH adjusted to 3 with dilute HCl, and any formed precipitate filtered off. The solution is poured over a 1 × 10 column of Dowex 50 W, 100- to 200-mesh, ammonium form, which should contain sufficient resin to retain any lysine or other strong bases originally present. The neutral, very water-soluble biocytin is washed through with a few hold-up volumes of water. The compound can be readily detected and judged free from lysine by thin-layer chromatography on silica gel sheets developed in ethanol–water (1:1), sprayed with 0.1% ninhydrin in butanol, and heat dried. Biocytin has an R_f value near 0.6 and lysine near 0.2. The aqueous effluent from the column is evaporated below 50° under partial vacuum to a small volume, the pH is adjusted to 7 with dilute LiOH, and poured over a 1 × 10 column of Dowex 1 X2, 100- to 200-mesh, chloride form, which should contain sufficient resin to retain any biotin or other acids originally present. Again biocytin

[1] J. Weijlard, G. Purdue, and M. Tishler, *J. Am. Chem. Soc.* **76**, 2505 (1954).

is washed through with a few hold-up volumes of water. Complete removal of biotin can be checked by thin-layer chromatography as before, but with iodine vapor to detect organic compounds as darker spots against a lighter background, or by the prior inclusion of a trace of radioactive biotin in the solution applied to the column. The aqueous effluent from the column is evaporated as before to a small volume, the pH is adjusted to 6 with a few drops of dilute LiOH, and biocytin is precipitated by the gradual addition with stirring of 10 volumes of acetone. The pure compound is collected on a sintered-glass funnel, rinsed with acetone, and dried *in vacuo*.

[73] Purification of Avidin

By N. M. GREEN

Introduction

Avidin has been purified from egg white (freshly separated or frozen) by a number of different procedures, most of them based on an initial adsorption step to separate it from the bulk of the other egg white proteins.[1-5] The following method is based on the use of carboxymethyl cellulose[2] followed by crystallization from ammonium sulfate.[3]

A more specific method has recently been published[5] using biocytin coupled to Sepharose. Although it was necessary to use 6 M guanidinium chloride, pH 1.5, to release the avidin, the yield and specific activity were excellent. Provided that the initial absorbent is conveniently available, this could eventually prove to be the method of choice. It has the advantage that the product will be relatively free of biotin since avidin molecules saturated with biotin will not be taken up by the adsorbent.

Assay

The colorimetric dye binding assay (this volume [74]) is used to assay column fractions. The titrimetric method is used for the final activity determination. All reagents are of analytical grade. All operations prior to crystallization may be performed at room temperature (23°).

Step 1. Carboxymethyl Cellulose. Approximately 100 g dry weight of

[1] H. Fraenkel-Conrat, N. S. Snell, and E. D. Ducay, *Arch. Biochem. Biophys.* **39,** 80 (1952).
[2] M. D. Melamed and N. M. Green, *Biochem. J.* **89,** 591 (1963).
[3] N. M. Green and E. J. Toms, *Biochem. J.*, in press, 1970.
[4] D. B. McCormick, *Anal. Biochem.* **13,** 194 (1965).
[5] P. Cuatrecasas and M. Wilchek, *Biochem. Biophys. Res. Commun.* **33,** 235 (1968).

CM-cellulose is used during the processing of each 8 lb of egg white. Whatman CM 11 or CM 22 grades are satisfactory provided that they are thoroughly washed. After removal of fines by decantation, the CM-cellulose is suspended in a mixture of equal volumes of 1 M NaOH and 1 M NaCl (6 liters/500 g of cellulose). The cellulose is filtered off, washed with 0.5 M NaCl (5 liters), suspended in water, and transferred to a column. The column is washed with water (30 liters), 1 mM EDTA (30 liters), and finally with deionized or glass-distilled water (30 liters). The CM-cellulose is then transferred as a thick slurry to a large Büchner funnel without any filter paper and allowed to drain. The filter cake is then sucked dry (3 kg containing 15% dry weight of CM-cellulose).

Step 2. Adsorption of Basic Egg White Proteins on CM-Cellulose. Frozen egg white (84 lb) is thawed, then gently homogenized in 5-liter batches with equal volumes of deionized water; the homogenate is partly clarified by centrifugation (1200 g, 20 minutes). After adjusting the pH to 7.0 with 2 N sulfuric acid, CM-cellulose filter cake (3 kg, 15% dry weight) is added and stirred for 10 minutes. After standing for several hours, the supernatant solution is siphoned from the CM-cellulose and then combined with further supernatant obtained by centrifuging the residual slurry (1200 g, 10 minutes). A further quantity of CM-cellulose (1.6 kg) is added, and the process is repeated. The combined CM-cellulose fractions are suspended in sufficient ammonium acetate (0.05 M, pH 7) to give a thick slurry, which is then poured onto a large Büchner funnel (30 cm) without any paper and allowed to drain without suction. The filter cake is washed under gravity with ammonium acetate (0.05 M, pH 7, 10 liters; 0.05 M, pH 9, 15 liters) followed by ammonium carbonate (0.3%, 5 liters). The cake is finally drained by gentle suction and suspended in ammonium carbonate (0.3%, 12 liters).

Step 3. Selective Elution of Avidin. The suspension is degassed and poured into a 9 × 100-cm column. During packing the downward flow rate is limited by reducing the overall head to 40 cm. The column is fitted with an upward flow adaptor and washed overnight with ammonium carbonate (0.4%, 15 liters). (The use of downward flow frequently leads to clogged columns.) Elution is then performed using a tenth percent stepwise gradient of 1.5 liters each of ammonium carbonate from 0.5 to 1.0% and 1.5 liters of 3%. Flow rate is 20 ml/min. Fractions (300 ml) can be collected manually. The A_{280} and avidin content of each is measured and the avidin-containing fractions are pooled. In order to regenerate the CM-cellulose for further use, the column is washed with 4% ammonium carbonate in 1 M ammonia (10 liters) followed by 0.5 M NaCl (10 liters). At this stage, the column is unpacked and the final washing with 0.05 M ammonium acetate (15 liters) is carried out on a large Büchner funnel, since swelling of the CM-cellulose in the dilute buffer prevents further washing on the column.

Lysozyme (40 mg/ml) is eluted by the 3% ammonium carbonate, and about 100 g is readily crystallized by raising the pH to 10 with NaOH (2 M) and adding NaCl (5% by weight).

Step 4. Chromatography on CM-Cellulose. The combined avidin fractions (7 liters) are acidified to pH 6.0 with acetic acid (50%), degassed, and run onto a column of CM-cellulose (6.5 × 80 cm). The column is eluted with a linear gradient of ammonium carbonate (0.5%–4.0% in 5 liters of total volume). The avidin emerges in the first peak (580 ml, 1.5 mg/ml) and is concentrated about 6-fold by vacuum dialysis.

Step 5. Crystallization of Avidin. The concentrated avidin solution is acidified to pH 5.0 with acetic acid (2 M), and ammonium sulfate (41.8 g/ 100 ml) is added to give a concentration of 2.5 M. The precipitate is centrifuged (12,000 g, 20 minutes) after standing overnight and rejected. Ammonium sulfate (9.2 g per 100 ml of solution) is added slowly with stirring (final concentration 3.0 M). Crystallization (thin plates) usually commences within a few hours, and at this stage a further 10.8 g of ammonium sulfate per 100 ml of solution is stirred in (3.5 M). The crystals are filtered after 24 hours at 4° and washed with a little 3.5 M ammonium sulfate. The avidin may be redissolved in water to give a concentration of 10–20 mg/ml, and the fractionation and crystallization may be repeated. The product is dissolved in 2.5 ml of water, dialyzed against deionized water, and freeze-dried.

A typical purification is given in the table.

PURIFICATION OF AVIDIN FROM FROZEN EGG WHITE (84 POUNDS)

Stage	Volume (ml)	A_{280}	Protein (g)	Avidin (g)	Specific activity (μg biotin/ mg protein)
Homogenized egg white	74,000	—	~4200	~1.2	~0.004
First column effluent	7,400	0.86	4.1	0.96	3.5
Second column effluent	580	2.4	0.90	0.83	14.1
First crystals[a]	40	29	0.75	0.75	15.1[b]

[a] It is possible to crystallize the avidin starting from material of specific activity 10, if seed crystals are available.

[b] Recrystallization gave no further increase in activity. A further 50–100 mg of slightly less pure avidin can be recovered from tail fractions from the columns and from the 2.5 M ammonium sulfate precipitate.

Properties

Avidin is a very stable, water-soluble glycoprotein. $E_{282} = 1.55$ mg/ml. The best preparations bind 15.1 μg of biotin per milligram of protein, corresponding to an equivalent weight of 16,200. Since the molecular weight is 68,000,[6] each molecule contains four subunits, into which it dissociates in guanidinium chloride $(5\ M)$[7] or HCl $(0.1\ M)$.[7,8] The lower activity of earlier preparations was probably due to the presence of a small amount of avidin–biotin complex.[8] If desired, this may be largely removed, with rather heavy losses, by the procedure described in the next section.

The avidin–biotin complex is even more stable than avidin, and does not completely release its biotin after 60 minutes at 100°.[9] It is necessary to autoclave at 120° (15 minutes) to obtain quantitative recovery. It has been reported[9,10] that the complex is less stable at low ionic strength, but this appears to have been found only with commercial preparations of avidin of low specific activity (2–3 units/mg). The effect of salt concentration on the stability of complexes of pure avidin both with biotin and with many weakly bound analogs is negligible at ordinary temperatures,[11] though it is possible that there may be some effect on the denaturation of avidin or avidin–biotin complex at 100°. It is also possible to release the biotin in 6–8 M guanidinium chloride at pH 1.5, where the avidin is dissociated into subunits. Removal of the biotin, followed by that of the guanidinium chloride and acid, allows the subunits to reassociate with good recovery of activity.[5,11]

Avidin Free of Biotin

It is possible to separate the avidin–biotin complex from avidin by taking advantage of the resistance of the complex to dissociation into subunits in 0.1 M HCl.[8] Avidin (20 mg) is dissolved in 0.1 M HCl (0.8 ml), and after standing for 30 minutes at room temperature, the solution is applied to a column of Sephadex G 100 (1 × 60 cm). Three peaks emerge. The first contains inactive aggregates, the second contains a mixture of some undissociated avidin with avidin–biotin complex, and the third contains avidin subunits which regain their activity after neutralization. About 50% of the starting activity is recovered in the subunit peak. It contains no tryptophan resistant to oxidation by N-bromosuccinimide and hence little or no avidin–biotin complex. Specific activity is 13–15 units/mg, depending on the starting material.

[6] N. M. Green, *Biochem. J.* **92,** 16c (1964).

[7] N. M. Green, *Biochem. J.* **89,** 609 (1963).

[8] N. M. Green and M. E. Ross, *Biochem. J.* **110,** 59 (1968).

[9] C. H. Pai and H. C. Lichstein, *Proc. Soc. Exptl. Biol. Med.* **116,** 197 (1964).

[10] R. Wei and L. D. Wright, *Proc. Soc. Exptl. Biol. Med.* **117,** 341 (1964).

[11] N. M. Green, unpublished experiments.

[74] Spectrophotometric Determination of Avidin and Biotin

By N. M. GREEN

Since avidin combines stoichiometrically with biotin, it is possible to use any physicochemical difference between avidin and the avidin–biotin complex as the basis of an assay method for either component. Two convenient techniques have been described. The first is based on the red shift of the absorption spectrum of the tryptophan residues of avidin, which accompanies combination with biotin, and requires measurement of the change in optical density at 233 nm.[1] The second is based on the use of the dye 4-hydroxyazobenzene-2'-carboxylic acid (HABA), which binds only to avidin and can therefore be used as an indicator for unoccupied binding sites.[2]

The method based on spectral shift is technically more demanding and has no advantages over the dye-binding method for routine estimations, but it is useful for the study of biotin analogs. Both methods are of lower sensitivity than the bioassays, but they are more precise, more convenient, and can be used over a wide range of pH and salt concentration. Methods for determination of avidin based on the use of radioactive biotin are described elsewhere (this volume [75]).

The unit of biotin binding activity is the amount of protein which binds 1 μg of biotin.

Reagents

    Avidin (Worthington, 12–13 units/mg; Nutritional Biochemicals, 2–3 units/mg) or biotin

    4-Hydroxyazobenzene-2'-carboxylic acid (HABA) (Eastman Kodak, Koch-Light). The dye should be recrystallized from aqueous methanol since impure preparations have given anomalous results at high salt concentration. One equivalent of base is required to dissolve it in water.

    Buffer, e.g., 0.1 M sodium phosphate, pH 7. Any other convenient buffer from pH 4–8 of any ionic strength can be used.

The binding of the dye HABA by avidin is accompanied by spectral changes summarized in Table I. More detailed spectral characteristics of the dye may be found elsewhere.[3] HABA is not bound by the avidin–biotin

---

[1] N. M. Green, *Biochem. J.* **89,** 585 (1963).
[2] N. M. Green, *Biochem. J.* **94,** 23C (1965).
[3] J. Baxter, *Arch. Biochem. Biophys.* **108,** 375 (1964).

complex, and since the dissociation constant of the latter is so low (10^{-15} M) the dye is stoichiometrically displaced by biotin. This can be made the basis of both colorimetric and titrimetric assays.

TABLE I

EXTINCTION COEFFICIENTS OF 4-HYDROXYAZOBENZENE-2'-CARBOXYLIC ACID (HABA) AND ITS COMPLEXES WITH AVIDIN AND STREPTAVIDIN[a]

	λ (nm)	ϵ_{282}	ϵ_{350}	ϵ_{500}
Avidin	282	25,000	0	0
HABA	350	2,800	20,500	480
Avidin–HABA complex	500	—	2,000	35,500
Streptavidin	280	$(57,000)^b$	—	0
Streptavidin–HABA complex	500	—	—	35,000

[a] All extinction coefficients are expressed per mole of biotin bound.

[b] This figure is corrected for slight opalescence of streptavidin solutions.

Colorimetric Assay for Avidin

1. To 2.0 ml of avidin solution (0.1–1 mg protein in 0.1 M phosphate or acetate buffer at any pH between 4 and 8) in a 1-cm cuvette, add HABA (10 mM, 50 μl).

2. Measure A_{500}.

3. Add biotin (2 mM, 50 μl).

4. Measure A_{500} to calculate

$$[\text{binding sites}] = \frac{\Delta A_{500}}{34} \text{ m}M$$

and

$$\text{avidin (mg/ml original solution)} = \frac{2.05}{2} \times \frac{16.2}{34} \Delta A_{500}$$
$$= 0.49 \Delta A_{500}$$

Colorimetric Procedure for Biotin

1. Pipette 2.0 ml of a solution of avidin–HABA complex (0.2–0.4 mg avidin/ml 0.25 mM HABA in 0.1 M phosphate buffer or acetate buffer, pH 4–8) into a 1-cm cuvette.

2. Measure A_{500} (A_1).

3. Add a known volume of biotin (v, ml) and again measure A_{500} (A_2) to calculate

$$[\text{biotin}] = \frac{A_1 - A_2(v + 2)/2}{34} \text{ m}M$$

Provided that $v < 0.2$ ml, the factor $(v + 2)/2$ has a negligible effect on ΔA_{500} and may be ignored. If $v > 1.0$ ml, correction is necessary for dilution of the dye as indicated in Table II. When v is negligible, micrograms of biotin per milliliter = $7.2 \Delta A_{500}$, or 1 μg biotin in 2 ml gives ΔA_{500} of 0.069.

It is possible to use a solution of pure biotin as a standard in place of the factors given.

Variations of Conditions

1. Use of microcuvettes increases the sensitivity.

2. The method can be used over a wide range of pH and salt concentration. Below pH 4 the dye begins to precipitate, since its carboxyl group is protonated (p$K \sim 4$), although binding to avidin is not much affected until the pH is below 2. Above pH 8 the hydroxyl group of HABA ionizes with a pK of 8.5, and the ionized form does not bind to avidin with accompanying spectral change. The apparent dissociation constant of the dye therefore increases rapidly above pH 7 (Table II). Between pH 4 and 8, the method can be used as described, since the extinction coefficient is independent of pH, and the dissociation constant varies only slightly (Table II).

The method has been used mostly with acetate or phosphate buffers, but it appears insensitive to the nature of the buffer or the concentration of salt.

TABLE II

EFFECT OF pH ON EXTINCTION COEFFICIENTS AND DISSOCIATION CONSTANTS OF THE AVIDIN DYE COMPLEX (AD)[a]

Protein	pH	$\Delta\epsilon_{500} \times 10^{-3}$ [b]	K $M \times 10^6$
Avidin	4	34	6
	5	33	13
	6	34	6
	7	34	7
	8	33	12
	9	23	65
Streptavidin	7	24	100

[a] The results for avidin at pH 7 are taken from N. M. Green, *Biochem. J.* **94**, 23C (1965). The remainder are from unpublished experiments.

[b] $\Delta\epsilon_{500}$ is an apparent $\Delta\epsilon$ calculated from the change in optical density following addition of excess biotin, when the concentration of dye is 0.25 mM. It is equal to ΔA_{500}/mole biotin bound per liter. Under these conditions, the binding sites are not quite saturated: [AD]/[A total] = $D/(D + K)$. The true

$$\epsilon_{500} \text{ (Table I)} = [\Delta\epsilon_{app}(D + K)/D + 480],$$

where 480 is the extinction coefficient of the free dye. This equation can be used to calculate $\Delta\epsilon_{app}$ at other dye concentrations. It is necessary to use a 15-fold greater dye concentration to achieve comparable saturation of streptavidin.

For example, 0.5 M ammonium sulfate has no effect on K or ϵ at pH 7.0, and even 3.5 M ammonium sulfate produces very small effects.

3. Lower concentrations of dye may be used provided that a correction is made for the smaller fraction of sites occupied by dye (Table II). It is inconvenient to use much higher dye concentrations because of the appreciable absorbance of the unbound dye (Table I).

Titration Method

The solutions used are the same as those employed above, but the biotin is added in small (1–10 μl) aliquots from a Hamilton microsyringe until no further change in A_{500} is observed. The end point, determined by plotting A_{500} against biotin added, is sharp, and the equivalence point between given solutions of avidin and biotin can be estimated within a few percent, so that the concentration of one may be determined if that of the other is known. The method is slightly more laborious than the colorimetric method, but it does not require any knowledge of the extinction coefficient or dissociation constant of the avidin–dye complex and can therefore be used over an even wider range of conditions without any corrections.

Specificity

The only other proteins that have been shown to give color changes with HABA are streptavidin[2] and serum albumin.[3] The binding to streptavidin is somewhat weaker (Table II), so that more dye is required to achieve a comparable saturation of the binding sites.[4] If this is taken into account it may be used in place of avidin for the estimation of biotin. Since the dye is not displaced from serum albumin by low concentrations of biotin it will not interfere with the assay, unless it is present in such large amounts that it binds a large fraction of the HABA.

Biotin Analogs

Combined forms of biotin, such as biocytin and various biotinyl enzymes (see below), are bound by avidin and titrate as biotin in this system. Biotin analogs will also displace HABA if the dissociation constants of their complexes are $<10^{-6}\ M$; however, they will not give sharp end points unless $K < 10^{-9}\ M$. Curvature in the neighborhood of the end point serves to distinguish weakly bound analogs from biotin, biotin sulfone, dethiobiotin, and biocytin.

Information obtained with bifunctional biotin compounds[5] and with biotin linked to an insoluble matrix[6] suggests that biotinyl proteins will

[4] N. M. Green, unpublished experiments.
[5] N. M. Green, *Biochem. J.* **104**, 64P (1967).
[6] P. Cuatrecasas and M. Wilchek, *Biochem. Biophys. Res. Commun.* **33**, 235 (1968).

react with avidin only when the biotin is separated from the surface of the macromolecule by at least five methylene groups. It is therefore possible that some forms of bound biotin will not be extimated by these methods. For example, the biotinyl groups in the polymeric form of acetyl coenzyme A carboxylase (activated by isocitrate) is not available to avidin.[7]

Estimation of Biotin Content of Biotinyl Enzymes[4,8]

When concentrated solutions of some biotinyl enzymes are added to avidin-HABA complex, the dye is displaced, but more slowly than by free biotin. When more than about 30% of the dye has been displaced, the reaction becomes even slower and does not go to completion. It appears that owing to the multi-subunit nature of both avidin and of the enzymes, it is sterically impossible to achieve a stoichiometric reaction. The magnitude of the anomaly varies with the enzyme and is, for example, much greater with pyruvate carboxylase than with transcarboxylase, which initially reacts fairly rapidly and completely. It is possible to eliminate these problems by digestion of the enzyme with pronase for 24 hours at 37° and assaying the digest by the colorimetric method described above, using concentrated solutions of the enzyme digest (2–5%) in order to obtain adequate optical density change. The presence of peptide digestion products does not affect the assay.

Spectral Shift Method for Assay of Biotin and Its Analogs[1]

The method requires avidin of high specific activity (>10 units/mg). The biotin solution should have low absorption in the UV down to 230 nm. Measurements of difference spectra are best made with a recording spectrophotometer. Some of the technical problems that may be encountered are considered in detail elsewhere.[9]

1. Two cuvettes are prepared with identical volumes (e.g., 2.0 ml) of a solution of avidin of $A_{280} < 0.4$ ($A_{233} < 1.6$). The difference spectrum (in this case a flat baseline) between 230 and 330 nm is determined with a recording spectrophotometer.

2. The wavelength is set to 233 nm (or 43,000 cm^{-1}) and a 2–10 µl sample of a solution of biotin (or analog) is added to one cuvette from a Hamilton microsyringe. An identical volume of buffer is added to the reference cuvette.

3. After adequate mixing (e.g., by a small magnetic stirrer in the cuvette), the A_{233} is recorded.

---

[7] E. Ryder, C. Gregolin, C. H. Chang, and M. D. Lane, *Proc. Natl. Acad. Sci. U.S.* **57**, 1455 (1967).

[8] M. C. Scrutton and A. S. Mildvan, *Biochemistry* **7**, 1490 (1968).

[9] T. T. Herskovits, Vol. XI, p. 771.

TABLE III

BINDING OF BIOTIN AND ANALOGS TO AVIDIN AT pH $7^{a,b}$

Compound	$\dfrac{\Delta A_{max}{}^c}{A_{282}}$	Dissociation constant of complex (M)
Biotin	0.94^d	$10^{-15\ e}$
Biotin sulfone	0.84^d	$<10^{-13\ f,g}$
N-Biotinyl-ϵ-aminohexanoate	0.76^d	$<10^{-13\ f,g}$
Biotin anilide	0.68^d	$<10^{-13\ f,g}$
Biotin methyl ester	0.74	$<10^{-13\ f,g}$
2′-Thiobiotin (pH 9.0)	1.09^i	$5 \times 10^{-13\ g,h}$
D-Dethiobiotin	0.47	$5 \times 10^{-13\ f,g}$
DL-Dethiobiotin methyl ester	0.39	$1 \times 10^{-11\ f,g}$
2′-Iminobiotin (free base)	0.76	$3.5 \times 10^{-11\ h}$
N-3′-Methoxycarbonyl biotin methyl ester	0.45	10^{-8} to $10^{-9\ d}$
N-1′-Methoxycarboxyl biotin methyl ester	0.52	$4 \times 10^{-7\ d}$
Diamine from biotin (free base) [δ-(3,4-diaminothiophan-2-yl)pentanoic acid]	0.73	$1.2 \times 10^{-7\ h}$
DL-3-n-Hexyl imidazolid-2-one	0.41	$1 \times 10^{-7\ f}$
DL-3-n-Butyl imidazolid-2-one	0.40	$1 \times 10^{-6\ f}$
DL-3-n-Propyl imidazolid-2-one	0.24	$5 \times 10^{-6\ f}$
DL-3-Ethyl imidazolid-2-one	0.22	$8 \times 10^{-6\ f}$
DL-3-Methyl imidazolid-2-one	0.23	$3 \times 10^{-5\ f}$
Imidazolid-2-one (ethylene urea)	0.26	$5 \times 10^{-4\ c}$
Urea	0.33	$4 \times 10^{-2\ d}$
DL-Lipoate	0.39	$7 \times 10^{-7\ d}$
n-Decanoate	0.11^i	$1 \times 10^{-5\ d}$
n-Hexanoate	0.10^i	$3 \times 10^{-4\ d}$
n-Hexyl alcohol	0.09^k	$4 \times 10^{-3\ f}$

[a] Information on the competition between biotin and a number of other analogs is given by L. D. Wright, H. R. Skeggs, and E. L. Cresson, *Proc. Soc. Exptl. Biol. Med.* **64**, 150 (1947).

[b] The difference extinction coefficients and dissociation constants do not depend strongly on pH between 4 and 10 unless the analog undergoes changes in ionization.

[c] Most complexes have $\lambda_{max} = 233.6$ nm ($43,000$ cm^{-1}). The values of $\Delta A_{max}/A_{282}$ were measured with a Unicam SP.700 recording spectrophotometer. Values some 10–15% higher are found with Cary spectrophotometers on account of their lower stray light level. $\Delta\epsilon$ per mole analog bound $= \Delta A_{max}/A_{282} \times 25,000$.

[d] N. M. Green, *Biochem. J.* **89**, 599 (1963).

[e] N. M. Green, *Biochem. J.* **89**, 585 (1963).

[f] N. M. Green, unpublished experiments.

[g] Approximate estimate from the rate of displacement of the analog by biotin. Procedure described for thiobiotin in footnote [h].

[h] N. M. Green, *Biochem. J.* **101**, 774 (1966).

[i] λ_{max} 238 nm.

[j] λ_{max} 229 nm.

[k] λ_{max} 231 nm.

4. The additions are repeated until no further change occurs.

5. The difference spectrum is recorded from 230 to 330 nm, as a check for anomalous spectral effects. When a particular system is thoroughly characterized, the recording of the difference spectrum may be omitted.

6. A_{233} is plotted against the amount of biotin or of analog added and the equivalence point is read from the graph. Sharp end points are obtained provided that K (the dissociation constant of the complex) is $< 10^{-7} M$.

Notes

1. The total A_{233} at the end point should be < 2.0 to avoid any errors from stray light. For this reason, avidin of high purity is required ($> 70\%$).

2. Volumes must be precisely controlled.

3. If the biotin analog contributes appreciably to A_{233}, this must be corrected either arithmetically or by use of tandem cells containing analog alone, at the same concentration as in the sample cuvette.[9]

4. The value of $\Delta\epsilon_{233}$, which is most readily expressed as $\Delta A_{233}/(A_{282}$ due to avidin), is characteristic of a particular analog, as is the profile of the difference spectrum in the 280 nm region. Table III gives values of $\Delta A_{233}/A_{282}$ and of the dissociation constants of a number of analogs to give some idea of the compounds that may affect assays involving avidin.

The main drawbacks of the method are the exacting technique required for good results and its sensitivity to interference by substances absorbing at 230 nm. Its precision can be nearly as good as that of the HABA methods, and it has advantages in that it can provide information about the dissociation constants of the complexes of avidin with biotin analogs and can serve as a method of characterizing such analogs.

[75] Assay of Avidin

By RU-DONG WEI

Definition of Unit and Specific Activity. A unit of avidin is defined as the amount that binds or inactivates 1 μg of (+)-biotin.[1] Specific activity is expressed as units of avidin per milligram of protein. On this basis, the

---

[1] R. E. Eakin, E. E. Snell, and R. J. Williams, *J. Biol. Chem.* **140,** 535 (1941).

purest preparation obtained by Melamed and Green[2] had a specific activity of 13.8 units per milligram of protein.

Assay Methods

In addition to microbiological methods,[1] which are extremely sensitive but require time-consuming manipulation for assay of avidin, several new methods have been developed recently in various laboratories.[2-8] Details of three simple and reliable assays are given below.

Method 1: Bentonite Adsorption[3,4]

Principle. The assay is based on the specific binding of the avidin to ^{14}C-labeled (+)-biotin, adsorption of the complex onto bentonite, trapping the complex on a cellulose acetate filter, and counting the radioactivity in a liquid scintillation counter.

Reagents

Ammonium carbonate, 0.2 M, pH 8.9
(+)-Biotin-carbonyl-[14]C (30–40 mCi/mmole), 0.1 μg/ml of 0.2 M ammonium carbonate (pH 8.9)
Bentonite
Bray's solution[9]

Procedure. The method is applicable to the determination of avidin in tissue homogenates. Tissue is homogenized in 6–10 times its weight of 0.2 M ammonium carbonate (pH 8.9) using a motor-driven Teflon pestle. The homogenate is centrifuged at 5000 g for 30 minutes, and the supernatant solution is centrifuged at 105,000 g for 2 hours. Avidin is measured in aliquots of the 105,000 g supernatant solution.

An aliquot of the 105,000 g supernatant solution, containing about 0.01–0.03 unit of avidin, is mixed with 0.5 ml of the ^{14}C-labeled biotin solution. After a short incubation at room temperature, 10 mg of bentonite is added. Five minutes later the mixture is transferred onto a Millipore filter under suction and washed with 0.2 M ammonium carbonate (pH 8.9). The filter containing the bentonite–avidin–biotin-^{14}C complex is treated in a

[2] M. D. Melamed and N. M. Green, *Biochem. J.* **89,** 591 (1963).
[3] S. G. Korenman and B. W. O'Malley, *Biochim. Biophys. Acta* **140,** 174 (1967).
[4] B. W. O'Malley, *Biochemistry* **6,** 2546 (1967).
[5] B. W. O'Malley and S. G. Korenman, *Life Sci.* **6,** 1953 (1967).
[6] R. D. Wei and L. D. Wright, *Proc. Soc. Exptl. Biol. Med.* **117,** 17 (1964).
[7] N. M. Green, *Biochim. Biophys. Acta* **59,** 244 (1962).
[8] N. M. Green, *Biochem. J.* **89,** 585 (1963).
[9] G. A. Bray, *Anal. Biochem.* **1,** 279 (1960).

counting vial with 10 ml of Bray's solution, leaving a bentonite sediment. Radioactivity is determined with a liquid scintillation counter.

Method 2: Sephadex Chromatography[6]

Principle. The method is based on measurement of the radioactivity of avidin–biotin complex which separates from free biotin when a reaction solution of avidin and excess radioactive biotin is applied to a Sephadex G 25 column.

Reagents

Ammonium carbonate, 0.2 M, pH 8.9
(+)-Biotin-carbonyl-[14]C (30–40 mCi/mmole), 0.1 μg/ml of 0.2 M ammonium carbonate (pH 8.9).
Sephadex G 25
Bray's solution.[9]

Procedure. To 0.6 ml of avidin sample solution, containing about 5 μg of avidin per milliliter of 0.2 M ammonium carbonate (pH 8.9), is added an equal volume of the ^{14}C-labeled biotin solution. After a brief mixing, a 1-ml aliquot is applied to a Sephadex G 25 column, 0.8 × 33 cm. The column is eluted with 0.2 M ammonium carbonate (pH 8.9) and fractions of 6 drops each are collected directly into counting vials with 10 ml of Bray's solution. The vials are counted by means of a liquid scintillation counter. The method gives a two-peak chromatogram. The area of the first peak corresponds to the avidin–biotin complex; in turn the amount of avidin can be calculated.

Method 3: Spectrophotometric Titration[7,8]

Principle. A shift of absorption spectrum to the red occurs during the avidin–biotin reaction. The change in extinction coefficient at 233 mμ is 25,000 per mole of biotin bound, which corresponds to a 25% increase in the absorbancy. This large increase provides the basis of the method for the estimation of avidin.

Reagents

(+)-Biotin, 0.01 M (24.4 mg/10 ml) in 0.2 M ammonium carbonate (pH 8.9).

Procedure. Into each of two cuvettes, pipette 3 ml of a sample solution containing approximately 0.2 mg (based on dry weight) of avidin per milliliter of 0.2 M ammonium carbonate. The cuvettes are placed in a double beam spectrophotometer. Zero the instrument at 233 mμ using one of the cuvettes as a blank (if total extinction is greater than 1.5, dilute the avidin solution further). Add 1 μl of the biotin solution to the sample cuvette, mix, and record the increase in absorbancy at 233 mμ. Continue to add 1-μl aliquots of biotin until no further increase in absorbancy is observed. Plot the results on graph paper to determine the equivalence point. The specific activity can then be calculated by

$$\text{units/mg} = \frac{(\mu\text{l biotin used at equivalence point})(\text{conc. in } \mu\text{g}/\mu\text{l})}{\text{mg avidin per 3 ml}}$$

[76] Newer Methods of Avidin Assay

By S. G. KORENMAN, and B. W. O'MALLEY

Recent renewed interest in avidin, the biotin binding protein of egg white, has been based both on its unusual chemical properties and on hormonal control of its synthesis by the chick oviduct. Study of the protein required assay procedures less cumbersome and more precise than the microbiological technique of Eakin, Snell, and Williams.[1]

Melamed and Green[2] devised a method of preparing relatively pure avidin. Using this material, Green determined the molecular weight and number of binding sites,[3] the kinetics of the binding,[4] and the nature of the binding site.[5] During the course of these studies, two spectrophotometric assays[6,7] and a radioligand binding assay[8] for avidin were used.

Because of our interest in the mechanisms of steroid hormone action, we sought a model system in which the synthesis of *specific* proteins is under the control of individual hormones. It was apparent that regulation of egg-white protein synthesis by sex hormones represents such a system, and particularly that avidin biosynthesis, which is regulated by both estrogens and progesterone,[9] would be of great interest. Accordingly, a convenient, specific, rapid, and precise radioligand binding assay system for avidin suitable for the study of tissue fractions has been developed.[10] By means of

---

[1] R. E. Eakin, E. E. Snell, and R. J. Williams, *J. Biol. Chem.* **140,** 535 (1941).

[2] M. D. Melamed and N. M. Green, *Biochem. J.* **89,** 591 (1963).

[3] N. M. Green, *Biochem. J.* **92,** 16c (1964).

[4] N. M. Green, *Biochem. J.* **89,** 585 (1963).

[5] N. M. Green, *Biochem. J.* **89,** 599 (1963).

[6] See footnote 5.

[7] N. M. Green, *Biochem. J.* **94,** 23c (1965).

[8] N. M. Green, *Biochem. J.* **89,** 599 (1963).

[9] R. M. Fraps, R. Hertz, and W. H. Sebrell, *Proc. Soc. Exptl. Biol. Med.* **52,** 140 (1943).

[10] S. G. Korenman and B. W. O'Malley, *Biochim. Biophys. Acta* **140,** 174 (1967).

this procedure, considerable information has been gathered on the mechanism of action of progesterone in stimulating avidin biosynthesis.[11–13] It has also been possible, employing a specific antiserum, to develop an immunoassay for avidin.[14]

Avidin Standards

Avidin of high specific activity is not always commercially available. Crude avidin preparations can be used as standards as long as their content of unsaturated biotin binding sites is carefully measured and their specific activity is determined. One unit of avidin activity is defined as that required to bind 1 μg of biotin. Pure avidin contains 13.8 U/mg.[15] By the method of Melamed and Green,[16] employing ion-exchange chromatography on carboxymethyl cellulose, material with a specific activity of 10 U/mg or greater can be prepared from egg whites. The ingenious procedure devised by McCormick[17] using a biotin-cellulose column is another means that can be used to enrich crude avidin preparations.

Spectrophotometric Assays

A. Titration of avidin concentration by measurement of the increase in optical density at 233 mμ produced by addition of biotin until all the binding sites of avidin are saturated.[18] This assay is insensitive and depends upon the accuracy of the titration procedure.

B. Binding of the anionic dye, 4'-hydroxyazobenzene-2-carboxylate, to avidin was shown to result in a new absorption band at 500 mμ.[19] The dye is bound at the same sites as biotin and can be displaced by biotin. For assay purposes, to ensure specificity, dye is added until the binding sites of a dilute avidin solution at pH 7.0 are just saturated, and then by titration the amount of biotin needed to eliminate the absorbance at 500 mμ is determined. The method is sensitive to about 25 μg/ml of avidin. It is not suitable for assay of tissue fractions containing only a small amount of avidin because dye binding to many proteins may occur and the effects of biotin titration may not be discernible.

[11] S. G. Korenman and B. W. O'Malley, *Endocrinology* **83,** 11 (1968).
[12] B. W. O'Malley, *Biochemistry* **6,** 2546 (1967).
[13] B. W. O'Malley, S. L. McGuire, P. O. Kohler, and S. G. Korenman, *Recent Progr. Hormone Res.* **25,** 105 (1969).
[14] B. W. O'Malley and S. G. Korenman, *Life Sci.* **6,** 1953 (1967).
[15] M. D. Melamed and N. M. Green, *Biochem. J.* **89,** 591 (1963).
[16] See footnote 15.
[17] D. B. McCormick, *Ann. Biochem.* **13,** 194 (1965).
[18] N. M. Green, *Biochem. J.* **89,** 599 (1963).
[19] N. M. Green, *Biochem. J.* **94,** 23c (1965).

Radioligand Binding Assays

In general, these procedures involve interaction of a labeled ligand, such as biotin, with a specific binding protein, such as avidin. By measurement of the amount of radioactivity either in the ligand–protein complex or unbound, under the appropriate conditions, the amount of either the ligand or the protein can be determined.

In these assays, to measure avidin content an excess of radioactive biotin is reacted with avidin, the avidin–biotin complex is separated from unbound biotin, and avidin is measured by the number of counts bound. The sensitivity of the method is limited only by the specific activity of the labeled biotin. In reality, the number of unsaturated biotin binding sites, rather than the total number of avidin molecules present, are measured. Because of the extremely low equilibrium constant of dissociation of the avidin-biotin complex,[20] the binding reaction may be considered irreversible. Green used both batch adsorption to carboxymethyl cellulose and ultrafiltration to separate bound from free biotin,[21] while Wei and Wright employed gel filtration on Sephadex.[22] Korenman and O'Malley[23] used adsorption of the avidin–biotin complex onto bentonite. Only the latter procedure, which is the most convenient, will be described in detail.

Water is distilled and deionized to ensure that it is iron free. Bentonite is used as obtained. Cellulose nitrate filters, 0.45-μ pore size, 25 mm in diameter, are obtained from the Millipore Corporation. Biotin ^{14}C with a specific activity of 50 μCi/mg is used. All of a sample must be shown to bind to an excess of avidin (radioactive biotin should always be tested for purity by demonstrating that it binds quantitatively to avidin).

Samples and standards are assayed in triplicate. Reagents and avidin samples are diluted in 0.2 M ammonium carbonate. The avidin and biotin solution are mixed, taken to 0.6 ml, and incubated for 10 minutes at 23°. One milliliter of 0.2 M ammonium carbonate containing 10 mg of bentonite is then added. The suspension is stirred, incubated for 5 minutes, and transferred to a Millipore filter under suction and rinsed once. Unbound biotin passes through the filter. The filter is transferred to a counting vial, wet with 1 drop of ethanol, and dissolved in Bray's solution,[24] leaving a bentonite sediment. Radioactivity is measured in a liquid scintillation spectrometer at high efficiency. Using the available biotin specific activity, avidin concentrations as low as 0.1 μg/ml can be measured with ease. The sensitivity

[20] N. M. Green, *Biochem. J.* **89**, 585 (1963).
[21] See footnote 20.
[22] R. D. Wei and L. D. Wright, *Proc. Soc. Exptl. Biol. Med.* **117**, 17 (1964).
[23] S. G. Korenman and B. W. O'Malley, *Biochim. Biophys. Acta* **140**, 174 (1967).
[24] G. A. Bray, *Ann. Biochem.* **1**, 279 (1960).

can be increased by utilizing tritiated biotin if it is available. This procedure is suitable for avidin assay in crude oviduct supernatant fractions.

Immunoassay

Avidin is quite antigenic.[25] By the use of rabbit antiavidin and sheep antirabbit γ-globulin, O'Malley and Korenman[26] were able both to assay avidin by biotin ^{14}C binding and to determine that portion which was newly synthesized by simultaneous measurement of incorporation of a radioactive amino acid into antibody-precipitable protein.

Specific antibodies are produced in rabbits by giving 4 weekly injections of avidin in Freund's adjuvant and then bleeding the animals. Sheep antirabbit γ-globulin is obtained commercially.

The reactions are carried out at 23°. Avidin is incubated with a small excess of specific antiserum for 2 hours. Then an excess of biotin ^{14}C is added and incubation is continued for 20 minutes. The avidin–antibody complex binds biotin just as well as free avidin. An excess of sheep antirabbit γ-globulin is added, and incubation is carried out for 3 hours. The resultant precipitate is centrifuged, washed, dissolved in 1 ml of NCS reagent (Nuclear Chicago Corp.), added to 10 ml of toluene phosphor, and counted in a liquid scintillation spectrometer. If the avidin specifically precipitated has been newly formed in the presence of a tritiated amino acid (i.e., as a result of induction by progesterone) the proportion of new protein synthesis due to avidin and the specific activity of the avidin can be determined by simultaneous measurement of ^{14}C and ^{3}H radioactivity.

Availability of radioactively labeled biotin has made it possible to develop a number of convenient assay procedures for avidin suitable for study of both its biochemical characteristics and regulation of its biosynthesis. The procedures outlined can be modified with ease to permit quantitative assay of streptavidin, biotin, and actively binding biotin analogs as well.

[25] D. V. Siva Sankar, B. J. Cossano, H. W. Theis, and C. R. Marks, *Nature* **181**, 619 (1958).

[26] B. W. O'Malley and S. G. Korenman, *Life Sci.* **6**, 1953 (1967).

Pyridoxine, Pyridoxamine, and Pyridoxal:
Analogs and Derivatives

[77] Analyzing Spectra of Vitamin B_6 Derivatives[1]

By ROBERT J. JOHNSON and DAVID E. METZLER

The 3-hydroxypyridine chromophore of vitamin B_6 is sensitively dependent upon structural factors and environment and provides a built-in indicator at the active sites of pyridoxal phosphate-dependent enzymes. Thus, the correct interpretation of subtle differences in the spectrum of the coenzyme may provide valuable information about events at the active site. This is one reason for the biochemist's interest in spectra of the various forms of vitamin B_6.

The spectrum of pyridoxol[2] (Fig. 1) is typical. Each of the three ionic

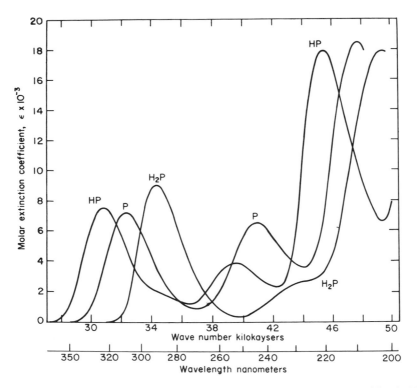

FIG. 1. Absorption spectra of the three individual ionic forms of pyridoxol. The forms are labeled H_2P, HP, and P according to the extent of protonation.

[1] This study was supported by grants from the U.S. Public Health Service (No. AM-1549) and from the National Science Foundation (No. GB-5378).
[2] Although the spectra are very similar, some discrepancies exist between our spectra shown in Fig. 1 and those reported by Bazhulina *et al.* (footnote 4) above 46 kK.

forms possesses three major absorption bands corresponding to transitions to three different excited states. The bands will be designated here simply as band I (lowest frequency; longest wavelength), band II, and band III. Both the peak positions and molar extinction coefficients vary with the state of dissociation of the ring nitrogen and of the phenolic hydroxyl groups. This article describes methods for obtaining information about dissociation constants and various other chemical equilibria from spectral data. The methods are equally applicable to the study of many other light-absorbing substances. In this paper spectra are described by giving absorbances and wave numbers of peaks in kilokaysers (kK) (1 kilokayser = 1000 cm^{-1}). This is desirable because of the direct proportionality of wave numbers with frequencies and energies. The wavelengths in nanometers (nm) are usually also given in parentheses, and figures carry both wave number and wavelength scales.

I. Digital Recording of Spectral Data

Collection of spectrophotometric data with direct digital output equipment offers several advantages over the usual methods. (1) Exact absorbances are recorded at precisely selected wavelengths and, at the same time, the spectra are recorded graphically in the usual way. (2) The "blank" or "baseline" absorbances can be subtracted automatically from the sample absorbances using a digital computer, and other corrections can be applied with equal ease. (3) Molar extinction coefficients, ϵ, can be computed with ease at all wavelengths, and precise plots of ϵ or log ϵ versus wavelength or wave number can be produced automatically and inexpensively at many existing computation centers. (4) Using a high-sp eed computer, it is possible to obtain least squares best fits for pK_a's and o ther equilibrium constants using data at *all* wavelengths and to make automatic comparison plots in which "theoretical" curves, using the best parameters found, are compared directly with the experimental spectra.[3] (5) The processed data in digital form, e.g., the spectra of individual ionic forms and pK_a's (as in Fig. 1) constitute a "library form" of spectral data which are available for future uses, such as studying spectra–structure relationships and comparing spectra obtained in different laboratories.

Many of the data shown in this article (e.g., Figs. 1, 8, and 10) were recorded with a Cary Model 15 spectrophotometer equipped with a Cary Datex digital output system and an IBM card punch. The so-called "wavelength interval command" was used to record absorbances at regular 2 nm intervals of wavelength. More closely spaced points can be obtained when necessary.

---

[3] K. Nagano and D. E. Metzler, *J. Am. Chem. Soc.* **89**, 2891 (1967).

II. Determination of pK_a's

The spectrophotometric determination of acid dissociation constants (as pK_a's, referred to hereafter simply as pK's) is simple and reliable for many aromatic substances. Specific advantages over the titration method are that very small amounts of sample are required and that pK's falling in the high and low pH ranges (above 10 and below 3) can be determined reliably. Dilute solutions (10^{-4} M) and cuvettes of path length 1.0 cm are usually employed; more concentrated solutions may be used by decreasing the light path. At concentrations from 10^{-5} to 10^{-2} M, Beer's law is obeyed to a high degree of accuracy for 3-hydroxypyridine and for pyridoxol, pyridoxal, pyridoxamine, and their phosphates.[4–7] This is true at all pH values, the spectra of the individual ionic forms being constant.

Procedures. Samples may be titrated and the pH measured in the cuvette between each spectral measurement,[8] a method which is desirable for use with small amounts of enzyme. Often it is simpler to prepare a series of dilutions of a single stock solution using buffers of selected pH. For example, an aqueous stock solution about 5 $\times$ 10^{-4} M in the absorbing substance may be diluted 5-fold into clean glass-stoppered volumetric flasks containing portions of appropriate stock buffers. Spectra (complete or at selected wavelengths) are recorded against solvent "blanks" containing the same buffers as the samples. The pH of each sample is measured with great care immediately before or after the spectrum is recorded.

For enzymes, a final concentration of about 2 mg of protein per milliliter is usually satisfactory. Dilution with a buffer may be made with precision pipettes or syringes directly into the cuvette or in a small test tube. The enzyme stock solution should be buffered only weakly, and the exact pH should be determined on each test sample. Use of a probe electrode permits measurement of the content of a single semimicro cuvette (0.5 ml or less) after transfer to a small test tube.

Selection and Preparation of Buffers. To determine a single pK, the spectra of each of the two ionic species must be established as well as the pK. An absolute minimum of three solutions of different pH is required, but at least 4–6 solutions are recommended, and at least 3 or 4 more

[4] N. P. Bazhulina, Yu. V. Morozov, M. Ya. Karpeisky, B. I. Ivanov, and A. I. Kuklin, *Biofizika* **11**, 42 (1966).

[5] Yu. V. Morozov, N. P. Bazhulina, V. I. Ivanov, M. Ya. Karpeisky, and A. I. Kuklin, *Biofizika* **10**, 595 (1965).

[6] Yu. V. Morozov, N. P. Bazhulina, M. Ya. Karpeisky, B. I. Ivanov, and A. I. Kuklin, *Biofizika* **11**, 228 (1966).

[7] Yu. V. Morozov, N. P. Bazhulina, L. P. Cherkashina, and M. Ya. Karpeisky, *Biofizika* **12**, 397 (1967).

[8] W. T. Jenkins and L. D'Ari, *J. Biol. Chem.* **241**, 5667 (1966).

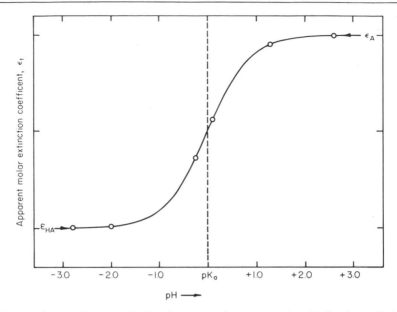

FIG. 2. Spectrophotometric titration curve of a monoprotic acid showing a desirable selection of pH values for experimental points for determining ϵ_{HA}, ϵ_A, and pK_a.

should be added for each additional pK of the compound. The pH values should be selected as follows (see also Fig. 2). Two should be near the pK (within ±0.3 unit) and one should be at least 2, and preferably 3 or more, units away from the pK in *each direction*. These extreme pH values are usually obtained with dilute (0.01–0.1 N) HCl or NaOH. Additional pH values near the extremes are desirable. Usually buffers of the total molar concentration of 0.02–0.2 are used; it is best to maintain a constant ionic strength. While Morozov *et al.*[4–7] have shown that the spectra of most forms of vitamin B_6 are not influenced by ionic strength, pK's will be affected because of interionic electrostatic effects. Most workers prefer to determine apparent pK's at relatively high ionic strength (0.1 or more) and to assume that activity coefficients are constant under these conditions.

Some suitable buffer acids are listed in Table I. Buffers of any desired pH can be prepared quickly by mixing together the component stock solutions in appropriate amounts according to Eq. (1), where the pK represents the *apparent* pK at the ionic strength of

$$pH - pK = \log \frac{[A]}{[HA]} = \log \frac{f_d}{1 - f_d} \tag{1}$$

the buffer and f_d is the degree of dissociation, i.e., the fraction of total

TABLE I

BUFFER ACIDS SUITABLE FOR USE WITH VITAMIN B$_6$ COMPOUNDS

Buffer acids	Apparent pK$_a$	pH Range	Suggested storage form^b
HCl	—	0–2.5	1 or 2 N HCl
Formic acid	3.7	2.7–4.7	1 or 2 N Formic acid^c
Acetic acid	4.6	3.6–5.6	1 or 2 N Acetic acid^c
Cacodylic acid	6.1	5.1–7.1	1 M Sodium cacodylate
Dihydrogen phosphate	6.8	5.8–7.8	0.5 M KH$_2$PO$_4$ and Na$_2$HPO$_4$
Triethanolammonium	7.9	6.9–8.9	1 M Triethanolamine hydrochloride^d
Bicarbonate	10.0	9.0–11.0	0.5 M NaHCO$_2$
Sodium hydroxide	—	11.5–13	1 or 2 N NaOH^a

^{a} For buffers from 0.02 to 0.2 M.

^{b} HCl, NaOH, formic and acetic acid solutions should be standardized to ±1%.

^{c} Stopper tightly; formic and acetic acid solutions lose acid readily and must be titrated quickly during standardization.

^{d} Triethanolamine hydrochloride may be prepared by mixing 250 g of triethanolamine in 300 ml of 8 N HCl, allowing to crystallize, collecting, and washing with cold methanol or ethanol. Recrystallize by dissolving in hot 0.5 N HCl and cooling to room temperature. Wash the crystals with cold methanol and air dry.

buffer acid in the dissociated form, A.[9] To save time the chart in Table II can be used. For example, to prepare a buffer having a pH 0.2 unit above the pK of the buffer acid (f_d = 0.61) one could either mix 39 parts of the acid component of the buffer with 61 parts of the conjugate base or mix 100 parts of the acid with 61 parts of NaOH. If the pH selected is within 0.5 unit of the pK$_a$, the solution prepared this way should have the desired pH to within 0.1 unit or less (with the exception that at very low pH, the observed pH's of formate buffers will be a little higher than expected, and at high pH the carbonate buffers will yield pH's a little lower than calculated).

Carboxylic acid and amine buffers absorb light significantly above 40 kK (below 250 nm). It is sometimes desirable to measure in HCl, phosphate and carbonate buffers and dilute NaOH (all very transparent) down to low wavelengths to obtain complete information about spectra, but to add data in less transparent buffers over a shorter wavelength range to obtain pK's. Buffers containing primary or secondary amine groups should usually be avoided when dealing with aldehyde forms of vitamin B$_6$; borate buffers should also be avoided.

Calculation of pK$_a$ Values. If Beer's law is obeyed by all ionic species and if no interaction occurs between species, the absorbance (A) of a

[9] Compositions of many buffers are also given in Vol. I, pp. 138–146.

TABLE II
The Degree of Dissociation, f_d, of a Monoprotic Acid versus pH;

$$\Delta = pH - pK = \log_{10} f_d/1 - f_d$$

Δ	f_d	Δ	f_d	1,6	2,7	3,8	4,9
0.00	0.500	0.00	0.500	6	12	18	23
−0.05	0.471	0.05	0.529	6	11	17	23
−0.10	0.443	0.10	0.557	6	11	17	22
−0.15	0.415	0.15	0.585	6	11	17	22
−0.20	0.387	0.20	0.613	5	11	16	21
−0.25	0.360	0.25	0.640	5	10	16	21
−0.30	0.334	0.30	0.666	5	10	15	20
−0.35	0.309	0.35	0.691	5	10	15	19
−0.40	0.285	0.40	0.715	4	9	14	18
−0.45	0.262	0.45	0.738	4	9	13	16
−0.50	0.240	0.50	0.760	4	8	12	16
−0.55	0.220	0.55	0.780	4	8	12	15
−0.60	0.201	0.60	0.799	3	7	11	14
−0.65	0.183	0.65	0.817	3	6	10	13
−0.70	0.166	0.70	0.834	3	6	9	12
−0.75	0.151	0.75	0.849	3	6	9	12
−0.80	0.137	0.80	0.863	3	5	8	11
−0.85	0.124	0.85	0.876	3	5	8	10
−0.90	0.112	0.90	0.888	2	5	7	9
−0.95	0.101	0.95	0.899	2	4	6	8
−1.00	0.091	1.00	0.909	2	4	6	8
−1.05	0.082	1.05	0.918	2	3	5	7
−1.10	0.074	1.10	0.926	2	3	5	6
−1.15	0.066	1.15	0.934	1	3	4	6
−1.20	0.059	1.20	0.941	1	2	3	5
−1.25	0.053	1.25	0.947	1	2	3	4
−1.30	0.048	1.30	0.952	1	2	3	4
−1.35	0.043	1.35	0.957	1	2	3	4
−1.40	0.038	1.40	0.962	1	2	2	3
−1.45	0.034	1.45	0.966	1	1	2	2
−1.50	0.031	1.50	0.969	1	2	2	3
−1.55	0.027	1.55	0.973	0	1	1	2
−1.60	0.025	1.60	0.975				
−1.70	0.020	1.70	0.980				
−1.80	0.016	1.80	0.984				
−1 90	0.012	1.90	0.988				
−2.00	0.010	2.00	0.990				
−2.10	0.008	2.10	0.992				
−2.20	0.006	2.20	0.994				
−2.30	0.005	2.30	0.995				
−2.40	0.004	2.40	0.996				
−2.50	0.003	2.50	0.997				
−2.70	0.002	2.70	0.998				
−3.00	0.001	3.00	0.999				

Proportional parts header spans columns: 1,6 | 2,7 | 3,8 | 4,9

solution will be given by Eq. (2) where C_j and ϵ_j are the molar concentration and molar extinction coefficient, respectively, of the jth species and l is the path length in centimeters.

$$A = \sum_j \epsilon_j C_j l \tag{2}$$

Dividing the left side of Eq. (2) by $C_t l$, where C_t is the total concentration of all species of a given substance which are in equilibrium together in the solution, we obtain Eq. (3), in which ϵ_t is the *apparent* molar extinction coefficient of the substance, and f_j is the

$$\frac{A}{C_t l} = \epsilon_t = \sum_j \epsilon_j C_j / C_t = \sum_j \epsilon_j f_j \tag{3}$$

fraction of the total represented by the jth species.

Computation is easiest for monoprotic acids and for polyprotic acids for which there is no significant overlapping of the dissociation steps. As illustrated in Fig. 2, the extinction coefficients of HA and A, ϵ_{HA} and ϵ_A, are obtained directly from samples of low and high pH. To ensure that 99% or more of a single ionic form is present, these pH's must be at least 2.0 units away from the pK. The apparent extinction coefficients, ϵ_t, of the samples with pH values near the pK are then used to determine the degrees of dissociation, f_d, for each sample according to Eq. (4), and Eq. (1) is used to obtain the pK. Table II contains the solutions

$$f_d = \frac{\epsilon_t - \epsilon_{HA}}{\epsilon_A - \epsilon_{HA}} \tag{4}$$

of Eq. (1) in convenient form; the values of pH $-$ pK_a (Δ) being given there for all possible values of f_d. Values of pK obtained for two or more points within ± 0.3 unit of the pK should be averaged. It is usually a good idea to construct the theoretical titration curve using the calculated pK value and to plot it, as in Fig. 2, together with all experimental points. This can be done quickly and easily with aid of Table II.

It is always desirable to compute the pK independently from the absorbance changes at another wavelength, preferably far removed from the first wavelength.

Usually the spectral curves of forms HA and A not only will differ, but will cross at one or more isosbestic points (points of constant extinction). At these wavelengths the absorbance of any mixture of HA and A of constant total concentration (of HA plus A) will be constant (Fig. 3). The presence of isosbestic points is often taken as an indication of a two-component system because, if a third component were present, it probably

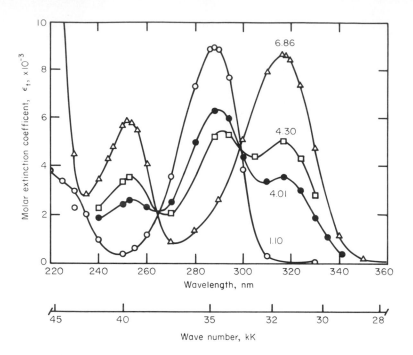

Fig. 3. Absorption spectra of pyridoxal at various pH values from 1.1 to 6.86 (footnote 11). The pH's are indicated by the numbers beside the curves.

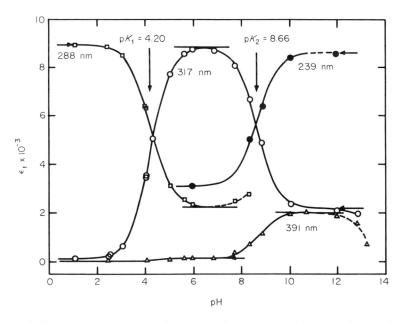

Fig. 4. Spectrophotometric titration curves for pyridoxal at four wavelengths.[11]

would have a spectrum which did not pass through the same points. This conclusion must be interpreted with care. A third component (e.g., a third ionic form) *may* be present with a spectrum identical to that of another form. Also, what appear by this criterion to be single ionic forms are often mixtures of two or more tautomers, hydrates, etc.

Consider now the dissociation of a diprotic acid,

$$H_2P \overset{K_1}{\rightleftharpoons} HP \overset{K_2}{\rightleftharpoons} P \tag{5}$$

H_2P (Eq. 5). If pK_1 and pK_2 are 5 units or more apart they may be considered independent, for at a pH halfway between them, 99.4% or more of the substance will be in the intermediate form, HP. For pyridoxol and most closely related materials, the separation of the two pK's is about 4.0, so that a slight "overlapping" occurs. The spectrophotometric titration curves of Fig. 4 for pyridoxal are typical. Since the points near pK_1, (which are of most importance in determining pK_1) are far removed from pK_2, the second dissociation step has no significant effect on these points. There is little difficulty in computing the pK's except for the fact that ϵ_{HP} cannot be obtained precisely by inspection. For example, in the curve for 317 nm in Fig. 4, ϵ_{HP} is just a little higher than the maximum value of ϵ_t obtained experimentally. A rapid procedure for establishing ϵ_{HP} and the pK's is to select a trial value of ϵ_{HP} a little higher than the maximum. Using this, a first approximation for the two pK's can be evaluated as in the nonoverlapping case. Then, at a pH midway between the two pK's the degree of dissociation for step 1 is ascertained (using Table II) and, *neglecting* the presence of P, the *fraction* of H_2P, f_{H_2P}, is obtained (this is simply $1 - f_d$ for the first step). Then, neglecting the presence of H_2P, the fraction of P, f_P, is estimated. (This is the same as the value of f_d for the second step.) The value of f_{HP} is obtained by difference, and a second value of ϵ_{HP} is computed by solving Eq. (6). Using the second trial value of ϵ_{HP}

$$\epsilon_t = \epsilon_{HP} f_{HP} + f_{H_2P} \epsilon_{H_2P} + f_P \epsilon_P \tag{6}$$

the final values of the pK's may be evaluated.

To construct a theoretical curve using the evaluated pK's, Eq. (13) is employed. At all values of pH except in the central overlapping region, this reduces to a simple form equivalent to that for the nonoverlapping case. The pK's of 3-hydroxypyridine, pyridoxol, and a number of related substances have been determined in this way.[10–12]

[10] A. K. Lunn and R. A. Morton, *Analyst* **77**, 718 (1952).
[11] D. E. Metzler and E. E. Snell, *J. Am. Chem. Soc.* **77**, 2431 (1955).
[12] K. Nakamoto and A. E. Martell, *J. Am. Chem. Soc.* **81**, 5857 (1959); *ibid.* **81**, 5863 (1959).

Now let us consider compounds with "overlapping" dissociations. Expressing the concentrations of H_2P, HP, and P in terms of the total concentration of all forms, C_t, the dissociation constants, K_1 and K_2, and the hydrogen ion activity, h, where

$$h = 10^{-pH} \tag{7}$$

we obtain the following:

$$C_{H_2P} = h^2/(K_1K_2) \cdot C_t/\alpha \tag{8}$$
$$C_{HP} = h/K_2 \cdot C_t/\alpha \tag{9}$$
$$C_P = C_t/\alpha \tag{10}$$

where

$$C_t = C_{H_2P} + C_{HP} + C_P \tag{11}$$
$$\alpha = 1 + h/K_2 + h^2/(K_1K_2) \tag{12}$$

The apparent extinction coefficient, ϵ_t, at any pH is given by Eq. (13). In deriving Eq. (13), we have

$$\epsilon_t = \frac{1}{\alpha}\left[\epsilon_P + \epsilon_{HP}\frac{h}{K_2} + \epsilon_{H_2P}\frac{h^2}{K_1K_2} \right] \tag{13}$$

expressed C_{H_2P} and C_{HP} in terms of h, the dissociation constants and C_P (i.e., C_t/α), but we could have equally well related the concentrations of all the individual forms to that of H_2P or of HP. In such a case the pH function, α, would have been defined differently.[13]

Thamer and Voigt[14] have determined overlapping dissociation constants by selecting a wavelength at which a decided maximum (or minimum) is seen in the plot of ϵ_t vs. pH (e.g., the 317 nm curve in Fig. 4). They point out that if no such wavelength occurs, data at different wavelengths can sometimes be combined into a composite, C, which does have a maximum or minimum (Eq. 14). Here i

$$C = \sum_i g_i(\epsilon_t)_i \tag{14}$$

represents the wavelengths used and the g_i's are arbitrary weighting factors used to give the most favorable maximum or minimum. The pH of the maximum must be located exactly. The values of ϵ_t, pH, and h at the maximum are $\epsilon_{t,m}$, pH_m, and h_m, respectively. At any point somewhat removed from the maximum and at distance $\Delta = pH - pH_m$ from the

[13] This has been done by Thamer and Voigt,[14] whose Eq. (4) is equivalent to our Eq. (13) except that the concentrations were all related to that of H_2P.

[14] B. J. Thamer and A. F. Voigt, *J. Phys. Chem.* **56**, 225 (1952).

maximum (Δ may be either positive or negative), Eq. (15) holds[15] and can be used to obtain the value of K_1. Here ϵ_t is the apparent extinction coefficient at the selected pH of distance, Δ, from the maximum. The value of K_2 is obtained from Eq. (16).

$$K_1 = \frac{h_m[(\epsilon_{t,m} - \epsilon_P)(\epsilon_t - \epsilon_{H_2P})10^{-\Delta} + (\epsilon_{t,m} - \epsilon_{H_2P})(\epsilon_t - \epsilon_P)10^{+\Delta} - 2(\epsilon_{t,m} - \epsilon_{H_2P})(\epsilon_t - \epsilon_P)]}{(\epsilon_{t,m} - \epsilon_P)(\epsilon_{t,m} - \epsilon_t)} \tag{15}$$

$$K_1K_2 = h_m{}^2 \frac{(\epsilon_{t,m} - \epsilon_{H_2P})}{(\epsilon_{t,m} - \epsilon_P)} \tag{16}$$

If $\epsilon_{H_2P} = \epsilon_P$ (the wavelengths for measurement can often be chosen deliberately so that this is true), a symmetrical bell-shaped curve of ϵ_t vs. pH results, and Eqs. (15) and (16) reduce to:

$$K_1 = \frac{2h_m}{(\epsilon_{t,m} - \epsilon_t)} [(\epsilon_t - \epsilon_P) \cosh(\Delta \ln 10) - (\epsilon_t - \epsilon_P)] \tag{17}$$

$$K_1K_2 = h_m{}^2 \tag{18}$$

As shown by Wigler and Wilson,[16] under these conditions Eq. (13) simplifies to:

$$(\epsilon_t - \epsilon_P) = \frac{\epsilon_{HP} - \epsilon_P}{1 + h/K_1 + K_2/h} \tag{19}$$

If the hydrogen ion activity at the two halfway points on the bell-shaped curve where $\epsilon_t - \epsilon_P = (\epsilon_{t,m} - \epsilon_P)/2$ are designated as h_a and h_b, it can be shown[16] that:

$$K_1 = h_a + h_b - 4h_m \tag{20}$$

Using Eqs. (20) and (18), Wigler and Wilson obtained the pK's of 3-hydroxypyridine.[16]

Nagano has devised an elegant procedure for determining up to 3 pK's automatically, using a high-speed computer.[2,17] In this method complete spectra are recorded. For each of the N different samples of various pH, the concentration of each of the ionic species is computed using trial pK values and equations analogous to Eqs. (8)–(12), but written to include 3 pK's and 4 ionic forms. The experimenter must provide the trial pK values, but they do not usually have to be very close to the true values.

[15] Our Eq. (15) is the same as Eq. (14) of Thamer and Voigt (footnote 14) with the symbols altered to conform to those in this article.

[16] P. W. Wigler and L. E. Wilson, *Anal. Biochem.* **15**, 421 (1966).

[17] D. E. Metzler and K. Nagano, *in* "Pyridoxal Catalysis: Enzymes and Model Systems" (E. E. Snell, A. E. Braunstein, E. S. Severin, and Yu. M. Torchinsky, eds.), p. 53. Wiley (Interscience), New York, 1968.

Using the computed concentrations of each ionic form, N equations like Eq. (2) can be written. These represent a set of linear equations in which the unknowns are the extinction coefficients of the 4 species. These extinction coefficients are determined by the least squares method at each wavelength. Then, at each wavelength and for each test solution, using the least squares values of the extinction coefficients, the expected absorbance is computed according to Eq. (2), and the squares of the differences between this expected absorbance and the experimentally observed absorbance at each wavelength and for each solution are computed and summed. If negative values of extinction coefficients are found, an additional term is added. The resulting sum of squares of deviations, designated as U_a, represents the error function which must now be minimized by adjusting the pK values and, at each stage of adjustment, recomputing U_a and comparing with its previous value.

If the pK's are only slightly overlapping, they may be adjusted one at a time beginning with pK_1. The whole procedure can then be repeated once to obtain a better fit. This is done by simply decreasing pK by an amount, Δ (0.4 is a convenient value for Δ). If a better fit is obtained with the new pK it is altered by a second decrease of amount, Δ. When a better fit is not found by subtraction of Δ, the pK is moved in the opposite direction by $\Delta/2$. This process continues until the value of Δ falls to below 0.004, about 12 cycles of adjustment and recalculation being required. Then pK_2 is adjusted in a similar way and finally pK_3.

When two pK's overlap extensively, serious problems in evaluation of pK's sometimes arise. Disregard of these difficulties may easily lead to false solutions, especially if the computer-assisted methods are applied uncritically.

Consider a compound of the type shown in Scheme 1, for which two orders of dissociation are possible in the first step to give either HP (A) or HP (B), or more likely, a mixture of the two. The successive dissociation constants, K_1 and K_2, are related to the "microscopic" dissociation constants, K_A, K_B, K_C, and K_D as follows:

$$K_1 = K_A + K_B \text{ and } 1/K_2 = 1/K_C + 1/K_D \tag{21}$$

The four "microscopic" constants are not independent, but

$$K_D = K_A K_C / K_B \tag{22}$$

Therefore, the system can be described by specifying three of these four constants. It is desirable (especially in more complex cases) to choose as the three independent constants K_A, K_C, and the tautomeric ratio, R, where

$$R = [\text{HP(B)}]/[\text{HP(A)}] = K_B/K_A = K_C/K_D \tag{23}$$

Expressing ϵ_t in terms of K_A, K_C, and R, Eq. (13) becomes:

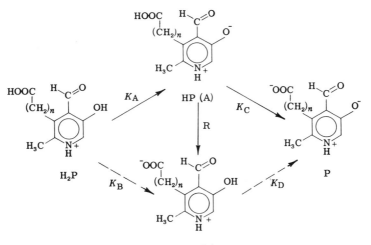

SCHEME 1

$$\epsilon_t = \frac{\epsilon_P K_A K_C + \epsilon_{HP} K_A (1 + R)h + \epsilon_{H_2P} h^2}{K_A K_C + K_A (1 + R)h + h^2} \tag{24}$$

The value of ϵ_{HP} depends on ϵ_A and ϵ_B, the extinction coefficients of sub-forms, A and B, of HP:

$$\epsilon_{HP} = (\epsilon_A + \epsilon_B R)/(1 + R) \tag{25}$$

Except when $\epsilon_{H_2P} - \epsilon_P = 0$, it is convenient to plot against pH not ϵ_t, but $y = (\epsilon_t - \epsilon_P)/(\epsilon_{H_2P} - \epsilon_P)$. From Eqs. (24) and (25) we obtain for y:

$$y = \frac{\epsilon_t - \epsilon_P}{\epsilon_{H_2P} - \epsilon_P} = \frac{a K_A h + h^2}{K_A K_C + K_A (1 + R)h + h^2} \tag{26}$$

where

$$a = \frac{(\epsilon_A - \epsilon_P) + (\epsilon_B - \epsilon_P)R}{\epsilon_{H_2P} - \epsilon_P} \tag{27}$$

Let us define the ratio S as

$$S = K_B/K_C = K_A R/K_C \tag{28}$$

Equation (26) can easily be transformed to:

$$y = \frac{1 + a(K_A/h)}{1 + (1 + R)(K_A/h) + R/S(K_A/h)^2} \tag{29}$$

Using Eq. (29) we can examine the effects on a spectrophotometric titration curve at a given wavelength of variations in the values of R, S, and a.

Consider first the case (Case I) in which the dissociation of the carboxyl group in Scheme 1 is assumed to have *no effect* on the spectrum, i.e., $\epsilon_B = \epsilon_{H_2P}$ and $\epsilon_A = \epsilon_P$. It follows from Eq. (27) that, in this case, $a = R$. Plots of y vs. $pH - pK_a$ according to Eq. (29) with $a = R$ are shown in Fig. 5 for several different values of R and S.[18] If the distance between the carboxyl and phenolic hydroxyl groups is so great that no significant electrostatic interaction takes place, $K_B = K_C$, $S = 1$, and Eq. (29) reduces to:

$$y = \frac{1}{1 + K_A/h} \tag{30}$$

This is the equation of a monoprotic acid of constant, K_a. In other words, if the carboxylate ion in the side chain of the compound neither exerts an electrostatic influence on the dissociation constant of the phenolic group nor affects the spectrum, we can determine only K_a and cannot learn anything from spectrophotometry about K_C or R, no matter what their values. The pH at the midpoint of the titration, for this case, is pK_a and the slope, $dy/d(pH) = -0.576$ at the midpoint. This is the extreme left-hand curve in Fig. 5.

If the charged carboxyl group of the side chain is assumed to have an electrostatic effect on the ease of dissociation of the phenolic hydroxyl group, $K_B \neq K_C$. In the absence of some special cooperative effect, as in thiamine,[19] $S > 1$, $K_B > K_C$ and $K_A > K_D$. The ratio, S, is a measure of the electrostatic spreading of the pK's.

$$K_1/K_2 = S \frac{(1 + R)^2}{R} \tag{31}$$

As S increases, the spread between K_1 and K_2 also increases; the K's no longer overlap as much, and the two dissociation steps can be detected spectrophotometrically even if one of the dissociations is accompanied by little or no change in the spectrum. This is especially clear in the right-hand curve of Fig. 5 for $S = 1000$ and for which the two titration steps are well separated. It is unlikely, however, that any real compound will resemble this case.

An example that approximates Case I with R near 1 and S about 3 (see curve in Fig. 5) is 5'-carboxymethyl-5-deoxypyridoxal (CMDPL), the compound of Scheme 1 when $n = 2$. Its spectra display a pair of rather sharp isosbestic points over the whole range of acidic pH values, but the titration curve at a fixed wavelength is broader than that of a monoprotic acid. From this it was clear that there must be two pK's separated by

---

[18] The curves in Fig. 5 were computed by Mr. William Baldwin, NSF research participant during the summer of 1968.

[19] G. D. Maier and D. E. Metzler, *J. Am. Chem. Soc.* **79**, 4386 (1957).

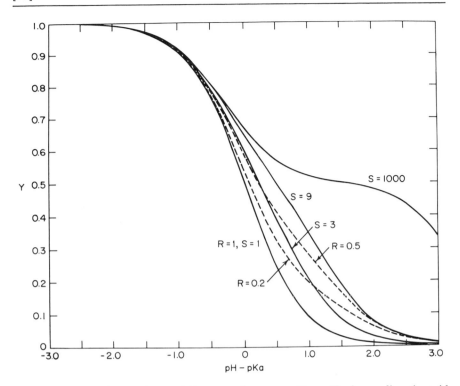

FIG. 5. Plot of $y = (\epsilon_t - \epsilon_P)/(\epsilon_{H_2P} - \epsilon_P)$ versus $pH - pK_a$ for a diprotic acid according to Eq. (29) for $a = R$ (Case I). Solid curves are for values of the tautomeric ratio, $R = 1$ and for four values of the electrostatic spreading factor, $S = K_B/K_C$, namely 1, 3, 9, and 1000. Dashed lines are for $S = 9$ and $R = 0.2$ and 0.5.

about 1.1 units but, because of the very small effect of dissociation of the carboxyl group on the spectrum, no unique values for the two pK's could be found using spectral data alone. Depending upon the trial values used, different pK's were found. The two sets, pK_1 = 3.41; pK_2 = 4.56 and pK_1 = 3.23; pK_2 = 4.35, were obtained in separate attempts. Both gave identically good fits and the comparison plots of theoretical vs. experimental titration curves (like Fig. 3) were excellent in both cases. The only differences were in the computed spectra of the intermediate form, HP. The two solutions indicated a different value of the ratio, R. The problem was solved by determining pK_2 by titration. The value so obtained (4.6) was used as a trial value in applying the computer program to evaluate both pK's. This led to the first of the two solutions given above; we believe it is substantially correct.

In Case II, the dissociation of either group in a diprotic acid causes a distinct spectral change. Careful analysis shows that even for CMDPL very small changes in the spectrum are caused by dissociation of the distant carboxyl group. When larger changes occur there is usually also a large electrostatic effect tending to spread K_1 and K_2. Both this effect and the existence of distinct spectral differences make these cases easier to solve by the methods already outlined.

III. Resolution of Tautomeric Mixtures

From a study of the spectrum versus pH, we obtain the successive dissociation constants, K_1, K_2, K_3, . . . and the spectra of the individual ionic forms H_3P, H_2P, HP, . . . In general each of these forms is a mixture of tautomers, hydrates, or other interconvertible forms. The equilibria between these forms are not affected by pH changes, and it is valid to lump all forms of the same net charge together in determining the pK's. The pH-independent equilibria may be represented by relating the concentration of each of them to that of a single form through an equilibrium constant, or tautomeric ratio, R (see Schemes 1 and 2). When several interconvertible forms exist, several tautomeric ratios must be defined.

If each of the subspecies of a single ionic form have distinctly different spectra, it is possible to resolve the experimentally observed spectrum into contributions of the individual species, to assume values of extinction coefficient for each form, to compute the fraction of each species present and, hence, to evaluate each of the ratios, R. Two methods are commonly used to obtain approximately the spectra of the individual species. The first is to change the solvent so that the equilibrium is shifted completely in favor of a single tautomer. The spectrum of that tautomer can then be measured directly; corrections for the effects of the solvent change on the band position and shape of the spectrum can be applied to improve the procedure. The second is to assume that the spectrum of a particular tautomer is similar to or identical to that of some methylated, or otherwise modified, derivative in which the tautomerism is blocked.

The spectrum of 3-hydroxypyridine in water–dioxane mixtures was studied by Metzler and Snell.[11] In 90% dioxane, two absorption bands at 36.1 kK (277 nm) and above 45.5 are present, but in pure water (at pH 6.8) there are two additional bands at 31.9 kK (313 nm) and 40.6 kK (246 nm). The former are attributed to the uncharged form, HP (B), of Scheme 2 and the latter to the dipolar ion, HP (A). The change from dioxane to water leads to a partial conversion of the uncharged into the dipolar ionic form. The spectra of 3-hydroxypyridine in a series of dioxane–water mixtures are intermediate between those in water and in 90% dioxane. Metzler and Snell estimated the tautomeric ratio R (Scheme 2) as 0.85

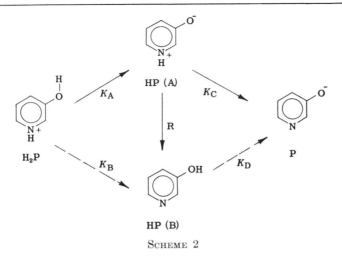

SCHEME 2

by assuming that ϵ for the uncharged form was the same in water as in 92% dioxane.

Pyridoxol displays the same type of solvent dependence of spectrum, but in water the dipolar ion constitutes a smaller fraction of the total and the presence of the uncharged form is indicated only by the small shoulder at about 35.0 kK (286 nm) in Fig. 1. In such cases an accurate estimate of R can be obtained only by using a reliable method of resolving individual spectral bands, such as that described in the following section on band shape analysis. It is considerably more correct to assume that the concentrations of the tautomers are proportional to the areas of the resolved bands than to the peak absorbancies.[19a]

The second method of establishing the spectrum of a single tautomer is illustrated by the investigations of Mason on spectra of hydroxypyridines and their N- and O-methylated derivatives[19b] and by that of Nakamoto and Martell on 2-hydroxymethyl-3-hydroxypyridine.[12] In the latter case the ϵ for the uncharged form in water was estimated from that in dioxane by assuming that the ratio ϵ(water):ϵ(dioxane) was the same for this compound as for its O-methyl derivative.

Spectral Band-Shape Analysis.[20] The absorption bands of 3-hydroxypyridine, pyridoxol, and many other substances are, in aqueous solutions, extremely smooth and very similar in shape. A comparison of shapes can be made, as shown in Fig. 6, by plotting ϵ/ϵ_{max} (where ϵ_{max} is the molar

[19a] D. B. Siano and D. E. Metzler, in preparation.
[19b] S. F. Mason, *J. Chem. Soc.* p. 5010 (1957).
[20] We are indebted to Donald B. Siano for the writing of this section.

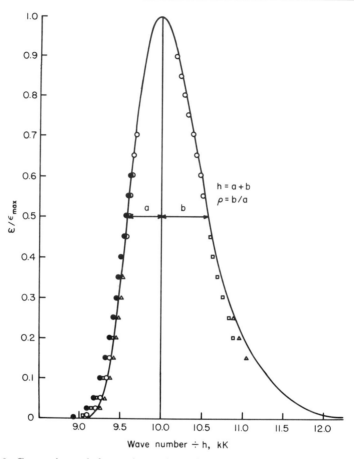

Fɪɢ. 6. Comparison of shapes for various absorption bands. ————, H₂P form of 3-hydroxypyridine; $\rho = 1.40$; ●, form P of pyridoxol; △, form H₂P of pyridoxol; ○, form P of 5-deoxypyridoxamine; □, form P of 5-deoxypyridoxal. All data are normalized both vertically and horizontally and shifted along the wave number axis to coincide at the peak.

extinction coefficient of a peak that does not overlap with other peaks significantly at the position of the maximum) against the wave number divided by the "half-width" (the band width in wave numbers at $\epsilon/\epsilon_{max} = 0.5$). The half-width must be corrected, when necessary, for overlapping of higher frequency absorption bands. Figure 6 shows the spectrum of form H₂P of 3-hydroxypyridine, a spectrum with a very low valley and hence little overlapping of bands. Several other spectra with low valleys have been superimposed by shifting them along the x-axis. The similarity is clear, but small differences exist, the principle one being in the degree of skewness, especially near the base. A small amount of the skewness can be removed by multiplying ϵ/ϵ_{max} by λ/λ_{max}, a procedure that is appealing for theoretical reasons.[21] However, only a small portion of the skewness is removed by this procedure. Some authors fit curves of this sort by using two or more Gaussian functions,[22] but there is absolutely no theoretical reason for believing that the basic shapes of the bands are always Gaussian. In fact, there are very good grounds for believing that the theoretical band shape is skewed.[23] Moreover, the utilization of several Gaussians to fit a skewed curve introduces an excessive number of parameters. Good fits result, but the parameters have almost no possible connection with theory.

A more reasonable approach is to fit each band by a function of $\bar{\nu}$ that is characterized by just four parameters. Reasonable choices for these parameters are: the position of the maximum, $\bar{\nu}_0$; the molar extinction coefficient at the maximum, ϵ_{max}; the half-width (i.e., the width at $\epsilon_{max/2}$), h; and the skewness, ρ, defined as the ratio of the distances of the curve at $\epsilon_{max/2}$ from $\bar{\nu}_0$ (see Fig. 6).

The choice of the fitting function is all-important. A pair of half-Gaussians with two different half-widths has sometimes been used,[22] but a log normal curve[23a] (Eq. 32) has been found to give much better fits.[23b] It is very easy to work with, and has many desirable properties.

$$\epsilon(\bar{\nu}) = \frac{\epsilon_{max}b}{\bar{\nu} - a} \cdot \exp[-c^2/2] \cdot \exp\left\{-\frac{1}{2c^2} \cdot \left[\ln\frac{(\bar{\nu} - a)}{b}\right]^2\right\}; \bar{\nu} > a;$$
$$\epsilon(\bar{\nu}) = 0; \bar{\nu} \leq a \tag{32}$$

The parameters a, b, and c are related to $\bar{\nu}_0$, h, and ρ by

$$c = (\ln \rho)/\sqrt{2 \ln 2}$$
$$b = h \cdot \exp[c^2/2] \cdot \rho/(\rho^2 - 1)$$
$$a = \bar{\nu}_0 - h\rho/(\rho^2 - 1)$$

and $\bar{\nu}$ is the wave number in kilokaysers. The Gaussian is the limiting form of $\epsilon(\bar{\nu})$ when c approaches zero.[23c]

The area, $\mathcal{Q}$, under $\epsilon(\bar{\nu})$ versus $\bar{\nu}$ is given by

$$\mathcal{Q} = a_s\epsilon_{max} h \tag{33}$$

[21] C. K. Jørgensen, "Absorption Spectra and Chemical Bonding in Complexes." Macmillan (Pergamon), New York, 1962.

[22] F. Quadrifoglio and D. W. Urry, J. Am. Chem. Soc. **90**, 2760 (1968).

[23] V. P. Klochov, Opt. Spectr. (USSR) **19**, 192 (1965) (English Transl.).

[23a] J. Aitcheson, "The Lognormal Distribution." Cambridge Univ. Press, London and New York, 1957.

[23b] D. B. Siano and D. E. Metzler, J. Chem. Phys. **51**, 1856 (1969).

[23c] P. Yuan, Ann. Math. Stat. **4**, 30 (1933).

where

$$a_s = \sqrt{2\pi}[\rho/\rho^2 - 1]c \exp(c^2/2)$$

and varies from 1.06 for $\rho = 1$ to 1.10 for $\rho = 1.5$.

We will designate the area under a peak as $\mathcal{C}^0$ (in units of liter mole^{-1} cm^{-2}) when a single molecular species is present. For tautomeric mixtures the value of $\mathcal{C}_j^0$ for a single species, j, is given by $\mathcal{C}_j/f_j$ where f_j is the fraction of the total represented by the jth species. Several functions of the refractive index and the dielectric constant have been suggested as proportionality constants to relate areas under bands to oscillator strengths, but in the absence of general agreement it is probably best to report experimental data in terms of areas. Further, in a series of careful measurements on 3-hydroxypyridine in water–dioxane and water–alcoholic solvent mixtures and in water at various temperatures, we find that the areas and skewness of bands are constant within experimental error while $\bar{\nu}_0$, ϵ_{max} and h for a given compound undergo easily measured changes of up to 10%.[19a] We note in passing that these and many other properties of the spectra are also observed in color centers of ionic crystals.[23d]

Once the fitting function has been chosen, one still is faced with the problem of resolving overlapping peaks so that each band can be studied separately. The usual method of moments is not appropriate, but the resolution can be accomplished either by analog methods (e.g., the DuPont curve analyzer[23e]) or by an iterative least squares method that utilizes a digital computer.[24] A similar procedure has been developed here and has been applied to spectra of pyridoxol and 5-deoxypyridoxal. The results are presented in Table III and in Fig. 7; the fits shown in Fig. 7 are typical.

Microscopic Dissociation Constants. It is often desired to know the "microscopic" or "intrinsic" dissociation constant for a single dissociable group in a particular structure. As in the cases of Schemes 1 and 2 discussed above, so in more complex cases, specification of a minimum number of microscopic constants together with the various tautomeric ratios permits easy evaluation of any of all the other microscopic constants. For example, in Scheme 3 four ionic forms, 1, 2, 3, and 4, each consisting of 2 or more tautomeric forms (or hydrates, etc.), 1a, 1b, 2a, 2b, 2c, etc. (circled) are in equilibrium. Three constants are labeled, as are the various tautomeric ratios, R_{1b}, R_{2b}, R_{2c}, etc. The stepwise constants, K_1, K_2, and K_3 can be related to the above constants as follows:

[23d] J. J. Markham and J. D. Konitzer, *J. Chem. Phys.* **29**, 673 (1958); *ibid.* **30**, 328 (1959).

[23e] F. W. Noble, J. E. Hodges, Jr., and M. Eden, *Proc. IEEE* **47**, 1952 (1959).

[24] J. Pitha and R. N. Jones, *Can. J. Chem.* **44**, 3031 (1966).

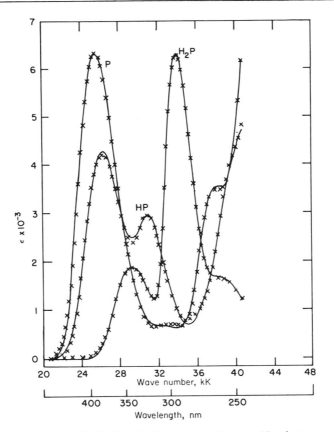

FIG. 7. Spectra of individual ionic forms of 5-deoxypyridoxal. ✕, experimental points; solid lines, sum of component bands (see Table III) found by fitting with log normal distribution curves.

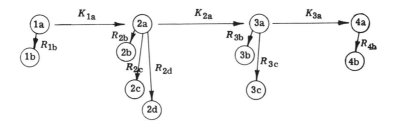

SCHEME 3

TABLE III

RESOLUTION OF SEVERAL SPECTRA INTO INDIVIDUAL BANDS

	Position, $\bar{\nu}_0$ (kK)	ϵ_{max} $\times 10^{-3}$	$\mathcal{C}^0$	h (kK)	ρ
1. Pyridoxol					
H$_2$P					
Band I	34.36	9.06	32.2	3.26	1.41
Band II	44.22	2.75	15.2	5.21	1.12
Band III	49.41	19.80	96.5	4.60	1.11
HP(A), dipolar ion					
(see Scheme 2)[a,b]					
Band I	30.82	7.56	38.8	3.40	1.32
Band II	39.62	3.87	25.2	4.35	1.33
Band III	45.59	17.67	107	4.05	1.39
HP(B), uncharged form					
(see Scheme 2)[a,b]					
Band I	35.09	1.18	14.2	3.12	1.44
P					
Band I	32.33	7.13	28.2	3.63	1.38
Band II	41.04	6.59	33.0	4.64	1.28
Band III	47.87	18.57	94.6	4.74	1.25
2. 5-Deoxypyridoxal[c,d]					
H$_2$P, aldehyde (see Scheme 4)					
Band I	29.20	1.90	29.3	4.76	1.40
Band II	39.42	1.30	17.7	4.21	1.35
H$_2$P, hydrate					
Band I	34.02	6.01	(32.2)[e]	3.24	1.51
HP, aldehyde					
Band I	26.41	4.33	29.4	4.89	1.30
HP, hydrate					
Band I	31.21	2.30	(38.8)[e]	3.36	1.31
P, aldehyde					
Band I	25.57	6.35	34.6	4.52	1.33
Band II	38.09	3.43	18.6	4.53	1.29
P, hydrate					
Band I	33.47	0.52	(28.2)[e]	5.03	1.41

[a] In addition to band I of the uncharged form, HP(B), bands II and III are expected at frequencies a few kilokaysers above those of the dipolar ion. These are well "buried" in the spectra of Fig. 1 and cannot be resolved at present. Their presence does distort somewhat the data for bands II and III of the dipolar ion.

[b] A tautomeric ratio of uncharged to dipolar ion of 0.385 was assumed and used in computing the values of $\mathcal{C}^0$ for bands I. Since bands II and III of the uncharged form could not be resolved from those of the dipolar ion, the values of $\mathcal{C}^0$ for bands II and III are only approximate.

[c] Additional bands at higher frequencies were fitted to obtain the curves in Fig. 7, but the data do not permit a unique solution for the parameters of these bands and they are not included in this table.

[d] The tautomeric ratios of hydrate to aldehyde for each form were estimated as 1.99 for the H$_2$P form and 0.276 for the HP form and 0.113 for the P-form by comparing areas of the resolved peaks and assuming the same values of $\mathcal{C}^0$ for the hydrates as for the corresponding forms of pyridoxol.

[e] Assumed values.

$$K_1 = K_{1a} \frac{(1 + R_{2b} + R_{2c} + R_{2d} \cdots)}{(1 + R_{1b} + \cdots)} \tag{34}$$

$$K_2 = K_{2a} \frac{(1 + R_{3b} + R_{3c} + \cdots)}{(1 + R_{2b} + R_{2c} + R_{2d} + \cdots)} \tag{35}$$

$$K_3 = K_{3a} \frac{(1 + R_{4b} + \cdots)}{(1 + R_{3b} + R_{3c} + \cdots)} \tag{36}$$

(The dots indicate that additional terms can be added when more tautomeric forms are present.) Other microscopic constants are related directly to these given, e.g.,

$$K_{1a \to 2b} = K_{1a} R_{2b} \tag{37}$$

$$K_{2c \to 3d} = \frac{K_{2a} R_{3d}}{R_{2c}} \tag{38}$$

$$K_{3c \to 4b} = \frac{K_{3a} R_{4b}}{R_{3c}}, \text{ etc.} \tag{39}$$

where $K_{1a \to 2b}$ represents the dissociation constant of species 1a dissociating to species 2b, etc.

In Scheme 1 and Table IV are shown the various microscopic dissociation constants of 5-deoxypyridoxal and of pyridoxal phosphate computed in this way.

The microscopic dissociation constants are important in understanding correlations of structure and acidity. They have been used with the Hammett equation[11,25,26] to compute "substituent constants" for the ring nitrogen (uncharged and protonated). The successful correlation of intrinsic pK's for various forms by the Hammett equation confirms the assignment of structures to the various tautomers.

IV. Spectra of Aldehydes

Band I of pyridoxal phosphate, 5-deoxypyridoxal, and related substances is strongly shifted to lower energies than in compounds lacking the aldehyde group in the 4-position and appears at about 25.6 kK (390 nm)[6,11,12] (Fig. 8). The band at this position is most certainly that of the free aldehyde with a dissociated phenolic hydroxyl group adjacent to it. Protonation of the phenolate ion shifts the band to about 29.4 kK (340 nm). Other bands are present at higher energies, the most prominent ones representing the hydrated aldehyde, or in the case of pyridoxal, the internal hemiacetal (Fig. 3).

Morozov *et al.* have examined the spectra of pyridoxal and pyridoxal phosphate in detail.[6,27] They report that the presence of the phosphate

[25] H. H. Jaffé and G. O. Doak, *J. Am. Chem. Soc.* **77**, 4441 (1955).

[26] H. H. Jaffé, *J. Am. Chem. Soc.* **77**, 4445 (1955).

[27] Yu. V. Morozov, N. P. Bazhulina, and M. Ya. Karpeisky, *in* "Pyridoxal Catalysis: Enzymes and Model Systems" (E. E. Snell, A. E. Braunstein, E. S. Severin, and Yu. M. Torchinsky, eds.), p. 53. Wiley (Interscience), New York, 1968.

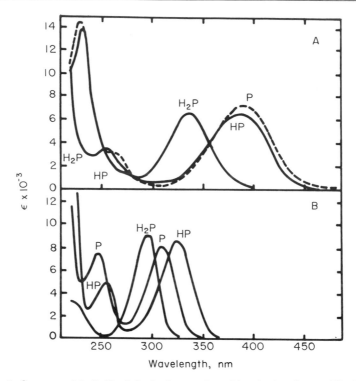

FIG. 8. Spectra of individual ionic forms of pyridoxal phosphate. (A) Aldehyde forms; (B) hydrated forms (see Scheme 4 for structures). Redrawn from Yu. M. Morozov *et al.*[6]

group has no discernible effect on the spectrum and that only the six species shown in Scheme 4 need be considered. This scheme is also appropriate for a consideration of the equilibria in 5-deoxypyridoxal.

By assuming that the positions, extinction coefficients, and half-widths of the bands of the hydrated forms of pyridoxal phosphate are similar to those of the corresponding forms of pyridoxol and pyridoxamine and their phosphates, Morozov *et al.*[6] were able to resolve the spectra into the independent contributions of the hydrated forms and of the aldehyde forms. These are shown in Fig. 8, and the ratios R_1, R_2, and R_3 taken from their work are given in Table IV. We have resolved the spectra of 5-deoxypyridoxal as described in Section III (Fig. 7). By assuming that extinction coefficients of the hydrated forms of 5-deoxypyridoxal are the same as those of corresponding forms of pyridoxol, the tautomeric ratios were obtained and are given, and microscopic dissociation constants were obtained (Table IV, Scheme 4).

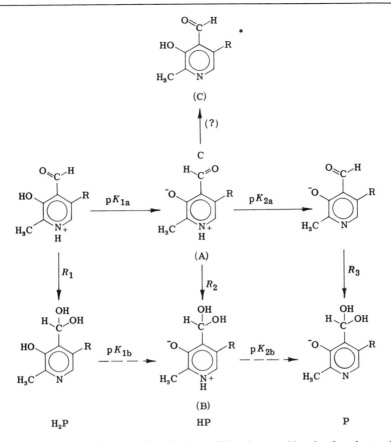

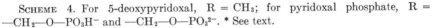

SCHEME 4. For 5-deoxypyridoxal, R = CH₃; for pyridoxal phosphate, R = —CH₂—O—PO₃H⁻ and —CH₂—O—PO₃²⁻. * See text.

Anderson and Martell, studying the infrared spectra of pyridoxal phosphate in D_2O arrived at a conflicting conclusion that form HP(C) of Scheme 4 (with a doubly charged phosphate anion in the "R-group") is a predominant form.[28] This conclusion is incompatible with our interpretations, and we believe it to be incorrect. Likewise, a recent claim based on NMR spectroscopy,[29] that the free aldehyde form of pyridoxal does not exist in aqueous solution, is doubtless in error (see also the article by Korytnyk and Ahrens in this volume[30]).

Most equilibrium constants vary with temperature. The pK's (especially those with high values) change, as do the tautomeric ratios. A

[28] F. J. Anderson and A. E. Martell, J. Am. Chem. Soc. 86, 715 (1964).
[29] O. A. Gansow and R. H. Holm, Tetrahedron 24, 4477 (1968).
[30] W. Korytnyk and H. Ahrens, this volume [79].

TABLE IV

EQUILIBRIUM CONSTANTS FOR PYRIDOXAL PHOSPHATE AND FOR 5-DEOXYPYRIDOXAL
(SEE SCHEME 4)

Parameter	Pyridoxal phosphate	5-Deoxypyridoxal
pK_1	4.14^a	4.17
pK_2	8.69^a	8.14
R_1	3.0^b	$1.99\ (3.4^c)$
R_2	0.275^b	$0.276\ (0.66^c)$
R_3	0.099^b	$0.113\ (0.06^c)$
pK_{1a}	3.64	3.80
pK_{2a}	8.63	8.08
pK_{1b}	4.68	4.46
pK_{2b}	9.07	8.50

[a] Data from V. R. Williams and J. B. Nielands, *Arch. Biochem. Biophys.* **53,** 56 (1954).
[b] Estimated from data given in footnote 6.
[c] Previous estimate given in footnote 11.

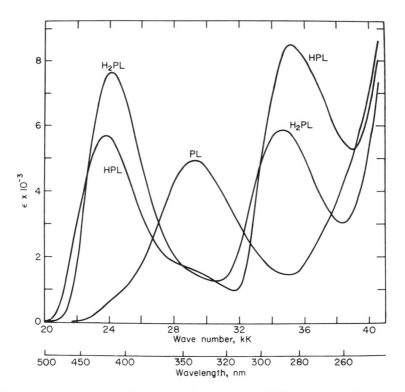

FIG. 9. Spectra of individual ionic forms of the Schiff's base of L-leucine with 5-deoxypyridoxal.

striking example is the reversible yellowing of pyridoxal solutions with increasing temperature, which results largely from the shift in equilibrium between the internal hemiacetal and free aldehyde, the proportion of the latter increasing with temperature.

V. Spectra of Schiff's Bases

Pyridoxal and related aldehydes react readily with α-amino acids and other amines in aqueous or alcoholic solutions to give Schiff's bases,

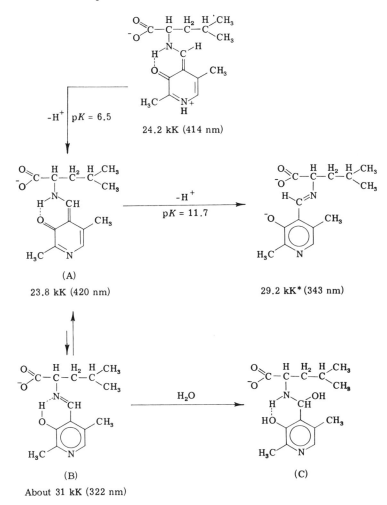

24.2 kK (414 nm)

(A)
23.8 kK (420 nm)

29.2 kK* (343 nm)

(B)
About 31 kK (322 nm)

(C)

SCHEME 5. Note that the value indicated by the asterisk (*) is atypically high. The corresponding form of the pyridoxal-valine Schiff's base absorbs at 27.2 kK (367 nm), and the pK is 10.5 rather than 11.7.

which are often intensely yellow.[31,32] The spectrum of the Schiff's base of 5-deoxypyridoxal plus leucine is typical (Fig. 9).[17] The lowest energy band appears at 23.2–24.0 kK (414–430 nm) depending on structure and solvent. As shown by Heinert and Martell,[33,34] this band represents the quinoid, or eneamine, structure (Scheme 5, structure A) rather than the tautomeric phenolic structure (Scheme 5,B) which absorbs at about 31 kK (322 nm). (However note that the latter may be responsible for the shoulder at 29.4 kK in Fig. 9.) Heinert and Martell measured spectra of many Schiff's bases of 3-hydroxypyridine 4-aldehyde, of salicylaldehyde, and of their O-methyl derivatives with amino acids in both the solid state and in dioxane solutions. Some of their data are given in Table V. They found mixtures of both tautomers (A) and (B), in both KBr pellets and in dioxane, whereas the O-methylated Schiff's bases gave spectra corresponding to those of only structure (B) in Scheme 5. The same conclusion had been reached earlier by the same authors on the basis of infrared spectra.[35] Recent NMR studies on related substances support these conclusions strongly.[36] The compound shown in Scheme 6 has been prepared recently and provides additional reference spectra. It is clear from Scheme 6 that it is not the possibility of internal hydrogen bonding, but rather the ability to form an orthoquinoid structure, which leads to the low-energy bands in the 23–24 kK region.

Protonation of Schiff's bases on the ring nitrogen has a minor effect on spectra (Fig. 9; Schemes 5 and 6), but dissociation of a proton from the "imine" nitrogen leads to a strong shift to about 27.2–27.5 kK (364–367 nm) (Schemes 5 and 6).[32,33] The Schiff's base of 5-deoxypyridoxal with leucine is atypical, giving a band at 29.2 kK (343 nm) in the dissociated form. This may indicate a nonplanar structure and suggests that great caution should be employed in speculation about structures of enzyme-bound Schiff's bases with similar spectra.

Spectra of Schiff's bases can sometimes be measured directly in aqueous solutions at very high concentrations of amino acid. Complete formation can be obtained in nonaqueous systems,[31,37] and spectra in water can sometimes be obtained by diluting nonaqueous (e.g., methanolic) stock solutions into water and measuring rapidly.[37] An equilibrium between

[31] Y. Matsuo, J. Am. Chem. Soc. **79**, 2011, 2016 (1957).

[32] D. Metzler, J. Am. Chem. Soc. **79**, 485 (1957).

[33] D. Heinert and A. E. Martell, J. Am. Chem. Soc. **85**, 183, 188 (1963).

[34] A. E. Martell, in "Chemical and Biological Aspects of Pyridoxal Catalysis" (E. E. Snell, P. M. Fasella, A. Braunstein, and A. Rossi Fanelli, eds.), p. 13. Macmillan (Pergamon), New York, 1963.

[35] D. Heinert and A. E. Martell, J. Am. Chem. Soc. **84**, 3257 (1962).

[36] G. O. Dudek and E. P. Dudek, Chem. Commun. p. 464 (1965).

[37] W. Brayneel, J. J. Charette and E. De Hoffmann, J. Am. Chem. Soc. **88**, 3808 (1966).

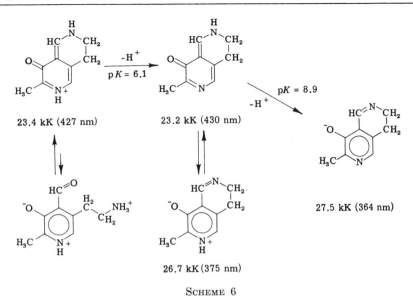

SCHEME 6

free aldehyde and amino acid and Schiff's base is usually attained rapidly. By measuring the spectra of such mixtures as a function of pH and at various concentrations of amino acids the spectra of the various ionic forms of the Schiff's base, their pK's and the formation constants of the Schiff's bases can be evaluated.

$$H_2P \xrightarrow{K_{1P}} HP \xrightarrow{K_{2P}} P$$
$$+$$
$$H_2L \xrightarrow{K_{1L}} HL \xrightarrow{K_{2L}} L$$
$$\updownarrow \ F_{PL}$$
$$H_3PL \xrightarrow{K_{1PL}} H_2PL \xrightarrow{K_{2PL}} HPL \xrightarrow{K_{3PL}} PL$$

SCHEME 7

Consider a typical scheme (Scheme 7). Here P represents the aldehyde, L the amino acid, and PL the Schiff's base. The K's are acid dissociation constants, and F_{PL} is the formation constant of form PL of the Schiff's base. It is often easiest to measure at a constant pH the apparent formation constant (which varies with pH), which we will designate K_{pH}. This constant is related to the total concentrations of Schiff's base, aldehyde, and amine, C_{tPL}, C_{tP}, and C_{tL}, by Eq. (40). It is related to the constants

$$K_{pH} = C_{tPL}/C_{tP} \cdot C_{tL} \tag{40}$$

in Scheme 7 by Eq. (41).[32]

TABLE V
SPECTRAL DATA FOR SELECTED COMPOUNDS

Structure	Band I			Band II			Band III		
	kK	nm	$\epsilon \times 10^{-3}$	kK	nm	$\epsilon_{max} \times 10^{-3}$	kK	nm	$\epsilon \times 10^{-3}$
	39.4 (38.9)[a]	254 (257)	0.204	49.1	204	7.4	54	185	80.0
	39.0	256	2.66	51.5	194	—	—	—	—
	39.0	256	5.32	—	—	—	—	—	—
	36.4	277	2.2	—	—	—	—	—	—
	35.0	286	5.9	—	—	—	—	—	—

Structure									
Pyridoxine cation (CH$_2$OH, HO, H$_3$C, N$^+$H)	34.4	291	8.9	43.5^b	230^b	2.4	49.5	202	19.4
Pyridoxine (CH$_2$OH, $^-$O, H$_3$C, N$^+$H)	30.8	324	7.5^c	39.4	254	3.8^c	45.4	220	17.8^c
Pyridoxine anion (CH$_2$OH, $^-$O, H$_3$C, N)	32.3	310	7.2	40.8	245	6.6	47.7	209	18.4
d — salicylaldehyde (CH=O, HO)	30.86	324	3.4	39.06	256	12.6	—	—	—
salicylaldehyde anion (HC=O, $^-$O)	26.53	377	6.7	37.81	264	7.5	—	—	—

(Continued)

TABLE V (*Continued*)

Structure	Band I			Band II			Band III		
	kK	nm	$\epsilon \times 10^{-3}$	kK	nm	$\epsilon_{max} \times 10^{-3}$	kK	nm	$\epsilon \times 10^{-3}$
(pyridoxal structure, N^+H, CH$_3$, H$_3$C, HC=O, $^-$O)	26.2	381	4.20	—	—	—	—	—	—
(structure *e*, CH$_2$—OPO$_3$H, N^+H, H$_3$C, HC=O, $^-$O)	25.8	388	4.90^c	—	—	—	—	—	—
(structure, CH$_2$—OPO$_3{}^{2-}$, N, H$_3$C, HC=O, $^-$O)	25.8	388	6.60^c	—	—	—	—	—	—

Structure							Solvent
Valine derivative (H$_3$CO-pyridine, HC=N)	32.2	310	6.2	39.2	255	6.9	Dioxane
Valine derivative (pyridine, N=CH, H—O)	30.9 / 31.0	323 / 322	2.5^c / —	39.7 / —	252 / —	7.0^c / —	Dioxane / KBr
Valine derivative (benzene, N=CH, H—O)	31.5	318	2.5^c	38.9	257	9.5^c	Dioxane
Valine derivative (CH, N, H---O, ketone)	24.3	411	4.6^c (9.7^g)	36.0	278	8.1^c	Dioxane

(Continued)

TABLE V (Continued)

Structure	Band I kK	nm	$\epsilon \times 10^{-3}$	Band II kK	nm	$\epsilon_{max} \times 10^{-3}$	Band III kK	nm	$\epsilon \times 10^{-3}$
Valine | N—CH (quinoid structure) f	23.5	425	3.9c	37.0	270	6.2c	Dioxane		
	23.7	422	(7.3h)	—	—	—	KBr		
Valine | HC=N (structure) f	26.5	377	6.6	38.0	263	7.1	Methanol		

a "Frank-Condon reference maximum" of 38.9 kK has been suggested by Stevenson (footnote 46) for use in comparing this band with those of other substances.

b Approximate values.

c These are extinction coefficients of the compound at the given maximum, (ϵ_t), but not those of the individual species shown. The latter are always higher.

d From Stevenson (footnote 46).

e From E. A. Peterson and H. A. Sober, J. Am. Chem. Soc. 76, 169 (1954). Data for pyridoxol, pyridoxamine, 4-deoxypyridoxol, and their phosphates are also given.

f From Heinert and Martell (footnote 33).

g Extinction coefficient of individual species assuming (Heinert and Martell[33]) that 50.6% of the Schiff's base is in the quinoid form in this case.

h Extinction coefficient of individual species assuming 53.3% in quinoid form (footnote 33).

$$K_{pH} = F_{PL} \cdot \frac{(1 + h/K_{3PL} + h^2/K_{2PL}K_{3PL} + h^3/K_{1PL}K_{2PL}K_{3PL})}{(1 + h/K_{2P} + h^2/K_{1P}K_{2P})(1 + h/K_{2L} + h^2/K_{1L}K_{2L})} \quad (41)$$

A common method of measuring K_{pH} is that of Ketelaar et al.[38] Spectra are measured for a series of solutions at a constant pH and aldehyde concentration, but with varying amino acid concentration. The reciprocal of $\epsilon_t - \epsilon_{tP}$, the apparent extinction coefficient at a selected wavelength minus that of the aldehyde alone at the same wavelength and pH, is plotted against $1/C_{tL}$. Equation (42) permits the calculation of ϵ_{tPL} (the apparent extinction of the Schiff's base at the chosen pH) from the intercept and K_{pH} from the slope of the resulting straight line. By measuring K_{pH} at several different values of pH the

$$\frac{1}{(\epsilon_t - \epsilon_{tP})} = \frac{1}{K_{pH}(\epsilon_{tPL} - \epsilon_{tP})} \cdot \frac{1}{C} + \frac{1}{(\epsilon_{tPL} - \epsilon_{tP})} \quad (42)$$

constant F_{PL} and the dissociation constants of the Schiff's base can be evaluated using Eq. (41). Likewise, from the values of ϵ_{tPL} the extinction coefficients of the individual ionic forms of the Schiff's base can be evaluated.

The constants in Eq. (41) and the spectra of individual ionic forms of the Schiff's base can also be evaluated from experimental data using a computer.[3,17] The spectra in Fig. 9 were determined in this way.[17] Important papers have been published by Bruice and co-workers[40] and by Christensen.[41]

VI. Spectra of Pyridoxal Phosphate-Dependent Enzymes

The methods described in preceding sections can be applied with little modification to enzymes. With highly purified enzymes, the absorption bands of the coenzyme are clearly visible below 33 kK (above 300 nm). The spectra of several forms of aspartate aminotransferase are shown in Fig. 10.[42] The extinction coefficients are based on the pyridoxal phosphate content and are plotted on a logarithmic scale to permit both the coenzyme peaks and the intense protein peak at 35.7 kK (280 nm) to be shown. The

[38] J. A. A. Ketelaar, C. Van De Stolpe, A. Goudsmit, and W. Dzcubas, Rec. Trav. Chim. **71,** 1104 (1952).

[39] F. Olivo, C. S. Rossi, and N. Siliprandi, in "Chemical and Biological Aspects of Pyridoxal Catalysis" (E. E. Snell, P. M. Fasella, A. Braunstein, and A. Rossi Fanelli, eds.), p. 13. Macmillan (Pergamon), New York, 1963.

[40] D. S. Auld and T. C. Bruice, J. Am. Chem. Soc. **89,** 2090, 2098 (1967) and preceding papers.

[41] H. N. Christensen, J. Am. Chem. Soc. **80,** 99 (1958).

[42] These have been reported previously by W. T. Jenkins et al. See W. T. Jenkins and L. D'Ari, J. Biol. Chem. **241,** 2845 (1966) and preceding papers. The spectra reproduced in Fig. 10 were measured on enzyme prepared by Dr. W. T. Jenkins.

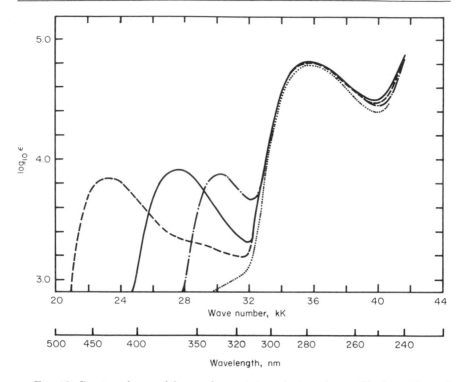

Fig. 10. Spectra of several forms of aspartate aminotransferase. The logarithms of extinction coefficients are plotted. ------, Low pH form of pyridoxal phosphate enzyme; ———, high pH form of pyridoxal phosphate enzyme; - · - · , pyridoxamine phosphate form (constant over a wide pH range); · · · · , apoenzyme. Enzyme was prepared and donated by Dr. W. T. Jenkins.

forms of pyridoxal-dependent enzymes absorbing maximally between 23 and 25 kK (400–430 nm) have been shown to be internal Schiff's bases with ϵ-amino groups of lysine residues of the proteins. Possible structures are forms (A) and (B) of Scheme 8. Some, but not all, of the enzymes dissociate to give absorption bands at about 27.6 kK (362 nm) or near 30 kK (330–340 nm). Possible structures for these forms are shown by the other structures in Scheme 8. Great care should be exercised in interpreting spectral bands near 30 kK (333 nm) because of the variety of structures that can give rise to absorption in this region. Care should also be taken in assuming that the state of dissociation of the pyridine ring can be deduced from spectral data, because for many forms the spectrum changes little with changes in the state of dissociation of the ring nitrogen.

An important aspect of the spectrophotometry of pyridoxal phosphate-

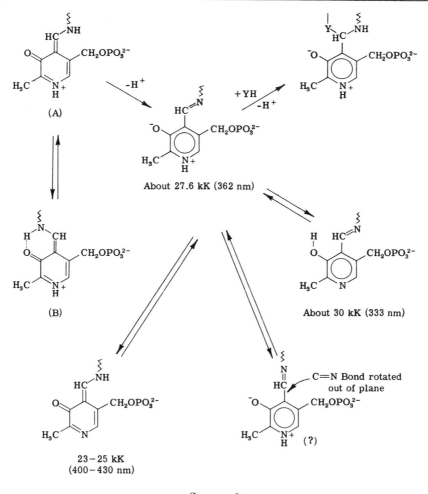

SCHEME 8

dependent enzymes is the fact that the chromophore is asymmetrically bound to the protein, giving rise to an induced optical activity. This is most conveniently studied by measurements of circular dichroism.[43]

Studies of absorption spectra and circular dichroism of pyridoxal phosphate-dependent enzymes have been reviewed by Fasella.[44]

[43] V. I. Ivanov, Yu. N. Breusov, M. Ya. Karpeisky, and O. L. Polianovsky, *Mol. Biol.* (*Russ.*) **4,** 588 (1967).

[44] P. Fasella, *Ann. Rev. Biochem.* **36,** 185 (1967).

VII. Relation to Spectra of Substituted Benzenes

The three bands of the spectra of 3-hydroxypyridines can be related directly to the three $\pi - \pi^*$ transitions of benzene[45] (see Table V). Band I of benzene is symmetry-forbidden, has a low intensity and displays a well-resolved fine structure. However, the $0 - 0$ band is missing, and, as suggested by Stevenson,[46] it is best to use a "reference Frank-Condon" maximum of 38.9 kK (in water) when comparing this peak position with those of other substances. The spectrum of pyridine is quite similar but is complicated somewhat by the presence of a weak $n - \pi^*$ transition (the $0 - 0$ frequency in vapor is 34.8 kK[47,48]) which overlaps with band I. This $n - \pi^*$ band is usually not easily discernible in the spectra of pyridine derivatives.

It is beyond the scope of this chapter to discuss the relationships between structure and spectra, but we will point out that impressively good correlations have been found between structures of benzene derivatives and shifts in peak positions from those of benzene itself. Doub and Vandenbelt[49] were among the first to develop such correlations; they have been improved by Petrushka[50] and by Stevenson[46] through the use of a molecular orbital-perturbation approach. Stevenson has tabulated the extensive data of Doub and Vandenbelt and has given the parameters for various substituents to be used in equations for predicting shifts in peak position for both bands I ($^1A - {}^1L_b$ transition) and band II ($^1A - {}^1L_a$ transition).

We have observed that the method of Stevenson can be applied quite well to the various ionic species of 3-hydroxypyridine regarded as disubstituted benzenes, but that application to the 4-aldehydes (regarded as trisubstituted benzenes) is not as satisfactory. In the 3-hydroxypyridine 4-aldehydes and their Schiff's bases, protonation on the ring nitrogen has little effect on the spectra; in fact in all cases the spectra are very similar to those of the corresponding ortho-disubstituted benzene compounds (Table V). The methods of Doub and Vandenbelt, Petrushka, and Stevenson predict that for two ortho groups, one electron-donating and one electron-accepting, the shifts in peak position will be greater, the stronger the electron-donating tendency of the donor and the stronger the electron-accepting ability of the acceptor. On this basis, for the following Schiff's

[45] H. H. Jaffé and M. Orchin, "Theory and Application of Ultraviolet Spectroscopy." Wiley, New York, 1962.

[46] P. E. Stevenson, *J. Mol. Spectr.* **15**, 220 (1965).

[47] J. E. Parkin and K. K. Innes, *J. Mol. Spectr.* **15**, 407 (1965).

[48] S. F. Mason, *J. Chem. Soc.* pp. 1240, 1247, 1253 (1959).

[49] L. Doub and J. M. Vandenbelt, *J. Am. Chem. Soc.* **69**, 2714 (1947); *ibid.* **71**, 2414 (1949); *ibid.* **77**, 4535 (1955).

[50] J. Petrushka, *J. Chem. Phys.* **34**, 1111, 1120 (1961).

bases the shifts (from 38.9 kK or 254 nm in benzene) should increase from

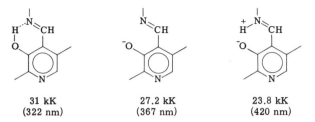

| 31 kK | 27.2 kK | 23.8 kK |
| (322 nm) | (367 nm) | (420 nm) |

left to right, as is observed. Note that the right-hand structure is a resonance form of the quinoid structure (A) of Scheme 5 drawn to show this structure as a disubstituted pyridine. All the structural assignments made in this chapter are consistent with the correlations discussed above.

Data for several compounds typical of those which have been discussed in this chapter are summarized in Table V. These, together with those in Schemes 5 and 6 provide reference structures sufficient to answer many of the questions that may arise about interpretation of spectra of compounds related to pyridoxal.

[78] Fluorometric Determination of Pyridoxal and Pyridoxal 5'-Phosphate in Biological Materials by the Reaction with Cyanide

By Z. TAMURA and S. TAKANASHI

Principle

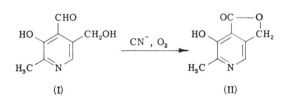

(I)   (II)

Pyridoxal (PAL) (I) is oxidized to 4-pyridoxolactone (II), which is a highly fluorescent compound, in the presence of oxygen in slightly alkaline solution by the catalytic action of cyanide.[1-3] PAL is previously separated

[1] V. Bonavita, *Arch. Biochem. Biophys.* **88**, 366 (1960).
[2] S. Oishi and S. Fukui, *Arch. Biochem. Biophys.* **128**, 606 (1968).
[3] S. Takanashi, Z. Tamura, A. Yoshino, and Y. Iidaka, *Chem. Pharm. Bull. (Tokyo)* **16**, 758 (1968).

by ion-exchange column treatment, treated with potassium cyanide at pH 7.5, and determined by measuring the fluorescence intensity at pH 10. Pyridoxal 5'-phosphate (PAL-P) is hydrolyzed to PAL by acid phosphatase and determined as PAL.[4]

Reagents

Acetate buffer, 0.2 M, pH 4.0

Phosphate buffer, 0.4 M, pH 7.5

Trichloroacetic acid solution, 20%

Sodium carbonate solution, 0.4 M

Potato acid phosphatase (0.4 unit/ml)

Dowex 1 X8, 100–200 mesh, acetate form

Amberlite CG 120, 100–200 mesh, bufferized to pH 4.0

PAL, standard solution. Dissolve 20.36 mg of PAL hydrochloride ($C_8H_9NO_3 \cdot HCl$) in 100 ml of 0.01 N HCl with protection from light (10^{-3} M solution). Nitrogen gas is bubbled into the solution for 20 minutes; the preparation is stored at below 5°. The solution is freshly prepared every week. Just before use, the stock solution is diluted with water to obtain a 10^{-5} to 10^{-6} M solution, which is diluted further with 0.4 M phosphate buffer (pH 7.5) to 10^{-6} to 10^{-7} M.

PAL-P, standard solution. Dissolve 26.52 mg of PAL-P monohydrate ($C_8H_{10}NO_6P \cdot H_2O$) in 100 ml of 0.01 N HCl with protection from light, then treat and store in the same way as PAL. The stock solution is freshly prepared every 4 days. Just before use, a 10^{-6} to 10^{-7} M solution in 0.2 M acetate buffer (pH 4.0) is prepared.

Preparation of Potato Acid Phosphatase.[5] Skinned potatoes (1 kg) are cut in pieces, homogenized in 500 ml of 0.1 M acetate buffer (pH 3.4) and filtered with gauze into a beaker. Celite (40 g) is added to the filtrate, mixed well, and filtered through a Büchner funnel by suction. The filtrate is cooled to 0°–4°, and 34 g of $(NH_4)_2SO_4$ per 100 ml of the filtrate is added in portions with stirring. The mixture is allowed to stand for 1 hour at 0°–4° and centrifuged at 8000 g for 15 minutes. The supernatant solution is discarded, and the precipitate is dissolved in 0.05 M Tris-HCl buffer (pH 7.2). The solution is dialyzed twice against 5 liters of water for 12 hours. The dialyzed solution is adjusted to pH 5 with 0.1 M acetate buffer (pH 3.4) and heated at 60° for 1 minute. The solution is cooled to 0°–4° and centrifuged. Supernatant solution is diluted with 0.2 M acetate

---

[4] Enzymatic assay of PAL and PAL-P; see Vol. II [109], Vol. III [142].

[5] N. Katsunuma, unpublished data (1967).

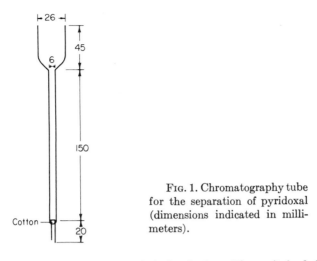

FIG. 1. Chromatography tube for the separation of pyridoxal (dimensions indicated in milli-meters).

buffer (pH 4.0) to obtain 0.4 unit/ml solution. The unit is defined as the amount of the enzyme which liberates 1 micromole of inorganic phosphate per 1 minute from 20 micromoles ($5 \times 10^{-3}\ M$) of β-glycerophosphate at pH 4.0.

Preparation of Ion-Exchange Columns. DOWEX COLUMN. Dowex 1 X8 (Cl^- form, 50 ml) is washed with water in a beaker until turbidity is removed and placed in a chromatography tube of 2×30 cm. The column is washed with 200 ml of 2 N NaOH, water, 50 ml of 2 M NaCl, 100 ml of 2 N HCl, water, and 200 ml of 2 N NaOH in this order. Finally 200 ml of 2 N acetic acid is passed through the column to obtain the acetate form, and the resin is pipetted into chromatography tubes 6 mm in diameter to form a packed layer 80 mm long (Fig. 1).

AMBERLITE COLUMN. Amberlite CG 120 (Na^+ form, 50 ml) is washed with water and placed in a tube of 2×30 cm. The column is washed with 200 ml of 2 N NaOH, water, 200 ml of 2 N HCl, 50 ml of 2 M NaCl, 200 ml of NaOH, water, and 200 ml of 2 N HCl in this order. Then 1 M acetate buffer (pH 4.0) is passed through the column until the pH of the effluent is 4, and finally the column is washed with 100 ml of 0.2 M acetate buffer (pH 4.0). The resin is pipetted into tubes 6 mm in diameter and up to 100 mm long (Fig. 1).

Pretreatment of Samples. TISSUES. Homogenize in 0.2 M acetate buffer (pH 4.0) at 0°–4° to make a 10% homogenate.

SAMPLES FOR PAL DETERMINATION. A sample (0.5–1 ml) is diluted to 6 ml with the acetate buffer, and 2 ml of 20% TCA is added. The mixture is heated at 50° for 15 minutes and centrifuged. The supernatant solution is taken.

SAMPLES FOR PAL-P DETERMINATION. A sample (0.5–1 ml) is diluted to 5 ml with the acetate buffer and heated at 80° for 5 minutes to inactivate enzymes in the sample. After cooling, 1 ml of the acid phosphatase (0.4 unit) is added and incubated at 37° for 1 hour with shaking. Two milliliters of 20% TCA is added to the incubation mixture, which is heated at 50° for 15 minutes and centrifuged.

For correction by the standard addition-extrapolation method, a certain volume of the standard solution of PAL-P is added to the sample and it is diluted to 5 ml with the acetate buffer, and treated in the same way.

Procedure. A 4-ml portion of the supernatant solution described above is added to a column of Dowex, to which a column of Amberlite is connected so as to let the effluent from the former flow into the latter. The combined columns are washed twice with 3 ml of the acetate buffer, and the Dowex column is removed. The Amberlite column is washed twice with 3 ml of water, and PAL adsorbed on the column is eluted with 0.4 M phosphate buffer (pH 7.5). The first fraction of nearly 10 ml is taken in a volumetric flask and made up to a volume of exactly 10 ml with the addition of the phosphate buffer. Portions of the effluent, 4 ml, are taken in two brown test tubes, and 0.1 ml of 0.1 M KCN is added to one and 0.1 ml of water is added to the other for a blank test. Similarly, each 4 ml of PAL standard solution in the phosphate buffer is placed in two test tubes, and KCN solution and water are added, respectively. All the test tubes are heated at 50° for 2 hours. After cooling, 2 ml of 0.4 M Na_2CO_3 is added to each of the test tubes and the fluorescence intensity is measured with 356 mμ for excitation and 432 mμ for emission.

The amount of PAL is calculated from the values obtained from the sample for PAL (unhydrolyzed sample) as follows:

$$PAL = (F - F')/(F_s - F'_s) \times A \times 2 \times 10/4$$

F, fluorescence of sample; F', blank fluorescence of sample; F_s, fluorescence of standard; F'_s, blank fluorescence of standard. A, amount of PAL in 4 ml of standard solution.

The amount of PAL-P is calculated by subtracting the amount of PAL from the value of the sample for PAL-P (hydrolyzed sample) and converting to PAL-P.

In the case of the PAL-P determination, it is desirable to correct the determined values by the standard addition extrapolation method.

Effect of pH on the Fluorescence Intensity. The fluorescence intensity of the reaction product, 4-pyridoxolactone, reaches a maximum that is stable at pH 9–10, and decreases markedly as the pH approaches the acidic region. The reaction product is unstable above pH 10.

Interfering Substances. The reaction is inhibited by borate in concentrations over 0.1 M, so borate buffer should not be used. The presence of 10^{-3} M glucose shows no effect. 4-Pyridoxic acid, which is a metabolite of the vitamin B$_6$ group, interferes with the fluorometry, but it is completely removed by the column treatment. Pyridoxine and pyridoxamine show no interference. Vitamin B$_2$ causes an increase of blank fluorescence and diminishes the accuracy of the determination if in high concentration. NAD becomes slightly fluorescent under the reaction conditions, but the interference by NAD is reduced to a negligible extent by the column treatment up to a concentration of 50 times as much as PAL.

Recovery of PAL and PAL-P. Recoveries of PAL and PAL-P (each 0.25–1 millimicromole) added to sera, plasma, urine, and tissues were 100 ± 5% and 90 ± 5%, respectively. The recovery of PAL-P added to whole blood (1 ml) was usually poor (65–70%).

Limit of the Determination. Limit of the determination is about 1 × 10^{-10} mole (20 mμg as PAL) in the original sample solution by the use of a Hitachi fluorescence spectrophotometer MPF-2A.

Application. The present method is applicable to sera, plasma, urine, tissues, and milk, but PAL and PAL-P in normal urine are not detectable.

[79] Nuclear Magnetic Resonance Spectroscopy of Vitamin B$_6$[1]

By W. KORYTNYK and H. AHRENS

Introduction

High-resolution nuclear magnetic resonance (NMR) spectroscopy is now recognized as being one of the most powerful and generally applicable physicochemical techniques for the study of organic molecules. Ever since it was first utilized in the vitamin B$_6$ group,[2] NMR spectroscopy has found increasing application in various phases of vitamin B$_6$ research as an invaluable tool for determining the structures of analogs and metabolites of vitamin B$_6$[3–6] and their derivatives.[7,8] The ring-chain tautomerism of

[1] This study was supported in part by a research grant (CA-08793) from the National Cancer Institute, U.S. Public Health Service.

[2] W. Korytnyk and R. P. Singh, *J. Am. Chem. Soc.* **85,** 2813 (1963).

[3] W. Korytnyk, E. J. Kris, and R. P. Singh, *J. Org. Chem.* **29,** 574 (1964).

[4] W. Korytnyk, *J. Med. Chem.* **8,** 112 (1965).

[5] W. Korytnyk and B. Paul, *J. Heterocyclic Chem.* **2,** 144 (1965).

[6] W. Korytnyk, B. Paul, A. Bloch, and C. A. Nichol, *J. Med. Chem.* **10,** 345 (1967).

[7] W. Korytnyk and B. Paul, *Tetrahedron Letters* **8,** 777 (1966).

[8] W. Korytnyk and B. Paul, *J. Org. Chem.* **32,** 3791 (1967).

pyridoxal[2,9] and some metabolites[10,11] has been demonstrated. The extent of the deuteration of various forms of vitamin B_6 can be readily followed by NMR, and this method has been used by Dunathan in his mechanistic and stereochemical enzyme studies.[12] More recently, the stereochemistry and the deuteration of Schiff bases have been studied by means of NMR,[9] and the acid-base equilibria of pyridoxamine and pyridoxal have been determined in D_2O solution.[13]

In the present chapter, NMR spectra of all forms of vitamin B_6 will be considered, in order to provide basic information regarding the conditions for determining and interpreting the spectra.

Methods

In the present study, NMR spectra were determined with a Varian A60A spectrometer, and were standardized according to the method of Jungnickel.[14]

Various solvents have been used in determining NMR spectra. D_2O is a very useful solvent for all forms of vitamin B_6, and was used in this study. The spectra are simple, since phenolic and alcoholic protons are immediately exchanged for deuterium. Positions of the peaks vary appreciably with the pD of the solution.[2,13] These shifts are characteristic of the anionic, zwitterionic, or cationic forms of the molecules.[2] The proton shifts can be rationalized in terms of the electronic properties of each form.[2]

Solutions of cations were obtained either by dissolving the hydrochlorides in D_2O, the pD then being around 3.0, or by dissolving the bases in $1 N$ D_2SO_4, which was prepared by adding H_2SO_4 to D_2O. Both methods gave the same spectra. Anions were determined in $1 N$ NaOD solution, which was prepared by dissolving sodium in D_2O under nitrogen. Zwitterionic forms of bases were determined around pD 7.0, which was obtained on careful addition of $1 N$ NaOD to an acid solution of the compound, or by dissolving the free base in D_2O. True pD values were calculated from the relationship pD = pH + 0.40.[15]

Whenever possible, a compound was used in the form of a 5–10% solution. A few compounds were not sufficiently soluble to permit 5%

[9] K. F. Turchin, V. F. Bystrov, M. Ya. Karpeisky, A. S. Olkhovoy, V. L. Florentiev, and Yu. N. Sheinker, in "Pyridoxal Catalysis: Enzymes and Model Systems" (E. E. Snell, A. E. Braunstein, E. S. Severin, and Yu. M. Torchinsky, eds.), p. 67. Wiley (Interscience), New York, 1968.

[10] B. Paul and W. Korytnyk, Chem. Ind. (London) p. 230 (1967).

[11] W. Korytnyk and B. Paul, same book as in footnote 9, p. 615.

[12] H. Dunathan, L. Davis, and M. Kaplan, same book as in footnote 9, p. 325.

[13] O. A. Gansow and R. H. Holm, Tetrahedron 24, 4477 (1968); ibid. J. Am. Chem. Soc. 91, 5984 (1969).

[14] J. L. Jungnickel, Anal. Chem. 35, 1985 (1963).

[15] P. K. Glasoe and F. A. Long, J. Phys. Chem. 64, 188 (1960).

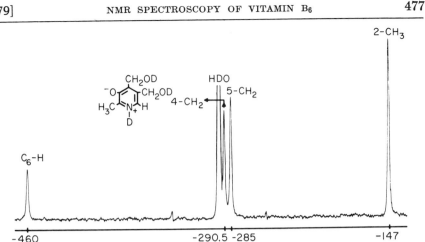

FIG. 1. 60 Mc/sec spectrum of pyridoxol zwitterion in D$_2$O. To move the HDO peak out of the methylene region, the NMR tube was cooled in ice before being inserted into the probe.

solutions, and hence were used in the form of saturated solutions. The exact concentration was not a critical factor in determining the positions of the peaks.

1,4-Dioxane and Tiers' salt, sodium 3-(trimethylsilyl) 1-propane-sulfonate, have been used as internal standards, and the virtues of the two substances have been discussed.[16] For the purposes of this chapter, spectra in D$_2$O with Tiers' salt rather than dioxane are reported.

Dimethyl-d_6 sulfoxide is also an excellent solvent, especially for some phosphorylated derivatives. Whenever the phenolic hydroxyl in pyridoxol derivatives is blocked, proton exchange is slowed down enough to permit observation of spin-spin splitting caused by the aliphatic hydroxyls.[16] On the other hand, if the phenolic hydroxyl is free, neither the expected splitting by the hydroxyl protons nor discrete peaks due to the various hydroxyls have been observed. This finding has been utilized for determining the substitution patterns of pyridoxol.[8] Another solvent of wide applicability is hexafluoroacetone hydrate-d_2, which has also been found suitable for some phosphorylated derivatives.[10] Solvents of limited applicability include CDCl$_3$ (for highly substituted analogs),[3] CD$_3$COOD (for pyridoxamine),[12] and CD$_3$OD (for some Schiff bases).[9] With all solvents other than D$_2$O, tetramethylsilane was used as the internal standard.

Spectra

Pyridoxol (Fig. 1). The assignment of peaks is obvious for 2-CH$_3$ and C$_6$-H protons. The two methylene protons have been assigned by com-

[16] W. Korytnyk and B. Paul, *J. Heterocyclic Chem.* **2**, 481 (1965).

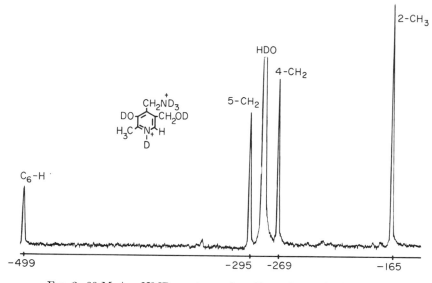

FIG. 2. 60 Mc/sec NMR spectrum of pyridoxamine cation in D_2O.

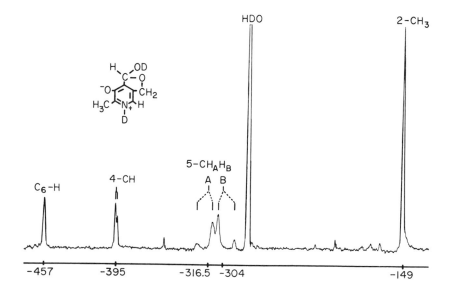

FIG. 3. 60 Mc/sec NMR spectrum of pyridoxal zwitterion in D_2O.

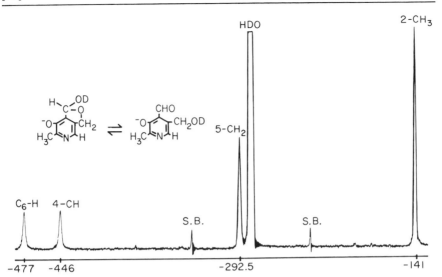

Fɪɢ. 4. 60 Mc/sec NMR spectrum of pyridoxal anion in 1 M NaOD at 35° probe temperature.

parison with close structural analogs and by deuterium labeling.[2,5] Irrespective of the pH and the ionic form of pyridoxol, the 5-CH₂ protons appear at a higher field than the 4-CH₂ protons.[5] Quaternization of pyridoxal as in N-methylpyridoxol labilizes the hydrogens in 2-CH₃, and they become very readily exchangeable with deuterons of D₂O under alkaline conditions.[2]

Pyridoxamine (Fig. 2). The methylene peaks have been assigned by comparison with pyridoxamine phosphate, in which the positions of the two peaks can be distinguished (see below).[2]

Pyridoxal (Figs. 3 and 4). It has been recognized for some time that pyridoxal can exist in both aldehyde (I) and hemiacetal (II) forms, and this ring-chain tautomerism has been studied by various physicochemical methods. In addition, pyridine aldehydes have a pronounced tendency to form hydrates of the general structure RCH(OH)₂, as shown in (III).

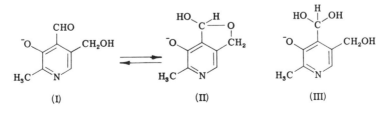

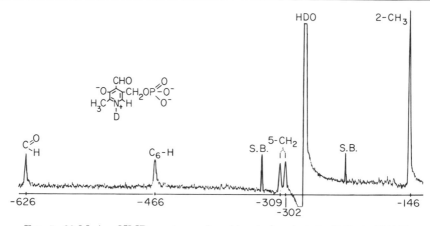

Fig. 5. 60 Mc/sec NMR spectrum of pyridoxal phosphate in D_2O at pD 7.25.

NMR spectroscopy has been used for following changes in the aldehyde group in solution. The aldehyde proton appears in a very low field (around -600 cps), and can be clearly distinguished from the hydrated and the hemiacetal forms.

Figure 3 shows the zwitterionic form of pyridoxal, which exists in the hemiacetal form. The 5-methylene protons are not equivalent, and they form an AB quadruplet. Their calculated positions are indicated in the graph. The hemiacetal (4-CH) proton appears as a doublet ($J = 1.5$ cps) and is coupled to one of the 5-methylene protons, as has been established by decoupling experiments. The pyridoxal cation exhibits a similar NMR pattern, because of its hemiacetal structure. In 1 N NaOD, the 5-methylene protons and the hemiacetal proton appear as sharp singlets (Fig. 4), whereas the corresponding ethyl hemiacetal still maintains all the features of a hemiacetal structure as noted for the zwitterion (Fig. 3).

This collapse of peaks can be interpreted as indicating the existence of a fast equilibrium between the "open" (I) and "closed" (II) forms of pyridoxal in the alkaline medium. This equilibrium is temperature-dependent and can be shifted toward the "open" form (I) with a temperature increase.[9] An alternative explanation for this behavior of pyridoxal in alkaline solution has recently been proposed, involving an equilibrium between the hemiacetal (II) and hydrate (III) forms of the compound.[13] This latter explanation, however, is not consistent with the following facts:

1. The position of the 4-CH peak of pyridoxal in the alkaline solution at -446 cps is not consistent with an equilibrium position between the hydrated aldehyde form (at -390 cps for pyridoxal phosphate) and the hemiacetal form (estimated at -383 cps). It can, in contrast, be reconciled with an equilibrium between the aldehyde form and either the hydrated or

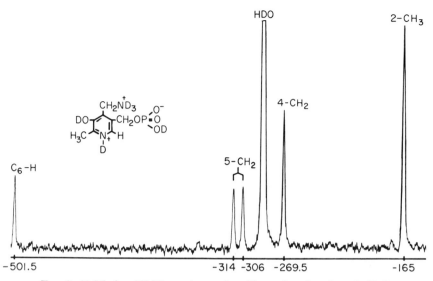

FIG. 6. 60 Mc/sec NMR spectrum of pyridoxamine phosphate in D₂O.

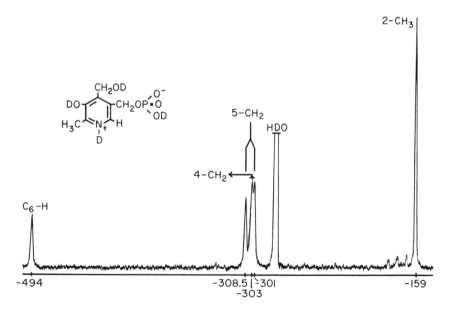

FIG. 7. 60 Mc/sec NMR spectrum of pyridoxol phosphate in D₂O.

hemiacetal form or both (the latter two forms cannot be distinguished by NMR alone).

2. Compounds related to pyridoxal, but which are devoid of the 5-hydroxymethyl group, such as 5-deoxypyridoxal and pyridoxal phosphate, exist in alkaline solutions as free aldehydes (singlets at around -610 cps).[2,17] Hydration of the aldehydes is pronounced under acidic but not alkaline conditions.

3. Other physical methods (ultraviolet and infrared) indicate the presence of a fraction of the aldehyde form in pyridoxal in alkaline solutions.

The UV spectrum of pyridoxal in alkaline solution exhibits a weak absorption at 390 mμ in addition to the absorption due to the hemiacetal form at 302 mμ. The low-intensity peak has been interpreted as being due to the free aldehyde form, since its position is similar to that of 5-deoxypyridoxal[18] and of pyridoxal phosphate in the same medium (Johnson and Metzler, this volume [77]).

The IR spectrum of pyridoxal in D_2O (pD 11.7) shows a free carbonyl band at 1647 cm^{-1}.[19]

From the position of the 4-CH peak at -446 cps (Fig. 4), the amount of the free aldehyde can be calculated as being 27% at 35° (probe temperature). The position of this peak varies regularly from -435.5 cps at $-1.5°$ to -452 cps at 101°, whereas the positions of other peaks remain constant.

Pyridoxal Phosphate (Fig. 5). The characteristic feature of the phosphorylated forms of vitamin B$_6$ is the ^{31}P-^{1}H coupling, which splits the methylene protons to give a doublet. At neutral and alkaline pD, pyridoxal phosphate exists entirely in the free aldehyde form, as is indicated by a sharp singlet (Fig. 5). In some model compounds (e.g., 5-deoxypyridoxal), considerable broadening of the aldehyde proton peak in an alkaline medium has been observed.[17] At acidic pD, the aldehyde group is hydrated. In contrast to the aldehyde ⇌ hemiacetal equilibria, which are very fast, the aldehyde ⇌ hydrate equilibria of pyridoxal phosphate and many model compounds are sufficiently slow for both forms to be observable by NMR spectroscopy. Thus, at pD 4.1, both the aldehyde (-630 cps) and hydrate (-392 cps) forms of pyridoxal phosphate can be observed. The hydration of the aldehyde at lower pD values is consistent with the conclusion reached by Anderson and Martell on the basis of their IR studies.[19]

Pyridoxamine Phosphate (Fig. 6). The NMR spectrum of pyridoxamine phosphate in acidic solution presents no problems, since the methylene

[17] C. Iwata and D. E. Metzler, *J. Heterocyclic Chem.* **4,** 319 (1967).
[18] D. E. Metzler and E. E. Snell, *J. Am. Chem. Soc.* **77,** 2431 (1955).
[19] F. J. Anderson and A. E. Martell, *J. Am. Chem. Soc.* **86,** 715 (1964).

protons can be readily distinguished by ^{31}P-^{1}H coupling in the 5-CH$_2$ group, which appears as a doublet.[2]

Pyridoxol Phosphate (Fig. 7). In the interpretation of the NMR spectrum of pyridoxol phosphate in acidic solution use has been made of the ^{31}P-^{1}H coupling constant of 7.5 cps to assign the peaks in the methylene region.

[80] Mass Spectrometry of Vitamin B$_6$[1]

By D. C. DeJongh and W. Korytnyk

Introduction

A mass spectrometer ionizes molecules in the gas phase and separates the ions formed, which correspond to the original molecule and to fragments of it, into groups according to their mass-to-charge ratio (m/e).[2] The most commonly used technique for ionization is bombardment with electrons of 70 eV energy, although other ionization methods are available. The sample pressure generally does not rise above 10^{-5} mm and the sample size can routinely be in the microgram range.

The ionized molecules are called molecular ions, or parent ions, and they can decompose by fragmentation or rearrangement. Because of the low sample pressure, these are unimolecular reactions. The presence of these ions is detected, and their abundances are plotted against their mass-to-charge ratio. The resulting graph is called a mass spectrum.

From the molecular-ion region of the mass spectrum, one can find an exact molecular weight and, with high-resolution mass spectrometers, elemental composition. From the fragmentations and rearrangements, it is possible to recognize structural features. Since cations are formed and their presence is recorded as a function of m/e, interpretation of a mass spectrum is based on stabilization of positive charge. Interpretation of the

[1] The work at Wayne State University was supported in part by a Wayne State University Faculty Research Fellowship awarded to D. C. DeJongh. The work at Roswell Park Memorial Institute was supported in part by U.S. Public Health Service Grant CA-08793 awarded to W. Korytnyk. We would like to thank Professor F. W. McLafferty of the Purdue Mass Spectrometry Center supported under USPHS Grant FR-00354 for the high-resolution data.
[2] J. H. Beynon, R. A. Saunders, and A. E. Williams, "The Mass Spectra of Organic Molecules," Elsevier, Amsterdam, 1968.

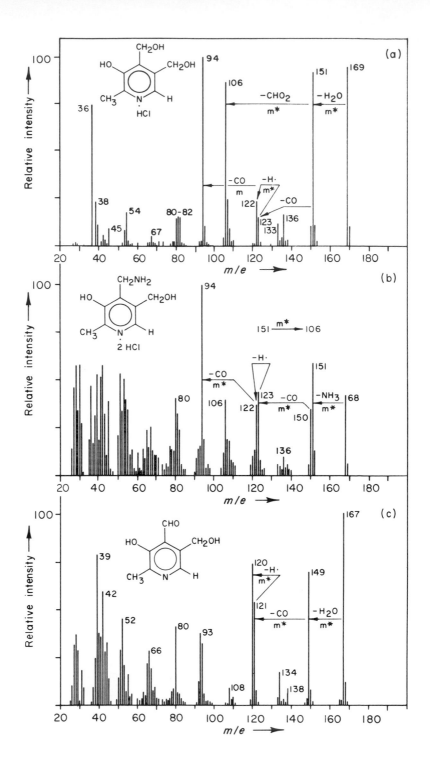

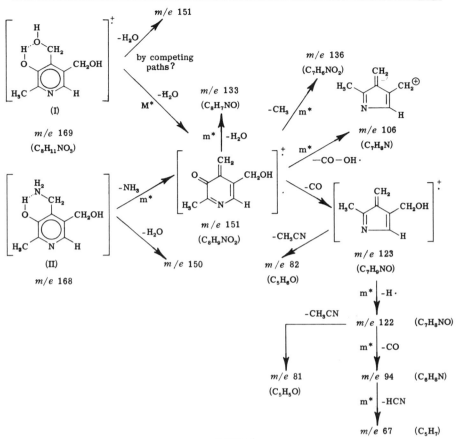

Scheme 1

mass spectra of vitamin B₆ compounds on this basis provides a valuable method for structural analysis and identification in that field.[3,4]

Mass spectrometry can be used in conjunction with chromatography, the latter being a separation technique and the former an identification technique. For example, the *O*-trimethylsilyl derivatives of pyridoxal have been studied by mass spectrometry after collection from a gas chro-

FIG. 1. (a) The 20 eV mass spectrum of pyridoxol, introduced as the hydrochloride. It is almost identical to the 70 eV mass spectrum (footnote 3), with the exception of the prominent HCl peak, m/e 36. (b) The 70 eV mass spectrum of pyridoxamine, introduced as the dihydrochloride. (c) The 70 eV mass spectrum of pyridoxal.

[3] D. C. DeJongh, S. C. Perricone, and W. Korytnyk, *J. Am. Chem. Soc.* **88,** 1233 (1966).
[4] D. C. DeJongh, S. C. Perricone, M. L. Gay, and W. Korytnyk, *Org. Mass Spectr.* **1,** 151 (1968).

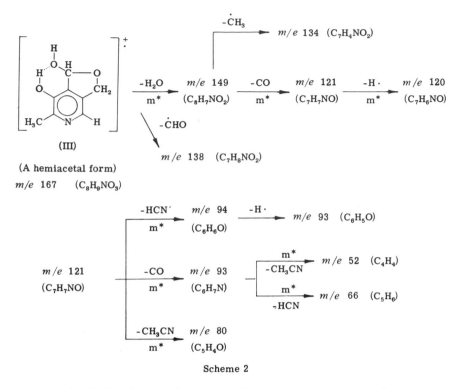

Scheme 2

matograph.[5] It is also possible to couple a mass spectrometer directly to a gas chromatograph and obtain mass spectra of components as they appear, without prior collection.[6]

Method

The mass spectra were determined with an Atlas CH4 mass spectrometer, ionizing potential 70 eV, ionizing current 18 μA. The solid samples were ionized by electron bombardment after sublimation directly into the electron beam from a small graphite crucible heated by a tungsten coil. A cathode with a tungsten wire of 0.15 mm diameter was used.

Above m/e 40 the mass spectra are identical whether the compounds are introduced into the mass spectrometer as hydrochlorides or free bases, indicating dissociation before electron bombardment.

The elemental compositions shown in the schemes were obtained from exact-mass measurements made with a CEC-110 double-focusing mass spectrometer. Microgram quantities of sample were used.

[5] W. Richter, M. Vecchi, W. Vetter, and W. Walther, *Helv. Chim. Acta* **50**, 364 (1967).
[6] F. A. J. M. Leemans and J. A. McCloskey, *J. Am. Oil Chemists' Soc.* **44**, 11 (1967).

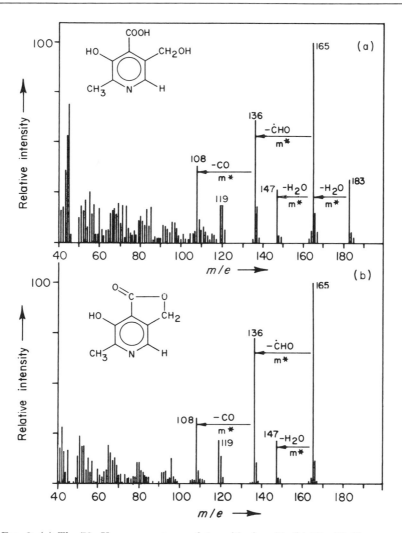

FIG. 2. (a) The 70 eV mass spectrum of 4-pyridoxic acid. (b) The 70 eV mass spectrum of 4-pyridoxic acid lactone.

Mass Spectra

Different Forms of the Vitamin

Mass spectra of pyridoxol (I), pyridoxamine (II), and pyridoxal (III) are presented in Fig. 1(a–c).

They are characterized by having a molecular ion peak, and hence they can be distinguished by this method on a microgram, or less, scale. Mixtures of the three forms could be analyzed both qualitatively and quantitatively.

The fragmentation paths for pyridoxol (I) and pyridoxamine (II) are similar and are presented in Scheme 1. The paths outlined are supported by metastable peaks, by shift peaks in the mass spectra of deuterated analogs, and by exact-mass measurements from which elemental compositions were calculated. The fragmentations are similar to those seen in the mass spectra of o-hydroxybenzyl alcohol, benzyl alcohol, and alkylpyridines.

Scheme 2 summarizes the major fragmentation paths for pyridoxal (III), which most likely exists in the hemiacetal form (see W. Korytnyk and H. Ahrens, this volume [79]). The fragmentation of pyridoxal differs from that of pyridoxol and pyridoxamine in that the fragmentation corresponding to m/e 106 is not seen, whereas m/e 80, arising from loss of CH_3CN from m/e 121, is more prominent.

4-Pyridoxic Acid and 4-Pyridoxic Acid Lactone

The most important metabolites of vitamin B_6, 4-pyridoxic acid and its lactone, can be readily identified by mass spectrometry (Fig. 2a and b) These mass spectra are strikingly similar below m/e 165. This suggests that the 4-pyridoxic acid loses H_2O to form an ion, m/e 165, which has the same structure as the molecular ion of its lactone. The loss of ·CHO, leading to m/e 136, is characteristic of phenolic compounds. Caution must be exercised in vaporizing the acid into the electron beam; if the temperature is too high, the lactone forms thermally and the molecular ion peak m/e 183 disappears.

It is interesting to note that 4-pyridoxic acid lactone can be readily distinguished from 5-pyridoxic acid lactone, which is the microbial metabolite of the vitamin B_6.[4]

Other metabolites, antimetabolites, and analogs of vitamin B_6 give characteristic mass spectra.[4]

Mass spectra of a number of derivatives have been determined, among which isopropylidene, acetyl, and trimethylsilyl derivatives are of special importance, because of their use in gas chromatography[6,7] (Korytnyk, this volume [83]). These three classes of relatively volatile derivatives of vitamin B_6 give rise to a prominent molecular ion peak, and hence are suitable for the application of mass spectrometry in combination with gas chromatography. The mass spectra of the isopropylidene derivatives of pyridoxol and some of the analogs have been discussed in some detail.[3] However, the 5-phosphate derivatives of vitamin B_6 were found insufficiently volatile for mass spectrometry.

---

[7] W. Korytnyk, G. Fricke, and B. Paul, *Anal. Biochem.* **17,** 66 (1966).

[81] Thin-Layer Chromatography and Thin-Layer Electrophoresis of Vitamin B_6[1]

By H. Ahrens and W. Korytnyk

The advantages of thin-layer chromatography and thin-layer electrophoresis are well documented, and excellent books and reviews (e.g., on thin-layer chromatography, this series, Vol. XII, p. 323) are available describing the technique. Thin-layer chromatography of vitamin B_6 has been the subject of several studies.[2-8] After reviewing the systems previously recommended, we have been able to develop some additional systems that have certain advantages.[8a] Furthermore, we have been able to apply thin-layer electrophoresis to the separation of compounds in the vitamin B_6 group.[8a]

Thin-Layer Chromatography

Preparation of Plates

I. Silica gel, 20 plates 20 × 5 cm, 300 μ thick: 30 g of silica gel HF_{254} (Merck A. G., Darmstadt, Germany) and 75 ml of water. Dried in air 30 minutes at room temperature and 30 minutes at 110°

II. Cellulose, 20 plates 20 × 5 cm, 300 μ thick: 15 g of MN 300 G cellulose (Macherey and Nagel, Dueren, Germany) and 90 ml of water, blended for 1 minute in a mixer; dried in air 30 minutes at room temperature and 20 minutes at 110°

III. Eastman Kodak "Chromagram 6065" cellulose sheets, used without pretreatment

Detection

When absorbents containing fluorescence indicators are used, all forms and derivatives of vitamin B_6 can be detected through quenching of

[1] This study was supported in part by a research grant (CA-08793) from the National Cancer Institute, U.S. Public Health Service.

[2] E. Stahl, "Thin Layer Chromatography," Academic Press, New York, 1965: (a) p. 240; (b) p. 489; (c) p. 496.

[3] E. Nuernberg, *Deut. Apotheker-Ztg.* **101**, 268 (1961).

[4] P. Gonnard, M. Camier, and N. Boigne, *Bull. Soc. Chim. Biol.* **64**, 407 (1964).

[5] R. Hakanson, *J. Chromatog.* **13**, 263 (1964).

[6] M. Yamada and A. Saito, *J. Vitaminol. (Kyoto)* **11**, 192 (1965).

[7] J. E. Gill, *J. Chromatog.* **26**, 315 (1967).

[8] E. N. Dementieva, N. A. Drobinskaya, L. V. Ionova, M. Ya. Karpeisky, and V. L. Florentiev, *Biokhimiya* **33**, 350 (1968); *Chem. Abstr.* **69**, 12902h (1968).

[8a] H. Ahrens and W. Korytnyk, *Anal. Biochem.* **30**, 413 (1969).

fluorescence in ultraviolet (UV) light (254 mμ). The limit of detection with UV light is 1 γ, which can be extended to approximately 0.1 γ by using either Gibbs' reagent or diazotized p-nitroaniline, as described here.

Gibbs' Reagent.[2b] N-2,6-Trichloro-p-benzoquinone imine (1% in ethanol or benzene) is sprayed on the plate, which is subsequently treated with ammonia vapor (blowing a stream of air over a bottle of conc. NH_4OH toward the plate) or sprayed with dilute NH_4OH or $NaOH$ solution to produce blue- or violet-colored spots.

Diazotized p-Nitroaniline.[2b] p-Nitroaniline (0.7 g) is dissolved in concentrated hydrochloric acid (9 ml), and the solution is made up to 100 ml with water. The reagent is prepared by adding 4 ml of this stock solution to 5 ml of 1% $NaNO_2$ solution cooled with ice, and making the mixture up to 100 ml with water. In some cases, treatment of the plate with ammonia vapor is necessary, depending on the compound used.

Both Gibbs' and the diazotized p-nitroaniline tests require the presence of the phenolic hydroxyl group. The orange-red color produced by diazotized p-nitroaniline is somewhat more stable. Blocking groups can readily be removed on the thin-layer plate by spraying with either 1 N NaOH or 1 N HCl (depending on the nature of the group) and subsequently heating the plate.[9]

Amino Groups. Amino groups, as in pyridoxamine and its phosphate, can be detected with a standard ninhydrin solution,[2c] 0.3 g of indan-1,2,3-trione in 100 ml of n-butanol plus 3 ml of glacial acetic acid.

Aldehydes. Aldehydes, such as pyridoxal and its phosphate, can be detected by spraying with phenylhydrazine hydrochloride (5% solution in water). Hakanson[5] has advocated using a semicarbazide spray and viewing the spots in UV light.

Separation

Separation of the nonphosphorylated forms of vitamin B_6 (pyridoxol, pyridoxal, pyridoxamine, and the most important metabolites, 4-pyridoxic acid and its lactone) involves problems different from those created by separation of the phosphorylated forms, and hence the two classes of compounds will be discussed individually.

Pyridoxal, Pyridoxol, Pyridoxamine, and 4-Pyridoxic Acid and Its Lactone. Although pyridoxamine can be separated easily from the other two nonphosphorylated forms of the vitamin in a number of systems (Table I), the separation of pyridoxal from pyridoxol presents difficulties. The most satisfactory method for separation of these two forms is the conversion of pyridoxal to its ethyl acetal, which can be readily separated from pyridoxol and pyridoxamine.[2a,3] For this purpose, a mixture of pyridoxal with one or

---

[9] W. Korytnyk and B. Paul, *J. Org. Chem.* **32**, 3791 (1967).

TABLE I

R_f VALUES ($\times 100$) OF VITAMIN B₆ COMPOUNDS[a]

Compound	Layer[a] Solvent[b]	I A	I B	I[d] B	I F	I[c] F	II[c] C	II[c] D	II[c] G	III[c] C	III[c] E
Pyridoxol		62	47	In the H₃BO₃ strip	—	—	—	—	—	—	—
Pyridoxal		68	56	36	—	—	—	—	—	—	—
Pyridoxamine		12	05	05	—	—	—	—	—	—	—
Pyridoxal ethyl acetal		54	84	83	—	—	—	—	—	—	—
4-Pyridoxic acid		91	49	—	—	—	—	—	—	—	—
4-Pyridoxic acid lactone		91	18	—	—	—	—	—	—	—	—
Pyridoxol phosphate		95	00	—	30	18	47	68	70	21	69[e]
Pyridoxal phosphate		95	00	—	54[f]	33[f]	64	80	78	37	79[e]
Pyridoxamine phosphate		86	00	—	41	28	07	36	62	01	57[e]
Pyridoxal phosphate phenylhydrazone		—	—	—	76	55	—	—	—	—	—

[a] *Layers:* (I) Silica gel HF₂₅₄; (II) cellulose MN 300G; (III) "Chromagram" cellulose sheets.

[b] *Solvents:* (A) 0.2% NH₄OH in water (1:139 v/v conc. NH₄OH:H₂O); (B) chloroform:methanol (75:25, v/v); (C) *n*-butanol saturated with 1 N HCl, upper layer; (D) *n*-butanol:conc. HCl:H₂O (25:5:10, v/v); (E) methanol:butanol:benzene: water:triethylamine (20:10:10:10:5, v/v); (F) methyl ethyl ketone:ethanol:conc. NH₄OH:H₂O (15:5:5:5, v/v); (G) *n*-butanol:acetic acid:water (15:10:10 v/v).

[c] Chamber saturation.

[d] H₃BO₃-treated plate.

[e] Spots show tailing.

[f] Phenylhydrazine test is negative.

both of the other two simple forms of the vitamin is heated in absolute ethanol on a steam bath for approximately 1 hour. Pyridoxamine and pyridoxol remain unchanged, but pyridoxal is transformed quantitatively into its ethyl acetal. After excess ethanol is removed *in vacuo*, the three components can be separated in one run, using solvent B (Table I). Previously a double-development technique was used.[3]

Pyridoxol forms a strong complex with boric acid across the phenolic OH and the 4-(hydroxymethyl) group,[10] whereas other forms of the vitamin do not. Thus plates treated with boric acid greatly retard pyridoxol, and this property can be used as the basis of the separation of pyridoxol from the other two simple forms of the vitamin. For this purpose, coated and activated silica gel plates are streaked with 2.5% boric acid in ethanol approximately 2 cm from the starting point. For a plate 5 cm wide, approxi-

[10] J. V. Scudi, W. A. Bastedo, and T. J. Webb, *J. Biol. Chem.* **136,** 399 (1940).

mately 50 μl of boric acid solution is required. After drying with a hair dryer for 5 minutes at approximately 50°, the plates are ready for use. Pyridoxol is completely retarded by the H_3BO_3 strip, but pyridoxal is only partially retarded. As well as providing greatly improved separation, this method is valuable in structural diagnosis, since compounds lacking the hydroxyl group in the 3- or α^4-position pass through the strip unhindered.[11]

Phosphorylated Forms of Vitamin B_6. Pyridoxamine phosphate and pyridoxol phosphate give sharp spots and can be readily separated in a number of systems (Table I). The analysis of pyridoxal phosphate presents special problems. It was found to give rise to several spots on thin-layer and paper chromatography.[4] We have now established that additional spots are due to the formation of photodecomposition products during chromatography, and that they can be avoided by working in the absence of light. A $5 \times 10^{-3} M$ solution of pyridoxal phosphate in water, when exposed to laboratory daylight and air, is decomposed more than 95% within 1 hour. The main products of decomposition[12] are 5,5'-bis(dihydroxyphosphinyloxymethyl)-3,3'-dihydroxy-2,2'-dimethyl-4,4'-pyridil (solvent D, R_f 0.40, quenches fluorescence) and 4-pyridoxic acid-5-(dihydrogen phosphate) (solvent D, R_f 0.70, strong blue fluorescence under UV light); some minor additional products have also been observed.

Pyridoxal phosphate has a tendency to give broad spots in some solvent systems. Nevertheless, for most of the systems listed in Table I, good separations were obtained, and the spots were compact in most cases. On application of Gibbs reagent, the color takes longer to develop and is weaker than in other phosphorylated derivatives.

Some commercial samples of pyridoxal phosphate were found to have impurities, but they were effectively purified by the method of Peterson and Sober.[13]

Considerable stabilization of pyridoxal phosphate can be achieved by allowing it to react with various carbonyl reagents, such as phenylhydrazine, semicarbazide,[5] or hydroxylamine. We have found phenylhydrazone to be the reagent of choice. A clear solution of the phosphates of pyridoxal, pyridoxamine, and pyridoxol in a minimum amount of 3 N HCl is treated with a few drops of a concentrated solution of phenylhydrazine hydrochloride in water. The resulting thick precipitate is heated in a steam bath for 30 seconds, and, after cooling, is treated dropwise with 1 N NH$_4$OH until a clear yellow solution is obtained, which is spotted immediately.

[11] H. Ahrens and W. Korytnyk, *J. Heterocyclic Chem.* **4**, 625 (1967).
[12] A. L. Morrison and R. F. Long, *J. Chem. Soc.* p. 211 (1958).
[13] E. E. Peterson and H. A. Sober, *J. Am. Chem. Soc.* **76**, 169 (1954).

Pyridoxamine phosphate, pyridoxol phosphate, and pyridoxal phosphate phenylhydrazone give sharp and well-separated spots on silica gel plates (solvent F, Table I). The excess phenylhydrazine migrates toward the solvent front, and does not interfere. For detection, diazotized p-nitroaniline solution is used (see above).

Substituted vitamin B₆ derivatives[9,14] and many metabolites[11] and antimetabolites have been conveniently separated by using organic solvents (ethyl acetate, chloroform, or methanol, or mixtures of those solvents). A 1:1 mixture of methanol (or ethanol) and chloroform appears to be the most useful. Mixtures consisting of acetic acid, butanol, and water (1:4:5) or ethanol, water, and butanol (1:4:5) have also been employed.[15]

Thin-Layer Electrophoresis

Thin-layer electrophoresis adds another important possibility to separation techniques, and can be utilized either in conjunction with thin-layer chromatography as a second dimension or independently of it in cases where thin-layer chromatography does not offer sufficient separation. We have investigated various buffer systems in the pH range 3.50 to 13.0, and have found sodium acetate buffers in the pH range 3.50 to 4.50 the most useful for separation (Table II). This system provides compact spots. Buffers of higher pH (formate, pH 4.80; borate, pH 7.80; acetate–triethylamine, pH 5.90; glycinate, pH 8.70; phosphate, pH 11.52, 12.50, and 13.00) were found often to give rise to broad and sometimes multiple spots,

TABLE II

ELECTROPHORETIC MOBILITIES OF VITAMIN B₆ COMPOUNDS[a]

Compound	Buffer pH 3.95[b] (mm)	Buffer pH 4.53[c] (mm)
Pyridoxol	−54	−19
Pyridoxal	−35	−9
Pyridoxamine	−56	−34
Pyridoxol phosphate	0	+12
Pyridoxal phosphate	+24	+21
Pyridoxamine phosphate	−11	−3

[a] Movements of compounds are given in millimeters from the origin. Plus indicates movement toward anode, and minus indicates movement toward cathode. Acetate buffers were used; the ionic strength was 0.05.

[b] Time of running was 55 minutes, at 500 V. The current increased from 19 to 30 mA.

[c] Time of running was 40 minutes, at 500 V. The current increased from 28 to 43 mA.

[14] W. Korytnyk, *J. Org. Chem.* **27**, 3724 (1962).

[15] I. Tomita, H. G. Brooks, and D. E. Metzler, *J. Heterocyclic Chem.* **3**, 178 (1966).

and insufficiently different rates of migration of useful separation. They might be suitable, however, in some special cases. Sodium acetate[16] and sodium butyrate[17] buffers have been used in paper electrophoresis of the compounds in the vitamin B_6 group, and this technique has been applied to a study of the nonenzymatic transamination reaction catalyzed by pyridoxal.[18]

Method

A WCLID Model E-800-2 Electrophoresis System (Warner Chilcott Laboratories Instrument Division) was used. The thin-layer electrophoresis plates (Silica gel, 10×20 cm) were prepared in the same way as the thin-layer chromatography plates (see above) and were sprayed with the appropriate buffer until they became transparent. Subsequently the compounds were spotted in the middle of each plate and were immediately subjected to electrophoresis. UV light and spray reagents (see thin-layer chromatography) were used for detection.

[16] N. Siliprandi, D. Siliprandi, and H. Lis, *Biochim. Biophys. Acta* **14**, 212 (1954).

[17] H. E. Wade and D. M. Morgan, *Biochem. J.* **60**, 264 (1955).

[18] P. Fasella, H. Lis, N. Siliprandi, and C. Baglioni, *Biochim. Biophys. Acta* **23**, 417 (1957).

[82] Gas Chromatography of Vitamin B_6

By A. J. Sheppard and A. R. Prosser

Principle

Analysis of vitamin B_6 analogs by gas–liquid partition chromatography (GLC) must be carried out with volatile derivatives of the substance under investigation. A few studies have been made of the chromatographic behavior of acetyl,[1–3] isopropylidene,[3] 3-O-benzyl,[3] trimethylsilyl,[3–6] and trifluoroacetyl[7] derivatives. Excellent separations can be obtained with any of these modified vitamin B_6 analogs, the volatilities of which are generally sufficient for gas chromatography under a variety of operating conditions.

[1] A. R. Prosser and A. J. Sheppard, *Federation Proc.* **25** (2, part 1), 669 (1966).

[2] A. R. Prosser, A. J. Sheppard, and D. A. Libby, *J. Assoc. Offic. Anal. Chemists* **50**, 1348–1353 (1967).

[3] W. Korytnyk, G. Fricke, and B. Paul, *Anal. Biochem.* **17**, 66 (1966).

[4] L. T. Sennello, F. A. Kummerow, and C. J. Argoudelis, *J. Heterocyclic Chem.* **4**, 295 (1967).

[5] Y. Ohnishi, Z. Horii, and M. Makita, *Yakugaku Zasshi* **87**, 747 (1967).

[6] B. E. Haskell, *Federation Proc.* **27**, 554 (March–April 1968) (Abstract No. 1899).

[7] T. Imanari and Z. Tamura, *Chem. Pharm. Bull.* (*Tokyo*) **15**, 896 (1967).

However, consideration should be given to the proper choice of derivative for routine analytical work. The reaction that yields the derivative must be rapid and quantitatively consistent at ambient temperatures and must be suitable for a wide range of concentrations of the initial starting material. In our experience, the acetyls are the most satisfactory general purpose derivatives for analysis of GLC vitamin B₆ analogs.

The potential practical applications of the GLC technique for measuring vitamin B₆ analogs have not been realized. Only one published paper[2] is available which indicates that the method is effective for measuring the vitamin B₆ analogs used in the preparation of pharmaceutical products. A paper[6] presented at the 52nd Annual meeting of the Federation of American Societies for Experimental Biology indicated that the GLC method is applicable to the determination of pyridoxal 5-phosphate in animal tissues. However, it was stated in the oral presentation that the technique has not been developed to a reliable quantitative procedure.

The techniques in this chapter are based on the authors' experience with pharmaceuticals. The method is very effective and reliable for pyridoxine (pyridoxol) hydrochloride, the vitamin B₆ analog used in pharmaceutical preparations. In the authors' laboratory the mean recovery of pyridoxine hydrochloride from pharmaceutical products has been 97.3% with a standard deviation of ±2.2%. The authors have had no experience in measuring the vitamin B₆ analogs present in biological materials; therefore, the method presented here is the one used to analyze pharmaceutical products for their vitamin B₆ content. Probably the only major development needed to measure biological vitamin B₆ analogs is a clean-up of the extracts prior to the preparation of the acetyl derivatives.

Preparation of Derivatives from Pharmaceutical Products

Tablets. Tablets previously weighed are ground to a fine powder in a mortar. A portion of the powdered tablets is weighed and placed in a 50-ml round-bottom flask. To this sample is added 1 ml of pyridine–acetic anhydride solution (25:75, v/v) per 10 mg of pyridoxol (based on label claim); a magnetic stirring bar is inserted, the flask is stoppered, and the mixture is stirred for 1 hour. The volume of the reaction mixture is then adjusted with chloroform[8] to obtain a concentration of 1 mg per milliliter of the expected pyridoxol content. The flask is stoppered, and the mixture is stirred for 15 minutes and then filtered through Whatman No. 42 paper into a volumetric flask of such size that the final concentration is 1 mg per milliliter of the expected pyridoxol content.

A duplicate sample containing a known amount of added vitamin B₆ analog(s) is prepared and processed in the same manner as described above. This duplicate containing added analog is necessary for control of

---

[8] Benzene should be used with the argon ionization detector.

the combined effects of reaction efficiency and compound recovery so that the values obtained from the unknown can be properly corrected if necessary.

Two microliters of each solution is injected onto the GLC column for vitamin B_6 analysis.[8]

Capsules. Slip capsules are opened and the contents emptied into a 50-ml round-bottom flask. Other types of capsules are cut open and the contents of the capsules and the capsule fragments are put into the flask. The sample is then processed in the same manner as for tablets.

Liquid Preparations. A portion of the liquid preparation is transferred by volumetric pipette to a 50-ml round-bottom flask. The volume transferred must contain enough vitamin B_6 analog, based on expected content, to give a final concentration of 1 mg per milliliter of the derivative for GLC analysis. The round-bottom flask is attached to a flash evaporator, the aqueous solution is removed under reduced pressure, and the procedure for tablets is followed. It is very important that all the water be removed before the reaction phase is started.

Preparation of Packed Columns for Gas Chromatography

The most satisfactory column packing evaluated by the authors for the analysis of the vitamin B_6 acetyl derivatives is a mixed immobile phase of 6.67% SE-30 and 0.33% neopentyl glycol succinate (NPGS)(w/w) coated on 100/120 mesh silanized Gas Chrom P. This packing completely resolves diacetyl pyridoxal, triacetyl pyridoxol, and triacetyl pyridoxamine. The inert support is silanized as described for gas chromatography of vitamin K_3 (this volume [235]). The immobile phase is coated on the silanized inert support as follows: 1.6675 g of SE-30 is weighed and dissolved in 50 ml of toluene, with heat if necessary. Then 0.0825 g of NPGS is weighed and dissolved in 25 ml of chloroform, with heat if necessary. Both of the immobile phase solutions are transferred to a Morton flask containing 23.25 g of the silane-treated inert support, and enough additional toluene: chloroform solution (2:1, v/v) is added to form a slurry. The mixture is allowed to stand for 1 hour with occasional shaking. The toluene is evaporated *in vacuo* with warming on a flash evaporator, and the prepared packing is dried at 80° for 1 hour. The finished product is stored in a dust-free container. The chromatographic column is filled with the prepared packing as described for gas chromatography of vitamin K_3 [this volume [235]).

Instrumentation

Any gas–liquid chromatograph fitted with either a hydrogen flame ionization detector or a β-argon ionization detector capable of using an

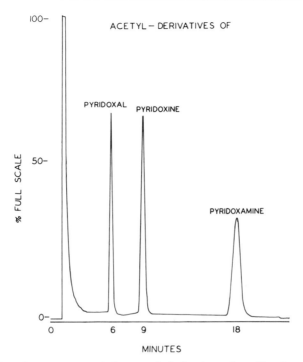

FIG. 1. Gas chromatogram of the acetyl derivatives of pyridoxal, pyridoxol, and pyridoxamine.

8-foot glass column at 250° can be used. The electrometer outputs should not be less than 3×10^{-7} A for the β-argon ionization detector and 1×10^{-9} A for the hydrogen flame ionization detector as expressed in a GLC system utilizing a 5-mV recorder. The recommended chart speed is 0.33 inch per minute.

The operating parameters for the GLC analysis of the vitamin B₆ acetyl derivatives are as follows: column, 230°; detector, 250°; flash heater, 240°; direct current voltage for β-argon detector, 900; carrier gas flow adjusted so that the triacetyl pyridoxol peak appears at about 9 minutes (carrier gas flow of approximately 45 ml/minute in the authors' experience). A typical chromatogram of the acetyl derivatives is shown in Fig. 1.

Standards and Calibration

Three standards of the three vitamin B₆ derivatives are prepared as follows. Weighed amounts of the individual vitamin B₆ acetyl derivatives are dissolved in chloroform[8] to provide solutions containing 0.5, 1.0, and 2.0 mg of each acetyl derivative per milliliter, respectively. The derivatives are prepared as above except that the starting materials are pure crystalline pyridoxal, pyridoxol, and pyridoxamine rather than sample materials. The standards are used to prepare a calibration plot for each acetyl derivative. A calibration plot covering a range of 1–4 μg has been the usual working range when either the hydrogen flame ionization detector or the β-argon ionization detector is used. Variation in response has been observed in GLC instruments having the same type of detectors; thus, the span of the calibration plot must be adjusted to suit the particular GLC system being used. The concentration of the standard solutions are adjusted so that a fixed 2-μl injection is used to obtain a minimum of three calibration points for each derivative. The response in square centimeters for each amount of the specified derivative injected is plotted on standard graph paper with the peak area in square centimeters as the ordinate and the amount of the derivative injected as the abscissa. A straight line is fitted to the calibration data of each individual derivative. A typical calibration plot for the three derivatives is given in Fig. 2. It is very important that the same volumes of standards and unknowns are injected to eliminate "needle cook-out effect."

Injection Technique

This technique is described for gas chromatography of vitamin K_3 (this volume [235]).

Example

The example that follows is based on a situation that was solved in the authors' laboratory. The product was a tablet with a claim of 50 mg of pyridoxine per tablet.

Three tablets were weighed and an average weight per tablet was established. The three tablets were ground in a mortar to a fine powder. Then the amount of powder equivalent to two tablets was weighed and transferred to a 50-ml round-bottom flask, and the triacetyl pyridoxol was prepared as previously described for tablets. The reaction mixture was filtered into a 100 ml volumetric flask and the volume was adjusted to the mark.

After the instrument was calibrated as shown in Fig. 2, using 2-μl injections, 2 μl of the sample preparation was injected onto the column. Both standards and samples were analyzed in duplicate.

The average response for the two sample injections was 835 digital readout units. The response of 835 on the ordinate corresponds to 2.05 μg on the abscissa of the calibration plot for triacetyl pyridoxol in Fig. 1. The amount of pyridoxine present per tablet is calculated by the following equation:

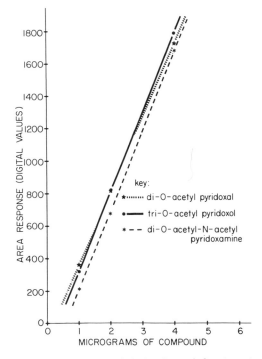

FIG. 2. Dose-response plots for the acetyl derivatives of the three forms of vitamin B₆.

$$P = \frac{(X/M)V}{N} \tag{1}$$

where P = milligrams of pyridoxine per tablet; X = micrograms of derivative found per injection; M = number of microliters injected; V = volume of extract in milliliters; N = number of tablets in extract.

Thus:

$$P = \frac{(2.05/2)100}{2}$$

$$P = 51.25 \text{ mg per tablet}$$

A duplicate control sample was weighed into a 50-ml round-bottom flask, and 100 mg of pyridoxol was added. The control sample was treated in the same manner as the sample for analysis. The reaction mixture was filtered into a 200-ml volumetric flask and adjusted to volume, to provide a concentration of 1 mg/ml, based on the claim plus the added standard. Injections into the instrument of 2 μl of the control sample preparations gave a response of 810 digital readout units. This corresponds to 2.00 μg on the abscissa of the calibration plot in Fig. 2 for triacetyl pyridoxol. The amount of pyridoxine present per tablet in the control sample is calculated in the same manner as the amount present in the unknown analysis sample:

$$P = \frac{(2.00/2)200}{2}$$

$$P = 100.0 \text{ mg per control tablet}$$

To calculate the true amount of pyridoxol in the product, the 51.25 mg of pyridoxol found in the unknown analysis sample must be corrected based on the recovery of added pyridoxol. The data correction factor is calculated as follows:

$$C = A/R \tag{2}$$

where A = milligrams of pyridoxol added; R = milligrams of pyridoxol recovered.

Thus:

$$C = 50.00/48.75$$

$$C = 1.0256 \text{ (97.5\% recovery)}$$

The amount of pyridoxine in the tablets is therefore:

$$51.25 \times 1.0256 \text{ or } 52.56 \text{ mg}$$

For liquids, the number of milliliters of preparation is substituted for the number of tablets in the example; similarly, for powders or biological materials, the number of grams used is substituted.

[83] Gas Chromatography of Vitamin B_6[1]

By W. KORYTNYK

Principle

A number of methods are available for the assay of vitamin B_6, but gas chromatography may offer unique advantages in speed of separation, sensitivity, and convenience in quantitation.[2] Although simple pyridine derivatives are amenable to gas chromatography, the presence of polar groups in vitamin B_6 compounds tends to keep them from being volatile enough for this method of analysis. Among the relatively volatile derivatives, the

[1] This study was supported in part by a research grant (CA 08793) from the Nationa Cancer Institute, U.S. Public Health Service.
[2] See Vol. VIII, p. 95.

acetates,[3,4] trimethylsilyl ethers,[3,5-7] isopropylidene derivatives,[3] and benzyl ethers[3] have shown promise for applications in gas chromatography.

In the present chapter, attention will be given chiefly to the gas chromatography of the fully acetylated derivatives of vitamin B₆, since they have been shown to be suitable for fast and quantitative separation of all three simple forms of the vitamin (but not the phosphates) and the most important metabolites.[3] As the technique is further developed, other derivatives will find extensive application. In this respect, the trimethylsilyl derivatives should be useful, particularly from the point of view of their application to the volatilization and gas chromatography of the phosphorylated forms of the vitamin; some preliminary experiments have been described.[3] Other derivatives, including the trifluoroacetyl esters, may offer special advantages, such as greater volatility and possible appropriateness for the more sensitive electron capture detectors; but they have not yet been investigated.

Peracetylated derivatives of vitamin B₆ have been utilized in quantitative determination of pyridoxol content in pharmaceutical preparations,[4] Mass spectrometry of these derivatives gives rise to molecular ion peaks,[7ª] and hence they should be suitable for a combination of gas chromatography with mass spectrometry (see this volume [80]).

Method

Apparatus. The gas chromatograph used by the present author was an F & M Model 400 Biochemical Gas Chromatograph equipped with a hydrogen flame detector and a direct injection port. The column consisted of a 4 ft × 1/4 in. glass column packed with 1.2% silicone gum rubber SE-30 on 90- to 100-mesh Chromsorb G, which was acid-washed and silanized.

GLC Conditions. The temperatures of the injection port and detector block were kept about 50° above the working column temperature, which was varied between 115° and 185°, depending on the individual sample. The flow rate was 75 ml/min (inlet pressure 30 psi) for helium, 50 ml/min for hydrogen, and 450 ml/min for air. The injected sample had a volume of 1–2 μl and contained 3–4 μg of any single component.

[3] W. Korytnyk, G. Fricke, and B. Paul, *Anal. Biochem.* **17**, 66 (1966).

[4] A. R. Prosser, A. J. Sheppard, and D. A. Libby, *J. Assoc. Offic. Agr. Chemists* **50**, 1348 (1967).

[5] L. T. Sennello, F. A. Kummerow, and C. J. Argoudelis, *J. Heterocyclic Chem.* **4**, 295 (1967); L. T. Sennello and C. J. Argoudelis, *Anal. Chem.* **41**, 171 (1969).

[6] Y. Ohnishi, Z. Horii, and M. Makita, *Yakugaku Zasshi* **87**, 747 (1967).

[7] W. Richter, M. Vecchi, W. Vetter, and W. Walther, *Helv. Chim. Acta* **50**, 364 (1967).

[7ª] D. C. DeJongh, S. C. Perricone, and W. Korytnyk, unpublished observations, 1966.

Preparation of Acetyl Derivatives. Peracetylated derivatives of the natural pyridoxine compounds have been described previously,[8,9] with the exception of the diacetate of pyridoxal, which has been prepared for the first time for the purposes of gas chromatography.[3] 3,α^4-O-Diacetylpyridoxal (I) was obtained by shaking pyridoxal hydrochloride with a mixture of pyridine and acetic anhydride overnight. It is a crystalline compound of

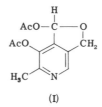

(I)

m.p. 108–110°. Its hemiacetal structure (I) was confirmed by NMR spectroscopy, including analogy with the NMR spectrum of the pyridoxal zwitterion (absence of aldehyde proton; the 5-CH_2 proton appears as an 10-AB quadruplet (see this volume [79]).

The retention times of the acetates are summarized in Table I. The

TABLE I[a]

RETENTION TIMES OF THE ACETATES AT 125°

Acetates	Time (min)	Relative to pyridoxol acetate
Pyridoxol	8.5	1.00
Pyridoxamine[b]	19.6	2.31
Pyridoxal	3.1	0.36
Pyridoxal ethyl acetal	2.4	0.27
4-Pyridoxic acid lactone	2.2	0.26
5-Pyridoxic acid lactone	2.8	0.33
4-Deoxypyridoxol	3.0	0.35

[a] From footnote 3.

[b] At 165°, the retention time was 2.3 minutes.

retention data indicate that the three nonphosphorylated forms of the vitamin and the most important metabolite, 4-pyridoxic acid lactone, can be separated as their acetates. For separating the three vitamers and 4-pyridoxic acid lactone, it was convenient to use temperature program-

[8] R. Kuhn and G. Wendt, *Chem. Ber.* **71**, 780 (1938); W. Korytnyk, E. J. Kris, and R. P. Singh, *J. Org. Chem.* **29**, 574 (1964).

[9] T. Sakuragi and F. A. Kummerow, *J. Org. Chem.* **22**, 825 (1957).

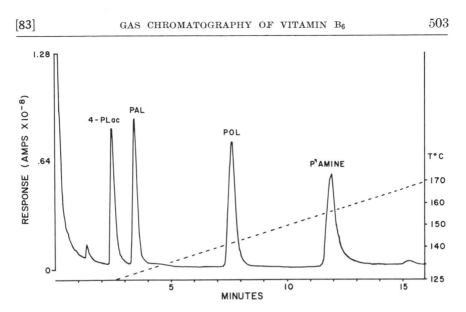

FIG. 1. Chromatograms of acetates of 4-pyridoxic acid lactone (4-PLac), pyridoxal (PAL), pyridoxol (POL), and pyridoxamine (P'AMINE) (from footnote 3).

ming, as indicated in Fig. 1. In this way, separation of the four compounds can be accomplished within 13 minutes.

Acetylation was carried out by treatment of the hydrochlorides or free bases of the vitamers or their analogs with a mixture of acetic anhydride and pyridine (1:1 by volume) at room temperature for 2 to 4 hours. The acetylation mixture was then injected directly into the column.

Pyridoxal may interfere with the analysis of 4-deoxypyridoxol and of 5-pyridoxic acid lactone, because of the similarity of the retention times (Table I). In such cases, it may be expedient to convert pyridoxal to its

TABLE II[a]

ANALYSIS OF SYNTHETIC MIXTURES OF PYRIDOXOL ACETATE
AND 4-PYRIDOXIC ACID LACTONE ACETATE

Sample No.	Pyridoxol acetate			4-Pyridoxic acid lactone acetate		
	Calcd. wt %	Calcd. mole %	Detd. area %	Calcd. wt %	Calcd. mole %	Detd. area %
1	60.9	51.9	57.9	39.1	48.1	42.2
2	75.8	68.6	73.5	24.2	31.4	26.4
3	81.4	75.4	80.3	18.6	24.6	19.7

[a] From footnote 3.

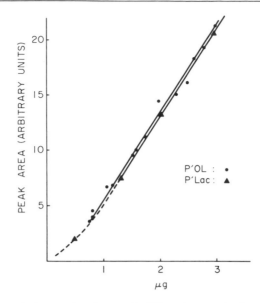

Fig. 2. Response in terms of peak areas to different amounts of acetates of pyridoxol (P'OL) and of 4-pyridoxic acid lactone (P'Lac) (from footnote 3).

hemiacetal, which can be accomplished quite readily and quantitatively by refluxing the reaction mixture with ethanol for 2 hours prior to acetylation.[3] Alternatively, pyridoxal may be reduced quantitatively to pyridoxol through the use of sodium borohydride.[3]

The quantitative aspects of the acetylation procedure have been studied, and the response to various amounts of the acetyl derivatives of pyridoxol and 4-pyridoxic acid lactone has been determined (Fig. 2). There is a linear and almost equal response to both acetyl derivatives in the range of 0.8 to 15.0 μg. This was borne out by analysis of a synthetic mixture of these two acetyl derivatives (Table II). It is to be noted that the values found correspond more closely to the actual weights of the acetyl derivatives than to the mole percentages. Under the conditions used, pyridoxol was found to be acetylated quantitatively within the limits of experimental error. This was also the case with other vitamin B_6 compounds, as was demonstrated by the use of internal standards.

[84] Microassay of Pyridoxal Phosphate Using Tryptophan-^{14}C with Tryptophanase

By Kiyoshi Okuda, Satoru Fujii, and Masahisa Wada

Hiroshi Wada and his co-workers[1] used apotryptophanase for the determination of pyridoxal phosphate because the enzyme reaction is dependent on the content of the coenzyme in the presence of excess apoenzyme. In this method, the amount of indole produced is measured colorimetrically by Ehrlich's aldehyde reaction. This method is not sensitive enough to determine the content of pyridoxal phosphate in small specimens, such as human liver tissue, obtained by needle biopsy. Therefore, the method of H. Wada *et al.* has been modified,[2,3] and radioactive tryptophan is used as the substrate of tryptophanase. The extent of the reaction is determined by measuring the radioactivity in the products formed.

Assay Method

Principle. The enzyme reaction catalyzed by tryptophanase is as follows:

$$\text{L-Tryptophan} + H_2O \rightarrow \text{pyruvate} + \text{indole} + NH_3$$

In the presence of excess apoenzyme and substrate, the reaction rate is dependent on the pyridoxal phosphate concentration. Tryptophan-methylene-3-^{14}C is used as the substrate, and the extent of the reaction is determined by measuring radioactivity in the pyruvate-^{14}C formed, which is extracted with ethyl acetate as pyruvate hydrazone. When tryptophan with ^{14}C in the benzene ring is used, the indole-^{14}C formed is measured after extraction by petroleum ether.

Thus, a plot of the rate of radioactive pyruvate or indole production vs. amount of pyridoxal phosphate may be used as a calibration curve to estimate the amount of this coenzyme in the sample solution.

The method with tryptophan-3-^{14}C is described.

Reagents

Reagent A: 1.0 M Tris hydrochloride, 0.1 M Na$_2$EDTA, 0.1 M ammonium sulfate, 0.1 M KCl, 0.01 M mercaptoethanol; at pH 8.1

[1] H. Wada, T. Morisue, Y. Sakamoto, and K. Ichihara, *J. Vitaminol. Japan* **3**, 183 (1957).

[2] K. Okuda, M. Wada, and S. Fujii, *in* "Symposium on Pyridoxal Enzymes" (K. Yamada, N. Katunuma, and H. Wada, eds.), p. 229. Maruzen, Tokyo, 1968.

[3] S. Fujii, *J. Osaka City Med. Center* **17**, 443 (1968).

Apotryptophanase (see below)

L-Tryptophan (methylene-^{14}C; specific activity: 0.15 μCi/micromole) 0.02 M

HCl, concentrated

Lithium (sodium) pyruvate, 0.01 M

Dowex 50 X8, H$^+$ form in a microcolumn (bed size is 5 × 0.4 cm with 200 to 400 mesh resin, preequilibrated with 1 N HCl)

HCl, 1 N

2,4-Dinitrophenyl hydrazine, 0.001 M in 1 N HCl

Ethyl acetate

Pyridoxal phosphate standard solutions:

  Stock standard: 0.1 mg per ml of 0.001 N KOH (keep in a colored bottle at -15 to $-20°$)

  Working standard: Dilute the stock standard solution with distilled water to make it 0.2 μg/ml. Discard after 1 day.

Procedure. The following procedure for the assay of pyridoxal phosphate is adopted. Reagent A (0.15 ml), enzyme solution (more than 0.8 unit in 0.25 ml), and pyridoxal phosphate solution (5–50 mμg of pyridoxal phosphate in 0.3 ml) are placed in a colored test tube.[4] After 10 minutes of the preincubation period at 37°, the tryptophanase reaction is started by the addition of radioactive tryptophan (0.3 ml), incubated for 15 minutes at 37°, and then the reaction is terminated by the addition of 0.2 ml of concentrated HCl. A small amount of pyruvate (1 micromole in 0.1 ml) is added as the carrier, and the mixture (total 1.3 ml) is applied to a column of Dowex 50 (H$^+$ form) to remove an excess of ^{14}C-tryptophan remaining. The test tube and the column are washed 3 times with 1.0 ml of 1 N HCl, and the effluent is collected in a glass-stoppered centrifuge tube. To 4.3 ml of combined solution (1.3 ml of reaction mixture and 3.0 ml of the washings), 2 ml of 2,4-dinitrophenyl hydrazine solution is added. The mixture is allowed to stand for 20 minutes at room temperature, and then the pyruvate hydrazone formed is extracted with 3.0 ml of ethyl acetate by vigorous shaking and centrifugation.[5] A 2-ml aliquot of the ethyl acetate layer is transferred to a planchet, and its radioactivity is measured by a gas flow counter after drying.

---

[4] It is unnecessary to add a toluene layer to the mixture because no inhibition is observed with 20 millimicromoles of additional indole in this assay system, and indole production with the usual amount of pyridoxal phosphate is calculated to be less than 10 millimicromoles.[3]

[5] The hydrazone formed is applied to paper chromatography using two solvent systems of 0.05 N NaOH and n-butanol saturated with 3% NH$_4$OH, and the radioactivity is found to be only in *cis* and *trans* isomers of pyruvate hydrazone by radiochromatoscanning technique.[2,3]

All the data from one lot of analysis should be corrected by the zero time incubation control value, which is obtained by the same procedure described above; to this, however, the concentrated HCl is added before the addition of radioactive tryptophan. For accuracy, a simultaneous control test using a known amount of pyridoxal phosphate is recommended.

Preparation of Sample Solution. The test solution can be prepared in various ways. The procedure described herein is one of the methods recommended for the extraction of pyridoxal phosphate contained in a tissue specimen.[6] The liver (or other) tissue is homogenized in 0.2 ml of 0.3 N H_2SO_4 per milligram of wet tissue in a small Potter-Elvehjem homogenizer. The homogenate is heated for 3 minutes in a boiling water bath, cooled rapidly in an ice–water mixture, neutralized with solid K_2CO_3 (adding a small drop of diluted bromothymol blue as indicator) and centrifuged at 800 g for 5 minutes. The supernatant solution is used as the test solution.

Definition of Unit. One unit of enzyme activity is defined as the amount of protein that produces 1 micromole of indole per hour at 37°. Specific activity is expressed in units per milligram of protein. Protein is determined by the method of Lowry *et al.*[8]

Purification Procedure

Although the preparation of highly purified apotryptophanase has been reported by W. A. Newton *et al.*[9] the following procedure is recommended for its simplicity.

Growth of Escherichia coli. The apotryptophanase preparation described herein is purified from cells of *E. coli* (strain K12) which is grown in a nutrient broth medium containing 1% polypeptone,[10] 1% meat extract and 0.5% NaCl. Culture is grown from a 1% inoculum at 37° with 5 liters per minute of aeration rate and 250 rpm of agitation in a 17-liter jar fermentor. The cells are harvested with a Sharples centrifuge at the end of the exponential phase of growth (about 20 hours) and are washed once with distilled water. Unless otherwise specified, the following operations are carried out at a temperature between 0° and 4°.

Step 1. Crude Extract. Seventy grams (wet weight) of washed cells of *E. coli* K12 is suspended in 350 ml of buffer A containing 0.1 M potassium phosphate, 5 mM mercaptoethanol, and 2 mM Na₂EDTA at pH 7.0.

[6] In this modified procedure of J. Arata and K. Tanioku[7] 94–103% of the additional pyridoxal phosphate is recovered in the sample solutions.[2,3]

[7] J. Arata and K. Tanioku, *Proc. Symp. Chem. Physiol. Pathol., Japan* **5**, 105 (1965).

[8] See Vol. III [73].

[9] W. A. Newton, Y. Morino, and E. E. Snell, *J. Biol. Chem.* **240**, 1211 (1965).

[10] Polypeptone is commercially available in powder form in Japan and is replaceable by Tryptone or Casamino acids.

They are disrupted in a cup-type sonic oscillator, 10 kc, 100 W, for 10 minutes and then centrifuged for 20 minutes at 12,000 g. The pellet is discarded.

Step 2. Ammonium Sulfate 30–70 Fraction. To the supernatant solution from step 1, solid ammonium sulfate is added gradually, with stirring over a 30-minute period, to 30% saturation. The precipitated protein is discarded. Then, to the supernatant, additional ammonium sulfate is added, as described above, to 70% saturation. The active precipitate is collected by centrifugation and is dissolved in 35 ml of buffer B (pH 7.5) which contains 0.05 M Tris hydrochloride, 2 mM Na$_2$EDTA, and 5 mM mercaptoethanol. The solution is then dialyzed, with mechanical stirring, against at least two changes of 1 liter of the same buffer.

Step 3. DEAE-Sephadex Chromatography. Forty-eight milliliters of the dialyzed protein solution (approximately 24 mg/ml) from step 2 is applied to a DEAE-Sephadex column (40 × 3.2 cm diameter; bed size 320 cm^3) which has been equilibrated with buffer B. The column is washed with 1 liter of the same buffer until almost no further protein is eluted. Apotryptophanase is developed out with a linear gradient elution established with 1 liter of buffer B in a mixer and 1 liter of 0.15 M ammonium sulfate (buffer B solution) in the reservoir, and the elution is continued with 0.15 M ammonium sulfate in buffer B until apotryptophanase is eluted.[11] Elution proceeds at an average flow rate of 1 ml per minute, and 10-ml fractions are collected by means of an automatic fraction collector.

Step 4. Hydroxylamine Treatment. No pyridoxal phosphate, in the step 3 enzyme preparation, is observed when the preparation is checked colorimetrically. Pyridoxal phosphate resolution from the apoenzyme is not completely satisfactory in the radiochemical procedure described here. Therefore, the following procedure is adopted to inactivate the coenzyme activity in the step 3 preparation. That is, the fractions from step 3 containing tryptophanase with activity in excess of 25 units are pooled, and 1 micromole of hydroxylamine hydrochloride per 2.5 units of tryptophanase is added with simultaneous stirring. After 20 minutes, the solution is brought to 80% saturation with ammonium sulfate and centrifuged, and the precipitate is dissolved in 2.0 ml of buffer B. The solution is passed through a Sephadex G 25 column (26 × 1.3 cm), which has been equilibrated with buffer B, and is eluted with the same buffer at an elution rate of 0.5 ml per minute. Thus the effluent containing apotryptophanase is pooled and used for the assay of pyridoxal phosphate.

---

[11] Tryptophanase activity is measured by the colorimetric procedure of D. B. McCormick *et al.*[12] or H. Wada *et al.*[1]

[12] D. B. McCormick, M. E. Gregory, and E. E. Snell, *J. Biol. Chem.* **236,** 2076 (1961).

Properties

Stability. Crude extract (step 1) and ammonium sulfate fraction (step 2) are fairly stable at least for 3 months in frozen form. The purified enzyme, particularly step 4 preparation is relatively unstable, but only a little loss of enzymatic activity is observed during 1 week of storage at 4°.

Other Properties. During a prolonged preincubation period (see Procedure), no change in the pyridoxal phosphate level or tryptophanase activity is observed within 30 minutes. Monoiodoacetate has no influence on the recovery of radioactive pyruvate formed in the reaction mixture. It seems likely that no extraneous reaction system concerned with pyridoxal phosphate or pyruvate itself is involved in this assay system during a suitable incubation period.

[85] Microassay of Pyridoxal Phosphate Using L-Tyrosine-1-^{14}C and Tyrosine Apodecarboxylase

By P. R. SUNDARESAN and D. B. COURSIN[1]

$$\text{L-Tyrosine-1-}^{14}\text{C} \xrightarrow[\substack{\text{tyrosine decarboxylase} \\ \text{apoenzyme}}]{\text{pyridoxal phosphate}} \text{tyramine} + {}^{14}\text{CO}_2$$

Assay Method

Principle. The assay is based on the pyridoxal phosphate (pyridoxal-P)-dependent enzymatic decarboxylation of L-tyrosine by tyrosine decarboxylase (L-tyrosine carboxy-lyase, EC 4.1.1.25). With L-tyrosine-1-^{14}C as the substrate, in the presence of tyrosine apodecarboxylase and pyridoxal-P, ^{14}CO$_2$ which is produced in the reaction is trapped and counted directly in a liquid scintillation spectrometer. The method is a modification of the one originally described by Umbreit *et al.*,[2] Hamfelt,[3] and Maruyama and Coursin.[4]

Reagents

L-Tyrosine-1-^{14}C solution: 0.008 M, in potassium acetate buffer, 0.8 M, pH 5.5, specific activity 10.1 μCi/millimole[5]

[1] These experiments were supported by National Institutes of Health, U.S. Public Health Service Grant No. NS 07361 and by Hoffmann-La Roche. The authors wish to express their thanks to Miss Lois Erb for her skillful technical assistance.

[2] W. W. Umbreit, W. D. Bellamy, and I. C. Gunsalus, *Arch. Biochem. Biophys.* **7,** 185 (1945).

Pyridoxal-P stock solution: $1.12 \times 10^{-4} M$, 3.031 mg dissolved in 100 ml of double distilled water. This solution is refrigerated in the dark and is stable for 2 days. In the assay, the stock solution is diluted 1000-fold with potassium acetate buffer, 0.01 M, containing 0.005 M ethylenediaminetetraacetate (EDTA), pH 5.5, to give a concentration of 27.0 ng/ml.

NCS solubilizer[6]

Cell-free extract of tyrosine apodecarboxylase, 15–18 units/ml (see tyrosine apodecarboxylase preparation). Aliquots of the apoenzyme are diluted 2-fold with potassium acetate buffer, 0.01 M, containing 0.005 M EDTA, pH 5.5, just before use.

Procedure. The incubation mixture contains 0.1 ml of tyrosine apo-decarboxylase (0.75 unit) in potassium acetate buffer, 0.01 M, pH 5.5, 0.1–0.3 ml (2.7–8.1 ng) of the standard pyridoxal-P solution or unknown, 0.6 ml of L-tyrosine-1-^{14}C solution, and water in a final volume of 1 ml. The tyrosine-1-^{14}C solution is added after the apoenzyme has been preincubated with pyridoxal-P or unknown for 15 minutes at 37° in 25-ml reaction flasks.[7] The flasks are then stoppered with septums containing polyethylene center wells[8] and incubated at 37° for 8 minutes. After incubation, the flasks are chilled in ice and the reaction is terminated by injecting 1 ml of 10% trichloroacetic acid through the septums. Then 0.3 ml of NCS is injected into the center well of each flask, and the flasks are shaken for 1 hour in a shaker at room temperature. The center well from each flask is carefully removed and transferred by cutting the polyethylene rod above the center well into a vial containing 10 ml of toluene solution containing 0.5% 2,5-diphenyloxazole (PPO) and 0.01% *p*-bis-[2-(5-phenyloxazoyl)]benzene (POPOP). The radioactivity is counted in a

---

[3] A. Hamfelt, *Scand. J. Clin. Lab. Invest.* **20**, 1 (1967).

[4] H. Maruyama and D. B. Coursin, *Anal. Biochem.* **26**, 420 (1968).

[5] Ten milliliters of L-tyrosine solution in dilute HCl (90.6 mg of L-tyrosine in 10 ml of 0.15 N HCl) is added to 80 ml of potassium acetate buffer, 1.0 M, pH 6.0, saturated with tyrosine (100 mg of L-tyrosine in 100 ml of buffer, warmed to 70° and filtered). This is followed by addition of 10 μCi of L-tyrosine-1-^{14}C (specific activity 19.1 mCi/millimole) in 200 μl of 1 N HCl. The pH of the solution is adjusted to 5.5 with 1 N HCl. Any precipitate that appears is filtered off on glass-fiber filter paper under vacuum. The solution is then made up to 100 ml with water. The radioactivity and concentration of tyrosine are determined to obtain the specific activity of tyrosine solution [$\epsilon = 2330$ at 293.5 mμ in 0.1 N NaOH; G. Weber *in* "Biochemists Handbook" (C. Long, ed.), p. 82. E. & F. N. Spon, London, 1961].

[6] Nuclear Chicago Corporation, Chicago, Illinois.

[7] Preincubation increases the reaction rate by 15%. Hence, samples are routinely preincubated with the apoenzyme before addition of tyrosine.

[8] Kontes Glass Company, Vineland, New Jersey.

liquid scintillation spectrometer. Two reaction blanks are run, one without pyridoxal-P to correct for the endogenous pyridoxal-P in the apoenzyme preparation and one without the enzyme. The assay is carried out in triplicate under yellow light, since pyridoxal-P is light sensitive.[9]

Standard Curve

A standard solution of pyridoxal-P is prepared as described (see Reagents). The concentration of pyridoxal-P in the standard solution is determined spectrophotometrically from its molar extinction coefficient of 6550 at 388 mμ in 0.1 N NaOH and 4900 at pH 7.0.[10] Dilute solutions of pyridoxal-P are prepared, and the reaction rates are determined by the method described above. A plot of disintegrations per minute in CO_2 versus amount of standard pyridoxal-P is used to determine the pyridoxal-P content of the unknown sample. The assay is linear between 1 and 8 ng of pyridoxal-P. Samples containing higher concentration of pyridoxal-P should be diluted with potassium acetate buffer, 0.01 M, containing 0.005 M EDTA, pH 5.5.

Definition of Unit and Specific Activity. One unit of L-tyrosine decarboxylase is defined as the amount of enzyme which catalyzes the decarboxylation of 1.0 micromole of tyrosine per minute under the conditions of the assay. Specific activity is defined as units per milligram of protein.

Preparation of Tyrosine Apodecarboxylase

All operations are carried out at 4° except where otherwise specified. One gram of dried *Streptococcus faecalis* cells[11] grown in pyridoxine-deficient medium is suspended in 20 ml of potassium acetate buffer, 0.01 M, pH 5.5, containing 0.005 M EDTA (buffer A) and 150 ml of saturated $(NH_4)_2SO_4$ prepared in dilute NH_4OH (3 ml of 30% NH_4OH, specific gravity 0.9, in 100 ml of H_2O). The suspension is incubated at 37° for 30 minutes with constant stirring. The cells are then recovered by centrifugation at 12,000 g for 10 minutes and washed twice in 40 ml of buffer A. The final washing is carried out in 40 ml of 0.01 M potassium acetate buffer, pH 5.5, containing 0.005 M EDTA, 0.005 M mercaptoethanol, and 0.0005 M L-tyrosine (buffer B). The washed cells are then suspended in 20 ml of buffer B and subjected to sonication. The sonication is carried out in a stainless steel vessel containing 1 g of glass beads, 5 μ in diameter, with a Branson Sonifier, 250 W, at 20 kc for 4 minutes. During sonication, the vessel is kept in a sodium chloride ice bath and the temperature of the suspension is main-

[9] E. A. Peterson and H. A. Sober, *J. Am. Chem. Soc.* **76**, 169 (1954).

[10] See Vol. III [142].

[11] Worthington Biochemical Company, Freehold, New Jersey.

tained at 5°. The broken cells are then centrifuged at 27,000 g for 10 minutes. The supernatant solutions are combined, the pH is adjusted to 5.5, and the solution is stored at $-20°$. The yield of the apoenzyme is about 600 units per gram of dried cells.

Stability. The tyrosine apodecarboxylase is stable for one year at $-20°$.

pH Optimum. The tyrosine decarboxylase has a sharp pH optimum at 5.5.[12]

Kinetics. The reaction follows a linear course for 8 minutes under the conditions of the assay. The K_m for tyrosine is 1.66 mM (cf. 1.5 mM footnote 3) and for pyridoxal-P, 1.66×10^{-7} M; in the presence of ATP, the K_m for pyridoxal is 3.0×10^{-7} M.[2,13] 3,4-Dihydroxyphenylalanine, which also serves as the substrate for tyrosine decarboxylase, has a K_m of 2.3 mM.[12]

Inhibitors. The following heavy metal ions: Ag^+, Cu^{2+}, Fe^{2+}, Fe^{3+}, and Pb^{2+}; and KCN, NH_2OH, and N_2H_4 inhibit tyrosine decarboxylase.[12] Sulfate ion[14] and trichloroacetic acid[2] also inhibit the reaction. The enzyme activity has been reported to be low in the following buffers as compared to potassium acetate buffer: citrate, phosphate, pyrophosphate, and phthalate.[12,14]

[12] H. M. R. Epps, *Biochem. J.* **38**, 242 (1944).
[13] I. C. Gunsalus and W. D. Bellamy, *J. Biol. Chem.* **155**, 357 (1944).
[14] G. H. Sloane-Stanley, *Biochem. J.* **44**, 567 (1959).

[86] Microbiological Determination of the Vitamin B_6 Group

By BETTY E. HASKELL and ESMOND E. SNELL

Principle

Microbiological assay remains one of the most widely used methods for determination of the vitamin B_6 group, pyridoxal, pyridoxamine, and pyridoxine, because of its sensitivity, its high degree of specificity, and its suitability for use in assaying crude extracts of biological materials. The method depends on the selection of a microorganism that requires vitamin B_6 for growth. (One or more forms of the vitamin may be active.) Such a microorganism is grown in media containing an excess of all nutrients required for growth except vitamin B_6. Vitamin B_6 is added in suboptimal amounts so that growth of the microorganism is limited by the vitamin B_6 content of the medium. Microbial growth is measured conveniently by turbidity. To measure the vitamin B$_6$ content of a sample, the turbidity of cultures grown with graded amounts of the sample is compared to the turbidity of cultures grown with known amounts of vitamin B$_6$ standard.

Although several test organisms have been used in the determination of vitamin B$_6$ (see Storvick *et al.*[1] for review), three are preferred: *Saccharomyces carlsbergensis* 4228 (ATCC 9080) for the determination of pyridoxal, pyridoxine, and pyridoxamine[2]; *Lactobacillus casei* (ATCC 7469) for the determination of pyridoxal[3]; and *Streptococcus faecium* ϕ51 (NIRD Culture 1229) for the determination of pyridoxal and pyridoxamine.[4,5] To determine how much of each of the three forms of the vitamin is present in a sample, the sample may be reassayed with test organisms of differing specificity.[6] Or, the sample may be chromatographed on ion exchange resins to separate the forms of vitamin B$_6$ prior to microbiological assay with *S. carlsbergensis*. Storvick *et al.*[1] have modified Toepfer and Lehmann's[7] procedure to obtain clean separation of pyridoxal, pyridoxine, and pyridoxamine without the use of hot buffers. Bain and Williams[8] have described a procedure for separating both phosphorylated and free forms of vitamin B$_6$ on ion-exchange resins.

Preparation of Sample

Biological samples must be hydrolyzed prior to microbiological assay to liberate free vitamin B$_6$ from its phosphorylated forms, which frequently occur in association with protein. To obtain maximum release of vitamin B$_6$ from most biological materials,[9] a finely divided sample containing about 2 μg of vitamin B$_6$ is autoclaved in 180 ml of 0.055 N HCl for 5 hours at 121°. It is cooled, adjusted to pH 4.5–5 with dilute KOH, diluted to a convenient volume, and filtered through Whatman 42 filter paper. The sample should be protected from light and stored at 4°. Unfortunately, no single hydrolytic procedure which gives optimal results with all types of

[1] C. A. Storvick, E. M. Benson, M. A. Edwards, and M. J. Woodring, *in* "Methods of Biochemical Analysis" (D. Glick, ed.), Vol. 12, p. 183. Wiley (Interscience), New York, 1964.

[2] L. Atkin, A. S. Schultz, W. L. Williams, and C. N. Frey, *Ind. Eng. Chem., Anal. Ed.* **15**, 141 (1943).

[3] J. C. Rabinowitz, N. I. Mondy, and E. E. Snell, *J. Biol. Chem.* **175**, 147 (1948).

[4] M. E. Gregory, *J. Dairy Res.* **26**, 203 (1959).

[5] J. C. Rabinowitz and E. E. Snell, *J. Biol. Chem.* **169**, 631 (1947).

[6] J. C. Rabinowitz and E. E. Snell, *J. Biol. Chem.* **176**, 1157 (1948).

[7] E. W. Toepfer and J. Lehmann, *J. Assoc. Offic. Agr. Chemists* **44**, 426 (1961).

[8] J. A. Bain and H. L. Williams, *in* "Inhibition in the Nervous System and Gamma-Aminobutyric Acid" (E. Roberts, ed.), p. 275. Macmillan (Pergamon), New York, 1960.

[9] J. C. Rabinowitz and E. E. Snell, *Anal. Chem.* **19**, 277 (1947).

MEDIA FOR MICROBIOLOGICAL ASSAY OF VITAMIN B$_6$

Ingredient	Amount per liter of double-strength medium		
	Saccharomyces carlsbergensis	Lactobacillus casei	Streptococcus faecium
Carbon source			
Glucose (g)	100	20	20
Nitrogen sources			
Ammonium sulfate (g)	3.75	—	—
Vitamin-free casamino acids (g)	5	15	10
DL-Alanine (g)	—	—	1
L-Asparagine (mg)	—	200	200
L-Cystine[a] (mg)	—	400	800
Glycine (mg)	—	400	400
L-Proline[b] (mg)	—	—	20
L-Phenylalanine[b] (mg)	—	—	20
L-Histidine[b] (mg)	—	—	20
L-Tyrosine[b] (mg)	—	—	20
DL-Tryptophan[a] (mg)	—	200	200
Salts			
Calcium chloride·2 H$_2$O (mg)	250	—	—
Citric acid (g)	2	—	—
Ferric sulfate·7 H$_2$O[c] (mg)	—	20	20
Ferric chloride (mg)	5	—	—
Magnesium sulfate·7 H$_2$O[c] (mg)	250	400	400
Manganese sulfate·4 H$_2$O[c] (mg)	5	20	20
Potassium chloride (mg)	850	—	—
Potassium phosphate, monobasic[d] (g)	1.1	1	6
Potassium phosphate, dibasic[d] (g)	—	1	6
Sodium acetate, anhydrous (g)	—	12	—
Sodium chloride[c] (mg)	—	20	20
Sodium citrate tribasic (g)	10	—	40
Vitamins			
p-Aminobenzoic acid (μg)	—	400	400
Biotin[e] (μg)	66	4	4
Calcium pantothenate[e] (μg)	500	800	800
Folic acid[b] (μg)	—	20	800
Inositol[e] (mg)	50	—	—
Nicotinic acid[e] (μg)	500	800	800
Riboflavin[a] (μg)	—	800	800
Thiamine·HCl[e] (μg)	500	400	400
Miscellaneous			
Adenine sulfate[a] (mg)	—	—	10
Guanine·HCl[a] (mg)	—	—	10
Uracil[a] (mg)	—	—	10
Tween 80 (ml)	—	—	2
pH adjusted to:	4.7	6.8	6.8

samples has yet been devised. For a discussion of enzymatic and other hydrolytic procedures which may be used, see Storvick et al.,[1] Sauberlich,[10] and Strohecker and Henning.[11]

Determination of Total Vitamin B_6

Maintenance of Stock Cultures. Saccharomyces carlsbergensis 4228 (ATCC 9080), the test organism for pyridoxal, pyridoxine, and pyridoxamine, is maintained on slants prepared from 2.5% agar and 10% malt extract. Cultures are transferred monthly. Fresh transfers are incubated about 24 hours at room temperature and stored at 4°.

Medium. A double-strength medium for microbiological assay of vitamin B_6 with *S. carlsbergensis* is given in the table. The medium is slightly modified from that of Atkin et al.[2] Some investigators supplement the medium with tryptophan[1,7,8] and other amino acids.[7,8] It is convenient to prepare a separate vitamin solution which should be renewed weekly (see notes to the table). The remainder of the medium ingredients may be combined in the quantities indicated in the table, diluted to 500 ml to make quadruple-strength medium, and stored under benzene at 4° for months. Double-strength medium is prepared fresh for each assay by combining suitable aliquots of the two solutions, adjusting the pH to 4.7 with KOH, and diluting with distilled water.

Standard. A solution containing 100 millimicrograms of pyridoxine·HCl per milliliter is prepared fresh for each assay. The standard curve is prepared by appropriate dilution into the basal medium to supply 0.1 to 1 millimicrograms of pyridoxine·HCl per milliliter of single-strength medium.

[10] H. E. Sauberlich, *in* "The Vitamins" (P. György and W. N. Pearson, eds.), Vol. 7, p. 169. Academic Press, New York, 1967.
[11] R. Strohecker and H. M. Henning, "Vitamin Assay," p. 143. Verlag Chemie, Weinheim, 1965.

Table Footnotes
  [a] To dissolve, use a few drops of dilute HCl.
  [b] To dissolve, use a few drops of dilute NaOH.
  [c] To prepare double-strength media for *L. casei* and *S. faecium*, use 10 ml per liter of the following solution: 10 g $MgSO_4$·7 H_2O, 0.5 g NaCl, 0.5 g $FeSO_4$·7 H_2O, 0.5 g $MnSO_4$·4 H_2O, and distilled water to make 250 ml.
  [d] To prepare double-strength media for *L. casei* and *S. faecium*, prepare the following solution: 25 g KH_2PO_4, 25 g K_2HPO_4, and distilled water to make 250 ml. For *L. casei*, use 10 ml per liter of double-strength medium; for *S. faecium*, use 60 ml per liter of double-strength medium.
  [e] For *S. carlsbergensis* medium, use 10 ml per liter of double-strength medium of the following solution: 200 μg of biotin, 1.5 mg of calcium pantothenate, 150 mg of inositol, 1.5 mg of nicotinic acid, 1.5 mg of thiamine·HCl, and distilled water to make 30 ml. Use a few drops of 95% ethanol, if necessary, to aid in dissolving biotin.

Inoculum. To prepare the inoculum culture, 10 ml of single-strength medium containing 1 millimicrogram of pyridoxine·HCl per milliliter is steamed for 10 minutes at 100° in a cotton-stoppered Erlenmeyer flask, cooled, and inoculated with one loopful of *S. carlsbergensis* cells from a malt agar slant. The flask is incubated with shaking for 18–24 hours at 30°. The cells are harvested by centrifuging in sterile, cotton-stoppered tubes. The medium is decanted. The cells are suspended in sterile, distilled water and are diluted to a concentration of 0.04 mg of dry cells per milliliter. A suspension of this concentration has an optical density of 0.08 when read in 0.5-inch tubes in a Bausch and Lomb Spectronic 20 colorimeter at 650 mμ. (For a procedure for the preparation of a turbidity–dry weight curve, see Pearson.[12]) The inoculum may be stored at 4° up to about 1 week in the inoculum medium.

Assay Procedure. Samples or standards to supply zero to 1 millimicrogram of vitamin B$_6$ per milliliter of the final, single-strength medium are placed in culture tubes or 50-ml Erlenmeyer flasks and diluted to 5 ml with distilled water. (An assay grown in 25 × 150 mm culture tubes may be read in a Bausch and Lomb colorimeter with a 1-inch adapter and light shield without transferring to colorimeter tubes. If the volume of sample and standards is 0.1 ml or less, the volume may be considered negligible; samples and standards may be diluted by adding 5 ml of water to each tube from an automatic pipette.) Five milliliters of double-strength medium is added to each tube. The tubes are capped, steamed at 100° for 10 minutes, cooled, and inoculated with 1 drop of diluted inoculum suspension. The assay is incubated at 30° for about 24 hours or until the difference in turbidity between tubes containing zero and 1 millimicrogram of vitamin B$_6$ per milliliter of medium is about 0.6 OD units. Before the assay is read, the tubes are shaken vigorously to suspend cells and allowed to stand for a few minutes to permit bubbles to rise. The turbidity of samples is read at 650 mμ against a water blank. The concentration of the vitamin is plotted against optical density to obtain a standard curve. The amount of vitamin B$_6$ in the sample is determined by reference to the standard curve.

Notes and Interpretation. Growth of *S. carlsbergensis* in response to 0–1 millimicrogram of vitamin B$_6$ per milliliter of medium yields a smooth standard curve which may be expected to vary somewhat from one assay to another. It is important, therefore, to run a full standard curve with each assay. Duplicate standard tubes usually agree within 2%. For samples, it is desirable to use aliquots of varying sizes. By determining the growth response per milliliter of sample for several aliquot sizes, one can check for growth depressing or growth-stimulating substances in the sample.[12,13] If

[12] W. N. Pearson, *in* "The Vitamins" (P. György and W. N. Pearson, eds.), Vol. 7, p. 1. Academic Press, New York, 1967.

the calculated amount of vitamin B$_6$ per milliliter of sample agrees within 10% for two or more aliquot sizes, one may average these values to obtain a reliable figure for the vitamin B$_6$ content of the sample.

Assays with *S. carlsbergensis* are more widely used than any other microbiological assay for vitamin B$_6$ because the microorganism grows vigorously on all three forms of vitamin B$_6$. The growth response to pyridoxamine frequently is slightly less than that to pyridoxal and pyridoxine.[4,6] In assaying samples that have been fractionated on ion-exchange resins, one can correct for the lower response to pyridoxamine by reading the pyridoxamine fraction against a standard curve prepared from pyridoxamine·2 HCl. High salt concentrations inhibit growth of *S. carlsbergensis*[14]; Scriver and Cullen[15] have reported low recoveries of pyridoxine· HCl standard added to acid-hydrolyzed urine. To prevent contamination of assay tubes with vitamin B$_6$ from previous assays, washing with ordinary detergents usually is adequate. If growth in assay tubes is erratic, they may be soaked in alcoholic potassium hydroxide prior to washing. Commercial sources of vitamin-free acid-hydrolyzed casein sometimes need to be treated with an adsorbent[16] before satisfactory blanks can be obtained in microbiological assays for vitamin B$_6$.

Determination of Pyridoxal and Pyridoxamine

Maintenance of Stock Cultures. Lactobacillus casei ATCC 7469, the test organism for pyridoxal, and *Streptococcus faecium* ϕ51, the test organism for pyridoxal and pyridoxamine, are maintained in stab cultures consisting of 1% yeast extract, 0.25% glucose, and 2% agar. The cultures are transferred monthly and are incubated at 37° for 24 hours. They are stored at 4°.

Media. The composition of double-strength media for assay of vitamin B$_6$ with *L. casei* and *S. faecium* is given in the table. The media are those of Rabinowitz *et al.*[3] and Rabinowitz and Snell[5] with minor modifications.[1,4] Ingredients that do not dissolve readily in water should be put into solution with the aid of a few drops of dilute HCl or NaOH prior to combining with other media ingredients. (For suggested procedures, see notes to the table.) Those who use microbiological assay routinely may find it advantageous to prepare media from stock solutions.[1,11] The assay media should be protected from light and stored frozen.

Standard. A freshly prepared solution containing pyridoxal·HCl is sterilized by passing it through a bacteriological filter. For assays carried

[13] E. E. Snell, *in* "Vitamin Methods" (P. György, ed.), Vol. 1, p. 327. Academic Press, New York, 1950.

[14] J. C. Rabinowitz and E. E. Snell, *Proc. Soc. Exptl. Biol. Med.* **70**, 235 (1949).

[15] C. R. Scriver and A. M. Cullen, *Pediatrics* **36**, 14 (1965).

[16] R. C. Raines and B. E. Haskell, *Anal. Biochem.* **23**, 413 (1968).

out in a final volume of 6 ml, a solution is prepared which contains 60 millimicrograms of pyridoxal·HCl per milliliter. Aliquots containing zero to 6 millimicrograms are diluted with medium to prepare the standard curve.

Inoculum. Single-strength assay medium, 6 ml, is placed in a 13 × 100 mm culture tube, capped, and autoclaved at 121° for 15 minutes. After cooling, 1 millimicrogram of pyridoxal·HCl per milliliter is added aseptically. The medium is inoculated with cells from a stab culture. *L. casei* is incubated for 24 hours at 37°; *S. faecium* for 24 hours at 30°. The cells are harvested by centrifuging, and the medium is decanted. The cells are suspended in sterile 0.9% sodium chloride solution and diluted with sterile saline to an optical density of 0.3, as determined in 0.5-inch tubes in a Bausch and Lomb Spectronic 20 colorimeter at 650 mμ. A freshly grown inoculum should be used for each assay.

Assay Procedure. *L. casei* and *S. faecium* are incubated without shaking in culture tubes about two-thirds full of medium. If one uses 13 × 100 mm culture tubes containing 6 ml of medium, one can read the assay directly in a Bausch and Lomb Spectronic 20 colorimeter with a 0.5-inch adapter without transferring the medium to colorimeter tubes. Standards and samples must be sterilized separately from the basal medium to prevent loss of pyridoxal.[3]

To set up the assay, enough water to bring the volume of sample or standard plus water to 3 ml is pipetted into 13 × 100 mm culture tubes. The tubes are capped, autoclaved at 121° for 15 minutes, and cooled. Double-strength medium is sterilized separately by autoclaving at 121° for 15 minutes in a cotton-stoppered Erlenmeyer flask. Samples or standard are sterilized by passing through a bacteriological filter and are added to each culture tube aseptically. The diluted inoculum is mixed with the double-strength assay medium, 1 ml of a cell suspension with an optical density of 0.3 being added for each 100 ml of double-strength medium. Three milliliters of the inoculated medium is added to each culture tube.

The assay is incubated for 22–24 hours at 37° for *L. casei* or at 30° for *S. faecium*. The tubes are stoppered with a rubber stopper and shaken vigorously; the turbidity is read at 650 mμ. The amount or pyridoxal (or pyridoxal plus pyridoxamine) in samples is determined by reference to a standard curve run in duplicate with each assay.

Notes and Interpretation. Growth of *L. casei* in a medium containing no vitamin B$_6$ is highly specific for pyridoxal.[3] Other forms of the vitamin have no activity if precautions are taken to prevent their partial conversion to pyridoxal. However, *L. casei* can grow in a rich medium containing no vitamin B$_6$ providing the medium contains relatively large amounts of D-alanine and peptides. About 10,000–25,000 μg of D-alanine is needed to replace 1 μg of pyridoxal for *L. casei* when this microorganism is grown in a peptide-enriched medium.[3] Since D-alanine is rare in biological materials, the possibility that it will interfere with microbiological assay of pyridoxal is slight. Only in samples consisting of hydrolyzed blood has D-alanine interference been observed.[17,18] Hydrolytic procedures used to prepare blood for microbiological assay partially racemize L-alanine present in blood and partially degrade blood proteins to peptides.[18] D-Alanine interference might be expected to occur only in samples with an unusually high ratio of protein to vitamin B$_6$. In blood, the ratio of protein to vitamin B$_6$ is about 10^6 to 1.

The growth response of *S. faecium* to pyridoxal and pyridoxamine is approximately equal. Pyridoxine is 6000 times less active in promoting growth than is pyridoxal.[19] *S. faecium* is the preferred test organism for microbiological assay of pyridoxal and pyridoxamine because it is relatively insensitive to stimulation by D- or L-alanine.[4,16] (*Streptococcus faecalis* R (ATCC 8043), another commonly used test organism for pyridoxal and pyridoxamine,[5] grows to near maximum in a vitamin B$_6$-free medium containing 0.1 mg DL-alanine per milliliter of medium[4,19]; the amounts of D- or L-alanine required for maximum growth of *S. faecium* in the absence of vitamin B$_6$ are 10–20 times greater.[16]) Growth of *S. faecium* in the absence of vitamin B$_6$ is stimulated by relatively large amounts of peptides containing L-alanine.[16] However, the only known instances of peptide interference are in assays of hydrolyzed blood[16,17] and of enzymatic casein hydrolyzate.[16] The reliability of values obtained for the vitamin B$_6$ content of samples with a high ratio of protein to vitamin B$_6$ can be checked by reassaying the sample with *S. carlsbergensis*. Neither peptides[16] nor D-alanine[4,18] appear to influence the response of *S. carlsbergensis* to vitamin B$_6$.

[17] C. A. Storvick and J. M. Peters, *in* "Vitamins and Hormones" (R. S. Harris, I. G. Wool, and J. A. Loraine, eds.), Vol. 22, p. 833. Academic Press, New York, 1964.

[18] B. E. Haskell and U. Wallnofer, *Anal. Biochem.* **19**, 569 (1967).

[19] P. Moller, *Acta Physiol. Scand.* **21**, 332 (1950).

[87] A Simplified Toepfer-Lehmann Assay for the Three Vitamin B$_6$ Vitamers

By MYRON BRIN

The innovative differential microbiological assay for the estimation of the 3 vitamin B$_6$ vitamers (pyridoxal, pyridoxine, and pyridoxamine) by Rabinowitz and Snell[1] has been followed by a variety of modifications in

[1] J. C. Rabinowitz and E. E. Snell, *J. Biol. Chem.* **176**, 1157 (1948).

attempts to increase specificity.[2-8] Of these the assay of Toepfer and Lehmann[7] proved to be particularly appropriate for the assay of food materials, and presented below is a simple adaptation of the latter to the routine assay of tissue extracts.[9,10]

Assay Method

Principle. An acid-hydrolyzed sample extract is transferred to an ion-exchange resin column from which each vitamin B_6 vitamer is eluted individually with appropriate buffers. The eluates are individually assayed microbiologically with *Saccharomyces carlsbergensis* against its own vitamer as reference standard.

Reagents

HCl, 0.055 N: 4.8 ml of concentrated HCl (11.7 N) per liter of solution

KOH, 6 N: 336 g of KOH pellets per liter of solution

HCl, 3 N: 256 ml of concentrated HCl (11.7 N) per liter of solution

Potassium acetate, 0.01 M: 0.9815 g per liter and adjusted to pH 4.5 with acetic acid

Potassium acetate, 0.02 M: 1.963 g per liter, and adjusted to pH 5.5 with acetic acid

Potassium acetate, 0.04 M: 3.926 g per liter and adjusted to pH 6.0 with acetic acid

Potassium acetate, 0.10 M: 9.815 g per liter and adjusted to pH 7.0 with KOH

KCl-K_2HPO_4 solution[7]: 74.6 g of KCl and 17.4 g of K_2HPO_4 are dissolved in 800 ml of water; the solution is adjusted to pH 8.0 with acetic acid, and water is added to 1 liter.

Wort agar slants: Dissolve 25 g of Bacto-wort agar in 500 ml of water with steaming. Pipette 10-ml portions into 16 × 150 mm test tubes, plug with cotton, autoclave for 15 minutes at 120°, and tilt while cooling to form slants. Excess heating results in liquefaction.[7] Store in dark refrigerator.

Bacto–Pyridoxine–Y-Medium[10,11] (P-Y-M): To make double-strength

[2] W. A. Winsten and E. Eigen, *Proc. Soc. Exptl. Biol. Med.* **67**, 513 (1948).

[3] D. B. Coursin and V. C. Brown, *Proc. Soc. Exptl. Biol. Med.* **98**, 315 (1958).

[4] V. Bonavita, *Arch. Biochem. Biophys.* **88**, 366 (1960).

[5] J. A. Bain and H. L. Williams, *in* "Inhibition in the Nervous System and Gamma-Aminobutyric Acid" (E. Roberts, ed.). Macmillan (Pergamon), New York (1960).

[6] E. W. Toepfer, M. J. MacArthur, and J. Lehmann, *J., Assoc. Offic. Agr. Chemists* **43**, 57 (1960).

[7] E. W. Toepfer and J. Lehmann, *J. Assoc. Offic. Agr. Chemists* **44**, 426 (1960).

[8] P. A. Hedin, *Agr. Food. Chem.* **11**, 343 (1963).

[9] V. F. Thiele and M. Brin, *J. Nutr.* **90**, 347 (1966).

[10] M. Brin and V. F. Thiele, *J. Nutr.* **93**, 213 (1967).

medium, dissolve in multiples of 5.3 g of powdered material in 100 ml of distilled water. Use immediately.

Liquid broth culture: To 20 ml of P-Y-M, add a solution containing 40 mμg each of pyridoxal, pyridoxine, and pyridoxamine hydrochlorides, and sufficient water to make the volume to 40 ml. Distribute 10-ml aliquots into culture tubes, plug with cotton, autoclave at 120° for 15 minutes, cool, and store in dark refrigerator.

Inoculum-rinse: Dilute P-Y-M with an equal volume of water, transfer 10 ml into culture tubes, plug with cotton, autoclave at 120° for 15 minutes, cool, store in dark refrigerator.

Pyridoxal standard solution: Dissolve 50 mg of carefully dried pyridoxal·HCl in 500 ml of 25% ethanol in water. Dilute 2 ml of this solution to 1 liter with 25% ethanol in water to make a stock solution containing 200 mμg of pyridoxine·HCl per milliliter. For the standard solution dilute 1 ml of the stock solution to 100 ml with distilled water, for a final concentration of 2 mμg of pyridoxine·HCl per milliliter.

Pyridoxine standard solution: Prepare with pyridoxine·HCl precisely as the pyridoxal solutions were prepared, and at the same concentrations.

Pyridoxamine standard solution: Dissolve 150 mg of carefully dried pyridoxamine·2 HCl in 500 ml of 25% ethanol in water. Dilute exactly as with the other two vitamers. The stock solution will contain 600 mμg of pyridoxamine·2 HCl per milliliter, and the standard solution will contain 6 mμg/ml.

In all cases, the stock solutions are stored in a dark refrigerator. Fresh standard solutions are made for each assay.

Preparation of Sample Extract. Tissue samples are homogenized in 0.055 N HCl in the proportion of 1 g of tissue per 200 ml of acid. The suspensions are autoclaved for 5 hours[9] at 120°. Upon cooling, the pH is adjusted to 4.5, 5 ml of 0.01 M KAc buffer, pH 4.5, is added, and distilled water is added to a final volume of 250 ml (per gram of sample).

Chromatography. (a) PREPARATION OF RESIN.[7] An excess of 6 N KOH is added to 400 g of Dowex AG 50W X8 (100–200 mesh) resin in the hydrogen form, until the supernatant is blue to litmus.[7] After settling, the supernatant is decanted, and the resin is rinsed until the washings are clear. To the resin is added 400 ml of 3 N HCl, and the suspension is heated for 0.5 hour in boiling water. The HCl is decanted, and the acid treatment is repeated twice more. The resin is rinsed until the washings are at pH 4–5. Again 6 N KOH is added until the supernatant is alkaline

---

[11] Purchased as code 0951 from Difco Laboratories, Detroit, Michigan.

(litmus), and after decantation the resin is rinsed until the washings are at pH 9.0 (pH meter). It is used in this form, and can be reused by beginning at the 3 N HCl step.

(b) PACKING OF RESIN COLUMNS. Standard thiamine columns, 160 mm long, with a terminal 2-mm capillary bore tube, an internal diameter of 11.5 mm, and an upper reservoir of 50-ml capacity are suitable for this work.[12] They are fitted with a short rubber tube and pinch clamp to control flow. Prepared resin, 10 ml, is washed into the column over a glass bead[9] in such fashion that all air bubbles are removed and the liquid level does not drop below the resin surface. A plug of glass wool may be placed above the resin to avoid disruption of its surface.[7] The packed column is rinsed with four 50-ml portions of water followed by two rinses with hot 0.01 M potassium acetate buffer, pH 4.5, and the effluent must be at pH 4.5 (or the last rinsing is repeated).[7]

(c) ABSORPTION OF SAMPLE EXTRACT. A volume of sample extract equivalent to 100–400 mg of tissue (kidney or liver, 100 mg; heart or brain, 300 mg; muscle, 400 mg[9]) is added to the column followed by three rinses of the column each with 12–15 ml of warm (75°) 0.02 M potassium acetate buffer, pH 5.5.[7] (To avoid air bubbles do not permit the liquid level to drop below the glass wool.)

(d) ELUTION OF FRACTIONS. Pyridoxal is eluted with 50 ml of boiling 0.04 M potassium acetate buffer, pH 6.0; pyridoxine with 50 ml of boiling 0.1 M potassium acetate buffer, pH 7.0; and lastly, pyridoxamine with 50 ml of boiling KCl–K_2HPO_4 buffer, pH 8.0,[7] in sequence. Forty milliliters of each eluate are collected in turn, the pH is adjusted to 5.0, and the volume is made up to 50 ml. The remaining 10 ml of each eluate are discarded. In each assay an aliquot of 0.055 N HCl is autoclaved and chromatographed as a sample blank. Also a standard mixture of 0.10 μg each of pyridoxal, pyridoxine, and pyridoxamine per 50 ml is hydrolyzed and chromatographed for percentage recovery calculations.

Microbiological Assay. (a) TEST ORGANISM.[7] *Saccharomyces carlsbergensis* (ATCC 9080) is maintained by weekly transfer on slants of wort agar. Incubate freshly inoculated (surface) slants at 30° for 24 hours and store in the refrigerator.

(b) PREPARATION OF INOCULUM. Transfer a loop of cells from a fresh agar slant to tubes of liquid broth culture, plug with cotton (surrounded with tape), and incubate on a shaker at 30° for 20–24 hours. Centrifuge the cells, decant the supernatant, and rinse twice with 10 ml of inoculum-rinse in similar fashion. The cells, when suspended in the third aliquot of inoculum-rinse, provide the inoculum. Aseptic technique is used throughout.

---

[12] Purchased from Kontes Glass Co., Vineland, New Jersey.

(c) ASSAY PROCEDURE. To duplicate tubes containing 0.5, 1.0, and 2.0 ml of column eluates obtained from the sample hydrolyzates water is added to make a total volume of 5 ml per tube. Standard curves are prepared by pipetting into duplicate tubes the following amounts of each vitamer (millimicrograms): pyridoxal·HCl, 1, 2, 3, 4, 6, 8; pyridoxine·HCl, 2, 3, 4, 5, 6, 8, 10; and pyridoxamine·2 HCl, 6, 9, 12, 15, 18, 24, 30, and water is added to make a total volume of 5 ml for each tube. Where a total vitamin B$_6$ assay is done, the proper aliquots of hydrolyzed samples (adjusted to pH 5.0) are pipetted directly into the assay tubes without chromatographic separation. The pyridoxine standard curve is prepared alongside the pyridoxine eluates, and so on. Because of the light sensitivity of the vitamin B$_6$, all procedures should be performed under gold light.[13] To the prepared tubes is added 5 ml of double-strength "Pyridoxine-Y-Medium," with shaking to mix the 2 phases, the tubes are capped or plugged, and steamed in the autoclave for 10 minutes at 100°. Upon cooling, 0.1 ml of inoculum is added aseptically to each tube. The tubes are incubated on a shaker (preferably rotary) at 30° for 20–24 hours. After steaming in the autoclave for 5 minutes, and cooling, they are read for optical density (OD) at 550 mμ with a spectrophotometer against the inoculated blanks, which are set at zero OD.

(d) CALCULATIONS. Average OD readings for duplicate tubes for each level of standard are plotted on linear graph paper against the appropriate vitamer concentration level to produce a standard reference curve. By interpolation of average OD values for duplicate tubes for each level of sample, the vitamer content of the sample tubes is determined, and the vitamer content of the sample, in micrograms of vitamer per gram of sample is extrapolated. Correction for the percentage recovery for each vitamer is made.

Comments. When 0.055 N HCl was compared with 0.44 N HCl, for both a 2- and a 5-hour hydrolysis period, the weaker acid was as effective with the longer time period, and is therefore recommended for general use.[9]

The vitamin B$_6$ contained in 100 mg of liver did not exceed the absorbing capacity of the resin, when used as suggested.[9] Total vitamin B$_6$ (total of 3 vitamers) of tissues assayed (in micrograms per gram of tissue): liver, 10.0; kidney, 5.2; brain, 2.5; muscle, 3.7; heart, 4.5.[10]

At the levels of vitamers in liver, the recovery percentages were: pyridoxal, 89; pyridoxine, 127; pyridoxamine, 85; with total recovery, 100.[9]

The use of "Pyridoxine-Y-Medium"[10] proved to be a more suitable procedure than the former vitamin-free yeast base[9] or than the total preparation of the assay medium.[7]

---

[13] E. W. Toepfer, personal communication, 1965.

[88] Synthesis of Vitamin B₆ Analogs[1]

By W. KORYTNYK and MIYOSHI IKAWA

Soon after the isolation of vitamin B_6 was independently reported by several laboratories in 1938, the vitamin and a number of related substances were synthesized. Some of these compounds were found to act as inhibitors of the vitamin.[2] Discovery that vitamin B_6 exists in multiple forms, and that pyridoxal phosphate has cofactor functions in various enzyme systems, has stimulated a renewed interest in this area, with the result that a great many additional compounds related to vitamin B_6 have been synthesized. Alterations have been made at every position of the pyridine ring, and pyrimidine and benzene analogs have also been prepared. The effects of these alterations on diverse biological and biochemical systems have been studied.

Biochemical interest in analogs of vitamin B_6 has been focused on the following topics[3]: the significance of various functional groups in reactions catalyzed by pyridoxal, in both nonenzymatic[3] and enzymatic[4] systems; the substrate specificity and inhibition of enzymes involved in vitamin B_6 metabolism, particularly pyridoxal phosphokinase[5] and pyridoxine oxidase[6]; the structure and synthesis of metabolites of vitamin B_6[7] (this volume, [98]) as well as the biosynthesis of the vitamin itself; the mode of binding of the cofactor to the apoenzyme and the general structural requirements

[1] The work at Roswell Park Memorial Institute was supported in part by U.S. Public Health Service Grant CA-08793, awarded to W. Korytnyk. We wish to thank Professor Esmond E. Snell for his comments and suggestions concerning this paper.

[2] Excellent accounts of earlier work are available: (a) R. J. Williams, R. E. Eakin, E. Beerstecher, and W. Shive, "The Biochemistry of B Vitamins," p. 652. Reinhold, New York, 1950; (b) W. H. Sebrell, Jr. and R. S. Harris, "The Vitamins," Vol. III, p. 219. Academic Press, New York, 1954; (c) E. E. Snell, *in* "Comprehensive Biochemistry" (M. Florkin and E. H. Stotz, eds.), Vol. 11, p. 48. Elsevier, Amsterdam, 1963.

[3] E. E. Snell, *Vitamins Hormones*, **16**, 77 (1958), summarizes earlier work on the vitamin B_6 analogs.

[4] E. E. Snell and J. E. Ayling, *in* "Symposium on Pyridoxal Enzymes" (K. Yamada, N. Katsunuma, and H. Wada, eds.), p. 5. Maruzen, Tokyo, 1968.

[5] D. B. McCormick and E. E. Snell, *J. Biol. Chem.* **236**, 2085 (1961).

[6] H. Wada and E. E. Snell, *J. Biol. Chem.* **236**, 2089 (1961).

[7] (a) E. E. Snell, R. W. Burg, W. B. Dempsey, E. J. Nyns, T. K. Sundarum, and D. Zäch, *in* "Chemical and Biological Aspects of Pyridoxal Catalysis" (E. E. Snell, P. M. Fasella, A. Braunstein, and A. Rossi Fanelli, eds.), p. 563. Macmillan, New York, 1963; (b) M. Ikawa, V. W. Rodwell, and E. E. Snell, *J. Biol. Chem.* **233**, 1555 (1958); (c) R. W. Burg, V. W. Rodwell, and E. E. Snell, *J. Biol. Chem.* **235**, 1164 (1960).

for its cofactor function[8,9] (this volume [89]–[91]); and the various functions of vitamin B$_6$ in microorganisms and higher animals, especially as demonstrated by blocking *in vivo* certain reactions dependent on the vitamin.[10]

For the medicinal chemist and pharmacologist, the involvement of pyridoxal phosphate in various enzyme systems offers an opportunity for rational design of inhibitors against various biological systems.[11] Numerous pharmacological effects have been observed with analogs of vitamin B$_6$, such as inhibition of the growth of microorganisms,[3,11,12] tumors,[13] and lymphoid tissues, including decreases in circulating lymphocytes.[13] Many observations have been reported concerning the effects of these agents on the central nervous system.[13,14]

With the enzymological information now available regarding a limited number of vitamin B$_6$ analogs, it is possible to recognize the significance of certain structural features. Substrates of pyridoxal phosphokinase must have at least an unchanged 5-(hydroxymethyl) group,[5,15] whereas inhibitors of this enzyme include various pyridoxal and isopyridoxal hydrazones.[5,16] A number of phosphorylated analogs and their derivatives can compete with the cofactor or replace it in suitable enzyme systems,[8,17] and are inhibitors of pyridoxine oxidase.[6] Schiff bases and their reduction products are structurally similar to appropriate reaction intermediates in the enzyme reaction, and there is at least one example of an effective inhibitor of this type.[18] Analogs containing reactive groups are being developed. Such compounds may react irreversibly with enzymes, and thus provide information regarding the nature of active sites.[19]

The synthetic procedures that have been utilized for the preparation of of vitamin B$_6$ analogs can be conveniently divided into three general categories:

[8] Y. Morino and E. E. Snell, *Proc. Natl. Acad. Sci. U.S.* **57**, 1692 (1967).

[9] E. H. Fischer, A. Pocker, S. Shaltiel, J. L. Hedrick, and S. D. Elsom, *in* "Symposium on Pyridoxal Enzymes" (K. Yamada, N. Katunuma, and H. Wada, eds.), p. 119. Maruzen, Tokyo, 1968.

[10] J. Olivard and E. E. Snell, *J. Biol. Chem.* **213**, 203, 215 (1955).

[11] W. Korytnyk and B. Paul, *in* "Pyridoxal Catalysis: Enzymes and Model Systems" (E. E. Snell, A. E. Braunstein, E. S. Severin, and Yu. M. Torchinsky, eds.), p. 615. Wiley (Interscience), New York, 1968.

[12] W. Korytnyk, B. Paul, A. Bloch, and C. A. Nichol, *J. Med. Chem.* **10**, 345 (1967).

[13] F. Rosen, E. Mihich, and C. A. Nichol, *Vitamins Hormones* **22**, 609 (1964).

[14] P. Holtz and D. Palm, *Pharmacol. Rev.* **16**, 113 (1964).

[15] J. Hurwitz, *J. Biol. Chem.* **217**, 513 (1955).

[16] W. Korytnyk, E. J. Kris, and R. P. Singh, *J. Org. Chem.* **29**, 574 (1964).

[17] P. F. Mühlradt, Y. Morino, and E. E. Snell, *J. Med. Chem.* **10**, 341 (1967).

[18] W. B. Dempsey and E. E. Snell, *Biochemistry* **2**, 1414 (1963).

[19] W. Korytnyk and B. Lachmann, work in progress.

1. Extensions of existing methods for the synthesis of vitamin B_6. These methods have found greatest application in the synthesis of 2- and 3-modified analogs.[17,20,21]

2. Reactions of readily available vitamin B_6 compounds with various reagents. A number of 4- and 5-modified analogs have been prepared by this method, using such compounds as pyridoxal, isopyridoxal, or the pyridoxic acids.

3. Blocking-deblocking procedures. One or two functional groups can be blocked with another group (e.g., isopropylidene, benzyl, or acyl), thus permitting the remaining functional groups in the vitamin B_6 molecule to be modified selectively; the blocking group can then be split off to give the desired compound.[11] The success of this approach depends on the ease with which the blocking group can be removed as well as introduced. Blocking-deblocking procedures have found extensive application in the preparation of 4- and 5-modified homologs and in the synthesis of certain metabolites.[11]

Tables I through VIII list a number of vitamin B_6 analogs, most of which have been found to have vitamin or antivitamin activity. These tables are arranged mainly according to the positions of modifications of the pyridoxol molecule, but phosphorylation of the 5-(hydroxymethyl) group and interconversion of such groups as 4-(hydroxymethyl), 4-formyl, and 4-(aminomethyl) are ordinarily ignored in the assignment of a specific compound to a specific table. References to the literature dealing with the synthesis of the compounds are provided, including references to articles in this volume. Methods for synthesis of certain compounds of particular interest are given in detail after the tables.

The synthetic procedures described also illustrate some reactions of particular interest in vitamin B_6 chemistry, such as oxidation and phosphorylation reactions. They also include methods for the synthesis of some key intermediates, such as α^4,3-O-isopropylidenepyridoxol and α^5,3-O-dibenzylpyridoxol, which have found extensive application in the modification of the 5- and 4-positions, respectively.

The tables also briefly summarize the biochemical activities of the analogs, in either enzymatic or more complex biological systems or in simulated enzymatic reactions. (A number of compounds of potential biochemical interest may have been omitted, for lack of biochemical data.) Although some analogs, such as 4-deoxypyridoxol and its 5-phosphorylated derivative, have been studied extensively, our information concerning the great majority of the analogs is fragmentary, and hence precludes, at least for the time being, a comprehensive systematic classification based on enzymological properties.

The current state of the nomenclature of compounds related to vitamin

[20] R. G. Jones and E. C. Kornfeld, *J. Am. Chem. Soc.* **73,** 107 (1951).
[21] J. M. Osbond, *Vitamins Hormones* **22,** 367 (1964).

B₆ leaves something to be desired.²² Accordingly, the nomenclature used in this paper is intended to be as simple as possible without being ambiguous. Trivial names in common use are adopted whenever they satisfy this principle, and semisystematic names derived from such trivial names are used in preference to more complicated systematic names, especially where systematic names would obscure relationships to vitamin B₆. Fully systematic names are used only where simpler names would be ambiguous. Alternatives, where they are of sufficient importance, are given in parentheses.

Synthetic Procedures²³

Pyridoxol N-Methyl Betaine (I-1)

This method was described by Harris *et al.*²³ᵃ To a solution of 0.5247 g of free pyridoxol base in a mixture of 240 ml of hot benzene and 20–30 ml of methanol, add 8 ml of methyl iodide, and reflux the mixture overnight. Concentrate the solution to a third of its volume. Filter off the crystals of pyridoxol methiodide, and wash with ether. The yield is 96.5%. Treat a solution of the methiodide in water with freshly prepared silver carbonate until it is free of iodide ion. Filter the iodide-free solution, and evaporate the filtrate to dryness *in vacuo*. Recrystallize the pyridoxol N-methyl betaine from absolute ethanol. The yield is 80%; melting point, 196°.

5-Deoxypyridoxol (V-1)

α⁴,3-O-Isopropylidenepyridoxol　(*I*)

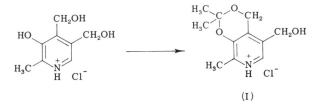

(I)

This procedure is a modification of that of W. Korytnyk and W. Wiedeman²⁴; other methods for the synthesis of this important intermediate have

---

²² W. Rennagel and W. Korytnyk, *Experientia* **24**, 304 (1968), and manuscript in preparation.

²³ A numerical designation consisting of a Roman numeral alone refers to a structure shown in this section; a numerical designation consisting of a Roman numeral and an Arabic numeral joined by a hyphen refers to the listing of the compound in the table indicated by the Roman numeral.

²³ᵃ S. A. Harris, T. J. Webb, and K. Folkers, *J. Am. Chem. Soc.* **62**, 3198 (1940).

²⁴ W. Korytnyk and W. Wiedeman, *J. Chem. Soc.* p. 2531 (1962).

TABLE I

VITAMIN B_6 COMPOUNDS MODIFIED IN THE 1-POSITION

No.	Compound	Structure	Synthesis[a]	Vitamin B_6 activity[a,b]
I-1	Pyridoxol N-methyl betaine		c, [88]	Inactive in rats;[c] slightly active in microorganisms[2a]
I-2	Pyridoxol N-oxide		d, e,	Active in rats[f]
I-3	Pyridoxal N-oxide		e, [90]	Active in model systems[g]
I-4	Pyridoxal 5′-phosphate N-oxide		h, [89]	Active in model systems, tryptophanase, and GOT[i,j]

[a] Articles in this volume are cited by article number enclosed in brackets [   ]. Superscript numbers refer to text footnotes.

[b] Unless otherwise indicated.

[c] S. A. Harris, T. J. Webb, and K. Folkers, *J. Am. Chem. Soc.* **62**, 3198 (1940).

[d] T. Sakuragi and F. A. Kummerow, *J. Org. Chem.* **24**, 1032 (1959).

[e] Y. Nakai, N. Onishi, S. Shimizu, and S. Fukui, *Bitamin* **35**, 213 (1967).

[f] T. Sakuragi and F. A. Kummerow, *Proc. Soc. Exptl. Biol. Med.* **103**, 185 (1960).

[g] N. Onishi, Y. Nakai, S. Shimizu, and S. Fukui, *Abstr. Intern. Symp. Pyridoxal Enzymes, Nagoya, Japan* Aug. 1967.

[h] Y. Nakai, N. Onishi, S. Shimizu, and S. Fukui, *Bitamin* **36,** 521 (1967).

[i] N. Onishi, Y. Nakai, S. Shimizu, and S. Fukui, *in* "Symposium on Pyridoxal Enzymes" (K. Yamada *et al.*, eds.), p. 43. Maruzen, Tokyo, 1968.

[j] Glutamic-oxaloacetic transaminase (aspartate aminotransferase).

TABLE II

Vitamin B6 Compounds Modified in the 2-Position

No.	Compound	Structure	Synthesis[a]	Vitamin B6 activity[a,b]
		(structure: R_4, HO–, CH_2OR_5, N, H)		
II-1	2-Norpyridoxol (2-demethyl-pyridoxol)	$R_4 = CH_2OH$; $R_5 = H$	17, c–e	—
II-2	2-Norpyridoxal	$R_4 = CHO$; $R_5 = H$	17, e	—
II-3	2-Norpyridoxamine	$R_4 = CH_2NH_2$; $R_5 = H$	e	—
II-4	2-Norpyridoxal 5'-phosphate	$R_4 = CHO$; $R_5 = PO_3H_2$	17, e	Replaces PLP[f] as a coenzyme[g,a,h]
II-5	2-Norpyridoxamine 5'-phosphate	$R_4 = CH_2NH_2$; $R_5 = PO_3H_2$	e	—
		(structure: R_4, HO–, CH_2OR_5, N, H_5C_2)		
II-6	α²-Methylpyridoxol (2-ω-methyl-pyridoxol, 2-ethyl-2-nor-pyridoxol)	$R_4 = CH_2OH$; $R_5 = H$	17, g, i	Little or no activity in lactic acid bacteria[j] and rats,[j] active on ex-cised tomato roots;[2a] oxidized by pyridoxol dehydrogenase[g]
II-7	α²-Methylpyridoxal	$R_4 = CHO$; $R_5 = H$	17, j	Partial activity in lactic acid bacteria[j]
II-8	α²-Methylpyridoxamine	$R_4 = CH_2NH_2$; $R_5 = H$	j	Slight activity in lactic acid bacteria[j]
II-9	α²-Methylpyridoxal 5'-phosphate	$R_4 = CHO$; $R_5 = PO_3H_2$	17, j	Slight activity in lactic acid bacteria;[j] replaces PLP as a coenzyme[8,h,i]
II-10	α²-Methylpyridoxamine 5'-phosphate	$R_4 = CH_2NH_2$; $R_5 = PO_3H_2$	j	Partial activity in lactic acid bacteria[j]

(Continued)

TABLE II (Continued)

No.	Compound	Structure	Synthesis[a]	Vitamin B_6 activity[a,b]
II-11	α^2,α^2-Dimethylpyridoxol	(pyridine ring: HO, CH_2OH, CH_2OH, $(CH_3)_2HC$, N)	g, m	Oxidized by pyridoxol dehydrogenase;[q] pyridoxine activity[m]
II-12	α^2-Propylpyridoxal 5′-phosphate	(pyridine ring: CHO, HO, $CH_2OPO_3H_2$, $H_3C(CH_2)_3$, N)	n [89]	Inhibits activation of arginine decarboxylase by PLP[l]
II-13	α^2-Hydroxypyridoxol	(pyridine ring: CH_2OH, HO, CH_2OH, HOH_2C, N)	o	Low activity in rats and yeast, no activity in *Lactobacillus casei*[p]
II-14	α^2-Hydroxypyridoxal 5′-phosphate	(pyridine ring: CHO, HO, $CH_2OPO_3H_2$, HOH_2C, N)	[91]	
II-15	α^2-Phenylpyridoxol	(pyridine ring: CH_2OH, HO, CH_2OH, ϕH_2C, N)	q	Slight activity in *Neurospora sitophila*;[q] slight inhibitory activity in *Staphylococcus aureus* and *Escherichia coli*[a]

II-16   2-Formyl-3-hydroxy-4,5-bis(hydroxymethyl)pyridine thiosemicarbazone

Inhibits the synthesis of DNA in Sarcoma 180 cells *in vitro*, has antitumor activity[r]

---

[a] Numbers correspond to text footnotes citing references to the literature; articles in this volume are cited by article number enclosed in brackets [ ]. Superscript numbers also refer to text footnotes.

[b] Unless otherwise indicated.

[c] S. M. Gadekar, J. L. Frederick, and E. C. DeRenzo, *J. Med. Pharm. Chem.* **5**, 531 (1962).

[d] B. Van der Wal, T. J. de Boer, and H. O. Huisman, *Rec. Trav. Chim.* **80**, 203 (1961).

[e] V. L. Florentiev, N. A. Drobinskaja, L. V. Ionova, and M. Ya. Karpeisky, *Tetrahedron Letters* p. 1747 (1967); *Dokl. Akad. Nauk SSSR* **177**, 617 (1967); *Chem. Abstr.* **69**, 27198g (1968).

[f] PLP, pyridoxal phosphate.

[g] P. Melius and D. L. Marshall, *J. Med. Chem.* **10**, 1157 (1967).

[h] A. L. Bocharov, V. I. Ivanov, M. Y. Karpeisky, O. K. Mamaeva, and V. L. Florentiev, *Biochem. Biophys. Res. Commun.* **30**, 459 (1968).

[i] S. A. Harris and A. N. Wilson, *J. Am. Chem. Soc.* **63**, 2526 (1941).

[j] M. Ikawa and E. E. Snell, *J. Am. Chem. Soc.* **76**, 637 (1954).

[k] R. Sandman and E. E. Snell, *Proc. Soc. Exptl. Biol. Med.* **90**, 63 (1955).

[l] S. L. Blethen, E. A. Boeker, and E. E. Snell, *J. Biol. Chem.* **243**, 1671 (1968).

[m] H. Davoll and F. B. Kipping, *J. Chem. Soc.* p. 1395 (1953).

[n] N. D. Doktorova, L. V. Ionova, M. Ya. Karpeisky, N. Sh. Padyukova, K. F. Turchin, and V. L. Florentiev, *Tetrahedron* **25**, 3527 (1969).

[o] G. R. Bedford, A. R. Katritzky, and H. M. Wuest, *J. Chem. Soc.* p. 4600 (1963).

[p] C. J. Argoudelis, *Federation Proc.* **27**, 553 (1953).

[q] A. Cohen and J. A. Silk, *J. Chem. Soc.* p. 4386 (1952).

[r] S. Clayman, K. C. Agrawal, and A. C. Sartorelli, *Abstr., 158th Meeting Am. Chem. Soc., New York, N.Y.*, Sept. 1969, Paper No. 77 of Med. Chem. Div.

TABLE III
VITAMIN B$_6$ COMPOUND MODIFIED IN THE 3-POSITION

No.	Compound	Structure	Synthesis[a]	Vitamin B$_6$ activity[a,b]
III-1	3-*O*-Methylpyridoxol	R$_3$ = OCH$_3$	c, d	Low activity in rats[2a]
III-2	3-*O*-Methylpyridoxal 5′-phosphate		e	Reactivates apophosphorylase b and apophosphorylase a[g,f]
III-3	3-Deoxypyridoxol	R$_3$ = H	g, h, i	Not oxidized by pyridoxol dehydrogenase[h]; substrate of pyridoxal phosphokinase,[15] arrests growth of Murphy lymphosarcoma,[i] causes convulsions in rats[j]

III-4 3-Amino-3-deoxypyridoxol	$R_3 = NH_2$	g	Substrate of pyridoxal phosphokinase[15]
III-5 3-Chloro-3-deoxypyridoxol	$R_3 = Cl$	k	B6 antagonist in rats, causes convulsive seizures[k]
III-6 3-Thio-3-deoxypyridoxol	$R_3 = SH$	l	Partial activity in *Saccharomyces carlsbergensis[l]*

[a] Numbers correspond to text footnotes citing references. Superscript numbers also refer to text footnotes.

[b] Unless otherwise indicated.

[c] R. Kuhn and G. Wendt, *Chem. Ber.* **71**, 1534 (1938).

[d] E. T. Stiller, J. C. Keresztesy, and J. R. Stevens, *J. Am. Chem. Soc.* **61**, 1237 (1939).

[e] A. Pocker and E. H. Fischer, *Biochemistry* **8**, 5181 (1969).

[f] S. Shaltiel, J. L. Hedrick, A. Pocker, and E. H. Fischer, *Biochemistry* **8**, 5189 (1969).

[g] R. G. Jones and E. C. Kornfeld, *J. Am. Chem. Soc.* **73**, 107 (1951).

[h] P. Melius and D. L. Marshall, *J. Med. Chem.* **10**, 1157 (1967).

[i] S. A. Harris, E. E. Harris, E. R. Peterson, and E. F. Rogers, *J. Med. Chem.* **10**, 261 (1967).

[j] F. Rosen, R. J. Milholland, and C. A. Nichol, *in* "Inhibition of the Nervous System and γ-Aminobutyric Acid," p. 338. Macmillan (Pergamon) New York, 1960.

[k] S. M. Gadekar, J. L. Fredrick, E. C. DeRenzo, *J. Med. Pharm. Chem.* **5**, 531 (1962).

[l] J. L. Greene, Jr. and J. A. Montgomery, *J. Med. Chem.* **7**, 17 (1964).

TABLE IV

VITAMIN B_6 COMPOUNDS MODIFIED IN THE 4-POSITION

No.	Compound	Structure	Synthesis[a]	Vitamin B_6 activity[a,b]
IV-1	Pyridoxol 5'-phosphate	R_4, CH_2OH, HO, H_3C, N (pyridine ring)	c, d, [88]	Substrate of pyridoxine oxidase,[6] inhibitor of glutamic–aspartic apotransaminase[e]
IV-2	α^4-O-Adamantoylpyridoxol	$R_4 = CH_2OCO-$ (adamantyl)	f	Inhibits *Saccharomyces carlsbergensis*[f]
IV-3	α^4-O-Methylpyridoxol (4-methoxy-pyridoxol)	$R_4 = CH_2OCH_3$	g	Active in chicks,[h] chick embryos,[i] and dogs[j]
IV-4	Pyridoxal 5'-phosphate oxime	$CH=NOH$, $CH_2OPO_3H_2$, HO, H_3C, N	k, [88]	Inhibits pyridoxine oxidase[6]
IV-5	4-Pyridoxic acid	$R_4 = COOH$	l, [88]	Metabolite of B_6 in man[m] and *Pseudomonas*[7e]
IV-6	4-Pyridoxic acid lactone (β-pyracin)	lactone ring structure, HO, H_3C, N	l	

Structure for IV-2: CH_2OH, $CH_2OPO_3H_2$, HO, H_3C, N (pyridine ring with adamantyl group)

	Name	Structure	Ref.	Activity
IV-7	4-Pyridoxic acid 5'-phosphate	COOH / CH$_2$OPO$_3$H$_2$; HO—, H$_3$C—, N (pyridine)	6	Inhibits pyridoxine oxidase[6]
IV-8	4-Pyridoxic acid hydrazide	R$_4$ = CONHNH$_2$	r	Antitubercular activity;[r] inhibits growth of S-180 cells *in vitro*[o]
IV-9	4-Deoxypyridoxol	R$_4$ = CH$_3$	g, p, q, [88]	Substrate of pyridoxal phosphokinase;[5,15] much studied potent antagonist of B$_6$ (reviewed[3,13])
IV-10	4-Deoxypyridoxol 5'-phosphate	CH$_3$ / CH$_2$OPO$_3$H$_2$; HO—, H$_3$C—, N	c	Inhibits tyrosine apodecarboxylase,[r] glutamic–aspartic apotransaminase,[e] tyrosine transaminase,[g] pyridoxine oxidase[6]
IV-11	5-Hydroxy-6-methyl-4-trifluoromethyl-3-pyridinemethanol (trifluoro-4-deoxypyridoxol)	CF$_3$ / CH$_2$OH; HO—, H$_3$C—, N	t	No anti-B$_6$ activity[t]
IV-12	3-Hydroxy-5-hydroxymethyl-2-methylpyridine [4-de(hydroxymethyl)pyridoxol]	R$_4$ = H	7c, u	Weak inhibitor of *S. carlsbergensis*[u]
IV-13	α^4-Methyl-4-deoxypyridoxol	CH$_2$CH$_3$ / CH$_2$OH; HO—, H$_3$C—, N	d	—
IV-14	α^4-Phenyl-4-deoxypyridoxol	CH$_2$φ / CH$_2$OH; HO—, H$_3$C—, N	w	—

(Continued)

TABLE IV (*Continued*)

No.	Compound	Structure	Synthesis[a]	Vitamin B_6 activity[a,b]
IV-15	α^4-Methylene-4-deoxypyridoxol		v	—
IV-16	7-Hydroxy-6-methyl-1-phenylfuro-[3,4-c]pyridine		w	—
IV-17	4-Demethylenepyridoxol (4-nor-pyridoxol)	$R_4 = OH$	x	Inhibits *S. carlsbergensis*;[y] inhibits growth of S-180 cells *in vitro*[o]
IV-18	α^4-Methylpyridoxol	$R_4 = CH(CH_3)OH$	11, w, [88]	Inhibits growth of S-180 cells *in vitro*[o]
IV-19	2-(4-Pyridoxyl)ethanol	$R_4 = (CH_2)_3OH$	11, w	Inhibits growth of S-180 cells *in vitro*[o]
IV-20	4-Vinylpyridoxol phosphate		z	Inhibits pyridoxine oxidase from rabbit liver;[z] also glutamate decarboxylase and apoaspartate aminotransferase[aa]
IV-21	3-Amino-8-methyl-2-oxo-2H-pyrano-(2,3-C)pyridine-5-methanol		bb	—

IV-22 2-(α^4-Pyridoxyl)-2-oxoacetic acid

[a] Numbers correspond to text footnotes citing references; articles in this volume are cited by article number enclosed in brackets [ ].

[b] Unless otherwise cited.

[c] E. A. Peterson and H. A. Sober, J. Am. Chem. Soc. 76, 169 (1954).

[d] W. Korytnyk and B. Lachmann, unpublished results, 1968.

[e] A. Meister, H. A. Sober, and E. A. Peterson, J. Biol. Chem. 206, 89 (1954).

[f] W. Korytnyk and G. Fricke, J. Med. Chem. 11, 180 (1968).

[g] S. A. Harris, J. Am. Chem. Soc. 62, 3203 (1940).

[h] W. H. Ott, Proc. Soc. Exptl. Biol. Med. 66, 215 (1947).

[i] D. A. Karnofsky, C. C. Stock, L. P. Ridgway, and P. A. Patterson, J. Biol. Chem. 182, 471 (1950).

[j] C. W. Mushett, R. B. Stebbins, and M. N. Barton, Trans. N.Y. Acad. Sci. 9, 291 (1947).

[k] D. Heyl, E. Luz, S. A. Harris, and K. Folkers, J. Am. Chem. Soc. 73, 3430 (1951).

[l] (a) H. Ahrens and W. Korytnyk, J. Heterocyclic Chem. 4, 625 (1967); (b) D. Heyl, J. Am. Chem. Soc. 70, 3434 (1948).

[m] J. W. Huff and W. A. Perlzweig, J. Biol. Chem. 155, 345 (1944).

[n] S. Emoto, J. Sci. Res. Inst. (Tokyo) 47, 37 (1953).

[o] M. Hakala, preliminary observations, 1968.

[p] R. G. Taborsky, J. Org. Chem. 26, 596 (1961).

[q] R. P. Singh and W. Korytnyk, J. Med. Chem. 8, 116 (1965).

[r] J. Hurwitz, J. Biol. Chem. 212, 757 (1955).

[s] F. Rosen, R. J. Milholland, and C. A. Nichol, in "Symposium on Pyridoxal Enzymes" (K. Yamada et al., eds.), p. 77. Maruzen, Tokyo, 1968.

[t] J. L. Green, Jr. and J. A. Montgomery, J. Med. Chem. 6, 294 (1963).

[u] L. A. Perez-Medina, R. P. Mariella, and S. M. McElvain, J. Am. Chem. Soc. 69, 2574 (1947).

[v] W. Korytnyk and B. Lachmann, to be published, 1969.

[w] W. Korytnyk and B. Paul, J. Med. Chem. 13, 187 (1970).

[x] D. Heyl, E. Luz, and S. A. Harris, J. Am. Chem. Soc. 73, 3437 (1951).

[y] W. Korytnyk and B. Paul, J. Heterocyclic Chem. 2, 144 (1965).

[z] W. Korytnyk and B. Lachmann, unpublished results, 1968.

[aa] M. Fonda, unpublished results, 1970.

[bb] W. Korytnyk and H. Ahrens, unpublished results.

TABLE V

Vitamin B_6 Compounds Modified in the 5-Position

No.	Compound	Structure	Synthesis[a]	Vitamin B_6 activity[a,b]
V-1	5-Deoxypyridoxol	$R_5 = CH_3$	12, c–e, [88]	Inhibits *Saccharomyces carlsbergensis*[l]
V-2	5-Deoxypyridoxal		d, e, g	Inhibits *S. carlsbergensis, Streptococcus faecalis, Lactobacillus helveticus*[l] and *Escherichia coli*;[h] weak competitive inhibitor of apotryptophanase;[i] substrate of pyridoxamine–pyruvate transaminase[j]
V-3	5-Deoxypyridoxamine		c	Substrate of pyridoxamine–pyruvate transaminase;[j] inhibits *S. carlsbergensis, S. faecalis*[l]
V-4	Isopyridoxal	$R_5 = CHO$	16, k, l	Microbial metabolite of pyridoxol;[m] active in *S. carlsbergensis*;[n] substrate of isopyridoxal hydrogenase[o]
V-5	4,5-Diformyl-3-hydroxy-2-methylpyridine (pyridoxdial)		p	—

V-6	5-Pyridoxic acid	R$_5$ = COOH	16	Microbial metabolite of pyridoxol[m]
V-7	5-Pyridoxic acid lactone (α-pyracin)		16	Microbial metabolite[m]
V-8	5-Pyridoxamide	R$_5$ = CONH$_2$	16	Inhibits *Escherichia coli*[h]
V-9	Isopyridoxamine	R$_5$ = CH$_2$NH$_2$	*l*	Slight activity in *S. carlsbergensis*[n]
V-10	5-Demethyleneisopyridoxamine (5-norisopyridoxamine)	R$_5$ = NH$_2$	*q*	Inhibits *S. carlsbergensis*[q]
V-11	C-(α^5-Pyridoxyl)methylamine	R$_5$ = CH$_2$CH$_2$NH$_2$	12	—
V-12	2-(α^5-Pyridoxyl)glycine	R$_5$ = CH$_2$CHCOOH, NH$_2$	12	—
V-13	5-Pyridoxthiol	R$_5$ = CH$_2$SH	*r*	Slight activity in rats[s]; radiation protective agent[t]
V-14	5-Pyridoxyl disulfide (pyrithioxin, pyritinol)	(CH$_3$S— ...)$_2$ structure	*r*	Slight activity in rats[s]; dynamizing neurotropic agent[u]
V-15	5-Pyridoxyl sulfide	(—CH$_2$— ...)$_2$ S structure	*v*	Analgesic activity[v]
V-16	*S*-Benzoyl-5-thiopyridoxal	CH$_2$SCOϕ structure	*w*	Inhibits *S. carlsbergensis*[z]

(*Continued*)

TABLE V (Continued)

No.	Compound	Structure	Synthesis[a]	Vitamin B6 activity[a,b]
V-17	5-Demethylenepyridoxol (5-nor-pyridoxol)	R_5 = OH	q	Inhibits S. carlsbergensis[q]
V-18	α5-Methylpyridoxol	R_5 = CH(CH3)OH	12	Inhibits S. carlsbergensis[12]
V-19	α5,α5-Dimethylpyridoxol	R_5 = C(CH3)2OH	12	Inhibits S. carlsbergensis[12]
V-20	ω-(5-Pyridoxyl)alkanols	R_5 = (CH2)2-4OH	12, y	Inhibit S. carlsbergensis[12]
V-21	3-Hydroxy-5-(2-hydroxyethyl)-2-methylpyridine-4-carboxaldehyde hemiacetal		w	Inhibits S. carlsbergensis[x]
V-22	ω-(5-Pyridoxyl)alkanoic acids	R_5 = (CH2)1-3COOH	y	Inhibit S. carlsbergensis[12]
V-23	3-(4-Formyl-3-hydroxy-2-methyl-5-pyridyl)propionic acid		z	Active in model systems[z] Competitive inhibitor of combination of PLP with apotryptophanase[i] and apoarginine decarboxylase[aa]
V-24	Ethyl 5-pyridoxyl acetate (α5-pyridoxylacetic acid ethyl ester)	R_5 = CH2CH2COOC2H5	12	Inhibits S. carlsbergensis[12]
V-25	α5-Ethyl-5-deoxypyridoxol	R_5 = CH2CH2CH3	12	Inhibits S. carlsbergensis[12]
V-26	3-Chloro-1-(5-pyridoxyl)-2-propanone	R_5 = CH2CH2COCH2Cl	bb	Inhibits apotryptophanase[bb]
V-27	α5-Methylene-5-deoxypyridoxol	R_5 = CH=CH2	cc	
V-28	5-(2-Nitrovinyl)-5-de(hydroxymethyl)pyridoxol	R_5 = CH=CHNO2	12	Inhibits S. carlsbergensis[12]
V-29	Pyridoxol 5'-sulfate	R_5 = CH2OSO3H	dd	

No.	Name	Structure	Ref.	Activity
V-30	Pyridoxal 5′-sulfate	CHO / CH₂OSO₃H pyridine ring HO, H₃C	dd	Inhibitor of glutamic acid decarboxylase[ee]
V-31	Pyridoxal 5′-(methyl phosphonate)	CHO / CH₂OP(O)CH₃, OH pyridine ring HO, H₃C	gg	Combines with apo-D-serine dehydrase[ff] and apoglutamate–aspartate transaminase[hh] but has negligible cofactor activity
V-32	Pyridoxal cyanoethyl 5′-phosphate	CHO / CH₂OPOCH₂CH₂CN, OH pyridine ring HO, H₃C	gg	Combines with apoglutamate–aspartate transaminase, and has substantial cofactor activity;[hh] combines with apo-D-serine dehydrase but has no cofactor activity[ff]
V-33	C-(5-Pyridoxyl)methylphosphonic acid	CH₂OH / CH₂CH₂PO(OH)₂ pyridine ring HO, H₃C	ii	Inhibitor of apoaspartate aminotransferase[ii]
V-34	C-(5-Pyridoxylidene)methylphosphonic acid	CH₂OH / CH=CHPO(OH)₂ pyridine ring HO, H₃C	ii	Same as V-33[ii]
V-35	2-(4-Formyl-3-hydroxy-2-methyl-5-pyridyl)ethylphosphonic acid	CHO / CH₂CH₂PO(OH)₂ pyridine ring HO, H₃C	ii	Same as V-33[ii]

(Continued)

TABLE V (Continued)

No.	Compound	Structure	Synthesis[a]	Vitamin B₆ activity[a,b]
V-36	2-(4-Formyl-3-hydroxy-2-methyl-5-pyridyl)ethenylphosphonic acid	(structure)	ii	Same as V-33[ii]
V-37	α⁵-Methylpyridoxal phosphate	(structure)	gg	Combines with apo-glutamate-aspartate transaminase[hh] and apo-D-serine dehydrase[ll] and has 3% and 24% cofactor activity, respectively
V-38	α⁵-Homopyridoxal phosphate	(structure)	gg	Combines with apo-glutamate decarboxylase and apo-glutamate-aspartate transaminase, and has 5% and 3% cofactor activity, respectively.[ii]
V-39	5-Thiopyridoxal	(structure)	w	—
V-40	7,8-Dihydro-3-methyl-2,6-naphthyridin-4-ol	(structure)	kk	—

[a] Numbers correspond to text footnotes citing references; articles in this volume are cited by article number in brackets [ ].
[b] Unless otherwise indicated.
[c] D. Heyl, S. A. Harris, and K. Folkers, *J. Am. Chem. Soc.* **75**, 653 (1953).

[d] P. F. Muhlradt and E. E. Snell, *J. Med. Chem.* **10**, 129 (1967).

[e] C. Iwata, *Biochem. Prep.* **12**, 117 (1968).

[f] J. C. Rabinowitz and E. E. Snell, *Arch. Biochem. Biophys.* **43**, 408 (1953).

[g] T. Kuroda, *Bitamin* **29**, 116 (1964); *Chem. Abstr.* **62**, 515g (1965).

[h] L. J. Areement, W. Korytnyk, and W. B. Dempsey, *Bact. Proc.* p. 121 (1968) and unpublished results (1968).

[i] W. A. Newton, Y. Morino, and E. E. Snell, *J. Biol. Chem.* **240**, 1211 (1965).

[j] J. E. Ayling and E. E. Snell, *Biochemistry* **7**, 1626 (1968).

[k] W. Korytnyk and E. J. Kris, *Chem. Ind. (London)* p. 1834 (1961).

[l] S. A. Harris, D. Heyl, and K. Folkers, *J. Am. Chem. Soc.* **66**, 2088 (1944).

[m] V. W. Rodwell, B. E. Volcani, M. Ikawa, and E. E. Snell, *J. Biol. Chem.* **233**, 1548 (1958).

[n] E. E. Snell and A. N. Rannefeld, *J. Biol. Chem.* **157**, 475 (1945).

[o] M. Fujioka, Y. Morino, and Y. Sakamoto, *J. Biochem. (Tokyo)* **49**, 333 (1961).

[p] P. D. Sattangi and C. J. Argoudelis, *J. Org. Chem.* **33**, 1337 (1968).

[q] W. Korytnyk and B. Paul, *J. Heterocyclic Chem.* **2**, 144 (1965).

[r] B. Paul and W. Korytnyk, *Tetrahedron* **25**, 1071 (1969).

[s] W. F. Korner and H. Nowak, *Z. Arzneimittelforsch.* **17**, 572 (1967).

[t] R. Koch, *Acta Chem. Scand.* **12**, 1873 (1958).

[u] Anon., *Angew. Chem. (Intern. Ed. Engl.)* **3**, 68 (1964).

[v] G. Schorre, U.S. Patent 3,086,023; *Chem. Abstr.* **59**, 9995c (1963).

[w] H. Ahrens, N. Angelino, W. Korytnyk, and B. Paul, *Abstr. 158th Am. Chem. Soc. Meeting, New York, Sept. 1969,* Paper No. 48 of Med. Chem. Div.

[x] A. Bloch, unpublished observations, 1969.

[y] W. Korytnyk, *J. Med. Chem.* **8**, 112 (1965).

[z] C. Iwata and D. E. Metzler, *J. Heterocyclic Chem.* **4**, 319 (1967).

[aa] S. L. Blethen, E. A. Boeker, and E. E. Snell, *J. Biol. Chem.* **243**, 1671 (1968).

[bb] W. Korytnyk and B. Lachmann, unpublished results, 1968.

[cc] W. Korytnyk and B. Lachmann, unpublished results, 1969.

[dd] T. Kuroda, *Bitamin* **28**, 21 (1963); *Chem. Abstr.* **62**, 515g (1965).

[ee] M. Matsuda and K. Makino, *Biochim. Biophys. Acta* **48**, 194 (1961).

[ff] W. Dowhan and E. E. Snell, to be published (1970).

[gg] W. Korytnyk and B. Lachmann, unpublished results, 1968.

[hh] F. S. Furbish, M. L. Fonda, and D. E. Metzler, *Biochemistry* **8**, 5169 (1969).

[ii] T. L. Hullar, *Tetrahedron Letters* p. 4921 (1967); *J. Med. Chem.* **12**, 58 (1969).

[jj] M. Fonda, unpublished results, 1970.

[kk] T. L. Fisher and D. E. Metzler, *J. Am. Chem. Soc.* **91**, 5323 (1969).

TABLE VI

VITAMIN B_6 COMPOUNDS MODIFIED IN THE 6-POSITION

No.	Compound	Structure	Synthesis[a]	Vitamin B_6 activity[a,b]
VI-1	6-Chloropyridoxol		c	Convulsant agent in rats[d]
VI-2	6-Methylpyridoxal 5'-phosphate		e	Replaces the cofactor in aspartate transaminase[f] and in D-serine dehydrase, tryptophanase, and arginine decarboxylase[g]
VI-3	6-Aminopyridoxol		h	Inactive against S-180 tumors in mice[h]

[a] Numbers correspond to text footnotes citing references to the literature. Superscript numbers also refer to text footnotes.

[b] Unless otherwise indicated.

[c] R. K. Blackwood, G. B. Hess, D. E. Larrabee, and F. J. Pilgrim, *J. Am. Chem. Soc.* **80**, 6244 (1958).

[d] F. Rosen, personal communication, 1962.

[e] N. D. Doktorova, L. V. Ionova, M. Ya. Karpeisky, N. Sh. Padyukova, K. F. Turchin, and V. L. Florentiev, *Tetrahedron* **25**, 3527 (1969).

[f] A. L. Bocharov, V. I. Ivanov, M. Ya. Karpeisky, O. K. Mamaeva, and V. L. Florentiev, *Biochem. Biophys. Res. Commun.* **30**, 459 (1968).

[g] E. E. Snell, personal communication, 1968.

[h] A. R. Katritzky, H. Z. Kucharska, M. J. Tucker, and H. M. Wuest, *J. Med. Chem.* **9**, 620 (1966).

been described.[25,26] A 5-liter Pyrex bottle (Standard taper 45) is provided with a two-hole rubber stopper. A gas dispersion tube is inserted into one hole and connected to a hydrogen chloride gas cylinder by means of dry Tygon or rubber tubing. The other hole contains a short piece of glass

[25] J. L. Greene, Jr., A. M. Williams, and J. A. Montgomery, *J. Med. Chem.* **7**, 20 (1964).

[26] J. Baddiley and A. P. Mathias, *J. Chem. Soc.* p. 2383 (1952); A. Cohen and E. G. Hughes, *J. Chem. Soc.* p. 4384 (1952).

tubing to which a drying tube filled with "Drierite" (anhydrous $CaSO_4$) is attached. The apparatus and reagents are carefully dried.

Next 100.0 g of dry pyridoxol hydrochloride is put into the bottle, and 1–2 liters of dry acetone is distilled in on top of the solid. The bottle is then placed in an ice bath for ca. 30 minutes. A rapid stream of hydrogen chloride gas is passed into the suspension until the solution turns bright yellow (about 20–30 minutes). At this point the acetone is saturated with hydrogen chloride,[27] and the gas escapes freely. The bottle is stoppered with a well-greased glass stopper and shaken for 60–75 minutes at room temperature in a box-carrying reciprocating shaker. The bottle can now be placed in a refrigerator or freezer overnight.

To the chilled suspension, 500 ml of ether is added; the product is filtered rapidly (sintered-glass filter) using suction. The crystalline material is washed with ether and dried in a vacuum desiccator; weight 113.6–114.5 g (95–96%), m.p. 209–211° (dec.).[28] It is free of the starting material, as shown by a negative Gibbs test and by thin-layer chromatography (silica gel "G"; 1:1 chloroform–methanol, as developing solvent).

The hydrochloride obtained by this procedure is converted to the free base by mixing it with an excess of a saturated solution of sodium bicarbonate. After all reaction has ceased, the mixture is filtered on a Büchner filter, and the solid is washed with cold water until the pH of the washings has fallen to about 6.

α⁴,3-O-Isopropylidene-α⁵-pyridoxyl Chloride Hydrochloride (II)

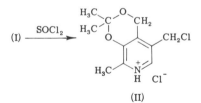

(II)

This procedure is an adaptation of that of Bennett *et al.*[28] as modified by Iwata.[29] In 220 ml of anhydrous benzene is suspended 12.3 g of iso-propylidenepyridoxol (I) hydrochloride; a solution of 6.5 g of thionyl chloride in 20 ml of dry benzene is added with stirring. The mixture is heated just to the boiling point of the solvent, then is cooled. The crude product is collected on a filter and washed with anhydrous ether. It can be

[27] At 0°C, 15–17% (w/w) of hydrogen chloride is taken up by acetone, but at room temperature, only 6–7%. α⁴, α⁵-isopropylidenepyridoxol is formed in the presence of insufficient HCl [W. Korytnyk, *J. Org. Chem.* **27**, 3724 (1962)].

[28] R. Bennett, A. Burger, and W. W. Umbreit, *J. Med. Pharm. Chem.* **1**, 213 (1959).

[29] C. Iwata, *Biochem. Prep.* **12**, 117 (1968).

TABLE VII
PYRIDOXAMINES AND ISOPYRIDOXAMINES MODIFIED ON THE α-NITROGEN

No.	Compound	Structure	Synthesis[a]	Vitamin B_6 activity[a,b]
VII-1	N-(4-Pyridoxyl)amino acids	R–CH$_2$NHCHCOOH; pyridine ring with CH$_2$OH, HO, H$_3$C, N	—	
a	-L-alanine		e, [88]	Low activity in rats[c] and microorganisms[d]
b	-DL-alanine		c	—
c	-β-alanine		c	Inhibits pyridoxamine–pyruvate transaminase;[18] weak inhibitor of growth of S-180 cells *in vitro*[f]
d	-L-arginine		e	—
e	-L-aspartic acid		e	—
f	-DL-aspartic acid		c	—
g	-L-asparagine		c	—
h	-L-glutamic acid		c, e	(as the pyrrolidone carboxylic acid[e])
i	-DL-glutamic acid		c	—
j	-glycine		c	—
k	-DL-isoleucine		c	—
l	-L-leucine		c	—
m	-DL-leucine		c	—
n	-DL-norleucine		c	—
o	-L-lysine		c	—
p	α-N-(4-Pyridoxyl)lysine		g	—
q	ε-N-(4-Pyridoxyl)lysine		g	Formed after reduction and hydrolysis of phosphorylase[h] cystathionase,[i] glutamic–aspartic transaminase[i,j]

	Compound	Reference		Notes
r	ε-N-(4-Pyridoxyl)-DL-lysine	k, l	—	
s	α,ε-N-Di(4-pyridoxyl)lysine	g	—	
t	-DL-methionine	c	—	
u	-DL-phenylalanine	e	—	
v	-L-serine	e	—	
w	-DL-serine	e	—	
x	-DL-threonine	c	—	
y	-L-tryptophan	e	—	
z	-DL-tryptophan	c	—	
aa	-L-tyrosine	c, e	—	
bb	-DL-valine	c	—	
VII-2	N-(5-Phospho-4-pyridoxyl) amino acids	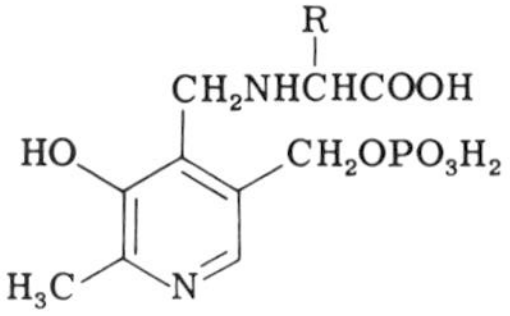 —	—	
a	-L-alanine	e, [88]	—	
b	-L-arginine	e	—	
c	-L-aspartic acid	e	—	
d	-L-glutamic acid	e	—	(As the pyrrolidone carboxylic acid[e]) Phosphorylates apo glutamic-aspartic transaminase[m]
e	ε-N-(5-Phospho-4-pyridoxyl)lysine	g		
f	-L-serine	e		
g	-L-tryptophan	e		Promotes conversion of tryptophanase dimer to tetramer[n]
h	-L-tyrosine	e	—	
VII-3	N-(4-Pyridoxyl)amines	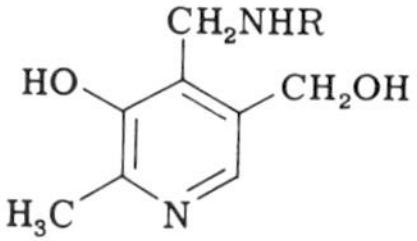 —		Most of the compounds show 50–100% activity in rats;[o,p] negligible activity in yeast and bacteria;[q] some show activity in *Neurospora*[q]

Structures:

VII-2 (R—CH₂NHCHCOOH substituent at 4-position; CH₂OPO₃H₂ at 5-position; 3-OH; 2-CH₃):

CH₂NHCHCOOH
|
R

CH₂OPO₃H₂ ... HO ... H₃C—

VII-3 (CH₂NHR at 4-position; CH₂OH at 5-position; 3-OH; 2-CH₃):

CH₂NHR ... CH₂OH ... HO ... H₃C—

(Continued)

TABLE VII (Continued)

No.	Compound	Structure	Synthesis[a]	Vitamin B_6 activity[a,b]
a	-methylamine		o	—
b	-ethylamine		o	—
c	-isobutylamine		o	—
d	-ethanolamine		o	—
e	-isopropanolamine		o	—
f	-aniline		o	—
g	-benzylamine		p	—
h	-β-phenylethylamine		p	—
i	-3-phenylpropylamine		o	—
j	-tryptamine		p	—
k	-tyramine		p	—
l	-histamine		p	—
m	-DL-arterenol		o	—
n	-3,4-dihydroxy-β-phenylethylamine		o	—
VII-4	Dipyridoxylamine		e	—
VII-5	N-(4-Pyridoxylidene)phenethylamine		r	Neurotoxic in mice; found in human phenylketonuric urine[r]; inhibitor of pyridoxal phosphokinase in brain[s]

VII-6	4-Pyridoxylhydrazine	(CH$_2$NHNH$_2$ pyridoxine structure)	t	Inhibits human neoplastic cells in vitro[t]
VII-7	sym-Di-(4-pyridoxyl)hydrazine	(CH$_2$NH— pyridoxine structure)$_2$	5, [88]	Inhibits pyridoxal phosphokinase[5]
VII-8	Pyridoxal hydrazone	—	5, [88]	—
VII-9	Pyridoxal azine	(CH=N— pyridoxine structure)$_2$	5, [88]	Inhibits pyridoxal phosphokinase[5]
VII-10	Pyridoxal methylhydrazone	—	u	Borderline retardation of tumor growth[u]
VII-11	Pyridoxal dimethylhydrazone	—	u	Borderline retardation of S-180 tumor growth[u]
VII-12	Pyridoxal phosphate dimethylhydrazone	—	u	—
VII-13	Pyridoxal nicotinoylhydrazone	—	v	Active against mammary cancer in mice and certain leukemia in mice[w]
VII-14	Pyridoxal isonicotinoylhydrazone	—	v	Same as for VII-13
VII-15	Pyridoxal phosphate isonicotinoylhydrazone	—	u	
VII-16	Pyridoxal semicarbazone	—		Inhibits pyridoxal phosphokinase[5]
VII-17	Pyridoxal thiosemicarbazone	—	x	Inhibits Mycobacterium tuberculosis[x]

(Continued)

TABLE VII (*Continued*)

No.	Compound	Structure	Synthesis[a]	Vitamin B_6 activity[a,b]
VII-18	Pyridoxal phenylthiosemicarbazone	—	x	Same as for VII-17
VII-19	Pyridoxal p-methoxyphenylthiosemicarbazone	—	x	Same as for VII-17
VII-20	Pyridoxal p-ethoxyphenylthiosemicarbazone	—	x	Same as for VII-17
VII-21	3-Hydroxy-2-methyl-5-(α-methylphenethylaminomethyl)-4-pyridinemethanol ("pyridoxyphen")	(structure: pyridine ring with CH_2OH, HO, H_3C, N, and $CH_2NHCHCH_2\phi$ / CH_3 substituents)	y	Hypotensive activity due to adrenolytic properties[a]
VII-22	Isopyridoxal hydrazone	—	16, [88]	Inhibits pyridoxal phosphokinase[16]
VII-23	Isopyridoxal azine	(structure: pyridine ring with CH_2OH, HO, H_3C, N, and CH=N— substituent, subscript 2)	16, [88]	Inhibits pyridoxal phosphokinase[16]
VII-24	Isopyridoxal dimethylhydrazone	—	16	Inhibits pyridoxal phosphokinase[16]
VII-25	Isopyridoxal phenylhydrazone	—	16	Inhibits pyridoxal phosphokinase[16]; inhibitor of growth of S-180 cells in *in vitro* cultures[f]
VII-26	Isopyridoxal phenylsemicarbazone	—	16	Inhibits pyridoxal phosphokinase[16]

[a] Numbers correspond to text footnotes citing references to the literature; articles in this volume are cited by article number enclosed in brackets [ ]. Superscript numbers also refer to text footnotes.
[b] Unless otherwise indicated.

c D. Heyl, S. A. Harris, and K. Folkers, J. Am. Chem. Soc. 70, 3429 (1948).

d E. E. Snell and J. C. Rabinowitz, J. Am. Chem. Soc. 70, 3432 (1948).

e M. Ikawa, Arch. Biochem. Biophys. 118, 497 (1967).

f M. Hakala, preliminary observations, 1968.

g E. H. Fischer, A. W. Forrey, J. L. Hedrick, R. C. Hughes, A. B. Kent, and E. G. Krebs, in "Chemical and Biological Aspects of Pyridoxal Catalysis" (E. E. Snell et al., eds.), p. 543. Macmillan, New York, 1963.

h E. H. Fischer, A. B. Kent, E. R. Snyder, and E. G. Krebs, J. Am. Chem. Soc. 80, 2906 (1958).

i E. H. Fischer and E. G. Krebs, Abstr. 136th Meeting ACS, September, 1959, p. 24c.

j C. Turano, P. Fasella, P. Vecchini, and A. Giartosio, Atti. Accad. Nazl. Lincei. Rend. Cl. Sci. Fis. Mat. Nat., 30, 532 (1961).

k O. L. Polyanovskii, Biokhimiya 28, 903 (1963); Biochemistry (USSR) (Engl. transl.) 28, 751 (1963).

l W. B. Dempsey and H. N. Christensen, J. Biol. Chem. 237, 1113 (1962).

m R. M. Khomutov, E. S. Severin, E. N. Khurs, and N. N. Gulaev, Biochim. Biophys. Acta 171, 201 (1969).

n Y. Morino and E. E. Snell, J. Biol. Chem. 242, 5591 (1967).

o D. Heyl, E. Luz, S. A. Harris, and K. Folkers, J. Am. Chem. Soc. 74, 414 (1952).

p D. Heyl, E. Luz, S. A. Harris, and K. Folkers, J. Am. Chem. Soc. 70, 3669 (1948).

q J. C. Rabinowitz and E. E. Snell, J. Am. Chem. Soc. 75, 998 (1953).

r Y. H. Loo, J. Neurochem. 14, 813 (1967).

s Y. H. Loo and V. P. Whittaker, J. Neurochem. 14, 997 (1967).

t E. Testa, A. Bonati, and G. Pagani, Chimia 15, 314 (1961).

u R. H. Wiley and G. Irick, J. Med. Pharm. Chem. 5, 49 (1962).

v P. P. T. Sah, J. Am. Chem. Soc. 76, 300 (1954).

w B. Freedlander and A. Furst, footnote in reference 120.

x P. P. T. Sah and C. T. Peng, Arch. Pharm. 293, 501 (1960).

y S. Ya. Arbuzov and S. M. Smirnova, Farmakol. Toksikol. 27, 420 (1964) [Chem. Abstr. 62, 2140e (1965)].

z S. Ya. Arbuzov, A. E. Aleksandrova, and S. M. Smirnova, Farmakol. Toksikol. 29, 521 (1966), Chem. Abstr. 66, 9764 (1967). S. Y. Arbuzuv, A. G. Gorodnik, and M. I. Nikiforov, Farmakol. Toksikol. 31, 152 (1968), Chem. Abstr. 69, 9594d (1968).

TABLE VIII

VITAMIN B_6 COMPOUNDS MODIFIED IN MULTIPLE POSITIONS AND IN THE RING

Pyridine Analogs with Multiple Modifications

No.	Compound	Structure	Synthesis[a]	Vitamin B_6 activity[a,b]
VIII-1	4-Formyl-1-methylpyridinium iodide		c	Active in model systems[c]
VIII-2	3-Hydroxypyridine-4-carboxaldehyde		d, e	Active in model systems;[f] active as substrate for pyridoxamine–pyruvate transaminase[g]
VIII-3	3-Hydroxypyridine-2-carboxaldehyde		d, h	Active model systems[h]
VIII-4	3,4-Dideoxypyridoxol		i	Weak vitamin B_6 antagonist in *Neurospora sitophila*[i]
VIII-5	3-O-Methyl-2-norpyridoxol		k	Inhibits *Escherichia coli*[j]

	Name	Structure	Ref.	Remarks
VIII-6	3-Hydroxy-6-methyl-2-pyridyl-methanol	HO—, CH$_3$, HOH$_2$C— pyridine	l	Inhibitor of *Saccharomyces cerevisiae* G.M.[l] Assignment of the structure of this compound, originally incorrect,[l] has been corrected[m]
VIII-7	4,5-Dideoxypyridoxol	CH$_3$, CH$_3$, HO—, H$_3$C— pyridine	n, [88]	Inactive or inhibitory in rats, tomato roots, and microorganisms[2a]
VIII-8	ω-(4-Deoxy-5-pyridoxyl)alkanols	CH$_3$, (CH$_2$)$_{2-3}$OH, HO—, H$_3$C— pyridine	o, [88]	Inhibits *S. carlsbergensis*[p]
VIII-9	5-Formyl-3-hydroxy-2-methylpyridine-4-carboxylic acid	COOH, CHO, HO—, H$_3$C— pyridine	q	Metabolite of pyridoxamine[7c]
VIII-10	3-Hydroxy-2-methylpyridine-4,5-dicarboxylic acid	COOH, COOH, HO—, H$_3$C— pyridine	q, r	Metabolite of pyridoxamine[7c]
VIII-11	3-Hydroxy-2-methylpyridine-5-carboxylic acid	COOH, HO—, H$_3$C— pyridine	$7c$, s, t	Metabolite of pyridoxamine[7c]

(Continued)

TABLE VIII (*Continued*)

No.	Compound	Structure	Synthesis[a]	Vitamin B_6 activity[a,b]
VIII-12	5-Chloromethyl-4-cyano-2-methyl-3-pyridinol		u	Potent *E. coli* growth inhibitor[j]; Inhibitor of growth of S-180 cells *in vitro*[v]
VIII-13	3-Amino-4-(ethoxymethyl)-2-ethyl-5-pyridylmethylamine		l	Pyridoxal phosphokinase inhibitor,[15] inhibitor of *S. cerevisiae* G.M.[l]
VIII-14	N^5,N^5-Bis(2-chloroethyl)-O^4-methylisopyridoxamine (methoxypyridoxyl nitrogen mustard)		w	Inhibits S-180 tumor growth[x]
VIII-15	5-(2-Chloroethylthiomethyl)-2,4-di-methyl-3-pyridinol		y	Inhibits *S. carlsbergensis*, no anti-tumor effect[y]
VIII-16	N^5,N^5-Bis(2-chloroethyl)-4-deoxy-isopyridoxamine (4-deoxy-pyridoxyl nitrogen mustard)		w	Weak antitumor effect[z]

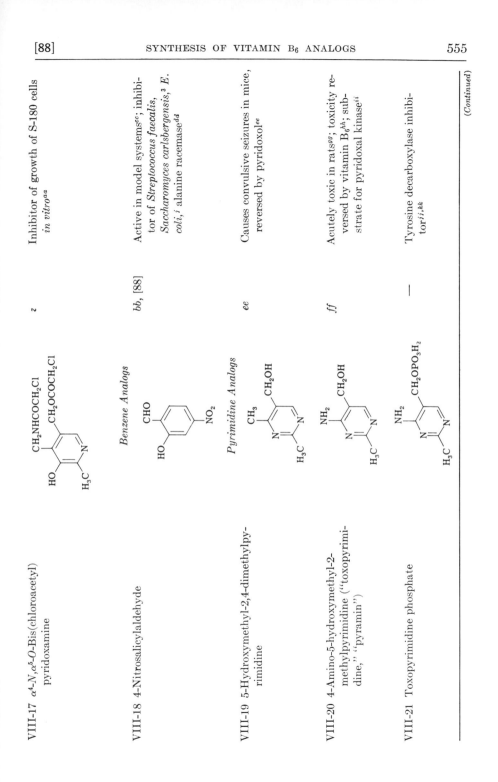

VIII-17 α^4-N,α^5-O-Bis(chloroacetyl) pyridoxamine		z	Inhibitor of growth of S-180 cells in vitro[aa]
Benzene Analogs			
VIII-18 4-Nitrosalicylaldehyde		bb, [88]	Active in model systems[cc]; inhibitor of *Streptococcus faecalis*, *Saccharomyces carlsbergensis*,[3] *E. coli*;[i] alanine racemase[dd]
Pyrimidine Analogs			
VIII-19 5-Hydroxymethyl-2,4-dimethylpyrimidine		ee	Causes convulsive seizures in mice, reversed by pyridoxol[ee]
VIII-20 4-Amino-5-hydroxymethyl-2-methylpyrimidine ("toxopyrimidine," "pyramin")		ff	Acutely toxic in rats[gg]; toxicity reversed by vitamin B₆[hh]; substrate for pyridoxal kinase[ii]
VIII-21 Toxopyrimidine phosphate		—	Tyrosine decarboxylase inhibitor[i,kk]

(Continued)

TABLE VIII (Continued)

No.	Compound	Structure	Synthesis[a]	Vitamin B_6 activity[a,b]
VIII-22	N-Methyltoxopyrimidine			B_6 antagonist[ll]
VIII-23	4-Amino-2,5-dimethylpyrimidine ("5-deoxypyramin")		mm	Active as B_6 antagonist[nn]
VIII-24	4-Amino-5-hydroxymethyl-2-(methylmercapto)pyrimidine ("methioprim")		oo	B_6 inhibitor in animals;[pp] antitumor agent[qq]
VIII-25	2-Chloro-4-(dimethylamino)-6-methylpyrimidine (Castrix)		rr	Potent vitamin B_6 antagonist in mice;[ss] inhibitor of E. coli

[a] Numbers correspond to text footnotes citing references; articles in this volume are cited by article number enclosed in brackets [ ].
[b] Unless otherwise indicated.
[c] J. R. Maley and T. C. Bruice, J. Am. Chem. Soc. 90, 2843 (1968).

[d] D. Heinert and A. E. Martell, Tetrahedron 3, 49 (1958).
[e] T. C. French, D. S. Auld, and T. C. Bruice, Biochemistry, 4, 77 (1965).
[f] D. S. Auld and T. C. Bruice, J. Am. Chem. Soc. 89, 2083, 2090, 2098 (1967).

g J. E. Ayling and E. E. Snell, *Biochemistry* **7**, 1626 (1968).

h D. E. Metzler, M. Ikawa, and E. E. Snell, *J. Am. Chem. Soc.* **76**, 648 (1954).

i R. P. Mariella and J. L. Leech, *J. Am. Chem. Soc.* **71**, 331 (1949).

j L. J. Arcement, W. Korytnyk, and W. B. Dempsey, *Bact. Proc.* p. 121 (1968) and unpublished results, 1968.

k W. Korytnyk, unpublished results, 1968.

l G. J. Martin, S. Avakian, and J. Moss, *J. Biol. Chem.* **174**, 495 (1948).

m A. Stempel and E. C. Buzzi, *J. Am. Chem. Soc.* **71**, 2969 (1949).

n S. A. Harris, *J. Am. Chem. Soc.* **62**, 3203 (1940).

o H. Ahrens, W. Korytnyk, and B. Lachmann, *Abstr. 157th Am. Chem. Soc. Meeting, April 1969*, MED-11.

p A. Bloch, unpublished observations, 1969.

q B. Paul and W. Korytnyk, *Chem. Ind. (London)*, p. 230 (1967).

r R. G. Jones and E. C. Kornfeld, *J. Am. Chem. Soc.* **73**, 107 (1951).

s C. J. Argoudelis and F. A. Kummerow, *J. Org. Chem.* **26**, 3420 (1961).

t D. Palm, A. A. Smucker, and E. E. Snell, *J. Org. Chem.* **32**, 826 (1967).

u D. Heyl, *J. Am. Chem. Soc.* **70**, 3434 (1948).

v M. Hakala, preliminary observations, 1968.

w E. Wilson and M. Tishler, unpublished results mentioned in footnote l.

x C. C. Stock, S. Buckley, K. Sugiura, and C. P. Rhoads, *Cancer Res.* **11**, 432 (1951).

y J. L. Greene, Jr., A. M. Williams, and J. A. Montgomery, *J. Med. Chem.* **7**, 20 (1964).

z W. Korytnyk and B. Paul, unpublished results, 1969.

aa M. Hakala, preliminary observations, 1968.

bb E. M. Bavin, R. J. W. Rees, J. M. Rosbon, M. Seiler, D. E. Seymour, and D. Suddaby, *J. Pharm. Pharmacol.* **2**, 764 (1950).

cc M. Ikawa and E. E. Snell, *J. Am. Chem. Soc.* **76**, 653 (1954).

dd J. Olivard and E. E. Snell, *J. Biol. Chem.* **213**, 203 (1955).

ee T. J. Schwan, H. Tieckelmann, J. F. Holland, and B. Bryant, *J. Med. Chem.* **8**, 750 (1965).

ff T. L. V. Ulbricht, *Progr. Nucleic Acid Res.* **4**, 189 (1965). (A review of various methods of synthesis of toxopyrimidine, methioprim, and analogous structures.)

gg R. Abderhalden, *Klin. Wochschr.* **18**, 171 (1939); S. Morii, *Biochem. Z.* **309**, 354 (1941).

hh K. Makino, T. Kinoshita, Y. Aramaki, and S. Shintani, *Nature* **174**, 275 (1954).

ii M. Tsubosaka, *Bitamin* **33**, 293 (1966); *Chem. Abstr.* **64**, 14507a (1966).

ii K. Makino and M. Koike, *Nature* **174**, 1056 (1954).

kk K. Makino and M. Koike, *Enzymologia* **17**, 157 (1954).

ll A. Schellenberger and K. Winter, *Z. Physiol. Chem.* **322**, 173 (1960).

mm S. Shintani, *J. Pharm. Soc. Japan* **77**, 746 (1957).

nn T. Sakuragi and F. A. Kummerow, *Arch. Biochem. Biophys.* **82**, 89 (1959).

oo T. L. V. Ulbricht and C. C. Price, *J. Org. Chem.* **21**, 567 (1956).

pp F. Rosen, J. F. Holland, and C. A. Nichol, *Proc. Am. Assoc. Cancer Res.* **2**, 243 (1957).

qq J. F. Holland, R. Guthrie, P. Sheele, and H. Tieckelmann, *Cancer Res.* **18**, 776 (1958).

rr K. Westphal, *Chem. Abstr.* **36**, 9113 (1942).

ss O. Karlog and E. Knudsen, *Nature* **200**, 790 (1963).

used as such for further reactions. If desired, it can be recrystallized from anhydrous acetone; m.p. 191–192°; yield 12 g (95%).

α⁴,3-O-Isopropylidene-5-deoxypyridoxol Hydrochloride (III)

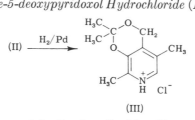

(III)

This procedure was originally described by Korytnyk *et al.*[12] A solution of 0.75 g of α^4,3-*O*-isopropylidene-α^5-pyridoxyl chloride hydrochloride (II) in 10 ml of cold water is made alkaline with $NaHCO_3$ and is extracted into ethyl acetate. The ethyl acetate solution is evaporated *in vacuo*, and the residue is dissolved in alcohol (40 ml) and is hydrogenated with H_2 at 2.1 kg/cm² for 1 hour in the presence of 10% Pd-C (250 mg) in a Parr hydrogenation apparatus. The catalyst is removed by filtration, the filtrate is evaporated to dryness *in vacuo*, and the residue is crystallized as the hydrochloride from an alcohol–ether mixture; yield, 0.58 g (91%); m.p. 215–216° (free base, m.p. 88–89°, from petroleum ether).

5-Deoxypyridoxol (V-1)

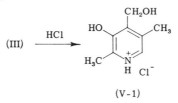

(V-1)

This method was described by Korytnyk *et al.*[12] A solution of 120 mg of α^4,3-*O*-isopropylidene-5-deoxypyridoxol hydrochloride (III) in 15 ml of 0.1 *N* HCl is heated on a steam bath for 1 hour. The acid solution is then cooled, filtered, and evaporated to dryness under reduced pressure. The residue is crystallized from an alcohol–ether mixture, and yields 90 mg (91%) of the desired compound, m.p. 143–144°.

α^4-Methylpyridoxol (IV-18)[30]

α^5-O-Benzyl-α^4,3-O-isopropylidenepyridoxol (IV)

α^4,3-*O*-Isopropylidenepyridoxol free base [(I) 10 g] is added to a suspension of sodium hydride (13.7 g of a 53% suspension in mineral oil, washed

[30] The method has been described by Korytnyk and Paul.[11]

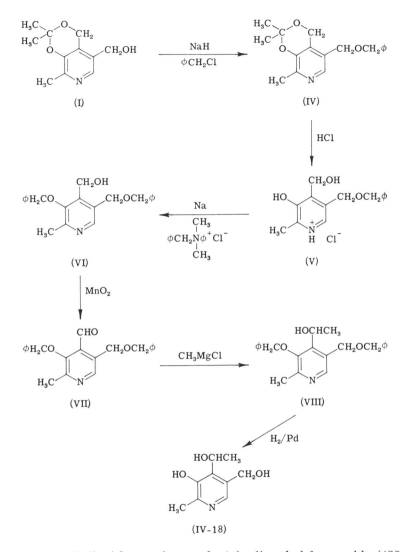

(IV-18)

free of mineral oil with petroleum ether) in dimethyl formamide (480 ml, purified by distillation over CaH$_2$) while the reaction mixture is being stirred and heated to 65°. The mixture is then gradually cooled to 45° over a 90-minute period, the heat is removed, and the flask is put into ice. Benzyl chloride (7.15 ml) is added dropwise, and the mixture is stirred overnight at 0°. After careful addition of water, the solution is extracted five times with petroleum ether. The petroleum ether extracts are dried and evaporated, yielding oily α⁵-O-benzyl-α⁴,3-O-isopropylidenepyridoxol (IV).

α^5-O-Benzylpyridoxol (V)

The oily α^5-O-benzyl-α^4,3-O-isopropylidenepyridoxol is dissolved in 100 ml of 1 N hydrochloric acid, and the mixture is heated on a steam bath for 1 hour. The aqueous layer is separated from any oily residue and is evaporated under reduced pressure. The residue from evaporating the aqueous layer is dissolved in ethanol, and crystallizes on the addition of ether; m.p. 152–153°. The free base is obtained by dissolving the hydrochloride in water, adding $NaHCO_3$ until the solution is basic, and extracting the aqueous solution with ethyl acetate. Evaporation and crystallization from ether provide the free base, m.p. 117–118°, in 78% yield (9.68 g).

3,α^5-O-Dibenzylpyridoxol (VI)

A solution of 14.25 g of benzyldimethylphenylammonium chloride in 30 ml of methanol is added to a solution of 1.65 g of sodium in 30 ml methanol, and then a solution of 9.6 g of α^5-O-benzylpyridoxol (V) in 100 ml of methanol is added. The mixture is allowed to stand for 20 minutes, and is then added over a period of 30 minutes to ca. 750 ml of hot (ca. 100°) toluene. During this time, volatile material slowly distills off (65–100°). When ca. 400 ml of residual toluene is left, the mixture is cooled, the toluene solution is decanted off, and the residue is washed with fresh toluene. The combined toluene solutions are evaporated *in vacuo* to an oil, which is taken up in a minimum volume of ether. α^5,3-O-Dibenzylpyridoxol (8.5 g, m.p. 69.5–72°) crystallizes out. The mother liquor is evaporated to an oil and subjected to steam distillation. The residue is then extracted with ether, and the extract is washed with water. After drying with $CaSO_4$, petroleum ether is added to the extract, precipitating additional dibenzylpyridoxol (2.0 g, m.p. 64–69°). The combined yield is 81%.

3,α^5-O-Dibenzylpyridoxal (VII)

To a stirred and cooled (ice bath) chloroform solution (500 ml, dry) of 3,α^5-O-dibenzylpyridoxol (10.06 g), freshly prepared active MnO_2 (60 g, prepared by heating $MnCO_3$ at 280–300° for 36–48 hours) suspended in dry chloroform (200 ml) is added. The mixture is stirred at room temperature for 17 hours; then thin-layer chromatography (TLC) (with ethyl acetate as solvent, the aldehyde has an R_f of 0.9, and the alcohol an R_f of 0.7) indicates the absence of the starting alcohol. The mixture is filtered using Celite filter aid, the residue is washed with chloroform, and the chloroform filtrates are evaporated *in vacuo* to a viscous oil. Addition of a few milliliters of ether results in crystallization, yielding 9.80 g (98.5%) of the hydrated aldehyde, m.p. 60–70° (72° after drying).

α^4-Methyl-$3\alpha^5$-O-dibenzylpyridoxol (VIII)

To a stirred suspension of 3,α^5-O-dibenzylpyridoxal (4.01 g) in anhydrous ether (50 ml), methylmagnesium chloride (12 ml of a 1.7 M solution in ether) is added dropwise, and the mixture is stirred overnight (ca. 12 hours) at room temperature. The reaction mixture is poured into 100–175 ml of an ice-water solution of ammonium chloride (24 g), allowed to stand for a few minutes, and extracted several times with ether. The combined extracts are washed with water, dried (over MgSO₄), and evaporated. The product crystallizes, yielding 3.82 g (90%), m.p. 72–74°. Recrystallization from acetonitrile raises the melting point to 83–84°.

α^4-Methylpyridoxol (IV-18)

α^4-Methyl-3,α^5-O-dibenzylpyridoxol (0.80 g) is dissolved in 67 ml of alcohol and is hydrogenolyzed in the presence of 1.33 g of palladium on charcoal. After 4 days the reaction is complete, and the solvent is evaporated *in vacuo*. The oily residue is taken up in a small amount of ethanol. The product precipitates as the hydrochloride on the addition of ether containing hydrogen chloride. The yield is 0.29 g (60%), m.p. 177–178°.

Pyridoxol 5'-Phosphate (IV-1)

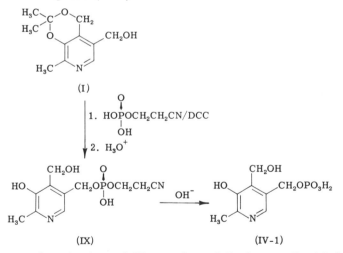

The procedure is that of Korytnyk and Lachmann.[31] α^4,3-O-Isopropylidenepyridoxol (I; 420 mg, 2 millimoles) and 2-cyanoethyl phosphate (520 mg, 3.5 millimoles, prepared by the method of Tener[32]) are dissolved

---

[31] W. Korytnyk and B. Lachmann, unpublished results.
[32] G. M. Tener, *J. Am. Chem. Soc.* **83**, 159 (1961).

in dry pyridine (8 ml). *N,N'*-Dicyclohexylcarbodiimide (1.25 g, 6 milli-moles) is added, and the solution is stirred at room temperature for 5 hours, moisture being excluded (drying tube). Addition of 2 ml of water, with stirring for another hour, produces dicyclohexylurea, which is filtered off, and the filtrate is evaporated *in vacuo*. The residue is dissolved in acetonitrile, leaving additional undissolved dicyclohexylurea, which is filtered off, and the filtrate is evaporated. The oily reaction product is dissolved in water and extracted with $CHCl_3$ in order to remove any un-reacted starting material. The aqueous layer is brought to pH 2 with 2 N HCl. After standing overnight, the solution is evaporated *in vacuo*. The residue is dissolved in 1 N LiOH (20 ml), and is refluxed for 1 hr while the pH is maintained well above 10. The solution is filtered, evaporated to a small volume, and applied to a Bio-Rad AG 50W × 8 column in the H+ form. Water elutes the phosphoric acid and 2 N HCl the phosphorylated compound which is contaminated with LiCl. Gibbs-positive fractions are combined, evaporated, and partially dissolved in ethyl alcohol. The un-dissolved LiCl is filtered off, the filtrate evaporated, and the residue dis-solved in a small amount of H_2O and added to an Amberlite CG-50 H+ column, which is eluted with water. The Gibbs-positive fractions are combined (except the first fraction), filtered, evaporated *in vacuo*, and crystallized from MeOH. The yield of pyridoxol phosphate is 350 mg (70%). This method appears to be general, and has been utilized for the synthesis of several phosphorylated pyridoxol analogs, including pyridoxol 5'-(meth-ylphosphonate) and 5'-homopyridoxol phosphate.

To obtain the intermediate pyridoxol 2-cyanoethyl 5'-phosphate (IX), the procedure is followed as described for pyridoxol 5'-phosphate through the chloroform extraction step. Then the aqueous layer is kept at room temperature for 2 days to hydrolyze the isopropylidene group. Water is removed *in vacuo*, and the oily residue is crystallized from MeOH-EtOH. The product, m.p. 188° (dec.) is obtained in 34% yield (300 mg from 630 mg of $\alpha^4,3$-*O*-isopropylidenepyridoxol).

4-Deoxypyridoxol and Some Homologs (IV-9, VIII-8)

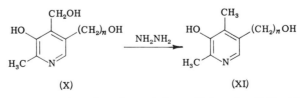

(X)   (XI)

The following method is adapted from Taborsky.[33] Pyridoxol hydro-chloride (X, $n = 1$; 1.09 g) is refluxed with hydrazine (7.2 g, dried by

[33] R. G. Taborsky, *J. Org. Chem.* **26,** 596 (1961).

standing over KOH, filtered, and distilled) for 17 hours, moisture being excluded. Excess hydrazine is removed *in vacuo* (oil pump, water bath temperature rising to 90°). The resulting crystalline product is dissolved in 7 ml of boiling methanol, and the solution is cooled in ice. Hydrazine hydrochloride (m.p. 91–92°) is removed by filtration. About two-thirds of the alcohol is distilled off, and additional hydrazine hydrochloride precipitates and is filtered off. The filtrate is treated with 2.0 ml of 10% HCl in methanol and is left at 40° to crystallize. Filtration gives colorless crystals (XI, $n = 1$; 67%, 0.673 g), which are recrystallized twice from methanol: m.p. 271° (lit.[33] m.p. 273°). Addition of ether to the mother liquor should give more crystalline fractions, which are contaminated with hydrazine hydrochloride and traces of pyridoxol. Synthesis of 4-deoxypyridoxol analogs (XI, $n = 2, 3$) can be accomplished by analogous procedures.[34]

4-Pyridoxic Acid (IV-5)

This method was described by Ahrens and Korytnyk.[35] Pyridoxol hydrochloride (0.991 g) is added to alcoholic potassium hydroxide (1.00 g of potassium hydroxide in 100 ml of absolute ethanol), and the mixture is stirred magnetically with 10.0 g of active manganese dioxide[36] for 3 hours at room temperature, moisture being excluded. Manganous salts are reoxidized by the addition of 5.0 ml of 30% hydrogen peroxide. The solution then becomes hot and is immediately filtered. The residue is washed with hot 0.1 N alcoholic potassium hydroxide, the wash liquid is combined with the filtrate, and the solution is neutralized to pH 7 with concentrated hydrochloric acid. The small amount of precipitated manganese dioxide is filtered off. Further acidification with concentrated hydrochloric acid to pH 4 precipitates 4-pyridoxic acid, which, after standing in a refrigerator for 6 hours, is filtered off and washed with ice water, ethanol, and ether. After thorough drying, the yield is 0.727 g (82%), m.p. 242° (dec.), raised to 256° (dec.) after recrystallization from boiling water.

4,5-Dideoxypyridoxol (VIII-7)

This method has been described by Harris.[37] The intermediate dibromo compound is prepared by the method of Ikawa.[38] Reflux 10 g of pyridoxol hydrochloride with 500 ml of 48% hydrobromic acid for 10 minutes, and then chill the mixture in an ice bath to crystallize the resulting 4,5-bis-(bromomethyl)-3-hydroxy-2-methylpyridine hydrobromide. Filter off the

[34] W. Korytnyk, H. Ahrens, and B. Paul, unpublished results, 1968.
[35] H. Ahrens and W. Korytnyk, *J. Heterocyclic Chem.* **4**, 625 (1967).
[36] O. Mancera, G. Rosenkranz, and F. Sondheimer, *J. Chem. Soc.* p. 2189 (1953).
[37] S. A. Harris, *J. Am. Chem. Soc.* **62**, 3203 (1940).
[38] M. Ikawa, unpublished results, 1960.

crystals and wash them with a small amount of 48% hydrobromic acid and then with a small amount of ice-cold water. Dry the product in a vacuum desiccator containing a dish of NaOH pellets. The yield is 14.05 g.

Dissolve the 4,5-bis(bromomethyl)-3-hydroxy-2-methylpyridine hydrobromide in ethanol and reduce it with hydrogen, using Pd-BaCO₃ catalyst. After the theoretical amount of hydrogen has been absorbed, neutralize the solution with sodium bicarbonate, and evaporate it to dryness. Extract the residue with acetone, and evaporate the extract to dryness. Dissolve the residue in ether. On concentration of the ether solution, 4,5-dideoxypyridoxol crystallizes out (m.p., 178°). The yield is 40% (from the bromo derivative). Additional amounts of the product as the hydrochloride may be obtained by passing hydrogen chloride into the ethereal mother liquor and recrystallizing the precipitate from absolute ethanol (m.p., 216°).

Pyridoxyl Amino Acids (VII-1)

N-Pyridoxyl-L-alanine (VII-1a)

The method described is that of Ikawa.[39] Add 5 millimoles each of L-alanine and pyridoxal free base to 25 ml of methanol and add 5 millimoles of 50% (w/v) KOH to effect solution. Next add 200 mg of PtO₂ catalyst, and hydrogenate the mixture at room temperature with H₂ at 1 atm until uptake of gas has ceased. The solution should turn from yellow to colorless during hydrogenation. Filter off the catalyst, add 5 millimoles of glacial acetic acid to the filtrate, and concentrate the filtrate *in vacuo* to a syrup, which should crystallize. Wash the crystals with ethanol and dry. The yield is 0.8 g.

N-(5-Phospho-4-pyridoxyl)-L-alanine (VII-2a)

The method described is that of Ikawa.[39] Dissolve 2 millimoles each of L-alanine and pyridoxal 5′-phosphate in 20 ml of water and add 6 millimoles of 50% (w/v) KOH. Hydrogenate the mixture with H₂ at 1 atm in the presence of 200 mg of PtO₂. After hydrogenation, filter the mixture and chromatograph the filtrate on a column of Amberlite CG-50 (H⁺-form, 1.6 × 22 cm). Develop the chromatogram with water, and collect 5-ml fractions. In order to detect the product, spot a drop from each fraction onto paper and, after drying, spray the paper with a solution of dichloroquinonechloroimide (0.1% in benzene). Those fractions (between 40 and 85 ml) giving a blue color changing to a shade of green (not those giving blue color that fades) are combined and evaporated *in vacuo* to dryness. Trituration of the residue with methanol yields the product (0.43 g), which

[39] M. Ikawa, *Arch. Biochem. Phys.* **118,** 497 (1967).

is purified by dissolving it in 2 ml of water and adding 3 ml of absolute ethanol. The resulting oil solidifies on washing with ethanol. The yield is 0.29 g.

Hydrazine Derivatives

Pyridoxal Hydrazone (VII-8)

The method described is that of Ikawa.[38] Dissolve 1 g of pyridoxal hydrochloride in 10 ml of water, and add the solution dropwise, with stirring, to a solution of 1 ml of 95% hydrazine in 20 ml of water. The solution should immediately turn yellow after the addition of each drop of pyridoxal solution, but quickly fade. During the addition, a white crystalline precipitate should begin to form. After the addition of the pyridoxal, allow the reaction mixture to stand at room temperature for an hour. Filter off the product, wash it with water, and dry. The yield is 0.85 g. The reaction must be carried out under basic conditions, since under acidic conditions the hydrazone is spontaneously converted into the azine.

Pyridoxal Azine (VII-9)

The method described is that of Ikawa.[38] Dissolve 1 g of pyridoxal hydrochloride in 20 ml of water. Acidify the solution with 1.2 ml of glacial acetic acid, and add 0.3 ml of 95% hydrazine. The copious yellow precipitate that results is allowed to stand for an hour, then is filtered, washed with water, and dried. The yield is 0.73 g.

sym-Di(4-pyridoxyl)hydrazine (VII-7)

The method described is that of Ikawa.[38] To 200 mg of pyridoxal azine suspended in 20 ml of 95% ethanol, add 100 mg of sodium borohydride, and stir. After 15 minutes, add an additional 100 mg of sodium borohydride, and allow the reaction mixture to stand for an hour. Filter off the white precipitate that results. Add 10 ml of water to the clear light-yellow filtrate, and acidify the solution to pH 3–4 with 2 M acetic acid. Concentrate the resulting solution to dryness *in vacuo*. To remove all the excess acetic acid, add a small amount of water to the resulting residue, and evaporate the solution again to dryness. The residue is recrystallized from water to provide *sym*-di(4-pyridoxyl)hydrazine. The yield is 120 mg.

Isopyridoxal Hydrazone (VII-22)

The method described is that of Korytnyk.[16] Dissolve 2.04 g of isopyridoxal hydrochloride in 18 ml of 0.1 M pH 4.5 sodium acetate buffer, and bring the pH to 8.5 with hydrazine. Heat the reaction mixture at 95–100° for 2–3 minutes, and then cool overnight in a refrigerator. The resulting crystals of isopyridoxal hydrazone are filtered off and are recrystallized from ethanol. The yield is 0.83 g; melting point, 175–176° (dec).

Isopyridoxal Azine (VII-23)

The method described is that of Korytnyk.[16] To a solution of 3.04 g of isopyridoxal hydrochloride in 20 ml of 0.1 M pH 4.5 sodium acetate buffer, add an equimolar amount of hydrazine. Boil the mixture for 2–3 minutes, then filter off the azine and wash it with water. The yield is 82%. The product can be recrystallized from dimethyl formamide.

4-Nitrosalicylaldehyde (VIII-18)

The method described is that of Ikawa.[38] Oxidation of the 2-hydroxy-4-nitrotoluene is carried out like that of 4-nitrotoluene.[40] To a 1-liter 3-neck flask equipped with a stirrer, reflux condenser, and thermometer, add 389 ml of acetic anhydride and 36 g (0.235 mole) of 2-hydroxy-4-nitrotoluene. Bring the mixture to refluxing, and then cool, and add 357 ml of glacial acetic acid. With the flask in an ice-salt bath, add 55 ml of concentrated sulfuric acid dropwise at such a rate that the temperature of the reaction mixture does not rise above 20°. Cool the reaction mixture to 5°, and add 65 g of CrO_3 in small portions at such a rate that the temperature does not rise above 10°.

When the temperature has dropped to 5° again, pour the reaction mixture into 3 liters of ice, and add a liter of water. Stir the mixture; when most of the ice has melted, filter the mixture, and wash the solid with cold water until the washings are colorless. To the solid add 50 ml of water, 50 ml of 95% ethanol, and 5 ml of concentrated sulfuric acid. Heat the mixture to refluxing, and add enough additional ethanol to dissolve all the solid. Continue the refluxing for 0.5 hour. Cooling should give a crystalline product, which is then filtered and washed with water. Additional amounts of solid can be obtained if the filtrate and washings are combined and the resulting oil is allowed to solidify.

Dissolve the product in ether, and extract the solution with 50 g of sodium metabisulfite ($Na_2S_2O_5$) dissolved in 250 ml of water. Reextract the ether layer with 25 g of sodium metabisulfite dissolved in 125 ml of water. Combine the sodium metabisulfite extracts and acidify by slowly adding, with stirring, 50 ml of concentrated sulfuric acid. On heating to boiling, a yellow solid is obtained. Cool the mixture; filter off the product and wash it with water. Recrystallize the product from aqueous ethanol. Yellow needles are obtained (m.p. 134–135°). The yield is 6.8 g.

[40] S. V. Lieberman and R. Conner, in "Organic Syntheses, Collective Vol. II" (A. H. Blatt, ed.), p. 441. Wiley, New York, 1943.

[89] Synthesis and Physicochemical and Coenzyme Properties of Alkyl-Substituted Analogs of the B₆ Vitamins and Pyridoxal Phosphate

By V. L. FLORENTIEV, V. I. IVANOV, and M. YA. KARPEISKY

Introduction: Analogs as a Tool for Investigation of Enzymatic Reaction Mechanisms and of Active Site Structure

The importance of studying the characteristics of enzymes reconstituted from the natural protein moiety and artificial coenzyme analogs is great, since such studies provide information on the significance of certain chemical groups of the cofactor in effecting the enzymatic reaction and in binding with the protein moiety. Alkyl analogs of pyridoxal phosphate (PLP) are of interest because they give a possibility of elucidating the puzzling role of the 2-methyl group of the coenzyme:

We refer to this group as "puzzling," since neither in some enzymatic[1]

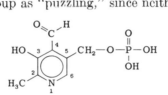

nor in model systems[2] is the methyl group essential for the reaction to occur. After the work of Snell and co-workers[1] and our studies[3] on the coenzyme properties of 2- and 6-alkyl analogs of pyridoxal phosphate, it became clear that the 2-methyl group does take some part in the binding of coenzyme to apoenzyme (apparently, by hydrophobic interaction). Although this bond is not strictly necessary for the enzyme-catalyzed reaction to proceed, it may assist in fine adjustment of the spatial interrelations between coenzyme and substrate, thus ensuring more favorable conditions for the fast initial steps of the reaction. This may be reflected in variation of Michaelis constants depending on the presence of a particular 2-alkyl analog.[3]

The subject of this paper is the synthesis of some analogs of pyridoxine (2-methyl-3-hydroxy-4,5-bis(hydroxymethyl)pyridine; PN), pyridoxamine (2-methyl-3-hydroxy-4-aminomethyl-5-hydroxymethylpyridine; PM), pyridoxal (2-methyl-3-hydroxy-4-formyl-5-hydroxymethylpyridine; PL) and

[1] Y. Morino and E. E. Snell, *Proc. Natl. Acad. Sci. U.S.* **57**, 1692 (1967).

[2] E. E. Snell, *Vitamins Hormones* **16**, 77 (1958).

[3] A. L. Bocharov, V. I. Ivanov, M. Ya. Karpeisky, O. K. Mamaeva, and V. L. Florentiev, *Biochem. Biophys. Res. Commun.* **30**, 459 (1968).

the 5'-phosphoric esters of the latter two (PLP and PMP). The latter compounds were assayed as coenzymes for aspartate aminotransferase (EC 2.6.1.1).

Synthesis of Analogs of the B_6 vitamins and Pyridoxal Phosphate

Principle of Method

A number of methods for the preparation of pyridoxine have been described in the literature. Among these a Diels-Alder condensation of 5-ethoxyoxazoles[4,5] seems to be the most appropriate. This "oxazole pathway" was used by us for a successful general procedure for the synthesis of 2- and 6-alkyl analogs of B_6 vitamins[6-9]:

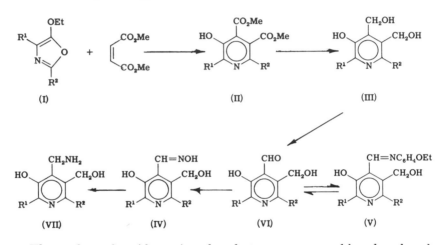

The analogs of pyridoxamine phosphate were prepared by phosphorylation of the pyridoxamine analogs with polyphosphoric acid[9,10]:

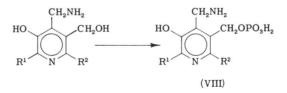

(VIII)

Syntheses of pyridoxal phosphate analogs were performed by two methods. The first method is the esterification of pyridoxal analog Schiff's

[4] E. E. Harris, R. A. Firestone, K. Pfister, R. R. Boettcher, F. I. Cross, R. B. Currie M. Monaco, E. R. Peterson, and W. Reuter, *J. Org. Chem.* **27**, 2705 (1962).
[5] R. A. Firestone, E. E. Harris, and W. Reuter, *Tetrahedron* **23**, 943 (1967).
[6] V. L. Florentiev, N. A. Drobinskaja, L. V. Ionova, and M. Ya. Karpeisky, *Tetrahedron Letters*, p. 1747 (1967).

bases with polyphosphoric acid followed by hydrolysis with an ion-exchange resin. In the second method, the analogs of pyridoxamine phosphate were transformed into pyridoxal phosphate analogs by means of transamination with glyoxylic acid in the presence of cupric acetate.

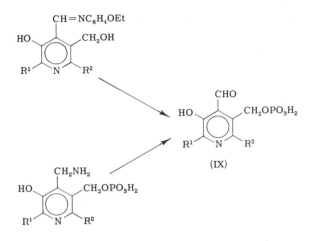

The letters a–e following a Roman numeral designating a structural formula indicate the meaning of R as follows:

(a) $R^1 = R^2 = H$

(b) $R^1 = H; R^2 = CH_3$

(c) $R^1 = R^2 = CH_3$

(d) $R^1 = iso\text{-}C_3H_7; R^2 = H$

(e) $R^1 = n\text{-}C_4H_9; R^2 = H$

Procedure

1. 5-Ethoxyoxazole (I a–e)

In a 2-liter three-necked flask fitted with a dropping funnel, mechanical stirrer, and reflux condenser are placed 340 ml of anhydrous chloroform (Note 1) and 142 g (1 mole) of phosphorus pentoxide. To the stirred mixture a solution, prepared from 0.5 mole of the required α-N-acylamino acid ethyl ester (Note 2) and 200 ml of anhydrous chloroform, is added in the

[7] N. A. Drobinskaja, L. V. Ionova, M. Ya. Karpeisky, K. F. Turchin, and V. L. Florentiev, *Dokl. Akad. Nauk SSSR* **177**, 617 (1967).

[8] N. A. Drobinskaja, L. V. Ionova, M. Ya. Karpeisky, and V. L. Florentiev, *Khim. Geterotsikl. Soedin.* p. 1028 (1969).

[9] N. A. Doktorova, L. V. Ionova, M. Ya. Karpeisky, N. Sch. Padiukova, K. F. Turchin, and V. L. Florentiev, *Tetrahedron* **25**, 3527 (1969).

[10] N. A. Drobinskaja, L. V. Ionova, M. Ya. Karpeisky, and V. L. Florentiev, *Khim. Geterotsikl. Soedin.* p. 1037 (1969).

course of 10 minutes. The stirrer is stopped and raised above the reaction mixture (Note 3). Then the mixture is heated on a steam bath under gentle reflux. The time of reaction is indicated in Table I. After completion of the heating, the flask is placed in an ice bath. With vigorous stirring, 750 ml of 20% aqueous solution of potassium hydroxide is added through the dropping funnel during 1.5 hours. After addition is completed, stirring is continued at room temperature for 30 minutes or until the solid mass has dissolved completely. The resulting solution is transferred to a separatory funnel, and the organic layer is separated. The aqueous layer is extracted with two 200-ml portions of chloroform. The combined chloroform extracts are washed with 100 ml of cold water and dried over anhydrous magnesium sulfate. Upon removal of solvent by distillation at atmospheric pressure, the residue is distilled at reduced pressure. The substances thus prepared are listed in Table I.

TABLE I
SYNTHESIS OF 5-ETHOXYOXAZOLES

Prepared 5-ethoxyoxazoles	Ethyl esters of	Time of reaction (hours)	Boiling point (°C)	Yield (%)
5-Ethoxyoxazole (I-a)	N-Formylglycine	4	74–76°/35 mm	16
2-Methyl-5-ethoxyoxazole (I-b)	N-Acetylglycine	8	83–84°/37 mm	60–64
2,4-Dimethyl-5-ethoxyoxazole (I-c)	N-Acetylalanine	6	89–90°/40 mm	56–60
4-Isopropyl-5-ethoxyoxazole (I-d)	N-Formylvaline	6	90–92°/40 mm	49–55
4-n-Butyl-5-ethoxyoxazole (I-e)	N-Formylnorleucine	6	94–96°/18 mm	44–49

Notes: 1. Technical grade chloroform is washed with concentrated sulfuric acid, then with water, dried over calcium chloride, and distilled over phosphorus pentoxide.

2. These N-acylamino acid esters were obtained by previously reported methods.[11,12]

3. When chloroform is refluxed, phosphorus pentoxide becomes a viscous scum, which sticks to the stirrer.

2. Dimethyl 5-Hydroxycinchomerates (II a–e)

A mixture of 28.8 g (0.2 mole) of dimethyl maleate (Note 1) and 0.1 mole of oxazole I (a–e) is placed in a round-bottomed flask fitted with a

[11] T. C. Sherhan and D. H. Yang, *J. Am. Chem. Soc.* **80**, 1154 (1958).
[12] R. C. Jones, *J. Am. Chem. Soc.* **71**, 644 (1949).

reflux condenser. The mixture is heated on an oil bath at 110–115°. The time of heating is indicated in Table II. The pale-yellow viscous mass is cooled then to room temperature, and 20 ml of a 25% solution of dry hydrogen chloride in absolute methanol is added. The hydrochlorides of dimethyl 5-hydroxycinchomerates are isolated by one of two alternative methods (A or B).

A. The resulting mixture is kept in a refrigerator for 2 hours. The crystalline product is filtered off with suction, washed with cold methanol and then with ether.

B. The mixture is shaken with 300 ml of anhydrous ether and allowed to crystallize in a refrigerator overnight. The crystalline product is collected by filtration and washed with ether.

The free bases II (b–e) are prepared by procedure C.

C. The carefully purified hydrochloride II (b–e) is dissolved in a minimum volume of water. To the resulting solution is added solid potassium carbonate to pH 6.5–7. The suspension is extracted with chloroform. The combined extracts are dried over magnesium sulfate, and solvent is removed by distillation at reduced pressure. The residue is dried in a vacuum desiccator over phosphorus pentoxide and paraffin.

TABLE II

SYNTHESIS OF DIMETHYL 5-HYDROXYCINCHOMERATE

Prepared dimethyl esters of	Oxazole	Time of heating (hours)	Method of isolation	Yield of hydrochloride (%)	Melting point of hydrochloride (°C)	Melting point of free base (°C)
5-Hydroxycinchomeronic acid (II-a)	5-Ethoxyoxazole	2	A	43–44	200–201	136–137
2-Methyl-5-hydroxycinchomeronic acid (II-b)	2-Methyl-5-ethoxyoxazole	6	B	28–30	123–124	57–58
2,6-Dimethyl-5-hydroxycinchomeronic acid (II-c)	2,4-Dimethyl-5-ethoxyoxazole	4	B	41–43	166–167	56–57
6-Isopropyl-5-hydroxycinchomeronic acid (II-d)	4-Isopropyl-5-ethoxyoxazole	2	B	72–76	144–145	66–67
6-n-Butyl-5-hydroxycinchomeronic acid (II-e)	4-n-Butyl-5-ethoxyoxazole	2	B	75–77	64–65	oil

The free base I (a) was obtained by procedure C only in poor yield. Procedure D is appropriate for the preparation of this substance.

D. To a suspension of the hydrochloride of I (a) (868 mg, 3.5 milli-moles) in 4 ml of anhydrous chloroform is added 0.4 g (0.55 ml, 4 milli-moles) of triethylamine. The mixture is warmed gently until the solid has dissolved completely. To this is added 40 ml of anhydrous ethyl acetate, and the mixture is kept in a refrigerator for 30 minutes. The solid is filtered off and washed with 10 ml of anhydrous ethyl acetate. Combined filtrates are evaporated *in vacuo* to dryness. The oily residue is mixed with 4 ml of water. Crystalline free base I (a) is filtered and dried. The yield is 715 mg (96.6%).

The substances prepared are listed in Table II.

Note: 1. Freshly distilled commercial dimethyl maleate was used.

3. Analogs of Pyridoxine (III a–e)

In a dry 250-ml three-necked flask, equipped with a reflux condenser, mechanical stirrer, and dropping funnel and protected from atmospheric moisture with drying tubes, are placed 1.14 g (30 millimoles) of lithium aluminum hydride and 50 ml of anhydrous ether (or anhydrous tetrahydro-furan, see Table III). The flask is cooled in an ice bath. A solution of 10 millimoles of free base of dimethyl cinchomerate II (a–e) in 50 ml of anhydrous ether (or tetrahydrofuran) is added, dropwise and with stirring, to the reaction mixture during about 10 minutes. The mixture is further stirred under gentle reflux for 6 hours and kept at room temperature over-night. The flask is placed in an ice bath, the complex is hydrolyzed, and the excess of lithium aluminum hydride is destroyed by the cautious addi-tion, dropwise and with stirring, of 100 ml of water (Note 1). This requires about 30 minutes. Carbon dioxide gas is passed through the resulting mix-ture for 30 minutes. The solid is collected by filtration with suction, trans-ferred to the same flask, and stirred with 100 ml of water–ethanol (1:1) mixture. Carbon dioxide gas is passed again through this suspension for 30 minutes. After filtration the solid is washed on the filter with three 50-ml portions of hot ethanol. Both filtrates and all washings are combined and evaporated *in vacuo* to dryness. The residue is mixed with 25 ml of ethanol, and the suspension is boiled for 5 minutes. The hot liquid is decanted onto a filter and filtered with suction. The remaining solid is mixed again with 25 ml of ethanol, and the above operation is repeated four times. The com-bined filtrates are evaporated *in vacuo* at 40–50° to dryness. Two alternative methods (A and B) are used for isolation of pyridoxine analogs.

A. The residue is mixed with 10 ml of anhydrous acetone. The solid is filtered, washed with 2 ml of anhydrous acetone, and dried. Free bases of pyridoxine analogs are obtained.

B. The residue is dissolved in 4 ml of a 12% solution of dry hydrogen chloride in absolute ethanol. Anhydrous ether is added to this solution until crystals begin to precipitate. The mixture is kept in a refrigerator overnight. The crystalline product is filtered, washed with anhydrous ether, and dried.

The substances obtained by this procedure are listed in Table III.

TABLE III
SYNTHESIS OF PYRIDOXINE ANALOGS

Substances prepared	Dimethyl esters of	Solvent	Method of separation	Yield (%)	Melting point (°C)
2-Norpyridoxine, hydrochloride (III-a)	5-Hydroxycinchomeronic acid	Tetrahydrofuran	A	71	125–126
2-Nor-6-methylpyridoxine (III-b)	2-Methyl-5-hydroxycinchomeronic acid	Ether	B	56–61	197–199 (decomp.)
6-Methylpyridoxine (III-c)	2,6-Dimethyl-5-hydroxycinchomeronic acid	Ether	B	60–64	174–177 (decomp.)
2′,2′-Dimethylpyridoxine, hydrochloride (III-d)	6-Isopropyl-5-hydroxycinchomeronic acid	Ether	A	72	190–191
2′-n-Propylpyridoxine, hydrochloride (III-e)	6-n-Butyl-5-hydroxycinchomeronic acid	Ether	A	74	187–188

Note: 1. The addition of water is accompanied by foaming, and care must be taken to avoid excessive loss of the solvent.

4. Oxidation of Analogs of Pyridoxine

In a 50-ml Erlenmeyer flask fitted with a magnetic stirrer is placed a solution of 3 millimoles of the pyridoxine analog in 0.3 M aqueous sulfuric acid (10 ml in the case of hydrochlorides or 15 ml in the case of free bases). To this solution is added 270 mg of active manganese dioxide (Note 1), and the mixture is stirred at room temperature for 3.5 hours. Near the end of this period, manganese dioxide dissolves completely or nearly completely, and the pH of the mixture is 5.5–6.

The pyridoxal analogs are isolated from reaction mixtures either in the form of oximes or of Schiff's bases by the following procedures.

A. To the mixture, obtained as above, is added 310 mg of hydroxylamine hydrochloride, and the solution is heated at 70° for 10 minutes. Then 820 mg of anhydrous sodium acetate (or 1.36 g of crystallohydrate) is added, the mixture is heated again at 70° for 10 minutes and then allowed to stand in a refrigerator for 2 hours. The precipitate is filtered, washed with cold water, and dried.

The oximes obtained are listed in Table IV.

TABLE IV

SYNTHESIS OF DERIVATIVES OF PYRIDOXAL (PL) ANALOGS

Substance	Yield (%)	Melting point (°C)
Oxime of		
2-Nor PL (IV-a)	75	201–203 (decomp.)
2-Nor-6-methyl PL (IV-b)	59	185–187 (decomp.)
6-Methyl PL (IV-c)	69	209–212 (decomp.)
2′,2′-Dimethyl PL (IV-d)	69	186–187 (decomp.)
2′-n-Propyl PL (IV-e)	71	173–175 (decomp.)
Schiff's base of		
2-Nor PL (V-a)	79	192–194 (decomp.)
2-Nor-6-methyl PL (V-b)	38	179–183 (decomp.)
6-Methyl PL (V-c)	66	177–180 (decomp.)
2′,2′-Dimethyl PL (V-d)	74	125–126
2′-n-Propyl PL (V-e)	76	126–127

B. The unreacted manganese dioxide is removed by filtration with suction and washed with 2 ml of water. To the combined filtrates is added 8 ml of 0.5 M aqueous p-phenetidine hydrochloride (Note 2) followed immediately by 12 ml of 2 N aqueous sodium acetate. The mixture is kept in a refrigerator for 2 hours. Orange needles of Schiff's bases are filtered off, washed with cold water, and dried.

The substances thus prepared are listed in Table IV.

Notes: 1. Satisfactory manganese dioxide (type "B") was obtained by the method described by Harnfest and co-workers.[13]

2. This solution may be prepared as follows: To freshly distilled p-phenetidine (1.3 ml) mixed with 15 ml of water is added 1 ml of concentrated hydrochloric acid. The volume is brought to 20 ml with water, and the mixture is shaken until the solid has dissolved completely. This solution must be used immediately.

5. Analogs of Pyridoxal (VI a–e)

The carefully purified Schiff's bases are dissolved in 1 N hydrochloric acid or ethanol (the solvent and its volume are indicated in Table V). The resulting solution is applied to the top of a column of Dowex 50 W-X4 (100–200 mesh) in acid form. The sizes of the columns are shown in Table V.

[13] M. Harnfest, A. Bavely, and W. A. Lazier, *J. Org. Chem.* **19,** 1608 (1954).
[14] H. Wada and E. E. Snell, *J. Biol. Chem.* **237,** 127 (1962).

The column is eluted with 1 N hydrochloric acid at a rate of 50 ml/hour. A typical elution curve is presented in Fig. 1. Fractions containing the pyridoxal analog are evaporated to dryness *in vacuo* at 40–45°. The oily residue is dried in a vacuum desiccator over potassium hydroxide. The hydrochloride VI (a–e) obtained in this way is analytically and chromatographically pure. The yield is nearly quantitative.

The substances prepared are presented in Table V.

TABLE V

SYNTHESIS OF PYRIDOXAL (PL) ANALOGS

Substances prepared	Amount of Schiff's base (mg)	Solvents and their volume	Size of column (cm)	Volume of previous fractions (ml)	Volume of fraction, containing PL-analog (ml)	Melting point (°C)
			Chromatography			
2-Nor PL, hydrochloride (VI-a)	200	1 N HCl, 3 ml	1.4 × 40	250	250	144–147 (decomp.)
2-Nor-6-methyl PL, hydrochloride (VI-b)	20	1 N HCl, 0.3 ml	1 × 20	250	100	170–175 (decomp.)
6-Methyl PL, hydrochloride (VI-c)	450	1 N HCl 7.5 ml	1.6 × 40	400	500	173–178 (decomp.)
2′,2′-Dimethyl PL, hydrochloride (VI-d)	200	EtOH, 4 ml	1.4 × 40	1300	900	119–121 (decomp.)
2′-*n*-Propyl PL, hydrochloride (VI-e)	470	EtOH, 15 ml	1.5 × 42	800	600	63–66

6. Analogs of Pyridoxamine (VII a–e)

In a 100-ml Erlenmeyer flask fitted with a magnetic stirrer is placed a solution of millimole of the pyridoxal analog oxime in 20 ml of water and 0.5 ml of concentrated hydrochloric acid. To this is added 150 mg of 5% palladium-on-charcoal catalyst. The flask is fitted to a hydrogen eudiometer, and the system is flushed three or four times with hydrogen. The hydrogenation is carried out with stirring at room temperature and atmospheric pressure. Hydrogenation is continued until about a 10% excess over the theoretical amount of hydrogen (2 millimoles) has been absorbed. This requires about 30 minutes. The catalyst is removed by filtration, using a hot water wash, and the clear filtrate is evaporated to dryness under reduced pressure at a temperature below 45°. The residue is dried in a vacuum desiccator over potassium hydroxide. The yields of pyridoxamine analogs are nearly quantitive.

The substances obtained are listed in Table VI.

TABLE VI
SYNTHESIS OF PYRIDOXAMINE (PM) ANALOGS

Substances prepared	Melting point (°C)
2-Nor PM, dihydrochloride (VII-a)	166–168 (decomp.)
2-Nor-6-methyl PM, dihydrochloride (VII-b)	234–240 (decomp.)
6-Methyl PM, dihydrochloride (VII-c)	182–184 (decomp.)
2′,2′-Dimethyl PM, dihydrochloride (VII-d)	169–170
2′-n-Propyl PM, dihydrochloride (VII-e)	163–164

7. Analogs of Pyridoxamine 5′-Phosphate (VIII a–e)

In a 50-ml pear-shaped flask equipped with a reflux condenser with a calcium chloride drying tube are mixed 0.52 g of 85% phosphoric acid and 0.40 g of phosphorus pentoxide. This mixture is cooled to room temperature. To this is then added 0.46 millimole of pyridoxamine analog hydrochloride. The mixture is allowed to stand at room temperature until completion of the evolution of hydrogen chloride (Note 1). The flask is then heated at 60° for 2 hours. After cooling of the mixture in an ice bath, 3 ml of ethanol is added, and this is stirred until the viscous mass has dissolved completely. Then the solution is mixed with 8 ml of ether and allowed to refrigerate for 1 hour. The liquid is decanted, and the residue is dissolved in 5 ml of 1 N aqueous hydrochloric acid and heated in a boiling water bath for 20 minutes. This solution is concentrated *in vacuo* to a volume of 1 ml, and brought to pH 5–6 by the addition of concentrated ammonia. The resulting syrup is applied to the top of a column of Amberlite CG-50 in the acid form. The column is eluted with water at a rate of 50 ml/hour (the sizes of the column are indicated in Table VII). A typical elution curve is presented in Fig. 2 (Note 2). The effluent, containing the pyridoxamine phosphate analog, is concentrated to a small volume *in vacuo* at 40–45° and mixed with ethanol until crystals begin to precipitate. After refrigeration for 2 hours, the solid is filtered, washed with ethanol, and dried. The pyridoxamine phosphate

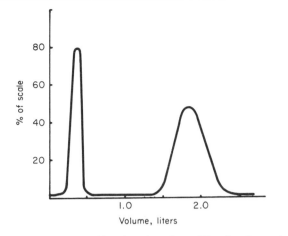

Fig. 1. Separation of 2-norpyridoxal and *p*-phenetidine by chromatography of the Schiff's base on ion-exchange resin.

analogs thus obtained (see Table VII) can be purified by recrystallization from water heated to 80°.

Notes: 1. Since the evolution of hydrogen chloride is accompanied by foaming, a flask of sufficient size should be used.

2. The effluent was controlled by measurement of its absorption at 295 mμ and by measurement of its conductivity.

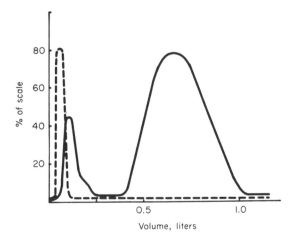

Fig. 2. Purification of the 2-norpyridoxamine phosphate: solid line, absorption at 295 mμ; broken line, conductivity.

TABLE VII

SYNTHESIS OF PYRIDOXAMINE PHOSPHATE (PMP) ANALOGS

Prepared substances	Purification				
	Amount of PM-analogs (mg)	Size of column (cm)	Volume of previous fraction (ml)	Volume of fraction, containing VIII (ml)	Yield (%)
2-Nor PMP, dihydrate (VIII-a)	100	1.6 × 50	400	600	65
2-Nor-6-methyl PMP, dihydrate (VIII-b)	290	1.7 × 65	1000	900	51
6-Methyl PMP, dihydrate (VIII-c)	250	1.6 × 65	1100	900	53
2′,2′-Dimethyl PMP, dihydrate (VIII-d)	400	2.1 × 50	2300	1400	63
2′-n-Propyl PMP, dihydrate (VIII-e)	350	2.1 × 50	1250	1000	64

8. Analogs of Pyridoxal 5′-Phosphate (IX a–e)

Method A. In a 25-ml pear-shaped flask fitted with a reflux condenser with a drying tube are mixed 3.72 g of 85% phosphoric acid and 2.86 g of phosphorus pentoxide. To the cooled mixture is added 2 millimoles of Schiff's base. The dark, reddish viscous mass is stirred cautiously and then heated as indicated in Table VIII. After heating, the mixture is cooled in an ice bath and mixed carefully with 6 ml of 0.1 N hydrochloric acid until a homogeneous solution is obtained. The resulting mixture is heated at 60° for 15 minutes, cooled to room temperature, and applied to the top of a column of Dowex 50 W-X4 (100–200 mesh) in the acid form (see Table VIII). The column is eluted with water (Note 1) at a rate of 50 ml/hour. A typical elution curve is presented in Fig. 3. The effluent containing the pyridoxal phosphate analog is concentrated *in vacuo* to a volume of 10 ml and applied on the same column of Dowex 50 W-X4. The chromatography is repeated under the conditions described above. The new effluent is concentrated *in vacuo* (Note 2) to a volume of 30–40 ml and lyophilized.

Method B. In a 20-ml round-bottomed flask, fitted with a magnetic stirrer, nitrogen-inlet tube, and dropping funnel with a pressure-equalizing side arm, is dissolved 1 millimole of pyridoxamine phosphate analog in 7 ml of water and 1 ml of 2 N aqueous sodium hydroxide. To the resulting solution 240 mg (2.5 millimoles) of sodium glyoxylate is added and the mixture is stirred at room temperature for 10 minutes. The solution is brought to pH 5 with glacial acetic acid and stirred for another 10 minutes. To this is

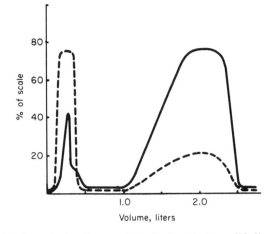

Fɪɢ. 3. Purification of the 2-norpyridoxal phosphate: solid line, absorption at 295 mμ; broken line, conductivity.

added dropwise and with stirring 3 ml of 0.25 M aqueous cupric acetate under a slow stream of nitrogen. Stirring is continued under a stream of nitrogen for 30 minutes. The resulting green solution is applied to the top of a column of Dowex 50 W-X4 (100–200 mesh) in acid form and eluted with water (Note 1) at a rate of 50 ml/hour. Further operations are carried out as in method A.

The substances obtained are listed in Table VIII.

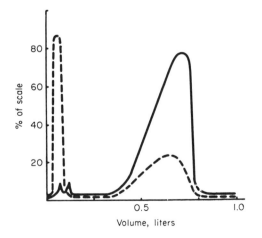

Fɪɢ. 4. Purification of 2-nor-6-methylpyridoxal phosphate: solid line, absorption at 295 mμ; broken line, conductivity.

TABLE VIII

SYNTHESIS OF PYRIDOXAL PHOSPHATE (PLP) ANALOGS

Prepared substances	Method of preparation	Reaction conditions	Amount of Schiff's base (g)	Purification			Yield (%)	Absorption in acid phenylhydrazine[a]
				Size of column (cm)	Volume of previous fraction (ml)	Volume of fraction, containing IX (ml)		
2-Nor PLP, monohydrate (IX-a)	A	40°, 4 hr	1.0	1.6 × 50	1000	1500	73	18,900
2-Nor-6-methyl PLP, monohydrate (IX-b)	B	—[b]	—[b]	—[b]	—[b]	—[b]	67	20,300
6-Methyl PLP, monohydrate (IX-c)	A	60°, 6 hr	0.25	1.4 × 25	500	500	53	21,000
2',2'-Dimethyl PLP, monohydrate (IX-d)	A	60°, 5 hr	0.5	1.4 × 40	400	600	69	21,200
2'-n-Propyl PLP, monohydrate (IX-e)	A	60°, 5 hr	1.0	1.6 × 50	2500	3000	71	20,900

[a] Absorption in acid phenylhydrazine was determined by the procedure of Wada and Snell.[14]
[b] See separation curve in Fig. 4.

TABLE IX
ABSORPTION MAXIMA AND MOLAR ABSORBANCE OF VITAMIN B$_6$ ANALOGS

| | λ_{max} mμ ($\epsilon \times 10^{-3}$) | | |
Substance[a]	0.1 N HCl	pH 7[b]	0.1 N KOH
2-Nor PN	289 (6.4)	251 (2.5)	242 (7.2)
		286 (3.0)	310 (5.6)
		324 (2.8)	
2-Nor-6-methyl PN	297 (6.2)	256 (3.6)	247 (7.2)
		294 (2.7)	317 (5.2)
		331 (3.3)	
6-Methyl PN	298 (10.3)	257 (6.3)	248 (8.5)
		332 (8.4)	317 (7.9)
2′,2′-Dimethyl PN	291 (8.7)	256 (4.0)	246 (6.5)
		328 (8.0)	310 (7.6)
2′-n-Propyl PN	293 (9.7)	254 (3.6)	245 (6.6)
		327 (7.5)	311 (1.9)
2-Nor PM	292 (6.8)	251 (2.6)	243 (7.5)
		287 (3.0)	307 (5.9)
		324 (3.2)	
2-Nor-6-methyl PM	299 (7.4)	253 (4.8)	245 (7.8)
		297 (2.9)	312 (5.8)
		328 (4.1)	
6-Methyl PM	302 (8.8)	255 (6.3)	248 (7.2)
		333 (8.7)	314 (7.2)
2′,2′-Dimethyl PM	296 (9.2)	254 (4.5)	246 (7.5)
		329 (8.1)	310 (8.1)
2′-n-Propyl PM	297 (9.2)	253 (4.9)	246 (6.7)
		328 (8.3)	309 (8.1)
2-Nor PL	284 (6.6)	249 (4.6)	240 (8.6)
		280 (2.1)	300 (5.0)
		314 (4.2)	390 (0.6)
		390 (0.08)	
2-Nor-6-methyl PL	292 (6.4)	247 (4.7)	240 (9.0)
		290 (2.1)	307 (5.6)
		323 (4.0)	
6-Methyl PL	295 (8.3)	250 (5.3)	242 (8.4)
		324 (7.3)	310 (7.8)
2′,2′-Dimethyl PL	289 (9.2)	253 (5.1)	238 (8.0)
		319 (8.3)	302 (6.3)
		390 (0.6)	390 (2.1)
2′-n-Propyl PL	292 (9.0)	254 (5.8)	243 (8.4)
		321 (8.2)	304 (6.5)
		390 (0.2)	390 (2.1)
2-Nor PMP	292 (7.1)	248 (3.7)	243 (7.3)
		283 (3.4)	307 (5.2)
		324 (3.6)	

(Continued)

TABLE IX (*Continued*)

ABSORPTION MAXIMA AND MOLAR ABSORBANCE OF VITAMIN B_6 ANALOGS

Substance[a]	λ_{max} mμ ($\epsilon \times 10^{-3}$)		
	0.1 N HCl	pH 7[b]	0.1 N KOH
2-Nor-6-methyl PMP	299 (8.1)	252 (4.9)	246 (8.6)
		297 (2.8)	313 (6.2)
		333 (4.6)	
6-Methyl PMP	302 (9.7)	252 (6.3)	248 (7.7)
		336 (8.9)	314 (8.0)
2,2'-Dimethyl PMP	294 (9.2)	252 (5.1)	244 (6.9)
		327 (8.1)	310 (7.9)
2'-n-Propyl PMP	293 (9.3)	253 (5.0)	245 (6.9)
		327 (8.2)	310 (7.9)
2-Nor PLP	295 (6.7)	284 (1.5)	305 (1.0)
	338 (1.4)	328 (2.4)	386 (5.6)
		384 (3.2)	
2-Nor-6-methyl PLP	295 (6.4)	290 (2.0)[c]	312 (3.9)
	340 (1.1)[c]	331 (4.3)	394 (2.9)
		380 (1.4)[c]	
6-Methyl PLP	302 (7.1)	332 (4.8)	313 (3.6)
	350 (1.0)	390 (2.5)	398 (4.3)
2',2'-Dimethyl PLP	297 (8.0)	329 (2.3)	304 (1.8)
	340 (1.2)[c]	390 (4.2)	393 (5.7)
2'-n-Propyl PLP	298 (7.9)	326 (2.5)	302 (2.7)
	350 (1.0)[c]	303 (4.3)	345 (4.2)

[a] PN, pyridoxine; PM, pyridoxamine; PL, pyridoxal; PMP, pyridoxamine phosphate; PLP, pyridoxal phosphate.

[b] Phosphate buffer, 0.1 M.

[c] Inflection point.

Notes: 1. It is necessary to use oxygen-free water. Satisfactory water may be prepared by boiling distilled water.

2. Since pyridoxal phosphate analogs are unstable, all concentration procedures are performed at a temperature not exceeding 30–35°.

Physicochemical Properties of 2- and 6-Alkyl Analogs
of Vitamins B_6 and Pyridoxal Phosphate

A. Ultraviolet (UV) spectra were taken on an CF-4 "Optica Milano" spectrophotometer. The data from UV spectra are presented in Table IX.

B. Infrared (IR) spectra were obtained on a UR-10 spectrophotometer for solid substances (pellets with potassium bromide). Correlation of frequencies in IR spectra is listed in Table X.

C. Proton magnetic resonance (PMR) spectra were determined on a JEOL JNM-4H-100 spectrometer. The chemical shifts are reported (see

TABLE X

INFRARED SPECTRA OF VITAMINS B6 ANALOGS

ν (cm^{-1})

Substances	3—OH	C=C pyridine	—NH$_3^+$	4—CHO	4—CH$_2$OH	5—CH$_2$OH	CH$_2$OPO$_3$H$_2$
Pyridoxine analogs, hydrochlorides	3310–3200 2850–2730	1645–1631 1557–1550 1510–1493 1421–1400	—	—	1006–992	1035–1007	—
Pyridoxal analogs, hydrochlorides	3390–3290 2770–2620	1662–1635 1583–1560 1509 1439–1420	—	—	—	—	—
Pyridoxamine analogs, dihydrochlorides	3370–3220 2980–2620	1642–1623 1596–1572 1510 1406–1400	1563–1540 1513–1488	—	—	1031–1019	—
Pyridoxamine phosphate analogs, dihydrate	3440–3400 3100–2650	1653–1643 — —	1547–1529	—	—	—	1294–1290 1174–1142 1084–1062
Pyridoxal phosphate analogs, monohydrate	3440–3410 2420–2550	1412–1400 1666–1646 1581–1560 1524–1520 1415–1409	— —	1720	—	—	1290–1274 1185–1169 1070–1061

TABLE XI

PROTON MAGNETIC RESONANCE SPECTRA OF VITAMIN B_6 ANALOGS

Substances	δ (ppm), D_2O								
	2—H	6—H	2—CH$_3$	6—CH$_3$	4—CH$_2$	4'—H	5—CH$_2$OH	2-*iso*-C$_3$H$_7$	2-*n*-C$_4$H$_4$
Pyridoxine analogs, hydrochlorides	—	8.14-8.12	2.60-2.61	—	4.97-4.99	—	4.75-4.76	1.31, 1.38	0.86, 1.29, 1.68, 2.98
Pyridoxal analogs, hydrochlorides	—	8.15-8.14	2.59	2.52	—	6.72-6.68	5.22-5.16	1.31, 1.38	0.86, 1.36, 1.69, 3.01
Pyridoxamine analogs, dihydrochlorides	8.18	8.21	2.65-2.69	2.68-2.70	4.46-4.42	—	4.81-4.85	1.31, 1.39	0.89, 1.34, 1.64, 3.05
Pyridoxal phosphate analogs	—	8.19	2.58-2.60	2.66	—	6.48-6.69	5.31-5.22 $J = 5.5$ cps	1.31, 1.38	0.85, 1.29, 1.82, 3.00

Table XI) in δ values in parts per million with the tetramethylsilane signal (0 ppm) or dioxane signal (3.27 ppm) as internal standard.

Superthin-Layer Chromatography of Analogs of Vitamins B₆ and Pyridoxal Phosphate

Superthin-layer chromatography is currently applied extensively for the analysis of mixtures.[15] This method, which has several advantages in comparison to other types of chromatography, was used for the separation of vitamin B₆ analogs.[16]

Procedure

Sorbent

Commercial silica gel is ground in a pebble mill with flint balls for 6 hours. The resulting powder is carefully stirred with 2.5 liters of water in a glass beaker of 15 cm diameter (height of suspension layer is also 15 cm) and allowed to stand for 40 minutes. The liquid is decanted into a similar beaker and allowed to stand for 2 hours. The liquid is removed by decantation, and the residue is stirred again with 2.5 liters of water in a beaker, as above. The sedimentation is repeated once more as above. The sorbent, obtained in this manner, is dried at 140° for 1 day. After cooling to room temperature, the sorbent is stirred carefully with freshly distilled chloroform to obtain a suspension of the consistency of thin sour cream. Glass slides (7.5 cm long and 2.5 cm wide) are dipped twice into the prepared mixture, excess of the silica gel suspension is allowed to drain off, and the slides are air-dried. After drying the sorbent is removed from the back surface of the slides.

Chromatography

Two major solvent systems are used for separation of vitamin B₆ analogs (system A) and pyridoxal phosphate and pyridoxamine phosphate analogs (system B).

System A. Ethyl acetate–acetone–25% aqueous ammonia (20:10:1.5).

System B. n-Butanol–ethanol–5% aqueous ammonia–glacial acetic acid (10:10:10:1).

In the case of chromatography of acid pyridine derivatives, good results were obtained with solvent system C: n-butanol–25% aqueous ammonia–water (40:9:1).

[15] B. G. Belenky, E. S. Gankina, and V. V. Nesterov, *Dokl. Akad. Nauk SSSR* **172,** 91 (1967).

[16] E. N. Dementieva, N. A. Drobinskaja, L. V. Ionova, M. Ya. Karpeisky, and V. L. Florentiev, *Biokhimiya* **33,** 350 (1968).

The chromatographs are developed in glass tanks with the plates inclined at an angle of approximately 90°. The position of areas is determined under ultraviolet light (365 mμ). All substances are clearly seen as spots with a bluish fluorescence.

Solutions of samples in methanol (in case of 5′-phosphoric esters, in water) are applied as spots with a micropipette in the usual manner. The convenient concentration of samples is 1–2 mg/ml. The distance from the border of the slide is 10–15 mm.

The areas may be detected easily following application of spots containing 2.0–2.5 μg of pyridoxine analogs, 1.0–1.5 μg of pyridoxamine analogs, and 2.0–2.5 μg of the 5′-phosphate esters.

Pyridoxal and its analogs are chromatographed in the form of their methyl acetals, prepared by heating of solution of samples in absolute methanol at 50° for 1 hour. In this case all developed aldehydes give one area and are detected easily with 1.5–2.0 μg of substances in the spots.

The R_f values are listed in Table XII.

TABLE XII

SUPERTHIN-LAYER CHROMATOGRAPHY OF VITAMIN B₆ ANALOGS

Substance[a]	Solvent system	R_f values
2-Nor PN	A	0.15 ± 0.02
2-Nor-6-methyl PN	A	0.27 ± 0.02
6-Methyl PN	A	0.52 ± 0.02
2′,2′-Dimethyl PN	A	0.54 ± 0.02
2′-n-Propyl PN	A	0.50 ± 0.02
2-Nor PM	A	0.67 ± 0.02
2-Nor-6-methyl PM	A	0.74 ± 0.02
6-Methyl PM	A	0.84 ± 0.02
2′,2′-Dimethyl PM	A	0.85 ± 0.02
2′-n-Propyl PM	A	0.82 ± 0.02
2-Nor PL	A	0.29 ± 0.01
2-Nor-6-methyl PL	A	0.44 ± 0.01
6-Methyl PL	A	0.65 ± 0.02
2′,2′-Dimethyl PL	A	0.68 ± 0.02
2′-n-Propyl PL	A	0.64 ± 0.02
2-Nor PMP	B	0.48 ± 0.02
2-Nor-6-methyl PMP	B	0.49 ± 0.02
6-Methyl PMP	B	0.56 ± 0.01
2-Nor PLP	B	0.59 ± 0.01
6-Methyl PLP	B	0.70 ± 0.01

[a] PN, pyridoxine; PM, pyridoxamine; PL, pyridoxal; PMP, pyridoxamine phosphate; PLP, pyridoxal phosphate.

Characteristics of the Pyridoxal Phosphate Analogs as Coenzymes for L-Aspartate:2-Oxoglutarate Aminotransferase (EC 2.6.1.1) (AAT)

L-Aspartate + α-ketoglutarate $\rightleftharpoons$ L-glutamate + oxaloacetate

The spectral properties of AAT at wavelengths above 300 mμ are due to the coenzyme, pyridoxal phosphate, which forms an aldimine bond with a lysyl ϵ-NH$_2$ group.[17] Optical activity in this spectral range is induced by the protein moiety of the enzyme.[18]

1. Application of the Circular Dichroism (CD) Method to Investigation of Apoenzyme–Coenzyme Interaction

The Principle of the CD Method

CD, like the phenomenon of optical rotation, is a manifestation of the optical activity of a substance. Both these effects are different consequences of the same cause—the absence of a plane and center of symmetry in electron density distribution in a chromophore.

Because the CD method is a relatively novel one in biochemical practice, let us consider briefly the meaning of the CD phenomenon.

Plane-polarized light can be considered as the geometric sum of two components circularly polarized in opposite directions. If the medium in which the light propagates is optically active, then the velocities of propagation of these components are different and, as a result a phase difference, will accumulate. Therefore the plane of polarization will be rotated for some angle (optical rotation). In the absorption band of an optically active substance, not only velocities of propagation of the circularly polarized components will be changed, but their amplitudes as well. This leads to elliptic polarization of the outgoing beam, which has its expression in anomalous dispersion of optical rotation. Optical rotatory dispersion is measured by spectropolarimetry. With a dichrograph one can measure directly the difference in absorption coefficients between the left- and right-circularly polarized components. The relation between CD and optical rotation for a single spectral transition is given by the Kronig-Cramers equation[19]:

$$[M(\lambda)] = \frac{2[\theta^0]}{\sqrt{\pi}} \left[e^{-(\lambda-\lambda_0)/\Delta_0} \int_0^{(\lambda-\lambda_0)/\Delta_0} e^{x^2} \, dx - \frac{\Delta_0}{2(\lambda+\lambda_0)} \right] \tag{1}$$

[17] R. C. Hughes, W. T. Jenkins, and E. H. Fisher, *Proc. Natl. Acad. Sci. U.S.* **48**, 1615 (1962).

[18] Yu. M. Torchinsky and L. G. Koreneva, *Biokhimiya* **28**, 1087 (1963); P. Fasella and G. Hammes, *Biochemistry* **3**, 530 (1964).

[19] C. Djerassi, "Optical Rotatory Dispersion," p. 165. McGraw-Hill, New York, 1960.

Here $[M(\lambda)]$ is molar rotation, connected with specific rotation by the equation:

$$[M] = M/100 \cdot \alpha \ (M, \text{ molecular weight}); \ [\theta^0] = 2.3 \cdot \frac{4500}{\pi} \Delta E \quad (2)$$

is the value of ellipticity; ΔE is the difference of the molar absorbancies of left- and right-circularly polarized components; λ_0 is the wavelength of the absorption maximum of a given absorption band; Δ_0 is the half-width of the absorption band.

Advantages of the CD Method in Comparison to the Determination of Optical Rotatory Dispersion

In spite of the existence of a simple interrelation between CD and optical rotation (Eq. 1), the measuring of CD has a number of essential advantages in practice. The main advantage is due to the fact that the CD effect is localized only within the absorption band of an optically active chromophore. In contrast to the CD bands, the optical rotation has also a substantial magnitude at a considerable distance from the absorption band. Therefore a distant band with strong rotation will overlap and mask effects due to a proximate band of less optical activity. This is the situation encountered in the case of enzymes. In particular, with aspartate aminotransferase the strong negative optical rotation due to the peptide chain makes it impossible to obtain quantitative information about some shortwave forms of the AAT coenzyme (PMP-form, for example) and about optical activity induced by the coenzyme in individual amino acid residues. The relation between CD and optical rotation of AAT is illustrated by Fig. 5.

An additional important advantage of measuring CD instead of absorption (in the cases when required information can be obtained using spectophotometry as well) is the independence of CD spectra from light scattering or slight turbidity of the sample. This is due to the fact that CD is the difference of absorptions between left- and right-circularly polarized beams. The screening effect of turbidity is thus automatically subtracted.

Quantitative Parameters Relevant to CD

A parameter, measured by a dichrograph, is the difference (D) of optical densities (ΔD) in left- and right-circularly polarized light,

$$D_L - D_R = \Delta D$$

Dividing this value by the length of the cell (c_l) in centimeters and the molar concentration of the sample, the molar CD is obtained,

$$\Delta E = \Delta D/c_l$$

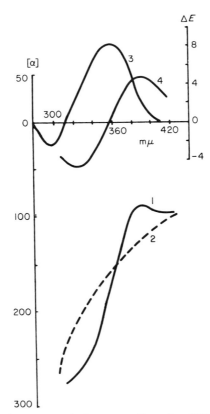

Fig. 5. Comparison between optical rotatory dispersion (ORD) and circular dichroism (CD) of aspartate aminotransferase (AAT) (pH 8.3): *1*, the experimental ORD spectrum of AAT; *2*, ORD of apo-AAT; *3*, the experimental CD spectrum of the AAT; *4*, ORD spectrum, calculated from curve *3* according to Eq. (1).

Experimental data concerning CD are sometimes given in terms of ellipticity (θ) instead of ΔE. Ellipticity is related to ΔE by Eq. (2). The ratio of dichroism to absorption, $\Delta D/D = \Delta E/E$ is another useful parameter. This value, designated as Kuhn's anisotropy factor, does not depend on concentration, and in some cases it allows one to identify enzyme–substrate complexes indistinguishable by their absorption spectra.[20]

The data here presented were obtained with a Roussel-Jouan dichrograph. Its sensitivity was about 0.5×10^{-4} unit of optical density. A detailed description of this apparatus is given by Velluz *et al.*[21]

[20] V. I. Ivanov, Yu. N. Breusov, H. Ya. Karpeisky, and O. L. Polyanovsky, *Moleculyarnaya Biologija (SSSR)* **1**, 588 (1967).

[21] L. Velluz, M. Legrand, and M. Grosjean, "Optical Circular Dichroism." Academic Press, New York, 1965.

2. Interactions of Pyridoxal Phosphate and Its Analogs with the Apoenzyme of AAT

Determining CD Spectra for Holoenzymes Reconstituted with Analogs of Pyridoxal Phosphate

Apoenzyme is prepared from the pyridoxamine phosphate, or amino, form of the enzyme according to the following procedure (based on footnote 22). To the enzyme solution, a saturated solution of ammonium sulfate (adjusted to pH 6.0) is added to achieve 0.6 saturation. Then the pH is adjusted to 4.5–5.0 with 1 N HCl and the solution is kept overnight in a refrigerator; the precipitated protein is centrifuged down, washed with 0.6 saturated ammonium sulfate solution and resedimented. The sediment thus obtained is dissolved in 0.2 M carbonate–bicarbonate buffer at pH 9.0, and the apoenzyme solution is dialyzed against distilled water for 15 hours. The apoenzyme thus prepared has residual enzymatic activity of about 5%. No visible absorption was present in the coenzyme range of spectra.

Apoenzyme concentration is determined in the following way: First, a standard sample of apoenzyme is prepared very carefully. Its concentration is determined from optical density at 280 mμ on the basis of the equation:

$$D_{280} \text{ for 1.0 mg/ml in a 1.0-cm cell} = 1.44$$

To this standard sample (the most appropriate concentration of apoenzyme is 3–5 mg/ml) at pH 5.2 (0.05 M acetate buffer) an excess of pyridoxal phosphate is added (1.5 $\times$ 10^{-4} M), and after a 10-minute interval, the CD spectrum is recorded. In this way ΔD_{430} was determined to be equal to 3.29 $\times$ 10^{-4} for 1.0 mg/ml per centimeter. Since pyridoxal phosphate is optically inactive per se, the excess of coenzyme in solution does not affect the CD spectrum of holoenzyme. In the experiments with artificial analogs of pyridoxal phosphate, the concentration of apoenzyme is first determined by adding an excess of pyridoxal phosphate to a separate aliquot of the protein solution and measuring $\Delta D_{430}^{1\,cm}$. Estimation of concentration based on ΔD_{430} has the advantage of determinating the concentration of active enzyme rather than total protein concentration, as in the case of measurement of D_{280}. This is of special importance in view of the easy denaturation of apotransaminase.

For obtaining CD spectra of holoenzyme reconstituted with artificial coenzyme analogs, an approximately 1.5 M excess of analog is added to an apoenzyme solution of concentration determined as described above. Holoenzyme formation is revealed by the CD spectrum (Fig. 6). To obtain the absorption spectra that are necessary for determination of anisotropy fac-

22 H. Wada and E. E. Snell, *J. Biol. Chem.* **193**, 45 (1951).

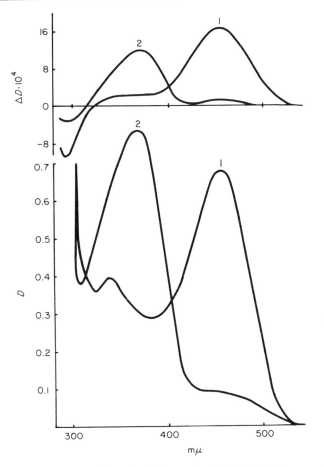

FIG. 6. Circular dichroism (above) and absorption (below) of holo-aspartate amino-transferase (holo-AAT) reconstituted with 6-methyl pyridoxal phosphate (PLP). *1*, pH 5.2; 0.05 *M* acetate buffer. *2*, pH 8.3, carbonate–bicarbonate buffer. Concentration of apo-AAT: 5.7 mg/ml; cuvettes: 1 cm.

tors $(\Delta D/D)$, the holoenzyme solution thus obtained is dialyzed 24 hours against the required buffer. The CD and absorption spectra obtained are shown in Fig. 6, and the anisotropy factors are listed in Table XIII. Using the mentioned type of dichrograph, the short-wave range of CD spectra $(\lambda < 300 \text{ m}\mu)$ can be recorded when the product protein concentration times length of cell $\simeq 1$ mg/ml per centimeter.

It should be noted that, because of increased absorption and, as a consequence, the necessity of having a high voltage on the photomultiplier, artifacts can arise in CD study of protein at wavelengths below 300 mμ.

TABLE XIII
PROPERTIES OF RECONSTITUTED TRANSAMINASES[3]

Property	PLP[a]	2'-Methyl PLP	2'-Propyl PLP	6-Methyl PLP	2-Nor PLP	2',2'-Di-methyl PLP
λ_{max}, pH 5.2	430	435	440	455	425	440
λ_{max}, pH 8.1	360	365	370	370	360	370
pK_a	6.25	6.5	6.4	6.3	5.8	7.1
$\Delta D/D \times 10^4$, 420–450 mμ	28	—	25	25	29	27
$\Delta D/D \times 10^4$ at 360 mμ	19	—	15	15	21	10
K_2, M^{-1} min^{-1}	1500	750	150	10,000	450	—
V_{max}, relative	1	0.32[b]	0.5	0.56	1.2	—
K_m for α-oxoglutarate (mM)	0.1	0.05[b]	0.08	0.14	0.24	—
K_m for L-aspartate (mM)	2.0	0.5[b]	0.6	1.0	1.8	—

[a] PLP, pyridoxal phosphate.
[b] Data from the study of Morino and Snell.[1]

Therefore, any effect detected should be checked for proportionality with optical density (by varying either concentration or cuvette width). In this way the negative dichroic band observed at 295–300 mμ in AAT (Figs. 5 and 6) was proved not to be an artifact. This band is absent in the spectrum of apoenzyme and arises upon recombination with coenzyme or its analog. We have assigned this band to a tyrosyl anion of the protein.[20]

Determination of the pK of Chromophores

For this purpose, recombination of an analog with apoenzyme is carried out in distilled water. Then a number of 2-ml aliquots with different pH's are prepared, and the CD spectra of these samples are recorded. The presence of a distinct isosbestic point at 390 mμ proved the occurrence of only one reaction of proteolytic dissociation involving the coenzyme chromophore (Fig. 7):

$$E + H^+ \overset{K}{\rightleftharpoons} EH$$

Here E and EH are the nonprotonated and protonated forms of the AAT. Hence

$$\log \frac{[E]}{[EH]} = pK - pH$$

Let the portion of protonated molecules [EH]/[E$_0$] be $\alpha(pH)$ (E$_0$ is the total enzyme concentration). Then

$$\log \frac{1 - \alpha}{\alpha} = pK - pH$$

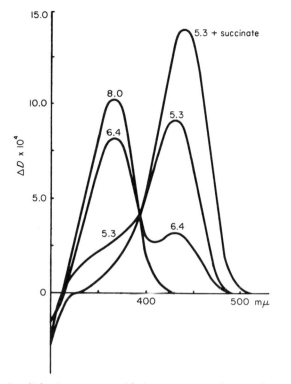

FIG. 7. Circular dichroism spectra of holo-aspartate aminotransferase reconstituted with 2-norpyridoxal phosphate (2-nor PLP) at different pH values. Concentrations: protein, 4 mg/ml; 2-nor PLP, $1.8 \times 10^{-4} M$.

In practice, it is more convenient to determine the value on the basis of ΔD for the nonprotonated form ($\lambda_{\max}$ at 360 mμ) rather than the protonated one ($\lambda_{\max}$ at 430 mμ), since denaturation of the protein at pH < 5.0 interferes with preparation of the fully protonated enzyme. Thus

$$\alpha(\text{pH}) = 1 - \frac{\Delta D_{360,\ \text{pH}}}{\Delta D_{360,\ \text{pH} = 8.3}}$$

Plotting $\log (1 - \alpha)/\alpha$ as a function of pH, one obtains pK as the intercept of the abscissa with this line (Fig. 8). The pK values determined in this way for a number of alkyl analogs of pyridoxal phosphate analogs, bound to the apoenzyme, are listed in Table XIII.

CD Titration of Apoenzyme with Coenzyme or a Coenzyme Analog

For the purpose of determinating the stoichiometric ratio between apoenzyme and coenzyme, CD titration is of great convenience. By the addi-

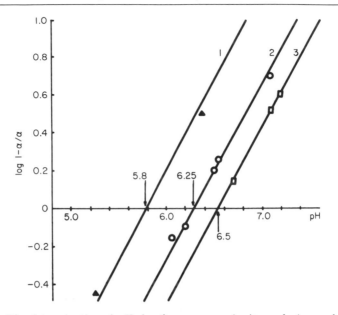

FIG. 8. The determination of pK_a for the coenzyme (or its analog) complexed with apo-aspartate aminotransferase. Explanation in the text. *1*, 2-nor PLP; *2*, PLP; *3*, 2'-methyl PLP. PLP = pyridoxal phosphate.

tion of calculated concentrations of an analog to the apoenzyme solution and recording of CD in the absorption band of the bound coenzyme, titration curves can be obtained wherein the breakpoint corresponds to completion of coenzyme binding. It follows from Fig. 9 that the 2-butyl analog of pyridoxal phosphate is bound to the apoenzyme with the same stoichiometry as the natural coenzyme, i.e., 1 mole of the analog is bound per 44,000 g of the protein. However, on titration with a racemic preparation of 5'-phosphate ester of the 5-(α-hydroxyethyl) analog of pyridoxal,[9] the inflection is reached on addition of the analog in twice the stoichiometric ratio (Fig. 10). This finding testifies to binding to the apoenzyme of only one of the two stereoisomers present in the racemic mixture.

Kinetics of Coenzyme Recombination with Apoenzyme, Studied by the CD Method

The kinetics of association of pyridoxal phosphate with apoenzyme of the AAT has previously been studied by plotting the increase in enzymatic activity, this activity being assumed to be proportional to the concentration of reconstituted holoenzyme.[23] Such a method requires sampling of aliquots

---

[23] B. E. C. Banks, A. J. Lawrence, C. A. Vernon, and J. F. Wootton, *in* "Chemical and Biological Asperty of Pyridoxal Catalysts," p. 147. Macmillan (Pergamon), New York, 1963.

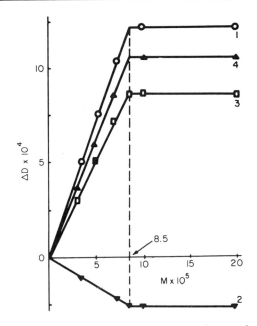

FIG. 9. Circular dichroism titration of apo-aspartate aminotransferase with pyridoxal phosphate (PLP) and 2′-*n*-propyl PLP. *1*, PLP, pH 5.2, λ = 430 mμ; *2*, PLP, pH 5.2, λ = 295 mμ; *3*, PLP, pH 8.3, λ = 360 mμ; *4*, 2′-*n*-propyl PLP, pH 5.2, λ = 430 mμ. The apo-AAT concentration is 3.8 mg/ml.

at fixed time intervals and the determination of initial reaction velocities for each sample.

The CD method has the important advantage of continuous recording and of circumventing complications connected with the presence of one more factor—the enzymatic process.

At pH 7.1 ($t = 20°$) and in the presence of 0.1 M phosphate buffer, the reaction is slow enough to be measured by the recorder of the dichrograph

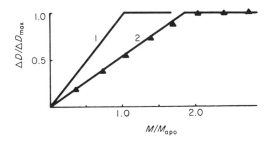

FIG. 10. Circular dichroism titration of apo-aspartate aminotransferase with pyridoxal phosphate (PLP) and 5′-methyl PLP. *1*, PLP, pH 5.2, λ = 430 mμ; *2*, 5′-methyl PLP, pH 5.2, λ = 430 mμ.

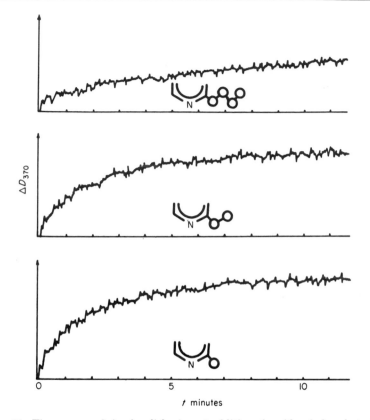

FIG. 11. Time course of circular dichroism at addition of pyridoxal phosphate (PLP) and its analogs to apoaspartate aminotransferase. Concentrations: protein, 4.3 mg/ml; analogs, 2.10^{-4} M; 0.1 M phosphate buffer, pH 7.1, $t = 20°$.

(inorganic phosphate is known to act as an inhibitor of the association reaction[23]). Under these conditions the characteristic times are of the order of minutes or hours, depending on the structure of the analog.

The procedure is as follows: to the cell containing the apoenzyme solution (protein concentration is varied from 1.0 to 5.0 mg/ml) of pH 7.1 in 0.1 M phosphate buffer and placed in the dichrograph, 0.01 ml of a suitably concentrated solution of analog is added. Final concentrations of the analog in the samples are varied from a 1.5-fold to a 3.0-fold excess over apoenzyme (calculated per mole of active sites). The sample is throughly mixed for 3–4 seconds, and the recording of CD at 360 mμ is started. The time elapsing between addition of cofactor and the start of recording is 10–15 seconds; for all analogs, with the only exception of 6-methylpyridoxal phosphate, it is negligibly low as compared with the characteristic times of the reaction.

The initial kinetic curves for pyridoxal phosphate and its 2-ethyl and 2-butyl analogs are shown in Fig. 11.

For a bimolecular reaction one can write:

$$k_2t = \frac{1}{C_0 - A_0} \ln \frac{A_0(C_0 - X)}{C_0(A_0 - X)} \tag{3}$$

Here k_2 is the bimolecular rate constant; C_0 is the total coenzyme concentration; A_0 is the total protein concentration, and X, the holoenzyme concentration,

$$X(t) = \frac{\Delta D_{360,\, t}}{\Delta D_{360,\, t\, =\, \infty}} A_0 \tag{4}$$

To express A_0 in moles of active sites, one should use the relation: 1 mg/ml of AAT = 2.2×10^{-5} M of active sites, which follows from data of Fig. 9. Substituting the X values for different periods of time from initial kinetic curves (Fig. 11) into Eq. 4, secondary plots are obtained, k_2 values being tangents of angles of slopes to abscissa of straightened lines (Fig. 12). If the rate-limiting step is, in fact, a bimolecular one, the same straight line must be obtained for an individual analog upon variation of apoenzyme and analog concentrations within sufficiently wide limits. It should be noted, that the k_2t values are the least reliable for period when X is approximating A_0, as well as in the initial period, when a "mixing effect" can exist. For the indicated type of dichrograph and the above-mentioned experimental conditions, the k_2t values are reliable within the range from $t \simeq 1$ minute to a stage corresponding to 70–80% completion of the reaction. The main cause

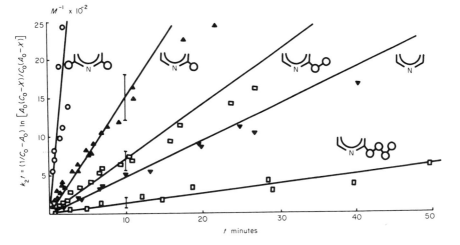

FIG. 12. Kinetics of association of pyridoxal phosphate and its analogs with apoaspartate aminotransferase plotted in the special coordinates for the bimolecular process.

of scatter of points in this working interval is due to uncontrolled variations in enzyme samples from different isolations.

The specific rates of association reaction for a number of alkyl analogs of pyridoxal phosphate are indicated in Table XIII.

Enzymatic Activity of the Reconstituted Holoenzymes

The enzymatic activity can be estimated spectrophotometrically by measuring the rates of increase in optical density at 280 mμ[24] (formation of the enol form of oxaloacetate produced in the enzymatic reaction).

The Michaelis parameters, K_m and V_{max}, for the individual systems were determined as described in footnote 25. The values thus obtained for holoenzymes reconstituted from apoenzyme and alkyl analogs of pyridoxal phosphate are listed in Table XIII.

[24] P. S. Cammarata and P. P. Cohen, *J. Biol. Chem.* **193,** 45 (1951).
[25] S. F. Velick and J. Vavra, *J. Biol. Chem.* **237,** 2109 (1962).

[90] Preparation and Catalytic Actions of Vitamin B$_6$ N-Oxides[1]

By SABURO FUKUI, NOBUKO OHISHI, and SHOICHI SHIMIZU

Preparation Procedure

Although PN N-oxide was first synthesized by Sakuragi and Kummerow[2] from triacetyl PN, oxidation of PN-free base with hydrogen perox-

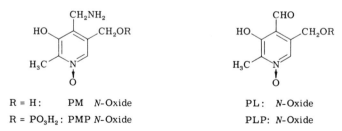

R = H :   PM *N*-Oxide
R = PO$_3$H$_2$: PMP *N*-Oxide

PL : *N*-Oxide
PLP: *N*-Oxide

FIG. 1. Structure of pyridoxal phosphate (PLP) N-oxide and pyridoxamine phosphate (PMP) N-oxide.

[1] Abbreviations used: PN, pyridoxine; PL, pyridoxal; PM, pyridoxamine; PLP and PMP, pyridoxal 5′-phosphate and pyridoxamine 5′-phosphate (Fig. 1).
[2] T. Sakuragi and F. A. Kummerow, *J. Org. Chem.* **24,** 1032 (1959).

ide in glacial acetic acid gives a better yield (37%). A solution of PN-free base (2 g) in a mixture of 20 ml of glacial acetic acid and 2 ml of 30% H_2O_2 is heated at 65°. Each 2-ml portion of H_2O_2 is added after 2 and 4 hours, respectively. The oxidation reaction is performed for 8 hours. After being cooled in an ice box overnight, the reaction mixture is evaporated to dryness *in vacuo*, and the residue is allowed to stand in a vacuum desiccator over KOH for complete removal of the remaining acetic acid. Recrystallization from water yields PN N-oxide as colorless needles (m.p. 170°, dec.).

Oxidation of PN N-Oxide to PL N-Oxide. This oxidation is performed as follows: Colloidal manganese dioxide is prepared by adding a saturated $NaHSO_4$ aqueous solution to 2.45 g of $KMnO_4$ dissolved in ca. 70 ml of water. After collection by centrifugation and thorough washing with water, the colloidal MnO_2 is suspended in about 120 ml of water. To the suspension is mixed 2.8 g of PN N-oxide with stirring, and 1.23 ml of conc. H_2SO_4 diluted with 30 ml of water is added gradually to the mixture. The resulting yellow transparent solution, after being allowed to stand overnight in an ice box, is concentrated *in vacuo* at below 35°. The resulting yellow precipitates are collected and recrystallized from water to yield colorless fibrous crystals containing 1 mole of water of crystallization (m.p. 120°, dec.). The yield of recrystallized PL N-oxide is 42%. This compound is positive to $FeCl_3$ and reacts with 2,4-dinitrophenylhydrazine or hydroxylamine. Reduction with palladium-black in ethanol yields PN N-oxide. Ultraviolet absorption maxima of PL N-oxide are 292 mμ in 2 N HCl, 259 and 297 mμ at pH 3.0, and 236, 310, and 395 mμ at pH 10.0.

PM N-Oxide. This oxide is prepared from triacetyl PM by oxidation with hydrogen peroxide in glacial acetic acid. A solution of triacetyl PM (7.67 g) in 65 ml of glacial acetic acid and 8.9 ml of 30% H_2O_2 is heated at 65° for about 2 hours. After the reaction mixture is evaporated to near dryness *in vacuo*, the residue is dissolved in a small amount of water, and the solution is allowed to stand in an ice box overnight. The yield of triacetyl PM N-oxide obtained by this procedure is 4.17 g. This compound is dissolved in a mixture of conc. HCl, ethanol, and water (7.5:60:32.5, v/v), and the solution is refluxed for 3 hours in an oil bath. After treatment with active charcoal, the reaction mixture is concentrated to near dryness under reduced pressure. The residue is kept in a vacuum desiccator over KOH, then dissolved in a small amount of water. On neutralization of the solution with NaOH, crude crystals of PM N-oxide precipitate (yield, 3.67 g). Recrystallization from methanol bubbled previously with HCl gas gives pure PM N-oxide·2HCl as colorless plates (m.p. 120°). This compound is positive to the ninhydrin test and to $FeCl_3$. It forms a yellow Schiff's base by reacting with *p-N*-dimethylaminobenzaldehyde, and converts to PM by catalytic reduction, PtO_2 being used as a catalyst. Absorption maxima in the UV spectra are at 296.5 mμ in 2 N HCl; 262 and 304 mμ at pH 3; 236, 257 (shoulder), and 325 mμ at pH 10.

PLP N-Oxide. This oxide is obtained by phosphorylation of PL N-oxide with P_2O_5 and H_3PO_4 after its aldehyde group is protected with *p*-toluidine according to the method of Murakami *et al.*[3]: The monohydrate form of PL N-oxide (2 g) is dissolved in a small amount of 2 N HCl. An aqueous solution of *p*-toluidine (1.8 g dissolved in a small amount of water) is added to the solution, and the pH of the mixture is adjusted to 5–6 by dropwise addition of 4 N NaOH under cooling. By this procedure the Schiff's base of PL N-oxide with *p*-toluidine precipitates as yellow crystals. This Schiff's base (2 g) is phosphorylated by reaction with a phosphorylating reagent prepared from 17.3 g of P_2O_5 and 19 g of 85% H_3PO_4 at 45° for 6 hours. In order to convert the resulting polyphosphates to monophosphate, the reaction mixture is heated with 0.05 N HCl for 10 minutes at 60° and adjusted with 2 N KOH to pH 3–4 under cooling in ice water. By this treatment, orange-red colored Schiff's base of PLP N-oxide is obtained. The precipitate (1.7 g) is dissolved in 8 ml of 2 N NaOH, and liberated *p*-toluidine is removed by extraction with ether. The pH of the aqueous layer is adjusted to 2–3, and the resulting dark-brown precipitate is discarded. After being adjusted to pH 7 with 2 N NaOH, the filtrate is subjected to column chromatography on Dowex 50-X8 (H$^+$ form) using water as a developer. Yellow fractions (absorption maximum, 390 mμ) eluted at an early stage of the elution are collected and concentrated *in vacuo* to concentrated syrup. On cooling this syrup in an ice box, crude PLP N-oxide separates out as light-yellow crystals. Recrystallization from water gives pure crystals of PLP N-oxide in a yield of 3 g (m.p. 176–178°, dec.).

This compound is positive to $FeCl_3$ and to the ammonium phosphomolybdate reagent and forms a 2,4-dinitrophenylhydrazone or osazone. The nuclear magnetic resonance (NMR) spectrum in D_2O consists of the following peaks: δ (in ppm from internal DSS reference) 2.40, 3H singlet of 2-CH$_3$; 5.08, 2H doublet (J, 6 Hz) of 5-CH$_2$-; 7.60, 1H singlet of 6-H; 10.17, 1H broad singlet of 4-CHO. Absorption maxima in the UV spectrum are 245 (shoulder), 304 and 350 mμ (shoulder) in 0.1 N HCl; 237, 296, and 393 mμ at pH 7.0; 238, 295, and 392 mμ in 0.1 N NaOH.

PMP N-Oxide. This oxide is prepared from PM N-oxide by phosphorylation with P_2O_5 and H_3PO_4 at 60–70° for 150 minutes, followed by hydrolysis of the resulting polyphosphates with water at 100° for 35 minutes. The reaction mixture is evaporated to dryness under a reduced pressure. To

[3] M. Murakami, M. Iwanami, and T. Murata, Japanese Patent 18,749 (1965); *Chem. Abstr.* **56**, 2069 (1966).

the residue is added absolute alcohol, and the mixture is allowed to stand in an ice box overnight. Crude PMP N-oxide is obtained as a white precipitate. After the crude crystals have been dissolved in water containing a small amount of pyridine, a large amount of acetone is added to the solution. Pure PMP N-oxide is obtained as white crystals (m.p. 210–213°, dec.). The yield is ca. 70%. This compound is positive to the ninhydrin test, to $FeCl_3$, and to ammonium phosphomolybdate. The NMR spectrum in D_2O shows the presence of 2-CH_3 (δ, 2.39 ppm), 4-CH_2- (4.02 ppm), 5-CH_2- (4.78 ppm, doublet), and 6-H (7.64 ppm). Absorption peaks in the UV spectra are 263 and 298 mμ in 0.1 N HCl; 257 (shoulder) and 327 mμ at pH 7.1; 257 (shoulder) and 323 mμ in 0.1 N NaOH.

Catalytic Activity in Nonenzymatic and Enzymatic Systems[4]

In the nonenzymatic α,β-elimination reactions of tryptophan[5] and serine,[6] the catalytic activity of PL N-oxide is approximately 80% of that of PL. Nonenzymatic transamination between PL N-oxide and glutamate occurs only slightly. On the other hand, the reaction between PM N-oxide and α-ketoglutarate takes place to an appreciable degree, but the reaction rate is significantly less than that between PM and the acid.[7]

The Schiff's base (imine) formed between PL N-oxide and valine exhibits absorption maxima at 319 and 412 mμ in weak acidic and neutral pH regions. The apparent equilibrium constant for the imine formation at pH 5.2 is 29 M^{-1}, which is markedly greater than that of PL-valine imine,[8] 0.5 M^{-1}, at the same pH. The imine chelate formed from PL N-oxide, valine, and Cu^{2+} shows absorption peaks at 305 and 382 mμ at pH 5.22. At that pH more than 80% of PL N-oxide converts to the imine chelate, while the conversion of PL calculated at 280 and 317 mμ has been reported as 75% by Davis et al.[9] These results indicate that PL N-oxide can form the imine as well as the imine chelate more easily than PL at the acidic pH region in spite of the lower catalytic activity of the analog in the nonenzymatic systems mentioned above.

PLP N-oxide serves as a coenzyme for E. coli tryptophanase.[10] The K_{Co} value, the concentration required for the half-maximum activity of the reconstituted enzyme, is ca. 5.0 μM. The value is a little larger than that of

[4] S. Fukui, N. Ohishi, Y. Nakai, and S. Shimizu, Arch. Biochem. Biophys. **130**, 584 (1969).
[5] E. McEvoy-Bowe, Arch. Biochem. Biophys. **113**, 167 (1966).
[6] D. E. Metzler and E. E. Snell, J. Biol. Chem. **198**, 353 (1952).
[7] D. E. Metzler and E. E. Snell, J. Am. Chem. Soc. **74**, 979 (1952).
[8] D. E. Metzler, J. Am. Chem. Soc. **79**, 485 (1957).
[9] L. Davis, F. Roddy, and D. E. Metzler, J. Am. Chem. Soc. **83**, 127 (1961).
[10] W. A. Newton, Y. Morino, and E. E. Snell, J. Biol. Chem. **240**, 1211 (1965).

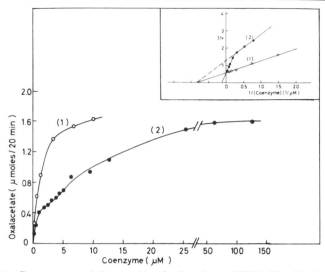

Fig. 2. Coenzyme activity pyridoxal phosphate (PLP) N-oxide in mammalian supernatant aspartate aminotransferase system. ○———○, PLP; ●---●, PLP N-oxide. (1) PLP. (2) PLPN-oxide.

PLP. V_{max} of the reaction catalyzed by the PLP N-oxide enzyme is about 65% of that of the PLP-dependent reaction. The low catalytic activity of PLP N-oxide in the tryptophanase reaction may be ascribed to the low electron-attracting ability of the pyridine nitrogen of the analog in a similar way to that of PL N-oxide in nonenzymatic systems. PLP N-oxide-bound apotryptophanase exhibits UV absorption maxima at 320 and 418 mμ.

Figure 2 shows the coenzyme activity of PLP N-oxide for supernatant aspartate aminotransferase (GOT) of pig heart muscle.[11] Although the reactivation of apo-GOT with PLP N-oxide requires a little longer time than that of PLP, 10 minutes' preincubation is sufficient for the complete reconstitution of the desired holoenzyme species. The analog-bound enzyme exhibits an obvious but significantly low activity in the enzymatic reaction. The shape of the reaction velocity–PLP N-oxide concentration curve is complicated and nonhyperbolic. The addition of a high concentration of PLP N-oxide to the reaction mixture exerts an allosteric activation effect. The N—O bond of the coenzyme analog is maintained during the enzyme reaction. The K_{Co} value obtained by the data of the 20-minute reaction is 0.56 μM, nearly equal to that of the native coenzyme, whereas the V_{max} value is about 26% of that catalyzed by the normal enzyme.

In the case of mitochondrial GOT, such complicated phenomena

[11] M. Martinez-Carrion, C. Turano, E. Chiancone, F. Bossa, A. Giartosio, F. Riva, and P. Fasella, *J. Biol. Chem.* **242**, 2397 (1967).

are not observed. PLP N-oxide-bound GOT shows two absorption bands (λ_{max} at 317 and 410 $\sim$ 420 mμ) varying in a reciprocal manner with pH as in the case of tryptophanase, and it exhibits a sharp CD peak at 420 mμ at pH 5.0. Optimal pH values of tryptophanase and GOT reactions dependent on PLP N-oxide are almost the same as those of the native enzymes.

[91] Preparation and Catalytic Actions of ω-Hydroxy Derivatives of Vitamin B₆[1]

By Saburo Fukui, Yoshiharu Nakai, and Fuminori Masugi

Although it has been known that the methyl group at position 2 of vitamin B₆ is not essential for its catalytic action in nonenzymatic model reactions, one should not preclude the possibility that a hydrophobic interaction between the 2-methyl group of PLP[1a] with apoprotein would have an interesting influence on the enzyme reactions dependent on the coenzyme. This paper deals with the preparation and catalytic actions of analogs of vitamin B₆ containing a hydroxymethyl group at position 2 instead of the methyl group (designated as ω-hydroxy PLP, etc.) (Fig. 1).

ω-Hydroxy PLP   ω-Hydroxy PMP

Fig. 1. Structure of ω-hydroxy pyridoxal phosphate (PLP) and ω-hydroxy pyridoxamine phosphate (PMP).

Preparation

Principle

ω-Hydroxy PM is obtained from PM N-oxide[1a] by heating with acetic anhydride, and phosphorylated to ω-hydroxy PMP with P_2O_5 and H_3PO_4. Conversion of ω-hydroxy PMP to ω-hydroxy PLP is carried out by transamination between ω-hydroxy PMP and pyridine 2-aldehyde.

---

[1] Abbreviations used: PN, pyridoxine; PL, pyridoxal; PM, pyridoxamine; PLP and PMP, pyridoxal 5′-phosphate and pyridoxamine 5′-phosphate.
[1a] See this volume [90].

Procedure

1. *ω-Hydroxy PM*. PM-free base (8.0 g) is dissolved in a mixture of 100 ml of acetic anhydride and 10 ml of pyridine, and allowed to stand at 37° for 24 hours. The volatile portion is removed *in vacuo*. Recrystallization of the resulting residue from water yields 8.7 g of 3,4,5-triacetyl PM in the form of white needles. A solution of 24 g of triacetyl PM in 180 ml of glacial acetic acid and 27 ml of 30% H_2O_2 is heated for 4.5 hours at 65°, and the reaction mixture is evaporated to dryness under reduced pressure. The resulting red-brown colored residue is, without further purification, dissolved in 200 ml of acetic anhydride, and the solution is heated for 40 minutes at 120–130°, then evaporated to dryness *in vacuo*. A dark-brown viscous residue is dissolved in 160 ml of a mixture of conc. HCl–water–ethanol (1:1:2), and the solution is heated for 2 hours at 110–120°, then the volatile portion is removed *in vacuo*. A solution of the resulting dark-brown residue in 20 ml of hot 50% ethanol is treated with active charcoal and the supernatant solution is allowed to stand at −20° overnight. Crude ω-hydroxy PM is obtained by this procedure, and recrystallization from 50% ethanol gives pure crystals in the form of colorless needles (m.p. 187–189°, dec.). The yield is 26.7%.

This compound is positive in the ninhydrin test, to Gibbs reagent and to $FeCl_3$. The result of elemental analysis coincides with $C_8H_{12}O_3N_2 \cdot 3\frac{1}{2}HCl$. Absorption maxima in UV spectra are at 294.0 mμ in 0.1 *N* HCl; 251 and 325 mμ at pH 7.0; 246 and 312 mμ in 0.1 *N* NaOH. The nuclear magnetic resonance (NMR) spectrum (in D_2O) shows the following peaks (in ppm from internal DSS reference): 4.42, 2H singlet of 4-CH_2-; 4.82, 2H singlet of 2-CH_2-; 5.05, 2H singlet of 5-CH_2-; 8.18, 1H singlet of 6H.

2. *ω-Hydroxy PMP*. Phosphorylation of ω-hydroxy PM (2 g) is performed at 70° for 2.5 hours using P_2O_5 and H_3PO_4 in the usual way. In order to convert the resulting polyphosphate to monophosphate, the reaction mixture is heated with water at 100° for 35 minutes, and evaporated *in vacuo* to a viscous concentrated syrup. To the residue is added 200 ml of anhydrous ethanol, and the mixture is allowed to stand at −20° overnight. White precipitates obtained from the solution are collected by centrifugation, washed successively with anhydrous ethanol and acetone, and dissolved in 10–15 ml of water. The pH of the solution is adjusted to 7.5 by the addition of saturated $Ba(OH)_2$, and the resulting $Ba_3(PO_4)_2$ is separated by centrifugation. After evaporation to about 5 ml and adjustment to pH 5.6 with NH_4OH, the supernatant solution is subjected to column chromatography on Amberlite CG 50 (H^+ form). Ninhydrin-positive fractions eluted with water are collected and evaporated to a concentrated syrup under reduced pressure. ω-Hydroxy PMP is crystallized in the form of

colorless needles from this syrup after being allowed to stand in an icebox (m.p. 135–137°). The yield is 62%.

The result of elemental analysis is consistent with $C_8H_{13}O_6N_2P \cdot 2\ H_2O$. This compound is positive to the ninhydrin test, to $FeCl_3$, and to the ammonium phosphomolybdate reagent. Absorption maxima in the UV spectrum are at 294.5 mμ in 0.1 N HCl, 253 and 327 mμ at pH 7.0, and 246 and 312 mμ in 0.1 N HCl. The UV peaks at 253 and 325 mμ in phosphate buffer (pH 7.4) shift to 298 mμ in borate buffer of the same pH. This phenomenon is considered to be due to the formation of a borate complex between the 2-hydroxymethyl group and the 3-hydroxy group of the compound. In the NMR spectrum, the peak at δ, 5.09 ppm assignable to the 2-CH$_2$- is a singlet, while the signal at 5.23 ppm assignable to the 5-CH$_2$- is a doublet. These results, together with the results of the enzymatic study mentioned later, strongly support the presence of the free 2-hydroxymethyl group.

3. *ω-Hydroxy PLP*. This compound is prepared by the transamination reaction between ω-hydroxy PMP and pyridine 2-aldehyde as described by Asano *et al.*[2] An aqueous solution of ω-hydroxy PMP and pyridine 2-aldehyde (molar ratio, 1:2) is adjusted to pH 6.5 with NaOH and allowed to stand at room temperature under a nitrogen atmosphere. After 1–3 hours, the reaction mixture is applied to a column of Dowex 50-X8 (H⁺ form). Yellow fractions (λ_{max}, 390 mμ) eluted with water are collected and evaporated to dryness *in vacuo*. By dissolving the residue in a small amount of methanol and scratching the flask, crude ω-hydroxy PLP is obtained as a light-yellow precipitate. Recrystallization from aqueous ethanol gives light-yellow crystals of ω-hydroxy PLP containing one mole of ethanol (m.p. 94–96°, dec.). The yield is 26%. The result of elemental analysis shows the composition as $C_8H_{10}O_7P \cdot C_2H_5OH \cdot 2H_2O$.

This compound is positive to $FeCl_3$ and to ammonium phosphomolybdate, and reacts with 2,4-dinitrophenylhydrazine or hydroxylamine. The absorption peaks in UV spectra are as follows: 294.5 and 338 mμ (shoulder) in 0.1 N HCl; 387 and 338 mμ (shoulder) at pH 7.0; 388 mμ in 0.1 N NaOH. The NMR spectrum in D_2O plus NaOD shows the following peaks: δ (in ppm from internal DSS reference): 4.77, 2H singlet of 2-CH$_2$-; 5.08, 2H doublet (J, 8 Hz) of 5-CH$_2$-; 7.80, 1H singlet of 6H; 8.70, 1H broad singlet of 4-CHO. The infrared (IR) spectrum of this compound is consistent with its structure. The paper electrophoretic behavior of ω-hydroxy PMP and ω-hydroxy PLP in $M/30$ phosphate buffer (pH 7.0) are almost identical with those of PMP and PLP, respectively. In ascending paper chromatography using a solvent system consisting of *t*-butyl alcohol (40), acetone

² K. Asano, T. Tanaka, M. Furukawa, and S. Shimada, Japanese Patent 1966–5095 (1966).

(35), diethylamine (5), and water (20), ω-hydroxy derivatives of B_6 exhibit lower R_f values than the corresponding vitamins: PN 0.56, ω-hydroxy PN 0.48; PM 0.78, ω-hydroxy PM 0.49; PMP 0.27, ω-hydroxy PMP 0.17; PLP 0.42, ω-hydroxyl PLP 0.28.

Catalytic Activity[3]

In the nonenzymatic transamination between α-ketoglutarate using Al^{3+} as catalyst,[4] ω-hydroxy PM is more effective than PM. The K_{Co} value of ω-hydroxy PLP for aspartate aminotransferase (GOT, EC 2.6.1.1) of pig heart muscle, the concentration required for a half-maximum activity of the reconstituted holoenzyme species, is 0.176 μM, nearly equal to that of PLP, and the V_{max} value of the ω-hydroxy PLP enzyme-dependent reaction is 1.3 times larger than that catalyzed by native GOT in Tris-HCl buffer (pH 8.0). In KPO_4 buffer (pH 7 to 8), however, the situation is markedly different: The K_{Co} value, ca. 25 μM, is about ten times as large as that of PLP and the V_{max} is approximately 60% of that of the native enzyme. In this case, phosphate anion acts as competitive inhibitor. On the other hand, the K_{Co} value for *Escherichia coli* tryptophanase is ca. 25 μM, markedly larger than that of PLP, and the V_{max} value is about 60% of that catalyzed by the native holoenzyme. Absorption maxima of ω-hydroxy PLP-bound apo-GOT are 429 mμ at pH 5.0 and 361 mμ at pH 8.0. The peaks are almost identical with those of native GOT.[5] ω-Hydroxy PLP-bound apotryptophanase shows absorption maxima at 332 mμ and 414 mμ at pH 7.0, corresponding to 337 mμ and 420 mμ of the PLP enzyme.[6]

[3] D. E. Metzler and E. E. Snell, *J. Am. Chem. Soc.* **74**, 979 (1952).
[4] S. Fukui, N. Ohishi, Y. Nakai, and F. Masugi, *Arch. Biochem. Biophys.* **132**, 1 (1969).
[5] Y. Morino and E. E. Snell, *Proc. Natl. Acad. Sci. U.S.* **57**, 1692 (1967).
[6] Y. Morino and E. E. Snell, *J. Biol. Chem.* **242**, 2800 (1967).

[92] Labeled Pyridoxine

By J. WURSCH, S. F. SCHAEREN, and H. FRICK

Synthetic procedures for the preparation of radioactive compounds should be straightforward and adaptable to small-scale work, especially if products of high specific activities are required. None of the numerous preparative methods for pyridoxine reported in the chemical literature[1] seems to have fulfilled these requirements before the principle of Diels-Alder condensations became widely applied in this series. Based on this approach, several research groups have successfully worked out simple

[1] For a review see J. M. Osbond, *Vitamins Hormones* **22**, 367 (1964).

syntheses of pyridoxine: A combination of these is presented here using readily accessible starting materials. A slight but not serious disadvantage of this method lies in the fact that one of the most advantageously applied radioactive starting materials, i.e., diethylfumarate, is always labeled in two positions, leading thus to pyridoxine labeled on two carbon atoms. On the other hand, products of correspondingly higher specific activities may be prepared for the same reason.

The reaction sequence is as follows: Diethylfumarate (II) is condensed with 4-methyl-5-ethoxyoxazole (I)[2] to give the intermediate (III) which is rearranged under controlled acidic conditions to diethyl 5-hydroxy-6-methylcinchomeronate (IV). Reduction of (IV) with lithium aluminum hydride affords pyridoxine (V). Either starting material may be labeled.

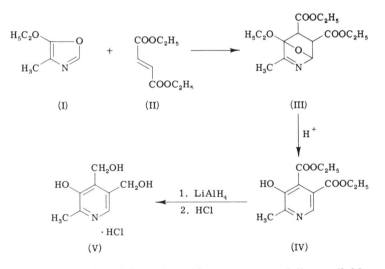

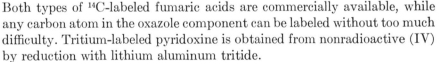

Both types of ^{14}C-labeled fumaric acids are commercially available, while any carbon atom in the oxazole component can be labeled without too much difficulty. Tritium-labeled pyridoxine is obtained from nonradioactive (IV) by reduction with lithium aluminum tritide.

5-Hydroxy-6-methylcinchomeronic Acid Diethyl Ester (IV)

A mixture of 870 mg (5 millimoles) of diethylfumarate and 635 mg (5 millimoles) of 4-methyl-5-ethoxyoxazole is kept under dry argon for 15 hours at 40°. With stirring 2.5 ml of water is added, the mixture is cooled to 10°, and 1.0 ml of 0.1 N HCl is dropped in over a period of 1 hour. Stirring is continued for 1 hour at 20°, after which time the mixture is cooled to 0° and treated dropwise with 1.0 ml of 1 N HCl. One-half hour after the

[2] E. E. Harris et al., J. Org. Chem. 27, 2705 (1962).

last drop, approximately 1.25 ml of 3 N HCl is added dropwise until a pH of 2.5 is reached. The mixture is now a suspension of crystals [hydrochloride of (IV)], which is dissolved by adjusting the pH to 8–9 with 3 N NH$_4$OH. Neutral impurities are removed by extraction of the ammonium salt solution with ether. The aqueous phase is readjusted to pH 5 with 3 N HCl and extracted 3 times with 20-ml portions of chloroform. The combined chloroform extracts are evaporated and the residue is distilled giving 1130 mg of (IV) boiling at 95°/0.01 mm of mercury, n_D^{20} 1.5229 (89% of theoretical yield). The purity of the material is checked by gas-liquid chromatography (GLC) on an Apiezon M column at 230°. This reaction may be performed on a much smaller scale.

Pyridoxine Hydrochloride (V)

Lithium aluminum hydride (180 mg, 4.7 millimoles) is placed into a stirring flask kept under dry argon. Three milliliters of anhydrous tetrahydrofuran free of peroxide is added and stirred with the hydride for 0.25 hour at room temperature. The mixture is now cooled to 10°, and 600 mg (2.4 millimoles) of (IV) dissolved in 4.5 ml of tetrahydrofuran is added dropwise during 0.7 hour. Stirring is continued for 2.5 hours at room temperature. With cooling in an ice bath, 2.5 ml of 96% ethanol, followed by 2.5 ml of 36% ethanolic HCl, is added dropwise, and stirring is continued for 0.5 hour at room temperature. The mixture is removed from the stirring flask with 15 ml of water and evaporated under reduced pressure. This evaporation is repeated twice with 15 ml-portions of water in order to remove all the organic solvent from the reaction mixture. The residue is taken up in 15 ml of water and adjusted to pH 7.2 with 25% NaOH. The solution is cooled to 15°, 3 ml of acetic anhydride is added, and the mixture is stirred for 1 hour at 15–20°. About 5 ml of water is distilled off under reduced pressure, and the residual solution is extracted 8 times with 10-ml portions of methylene chloride. The combined extracts are washed with 5 ml of water and evaporated to dryness. The residue is dissolved in 10 ml of 10% HCl and warmed to 90° for 0.5 hour. After that time the hydrochloric acid is evaporated at 50°, last traces being removed by several evaporations with 10-ml portions of anhydrous ethanol. Upon trituration with methanol, the residue crystallizes completely and after drying in a desiccator weighs 450 mg. It is recrystallized from 6 ml of methanol containing one drop of concentrated HCl. A first crop of 310 mg is followed, after concentration of the mother liquor, by further crops of 60 and 15 mg. Colored material is effectively decolorized with small amounts of Norite SX-1 in very dilute aqueous solution (25 ml for 50 mg). A total yield of 385 mg is obtained corresponding to 79.5% of the theoretical based on (IV). The purity of the material is better than 99% as shown by paper electrophoresis and radioscanning.

On this scale ^{14}C-labeled pyridoxines were prepared with both types of ^{14}C-labeled diethylfumarate as starting materials. The specific activities obtained were: Pyridoxine hydrochloride labeled in the ring: 142 μCi/mg, 29 mCi/millimole; labeled in the hydroxymethyl groups: 53 μCi/mg, 11 mCi/millimole.

[93] Preparation and Properties of Antigenic Vitamin and Coenzyme Derivatives

By Jean-Claude Jaton and Hanna Ungar-Waron

The vitamin B₆ family plays vital roles as coenzymes in a wide variety of metabolic transformations of amino acids, such as decarboxylation, transamination, and racemization.[1] The possibility that antibodies to vitamin B₆ may prove to be helpful in the elucidation of various enzymatic steps in the metabolism governed by the active forms of the vitamin makes it desirable to produce such antibodies against pyridoxal.

The covalent attachment of small molecules of biological interest to synthetic polypeptides or proteins has been shown to result in conjugates capable of eliciting antibodies with specificity directed toward the determinant groups coupled to the carrier. Thus, the production of antibodies specific for monosaccharides,[2] nucleosides,[3] hormones,[4–6] and drugs[7,8] has been reported.

Similarly, pyridoxal can confer both immunogenicity and antigenic specificity, when properly attached to the essentially nonimmunogenic multichain poly-DL-alanine.[9] The coenzyme can react with the free amino groups of the poly-DL-alanine side chains by forming a Schiff base at neutral pH (Fig. 1). The double bond thus formed is subsequently reduced by means of sodium borohydride, which stabilizes the linkage between pyridoxal and the polypeptide. Rabbit antibodies, provoked upon immunization with

[1] W. W. Umbreit, *in* "The Vitamins" (W. H. Sebrell, Jr. and R. S. Harris, eds.), Vol. III, p. 234. Academic Press, New York, 1954.

[2] E. Rüde, O. Westphal, E. Hurwitz, S. Fuchs, and M. Sela, *Immunochemistry* **3**, 137 (1966).

[3] H. Ungar-Waron, E. Hurwitz, J. C. Jaton, and M. Sela, *Biochim. Biophys. Acta* **138**, 513 (1967).

[4] T. L. Goodfriend, L. Levine, and G. D. Fasman, *Science* **144**, 1344 (1964).

[5] J. McGuire, R. McGill, S. Leeman, and T. Goodfriend, *J. Clin. Invest.* **44**, 1672 (1965).

[6] S. J. Gross, D. H. Campbell, and H. H. Weetall, *Immunochemistry* **5**, 55 (1968).

[7] R. N. Hamburger, *Science* **152**, 203 (1966).

[8] J. C. Jaton and H. Ungar-Waron, *Arch. Biochem. Biophys.* **122**, 157 (1967).

[9] H. Ungar-Waron and M. Sela, *Biochim. Biophys. Acta* **124**, 147 (1966).

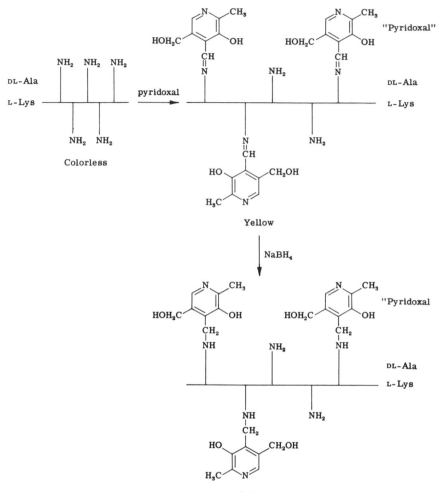

Fig. 1. Schematic representation of the synthesis of multichain pyridoxal-poly-peptide conjugate, pyridoxal-polyDLAla–polyLLys. From H. Ungar-Waron and M. Sela, *Biochim. Biophys. Acta* **124**, 147 (1966).

the resulting pyridoxal-polyDLAla–polyLLys conjugate, are mainly pyri-doxamine-specific and exert a marked inhibitory effect in the glutamic–oxaloacetic transaminase system.[9]

Synthesis of Pyridoxal Conjugate with Multichain Poly-DL-alanine

Multichain poly-DL-alanine(polyDLAla–polyLLys), a branched polymer in which side chains of poly-DL-alanine are attached to a poly-L-lysine core, is synthesized as previously reported.[10]

Pyridoxal hydrochloride (600 mg) is dissolved in 5 ml of water, and the resulting solution is neutralized with 1 N NaOH (pH 7.0) and added to polyDLAla–polyLLys (1 g) in a final volume of 20 ml of water. The reaction mixture is stirred for 15 minutes at room temperature and then cooled to 4°. Sodium borohydride (500 mg) is subsequently added; the reaction mixture is left for another 15 minutes in the cold and then dialyzed against 6 liters of distilled water for 3 days with two daily changes of water. The resulting pyridoxal-polyDLAla–polyLLys conjugate is recovered by lyophilization and stored at −20°.

The pyridoxal content of the polymer, as determined from the extinction at 325 mμ, at pH 7.0, and colorimetrically (dichloroquinone–chloroimide reagent),[11] ranges for various preparations, between 7 and 9% (weight ratio). The molecular weight of the conjugate, as determined from sedimentation and diffusion data, depends on the molecular weight of the multichain poly-DL-alanine used as a "carrier" for the synthesis of the pyridoxal conjugate, and is usually in the range of 90,000–150,000.

The immunization procedure with the antigen, pyridoxal-polyDLAla–polyLLys, as well as the immunospecific isolation of rabbit antipyridoxal antibodies are similar to those described in the section on "Antibodies to Folic Acid." Amounts of 100–300 μg of pyridoxal specific antibodies per milliliter of antiserum are usually obtained.

---

[10] M. Sela and S. Fuchs, in "Methods in Immunology and Immunochemistry" (M. W. Chase and C. A. Williams, eds.), Vol. 1, p. 167. Academic Press, New York, 1967.
[11] M. Hochberg, D. Melnick, and B. L. Oser, J. Biol. Chem. 155, 109 (1944).

[94] Pyridoxal Kinases (ATP:Pyridoxal 5-Phosphotransferase, EC 2.7.1.35) from Bacteria and from Mammalian Tissues

By DONALD B. MCCORMICK and ESMOND E. SNELL

Pyridoxal, -ine, -amine + ATP → pyridoxal, -ine, -amine 5-phosphate + ADP

Assay Method

Principle. The pyridoxal 5-phosphate formed is measured by a modification[1] of the method of Wada *et al.*[2] in which apotryptophanase from *Escherichia coli* is activated by the coenzyme to degrade tryptophan to

[1] D. B. McCormick, M. E. Gregory, and E. E. Snell, J. Biol. Chem. 236, 2076 (1961).
[2] H. Wada, T. Morisue, Y. Sakamoto, and K. Ichihara, J. Vitaminol. (Kyoto) 3, 183 (1957).

indole, the quantity of which is determined by colorimetric reaction with
p-dimethylaminobenzaldehyde in acid–ethanol. When the 5-phosphate of
pyridoxine or of pyridoxamine is the product from the kinase reaction, the
filtrate is first incubated with an oxidase[3] from rabbit liver to form pyridoxal
5-phosphate.

Reagents

> Pyridoxal, -ine, -amine, 0.001 M. Fresh solutions are made by neu-
> tralizing the hydrochloride salts.
> ATP, 0.01 M, pH 8
> MgSO$_4$, 0.01 M
> ZnSO$_4$, 0.001 M
> Potassium acetate buffer, 0.75 M, pH 5.25
> Potassium phosphate buffers, 0.75 M, pH 6.0, 6.5, 7.0, and 8.3
> Pyridoxal 5-phosphate, 10 mg neutralized in 10 ml. The stock solu-
> tion is kept refrigerated in the dark; standard solutions are made
> by diluting 0.1 ml to 100 ml for 1 μg/ml.
> Potassium bicarbonate, 0.5 M
> Toluene
> Reduced glutathione, 0.1 M
> L-Tryptophan, 0.05 M
> Trichloroacetic acid, 100% (w/v) in water
> p-Dimethylaminobenzaldehyde reagent, 5% (w/v) in 95% ethanol
> Sulfuric acid–ethanol, 80 ml of H$_2$SO$_4$ per liter of 95% ethanol

Enzymes

> Pyridoxine (pyridoxamine) phosphate oxidase from rabbit liver is
> purified through the alumina C$_\gamma$ gel step according to Wada and
> Snell (see this volume [96]).
> Apotryptophanase from $E.$ $coli$ is prepared by growing the Crookes
> strain for 24 hours at 37° in a medium containing 1% tryptone, 1%
> yeast extract, 0.5% potassium phosphate, and 0.1% glucose at an
> initial pH of 7. The cells are harvested by centrifugation, rinsed
> with cold water, lyophilized, and stored at −15°. Subsequent oper-
> ations are performed at 0–4°. Two grams of lyophilized cells are
> suspended in 50 ml of 0.02 M potassium phosphate buffer, pH 7.0,
> and treated for 30 minutes with a Raytheon 250-W, 10-kc magneto-
> striction apparatus. Debris is removed by centrifuging at 18,500 g
> for 30 minutes. The supernatant liquid is suspended in 50 ml of
> water and dialyzed overnight against 5 liters of distilled water.

---

[3] H. Wada and E. E. Snell, $J.$ $Biol.$ $Chem.$ **236**, 2089 (1961).

This dialyzed preparation is stored in convenient volumes at -10 to $-15°$. Immediately before use, a tube of the frozen preparation is thawed and any precipitated protein is homogenized briefly in a TenBroeck apparatus.

Procedure. Optimal conditions for the kinase vary with the origin, degree of purification, and amount of the enzyme used. Satisfactory mixtures contain 0.1–0.5 mM pyridoxal, -ine, or -amine, 0.1–0.5 mM ATP, 0.1 mM Mg^{2+} (for bacterial kinases) or 0.01 mM Zn^{2+} (for mammalian kinases), 75 mM potassium acetate, pH 5.25 (for bacterial kinases), or phosphate, pH 6.0 (for liver kinase) or 6.5 (for brain kinase), and enzyme (0.2–2.0 mg of protein) per total volume of 2.5 ml. Incubation is carried out for 1 hour at 37°. The reaction is stopped by heating the mixture in a boiling water bath for 3 minutes.

When pyridoxine or pyridoxamine is the substrate used in the kinase reaction, an aliquot of the boiled mixture containing 0–10 μg of pyridoxine 5-phosphate or 0–5 μg of pyridoxamine 5-phosphate is mixed with 0.4 ml of 0.5 M potassium bicarbonate and 0.1 ml (0.5–1.0 mg of protein) of pyridoxine (pyridoxamine) phosphate oxidase and diluted to 2.5 ml. After incubation for 1 hour at 37°, this mixture is heated in a boiling water bath for 3 minutes to stop enzymatic action, and an aliquot of 1.5 ml is taken for determination of pyridoxal 5-phosphate. Conversion to pyridoxal 5-phosphate is incomplete,[3] so that a standard curve should be established with each set of samples.

Graded volumes of a standard solution of pyridoxal 5-phosphate (to supply 0–1 μg) and samples to be assayed are pipetted into separate 50-ml Erlenmeyer flasks and diluted to 1.5 ml. To each flask is then added 5 ml of toluene followed by 0.4 ml of a solution containing 0.25 ml of 0.75 M potassium phosphate, pH 8.3, 0.1 ml of 0.1 M reduced glutathione, and 0.1 ml of the apotryptophanase preparation. The vessels are stoppered, and the contents are incubated for 10 minutes with moderate shaking in a water bath at 37° to permit association of apoenzyme and coenzyme. L-Tryptophan (0.2 ml of a 0.05 M solution) is then added to each flask, and the contents are incubated for another 10-minute period with shaking. The reaction is terminated by the addition of 0.2 ml of 100% trichloroacetic acid. The flasks are shaken vigorously for several seconds to ensure complete extraction of indole into the toluene phase. After separation of the phases, 0.5 ml of the upper toluene layer is mixed with 0.5 ml of the p-dimethylaminobenzaldehyde reagent and 5 ml of the sulfuric acid–ethanol solution. Color is allowed to develop for 10 minutes at room temperature before reading the absorbancy at 540 mμ in a spectrophotometer.

Definition of Unit and Specific Activity. One unit of enzyme is that amount of kinase which catalyzes the synthesis of 1 millimicromole of pyridoxal 5-phosphate in 1 hour at 37° in the above assay. Specific activity is defined as units per milligram of protein determined by the method of Lowry et al.[4]

Comments. The sensitivity of the apotryptophanase method for assay of pyridoxal 5-phosphate can be substantially increased or decreased by varying the size of the aliquot of toluene extract. The manometric procedure of Gunsalus and Razzell (see Vol. II [109]) as modified by Hurwitz[5] uses tyrosine apodecarboxylase from *Streptococcus faecalis*. Other modifications which use these apoenzymes and ^{14}C-substrates for satisfactory assay of pyridoxal 5-phosphate are given in other sections (see this volume [84] and [85]).

Purification Procedure

Pyridoxal kinase has been detected in numerous organisms, but the activity of similarly prepared extracts varies enormously from one organism to another.[1] Both *Streptococcus faecalis* 8043[6] and especially *Lactobacillus casei* 7469[7] afford good sources of the bacterial enzyme, which appears to be partially inducible. The kinase from mammalian cells is located in the true supernatant solution and can be readily purified from liver and brain.[1,8]

From Lactobacillus casei and Streptococcus faecalis

Step 1. Sonic Rupture of Cells. Broken-cell suspensions of bacteria are prepared by treatment of 4% cell suspensions in cold, 0.02 M potassium phosphate, pH 6.0, for 30–60 minutes in a Raytheon 10-kc sonic oscillator. All subsequent operations are performed at 0–4°.

Step 2. Preparation of Supernatant Solution. The broken-cell suspension is centrifuged at 25,000 g for 30 minutes, and the debris is discarded.

Step 3. Treatment with Protamine Sulfate. To the supernatant solution, one-fifth volume of a 2% protamine sulfate solution, pH 5.5, is added. After stirring for several minutes, the inactive precipitate is removed by centrifugation at 18,500 g for 15 minutes.

Step 4. Ammonium Sulfate Precipitation and Dialysis. The protamine sulfate-treated supernatant solution is brought to 40% saturation by addition of solid ammonium sulfate over a 1-hour period. The mixture is centrifuged at 18,500 g for 15 minutes, and the precipitate is discarded. The

[4] O. H. Lowry, N. J. Rosebrough, A. L. Farr, and R. J. Randall, *J. Biol. Chem.* **193**, 265 (1951).

[5] J. Hurwitz, *J. Biol. Chem.* **205**, 935 (1953).

[6] J. C. Rabinowitz and E. E. Snell, *J. Biol. Chem.* **169**, 631 (1947).

[7] J. C. Rabinowitz, N. I. Mondy, and E. E. Snell, *J. Biol. Chem.* **175**, 147 (1948).

[8] D. B. McCormick and E. E. Snell, *Proc. Natl. Acad. Sci. U.S.* **45**, 1371 (1959).

supernatant solution is then brought to 60% saturation with ammonium sulfate over a similar period of time and the active precipitate is collected by centrifugation. The precipitate is dissolved in a small volume of 5 mM potassium phosphate, pH 6.0, containing 0.1 mM reduced glutathione, and dialyzed overnight against a large volume of this solution; the dialyzate is clarified by centrifuging at 12,500 g for 15 minutes.

Typical partial purification of pyridoxal kinases from *L. casei* and *S. faecalis* is summarized by the data in Table I.

TABLE I

PURIFICATION OF PYRIDOXAL KINASES FROM *Lactobacillus casei* AND *Streptococcus faecalis*

	L. casei		*S. faecalis*	
Fraction	Total[a] activity (units)	Specific activity (units/mg)	Total[a] activity (units)	Specific activity (units/mg)
1. Sonicate	38,200	44.1	2120	2.4
2. Supernatant solution	37,600	54.2	2420	4.1
3. Protamine-treated solution	39,800	156	2800	59.0
4. Dialyzed (NH$_4$)$_2$SO$_4$ fraction[b]	—	—	3020	216

[a] From 2 g of lyophilized cells.

[b] This step causes considerable loss of activity with preparations from *L. casei*, but it can be used for more complete resolution of metal ion.

Comments. These procedures achieve a 90-fold purification with an apparent complete recovery of kinase activity from *S. faecalis*. The enzyme from *L. casei* behaves differently in that higher activities are seen in crude extracts, but this kinase is partially destroyed by the conventional precipitation with ammonium sulfate.

From Liver and Brain

Step 1. Homogenization of Tissue. Animals are sacrificed, and the appropriate organs are extirpated, rinsed, blotted, and homogenized with 4 volumes of cold, 0.05–0.1 M potassium phosphate buffer, pH 6.5, with the TenBroeck apparatus. All subsequent operations are performed at 0–4°.

Step 2. Preparation of Supernatant Solution. The 20% homogenate is centrifuged at 25,000 g for 30 minutes, and the debris is discarded.

Step 3. Heat Treatment (for Liver Kinase). Pyridoxal (1 mM) is added to the supernatant solution, which is warmed to 55° with stirring and held at this temperature for 15 minutes. The mixture is centrifuged at 18,500 g for 15 minutes, and the precipitate is discarded.

Step 4. First Ammonium Sulfate Precipitation and Dialysis. The supernatant solution with (liver) or without (brain) heat treatment is brought to 40% saturation by addition of solid ammonium sulfate over a 1-hour period. The mixture is centrifuged at 18,500 g for 15 minutes, and the precipitate is discarded. The supernatant solution is then brought to 60% saturation with ammonium sulfate over a similar period of time and the active precipitate is collected by centrifugation. The precipitate is dissolved in a small volume of 5 mM potassium phosphate, pH 6.5, containing 0.1 mM reduced glutathione, dialyzed overnight against a large volume of this solution, and the dialyzate clarified by centrifuging at 12,500 g for 15 minutes.

Step 5. Acetic Acid Treatment (for Liver Kinase). The dialyzed solution is adjusted to pH 5.0 by the addition of 1 M acetic acid, the turbid solution is centrifuged at 12,500 g for 15 minutes, and the precipitate is discarded.

Step 6. Alumina Gel Fractionation. The supernatant solution with (liver) or without (brain) acetic acid treatment is treated stepwise with alumina C_γ gel (7.5 mg/ml) in amounts sufficient to absorb most of the protein within 6 to 8 equal additions. After each gel addition, the mixture is stirred for 15 minutes and centrifuged at 10,000 g for 5 minutes; the supernatant fluid is subjected to the next gel treatment. Protein is eluted from the gel by stirring for 15 minutes with 5 volumes (buffer to gel) of 0.2 M potassium phosphate, pH 8.0. The most active eluates are combined for approximately one-half of the protein subjected to the fractionation.

Step 7. Second Ammonium Sulfate Precipitation and Dialysis. The combined eluates are brought to 40% saturation by the addition of solid ammonium sulfate over a 1-hour period. The mixture is centrifuged at 18,500 g for 15 minutes, and the precipitate is discarded. The supernatant solution is then brought to 55% saturation with ammonium sulfate over a similar period of time, and the active precipitate is collected by centrifugation. The precipitate is dissolved in a small volume of water and dialyzed for 6–15 hours against several liters of 5 mM potassium phosphate, pH 7.5; the dialyzate is clarified by centrifuging at 12,500 g for 15 minutes.

Step 8. Chromatography on DEAE-Cellulose. The clear supernatant solution is poured over a column of DEAE-cellulose (1 g per 50 mg of protein) which has been previously equilibrated with a buffer containing 0.02 M glycine in 0.005 M potassium phosphate pH 6.8. Protein is eluted fractionally in a linear gradient established between 0.005 M and 0.1 M potassium phosphate buffers, pH 6.8, each 0.02 M with respect to glycine. Fractions which contain most pyridoxal kinase are combined. The enzyme is precipitated by 60% saturation with ammonium sulfate, collected by centrifugation at 12,500 g for 15 minutes, and dissolved in a small volume of 5 mM potassium phosphate buffer, pH 7.0, containing 0.1 mM reduced glutathione. The solution is dialyzed against this buffer to remove ammonium sulfate.

TABLE II
PURIFICATION OF PYRIDOXAL KINASES FROM LIVER AND BRAIN OF RATS

	Liver		Brain	
Fraction	Total[a] activity (units)	Specific activity (units/mg)	Total[b] activity (units)	Specific activity (units/mg)
1. Homogenate	5,540	0.62	412	0.70
2. Supernatant solution	18,000	5.6	2140	18.7
3. Heat-treated solution	45,500	42.6	—	—
4. First $(NH_4)_2SO_4$ fraction	23,700	53.7	1520	80.1
5. Acid-treated solution	25,000	74.3	—	—
6. Alumina gel eluate	14,700	123	963	126
7. Second $(NH_4)_2SO_4$ fraction	9,790	178	398	240
8. DEAE-cellulose eluate	3,020	750	—	—

[a] From 36 g (wet weight) of tissue.
[b] From 4.9 g (wet weight) of tissue.

Typical partial purification of pyridoxal kinases from liver and brain of rats is summarized by the data in Table II.

Comments. These procedures achieve a purification of several 100-fold. Kinase from beef brain has also been obtained with a specific activity of 750 after chromatography on DEAE-cellulose.[1] The enzyme from human brain is similarly enriched.[8] By reapplication of the most active fractions to DEAE-cellulose, mammalian kinase preparations with specific activities of 1000 and greater have occasionally been obtained. These latter are still polydisperse in the analytical ultracentrifuge.

Properties

Optima for Temperature and pH. With the 1-hour assay time described above, apparent temperature optima for kinase preparations from *S. faecalis*, *L. casei*, liver, and brain were 63, 55, 55, and 40°, respectively.[1] The preparation from human brain which has a optimum of 40° with Zn^{2+} as activating ion shows an optimum of 50° when Mg^{2+} is substituted for Zn^{2+}. Temperature coefficients of reaction for a 10° increase in temperature below the optima are in the expected range of 1.5–2.0 for all preparations. Optima for pH are approximately 5.0, 5.2, 5.7, and 6.0 for *L. casei*, *S. faecalis*, liver, and brain, respectively.[1] The pH optima vary somewhat with the means and extent of purification and sometimes the nature and concentration of activating cation. For example, the kinase from *L. casei* shows a pH optimum of 5.5 when determined with a crude autolyzate, of 5.0 after protamine treatment, and of 4.5 after further precipitation with ammonium sulfate. A shift toward more acidic values is exhibited by both the kinase from beef brain and that from rat liver as the concentration of Zn^{2+} is increased to the optimum. No such shift is observed when Mg^{2+} replaces Zn^{2+} as the activating ion.

Requirements for Metal Ions and Nucleoside Triphosphates. Mg^{2+} is superior to Zn^{2+} as an activator for the purified kinases of *L. casei* and *S. faecalis.*[1] However, Zn^{2+} is markedly superior to Mg^{2+} for the kinase from rat or turkey liver[1] and is also the preferred activator of the enzyme from brains of rat, bovine, and human.[1,8] In the presence of optimal amounts of Zn^{2+}, an activating effect of K^+ is observed with the kinase from beef brain.[1] The unexpected function of Zn^{2+} with certain of the B_6 kinases that we showed earlier[1,9] has now been found with the highly purified enzyme from *E. coli.*[10] ATP is the only effective phosphorylating agent, though weak activity is seen with GTP, UTP, and CTP and the kinase from *S. faecalis.* ADP can replace ATP in crude liver preparations which contain adenylate kinase.

Specificity for Pyridoxal, Pyridoxine, and Pyridoxamine. Pyridoxal appears to be the preferred substrate for all the kinases studied except that from yeast.[1] The comparative affinities of some of the kinases for the three forms of vitamin B_6 are given in Table III.

TABLE III
COMPARATIVE AFFINITIES OF PYRIDOXAL KINASES FOR
PYRIDOXAL, PYRIDOXINE, AND PYRIDOXAMINE

Source of kinase	Apparent K_s values $(M \times 10^4)$[a]		
	Pyridoxal	Pyridoxine	Pyridoxamine
Lactobacillus casei	0.30	40	50
Streptococcus faecalis	0.15	15	25
Rat liver	0.15	0.25	1.5
Beef brain	0.50	2.0	5.0

[a] Values are given to the nearest half unit.

Effects of Inhibitors. A variety of vitamin B_6 analogs and variously substituted oximes and hydrazones of pyridoxal have been examined as inhibitors of pyridoxal kinases from *L. casei*, *S. faecalis*, rat liver, beef brain, and *Saccharomyces carlsbergensis.*[9] Inhibitory analogs containing a 4-formyl group (5-deoxypyridoxal, ω-methylpyridoxal) resemble pyridoxal in their affinity for the kinases; similarly, those lacking the 4-formyl group (4-deoxypyridoxine, 2-methyl-3-hydroxy-5-hydroxymethylpyridine) resemble pyridoxine. ω-Methylpyridoxal is phosphorylated by *S. faecalis* kinase

[9] D. B. McCormick and E. E. Snell, *J. Biol. Chem.* **236**, 2085 (1961).
[10] See this volume [95].

much more slowly than pyridoxal. Several other 2-methylpyridines containing altered substituents at other positions also inhibit the kinase.

Condensation products formed from pyridoxal and hydroxylamine, *O*-substituted hydroxylamines, hydrazine, and substituted hydrazines are extremely potent inhibitors for all the pyridoxal kinases tested. Affinities for these inhibitors range from 100 to 1000 times those for pyridoxal. The corresponding derivatives of pyridoxal 5-phosphate are not inhibitory. The free carbonyl reagents are ineffective as inhibitors as shown by their failure to inhibit kinase when pyridoxine is the substrate.

[95] Pyridoxine Kinase (ATP:Pyridoxal 5-Phosphotransferase, EC 2.7.1.35) from *Escherichia coli*

By ROSEANN WHITE and WALTER B. DEMPSEY

$$\text{Pyridoxol} + \text{ATP} \xrightarrow[\text{Zn}^{2+}]{\text{Mg}^{2+} \text{ or}} \text{pyridoxol 5'-phosphate} + \text{ADP}$$

Assay Methods

Principle. Pyridoxine kinase activity is measured by determining the amount of pyridoxal 5'-phosphate or ADP produced with pyridoxal as the substrate. The former is determined by either the apotryptophanase method of Wada,[1] as modified by McCormick,[2] or by direct spectrophotometry at 388 mμ. ADP is measured by coupling its formation in the pyridoxine kinase reaction to NADH oxidation in a pyruvate kinase–lactic dehydrogenase system.

Application of Assay Methods. Because of its great sensitivity and relative freedom from interfering reactions, the apotryptophanase method, in which tryptophan is enzymatically converted to indole, pyruvate, and ammonia as a function of pyridoxal 5'-phosphate concentration, is used in early stages of kinase purification. This assay, however, has a major drawback, which stems from the necessity of stopping the kinase reaction by heat denaturation, a method that will prevent further kinase activity and yet not inhibit the apotryptophanase reaction. For *E. coli* pyridoxine kinase, stopping the reaction with heat takes approximately 5 minutes. A linear dependence of enzyme activity on enzyme concentration can be obtained with this assay if the enzyme is limited to 0.06 to 0.4 units.

[1] H. Wada, T. Morisue, Y. Sakamoto, and K. Ichihara, *J. Vitaminol. (Kyoto)* **3**, 183 (1957).
[2] D. B. McCormick, M. E. Gregory, and E. E. Snell, *J. Biol. Chem.* **236**, 2076 (1961).

The spectrophotometric, initial rate assays, are used only with the 200 or more fold purified enzyme because of the extremely high protein concentrations required to measure activity. At these high protein concentrations, excessive interference from side reactions is seen.

When used to assay partially purified enzyme, the spectrophotometric assays have a number of advantages over the kinase–apotryptophanase assay. Both spectrophotometric assays have the advantage of being able to measure initial velocities, which give a strictly linear relationship between enzyme activity and enzyme concentration. In addition, the kinase-ADP coupled assay has the advantage of measuring kinase activity independent of the nature of the pyridoxine substrate used. Accordingly, the spectrophotometric assays are used for rate studies.

Reagents

1. PYRIDOXINE KINASE–APOTRYPTOPHANASE
   a. KINASE
   ATP, 0.01 M
   Pyridoxal-HCl, 0.01 M
   ZnSO$_4$, 0.01 M
   Potassium phosphate buffer, 1 M, pH 7.0
   Pyridoxine kinase enzyme preparation
   b. APOTRYPTOPHANASE ASSAY
   Trichloroacetic acid (TCA), 100 g/100 ml
   L-Tryptophan, 10.2 mg/ml
   Potassium phosphate buffer, 1 M, pH 8.3
   Apotryptophanase enzyme; see preparation of apotryptophanase
      enzyme
   Toluene, reagent grade
   Ethanol–sulfuric acid reagent, 8 ml of H$_2$SO$_4$ in 100 ml of ethanol
   p-Dimethylamino benzaldehyde, 5 g/100 ml of 95% ethanol
2. PYRIDOXINE KINASE-DIRECT SPECTROPHOTOMETRY AT 388 mμ. Same
   reagents as kinase substrates in 1a, but at pH 6.0.
3. PYRIDOXINE KINASE–ADP ASSAY
   Potassium phosphate buffer, 1 M, pH 6.0
   KCl, 2 M
   Tricyclohexylammonium phosphoenol pyruvate, 0.1 M
   MgCl$_2$, 0.1 M
   NADH, 0.01 M
   Lactic dehydrogenase, and pyruvate kinase, in excess
   ATP, 0.01 M
   Pyridoxal, pyridoxol, or pyridoxamine, 0.01 M
   Pyridoxine kinase, 200-fold or greater purified

Procedures

1. The following amounts of reagents are incubated in small stoppered test tubes at 37° for 15 minutes: 0.01 ml ATP, 0.01 ml pyridoxal–HCl, 0.1 ml ZnSO$_4$, 0.01 ml potassium phosphate buffer, pH 7.0, varying concentrations of enzyme, and water to a final volume of 1 ml. Controls measuring endogenous pyridoxal 5′-phosphate lack only pyridoxal. The reactions are started by the addition of enzyme to each tube and terminated after 15 minutes by heating in a boiling water bath for 5 minutes. The previously published boiling time of 3 minutes is insufficient.

Apotryptophanase Assay. Aliquots from each kinase reaction tube and a series of pyridoxal-phosphate standards from 0 to 6 nanomoles are added to 25-ml flasks, and the total volumes are brought to 1.5 ml. To each flask is added 0.4 ml of an apotryptophanase preparation consisting of one volume of 0.1 *M* glutathione, one volume of apotryptophanase (see preparation of apotryptophanase), and two volumes of 1 *M* potassium phosphate, pH 8.3. Toluene, 5 ml, is also added, and the stoppered flasks are shaken gently for 10 minutes in a 37° rotary water bath shaker. The reaction is started by the addition of 0.2 ml of tryptophan solution. After incubation with shaking for 10 minutes, the reaction is stopped by the addition of 0.2 ml of TCA. From each flask 0.5 ml of the upper toluene layer is removed and added to colorimeter tubes; 0.5 ml of *p*-dimethylaminobenzaldehyde and 5 ml of ethanol-H$_2$SO$_4$ are added to each tube. The color is allowed to develop for 10 minutes, and the absorbance at 560 mμ is read.

2. An aliquot containing 0.001–0.010 unit of a 200-fold or greater purified pyridoxine kinase preparation is added to a cuvette containing 0.1 ml of ATP, 0.1 ml of pyridoxal, 0.01 ml of potassium phosphate, 0.05 ml of ZnSO$_4$, and water to a final volume of 1.0 ml. The increase in absorbance is followed with time at 388 mμ in a recording spectrophotometer with the scale adjusted from 0 to 0.1 OD units. Because incubation of the enzyme with pyridoxal causes enzyme inactivation with time, only early optical density changes are recorded for initial rates.

3. An aliquot containing 0.001–0.010 unit of a 200-fold or greater purified pyridoxine kinase preparation is added to a cuvette containing 0.015 ml of potassium phosphate, 0.0125 ml of KCl, 0.01 ml of phosphoenol pyruvate, 0.05 ml of MgCl$_2$, 0.1 ml of ATP, 0.01 ml of pyridoxine substrate, 0.01 ml of NADH, excess pyruvate kinase–lactic dehydrogenase, and water to make a final volume of 1.0 ml. The decrease in absorbance at 340 mμ due to NADH oxidation is followed with time in a double-beam spectrophotometer (full scale from 0 to 0.1 OD), and the difference between the rate prior to and that subsequent to enzyme addition is equal to the net pyridoxine kinase rate.

Preparation of Apotryptophanase

GROWTH OF CELLS. Fernbach flasks containing 1 liter of a medium composed of 1% tryptone, 1% yeast extract, 0.5% potassium phosphate (dibasic salt), and 0.1% glucose are inoculated with 50 ml of a culture of *E. coli* (Crookes strain) grown in medium of the same composition. The Fernbach flasks are incubated at 30° for 4 hours without shaking and for 28 hours with shaking. The cells are harvested in a Sharples centrifuge and washed with 200 ml of water before lyophilization.

PREPARATION OF ENZYME EXTRACTS. Two grams of lyophilized Crookes strain are suspended in 50 ml of 0.02 M potassium phosphate, pH 7.0, and disrupted by sonic oscillation for 5 minutes at 2×10^4 Hz or 30 minutes at 1×10^4 Hz. The cells are subsequently centrifuged at 18,500 g for 30 minutes. The supernatant solution, which is maintained at 0–5°, is adjusted to a pH of 4.7 by the addition of 1 M acetic acid (approximately 1.5–2.0 ml). The precipitate is suspended in 45 ml of H_2O and dialyzed overnight against 5 liters of water. This enzyme preparation is stored frozen in 3-ml aliquots. Each tube is considered as 1 volume of apotryptophanase preparation.[3]

Definition of Unit of Specific Activity. With all three pyridoxine kinase assays, the specific activities of the kinase are expressed as units per milligram of protein. In all cases protein concentration is determined by the method of Lowry.[4]

1. One unit is the amount of enzyme that forms 1 micromole of pyridoxal 5'-phosphate per minute at 37°. The amount of pyridoxal 5'-phosphate formed per minute is determined from the micromoles formed in a 15-minute incubation time, although the increase in pyridoxal 5'-phosphate is not linear over the entire time interval.

2. One unit is the amount of enzyme that forms 1 micromole of pyridoxal 5'-phosphate per minute at room temperature. The micromoles of pyridoxal 5'-phosphate formed are determined from the extinction coefficient of 4.9 OD units for a millimolar solution of pyridoxal phosphate in a 1-cm cuvette.

3. One unit is the amount of enzyme that forms 1 micromole of NAD per minute at room temperature (25°). The extinction coefficient for NAD at 340 mμ is 6.2 OD units for a millimolar solution of NAD in a 1-cm cuvette.

Purification Procedure

Step 1. Preparation of Crude Extract. Approximately 260 g of lyophilized *E. coli* B, ¾ grown in enriched medium, obtained from Grain Processing

---

[3] D. B. McCormick, M. E. Gregory, and E. E. Snell, *J. Biol. Chem.* **236**, 2076 (1961).

[4] O. H. Lowry, N. J. Rosebrough, A. L. Farr, and R. J. Randall, *J. Biol. Chem.* **193**, 265 (1951).

Company, Muscatine, Iowa, are suspended in 0.01 M potassium phosphate buffer, pH 7.0, and disrupted ultrasonically for 2.5 hours at 0–5°. Sonication is carried out in a 2000-ml beaker containing a stirring bar and refrigerated coil which surrounds the probe. General Electric Antifoam 60 is added to prevent excessive foaming. Either the maximum amount of Lowry reactive material or OD of 260/280 readings are followed with time to determine when the cells are maximally disrupted. The broken cells are then centrifuged at 25,000 g for 3 hours. The enzyme is present in the supernatant fluid.

Step 2. Ammonium Sulfate Precipitation. A room-temperature-saturated ammonium sulfate solution, adjusted to pH 7.2 at a 1:50 dilution by the addition of NaOH, is added dropwise to the 25,000 g supernatant at 2° with continual stirring until 30% saturation is achieved (calculated as percent saturation at 25°). The enzyme solution is stirred and allowed to equilibrate overnight at 2–4° and then centrifuged 45 minutes at 25,000 g. The precipitate is discarded and the supernatant fluid is brought to 52% saturation and stirred 2 hours at 2–4°. The precipitate is collected by centrifugation for 45 minutes at 25,000 g, and suspended in approximately 300 ml of 0.01 M potassium phosphate buffer at pH 7.0. Step 1 and Step 2 are repeated until five 260 g-preparations have been brought to this point.

Step 3. DEAE-Cellulose Column Chromatography at pH 7.0, Stepwise Elution. The pooled 30–52% ammonium sulfate precipitates are dialyzed against 10 liters of 0.01 M potassium phosphate buffer at pH 7.0. The buffer is repeatedly changed, and dialysis is allowed to continue until the ammonium ion concentration in the 8-hour dialyzate is less than 0.01 M as determined by Nessler's reaction. The retentate is loaded onto an 8 × 100 cm DEAE-cellulose column equilibrated with 0.01 M potassium phosphate buffer at pH 7.0, and the column is washed with approximately 2 liters of that buffer or until the first protein peak, as measured by OD at 260/280 readings, washes off. The column is then washed with approximately 3 liters of 0.01 M potassium phosphate buffer, 0.08 M in KCl, until all the kinase peak washes off. Fractions of 20 ml per tube are collected, and the peak 1 liter of kinase enzyme of specific activity 0.0008 or greater is pooled for the next step.

Step 4. DEAE-Cellulose Column Chromatography at pH 7.0, Gradient Elution. The pooled fractions from the previous DEAE-cellulose column are dialyzed overnight against 10 liters of 0.01 M potassium phosphate, pH 7.0, and loaded on a 2.5 × 100 cm DEAE-cellulose column equilibrated with that same buffer. The column is washed with approximately 1 liter of that buffer, and a 1.6-liter gradient is applied from 0 to 0.5 M KCl in potassium phosphate buffer at pH 7.0. Fractions of specific activity 0.002 or greater are collected and pooled.

Step 5. DEAE-Cellulose Column Chromatography at pH 6.0, Gradient Elution. The pooled fractions from the previous DEAE-cellulose column are dialyzed overnight at 4° against 6 liters of 0.01 M potassium phosphate, pH 6.0, and loaded on a 2.5 × 100 cm DEAE-cellulose column equilibrated with that same buffer. The column is washed with 500 ml of that buffer, and a 1.6-liter gradient is applied from 0 to 0.5 M KCl in 0.01 M potassium phosphate buffer at pH 6.0. Fractions (11.0 ml per tube), of specific activity 0.0035 or greater contained in a total volume of approximately 100 ml, are collected and pooled.

Step 6. Sephadex G-100 Column Chromatography. The active enzyme pooled from the previous column is concentrated to 80 ml, and 40 ml of the preparation is loaded on a 5 × 100 cm Sephadex G-100 column equilibrated with 0.01 M potassium phosphate at pH 7.0. The protein is eluted by reverse flow of the buffer, and 11.0-ml fractions are collected. The other half of the enzyme pool is also subjected to the same procedure, and the most active 77 ml from each column are pooled.

Step 7. DEAE-Cellulose Column Chromatography at pH 8.0, Gradient Elution. The active eluent pooled from the two Sephadex G-100 columns is loaded on a 0.9 × 80 cm DEAE-cellulose column equilibrated with 0.01 M potassium phosphate buffer at pH 8.0. The column is washed with 200 ml of that buffer, and a 1-liter gradient in the equilibration buffer is applied from 0 to 0.5 M KCl. The most active fractions of specific activity 0.043 or greater are pooled.

Step 8. DEAE-Cellulose Column Chromatography at pH 6.0, Gradient Elution. The pooled fractions from the previous column are dialyzed overnight against 2 liters of 0.01 M potassium phosphate buffer at pH 6.0 and are loaded on a 0.9 × 80 cm DEAE-cellulose column equilibrated with 0.01 M potassium phosphate buffer at pH 6.0. The column is washed with 200 ml of that buffer, and a 1-liter linear gradient in the equilibration buffer is applied from 0 to 0.3 M KCl. The most active fractions (6.0 ml per tube), of specific activity 0.13 or greater contained in a total volume of 50 ml, are pooled.

Step 9. DEAE-Cellulose Column Chromatography at pH 8.0, Gradient Elution. The pooled fractions from the previous column are dialyzed overnight against 2 liters of 0.01 M potassium phosphate buffer at pH 8.0 and are loaded on a 0.9 × 45 cm DEAE-cellulose column at pH 8.0. The column is washed with 200 ml of that buffer, and a 800-ml linear gradient in the equilibration buffer is applied from 0 to 0.4 M KCl. The most active fractions from this column exhibit a constant specific activity over a number of fractions; however, the majority of the kinase units are of approximately 20% lower specific activity. Polyacrylamide gel electrophoresis by the method of Davis[5] yields a single band for the most active fractions. If the unstained gels are scanned at 280 mμ, the peak obtained indicates that the sample is at least 92% pure.

The purification at each stage is indicated by the data in the table. Specific activities shown in this table were determined by assay method 1, exclusively. A specific activity of 0.27 by assay 1 is equivalent to a specific activity of 4 at room temperature or a specific activity of 8.0 at 37° when assay 2 is used.

PURIFICATION OF PYRIDOXINE KINASE

Fractionation stage	Total units	Total protein (mg)	Volume (ml)	Specific activity $\times 10^3$	Yield (%)
Sonicated cells	80	840,000	8000	0.095	100
Centrifuged cells	73	510,000	7800	0.15	92
30–52% $(NH_4^+)_2SO_4^{2-}$ ppt	57	190,000	2000	0.30	71
DEAE-cellulose, pH 7.0, stepwise	33	38,000	1000	0.97	41
DEAE-cellulose, pH 7.0, gradient	13	8,400	160	2.7	29
DEAE-cellulose, pH 6.0, gradient	19	4,200	100	4.4	23
Sephadex G-100 column	10	340	150	29.0	13
DEAE-cellulose, pH 8.0	4.9	60	30	75	6.1
DEAE-cellulose, pH 6.0	3.0	16	35	200	3.8
DEAE-cellulose, pH 8.0	2.3	8.5	35	270	2.9

Properties

Stability. The pyridoxine kinase enzyme purified through at least the first DEAE-cellulose column is stable for storage in a refrigerator or freezer for at least a month. The 30–52% $(NH_4^+)_2SO_4$ preparations will lose 20% of the total kinase units in 2 weeks. In addition, the enzyme seems somewhat heat stable since it requires boiling for at least 5 minutes to ensure complete denaturation.

Specificity. Apparent K_m and K_i determinations for pyridoxine compounds yielded an apparent K_m for pyridoxal of 3×10^{-4} and apparent K_i values for pyridoxol, pyridoxamine, and 5-deoxypyridoxal of 8×10^{-6}, 5×10^{-4}, and 1.5×10^{-4} M, respectively.

Effect of pH. The pH optimum of the pyridoxine kinase is 6.0 with potassium phosphate buffer at 0.01 M concentration. This optimum was determined using pyridoxal as the substrate. The kinase is also active at pH 5, 7, and 8.

[5] B. J. Davis, *Ann. N.Y. Acad. Sci.* **121**, 404 (1964).

Activators and Inhibitors. The pyridoxine kinase enzyme requires ATP, one of the pyridoxine substrates, and divalent cation for activity. The optimal concentrations for Zn^{2+} and Mg^{2+} divalent cations as determined by direct spectrophotometry at 388 mμ are $7 \times 10^{-4}\ M$ and $1 \times 10^{-3}\ M$, respectively. The velocity of the reaction at the optimal Zn^{2+} concentration is twice the velocity at the optimal Mg^{2+} concentration, and the Zn^{2+} and Mg^{2+} effects appear to be competitive. The approximate turnover number for the kinase as determined by direct spectrophotometry at 388 mμ with pyridoxal as substrate is 350 moles of pyridoxal per minute per mole kinase at 37°. The kinase is rapidly inactivated by pyridoxal but comparatively uninhibited by the other substrates or products; because of this the rates of phosphorylation must be measured within the first 30 seconds of reaction to ensure that true initial velocities of reaction are measured.

Molecular Weight Determinations. The molecular weight of the purified pyridoxine kinase enzyme has been determined by the sucrose density gradient centrifugation method of Martin and Ames.[6] The average molecular weight of the pyridoxine kinase enzyme utilizing malic dehydrogenase and peroxidase as standards is $47,000 \pm 4000$. In addition, the molecular weight has been determined to be 35,000 from 18 hour sedimentation equilibrium patterns and assuming the partial specific volume is 0.74.

[6] R. G. Martin and B. N. Ames, *J. Biol. Chem.* **236,** 1372 (1961).

[96] Pyridoxine Phosphate Oxidase

By HIROSHI WADA

$$\text{Pyridoxine 5-phosphate} + O_2 \rightarrow \text{pyridoxal 5-phosphate} + H_2O_2 \tag{1}$$

$$\text{Pyridoxamine 5-phosphate} + O_2 \rightarrow \text{pyridoxal 5-phosphate} + NH_3 + H_2O_2 \tag{2}$$

Pyridoxine phosphate oxidase catalyzes reactions (1) and (2). It has been observed in crude extracts of *Escherichia coli* and in rabbit liver and brain that pyridoxine 5-phosphate is oxidized to pyridoxal 5-phosphate much faster than pyridoxine is to pyridoxal.[1,2] This observation suggested that this enzyme has an important physiological function in the formation of pyridoxal phosphate in living cells. It was shown previously that the oxidation of pyridoxamine 5-phosphate is catalyzed by an enzyme present

[1] H. Wada, T. Morisue, Y. Nishimura, Y. Morino, Y. Sakamoto, and K. Ichihara, *Proc. Japan Acad.* **35,** 299 (1959).
[2] T. Morisue, Y. Morino, Y. Sakamoto, and K. Ichihara, *J. Biochem.* (*Tokyo*) **48,** 18 (1960).

in rabbit liver, and that the enzyme is distinct from both monoamine oxidase and diamine oxidase.[3] Comparison of the two activities represented by Eqs. (1) and (2) at various steps of the purification procedure indicates that a single enzyme catalyzes both reactions.[4] The following method of assay and purification is taken from Wada and Snell.[4]

Assay Method

Principle. Pyridoxine phosphate oxidase is assayed by measuring the rate of formation of pyridoxal phosphate as its phenylhydrazine derivative which absorbs at 410 mμ.[5]

Reagents

Phenylhydrazine reagent, phenylhydrazine hydrochloride (2 g) is dissolved in 100 ml of 10 N H$_2$SO$_4$. The reagent can be kept in the refrigerator for several weeks but should be renewed when any brown color appears.

Pyridoxine 5-phosphate solution, 0.01 M, kept frozen

Trichloroacetic acid, 100% (w/v)

Procedure. The reaction mixture contains 0.7 ml of buffer, 0.1 ml of substrate (1 micromole), and enzyme to a final volume of 3.5 ml. The reaction is carried out at 37° for 30 minutes with vigorous shaking and stopped by the addition of 0.3 ml of 100% trichloroacetic acid. After a brief centrifugation to remove the precipitated protein, 0.2 ml of the phenylhydrazine reagent is added to 3.0 ml of the supernatant solution. The samples are allowed to stand 10 minutes at room temperature, and read at 410 mμ. When this method is applied to a crude extract, an appropriate blank, containing the enzyme extract but no substrate, is required for the correction.

Definition of Specific Activity. Specific activity of this enzyme is expressed as the millimicromoles of pyridoxal phosphate formed per milligram of protein per hour.

Purification Procedure

The enzyme is extracted from rabbit liver, which contains a much larger amount than either rat or beef liver. The livers of freshly slaughtered

[3] B. M. Pogell, *J. Biol. Chem.* **232**, 761 (1958).

[4] H. Wada and E. E. Snell, *J. Biol. Chem.* **236**, 2089 (1961).

[5] Both pyridoxal and pyridoxal phosphate form highly colored hydrazones with phenylhydrazine. Pyridoxal phosphate produces a stable, intense yellow color ($\lambda_{max} = 410$ mμ; molecular extinction coefficient, 24,300). Pyridoxal reacts more slowly but gives the same high extinction on heating. This slow color development with pyridoxal is due to the fact that this compound exists primarily as the hemiacetal in acid solution, and this difference in behavior permits the determination of pyridoxal phosphate in the presence of pyridoxal. See footnote 4.

animals are held at $-15°$ until used. Brain and kidney are much less active than liver. All preparative steps are carried out at $0-5°$.

Step 1. Extraction and Acid Treatment. Frozen rabbit liver (150 g) is thawed, diced, homogenized for 5 minutes in a Waring blendor with 600 ml of 0.02 M potassium phosphate buffer, pH 7.0, and centrifuged for 30 minutes at 18,000 g. The supernatant solution (fraction I, see table) is adjusted to pH 5.0 by dropwise addition with constant stirring of 2 N acetic acid. After 10 minutes, the precipitate is removed by centrifugation and the supernatant solution is brought to pH 7.0 with 2 N NaOH. A slight precipitate sometimes appears, which is removed by centrifugation.

Step 2. Heat Treatment. The clear supernatant solution (fraction II) is warmed to $50°$ by swirling in a $60°$ water bath, held at $50°$ for 8 minutes, cooled rapidly to approximately $20°$ and centrifuged.

Step 3. Ammonium Sulfate Fractionation. To the supernatant solution (fraction III), solid ammonium sulfate (22.8 g/100 ml, 30% saturation) is added. After 30 minutes at $2-5°$, the enzymatically inactive precipitate is removed by centrifugation and additional ammonium sulfate (15.2 g/100 ml, 50% saturation) is added gradually to the supernatant solution. After 60 minutes, the active precipitate is collected by centrifugation and dissolved in approximately one-tenth of the initial volume of water. The clear solution is dialyzed against distilled water overnight and centrifuged to remove precipitated protein.

Step 4. Alumina C_γ Gel Treatment. The dialyzed solution (fraction IV) is adjusted to pH 6.0 with 1 N acetic acid and centrifuged. To each 10-ml portion of the supernatant solution is added 2.0 ml of alumina C_γ gel (7.0 mg of alumina). After 20 minutes of constant stirring, the suspension is centrifuged, and the supernatant solution is treated a second time with the same amount of alumina gel. This procedure is repeated four more times. For elution, each gel precipitate is suspended separately in 10 ml of 0.5 M potassium phosphate buffer, pH 8.0, stirred occasionally for 20 minutes, and centrifuged. Most of the oxidase activity appears in the eluates from the third and fourth gel treatments. These are combined (fraction V) and used for further purification.

Step 5. Fractionation on DEAE-Cellulose. Fraction V (160 mg of protein) is dialyzed against distilled water overnight and poured over a 1.7×12 cm column of DEAE-cellulose, which had been equilibrated with 0.01 M potassium phosphate buffer, pH 8.0. Elution is carried out by a linear gradient established between 0.01 M potassium phosphate buffer (pH 8.0) and this buffer plus 0.2 M NaCl using 300 ml of each solution. The elution pattern is quite reproducible. The most active fraction (fraction VI) contains protein of specific activity 432, purified approximately 66-fold over the protein of the solution from step 1.

A typical purification of the enzyme from 150 g of frozen rabbit liver is summarized in the table.

PURIFICATION OF PYRIDOXINE PHOSPHATE OXIDASE FROM RABBIT LIVER

Fraction no.	Volume (ml)	Protein (mg/ml)	Specific activity	Yield (%)	Mole ratio oxidized, PNP:PMP[a]
I	610	34.0	6.5	100	1.07
II	520	16.4	13.0	83	1.16
III	485	10.8	18.0	70	1.14
IV	69	32.5	36.4	59	1.08
V	20	11.0	166.0	27	1.05
VI	—	—	431.5	—	1.10

[a] PNP, pyridoxine phosphate; PMP, pyridoxamine phosphate.

Properties

Substrate Specificity. As shown in the table, oxidase activity toward pyridoxine phosphate exactly parallels that toward pyridoxamine phosphate. Michaelis constants for pyridoxine phosphate and pyridoxamine phosphate are $3.1 \times 10^{-5} M$ and $1.4 \times 10^{-4} M$, respectively. The maximal velocities for the two substrates are identical. The purified preparation slowly oxidizes pyridoxamine at pH 10.0, but not at pH 8.0. Pyridoxine is not oxidized at any pH tested.

pH Optimum. The optimal pH lies around 9.0–10.0, irrespective of substrate. Below pH 9.0, the activity toward pyridoxine phosphate gradually decreases. On the other hand, the decrease of the activity toward pyridoxamine phosphate is much more rapid. This is probably due to the fact that the aminomethyl group of pyridoxamine phosphate ($pK \approx 10.9$)[6] is largely protonated below pH 9.0, and the unprotonated group is attacked preferentially by the enzyme.

Product Inhibition of the Oxidase and Its Alleviation by Amines. The reaction product, pyridoxal phosphate, strongly inhibits oxidation of both pyridoxine phosphate and pyridoxamine phosphate. Tris partially overcomes this inhibition through the formation of a less inhibitory Schiff's base. This fact could explain the observation that although the initial rate of pyridoxal phosphate formation is the same in both phosphate and Tris buffers, subsequently the rate decreases much more rapidly in phosphate than in Tris buffer. Besides Tris, other amino compounds such as alanine, valine, or α-aminobutyric acid exert a similar effect. Cysteine, which readily forms a thiazolidine derivative with pyridoxal phosphate, is most effective.

[6] E. E. Snell, *Vitamins Hormones* **16,** 77 (1958).

Coenzyme. FMN (flavin mononucleotide) has no effect on the activity of the crude enzyme. The enzyme can be resolved by a modification of the method of Warburg and Christian[7] (pH 2.8, 50% saturated ammonium sulfate solution). The resolved enzyme is essentially inactive in the absence of added FMN. The Michaelis constant for FMN is $3.1 \times 10^{-8}\,M$. Flavin adenine dinucleotide is about 0.1% as active as FMN. Riboflavin and isoriboflavin have no effect.

[7] O. Warburg and W. Christian, *Biochem. Z.* **298,** 150 (1938).

[97] Acid Phosphatase Having Pyridoxine-Phosphorylating Activity[1]

By Yoshiki Tani and Koichi Ogata

$$\text{Pyridoxine} + \text{R—O—P} \rightarrow \text{pyridoxine-P} + \text{R—OH}$$
$$\text{R—O—P} + \text{H}_2\text{O} \rightleftharpoons \text{R—OH} + \text{P}_i$$

Assay Method

Principle. The assay is based on the formation of pyridoxine 5′-phosphate in the transphosphorylation between pyridoxine and *p*-nitrophenyl phosphate, and of *p*-nitrophenol in the hydrolysis of *p*-nitrophenyl phosphate. The colorimetric assay measuring *p*-nitrophenol may be conveniently used.[2]

Reagents

Tris-maleate buffer, 0.2 *M*, pH 6.0
Pyridoxine, 0.02 *M*
p-Nitrophenyl phosphate, 0.1 *M*
Sodium carbonate, saturated solution

Procedure. The following reagents are present in a total volume of 3 ml: 200 micromoles of Tris-maleate buffer, pH 6.0, 10 micromoles of pyridoxine, 50 micromoles of *p*-nitrophenyl phosphate, and the enzyme solution to be tested. After incubation at 37° for 30 minutes, the reaction is stopped with 50% trichloroacetic acid, and a suitable aliquot of the supernatant solution is diluted. *p*-Nitrophenol is determined with addition of 2 ml of saturated sodium carbonate to 2 ml of the diluted sample, and the absorption at 430 mμ is read in a photoelectric photometer.[2] For the measurement of

[1] Y. Tani, T. Tochikura, H. Yamada, and K. Ogata, *Agr. Biol. Chem. (Tokyo),* **32,** 1220 (1968).
[2] S. Omori, *Enzymologia* **4,** 217 (1937).

pyridoxine 5′-phosphate, the reaction must be stopped by heating for 5 minutes in a boiling water bath and then oxidized by pyridoxine 5′-phosphate oxidase from *Alcaligenes faecalis*.[3] The oxidized product, pyridoxal 5′-phosphate, is assayed with the phenylhydrazine method.[4]

In routine assays required in the purification procedure, the measurement of *p*-nitrophenol is useful.

Protein is estimated by absorbancy or by the Lowry procedure.[5]

Units and Specific Activity. One unit of enzyme activity is defined as the amount of enzyme that forms 1 micromole of *p*-nitrophenol or pyridoxine 5′-phosphate under standard assay conditions. Specific activity is the number of enzyme units per milligram of protein.

Purification Procedure

Acid phosphatase may be purified from *Escherichia freundii* K1, which was isolated from soil and preserved in the Laboratory of Applied Microbiology, Kyoto University, Kyoto. The organism is grown in 2-liter flasks containing 500 ml of peptone–glucose–salts medium with 5 ml of the exponential growth culture, at 28° for 20–24 hours. Unless otherwise stated, all procedures are carried out in the cold and all centrifugations are performed at 12,000 *g* in a refrigerated centrifuge at 0°.

Step 1. Preparation of Crude Extract. The cells of *E. freundii* cultivated on 300 liters of the culture medium are harvested by centrifugation and washed twice with deionized water. The washed cells are suspended in 0.01 *M* potassium phosphate buffer, pH 7.0, and disrupted with a Kaijo-Denki 19 Hz ultrasonic oscillator. The cell debris is centrifuged off.

Step 2. Heat Treatment. The cell-free extract is immediately heated for 10 minutes at 60° in a water bath under gentle stirring. The resulted precipitates are centrifuged off. The supernatant solution is then dialyzed overnight against deionized water.

Step 3. Ammonium Sulfate Fractionation I. Solid ammonium sulfate is added to the dialyzed enzyme solution to 35% saturation (26.6 g/100 ml). The precipitated protein is removed by centrifugation and discarded. The ammonium sulfate concentration is then increased to 65% saturation by the further addition of solid ammonium sulfate, 20.3 g/100 ml. The precipitate is collected by centrifugation. This process is repeated separately until 340 g as dry weight of the cells are treated, and the active precipitates are combined for the further purification. The combined precipitate is dissolved in 0.01 *M* potassium phosphate buffer, pH 7.0, resulting in a total

[3] K. Ogata, T. Tochikura, Y. Tani, and S. Yamamoto, *Agr. Biol. Chem. (Tokyo)* **30**, 829 (1966).

[4] H. Wada and E. E. Snell, *J. Biol. Chem.* **236**, 2089 (1961).

[5] O. H. Lowry, N. J. Rosebrough, A. L. Farr, and R. J. Randall, *J. Biol. Chem.* **193**, 265 (1951).

volume of 1 liter. The solution is then dialyzed overnight against the same buffer. The inactive precipitate that appears is centrifuged off.

Step 4. Protamine Sulfate Treatment. Freshly prepared 2.0% protamine sulfate solution neutralized with sodium hydroxide, 100 ml, is slowly added under stirring to 1100 ml of the dialyzed enzyme solution from step 3 and allowed to stand for 15 minutes. The precipitate formed is removed by centrifugation, and the resultant supernatant solution is dialyzed overnight under constant stirring against three changes of 0.01 M potassium phosphate buffer, pH 7.0, 12 liters each.

Step 5. DEAE-Cellulose Column Chromatography. The absorbent equilibrated with 0.01 M potassium phosphate buffer, pH 7.0, is packed into a column (3 × 40 cm). The enzyme solution is placed on the column and eluted with 2.5 liters of 0.01 M potassium phosphate buffer, pH 7.0. The active fractions are combined to give 860 ml of solution to which solid ammonium sulfate is added to 80% saturation, 60.8 g/100 ml. The precipitate obtained by centrifugation is dissolved in 0.01 M potassium phosphate buffer, pH 7.0, and dialyzed overnight against 10 liters of the same buffer.

Step 6. Ammonium Sulfate Fractionation II. To 90 ml of the dialyzed enzyme solution from step 5, solid ammonium sulfate, 30.8 g, is added. The precipitate is centrifuged off. The ammonium sulfate concentration is then increased to 55% saturation by the addition of 6.8 g of solid ammonium sulfate. The precipitate is collected by centrifugation and dissolved in a small amount of 0.01 M potassium phosphate buffer, pH 7.0. The fractionated solution is dialyzed overnight against three changes of the same buffer, 5 liters each.

Step 7. Hydroxylapatite Column Chromatography. The dialyzed enzyme solution (25 ml) is subjected to hydroxylapatite[6] column chromatography. The absorbent equilibrated with 0.01 M sodium phosphate buffer, pH 6.8, is used to pack a column (5 × 5 cm). The dialyzed enzyme is placed on the column, and the column is washed with 650 ml of 0.03 M of the same buffer. The enzyme is subsequently eluted with 0.1 M of the same buffer containing 0.1 M sodium chloride, and active fractions are combined to give 150 ml of solution. The active fraction is concentrated by the addition of solid ammonium sulfate to 80% saturation. The precipitate obtained by centrifugation is dissolved in 0.01 M potassium phosphate buffer, pH 6.6, and dialyzed overnight against two changes of the same buffer, 5 liters each.

Step 8. CM-Sephadex Column Chromatography. The dialyzed enzyme solution (9.9 ml) is subjected to CM-Sephadex column chromatography. The absorbent equilibrated with 0.01 M potassium phosphate buffer, pH 6.6, is used to pack a column (2.2 × 90 cm). The enzyme solution is placed

[6] A. Tiselius, S. Hjertén, and Ö. Levin, Arch. Biochem. Biophys. **65**, 132 (1956).

on the column and eluted with 0.03 M of potassium phosphate buffer, pH 6.6. The active fractions which are enriched by approximately 3000-fold are combined to give 63 ml of solution, and enzyme is precipitated by the addition of solid ammonium sulfate to 80% saturation.

Step 9. Crystallization. The purified enzyme preparation obtained from CM-sephadex column chromatography is used for the crystallization of the enzyme. The precipitate is collected by centrifugation at 15,000 g for 30 minutes and dissolved in a small volume of 0.03 M potassium phosphate buffer, pH 7.0. The insoluble precipitate is centrifuged off. Solid ammonium sulfate is added gradually to the supernatant solution until the solution becomes faintly turbid. The turbid suspension is left in the refrigerator for 1 week. The crystals will be observed microscopically as highly refractive needles.

The results of the purification procedure are summarized in the table.

PURIFICATION OF ACID PHOSPHATASE HAVING PYRIDOXINE-PHOSPHORYLATING ACTIVITY

Step	Total protein	Total units	Specific activity	Ratio of p-nitro-phenol: pyri-doxine-P activity
Cell-free extract	343,300	308,970	0.9	33
Heat treatment	58,200	302,640	5.2	35
Ammonium sulfate	27,500	299,750	10.9	35
Protamine treatment	18,200	305,760	16.8	35
DEAE-cellulose	4,000	228,400	57.1	32
Ammonium sulfate	1,420	116,080	81.7	33
Hydroxylapatite	113	133,815	1184.2	32
CM-Sephadex	18	37,803	2716.1	31
Crystallization	12	33,602	2800.2	32

Properties

Specificity. The acid phosphatase preparation hydrolyzes a large variety of phosphomonoesters and transfers the phosphoryl moiety of these phosphates to pyridoxine. Among the three forms of free vitamin B_6, pyridoxine is the best acceptor substrate, followed by pyridoxamine. Pyridoxal has lower activity.

pH Optimum and Heat Stability. The enzyme has a pH optimum of pH 6.0 for transphosphorylation and hydrolysis. The purified enzyme is stable toward heating at 65° for 10 minutes.

Inhibitor and K_m. Fluoride and mercuric ions are inhibitory at $10^{-3} M$. The K_m is $1.7 \times 10^{-4} M$ for p-nitrophenyl phosphate and $2.6 \times 10^{-1} M$ for pyridoxine under the standard assay conditions.

Stability. The crystalline enzyme can be stored at least a few months as a suspension in 0.03 M potassium phosphate buffer, pH 7.0, containing ammonium sulfate.

Physical Constants. The enzyme shows an absorption maximum at 279 mμ. The sedimentation coefficient ($s_{20,w}^0$) of the enzyme is 7.5 S and the diffusion coefficient ($D_{20,w}$) is 6.15 × 10^{-7} cm^2/sec. From these values, the molecular weight of the enzyme was calculated to be about 120,000 according to the equation of Svedberg and Erikson.[7]

[7] A. Ehrenberg, *Acta Chem. Scand.* **11**, 1257 (1957).

[98] The Catabolism of Vitamin B$_6$

By RICHARD W. BURG

Little is known concerning the catabolism of vitamin B$_6$ by higher organisms. The only known catabolite in mammals is 4-pyridoxic acid, which is a major excretory product. This compound probably arises through the oxidation of pyridoxal by liver aldehyde oxidase.[1]

The most detailed account of vitamin B$_6$ catabolism has come from studies of a group of microorganisms that were isolated by enrichment culture on the three forms of vitamin B$_6$.[2] Two pseudomonads have been studied extensively. One of these, designated *Pseudomonas* sp. IA, oxidizes pyridoxol via the pathway[2,3]:

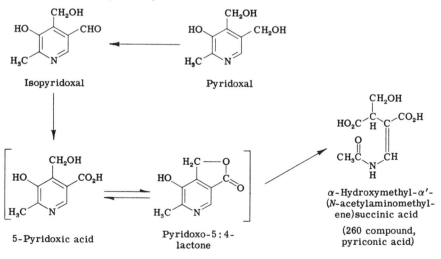

Isopyridoxal

Pyridoxal

5-Pyridoxic acid

Pyridoxo-5:4-lactone

α-Hydroxymethyl-α'-(*N*-acetylaminomethyl-ene)succinic acid

(260 compound, pyriconic acid)

The second isolate designated *Pseudomonas* sp. MA oxidizes pyridoxamine via the following pathway[4]:

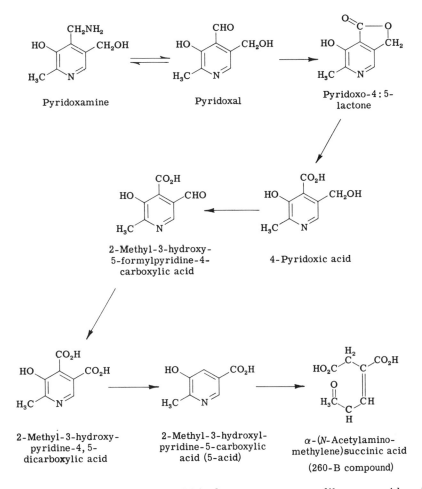

Pyridoxamine   Pyridoxal   Pyridoxo-4:5-lactone

2-Methyl-3-hydroxy-5-formylpyridine-4-carboxylic acid   4-Pyridoxic acid

2-Methyl-3-hydroxy-pyridine-4,5-dicarboxylic acid   2-Methyl-3-hydroxyl-pyridine-5-carboxylic acid (5-acid) α-(N-Acetylamino-methylene)succinic acid (260-B compound)

Although *Pseudomonas* sp. MA does not grow readily on pyridoxol, repeated transfer on that substrate gave rise to a variant that grew quite well. This strain, designated *Pseudomonas* sp. MA-I, oxidizes pyridoxol to pyridoxal which is degraded via the above pathway. *Pseudomonas* sp. MA-I

---

1 Y. Morino, H. Wada, T. Morisue, Y. Sakamoto, and K. Ichihara, *J. Biochem.* (*Tokyo*) **48**, 18 (1960).

2 V. W. Rodwell, B. E. Volcani, M. Ikawa, and E. E. Snell, *J. Biol. Chem.* **233**, 1548 (1958).

3 M. Ikawa, V. W. Rodwell, and E. E. Snell, *J. Biol. Chem.* **233**, 1555 (1958).

4 R. W. Burg, V. W. Rodwell, and E. E. Snell, *J. Biol. Chem.* **235**, 1164 (1960).

still produces the enzyme for the conversion of pyridoxamine to pyridoxal. Since pyridoxol is a more readily available substrate than pyridoxamine, this variant has been used for most enzyme studies.

The relevance of the catabolic pathways utilized by these microorganisms to the catabolic pathways in higher organisms is problematical. However, the biochemical reactions involved have their own intrinsic interest, and a number of the enzymes involved have been purified and studied extensively.

Separation and Characterization of Vitamin B_6 Catabolites

Paper Chromatography

The catabolites of vitamin B_6 may be separated and identified by ascending chromatography on Whatman No. 1 paper. The R_f values and colors with spray reagents are summarized in Table I.

Reagents

> Solvent A. *tert*-amyl alcohol–acetone–water–diethylamine
> (40:35:20:5) (v/v)
> Solvent B. *tert*-amyl alcohol–acetone–water–glacial acetic acid
> (40:35:20:5) (v/v)
> Solvent C. *tert*-amyl alcohol–acetone–water–benzylamine
> (40:35:20:5) (v/v)
> 2,6-Dichloroquinone chlorimide,[3] 1% solution in toluene
> Ammonium hydroxide, 1 M
> Sulfanilic acid, 0.9%. Dissolve 4.5 g of sulfanilic acid by warming
> with 45 ml of 12 N HCl. Dilute with water to 500 ml. Store at 5°.
> $NaNO_2$, 4.5%
> Na_2CO_3, 10%

The chromatogram may be sprayed with 2,6-dichloroquinone chlorimide followed by dilute ammonia. The diazotized sulfanilic acid spray reagent is prepared by mixing the sulfanilic acid solution with an equal volume of $NaNO_2$ solution. Cool the mixture in an ice bath, and add an equal volume of Na_2CO_3 solution.[5]

Ion-Exchange Chromatography[3,4]

The catabolites of pyridoxol and pyridoxamine can be isolated by ion-exchange column chromatography.

Procedure. A 4.5 × 50 cm column of Dowex 1 chloride (8% cross-linked, 200–400 mesh) is converted to the formate form by washing with 3 M

---

[5] H. K. Berry, H. E. Sutton, L. Cain, and J. S. Berry, University of Texas, Publ. No. 5109 (Biochemical Institute Studies IV), University of Texas, Austin, p. 30, 1951.

TABLE I
CHROMATOGRAPHIC PROPERTIES OF VITAMIN B₆ CATABOLITES

Compound	Fluorescence[a]	Color with 2,6-dichloroquinone chlorimide[b]	Color with diazotized sulfanilic acid[c]	R_f Solvent A	Solvent B	Solvent C
Pyridoxol	B	B	Y-O	0.36	0.57	0.58
Pyridoxamine	DB	B	O	0.73	0.20	0.72
Pyridoxal	Y	B	Y	0.68	0.60	0.87
Isopyridoxal	F	B → Br	P	0.4–0.6	0.66	—
Pyridoxo-5:4-lactone	F	B	Y	0.56	0.87	—
5-Pyridoxic acid	F	B	Y	0.15	0.85	—
α-Hydroxymethyl-α'-(N-acetylaminomethylene)succinic acid	Q	None	None	0.10	0.70	—
Pyridoxo-4:5-lactone	BB	B	Y-O	0.51	0.68	0.46
4-Pyridoxic acid	BB	B → Br	O	0.71	0.52	0.66
2-Methyl-3-hydroxy-5-formylpyridine-4-carboxylic acid	B	B		0.77	—	0.78
2-Methyl-3-hydroxypyridine-4,5-dicarboxylic acid	BB	B → Br	—	0.28	—	0.31
2-Methyl-3-hydroxypyridine-5-carboxylic acid	DB	B → Br	—	0.18	—	0.47
α-(N-Acetylaminomethylene)succinic acid	Q	None	—	0.14	—	0.15
Pyridoxine dimer	F	B	Y-O	0.18	0.65	—
Pyridoxal oxime	F	B	Y	0.55	0.85	—
Isopyridoxal oxime	F	B	Y-O	0.45	0.72	—
2-Methyl-3-hydroxy-5-hydroxymethylpyridine hydrochloride	B	B	—	0.41	—	0.69

[a] Under ultraviolet light, all the compounds fluoresce (F) except two which quench (Q). Fluorescent colors are designated as blue (B), dark blue (DB), bright blue (BB), and yellow (Y). 2-Methyl-3-hydroxy-5-formylpyridine-4-carboxylic acid fluoresces yellow when chromatographed in solvent C.

[b] All the 3-hydroxypyridine compounds give a blue color when sprayed with 2,6-dichloroquinone chlorimide. Those designated B → Br fade rapidly to brown.

[c] The colors formed with diazotized sulfanilic acid are yellow (Y), orange (O), yellow-orange (Y-O), and purple (P).

ammonium formate. When chloride is no longer eluted, the column is washed with water to free it of ammonium formate. Six liters of fermentation broth is concentrated to 300 ml by rotary evaporation at 50° and applied to the top of the column.

The catabolites of pyridoxol can be eluted stepwise by the addition of 3 liters of distilled water, 4 liters of 0.1 M formic acid, and 2.5 liters of 3 M formic acid. The water effluent contains pyridoxol and isopyridoxal which

TABLE II

ULTRAVIOLET ABSORPTION SPECTRA OF VITAMIN B_6 AND CATABOLITES

| | Ultraviolet absorption | | | | | |
| | 0.1 M HCl | | pH 7.0 | | 0.1 M NaOH | |
Compound	λ_{max} (mμ)	a_M ($\times 10^{-3}$)	λ_{max} (mμ)	a_M ($\times 10^{-3}$)	λ_{max} (mμ)	a_M ($\times 10^{-3}$)
Pyridoxol[a]	232	2.1	254	3.9	245	6.3
	291	8.6	324	7.2	310	6.8
Pyridoxamine[a]	226	2.0	252	4.5	245	6.2
	292	8.2	326	7.9	310	7.2
Pyridoxal[a]	—	—	352	5.8	240	8.9
	288	9.0	317	8.9	301	6.0
	—	—	390	0.2	394	1.7
Isopyridoxal[b]	230[d]	2.8	256	5.9	244	7.4
	284	7.2	312	7.4	296	6.1
5-Pyridoxic acid[b]	—	—	267	3.5	250[d]	5.1
	296	8.4	327	6.0	315	6.5
Pyridoxo-5:4-lactone[b]	252	4.4	277	6.1	—	—
	292	7.8	321	5.6	—	—
α-Hydroxymethyl-α'-(N-acetylaminomethylene) succinic acid[c]	265	21	262	17	266	16
Pyridoxo-4:5-lactone[a]	—	—	253	3.6	—	—
	316	7.6	356	8.0	—	—
4-Pyridoxic acid[a]	—	—	—	—	247	6.4
	317	6.1	316	6.0	308	7.2
2-Methyl-3-hydroxy-5-formyl-pyridine-4-carboxylic acid[a]	—	—	—	—	249	15.7
	316	5.9	320	5.3	364	5.3
2-Methyl-3-hydroxypyridine-4,5-dicarboxylic acid[a]	303	6.6	314	5.7	318	6.9
2-Methyl-3-hydroxypyridine-5-carboxylic acid[a]	241	3.7	256	4.2	256	6.2
	298	8.6	289	4.0	317	6.4
	—	—	327	4.2	—	—
α-(N-Acetylaminomethylene) succinic acid[a]	265	21.3	261	17.5	261	17.5

[a] These data are taken from footnote 4.
[b] These data are taken from footnote 2.
[c] These data are taken from footnote 3.
[d] Shoulder.

are followed by their absorbance at 320 mμ; the 0.1 M formic acid effluent contains pyridoxo-5:4-lactone and 5-pyridoxic acid measured at 300 mμ; the 3 M formic acid effluent contains α-hydroxymethyl-α'-(N-acetylaminomethylene)succinic acid measured at 260 mμ.

The catabolites of pyridoxamine which are more acidic can be eluted by the stepwise addition of 3 liters of distilled water, 6 liters of 0.1 M formic acid, 4 liters of 2 M formic acid, 2 liters of 3 M formic acid, and 4 liters of 5 M formic acid. Because of the numerous compounds and the variations of their ultraviolet spectra with changing pH, it is more convenient to follow the elution by scanning the spectra rapidly, using a recording spectrophotometer. Pyridoxamine and pyridoxal are eluted by water, pyridoxo-4:5-lactone and 2-methyl-3-hydroxypyridine-5-carboxylic acid by 0.1 M formic acid, 4-pyridoxic acid and α-(N-acetylaminomethylene)succinic acid by 2 M formic acid, 2-methyl-3-hydroxy-5-formylpyridine-4-carboxylic acid by 3 M formic acid, and 2-methyl-3-hydroxypyridine-4,5-dicarboxylic acid by 5 M formic acid.

Thin-Layer Chromatography

Thin-layer chromatography of vitamin B₆ and its derivatives is described in this volume [81].

Gas Chromatography

Gas chromatography of vitamin B₆ and its derivatives is described in this volume [82, 83].

Ultraviolet Absorption Spectra

The ultraviolet absorption spectra are summarized in Table II.

Fluorescence

The application of fluorometry to the assay of vitamin B₆ is presented in this volume [78].

Derivatives

Melting or decomposition points of vitamin B₆ catabolites and some derivatives are given in Table III.

Preparation of Catabolites of Vitamin B₆

Direct Fermentation

A number of catabolites of vitamin B₆ can be produced by fermentation. *Pseudomonas* sp. IA can be used to produce isopyridoxal, 5-pyridoxic acid, and α-hydroxymethyl-α'-(N-acetylaminomethylene)succinic acid from pyridoxol. *Pseudomonas* sp. MA can be used to produce 2-methyl-3-hydroxy-5-formylpyridine-4-carboxylic acid, 2-methyl-3-hydroxypyridine-4,5-dicarboxylic acid, and 2-methyl-3-hydroxypyridine-5-carboxylic acid.

TABLE III
Derivatives of Vitamin B$_6$ and Catabolites

Compound	Reagent	Derivative	Melting point
Pyridoxol	—	—	160°
	HCl	Pyridoxol hydro-chloride	205–212° (dec.)
Pyridoxamine	—	—	193°
	HCl	Pyridoxamine dihy-drochloride	226–227° (dec.)
Pyridoxal	—	—	—
	HCl	Pyridoxal hydro-chloride	165°
	Sodium borohydride	Pyridoxol	160°
	Hydroxylamine	Pyridoxal oxime	225–226° (dec.)
Isopyridoxal	—	—	185–187° (dec.)
	Sodium borohydride	Pyridoxol	160°
	Hydroxylamine	Isopyridoxal oxime	192–195° (dec.)
	Silver oxide	5-Pyridoxic acid	273° (dec.)
5-Pyridoxic acid	—	—	273° (dec.)
	HCl	Pyridoxo-5:4-lactone hydrochloride	242–246° (dec.)
Pyridoxo-5:4-lactone	—	—	272° (dec.)
	HCl	Pyridoxo-5:4-lactone hydrochloride	242–246° (dec.)
α-Hydroxymethyl-α'-(N-acetylamino-methylene)-succinic acid	—	—	152–153° (dec.)
Pyridoxo-4:5-lactone	—	—	263–265°
4-Pyridoxic acid	—	—	247–248°
2-Methyl-3-hydroxy-5-formylpyridine-4-carboxylic acid	—	—	245° (dec.)
	Semicarbazide	Semicarbazone of 2-methyl-3-hydroxy-5-formylpyridine-4-carboxylic acid	>300° (dec.)
	Sodium borohydride	4-Pyridoxic acid	247–248° (dec.)
2-Methyl-3-hydroxy-pyridine-4,5-dicarboxylic acid	—	—	269–270° (dec.)
2-Methyl-3-hydroxy-pyridine-5-carboxylic acid	—	—	325° (sublimes)
	Methanolic HCl	Methyl ester of 2-methyl-3-hydroxy-pyridine-5-carboxylic acid	239–240°
	Lithium aluminum	2-Methyl-3-hydroxy-5-hydroxymethyl-pyridine hydro-chloride	169–172°

The medium is prepared by mixing the following three solutions:

Solution A: Combine KH$_2$PO$_4$, 1.0 g; CaCl$_2$, 3.0 mg; FeSO$_4$, 3.0 mg; and distilled water, 1 liter. Adjust to pH 7.0 with KOH, and autoclave.

Solution B: Combine MgSO$_4$·7 H$_2$O, 0.5 g, and H$_2$O, 10 ml. Autoclave.

Solution C: To pyridoxol hydrochloride, 2.0 g, or pyridoxamine dihydrochloride, 1.5 g, add distilled water, 20 ml. Neutralize with KOH, and sterilize by filtration.

Cultures are maintained by monthly transfer on slants of the above medium containing 2.5% agar. Inocula are prepared in Roux bottles containing 100 ml of agar medium. Cells are washed from a Roux bottle with distilled water and added to 1 liter of liquid medium in a 2.8-liter Fernbach flask. The flask is covered with aluminum foil to prevent light-catalyzed degradation of the vitamin B$_6$, and the flask is shaken at 30° for 2–7 days. The progress of the fermentation is followed by determining the ultraviolet absorption spectrum and by chromatographing the medium. The products are isolated by ion-exchange chromatography as described above.

Resting-Cell Fermentation

2-Methyl-3-hydroxy-5-formylpyridine-4-carboxylic acid can be produced in good yield by incubation of lyophilized cells of *Pseudomonas* sp. MA with 4-pyridoxic acid in the presence of sodium bisulfite. Lyophilized cells (4.4 g) from 18 liters of a 48-hour culture in pyridoxamine medium are suspended in a solution containing 624 mg of sodium bisulfite and 1.1 g of 4-pyridoxic acid in 600 ml of 0.1 M potassium phosphate buffer, pH 7.0. The suspension is shaken at 30° in a 2.8-liter Fernbach flask. The progress of the oxidation is followed by determining the optical density at 308 mμ and 364 mμ of aliquots diluted in 0.1 M NaOH. After 12 hours when the absorption at 364 mμ exceeds that at 308 mμ, the cells are removed by centrifugation. The cells can be used at least twice more before they lose their activity. The supernatant solutions are combined, evaporated to 300 ml by rotary evaporation at 50° and chromatographed on a Dowex 1 formate column as described above. The bisulfite addition complex is eluted from the column with 1 N HCl. The fractions containing the 2-methyl-3-hydroxy-5-formylpyridine-4-carboxylic acid are combined and evaporated to dryness. The compound is recrystallized from boiling water to which a little decolorizing charcoal is added.

Enzymatic Syntheses

Purified enzymes can be used for synthesis of some catabolites that are not accumulated to any significant extent by fermentation. For example, α-(*N*-acetylaminomethylene)succinic acid is found in trace amounts in normal fermentation, but it has been produced from 2-methyl-3-hydroxypyridine-5-carboxylic acid using the oxygenase described in Part VI of this article.

Chemical Syntheses

Space does not permit a detailed description of the chemical syntheses of vitamin B_6 catabolites. Table IV lists those compounds for which syntheses are published. The syntheses from vitamin B_6 are especially convenient since they involve fewer steps and give high yields.

TABLE IV

CHEMICAL SYNTHESES OF VITAMIN B_6 CATABOLITES

	Literature reference	
Compound	Complete synthesis	Synthesis from vitamin B_6
Isopyridoxal	6	11
5-Pyridoxic acid and lactone	7	3
4-Pyridoxic acid and lactone	—	12
2-Methyl-3-hydroxy-5-formylpyridine-4-carboxylic acid	—	13
2-Methyl-3-hydroxypyridine-4,5-dicarboxylic acid	8	13
2-Methyl-3-hydroxypyridine-5-carboxylic acid	9	—
α-(N-Acetylaminomethylene)succinic acid	10	—

Enzymes of Vitamin B_6 Catabolism

Unless otherwise stated, the enzymes discussed below were extracted from *Pseudomonas* sp. MA-I grown on the growth medium.

Growth Medium: Combine KH_2PO_4, 1.0 g; $CaCl_2$, 3.0 mg; $FeSO_4$, 3.0 mg; $MnSO_4$, 0.5 mg; $ZnSO_4$, 0.5 mg; yeast extract, 0.2 g; and H_2O, 1 liter. Adjust to pH 7.0 with KOH, and autoclave. Add $MgSO_4\cdot7\ H_2O$, 0.5 g, and H_2O, 10 ml. Autoclave. Then add pyridoxol hydrochloride, 2.0 g, and H_2O, 20 ml. Adjust to pH 7.0 with KOH. Sterilize by filtration.

Culture Maintenance and Cell Production

The growth medium is prepared by mixing the three sterile solutions. *Pseudomonas* sp. MA-I is maintained by monthly transfer on 2.5% agar

[6] S. A. Harris, D. Heyl, and K. Folkers, *J. Am. Chem. Soc.* **66,** 2088 (1944).
[7] S. A. Harris, E. T. Stiller, and K. Folkers, *J. Am. Chem. Soc.* **61,** 1242 (1939).
[8] R. G. Jones and E. C. Kornfeld, *J. Am. Chem. Soc.* **73,** 107 (1951).
[9] C. J. Argoudelis and F. A. Kummerow, *J. Org. Chem.* **26,** 3420 (1961).
[10] F. Lingens and R. Hankwitz, *Ann. Chem.* **670,** 31 (1963).
[11] W. Korytnyk, E. J. Kris, and R. P. Singh, *J. Org. Chem.* **29,** 574 (1964).
[12] H. Ahrens and W. Korytnyk, *J. Heterocyclic Chem.* **4,** 625 (1967).
[13] B. Paul and W. Korytnyk, *Chem. Ind. (London)* p. 230 (1967).

slants of this medium. Liquid cultures for inocula may be kept up to 6 months at 4° or frozen at −20°. Twenty milliliters of liquid inoculum is used to inoculate 1 liter of medium in a 2.8-liter Fernbach flask which is incubated at 30° on a rotary shaker for 3–4 days. This can be used to inoculate 15 liters of medium in a 20-liter carboy. The carboy culture is aerated vigorously through a glass sparger.

Cell growth is followed by turbidity at 650 mμ, and cells are harvested by centrifugation when the cell density reaches 1.0–1.5 g (dry weight) per liter. The cells are washed twice with distilled water. The cell paste may be used directly or be lyophilized or frozen and stored at −20° until needed.

Protein Assay

All protein determinations for the following enzyme preparations were made by the method of Lowry *et al.*[14] using bovine serum albumin as standard.

I. Pyridoxamine–Pyruvate Transaminase

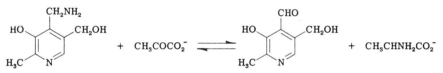

Assay Method

Pyridoxal formed in the reaction is determined by the phenylhydrazine method of Wada and Snell[15] as modified by Dempsey and Snell.[16] The purified enzyme can also be assayed spectrophotometrically by following the appearance or disappearance of pyridoxal at 400 mμ.

Reagents

Tris (hydroxymethylamino)methane-HCl buffer, 1 M, pH 8.5
Pyridoxamine dihydrochloride, 0.1 M, neutralized
Sodium pyruvate, 0.1 M
EDTA, 0.2 M
H₂SO₄, 9 M
Phenylhydrazine reagent: Dissolve 2 g of phenylhydrazine hydrochloride in 100 ml of 10 N H₂SO₄. This solution may be stored at 5° until it begins to darken.

[14] O. H. Lowry, N. J. Rosebrough, A. L. Farr, and R. J. Randall, *J. Biol. Chem.* **193**, 265 (1951).

[15] H. Wada and E. E. Snell, *J. Biol. Chem.* **236**, 2089 (1961).

[16] W. B. Dempsey and E. E. Snell, *Biochemistry* **2**, 1414 (1963).

Procedure. The reaction mixture contains 0.3 ml of Tris buffer, 0.1 ml of pyridoxamine, 0.1 ml of sodium pyruvate, 0.05 ml of EDTA, and 2.45 ml of deionized water. Warm to 37° and add the enzyme to start the reaction. When a yellow color appears, stop the reaction by adding 0.5 ml of H_2SO_4. Centrifuge to remove any precipitate and add 0.2 ml of phenylhydrazine reagent to 2.8 ml of the supernatant solution. Heat at 60° for 20 minutes, cool to room temperature, and measure the optical density at 410 mμ. This assay will measure from 0.01 to 0.1 micromole of pyridoxal.

Definition of Unit. One unit of enzyme is that amount which produces 1 micromole of pyridoxal per minute at 37°.

Purification Procedure

The enzyme was first crystallized by Wada and Snell,[17] and an improved procedure was developed by Dempsey and Snell.[16] It has been the subject of a number of detailed studies.[18-20]

Step 1. Preparation of Cell-Free Extract. Suspend 2 g of lyophilized cells in 50 ml of 0.02 M potassium phosphate buffer, pH 7.0, and treat for 20 minutes in a Raytheon 10-kc oscillator. Prepare several such batches. Centrifuge the sonicates at 15,000 g for 15 minutes. Extract the residue with 17 ml of buffer, centrifuge, and combine the supernatant fractions. This and all succeeding steps are carried out at 5°.

Step 2. Protamine Fractionation. Add freshly prepared 4% protamine sulfate in 0.1 M potassium phosphate buffer, pH 7.0, to the pooled extracts to supply 30 mg of protamine sulfate for each 100 mg of protein. Stir for 15 minutes, centrifuge, and discard the precipitate.

Step 3. Heat Treatment. Add 0.2 volume of 0.5 M potassium phosphate buffer, pH 6.5, containing 60 micromoles of pyridoxol per milliliter to the supernatant solution from the protamine fractionation. Stir the solution in an 85° water bath until its temperature reaches 70°, then transfer to a 70° water bath for 10 minutes. Cool in ice to 20° and centrifuge. Discard the precipitate.

Step 4. Ammonium Sulfate Fractionation. Cool the enzyme solution in an ice bath. Add a saturated solution of ammonium sulfate (adjusted to pH 7.0 with ammonia) to bring to 45% saturation. Centrifuge and discard the precipitate. Add additional ammonium sulfate solution to bring to 60% saturation. Centrifuge and discard the supernatant solution. Dissolve the precipitate in 0.02 M potassium phosphate buffer, pH 7.0, and dialyze against 100 volumes of the same buffer for 12 hours.

[17] H. Wada and E. E. Snell, *J. Biol. Chem.* **237**, 133 (1962).
[18] M. Fujioka and E. E. Snell, *J. Biol. Chem.* **240**, 3044, 3050 (1965).
[19] J. E. Ayling and E. E. Snell, *Biochemistry* **7**, 1616, 1626 (1968).
[20] H. Kolb, R. D. Cole, and E. E. Snell, *Biochemistry* **7**, 2946 (1968).

Step 5. Acetone Fractionation. Place the dialyzed enzyme preparation in a −20° bath and add one-half volume of acetone (−20°) in 1 minute with rapid stirring. Stir for an additional 2 minutes and then centrifuge at −15° at 15,000 g for 10 minutes. Suspend the precipitate in 0.02 M potassium phosphate buffer, pH 7.0, and dialyze overnight against the same buffer containing 0.001 M pyridoxal. Repeat the above procedure three more times on the supernatant fraction using the same volume of acetone as the first time. The transaminase precipitates during the third addition of acetone and can be identified by its bright yellow color.

Step 6. Crystallization. Add saturated ammonium sulfate (pH 7.0) to the dialyzed solution until the first permanent turbidity is noted. The enzyme crystallizes within 12 hours. The enzyme may be recrystallized from 0.02 M potassium phosphate buffer, pH 7.0, containing 0.001 M pyridoxal by repeating this ammonium sulfate treatment. At least 5 to 7 recrystallizations are required to attain constant specific activity.

A typical purification is summarized in Table V.

TABLE V
PURIFICATION OF PYRIDOXAMINE–PYRUVATE TRANSAMINASE

Fraction	Volume (ml)	Units	Protein (mg)	Specific activity (units/mg)
I. Cell-free extract	675	1290	10100	0.128
II. Protamine supernatant	705	1260	5640	0.224
III. Heat treatment	793	1230	1660	0.740
IV. Ammonium sulfate	147	1500	235	6.37
V. Acetone precipitate 1				1.17
2				2.00
3				23.1
4				1.75
VI. Recrystallized 7 times				40.0[a]

[a] Data of Dempsey and Snell;[16] a lower specific activity is usually obtained.

Stability. The purified enzyme must be kept in the presence of pyridoxal to prevent denaturation. The crystalline enzyme (30 mg/ml) can be stored at 5° in 60% saturated ammonium sulfate, pH 7.2, containing 0.001 M pyridoxal where it is stable for at least 6 months.

Properties

Molecular Weight. The transaminase has a molecular weight of 150,000 as determined from sedimentation and diffusion coefficients and by sedimentation equilibrium.[20] The enzyme dissociates into 4 subunits of equal size in 8 M urea.

Specificity. The transaminase is highly specific for pyruvate and L-alanine although α-aminobutyrate, DL-alanine ethyl ester, and DL-alanine amide react at about 5% of the rate of L-alanine. The specificities for pyridoxal and pyridoxamine are less exacting and a number of analogs containing the basic 3-hydroxypyridine-4-aldehyde structure are active. The 5-phosphate esters are inactive.

Effect of pH. The optimal pH for the transaminase is 9.2 in Tris buffer. Reaction rates are similar in pyrophosphate, borate, and phosphate buffers.

Inhibitors. The transaminase contains 6 sulfhydryl groups. Two of them react rapidly with *p*-chloromercuribenzoate with the release of the 2 moles of pyridoxal which are bound to each mole of enzyme, resulting in loss of activity.[18] The enzyme is inhibited by pyridoxol (K_i 5.2×10^{-5} M) and a number of pyridoxyl amino acids. Pyridoxyl-L-alanine is an especially potent inhibitor with a K_i of 1.8×10^{-7} M.

Kinetic Constants. The K_m for L-alanine is 1.6×10^{-3} M and that for pyruvate is 3.5×10^{-4} M. The K_m for pyridoxal is 1.2×10^{-5} M and that for pyridoxamine is 1.3×10^{-5} M. The $V_{\max}$ for the forward reaction is 10.1 micromoles/minute per milligram of enzyme, and that for the reverse reaction is 8.4 micromoles/minute per milligram of enzyme. In sodium pryophosphate buffer, pH 8.85,[19] at 25°

$$K_{eq} = \frac{[\text{pyridoxal}][\text{alanine}]}{[\text{pyridoxamine}][\text{pyruvate}]} = 1.21$$

II. Pyridoxol 4-Dehydrogenase[21]

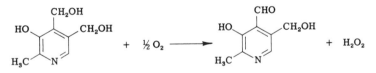

Assay Method

The dye 2,6-dichlorophenolindophenol acts as an electron acceptor for the oxidation of pyridoxol by pyridoxol 4-dehydrogenase. The rate of decolorization of the dye is followed at 620 mμ using a recording spectrophotometer. The enzyme may also be assayed by following the uptake of oxygen manometrically in a Warburg apparatus.

Reagents

Potassium phosphate buffer, 0.5 M, pH 8.0
2,6-Dichlorophenolindophenol, 100 μg/ml
KCN, 0.01 M

[21] T. K. Sundaram and E. E. Snell, *J. Biol. Chem.* **244**, 2577 (1969).

Flavin adenine dinucleotide, 3×10^{-4} M
Pyridoxol hydrochloride, 0.05 M. Adjust to pH 7.

Procedure. The reagents are mixed in a 1-cm cuvette in the following order: phosphate buffer, 0.1 ml; 2,6-dichlorophenolindophenol, 0.1 ml; KCN, 0.1 ml; FAD, 0.1 ml; H$_2$O, 0.5 ml, and enzyme. The reaction is started by the addition of 0.1 ml of pyridoxol.

Definition of Unit. A unit of enzyme is that amount which causes a decrease in absorbancy of 0.1 per minute at 26°. This corresponds to the reduction of 0.005 micromole of the dye.

Purification Procedure

Step 1. Preparation of Cell-Free Extract. Suspend 20 g (wet weight) of frozen packed cells in 80 ml of 0.02 M potassium phosphate buffer, pH 8.0, and treat for 20 minutes in a Raytheon 10-kc oscillator. Centrifuge for 1 hour at 35,000 g. Discard the residue. All steps are carried out at 5° unless otherwise stated.

Step 2. Protamine Fractionation. Add a freshly prepared 2% solution of protamine sulfate, pH 5, to supply 15 mg of protamine sulfate per 100 mg of protein. Stir for 15 minutes and then centrifuge at 35,000 g for 30 minutes. Discard the precipitate.

Step 3. Heat Treatment. Add pyridoxol to the supernatant fraction to bring the concentration to 5×10^{-3} M. Heat in a 50° water bath for 20 minutes and then cool in an ice bath. Centrifuge and discard the precipitate.

Step 4. Ammonium Sulfate Fractionation. Add saturated ammonium sulfate solution (pH 7) to 55% saturation. Stir for 45 minutes. Centrifuge and discard the supernatant solution. Dissolve the precipitate in 0.02 M potassium phosphate buffer, pH 8.0. Dialyze for 6 hours against 200 volumes of the same buffer.

Step 5. Calcium Phosphate Gel Absorption. Adjust the protein concentration of the dialyzed enzyme to 12 mg of protein per milliliter by the addition of 0.02 M potassium phosphate buffer, pH 8.0. Add an amount of calcium phosphate gel equal to twice the weight of the protein. Stir for 15 minutes and then centrifuge and discard the gel.

A typical purification is summarized in Table VI.

Properties

Specificity. The enzyme reacts maximally with pyridoxol but ω-methyl pyridoxol is oxidized at 50% and isopyridoxal and pyridoxol phosphate at 10% of the rate of pyridoxol.

Effect of pH. The activity of the enzyme is highly dependent upon the buffer used. Activity decreases in the following order: phosphate > pyro-

TABLE VI
PURIFICATION OF PYRIDOXOL 4-DEHYDROGENASE

Fraction	Volume (ml)	Units	Protein (mg)	Specific activity (units/mg)
I. Cell-free extract	68	19,000	1020	18.6
II. Protamine	71	15,000	740	20.3
III. Heat treatment	71	21,000	400	52.5
IV. Ammonium sulfate	9.1	13,000	130	100
V. Calcium phosphate gel	16.2	13,000	58	224

phosphate > Tris. The pH optima are 8 and 7.5 in pyrophosphate and Tris buffers, respectively. In phosphate buffer, the activity increases up to pH 8.

Cofactors. The enzyme requires FAD for activity. Both molecular oxygen and 2,6-dichlorophenolindophenol can act as electron acceptors.

Inhibitors. The enzyme is inhibited by p-chloromercuribenzoate when cyanide is omitted from the assay mixture. EDTA does not inhibit the enzyme but α,α'-dipyridyl, o-, m-, and p-phenanthroline do. Inhibition by m- and p-phenanthroline indicate that this inhibition is not due to chelation of a metal ion cofactor.

Substrate Affinities. The K_m for pyridoxol is 4.3×10^{-5} M and that for FAD is 2.4×10^{-6} M. The latter is unusually high compared with other FAD-requiring enzymes. Maximum velocity is attained with 8.3×10^{-6} M 2,6-dichlorophenolindophenol.

III. Pyridoxal Dehydrogenase[22]

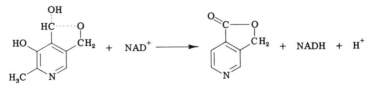

Assay Method

Pyridoxal dehydrogenase is assayed by following the increase in absorbancy at 340 mμ due to the formation of NADH and pyridoxo-4:5-lactone. The extinction coefficients are 6.2×10^3 and 6.7×10^3, respectively.

[22] R. W. Burg and E. E. Snell, *J. Biol. Chem.* **244**, 2585 (1969).

Reagents

Sodium pyrophosphate buffer, 0.1 M, pH 9.0
NAD, 0.01 M
Pyridoxal, 0.01 M

Procedure. In a 1-cm cuvette, mix 0.5 ml of sodium pyrophosphate buffer, 0.35 ml of deionized water, 0.1 ml of NAD, and 5 μl of enzyme preparation (diluted with deionized water so that the optical density change at 340 mμ does not exceed 0.2 per minute). Start the reaction by the addition of 0.05 ml of pyridoxal and follow the course of the reaction for 1 minute using a recording spectrophotometer. The reaction rate is determined from the slope at zero time.

Definition of Unit. A unit of enzyme is that amount which causes an optical density change of 0.1 per minute at room temperature.

Purification Procedure

Pyridoxal dehydrogenase has not been extensively purified, but it can be separated from pyridoxo-4:5-lactonase by the following procedure.

Step 1. Preparation of Cell-Free Extract. Packed cells of *Pseudomonas* sp. MA grown on the pyridoxamine medium given on page 641 which have been stored at $-20°$ are broken in a Hughes press. The broken cell paste is mixed with two volumes of deionized water and treated in a Raytheon 10-kc oscillator for 1–2 minutes to reduce the viscosity of the extract. Centrifuge at 10,000 g for 15 minutes, and discard the residue. This and all subsequent steps are carried out at 5°. Freezing and thawing of the extract may produce some insoluble material which is removed by centrifuging. The dehydrogenase is quite stable to freezing and thawing if the protein concentration is kept above 7 mg/ml.

Step 2. Protamine Fractionation. Add 0.1 to 0.12 ml of 2% protamine sulfate solution per milliliter of extract. No adjustment of pH is made. Centrifuge and discard the precipitate.

Step 3. Ammonium Sulfate Fractionation. Add solid ammonium sulfate to bring the supernatant solution to 20% saturation. Centrifuge and discard the precipitate. Add additional solid ammonium sulfate to bring to 60% saturation. Centrifuge, discard the supernatant solution, and dissolve the precipitate in 3 ml of deionized water. Dialyze against deionized water.

Step 4. Second Ammonium Sulfate Fractionation. To the dialyzed solution from step 3, add sufficient solid ammonium sulfate to bring it to 30% saturation. Centrifuge, and discard the precipitate. Bring the supernatant solution to 50% saturation with ammonium sulfate. Centrifuge, discard the supernatant solution, and dissolve the precipitate in 0.01 M potassium phosphate buffer, pH 7.0. Dialyze against the same buffer. This preparation can be stored at $-20°$ and is stable to repeated freezing and thawing. It is free of detectable pyridoxo-4:5-lactonase.

A typical purification is summarized in Table VII.

TABLE VII
PURIFICATION OF PYRIDOXAL DEHYDROGENASE

Fraction	Volume (ml)	Units	Protein (mg)	Specific activity (units/mg)
I. Cell-free extract	20	2630	178	14.8
II. Protamine supernatant	20	2240	128	17.5
III. First ammonium sulfate fraction	3.3	2290	52	44.0
IV. Second ammonium sulfate fraction	2.7	1360	22	62.0

Properties

Specificity. Pyridoxal and ω-methylpyridoxal are dehydrogenated at comparable rates. 5-Deoxypyridoxal is not a substrate, but it does stimulate the reaction with pyridoxal during which it is reduced to 5-deoxypyridoxol. Slight activity was observed with isopyridoxal, 2-methyl-3-hydroxy-5-formylpyridine-4-carboxylic acid, and pyridoxal phosphate. Acetaldehyde, benzaldehyde, and salicylaldehyde are not substrates. The enzyme requires NAD, and NADP is neither active nor inhibitory.

Effect of pH. The pH optimum is near 9.4. The activity is somewhat less in sodium carbonate buffer than in sodium pyrophosphate buffer. The K_m for pyridoxal increases with increasing pH while that of NAD remains constant.

Substrate Affinities. The K_m for pyridoxal is $7.6 \times 10^{-5}\,M$ at pH 9.3 and that of NAD is $2.9 \times 10^{-4}\,M$.

Inhibitors. Pyridoxo-4:5-lactone, pyridoxol, 4-pyridoxic acid, and pyridoxamine are mild competitive inhibitors with K_i's of $7 \times 10^{-4}\,M$, $8 \times 10^{-4}\,M$, $1 \times 10^{-3}\,M$, and $1 \times 10^{-3}\,M$, respectively. The enzyme is inhibited by o-phenanthroline (50% inhibition at $2.6 \times 10^{-4}\,M$), 8-hydroxyquinoline, and α,α'-dipyridyl. However, this inhibition does not appear to be due to chelation of a metal ion since quinoline and m- and p-phenanthroline are also inhibitors. The latter two are extremely potent, inhibiting 50% at $2 \times 10^{-6}\,M$.

Effect of Order of Addition of Reactants. Only 50% of the maximum activity is obtained when the enzyme is mixed with pyridoxal in the absence of NAD. The enzyme loses activity when diluted, but both sodium pyrophosphate buffer, pH 9, and NAD serve to stabilize it.

IV. Pyridoxo-4:5-lactonase[22]

Assay Method

Pyridoxo-4:5-lactone has an ultraviolet absorption maximum near 350 mμ while that of 4-pyridoxic acid is at 316 mμ. The reaction may be followed by the decrease in absorbancy at 350 mμ. Since the wavelength and extinction coefficient are pH dependent, the reaction may also be followed at 338 mμ, the isosbestic point (extinction 6.7 for 1 micromole/ml).

Reagents

Sodium pyrophosphate buffer, 0.1 M. Adjust to pH 7.7 with HCl.
Pyridoxo-4:5-lactone, 0.001 M. Adjust to pH 4.

Procedure. To a 1-cm cuvette, add 0.5 ml of sodium pyrophosphate buffer, 0.3 ml of deionized water, 0.1 ml of pyridoxo-4:5-lactone, and 0.1 ml of cell extract. Follow the reaction at either 340 or 350 mμ. The reaction rate is linear to completion.

Purification

This enzyme has not been purified to any significant extent. The purification procedure for pyridoxal dehydrogenase may be used. The lactonase remains in solution in the 60% ammonium sulfate solution of step 3. It is precipitated by adding solid ammonium sulfate to 75% saturation. The precipitate is collected by centrifugation, dissolved in 0.02 M potassium phosphate buffer, pH 7.0, and dialyzed against the same buffer. This preparation is only 2-fold purified over the crude extract.

Effect of pH. The optimal pH is near 7.6 in sodium pyrophosphate buffer.

Substrate Affinity. The K_m is 3×10^{-6} M for pyridoxo-4:5-lactone. There is substrate inhibition above 2×10^{-5} M.

Inhibitors. The phenanthrolines inhibit this enzyme, but not so strongly as they inhibit pyridoxal dehydrogenase. Again m- and p-phenanthroline are more effective than o-phenanthroline.

V. 2-Methyl-3-hydroxypyridine-4,5-dicarboxylic Acid 4-Decarboxylase[23]

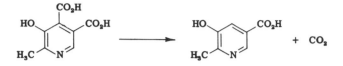

Assay Method

At pH 3.8, 2-methyl-3-hydroxypyridine-4,5-dicarboxylic acid has an absorption maximum at 318 mμ while that of 2-methyl-3-hydroxypyridine-5-carboxylic acid is at 292 mμ.

Reagents

> Tris buffer, 0.2 M, pH 8.0
> KCN, 0.1 M
> MnSO$_4$, 0.001 M
> 2-Methyl-3-hydroxypyridine-4,5-dicarboxylic acid, 0.1 M
> HClO$_4$, 70%
> Citrate buffer, 0.5 M, pH 3.8

Procedure. Mix 0.1 ml of 2-methyl-3-hydroxypyridine-4,5-dicarboxylic acid, 0.3 ml of MnSO$_4$, 1.5 ml of Tris buffer, 0.6 ml of KCN, and 0.4 ml of deionized water. Warm to 37° and add 0.1 ml of enzyme preparation. After 12 minutes, add 0.1 ml of 70% HClO$_4$ to stop the reaction. Centrifuge to remove the precipitated protein and dilute 0.4 ml of the supernatant solution with 9.2 ml of citrate buffer. Determine the optical density at 318 mμ. The molar extinction coefficient of 2-methyl-3-hydroxypyridine-4,5-dicarboxylic acid is 6.3 × 10^3 at 318 mμ and that of 2-methyl-3-hydroxypyridine-5-carboxylic acid is 0.4 × 10^3.

Definition of Unit. One unit of enzyme is that amount which will decarboxylate 1 micromole of substrate in 1 minute at 37°.

Purification Procedure

Step 1. Preparation of Cell-Free Extract. Either lyophilized or fresh cells of *Pseudomonas* sp. MA-I may be used. Suspend 5 g of lyophilized cells in 100 ml of deionized water. Treat 20-ml portions of the suspension for 15 minutes with a Branson 9-kc oscillator. Remove insoluble material by centrifuging at 10,000 g for 30 minutes. This and all subsequent steps are carried out at 5°.

Step 2. Protamine Fractionation. Add sufficient 1% protamine sulfate solution to provide 0.3 mg per milligram of protein in the cell-free extract.

[23] E. E. Snell, A. A. Smucker, E. Ringelman, and F. Lynen, *Biochem. Z.* **341**, 109 (1964).

Centrifuge and discard the supernatant solution. Extract the enzyme from the precipitate by stirring twice with 60 ml of 0.25 M sodium pyrophosphate, 0.35 M potassium phosphate buffer, pH 8.4. The insoluble residue is removed by centrifugation. The enzyme does not precipitate with the protamine if the extract is prepared in 0.1 M potassium phosphate buffer, pH 7.0, instead of water.

Step. 3. Manganese Fractionation. Add dropwise with stirring 65 ml of 1 M manganese sulfate solution (Mn^{2+}/pyrophosphate = 2:1) to the pyrophosphate–phosphate extract. Remove the copious precipitate by centrifugation and wash it once with a little water. Dialyze the combined supernatant solutions against 4 liters of deionized water.

Step 4. Calcium Phosphate Gel Adsorption. Adjust the pH of the dialyzed enzyme to 5.8–6.0 by the addition of 0.1 M acetic acid. Stir with calcium phosphate gel (0.1 mg of gel per milligram of protein) for 5 minutes. Remove the gel by centrifuging. Repeat the gel adsorption three more times, using the same amount of gel each time. Extract the enzyme from the gel by stirring with 8–10 ml of 0.01 M potassium phosphate buffer, pH 8.0.

A typical purification is summarized in Table VIII.

TABLE VIII
PURIFICATION OF 2-METHYL-3-HYDROXYPYRIDINE-4,5-DICARBOXYLIC
ACID 4-DECARBOXYLASE

Fraction	Volume (ml)	Units	Protein (mg)	Specific activity (units/mg)
I. Cell-free extract	100	153	900	0.17
II. Protamine extract	112	158	599	0.26
III. Manganese supernatant	220	128	176	0.73
IV. Gel eluate 1	8	12	7.8	0.28
2	9	5	11.5	1.30
3	9	24	9.2	2.56
4	10	23	11.5	2.03

Properties

Effect of pH. The decarboxylase has a broad pH optimum from 7.4 to 8.4.

Substrate Affinity. The K_m for 2-hydroxy-3-methylpyridine-4,5-dicarboxylic acid is $3.6 \times 10^{-5} M$. There is substrate inhibition above $2 \times 10^{-3} M$ which can be partially reversed by increasing the concentration of Mn^{2+}.

Cofactors. The decarboxylase requires a divalent cation for activity. Mn^{2+} is most active but can be partially replaced by high concentration of Ca^{2+} or Mg^{2+}. A reducing agent is also required. Cyanide is most effective, but thioglycol and cysteine are also active.

VI. 2-Methyl-3-hydroxypyridine-5-carboxylate Oxygenase[24]

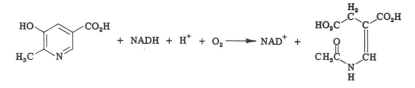

$$+ \text{ NADH } + \text{ H}^+ + \text{O}_2 \longrightarrow \text{NAD}^+ +$$

Assay Method

The enzyme is assayed by following the disappearance of NADH and 2-methyl-3-hydroxypyridine-5-carboxylate (5-acid) at 340 mμ. The extinction coefficients are 6.2×10^3 and 17.5×10^3, respectively.

Reagents

    Potassium phosphate buffer, 0.1 M, pH 8.0
    2-Mercaptoethanol, 0.4 M
    Flavin adenine dinucleotide, $3 \times 10^{-4} M$
    NADH, $2 \times 10^{-3} M$
    2-Methyl-3-hydroxypyridine-5-carboxylic acid, $2 \times 10^{-3} M$

Procedure. In a 1-cm cuvette, mix 0.5 ml of potassium phosphate buffer, 0.1 ml of 2-mercaptoethanol, 0.1 ml of FAD, 0.1 ml of NADH, and 0.1 ml of 5-acid. Start the reaction by the addition of 0.1 ml of the enzyme preparation, and follow the change in absorbancy at 340 mμ.

Definition of Unit. One unit of enzyme is that amount causing a decrease in absorbancy at 340 mμ of 0.1 per minute at room temperature.

Purification Procedure

Step 1. Preparation of Cell-Free Extract. *Pseudomonas* sp. MA-I grown on pyridoxol as previously described was used as the source of the enzyme. Packed cells can be stored at $-20°$ for at least 6 months. Suspend 30 g of frozen cells in 150 ml of 0.1 M potassium phosphate buffer, pH 8.0, containing 5 mM EDTA and 5 mM Versene-Fe^{3+}. Treat 50-ml batches for 20 minutes in a Raytheon 10-kc oscillator. Centrifuge for 1 hour at 105,000 g, and discard the residue. This and all subsequent steps are carried out at 5°.

Step 2. Protamine Fractionation. Add 2-mercaptoethanol to the supernatant solution to a final concentration of 0.1%, and then add sufficient

[24] L. G. Sparrow, P. T. Ho, T. K. Sundaram, D. Zäch, E. J. Nyns, and E. E. Snell, *J. Biol. Chem.* **244**, 2590 (1969).

1% protamine sulfate solution to provide 11 mg per 100 mg of protein. Stir for 15 minutes, centrifuge, and discard the precipitate.

Step 3. Ammonium Sulfate Fractionation. While maintaining the pH at 8.0 by the addition of ammonium hydroxide solution, add sufficient solid ammonium sulfate to bring the solution to 40% saturation. Centrifuge and discard the precipitate. Bring to 55% saturation by adding ammonium sulfate, again maintaining the pH at 8.0. Centrifuge, discard the supernatant solution, and dissolve the precipitate in 10 ml of 1 mM potassium phosphate buffer, pH 6.8, containing 0.1% 2-mercaptoethanol. Dialyze for 4 hours against 0.1 mM potassium phosphate buffer containing 0.1% 2-mercaptoethanol.

Step 4. Hydroxylapatite Chromatography. Pack hydroxylapatite (prepared according to Tiselius *et al.*[25]) by gravity to give a 3 × 16 cm column and equilibrate overnight with 0.1 M potassium phosphate buffer, pH 6.8, containing 0.1% 2-mercaptoethanol. Add the dialyzed protein solution from step 3, containing 300 mg of protein, to the column. Elute the column with a linear gradient in which the mixing chamber contains 500 ml of 0.01 M potassium phosphate buffer, pH 6.8, and 0.1% 2-mercaptoethanol and the reservoir contains 0.15 M potassium phosphate buffer, pH 6.8, and 0.1% 2-mercaptoethanol, using a flow rate of 30–35 ml per hour. Collect 7-ml fractions. The transaminase and 260-B hydrolase (see the following section) are eluted early, followed by the oxygenase in fractions 85–100, which is recognized by its faint yellow color. Combine the active fractions, and precipitate the enzyme by adding solid ammonium sulfate to 60% saturation. Centrifuge and discard the supernatant solution.

Step 5. Crystallization. Dissolve the precipitated enzyme from step 4 in 3 ml of 25% ammonium sulfate solution containing 0.1% 2-mercaptoethanol. Centrifuge and discard the insoluble material. Add saturated ammonium sulfate solution dropwise with stirring until a slight permanent turbidity appears. Clarify by centrifuging and add 1 drop of saturated ammonium sulfate. A birefringent precipitate appears. Allow the solution to stand in ice overnight, collect the crystals by centrifugation, and dissolve in 1.5 ml of 0.01 M potassium phosphate buffer, pH 8.0, containing 50% glycerol and 0.1% 2-mercaptoethanol. The enzyme can be stored at −20° in this solution for one month with a 20% loss of activity. The oxygenase is unstable in the absence of 2-mercaptoethanol.

Modified Purification

When the specific activity of the oxygenase in crude extracts is lower than that given in the example in Table IX, the following modifications are necessary in order to obtain a crystalline enzyme.

²⁵ A. Tiselius, S. Hjerten, and Ö. Levin, *Arch. Biochem. Biophys.* **65,** 132 (1956).

TABLE IX

PURIFICATION OF 2-METHYL-3-HYDROXYPYRIDINE-5-CARBOXYLATE OXYGENASE

Fraction	Units	Protein (mg)	Specific activity (units/mg)
I. Cell-free extract	15,600	1690	9
II. Protamine supernatant	17,300	1440	13
III. Ammonium sulfate	14,500	297	49
IV. Hydroxylapatite chromatography	14,300	63	226
IVa. Ammonium sulfate	13,500	50	270
V. Crystalline enzyme	6,750	22	300

Step 5. Sephadex Chromatography. Instead of dissolving the active fraction from step 4 above in 25% ammonium sulfate, dissolve it in 3 ml of 0.05 *M* potassium phosphate buffer, pH 7.0, containing 0.1% 2-mercaptoethanol. Apply this solution to a 2.5 × 90 cm column of Sephadex G-200 which has been equilibrated with the same buffer. Elute the column with this buffer at a flow rate of 12 ml per hour, collecting 2-ml fractions. Combine the active fractions and precipitate the enzyme by adding ammonium sulfate to 60% saturation. Collect the precipitate by centrifuging, and dissolve it in 0.1% 2-mercaptoethanol to give 15–25 mg of protein per milliliter.

Step 6. Crystallization. Add solid ammonium sulfate, maintaining the pH at 7.5–8.0 by the addition of ammonium hydroxide. When the first permanent turbidity appears, clarify the solution by centrifuging, then slowly add saturated ammonium sulfate solution to the supernatant solution until crystallization occurs.

Properties

Effect of pH. The oxygenase exhibits a broad optimum from pH 6.5 to 8.0 in both phosphate and pyrophosphate buffers. It is considerably less active in Tris buffer.

Substrate Specificity. The enzyme is highly specific for 2-methyl-3-hydroxypyridine-5-carboxylate. Of a number of analogs tested, 5-pyridoxic acid reacted at 3% of the rate of the "5-acid." Molecular oxygen is required and cannot be replaced by 2,6-dichlorophenolindophenol.

Cofactors. The oxygenase is a flavoprotein which requires FAD specifically. A reducing agent is required for reduction of the FAD–enzyme complex. Either NADH or NADPH are effective, and sodium borohydride and sodium hydrosulfite may also be used. There is no demonstrable metal ion requirement.

Substrate Affinities. The K_m for 2-methyl-3-hydroxypyridine-5-carboxylate is $4.8 \times 10^{-5}\ M$ and that for NADH is $1.0 \times 10^{-4}\ M$.

Inhibitors. A number of substrate analogs are mild inhibitors of the oxygenase. 5-Pyridoxic acid and 6-methylnicotinic acid are competitive inhibitors with K_i's of 6.0×10^{-5} and $2.2 \times 10^{-4}\ M$, respectively. PCMB inhibits at $5 \times 10^{-5}\ M$, and this inhibition is reversed by 4 mM 2-mercaptoethanol.

VII. α-(N-Acetylaminomethylene)succinic Acid Hydrolase[26]

$$CH_3CONHCH{=}\overset{\displaystyle CO_2^-}{\overset{|}{C}}{-}CH_2{-}CO_2^- + 2\ H_2O \rightarrow$$
$$CH_3{-}CO_2^- + NH_3 + CO_2 + OHCCH_2CH_2CO_2^-$$

Assay Method

The disappearance of α-(N-acetylaminomethylene) succinate is followed by the decrease in absorbancy at 260 mμ. The molar extinction coefficient is 17.5×10^3.

Reagents

> Sodium pyrophosphate buffer, pH 7.0. Adjust 0.5 M sodium pyrophosphate to pH 7.0 by adding 0.2 M phosphoric acid.
> α-(N-Acetylaminomethylene) succinic acid, $5 \times 10^{-4}\ M$

Procedure. Mix 0.5 ml of pyrophosphate buffer, 0.05 ml of enzyme, and 0.40 ml of deionized water in a 1-cm cuvette. Start the reaction by adding 0.05 ml of α-(N-acetylaminomethylene)succinic acid (260-B compound).

Definition of Unit. A unit of enzyme is that amount which brings about a decrease in absorbancy at 260 mμ of 0.1 per minute at room temperature.

Purification

The purification procedure is identical with that for the 5-acid oxygenase through step 4, hydroxylapatite chromatography. The 260-B hydrolase is eluted earlier than the oxygenase.

Step 5. Sephadex Chromatography. Apply 15 ml of the enzyme fraction from the hydroxylapatite column containing about 100 mg of protein to a 3×120 cm column of Sephadex G-75. Elute the column with $10^{-3}\ M$ potassium phosphate buffer, pH 7.3, using a flow rate of 18 ml per hour. This step separates the hydrolase from pyridoxamine transaminase and results in a 20-fold purification over the crude cell-free extract.

---

[26] E. J. Nyns, D. Zäch, and E. E. Snell, *J. Biol. Chem.* **244**, 2601 (1969).

Properties

Effect of pH. The optimal pH depends upon the buffer used. It is 6.0 in phosphate buffer, 7.0 in pyrophosphate buffer, and 7.5 in Tris buffer.

Substrate Specificity. The enzyme will not hydrolyze α-hydroxymethyl-α'-(N-acetylaminomethylene)succinic acid.

Cofactors. No cofactor requirement has been demonstrated.

Substrate Affinity. The K_m for α-(N-acetylaminomethylene)succinic acid is 5.3×10^{-5} M.

Author Index

Numbers in parentheses are reference numbers and indicate that an author's work is referred to although his name is not cited in the text.

Subject Index